Genome Mapping and Molecular Breeding in Plants
Volume 5

Series Editor: Chittaranjan Kole

Volumes of the Series
Genome Mapping and Molecular Breeding in Plants

Volume 1
Cereals and Millets

Volume 2
Oilseeds

Volume 3
Pulses, Sugar and Tuber Crops

Volume 4
Fruits and Nuts

Volume 5
Vegetables

Volume 6
Technical Crops

Volume 7
Forest Trees

Chittaranjan Kole (Ed.)

Vegetables

With 55 Illustrations, 5 in Color

Springer

Chittaranjan Kole
Department of Horticulture
316 Tyson Building
The Pennsylvania State University
University Park, PA 16802
USA

e-mail: cuk10@psu.edu

Library of Congress Control Number: 2006933735

ISBN-13 978-3-540-34535-0 Springer Berlin Heidelberg New York

Springer is a part of Springer Science+Business Media
springer.com

Editor: Dr. Sabine Schreck, Heidelberg, Germany
Desk Editor: Dr. Jutta Lindenborn, Heidelberg, Germany
Cover design: WMXDesign GmbH, Heidelberg, Germany
Typesetting and production: LE-TEX Jelonek, Schmidt & Vöckler GbR, Leipzig, Germany
39/3100/YL 5 4 3 2 1 0 – Printed on acid-free paper SPIN 11746096

Preface to the Series

Genome science has emerged unequivocally as the leading discipline of this new millennium. Progress in molecular biology during the last century has provided critical inputs for building a solid foundation for this discipline. However, it has gained fast momentum particularly in the last two decades with the advent of genetic linkage mapping with RFLP markers in humans in 1980. Since then it has been flourishing at a stupendous pace with the development of newly emerging tools and techniques. All these events are due to the concerted global efforts directed at the delineation of genomes and their improvement.

Genetic linkage maps based on molecular markers are now available for almost all plants of significant academic and economic interest, and the list of plants is growing regularly. A large number of economic genes have been mapped, tagged, cloned, sequenced, or characterized for expression and are being used for genetic tailoring of plants through molecular breeding. An array of markers in the arsenal from RFLP to SNP; tools such as BAC, YAC, ESTs, and microarrays; local physical maps of target genomic regions; and the employment of bioinformatics contributing to all the "-omics" disciplines are making the journey more and more enriching. Most naturally, the plants we commonly grow on our farms, forests, orchards, plantations, and labs have attracted emphatic attention, and deservedly so. The two-way shuttling from phenotype to genotype (or gene) and genotypte (gene) to phenotype has made the canvas much vaster. One could have easily compiled the vital information on genome mapping in economic plants within some 50 pages in the 1980s or within 500 pages in the 1990s. In the middle of the first decade of this century, even 5,000 pages would not suffice! Clearly genome mapping is no longer a mere "promising" branch of the life science; it has emerged as a full-fledged subject in its own right with promising branches of its own. Sequencing of the *Arabidopsis* genome was complete in 2000. The early 21st century witnessed the complete genome sequence of rice. Many more plant genomes are waiting in the wings of the national and international genome initiatives on individual plants or families.

The huge volume of information generated on genome analysis and improvement is dispersed mainly throughout the pages of periodicals in the form of review papers or scientific articles. There is a need for a ready reference for students and scientists alike that could provide more than just a glimpse of the present status of genome analysis and its use for genetic improvement. I personally felt the gap sorely when I failed to suggest any reference works to students and colleagues interested in the subject. This is the primary reason I conceived of a series on genome mapping and molecular breeding in plants.

There is not a single organism on earth that has no economic worth or concern for humanity. Information on genomes of lower organisms is abundant and highly useful from academic and applied points of view. Information on higher animals including humans is vast and useful. However, we first thought to concentrate only on the plants relevant to our daily lives, the agronomic, horticultural and technical crops, and forest trees, in the present series. We will come up soon with commentaries on food and fiber animals, wildlife and companion animals, laboratory animals, fishes and aquatic animals, beneficial and harmful insects, plant- and animal-associated microbes, and primates including humans in our next "genome series" dedicated to animals and microbes. In this series, 82 chapters devoted to plants or their groups have been included. We tried to include most of the plants in which significant progress has been made. We have also included preliminary works on some so-called minor and orphan crops in this series. We would be happy to include reviews on more such crops that deserve immediate national and international attention and support. The extent of coverage in terms of the number of pages, however, has nothing to do with the relative importance of a plant or plant group. Nor does the sequence of the chapters have any correlation to the importance of the plants discussed in the volumes. A simple rule of convenience has been followed.

I feel myself fortunate to have received highly positive responses from nearly 300 scientists of some 30-plus countries who contributed the chapters for this series. Scientists actively involved in analyzing and improving particular genomes contributed each and every chapter. I thank them all profoundly. I made a conscientious effort to assemble the best possible team of authors for certain chapters devoted to the important plants. In general, the lead authors of most chapters organized their teams. I extend my gratitude to them all.

The number of plants of economic relevance is enormous. They are classified from various angles. I have presented them using the most conventional approach. The volumes thus include cereals and millets (Volume I), oilseeds (Volume II), pulse, sugar and tuber crops (Volume III), fruits and nuts (Volume IV), vegetables (Volume V), technical crops including fiber and forage crops, ornamentals, plantation crops, and medicinal and aromatic plants (Volume VI), and forest trees (Volume VII).

A significant amount of information might be duplicated across the closely related species or genera, particularly where results of comparative mapping have been discussed. However, some readers would have liked to have had a chapter on a particular plant or plant group complete in itself. I ask all the readers to bear with me for such redundancy.

Obviously the contents and coverage of different chapters will vary depending on the effort expended and progress achieved. Some plants have received more attention for advanced works. We have included only introductory reviews on fundamental aspects on them since reviews in these areas are available elsewhere. On other plants, including the "orphan" crop plants, a substantial amount of information has been included on the basic aspects. This approach will be reflected in the illustrations as well.

It is mainly my research students and professional colleagues who sparked my interest in conceptualizing and pursuing this series. If this series serves its purpose, then the major credit goes to them. I would never have ventured to take up this huge task of editing without their constant support. Working and interacting with many people, particularly at the Laboratory of Molecular Biology and Biotechnology of the Orissa University of Agriculture and Technology, Bhubaneswar, India as its founder principal investigator; the Indo-Russian Center for Biotechnology, Allahabad, India as its first project coordinator; the then-USSR Academy of Sciences in Moscow; the University of Wisconsin at Madison; and The Pennsylvania State University, among institutions, and at EMBO, EUCARPIA, and Plant and Animal Genome meetings among the scientific gatherings have also inspired me and instilled confidence in my ability to accomplish this job.

I feel very fortunate for the inspiration and encouragement I have received from many dignified scientists from around the world, particularly Prof. Arthur

Kornberg, Prof. Franklin W. Stahl, Dr. Norman E. Borlaug, Dr. David V. Goeddel, Prof. Phillip A. Sharp, Prof. Gunter Blobel, and Prof. Lee Hartwell, who kindly opined on the utility of the series for students, academicians, and industry scientists of this and later generations. I express my deep regards and gratitude to them all for providing inspiration and extending generous comments.

I have been especially blessed by God with an affectionate student community and very cordial research students throughout my teaching career. I am thankful to all of them for their regards and feelings for me. I am grateful to all my teachers and colleagues for the blessings, assistance, and affection they showered on me throughout my career at various levels and places. I am equally indebted to the few critics who helped me to become professionally sounder and morally stronger.

My wife Phullara and our two children Sourav and Devleena have been of great help to me, as always, while I was engaged in editing this series. Phullara has taken pains ("pleasure" she would say) all along to assume most of my domestic responsibilities and to allow me to devote maximum possible time to my professional activities, including editing this series. Sourav and Devleena have always shown maturity and patience in allowing me to remain glued to my PC or "printed papers" ("P3" as they would say). For this series, they assisted me with Internet searches, maintenance of all hard and soft copies, and various timely inputs.

Some figures included by the authors in their chapters were published elsewhere previously. The authors have obtained permission from the concerned publishers or authors to use them again for their chapters and expressed due acknowledgement. However, as an editor I record my acknowledgements to all such publishers and authors for their generosity and good will.

I look forward to your valuable criticisms and feedback for further improvement of the series.

Publishing a book series like this requires diligence, patience, and understanding on the part of the publisher, and I am grateful to the people at Springer for having all these qualities in abundance and for their dedication to seeing this series through to completion. Their professionalism and attention to detail throughout the entire process of bringing this series to the reader made them a genuine pleasure to work with. Any enjoyment the reader may derive from this books is due in no small measure to their efforts.

Pennsylvania, Chittaranjan Kole
10 January 2006

Preface to the Volume

Vegetables include a versatile group of crop plants. Tomato, eggplant or Capsicums are botanically fruits but traditionally they are treated as vegetables and with due respect to the convention, we have included them in Volume 5 dedicated to vegetables. On the other hand, we could have included crops such as cowpea, pea and potato in this volume; however, they found place in Volume 3 representing pulses, sugar and tuber crops. Beets could also have appeared in Volume 3 for they include sugar beets, contributing to one quarter of the sugar production of the world. But, in this volume deliberations are needed on other beets. This is also true for *Brassica rapa*. The vegetable *B. rapa* includes some crops widely used as vegetables, however genetically they hardly differ from the oilseed *B. rapa* crops and have highly homologous genomes to other oilseed *Brassica* species. This is the main reason for detailing genomics and breeding of this species in Volume 2 on oilseeds, and presenting only a brief on molecular works but depicting the various vegetable *Brassica rapa* crops in detail in this volume. Green papaya and plantain are highly popular items in many curries, particularly in Asian countries and could have been included in this volume. However, papaya has been included independently as a fruit crop, and plantain has been discussed alongside banana in Volume 4 devoted to fruits and nuts.

The relative attention drawn by a crop of the scientific community depends not on its relative global importance but on its acreage in the developed nations. Tomato is a prime vegetable crop in the United States and the leading one slightly ahead of lettuce. It could win a place in several laboratories and has been well supported by funding agencies for works on genome mapping, molecular breeding and advanced genomics. The International Tomato Genome Project is progressing nicely and a stupendous amount of sequence data has already been accumulated. Tomato has a significant place in classical as well as modern genetics as a model plant. It can boast of its ranking among a few plant species possessing the earliest chromosome maps based on morphological, cytological, biochemical and/or molecular markers. Tomato is the proud crop plant to witness emergence of many concepts and strategies including molecular mapping of a gene; gene isolation through positional cloning and transposon tagging; comparative mapping; Mendelization, fine-mapping and cloning of QTLs; besides the homely terms of chromosome landing or reverse genetics amongst an array of others. Recent welcoming of the tomato into the *Solanum* genus has facilitated concerted genomic efforts along with other vegetable crops such as potato, Capsicums and eggplant.

Lettuce is grown and consumed in limited countries of the world but witnessed significant progress due to its horticultural importance in developed countries particularly in the United States. The most popular among all single genetic strategies practiced today, bulked segregant analysis, was devised originally for this crop.

Brassica oleracea includes many globally grown vegetables and appreciable progress has been made in this species. Again this species could have been better dealt with alongside other *Brassica* species in Volume 2 from a genetic perspective. The conventional classification of crop plants based on their utility was the deciding factor for the placement of this *Brassica* species in Volume 5.

Extensively grown crops like eggplant, Capsicums or Alliums have attracted relatively less attention for advanced genomics works. We feel Cucurbits should have been addressed more since they are used so widely in the countries with the highest populations on Earth. Included are the popularly used vegetable crops such as melon, watermelon, squash and cucumber with considerable progress, but "orphan crops" like bitter gourd, sponge gourd, snake gourd, bottle gourd, spine gourd, pointed gourd, pumpkin, etc., have no molecular story to tell.

People in developing nations do not consume a large amount of salad in their diets, and fewer pages on radish, beets or carrot could therefore have been justified. Still, appreciable progress has been made in these crops due to their popularity in developed countries.

Some "poor man's vegetable crops" with modified roots and stems, such as *Alocasia* and *Colocasia*, fruits like ladies' finger, country bean, drum stick, fig and jackfruit must be discussed in the near future. Several herbs consumed in many countries under the genera *Spinacea*, *Corchorus*, *Ipomoea*, *Amaranthus*, *Chenopodium*, etc., also deserve attention.

Considerable progress has been made with other vegetables, particularly asparagus and faba bean and should have been included; this is also true of some leafy vegetables. We are hopeful to see a sea-change in the table of contents of this volume in future editions.

The authors of two chapters of this volume, lettuce and onion, took the pain and pleasure to draft their chapters single-handedly. The lead authors of the remaining chapters have arranged themselves into multi-lab or even multi-national groups. Our thanks and gratitude go to all of them and all the co-authors for their useful contributions.

I am grateful to Springer for the work done in achieving this volume, like the preceding ones. All the shortcomings and flaws related to the content and its presentation are fully mine, and I will be looking forward to suggestions on how to address them in the future.

Pennsylvania, Chittaranjan Kole
18 June 2006

Contents

[1] TILLING is a registered trademark of Arcadia Biosciences

Contributors

Anna Amen
Arcadia Biosciences
410 W. Harrison
Ste 150
Seattle, WA 98119, USA
anna.amen@arcadiabio.com

Koh Aoki
Kazusa DNA Research Institute
Kazusa-Kamatari 2-6-7
Kisarazu, 292-0818, Japan
kaoki@kazusa.or.jp

Danuta Babula
Institute of Plant Genetics
Polish Academy of Sciences
Strzeszynska 34, 60-479 Poznań
Poland
dbab@igr.poznan.pl

Angela M. Baldo
USDA-ARS Plant Genetic
Resources Unit and Grape
Genetics Research Unit
Geneva, NY 14456, USA
angela.baldo@ars.usda.gov

Sang Woo Bang
Laboratory of Plant Breeding
Faculty of Agriculture
Utsunomiya University
350 Minemachi
Utsunomiya 321-8505, Japan
bang@cc.utsunomiya-u.ac.jp

Nicholas E. Blanchard
Department of Biology
California State University-Fresno
Fresno, CA 93740, USA
nblanchard01@csufresno.edu

Ana L. Caicedo
Biology Department
University of Massachusetts
Amherst, MA 01003, USA
caicedo@bio.umass.edu

Arnon Ben-Chaim
Department of Plant Breeding
Cornell University, Ithaca
NY 14853, USA
ab357@cornell.edu

Enrico Biancardi
CRA – Istituto Sperimentale
per le Colture Industriali
Sezione di Rovigo, viale Amendola 82
45100 Rovigo, Italy
enrico.biancardi@entecra.it

James M. Bradeen
Department of Plant Pathology
University of Minnesota
495 Borlaug Hall/1991 Upper
Buford Circle
St. Paul, MN 55108, USA
jbradeen@umn.edu

Mathilde Causse
Laboratory of Vegetable
and Fruit Genetics and Breeding
INRA, UR 1052
BP94, 84143
Montfavet, France
mcausse@avignon.inra.fr

Pablo F. Cavagnaro
INTA, EEA La Consulta, CC8 (5567)
La Consulta, San Carlos, Mendoza
Argentina
and
Consejo Nacional de Investigaciones
Científicas y Tecnológicas (CONICET)
pcavagnaro@laconsulta.inta.gov.ar

Roger T. Chetelat
C.M. Rick Tomato Genetics Resource
Center
Department of Plant Sciences
University of California
One Shields Avenue
Davis, CA 95616, USA
chetelat@plantsciences.ucdavis.edu

Marie-Christine Daunay
INRA, Unité de Génétique
& Amélioration des Fruits et Légumes
Domaine St. Maurice, BP 94
84143 Montfavet cedex, France
marie-christine.daunay
@avignon.inra.fr

Ralph A. Dean
Fungal Genomics Laboratory
North Carolina University
Campus Box 7251, Raleigh
NC 27695, USA
ralph_dean@ncsu.edu

Sami Doganlar
Izmir Institute of Technology
Department of Biology
Gulbahcekoyu Campus
Urla 35430, Izmir, Turkey
samidoganlar@iyte.edu.tr

Daniel Facciotti
Arcadia Biosciences
202 Cousteau Place
Ste 200
Davis, CA, 95616 USA
daniel.facciotti@arcadiabio.com

David Francis
Ohio Agricultural Research
and Development Center
Department of Horticulture
and Crop Science
The Ohio State University
Wooster, OH 44691, USA
francis.77@osu.edu

Anne Frary
Izmir Institute of Technology
Department of Biology
Gulbahcekoyu Campus
Urla 35430, Izmir, Turkey
annefrary@iyte.edu.tr

Carrie Freeman
Arcadia Biosciences
202 Cousteau Place
Ste 200
Davis, CA 95616, USA
carrie.barrios@arcadiabio.com

Theresa Fulton
Institute for Genomic Diversity
Cornell University
Ithaca, NY 14853, USA
tf12@cornell.edu

Claudio R. Galmarini
INTA, EEA La Consulta, CC8 (5567)
La Consulta, San Carlos, Mendoza
Argentina
and
FCA, UNCuyo, Alte. Brown 500
Chacras de Coria (5505) Mendoza
Argentina
and
Consejo Nacional de Investigaciones
Científicas y Tecnológicas (CONICET)
crgalmarini@mendoza.inta.gov.ar

Maria Jose Gonzalo
Ohio Agricultural Research
and Development Center
Department of Horticulture
and Crop Science
The Ohio State University
Wooster, OH 44691, USA
gonzalo.1@osu.edu

John Goodstal
Arcadia Biosciences
202 Cousteau Place
Ste 200
Davis, CA 95616, USA
john.goodstal@arcadiabio.com

Silvana Grandillo
CNR-IGV
Institute of Plant Genetics
Portici
Via Università 133
80055 - Portici (NA), Italy
grandill@unina.it

Masashi Hirai
Kyoto Prefectural University
Kita-Inayazuma, Seika
Kyoto 619-0244, Japan
mhirai@kab.seika.kyoto.jp

Susan R. Hurst
Arcadia Biosciences
410 W. Harrison
Ste 150
Seattle, WA 98119, USA
susan.hurst@arcadiabio.com

Yoko Iijima
Kazusa DNA Research Institute
Kazusa-Kamatari 2-6-7
Kisarazu, 292-0818, Japan
yiijima@kazusa.or.jp

Molly Jahn
Department of Plant Breeding
Cornell University, Ithaca
NY 14853, USA
mmj9@cornell.edu

Yuanfu Ji
Gulf Coast Research
& Education Center
University of Florida
14625 CR 672
Wimauma, FL 33598, USA
yji@ufl.edu

Tarek Joobeur
Fungal Genomics Laboratory
North Carolina University
Campus Box 7251, Raleigh
NC 27695, USA
tarek_joobeur@ncsu.edu

Małgorzata Kaczmarek
Institute of Plant Genetics
Polish Academy of Sciences
Strzeszynska 34, 60-479 Poznań
Poland
mrum@igr.poznan.pl

Yukio Kaneko
Laboratory of Plant Breeding
Faculty of Agriculture
Utsunomiya University
350 Minemachi, Utsunomiya 321-8505
Japan
kaneko@cc.utsunomiya-u.ac.jp

Byoung-Cheorl Kang
Department of Plant Breeding
Cornell University, Ithaca
NY 14853, USA
bk54@cornell.edu

Chiaki Kimizuka-Takagi
Mikado Seed Growers Co, Ltd
2789-1 Shimo-Ohtaki, Ohtaki
Chiba Pref. 298-0202, Japan
c-kimizuka@mikadokyowa.com

Esther van der Knaap
Ohio Agricultural Research
and Development Center
Department of Horticulture
and Crop Science
The Ohio State University
Wooster, OH 44691, USA
vanderknaap.1@osu.edu

Joanne A. Labate
USDA-ARS Plant Genetic
Resources Unit
Geneva, NY 14456, USA
joanne.labate@ars.usda.gov

Barbara E. Liedl
Gus R. Douglass Institute
Agricultural and Environmental
Research Station
West Virginia State University
Institute, WV 25112, USA
liedlbe@wvstateu.edu

Dayna Loeffler
Arcadia Biosciences
410 W. Harrison
Ste 150
Seattle, WA 98119, USA
dayna.loeffler@arcadiabio.com

Satoru Matsumoto
National Institute of Vegetable
and Tea Science
National Agriculture and Bio-oriented
Research Organization
Ano, Mie 514-2392, Japan
ssmats@affrc.go.jp

Yasuo Matsuzawa
Laboratory of Plant Breeding
Faculty of Agriculture
Utsunomiya University
350 Minemachi, Utsunomiya 321-8505
Japan
yasuom@cc.utsunomiya-u.ac.jp

John A. McCallum
New Zealand Institute
for Crop & Food Research
Private Bag 4704, Christchurch
New Zealand
mccallumj@crop.cri.nz

J. Mitchell McGrath
USDA-ARS Sugarbeet
and Bean Research
494 PSSB Michigan State University
East Lansing, MI 48824-1325, USA
mitchmcg@msu.edu

Cate McGuire
Arcadia Biosciences
202 Cousteau Place
Ste 200
Davis, CA 95616, USA
cate.mcguire@arcadiabio.com

Lukas A. Mueller
Department of Plant Breeding
and Genetics
Cornell University
Ithaca, NY 14853, USA
lam87@cornell.edu

Stéphane Muños
Laboratory of Vegetable
and Fruit Genetics and Breeding
INRA, UR 1052
BP94, 84143
Montfavet, France
Stephane.Munos@avignon.inra.fr

Mary A. O'Connell
Department of Plant
and Environmental Sciences
New Mexico State University
Las Cruces, NM 88003, USA
moconnel@nmsu.edu

Ilan Paran
Department of Plant Genetics
and Breeding
Agricultural Research Organization
The Volcani Center, P. O. Box 6
Bet Dagan, 50250, Israel
iparan@volcani.agri.gov.il

Iris Peralta
Department of Agronomy
National University of Cuyo
Almirante Brown 500, 5505
Chacras de Coria, Luján
Mendoza, Argentina; and IADIZA-CONICET
C.C. 507, 5500 Mendoza, Argentina
iperalta@fca.uncu.edu.ar

James P. Prince
Department of Biology
California State University-Fresno
Fresno, CA 93740, USA
jamespr@csufresno.edu

Matthew D. Robbins
Department of Horticulture
and Crop Science
Ohio Agricultural Research
and Development Center
Wooster, OH 44691, USA
robbins.184@osu.edu

Massimo Saccomani
Dipartimento di Biotecnologie Agrarie
Università di Padova
viale dell'Università 16
35020 Legnaro (Pd), Italy
massimo.saccomani@unipd.it

Jan Sadowski
Department of Biotechnology
Adam Mickiewicz University
Międzychodzka 5, 60-371 Poznań
Poland
jsad@amu.edu.pl

J. W. Scott
Gulf Coast Research
& Education Center
University of Florida
14625 CR 672
Wimauma, FL 33598, USA
jwsc@ufl.edu

Philipp W. Simon
USDA-ARS & Department
of Horticulture
University of Wisconsin-Madison
1575 Linden Drive
Madison
WI 53706, USA
psimon@wisc.edu

Ann J. Slade
Arcadia Biosciences
410 W. Harrison
Ste 150
Seattle, WA 98119, USA
ann.slade@arcadiabio.com

Brian Smith-White
National Center
for Biotechnology Information
USA National Library of Medicine
NIH, Bethesda
MD 20894, USA
smtwhite@ncbi.nlm.nih.gov

Jack E. Staub
USDA-ARS
Department of Horticulture
University of Wisconsin
1575 Linden Drive
Madison
WI 53703, USA
jestaub@wisc.edu

Michael N. Steine
Arcadia Biosciences
410 W. Harrison
Ste 150
Seattle, WA 98119, USA
mike.steine@arcadiabio.com

Piergiorgio Stevanato
Dipartimento di Biotecnologie Agrarie
Università di Padova
viale dell'Università 16
35020 Legnaro (Pd), Italy
p.stevanato@unipd.it

Mikel R. Stevens
Department of Plant
and Wildlife Sciences
Brigham Young University
287 Widstoe Building
Provo, UT 84602, USA
mikel_stevens@byu.edu

David W. Still
Department of Plant Sciences
& Technology
California State Polytechnic University
3801 West Temple Avenue, Pomona
CA 91768, USA
dwstill@csupomona.edu

John R. Stommel
USDA-ARS
Vegetable Laboratory
Bldg. 010A
BARC-West
10300 Baltimore Avenue
Beltsville, MD 20705, USA
john.stommel@ars.usda.gov

Dylan B. Storey
Department of Biology
California State University-Fresno
Fresno, CA 93740, USA
dylan.storey@gmail.com

Dionne Vafeados
Arcadia Biosciences
410 W. Harrison
Ste 150
Seattle, WA 98119, USA
dionne.vafeados@arcadiabio.com

Joyce Van Eck
Boyce Thompson Institute
for Plant Research
Tower Road
Ithaca, NY 14853, USA
jv27@cornell.edu

Jaw-Fen Wang
AVRDC - The World Vegetable Center
P.O. Box 42
Shanhua
Tainan
Taiwan 74199, ROC
jfw@netra.avrdc.org.tw

Yi-Hong Wang
School of Science
Behrend College
Penn State University
Erie, PA 16563, USA
yxw23@psu.edu

Wencai Yang
Ohio Agricultural Research
and Development Center
Department of Horticulture
and Crop Science
The Ohio State University
Wooster, OH 44691, USA
yangwencai@cau.edu.cn

Piotr A. Ziółkowski
Department of Biotechnology
Adam Mickiewicz University
Międzychodzka 5
60-371 Poznań
Poland
pzio@amu.edu.pl

Abbreviations

AB	Advanced backcross
ABA	Abscisic acid
ACSOs	S-alk(en)yl cysteine sulfoxides
AB-QTL	Advanced backcross QTL
AFLP	Amplified fragment length polymorphism
AIS	Alcohol insoluble solids
ANOVA	Analysis of variance
AP1	*APETALA1*
ARS	Agriculture Research Service
ASPE	Allele-specific primer extension
atpB	ATP synthase B
BAC	Bacterial artificial chromosome
BC	Backcross
BC1	First backcross generation
BIL	Backcross inbred line
BLAST	Basic local alignment search tool
BNYVV	Beet necrotic yellow vein virus
BOCC	Core collection of the *Brassica oleracea* gene pool
BSA	Bulked segregant analysis
Bt	*Bacillus thuringiensis*
cab	chlorophyll a-b binding protein
CAL	CAULIFLOWER (gene)
CaMV	Cauliflower mosaic virus
CAPS	Cleaved amplified polymorphic sequence
CE	Capillary electrophoresis
CGP	Composite genome project
CHS	Chalcone synthase
CIM	Composite interval mapping
CMS	Cytoplasmic male-sterility
CMV	*Cucumber mosaic virus*
CO	CONSTANS (gene)
COD	Clade-oriented database
COS	Conserved ortholog set
DBM	Diamondback moths
dCAPS	Derived cleaved amplified polymorphisms
DFFS	Diversity fixed foundation sets
DH	Doubled haploid
Dm	Downy mildew (gene)
ds	Double stranded
DTB	Days-to-bolting
EMS	Ethyl-methane sulfonate
ENU	N-ethyl-N-nitrosourea
EST	Expressed sequence tag
FAO	Food and Agriculture Organization
FISH	Fluorescent in situ hybridization

FLC	Flowering Locus C (gene)
FR-D	Far-red to dark
FRI	FRIGIDA (gene)
FTICR	Fourier transform ion cyclotron resonance
G:F	Glucose:fructose
GBSSI	Granule bound starch synthase I
GC	Gas chromatography
GFP	Green fluorescent protein
GISH	Genomic in situ hybridization
GME	GDP-mannose epimerase
GSL	Glucosinolates
GST	Genomic sequence tag
HCA	Hierarchical clustering analysis
HPLC	High-performance liquid chromatography
HTG	High-temperature germination
IBL	Inbred backcross line
IL	Introgression line
ILH	Introgression line hybrid
IMA	Inter microsatellite amplification
indel	insertion/deletion
IRD	Infra-red fluorescent dye
ISH	In situ hybridization
ISSR	Inter-simple sequence repeat
ITC	Isothiocyanate
ITS	Internal transcribed spacer
LC	Liquid chromatography
LD	LUMINIDEPENDES (gene)
L-DOPA	L-3,4-Dihydroxyphenylalanine
LG	Linkage group
LOD	Logarithm of odds
LRR	Leucine-rich repeat
LTG	Low-temperature germination
LTR	Long terminal repeat
MABB	Marker-assisted backcross breeding
MAS	Marker-assisted selection
MBGP	Multinational Brassica Genome Project
MCSO	S-Methyl-L-cysteine sulfoxide
MDHAR	Monodehydroascorbate reductase
MF	Methyl filtrated
Mf	Male-fertility
MI	First metaphase
MIAME	Minimum information about a microarray experiment
MLB	Multiple lateral branching
mRNA	Messenger ribonucleic acid
MS	Male-sterility
MS	Mass spectrometry
My	Million years
Mya	Million years ago
NBR	Nucleotide binding region
NBS	Nucleotide binding site
NCBI	National Center for Biotechnology Information

Ne	Effective population size
NIL	Near-isogenic line
ODO	Overdominant
ORF	Open reading frame
PCA	Principle component analysis
PCR	Polymerase chain reaction
PCSO	S-Propyl-L-cysteine sulfoxide
PDA	Photodiode array
PECSO	S-(1/2-Propenyl)-L-cysteine sulfoxide
PERL	Practical extraction and report language
PFGE	Pulse-field gel electrophoresis
PGD	6-Phosphogluconic dehydrogenase
PGI	Phosphoglucoseisomerase
PGM	Phosphoglucomutase
PMC	Pollen mother cell
PPR	Pentatricopeptide-repeat
PSSM	Position specific scoring matrix
PVMV	Pepper veinal mottle virus
QML	Quantitative metabolic loci
QR	Quinone reductase
QTL	Quantitative trait loci
QTN	Quantitative trait nucleotide
RAPD	Random(ly) amplified polymorphic DNA
rbcL	Ribulose-bisphosphate carboxylase Large subunit
rbcS	Ribulose-bisphosphate carboxylase Small subunit, Rubisco
rep-PCR	Repetitive-sequence-based-PCR
RFLP	Restriction fragment length polymorphism
RGA	Resistance gene analog
RIL	Recombinant inbred line
RNAi	RNA interference
RT-PCR	Reverse transcription PCR
SAGE	Serial analysis of gene expression
SBP	SQUAMOSA promoter binding protein
SC	Self-compatible
SCAR	Sequence characterized amplified region
SFP	Single feature polymorphism
SGN	Solanaceae Genomics Network
SI	Self-incompatibility
SI	Similarity index
SL	Substitution line
SLG	S-Locus-specific glycoprotein
SNP	Single nucleotide polymorphism
SRAP	Sequence-related amplified polymorphism
SS	Soluble solids
SSAP	Sequence-specific amplification polymorphism
SSC	Soluble solids content
SSCP	Single-strand conformation polymorphism
SSH	Suppressive subtractive hybridization
SSR	Simple-sequence repeat
SST-1	1-Fructosyltransferase
STS	Sequence tagged site

TILLING	Targeting induced local lesions in genomes
TIR	Toll and interleukin-1 receptor
TMRD	Tomato Mapping Resource Database
TMV	*Tobacco mosaic virus*
TOF	Time of flight
TSWV	*Tomato spotted wilt virus*
TuMV	*Turnip mosaic virus*
UPGMA	Unweighted pair-group method with arithmetic mean
USDA	United States Department of Agriculture
VIGS	Virus induced gene silencing
VNTR	Variable number tandem repeats
WAE	Weeks after emergence
WIS	Water insoluble solids
wt	Wild-type
WUE	Water-use efficiency

1 Tomato

Joanne A. Labate[1], Silvana Grandillo[2], Theresa Fulton[3], Stéphane Muños[4], Ana L. Caicedo[5], Iris Peralta[6], Yuanfu Ji[7], Roger T. Chetelat[8], J. W. Scott[7], Maria Jose Gonzalo[9], David Francis[9], Wencai Yang[9], Esther van der Knaap[9], Angela M. Baldo[1,10], Brian Smith-White[11], Lukas A. Mueller[12], James P. Prince[13], Nicholas E. Blanchard[13], Dylan B. Storey[13], Mikel R. Stevens[14], Matthew D. Robbins[9], Jaw-Fen Wang[15], Barbara E. Liedl[16], Mary A. O'Connell[17], John R. Stommel[18], Koh Aoki[19], Yoko Iijima[19], Ann J. Slade[20], Susan R. Hurst[20], Dayna Loeffler[20], Michael N. Steine[20], Dionne Vafeados[20], Cate McGuire[21], Carrie Freeman[21], Anna Amen[20], John Goodstal[21], Daniel Facciotti[21], Joyce Van Eck[22], and Mathilde Causse[4]

[1] USDA-ARS, Plant Genetic Resources Unit, Geneva, NY 14456, USA, *e-mail*: joanne.labate@ars.usda.gov
[2] CNR-IGV, Institute of Plant Genetics, Portici, Via Università 133, 80055, Portici, (NA), Italy
[3] Institute for Genomic Diversity, Cornell University, Ithaca, NY 14853, USA
[4] Laboratory of Vegetable and Fruit Genetics and Breeding, INRA, UR 1052, BP94, 84143, Montfavet, France
[5] Biology Department, University of Massachusetts, Amherst, MA 01003, USA
[6] Department of Agronomy, National University of Cuyo, Almirante Brown 500, 5500, Chacras de Coria, Luján, Mendoza, Argentina and IADIZA-CONICET, C.C. 507, 5500, Mendoza, Argentina
[7] Gulf Coast Research and Education Center, University of Florida, 14625 CR 672, Wimauma, FL 33598, USA
[8] C.M. Rick Tomato Genetics Resource Center, Department of Plant Sciences, University of California, One Shields Avenue, Davis, CA 95616, USA
[9] Ohio Agricultural Research and Development Center, Department of Horticulture and Crop Science, The Ohio State University, Wooster, OH 44691, USA
[10] USDA-ARS, Grape Genetics Research Unit, Geneva, NY 14456, USA
[11] National Center for Biotechnology Information, USA National Library of Medicine, NIH, Bethesda, MD 20894, USA
[12] Department of Plant Breeding and Genetics, Cornell University, Ithaca, NY 14853, USA
[13] Department of Biology, California State University-Fresno, Fresno, CA 93740, USA
[14] Department of Plant and Wildlife Sciences, Brigham Young University, 287 Widstoe Building, Provo, UT 84602, USA
[15] AVRDC - The World Vegetable Center, P.O. Box 42, Shanhua, Tainan, Taiwan 74199, ROC
[16] Gus R. Douglass Institute, Agricultural and Environmental Research Station, West Virginia State University, Institute, WV 25112, USA
[17] Department of Plant and Environmental Sciences, New Mexico State University, Las Cruces, NM 88003, USA
[18] USDA-ARS, Vegetable Laboratory, Bldg. 010A, BARC-West, 10300 Baltimore Avenue, Beltsville, MD 20705, USA
[19] Kazusa DNA Research Institute, Kazusa-Kamatari 2-6-7, Kisarazu, 292-0818, Japan
[20] Arcadia Biosciences, 410 W. Harrison, Ste 150, Seattle, WA 98119, USA
[21] Arcadia Biosciences, 202 Cousteau Place, Ste 200, Davis, CA 95616, USA
[22] Boyce Thompson Institute for Plant Research, Tower Road, Ithaca, NY 14853, USA

1.1 Introduction

Tomatoes (*Solanum lycopersicum*) are consumed as either fresh fruit by themselves, in salads, as ingredients in many recipes, or in the form of various processed products such as paste, whole peeled tomatoes, diced products, and various forms of juices and soups. The tomato is a favorite garden plant in many parts of the world, an important source of vitamins and nutrients (see Sect. 1.14), and an economically important agricultural commodity (see Sect. 1.3.3).

Tomato was among the first crops for which molecular markers (isozymes) were suggested for marker-assisted selection (MAS) in breeding (Rick and Fobes 1974; Tanksley and Rick 1980). Tanksley (1983) discussed the viability of using isozymes for MAS in tomato and concluded that DNA-based markers would probably be utilized for the next iteration of MAS in tomato.

The most daunting challenge of effectively implementing MAS in cultivated tomato has been the low frequency of easily identifiable molecular polymorphisms within *S. lycopersicum* (Stevens and Robbins 2007) (see Sect. 1.4). This impediment was rec-

Genome Mapping and Molecular Breeding in Plants, Volume 5
Vegetables
C. Kole (Ed.)

ognized as quickly as the theoretical concepts of using MAS were developed (Tanksley 1983; Helentjaris et al. 1985). In 1985, Helentjaris et al. demonstrated that DNA-based molecular markers in the form of restriction fragment length polymorphisms (RFLPs) could effectively identify differences between cultivated tomato and wild tomato species. Two years later, Nienhuis et al. (1987) demonstrated that MAS could identify quantitative trait loci (QTL) associated with insect resistance derived from *Solanum habrochaites* in an interspecific cross.

The clear demonstration that polymorphic markers were relatively abundant between cultivated tomato and its wild relatives opened a new line of MAS utilizing DNA-based markers. Young et al. (1988) exploited the abundance of polymorphisms derived from linkage drag surrounding genes introgressed from *S. peruvianum* into tomato. They utilized near-isogenic lines (NIL) to identify two RFLP markers tightly linked to the *Tm-2a* viral resistance gene.

Tomato's importance as a crop and role as a model for genetics, fleshy fruit development, secondary metabolism, disease resistance, domestication, and evolution, has led to concerted efforts to develop genetic and genomic resources for this species. These efforts have rapidly advanced tomato genome mapping and MAS, and have culminated in the adoption of tomato as the model genome (see Sect. 1.6, 1.18) for the commercially important (e.g., potato, pepper, eggplant) Solanaceae family.

1.2 Tomato Evolution and Taxonomy

The tomato clade is an evolutionarily young group that has diversified to occupy a great variety of habitats. The age of the genus *Solanum* is estimated at ~12 million years (My) based on nuclear (18S rDNA) and chloroplast markers ribulose-bisphosphate carboxylase large subunit (*rbcL*) and ATP Synthase B (*atpB*) (Wikström et al. 2001), and the radiation of the tomato clade has been estimated as ~7 My based on four nuclear genes (Nesbitt and Tanksley 2002). In this amount of time, tomato species have evolved to occupy various habitats along the western coast of South America, from central Ecuador to northern Chile, as well as the Galápagos Islands, ranging from sea level to above 3,000 m in altitude, within various grades of xeric to mesic environments (Taylor 1986). For evolutionary biologists, the group is a prime example of rapid evolution and adaptation to diverse environments and environmental stresses. For breeders, wild tomato species contain useful traits that can be introgressed into cultivated tomato, ranging from resistance to multiple pathogens to tolerance to drought, salinity, etc. (Bohn and Tucker 1940; Stevens and Rick 1986). Wild germplasm has played an important role in the modern breeding of cultivated tomato (Stevens and Rick 1986), fueling interest in the study of wild tomatoes and the evolution of the group as a whole.

1.2.1 Taxonomic Placement

Traditionally, wild and cultivated tomatoes have been considered within the genus *Lycopersicon* in the Solanaceae family, mainly based on the typical androceum where the anthers are connivent laterally to form a flask-shaped cone with an elongated sterile tip at the apex. However, there has been considerable flux in the taxonomy of tomatoes in the last ~450 years.

Early European botanists recognized the close relationship of tomatoes with the genus *Solanum*, and commonly referred to them as *S. pomiferum* (Luckwill 1943b). Tournefort (1694) was the first to name cultivated tomatoes as *Lycopersicon* ("wolf peach" in Greek). Tournefort placed forms with large multilocular fruit in *Lycopersicon*, but included forms with bilocular fruit in *Solanum*. In *Species Plantarum*, Linnaeus (Linnaeus 1753) began to consistently use Latin binomials to name species; he included tomatoes in the genus *Solanum* and described *S. lycopersicum* (the cultivated tomato) and *S. peruvianum*. One year later, in the 4th edition of The Gardener's Dictionary, Miller (1754) followed Tournefort and formally described the genus *Lycopersicon*. Later, Miller (1768) began to use Linnaeus' binomial system and published descriptions under *Lycopersicon* for several species, among them were three tomato species *L. esculentum*, *L. peruvianum*, *L. pimpinellifolium*, and also potatoes as *L. tuberosum* ("Lycopersicon radice tuberose, esculentum"). He supported the inclusion of potatoes based on the argument that "This Plant was always ranged in the Genus of *Solanum*, or Nightshade, and is now brought under that title by *Dr. Linnaeus*; but as *Lycopersicon* has now been established as a distinct Genus, on account of the Fruit being divided into several Cells, by intermediate Partitions, and as the Fruit of this Plant [the potato] exactly agrees with the Characters of the other species of this Genus, I have inserted it here." In the posthumously published edition of The Gardener's and Botanist's Dictionary (Miller 1807) the editor, Thomas Martyn, followed Linnaeus criterion and included *Lycopersicon* back into *Solanum*. Nevertheless, following Miller's early circumscription, tomatoes have been traditionally recognized under *Lycopersicon* by the majority of taxonomists.

Today, the treatment of tomatoes in the genus *Solanum* has gained wide acceptance supported by evidences from phylogenetic studies of the Solanaceae family based on molecular and morphological characters. These phylogenetic results have unambiguously shown tomatoes to be deeply nested within *Solanum* (Spooner et al. 1993, 2005; Bohs and Olmstead 1997; Olmstead and Palmer 1997; Olmstead et al. 1999; Peralta and Spooner 2001; Bohs 2005).

1.2.2 Phylogenetic Relationships

Based on morphological characters, phylogenetic relationships, and geographic distribution, 13 species of wild tomatoes, including the cultivated tomato (*Solanum lycopersicum*), and four closely related species have been recognized as part of the tomato clade. Their distinguishing characteristics, distribution, and habitats, are described in Table 1. Four species have been segregated from the highly polymorphic green-fruited species *S. peruvianum* sensu lato (sensu lato refers to a broad concept of a species): *S. arcanum*, *S. huaylasense*, *S. peruvianum*, and *S. corneliomulleri*. The first two have been described as new species (Peralta et al. 2005) from Perú, while the latter two had already been named by Linnaeus (1753) and MacBride (1962), respectively. Another yellow- to orange-fruited species, *S. galapagense*, was segregated from *S. cheesmaniae*; both are endemic to the Galápagos Islands (Darwin et al. 2003; Knapp and Darwin 2007). Peralta et al. (2007) are treating these 13 species, in addition to four closely related species (*S. juglandifolium*, *S. lycopersicoides*, *S. ochranthum*, *S. sitiens*), in the taxonomic series *Systematic Botany Monographs*.

Resolving evolutionary relationships within the tomato group has been difficult due to the young age of the group, which increases the probability of shared polymorphism among groups, and the low levels of variation within and between the self-compatible taxa. Morphologically, classification has also been made difficult by the unfamiliarity of initial systematists with the intraspecific variation (Warnock 1988). A wide variety of studies, using morphological (Peralta and Spooner 2005) and molecular characters such as chloroplast and mitochondrial restriction sites, nuclear RFLPs, isozymes, internal transcribed spacer (ITS), and granule bound starch synthase (GBSSI or *waxy*) sequences, simple sequence repeats (SSRs), and amplified fragment length polymorphisms (AFLPs) have been carried out to determine the wild tomato phylogeny. Although evolutionary relationships are not completely resolved, consensus has been emerging on several aspects of the group. Some recent results based on sequences of the structural gene GBSSI (Peralta and Spooner 2001), AFLPs (Spooner et al. 2005), morphology (Peralta and Spooner 2005), and more recently with multiple single-copy genes (Rodriguez et al. 2006) are highlighted below, illustrating our current understanding of phylogenetic relationships within the group (Fig. 1):

1. The tomatoes s.l. (sects. Lycopersicoides, Juglandifolia and Lycopersicon) are clearly monophyletic and sister to the potatoes, with sect. Etuberosum clearly monophyletic and sister to potatoes + tomatoes s.l.
2. Section. Lycopersicoides (formerly recognized as a subsection of sect. Lycopersicon) is clearly monophyletic and sister to sect. Juglandifolia + sect. Lycopersicon.
3. Section. Juglandifolia is clearly monophyletic and sister to sect. Lycopersicon.
4. Within sect. Lycopersicon, *S. pennellii* in most cases appears at the base of the trees as a polytomy with *S. habrochaites*, or sometimes forms a clade with this species. We consider this relationship unresolved, but the morphological data suggest that *S. pennellii* is sister to the rest of the tomatoes (sect. Lycopersicon); it is the only species that lacks the sterile anther appendage character shared with the outgroups.
5. *Solanum chilense*, *S. corneliomulleri*, *S. huaylasense*, *S. peruvianum*, *S. habrochaites* and *S. pennellii* appear as a polytomy at the base of the GBSSI tree, and the first four clustered with morphological characters, but there is a conflict with the AFLP and morphological data regarding the

Table 1. Species list for tomatoes and wild relatives (with equivalents in the previously recognized genus *Lycopersicon*, now part of a monophyletic *Solanum*), along with characteristic fruit color, breeding system, and distribution

Name in Peralta et al. (2007)	*Lycopersicon* equivalent	Fruit color	Breeding system[1]	Distribution and habitats
S. lycopersicoides Dunal	*L. lycopersicoides* (Dunal in DC.) A. Child ex J.M.H. Shaw	Green-yellow when maturing, black when ripe	SI, allogamous	Southern Peru to northern Chile on the western slopes of the Andes on dry rocky hillsides, 1,500–3,700 m
S. sitiens I.M. Johnst.	*L. sitiens* (I.M. Johnst.) J.M.H. Shaw	Green-yellow when maturing, brown and dry when ripe	SI, allogamous	Northern Chile, western Andean slopes on rocky hillsides and dry quebradas, 2,500–3,500 m
S. juglandifolium Dunal	*L. juglandifolium* (Dunal) J.M.H. Shaw	Green to yellow-green	SI, allogamous	Northeastern Colombia to southern Ecuador, on the edges of forest clearings, open areas and roadsides, 1,200–3,100 m
S. ochranthum Dunal	*L. ochranthum* (Dunal) J.M.H. Shaw	Green to yellow-green	SI, allogamous	Central Colombia to southern Peru, in montane forests and riparian sites, 1,400–3,660 m
S. pennellii Correll	*L. pennellii* (Correll) D'Arcy	Green	Usually SI, some SC in South of species range	Northern Peru to northern Chile, in dry rocky hillsides and sandy areas, from sea level to 2,850 m
S. habrochaites S. Knapp and D.M. Spooner	*L. hirsutum* Dunal	Green with darker green stripes	Typically SI, with SC populations in N and S of species range	Central Ecuador to Central Peru. In premontane forests to dry forests on the western slopes of the Andes, occasionally in lomas formations in northern Peru, 400–3,600 m
S. chilense (Dunal) Reiche	*L. chilense* Dunal	Green to whitish green with purple stripes	SI, allogamous	Southern Peru to northern Chile. On western slopes of the Andes, hyper-arid rocky plains, dry river beds, and coastal deserts, from sea level to 3,000 m
S. huaylasense Peralta	Part of *L. peruvianum* (L.) Miller	Typically green with dark green stripes	Typically SI, allogamous	Northern Perú (Department of Ancash). On the rocky slopes along rivers, 1,700–3,000 m
S. peruvianum L.	*L. peruvianum* (L.) Miller	Typically green to greenish-white, sometimes flushed with purple	Typically SI, allogamous	Central Peru to northern Chile. In lomas formations and occasionally in coastal deserts from sea level to 600 m, sometimes growing as a weed at field edges in coastal river valleys
S. corneliomuelleri J.F. Macbr. (one geographic race: Misti nr. Arequipa)	Part of *L. peruvianum* (L.) Miller; also known as *L. glandulosum* C.F. Müll.	Typically green with dark green or purple stripes, sometimes flushed with purple	Typically SI, allogamous	Central to southern Perú. On western slopes of the Andes, (400) 1,000–3,000 m, and on lower slopes on the edges of landslides (huaycos) towards the southern range of the species distribution

[1] SI Self-incompatible; SC Self-compatible

Table 1. (continued)

Name in Peralta et al. (2007)	*Lycopersicon* equivalent	Fruit color	Breeding system[1]	Distribution and habitats
S. arcanum Peralta (four geographic races: "humifusum", lomas, Marañon, Chotano-Yamaluc)	Part of *L. peruvianum* (L.) Miller	Typically green with dark green stripes	Typically SI, allogamous, rare populations SC, autogamous, facultative allogamous	Northern Peru. Coastal and inland Andean valleys, on dry rocky slopes, 100 to 2,500 m
S. chmielewskii (C.M. Rick, Kesicki, Fobes and M. Holle) D.M. Spooner, G.J. Anderson and R.K. Jansen	*L. chmielewskii* C.M. Rick, Kesicki, Fobes and M. Holle	Typically green with dark green stripes	SC, facultative allogamous	Southern Peru to northern Bolivia (Sorata). In high dry Andean valleys, 2,300–3,000 m
S. neorickii D.M. Spooner, G.J. Anderson and R.K. Jansen	*L. parviflorum* C.M. Rick, Kesicki, Fobes and M. Holle	Typically green with dark green stripes	SC, highly autogamous	Southern Ecuador to southern Perú. In dry Andean valleys, 1,950–3,000 m, often growing over rocky banks and roadsides. Sometimes found in sympatry with *S. chmielewskii*
S. pimpinellifolium L.	*L. pimpinellifolium* (L.) Miller	Red	SC, autogamous, facultative allogamous	Apparently native to coastal areas from Central Ecuador to southern Peru, 0–500 m, parts of Chile[2], adventive in N. America. Grows in humid places and on the edges of cultivated fields throughout its native range and has apparently escaped from cultivation in the Galápagos
S. lycopersicum L.	*L. esculentum* Miller	Red	SC, autogamous, facultative allogamous	Apparently native to Perú; the domesticated form of *S. lycopersicum* now occurs worldwide. The cherry tomato, commonly known as *S. lycopersicum* var. *cerasiforme*, has been suggested as the ancestor of cultivated tomato and can often be found growing as a weed in temperate habitats and the edges of cultivated fields, where it is not necessarily native; recent studies suggest *S. lycopersicum* var. *cerasiforme* may be a mixture of wild and cultivated tomatoes, rather than an ancestor (Nesbitt and Tanksley 2002)
S. cheesmaniae (L. Riley) Fosberg	*L. cheesmaniae* L. Riley	Yellow, orange	SC, exclusively autogamous	Endemic to the Galápagos Islands (Ecuador) from sea level to 1,300 m
S. galapagense S.C. Darwin and Peralta	Part of *L. cheesmaniae* L. Riley	Yellow, orange	SC, exclusively autogamous	Endemic to the Galápagos Islands, particularly the western and southern islands, mostly occurring on coastal lava and on volcanic slopes, sea-level to 650 m, exceptionally up to 1,500 m in Fernandina and Santiago Islands

[2] It is uncertain wether the populations are native or adventive in Chile

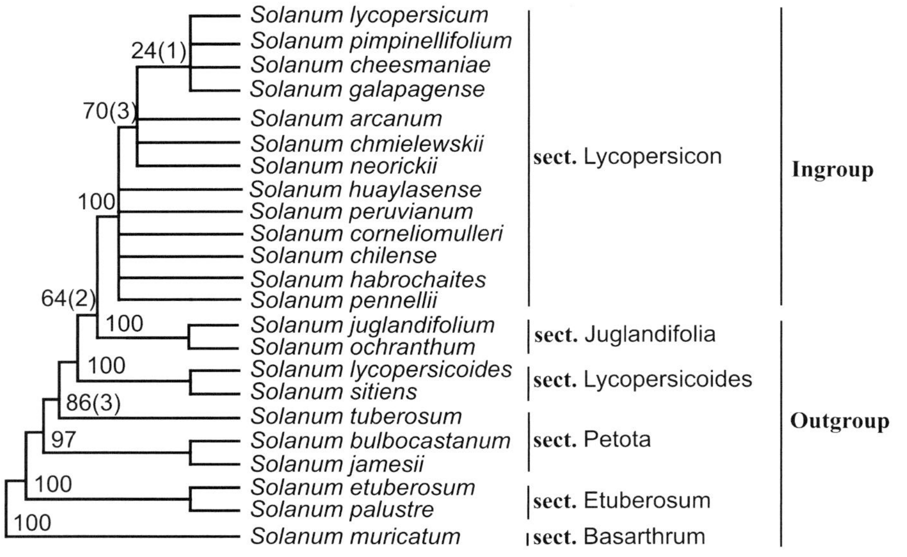

Fig. 1. Abstracted cladistic results of the 65 accessions of the 13 tomato species (ingroup sect. Lycopersicon) and ten outgroup taxa (outgroups sect. Juglandifolia, sect. Lycopersicoides, sect. Petota, sect. Etuberosum and sect. Basarthrum) examined in the phylogenetic analysis of the GBSSI gene sequences by Peralta and Spooner (2001). *Numbers* indicate bootstrap values, and decay values are indicated *between parentheses*

relationships of *S. huaylasense* (only one accession of *S. huaylasense* was examined with GBSSI). AFLPs place *S. huaylasense* with *S. arcanum*, but with bootstrap values below 50%; morphological phenetics places *S. huaylasense* with *S. chilense*.

6. The self-compatible green-fruited species *S. chmielewskii* and *S. neorickii* are closely related to *S. arcanum*, comprising a monophyletic group supported in almost all data sets.
7. The four species with brightly colored fruit (*S. cheesmaniae*, *S. galapagense*, *S. lycopersicum*, *S. pimpinellifolium*) unambiguously form a closely related monophyletic group.

1.2.3 Character Evolution and Genetic Diversity in Wild Tomatoes

Evolution of Fruit Color

Three variable characters within the tomato clade have garnered much attention in attempts to understand evolutionary dynamics of the group. The most obvious of these characters is the evolution of carotenoid pigments affecting fruit color. Fruit color of wild tomatoes varies from green to red, orange, and yellow. Various studies have demonstrated that green fruit is the ancestral character and that colored fruit arose once within the clade (e.g., Palmer and Zamir 1982; Miller and Tanksley 1990; Breto et al. 1993; Alvarez et al. 2001; Marshall et al. 2001; Peralta and Spooner 2001), and their origin is believed to be recent, about ~1 million years ago (mya) (Nesbitt and Tanksley 2002). The four most derived tomato species bear colored fruit, including the cultivated tomato (*S. lycopersicum*), *S. pimpinellifolium*, *S. cheesmaniae*, and *S. galapagense*. Within *S. cheesmaniae* and *S. galapagense*, fruit color can vary from yellow to deep orange; the other species in the group produce red fruit. Limited levels of genetic variation occur within the tomato clade containing colored fruit, due to its recent origin, which has made resolution of phylogenetic relationships between species difficult. The reason behind a transition to colored fruit in the evolution of the group is unknown, but may be related to seed dispersal.

Mating System Evolution

Two additional characters of wild tomatoes, self-incompatibility (SI) and allogamy (i.e., mating system), have played an important role in the evolution of the clade. SI is an ancestral character in the tomato clade (Rick et al. 1977; Miller and Tanksley 1990), and loss of SI has occurred independently several times. Notably, SI has been lost in some populations of *S. pennellii* (Rick and Tanksley 1981), some populations of *S. habrochaites* (Rick et al. 1979), and in the clade containing colored-fruited tomatoes and the species *S. neorickii* and *S. chmielewskii* (Rick et al. 1976; Taylor 1986). In this latter group there is support for loss of SI being of monophyletic origin (Miller and Tanksley 1990; Marshall et al. 2001).

Lower levels of genetic variation are expected in self-compatible (SC) species as compared to their SI relatives, primarily because of the reduction in N_e due to the increase in inbreeding (Charlesworth 2003). Various molecular markers, including allozymes (Breto et al. 1993), SSRs (Alvarez et al. 2001), RFLPs (Miller and Tanksley 1990), random amplified polymorphic DNAs (RAPDs) (Egashira et al. 2000), and nuclear DNA sequences (Baudry et al. 2001; Roselius et al. 2005) have consistently documented lower levels of genetic variation in SC wild tomato species.

Among SC wild tomato taxa, variation exists in the degree of allogamy, ranging from the almost entirely selfing species, *S. cheesmaniae*, to the facultative outcrossers, *S. chmielewskii* and *S. pimpinellifolium* (Taylor 1986). *S. pimpinellifolium* is notable in that variation in the degree of allogamy occurs among populations, with levels of outcrossing varying between 0 to 40% (Rick et al. 1977). Also of interest are the species *S. chmielewskii* and *S. neorickii*, which, though sympatric, differ in that *S. chmielewskii* is a facultative outcrosser, while *S. neorickii* is autogamous (Rick et al. 1976). The two species often grow in contact with each other, leading to the speculation that *S. neorickii* may have evolved from *S. chmielewskii* by acquiring autogamous reproduction (Rick et al. 1976); predictably, greater levels of allozyme diversity are found in *S. chmielewskii* (Rick 1984). Both within and between SC species, levels of outcrossing have been found to be correlated with various morphological features, including larger corolla size and greater stigma exsertion, which aid in cross-pollination (Rick et al. 1976, 1977; Rick 1984).

1.3 Domestication, Modern Breeding, and Genetic Resources

1.3.1 Domestication

Although the natural distribution of the wild species is restricted to the Andean region, the site(s) of domestication remain uncertain. Two alternative hypotheses have been proposed: southern Mexico (Jenkins 1948; Rick and Fobes 1975) versus Peru (De Candolle 1886). The Mexican domestication hypothesis states that the populations of wild cherry tomatoes (*S. lycopersicum* var. *cerasiforme*[1]) migrated from Peru into Meso-America, and became domesticated in Mexico (Jenkins 1948; Rick 1995). There is scant information on when domestication occurred, however; by the time of the Spanish conquest of Mexico in 1521, large-fruited types – a sign of human selection – were already being grown for food (Hancock 2004). Domestication and use of tomato as food probably first occurred in Central America as evidenced by cultural, linguistic, and historical records as well as genetic findings (Rick 1995; Cox 2000). With respect to the cultural evidence, decoration of textiles and ceramics with depictions of crop plants was a common practice of Pre-Columbian cultures in Peru. Important food plants commonly represented in this way include maize, potato, pepino, and others, but, conspicuously, not tomato. Putative archaeological evidence of tomato is decorated functional ceramics ("spindle whorls") produced by the native Quimbaya culture (500–1000 AD) of Colombia (McMeekin 1992). However, these ceramics could also be interpreted as representations of other Solanum flowers (possibly potato). Furthermore, there have been no discoveries of preserved tomato fossils or other archaeological remains in Peru, suggesting it is unlikely that tomato had undergone domestication there (Rick 1995). However, soft plant tissues generally do not preserve well. Studies of microfossils (starch grains, phytoliths, etc.) have revealed evidence of the early history of pepper in South America. Linguistic records also support the theory of domestication in Central rather than South America. The modern name "tomato" is derived from "tomatl", the word for this plant in the native language of the Aztecs (Gould 1983). The Aztecs mixed tomatoes, chilies and ground squash seeds into a concoction that seems likely to be the original salsa recipe (Cutler 1998). However, no writings or records from ancient Peruvian tribes ever mentioned a tomato-like fruit as being an important part of the diet or even a word meaning tomato. Supporting evidence also comes from the relationship between the introductions of the tomato to Europe relative to the conquests of Mexico and Peru. In possibly the first Italian herbal to mention tomato, Matthiolus (1544) implies a considerably earlier introduction of the tomato by his statement that it was already "eaten in Italy with oil, salt, and pepper". The earliness of this date favors a Mexican origin considering the capture of Mex-

[1] The correct name for the cherry tomato under the International Code of Botanical Nomenclature is *S. lycopersicum* var. *leptophyllum*. The more widely applied name, *S. lycopersicum* var. *cerasiforme*, is used in this chapter to avoid confusion.

ico City in 1521 vs. the conquest of Peru in 1531 (Rick 1995).

In addition to these cultural considerations, genetic evidence also supports Central American domestication. Data from allozymes demonstrated that older European cultivars, i.e., descendents of the tomatoes introduced by the Spanish explorers, are extremely homogeneous at most loci and are closely related to cultivars and wild cherry tomatoes from southern Mexico and Central America, but differ from those of South America (Rick and Holle 1990). In addition, genetic variability in wild cherry tomato from the Andean region is greater than in accessions collected outside of this region (Rick and Holle 1990). This is consistent with the model of migration beyond the center of origin, which would entail founder effects, selection, and the consequent loss of variation associated with those processes. In many crops, a reduction in genetic diversity has been one of the major features accompanying domestication, and this seems to be true for tomato.

Arguments supporting the Peruvian domestication hypothesis were presented by Moore (1935), Müller (1940a, b), and Luckwill (1943a, b). These arguments trace back to botanical (Bauhin 1623; Ruiz and Pavón 1797), historical (Hernández 1651), and linguistic (Roxburgh 1832) evidence summarized and interpreted by De Candolle (1886). De Candolle (1886) found that there were no definitive original records of tomato outside of the Americas before its European discovery there. Initial names for tomato cited by early 16th century botanists, "mala peruviana" and "pomi del Peru" (Bauhin 1623), suggested Peruvian origin and subsequent dispersal of tomato from Peru to Europe. De Candolle (1886) also argued that (i) its evolutionary origin was from the wild cherry tomato, that by the mid- to late 19th century was spread from coastal Peru, to Mexico, and to southwestern USA (California), and also reported in Asia, (ii) the crop had only recently been domesticated before the discovery of the Americas, and (iii) the distribution of cultivated tomato and its progenitor outside of Peru originated by garden escapes. Contrary to Jenkins' (1948) statements that there are no indigenous Peruvian names for tomato, Horkheimer (1973) documented a Quechua name for tomato (pirca), and Yakovleff and Herrera (1935) cited another Quechua name (pescco) possibly referring to the cherry tomato. Peralta and Spooner (2007) reviewed evidence for a Mexican versus Peruvian origin of domesticated tomato, as well as methods applied for inferring geographical location of crop origins. They concluded that the original site of tomato domestication remains uncertain, and may be unfeasible to resolve.

De Candolle's (1886) assertion that wild cherry tomato is the immediate progenitor of cultivated tomato also continues to be actively debated (Nesbitt and Tanksley 2002) (Table 1). This hypothesis is supported by *S. lycopersicum* var. *cerasiforme's* exceedingly close genetic resemblance to the cultivated forms of tomato, and its common occurrence in Central America. In terms of morphology, the cherry and modern domesticated types differ primarily by the larger fruit size of the latter. Like cultivated tomato, the wild cherry tomato is naturally self-pollinated. In contrast, many of the other wild species are either obligate outcrossers (SI) or have a mixed mating system (see Sect. 1.2.3). Self-pollination in the cherry tomato is promoted by a stigma and style that protrudes only a short distance or not at all, beyond the end of the anther cone (Müller 1940a; Luckwill 1943b; Rick 1995; Cox 2000). Aside from increased fruit size, the gradual shortening of style length and other factors favoring self-pollination are believed to be a major feature of domestication. Besides the uniformity associated with inbreeding, a shorter style and recessed stigma position tend to favor fruit-setting ability, hence yield (Rick 1995; Scott and Angell 1998). Early North American cultivars tended to have slightly exserted stigmas. This resulted in diminished fruit setting, especially in the absence of appropriate pollinating insects or under conditions of excessive temperature (which tends to reduce pollen fertility). Selection for less exserted stigmas resulted in stigmas approximately flush with the mouth of the anther tube, as seen in many early European cultivars. Full enclosure of the pistil by the anthers was inadvertently selected during breeding of mechanically harvested processing tomatoes for California conditions, apparently because it ensured a reliable, concentrated fruit set. This improvement virtually guarantees self-pollination and has been bred into many modern cultivars (Rick 1995).

1.3.2 Dispersal

The Spanish distributed the most desirable vegetables and fruit throughout their empire. From Central America and Mexico, the Spanish introduced the tomato into the Caribbean and the Philippines. From the Philippines, use of tomato spread to Southeast Asia and ultimately the rest of Asia (Smith 1994). Through the Spaniards the tomato was also taken back to Spain and disseminated throughout Europe. The earliest mention of tomato in European literature appeared in a document written by an Italian herbalist, Petrus Andreas Matthiolus (1544). He described tomatoes as pomi d'oro (golden apple), indicating the first tomatoes used as food by Europeans were yellow fruited. Red tomatoes were introduced into Italy by two Catholic priests several years later, and were documented in 1554 by Matthiolus (McCue 1952).

In Spain the tomato was called "pome dei Moro" (Moor's apple) (Cutler 1998). The tomato then became widespread in Spain, Italy, and France in the following decades. The French referred to it affectionately as "pomme d'amour" (love apple), perhaps because of its suspected aphrodisiac properties (Gould 1983). Despite its use as a food source in southern Europe, especially Italy, in the northern European countries tomato was regarded as a garden curiosity for over a century (Rick 1995). This was mainly due to fears of toxicity, a notion based upon the presence of poisonous glycoalkaloids in the foliage and fruit of other familiar members of the nightshade family such as henbane, mandrake and deadly nightshade, to which tomato bore some morphological resemblance (Cox 2000). English authors spoke of the tomato as an ornamental plant as early as 1578 (Gould 1983). One English gardener wrote in 1596 that "these love apples are eaten abroad", but he considered the entire plant to be "of ranke and stinking savour" (Cox 2000). By 1623 four types of tomatoes were known: red, yellow, orange and golden (Gould 1983). The first cookbook containing tomato recipes was published in Naples in 1692, but suspicion of tomatoes persisted into the 19th century in both England and the USA (Cutler 1998). The reputation of the tomato was somewhat improved by English authors in the 1750's when *Lycopersicon esculentum*, which means "edible wolf peach", was given as its botanical name by Miller. In 1752 tomatoes were used in England for flavoring soups. In 1758 tomato recipes appeared in a popular British cookbook, "The Art of Cookery" by Hannah Glass (Cutler 1998). The cultivation of tomato for marketing dates from about 1800 in Europe but its true value was not realized until 1822 when details for its cultivation were described (Gould 1983).

Tomato plants were brought from Britain to North America as horticultural ornamentals by the early colonists either when they emigrated from Europe or after the Declaration of Independence was signed. First mention of tomato cultivation in the USA was made by Thomas Jefferson, who grew tomato at Monticello in 1781 (Gould 1983). Later it was introduced to Philadelphia in 1798, Massachusetts in 1802, and other parts of the country. However, tomato was still considered by most people at that time to be of questionable safety as a food, and was grown mainly for ornamental or medicinal purposes. Tomato was first reported as food in the USA in New Orleans in 1812, probably due to the strong French influence in that region. Another two decades passed before tomatoes were widely cultivated as an edible vegetable during 1830 to 1840. Tomatoes were included in American cookbooks, such as The Cook's Own Book (Lee 1832), and in garden books, such as the 1843 Shaker Gardener's Manual (Harrison 2004) and The Gardener's Text-Book (Schenk 1851).

The rising popularity of the tomato as a food source encouraged the production of new cultivars. Burr listed 23 tomato cultivars in 1863, among which was "Trophy", the first large, fairly early, red, smooth, apple-shaped variety (Gould 1983). The tomato's popularity was further elevated by debate about its status as a vegetable or a fruit. The issue went all the way to the USA Supreme Court which ruled in 1887 to classify tomato as a vegetable and subject to import quotas and taxation (Cox 2000). Large scale breeding for economic traits took place in both Europe and the USA, beginning in the early 1870s. By the late 19th century tomato had firmly implanted itself in western culture. Production began to soar in the early 1920s with the advent of mass canning. Disease tolerant cultivars became increasingly prevalent in the USA but the development of truly resistant types would depend on resistance genes found in the closely related wild species of tomato, beginning in the late 1930s to early 1940s (Langford 1937; Stevens and Rick 1986). Wild relatives continue to be the main source of new resistance and other novel traits for tomato breeders today.

1.3.3 Economic Importance

Tomato is a major vegetable crop, has gained in popularity over the last century, and is now grown in almost every country of the world (Robertson and

Labate 2007). Worldwide production of fresh market and processing tomatoes has steadily increased during the last decade and reached an annual production of 124.4 million tons in 2004, with an average yield of 27.5 ton/hectare (Table 2; FAOSTAT 2004). The tomato growing area has increased by 38% and production by 42% worldwide in the past 10 years. Most of the increase came from China, where the growing area nearly tripled from 0.47 million ha in 1995 to 1.26 million ha in 2004, and production more than doubled from 13.2 million ton to 30.1 million ton. More than half of the total tomato production was from the six top producing countries: China, USA, Turkey, India, Egypt, and Italy (USDA-FAS 2005). Mexico ranks as number one in tomato exports, followed by Turkey, USA, European Union, Canada, and China, while USA ranks first in tomato imports, followed by Russia, European Union, and Canada.

The USA is one of the world's leading producers of tomatoes, second only to China. Annual per capita use of fresh-market tomatoes increased 15% between the early 1990s and the early 2000s to nearly 8 kg per person, while use in processed products declined 9% to 31 kg (fresh-weight basis) (USDA-ERS 2006). Tomato harvested area for fresh market tomatoes in the USA has declined from an average of ~62,000 ha in the 1960s to an average of ~54,000 ha in the 1970s, and remained at an average of ~52,000 in the 1980s, 1990s, and 2000s. However, the average production and value have increased steadily during the past five decades (Table 3). Similarly, although the harvested area for processing tomatoes has not changed greatly in the last half century, the average production and value have gradually increased, with the crop value remaining relatively stable in the 1990s and 2000s. Mexico and Canada are important suppliers of fresh market tomatoes to the USA, and Canada is the leading USA export market for fresh and processed tomatoes. Overall, the USA has maintained a trade deficit in fresh market production, but a surplus in processed tomato (USDA-FAS 2005).

Florida and California are the two leading states for fresh tomato production. The average production and crop value from the two states in the 2000s account for ~70% of that for the entire USA (Table 3). California is the top producer of processing tomatoes and has produced nearly 95% of the total for the USA. The second leading producer of processing tomato, Indiana, accounts for only 3% of the production value.

1.3.4 Germplasm Collections

A wide range of germplasm resources are available for breeding and research on tomato (see recent reviews

Table 2. World production of tomatoes in 1995 and 2004 (FAOSTAT 2004)

Location	1995 Area (ha × 10^3)	Yield (ton ha^{-1})	Production (ton × 10^3)	2004 Area (ha × 10^3)	Yield (ton ha^{-1})	Production (ton × 10^3)
World	3,286	26.7	87,722	4,530	27.5	124,422
Africa	505	20	10,090	703	20	12,428
North and Central America	359	42.7	15,341	333	52.8	15,838
South America	156	37	5,769	147	43.4	6,481
Asia	1,538	23.9	36,778	2,655	23.8	53,290
Europe	716	26.9	19,264	682	29.8	19,969
Oceania	12	39.3	481	10	50	492
Leading Countries						
China	474	27.8	13,172	1,255	24	30,144
USA	192	61.3	11,784	175	73.7	12,867
Turkey	175	41.4	7,250	255	37	9,440
India	350	15	5,260	540	14.1	7,600
Egypt	149	33.7	5,034	195	39.2	7,640
Italy	114	45.1	5,182	145	53	7,682

by Chetelat and Ji 2007; Ji and Scott 2007; Robertson and Labate 2007). Tomato germplasm collections in the USA are currently maintained at two major genebanks, the C.M. Rick Tomato Genetics Resources Center (TGRC), at the University of California – Davis, and the United States Department of Agriculture, Agricultural Research Service (USDA-ARS) Plant Genetic Resources Unit (PGRU) in Geneva, NY. The former consist mainly of wild species, mutants, prebreds, and other types of genetic stocks (Table 4), the latter emphasizes open-pollinated cultivars and some wild relatives. In addition, the Seed Savers Exchange in Decorah, Iowa, maintains a large set of heirloom and antique varieties through its network of growers. The major international collection of tomato germplasm is preserved at the Asian Vegetable Research and Development Center (AVRDC), now referred to as The World Vegetable Center, located at Taiwan. Other significant collections are housed in The Netherlands, Germany, France, and Russia. Worldwide, over 75,000 accessions of *Solanum* sect. *Lycopersicon* germplasm are maintained in more than 120 countries in a number of national institutions; for detail see review by Robertson and Labate (2007).

Germplasm collections in these genebanks have been intensively utilized as genetic resources for many types of basic and applied research involving tomato (Rick and Chetelat 1995; Ji and Scott 2007). Uses include searching the collections for traits of economic interest, such as disease and insect resistances, stress tolerances, and improved horticultural and fruit characteristics. Research topics of more fundamental nature include studies in physiology and development (e.g., leaf development, fruit ripening and self-incompatibility), genetics (e.g., wide hybridization, linkage and QTL mapping), and genomics (e.g., map-based cloning and comparative sequence analysis).

1.3.5 Germplasm Stocks

Genetic stocks such as those maintained by the TGRC are categorized into a number of groups based on taxonomy, monogenic characters, chromosomal modifications, ploidy, and other criteria (Table 4; Chetelat 2004, 2005, 2006).

Wild Species

The TGRC collections include over 1,000 accessions of wild species representing 9 to 13 species – depending on the taxonomic system used – in *Solanum* sect. *Lycopersicon*, and four related *Solanum* species (Chetelat 2004). Detailed "passport" information is available for each accession at the TGRC website (http://tgrc.ucdavis.edu). In addition, 458 and 659 accessions of wild species, the majority (~71%) of which are *S. peruvianum* and *S. pimpinellifolium*, are preserved at USDA-PGRU (http://www.ars.usda.gov) and AVRDC (http://www.avrdc.org), respectively. Together, approximately 2,200 accessions of

Table 3. Average fresh market and processing tomato harvested area, production, and crop value in the USA from 1960s to 2000s, and in the leading states in 2000s[a]

Years	Fresh market Area (ha × 10^3)	Production (ton × 10^3)	Value ($ × 10^6)	Processed Area (ha × 10^3)	Production (ton × 10^3)	Value ($ × 10^6)
1960s	62.2	914	183	119.7	4,401	164
1970s	53.6	926	359	123.4	5,967	359
1980s	51.8	1,356	758	112	6,624	479
1990s	52.2	1,591	1,054	131.5	9,628	679
2000s[b]	51.8	1,718	1,333	119.6	9,869	682
Florida	17.2	674	580			
California	16.4	533	374	112.4	9,333	629
% CA				93.7	94.5	92.2
% (FL + CA)	64.7	70.2	70.1			

[a] Summarized from data at USDA Economic Research Service (USDA-ERS 2003, 2006)

[b] Includes years 2000–2007 with forecasted data for 2006–2007

Table 4. Genetic stocks maintained by the C. M. Rick Tomato Genetics Resource Center (University of California – Davis; http://tgrc.ucdavis.edu) and the USDA-ARS Plant Genetic Resources Unit (Geneva, NY; http://www.ars.usda.gov)

Category	TGRC	USDA
WILD SPECIES STOCKS	1,015	458
S. pimpinellifolium (*L. pimpinellifolium*)	307	230
S. cheesmaniae (*L. cheesmanii*)	44	7
S. galapagense (*L. cheesmanii* f. *minor*)	29	5
S. chilense (*L. chilense*)	112	1
S. chmielewskii (*L. chmielewskii*)	37	1
S. habrochaites (*L. hirsutum*)	118	60
S. neorickii (*L. parviflorum*)	59	6
S. pennellii (*L. pennellii*)	68	10
S. peruvianum (*L. peruvianum*)	129	124
S. huaylasense (*L. peruvianum*)	8	
S. corneliomulleri (*L. peruvianum* f. *glandulosum*)	13	12
S. arcanum (*L. peruvianum* var. *hunifusum*)	38	2
S. lycopersicoides	23	–
S. sitiens	13	–
S. juglandifolium	8	–
S. ochranthum	9	–
S. lycopersicum (*L. esculentum*)		
CULTIVARS AND LANDRACES	424	4,914
Modern and Vintage Cultivars	198	na[a]
Latin American Cultivars	226	na
Wild Cherry Tomato (*L. esculentum* var. *cerasiforme*)	271	267
S. lycopersicum Hybrids	–	158
MONOGENIC STOCKS	1,017	–
PREBRED STOCKS	376	–
Introgression Lines	254	
Backcross Recombinant Inbreds of *S. pimpinellifolium*	99	
Alien Substitution Lines	11	
Monosomic Addition Lines	10	
Interspecific Hybrids	2	
STRESS TOLERANT STOCKS	52+	–
Drought Tolerance	7+	
Flooding Tolerance	3+	
High Temperature Tolerance	3	
Chilling Tolerance	10	
Aluminum Tolerance	1	
Salinity and/or Alkalinity Tolerance	26	
Arthropod Resistance	2+	
CYTOGENETIC STOCKS	89	–
Translocations	37	
Trisomics	35	
Autotetraploids	17	

[a] Not available

Table 4. (continued)

Category	TGRC	USDA
CYTOPLASMIC VARIANTS	3	–
GENETIC MARKER COMBINATIONS	483	–
Chromosome Marker Stocks	182	
Linkage Screening Testers	13	
Miscellaneous Marker Combinations	288	
PROVISIONAL MUTANTS	107	–
TOTAL NUMBER OF ACCESSIONS	3,837	5,797

wild species are maintained in these three gene banks.

Mutants

The TGRC collection includes more than 1,000 monogenic mutants at over 600 loci, including spontaneous and induced mutations affecting many aspects of plant development and morphology, disease resistance genes, protein marker stocks, and other traits of economic importance (Chetelat 2005). Genetic data on individual stocks, including phenotypes, images, chromosome locations, etc. are available at the TGRC website. An additional series of provisional (i.e., less well characterized) mutants are also available. The Hebrew University of Jerusalem offers an isogenic mutant library in the genetic background of cv. M82 (http://zamir.sgn.cornell.edu/mutants/index.html). A total of 13,000 M_2 families, generated by ethylmethane sulfonate (EMS) and fast-neutron mutagenesis, were phenotypically analyzed and catalogued into at least 3,417 mutations (Menda et al. 2004). This series of mutations includes many previously described mutant phenotypes as well as many novel mutants, and multiple alleles per locus. Together these mutant collections provide important tools for analyses of gene function.

Cultivars and Landraces

The TGRC maintains a collection of 198 modern and vintage cultivars, inbreds, and breeding lines for various purposes, kept mainly to provide isogenic (or nearly isogenic) stocks for specific mutants, standards for genetic comparison, sources of disease resistances, or other purposes. Another 226 cultivated tomato accessions are Latin American landraces assembled from various sources, mainly collected from local markets in Latin American. Much larger collections of cultivars are preserved at USDA-PGRU and AVRDC, with 4,914 and 5,311 accessions, respectively (Table 4; Robertson and Labate 2007).

Prebred Stocks

Collections of prebred stocks are maintained by the TGRC and are popular for mapping and breeding purposes. These include introgression lines (ILs), alien substitution lines, monosomic alien addition lines, and backcross (BC) recombinant inbreds. The introgression libraries were derived from *S. pennellii* (Eshed and Zamir 1994a; Liu and Zamir 1999), *S. habrochaites* (Monforte and Tanksley 2000a), or *S. lycopersicoides* (Canady et al. 2005). An entire genome of *S. pennellii* is covered by 50 overlapping introgressions; 26 additional sublines provide increased mapping resolution in some regions (see Sect. 1.8.3). Each line is homozygous for a single introgression from *S. pennellii* (LA0716) in the background of *S. lycopersicum* cv. M82. The *S. habrochaites* and *S. lycopersicoides* prebreds have a similar genetic makeup but with less complete genome coverage and in some cases more than one introgressed region per line. Because of sterility factors, some of the *S. lycopersicoides* lines are maintained via heterozygotes. Introgression line libraries are also being developed for other wild species, such as *S. chmielewskii* (http://www.keygene.com/services/plants/services_plants_line.htm).

The second type of prebred stocks maintained at the TGRC is a group of BC recombinant inbred lines (RILs) originating from the cross *S. lycopersicum* × *S. pimpinellifolium* (Doganlar et al. 2002c). Two generations of backcross followed by at least 6 generations of inbreeding via single-seed descent resulted in a high level of homozygosity in these RILs, whose residual heterozygosity is averaged at ~3%. The population has been genotyped at 127 marker loci, and the corresponding maps and relevant data files are available from the Solanaceae Genome Network (SGN) (http://www.sgn.cornell.edu). A set of 99 RILs has been selected for optimum mapping resolution. This provides a permanent, high-resolution mapping population. Numerous other types of unbalanced populations have also been developed to improve mapping resolution; see review by Ji and Scott (2007) for more details.

A few alien substitution and alien addition lines, each containing an intact alien chromosome from one of the wild relatives in the background of cultivated tomato, are also available. The substitutions represent seven of the 12 *S. pennellii* (LA0716) chromosomes (Rick 1969, 1972; Weide et al. 1993), and four *S. lycopersicoides* (LA2951) chromosomes (Chetelat and Meglic 2000; Ji and Chetelat 2003). The monosomic alien addition lines, of which there are ten, each contains one extra chromosome (i.e., $2n + 1$) from *S. lycopersicoides* LA1964 added to the tomato genome (Chetelat et al. 1998).

Stress Tolerant Stocks

The TGRC maintains a group of selected wild and cultivated accessions with known or inferred tolerances to various environmental (abiotic and biotic) stresses. Tolerances include drought, flooding, high temperature, chilling injury, aluminum toxicity, salinity and/or alkalinity, and damage from arthropods. These and other stress tolerant stocks, including the wild species, provide useful starting material for breeding, genetic mapping, and other uses.

Cytogenetic Stocks

The tomato genome is well represented by various types of cytogenetic stocks. The TGRC maintains autotetraploids, sesquidiploids, translocations, and several types of trisomics. These stocks contain cytologically detectable changes in chromosome numbers or chromosome structure. The autotetraploids and sesquidiploids are euploids (i.e., have changes in whole sets of chromosomes), while the trisomics are aneuploids, having a single extra chromosome which may be rearranged in various ways. The autotetraploids arose from spontaneous or induced chromosome doubling, the sesquidiploids (i.e., interspecific triploids) from artificial hybridization between tomato and related wild species (Rick et al. 1986). The trisomics were isolated as spontaneous unfruitful plants in fields of tomato, or were selected in progeny of autotriploids or following mutagenic treatments. A complete set of primary trisomics, containing one extra, nonrearranged tomato chromosome, is available. In addition, there are partial sets of secondary trisomics, which contain an extra isochromosome (i.e., two identical chromosome arms on either side of a centromere), tertiary trisomics, which contain a translocated chromosome (one arm from each of two different chromosomes), telo-trisomics, with an extra chromosome arm, and compensating trisomics, with the loss of a normal chromosome compensated by the presence of two arms in new, translocated associations (Khush and Rick 1968a). All types of trisomics are maintained via seed and have been utilized for gene mapping and other cytogenetic studies. For example, the assignment of certain linkage groups of morphological markers to their respective chromosomes was based on distortion in segregation ratio in the F_2 populations of primary trisomics (Lesley 1932; Rick and Barton 1954). Other trisomic types can be used in finding information regarding arm location, position of centromeres and orientation of linkage markers (Frary et al. 1996). In addition, trisomics have been used for chromosomal assignment of molecular markers by dosage analysis (Fobes 1980; Young et al. 1987), and for identification of individual chromosomes in synaptonemal complex spreads (Sherman and Stack 1992).

Changes in chromosome structure constitute another major type of chromosomal variation in tomatoes. Various kinds of structural alterations, including duplications, deficiencies, inversions, and translocations, have been identified. Only the reciprocal translocations have been maintained, probably because they are transmissible to the next generation (unlike deficiencies), fertile, and easily applied to mapping studies. They have been used for assignment of gene loci to their respective chromosomal arms and as a source of tertiary trisomics (Khush and Rick 1967a). The TGRC collections maintain 37 such lines, eight of which consist of a tester set involving all 12 tomato chromosomes (Gill 1983).

Cytoplasmic Variants

Three lines of cytoplasmically inherited chlorotic variants are maintained at TGRC. These lines were induced by mutagens and are inherited in a strictly maternal fashion. They are not transmitted by pollen, but when pollinated with any male parent the progeny are 100% variant.

Multiple Marker Combinations

There are a large number of genetic stocks containing combinations of morphological markers. These include chromosome marker stocks in which a series of marker genes has been assembled for a single chromosome, linkage-screening testers, each combining two pairs of strategically situated markers on two different chromosomes, and miscellaneous marker combinations in which various mutant genes have been combined for various purposes.

1.4 Marker Development: Leveraging Public Databases and Software Tools

Despite extensive availability of tomato molecular markers, many markers were developed based on polymorphism in wide crosses and are not informative when used within closely related germplasm. For example, whereas a saturated linkage map of tomato is available from a wide cross, *S. lycopersicum* × *S. pennelli* (Fulton et al. 2002b), populations derived from crosses of closely related species and crosses within *S. lycopersicum* require substantial additional effort to construct molecular linkage maps and to associate a marker with a phenotype. For this reason the development of markers continues to be important in tomato despite the availability of high-density linkage maps and the emerging genome sequence (Mueller et al. 2005b). In this section methods to exploit the available genetic and genomic resources of tomato to improve mapping and marker development for populations derived from closely related parents are presented. How these available resources were used to develop a polymerase chain reaction (PCR) based linkage map of an F_2 population derived from a cross between *S. pimpinellifolium* LA1589 and *S. lycopersicum* cv Rio Grande is then highlighted.

1.4.1 Markers

RFLPs were used to generate the first high-density linkage maps in tomato (Bernatzky and Tanksley 1986; Tanksley et al. 1992) (see Sect. 1.5) but are not amenable to high-throughput and automated genotyping analyses. AFLP markers alleviated the throughput limitations of RFLPs, but are largely anonymous, population-specific, and dominant (Haanstra et al. 1999). Sequence-dependent PCR based markers are preferred over RFLP and AFLP, and are now exploited in many plant species. These markers display important characteristics including codominance of alleles and robustness of the assay, features that are particularly attractive for high throughput MAS as well as high-resolution fine-mapping. SSRs, insertion and deletion polymorphisms (indels), and nucleotide substitutions provide the basis for sequence-dependent PCR-based markers. SSRs consist of two, three or four base-pair units that are repeated in a sequence of interest. A polymorphism arises when a difference in the number of repeats is found between alleles. Insertion or deletions that are not SSR are designated indel. Single base changes or single base indels define the single nucleotide polymorphisms (SNPs) class. These are valuable DNA markers because of their high frequency, widespread distribution in the genome, and their suitability for high-throughput, automated genotyping (Shi 2001). Furthermore, SNPs located in coding and regulatory regions may be the direct cause of phenotypic variation. Therefore, SNP markers are very promising in association mapping studies which aim to link phenotype to genotype without the development of structured mapping populations.

1.4.2 SSR, Indel and SNP Discovery

Ab initio and in silico Discovery of SSR, Indel and SNP Markers Without a priori Knowledge of Map Position

The identification of tomato SSRs and SNPs was greatly improved once sufficient sequence database resources became available. In particular, large collections of expressed sequence tags (ESTs) hosted in various databases greatly facilitated molecular marker development (Table 5). In the first attempt at in silico marker discovery for tomato, the search for SSRs by computational screening of genomic sequences from EMBL and GenBank databases resulted in the identification and confirmation of 36 polymorphic SSR markers between tomato cultivars and species (Smulders et al. 1997). Subsequently using different search parameters, putative SSRs were detected in the ESTs and genomic sequences (He et al. 2002; Frary et al. 2005). Tomato EST databases contained over 600 unique sequences with SSR motifs, of which a set of 109 were confirmed and mapped in a reference population derived from a *S. lycopersicum* × *S. pennellii* cross (Frary et al. 2005). These confirmed SSRs were used to detect size polymorphisms within and between species (He et al. 2002; Frary et al. 2005; Yang et al. 2005b) (Fig. 2). The same EST databases were also mined for SNPs. Since tomato EST collections are derived from more than one cultivar, this approach has been practical (Yang et al. 2004; Labate and Baldo 2005). Putative SNPs were verified via restriction enzyme digests or resequencing and mapped in different tomato populations and cultivars (Yang et al. 2004) (Fig. 2). ESTs from other varieties and putative SNP information are available at other websites (Tsugane et al. 2005), in addition to bacterial artificial chromosome (BAC) and BAC-end sequences which can be mined for additional putative SNPs.

Ab initio Discovery of Indel and SNP Markers Without a priori Knowledge of Map Position

One disadvantage to developing SSR or SNP markers in transcribed sequences is their low polymorphism between closely related tomato species and within cultivated germplasm. The sequence of coding regions is more conserved than non-coding regions. Therefore, the design of primers corresponding to the 5' or 3' UTR provides one approach to increase the chance for polymorphic SSR, indel or SNP detection (Cato et al. 2001). Another approach for detection of polymorphisms is to mine intron sequences. This approach is feasible because intron position is conserved between diverged species, especially for highly conserved genic sequences. Using conserved orthologous sequences (COS) derived for tomato (Fulton et al. 2002b; Wu et al. 2006) the position of introns can sometimes be predicted by comparison to the *Arabidopsis* genome (Timms et al. 2006; Van Deynze et al. 2006; Wu et al. 2006). Tools to find the position of introns are available from several genome projects including the Compositae Project Python blast viewer and the SGN Intron Finder (Table 5). For tomato, the intron mining approach resulted in the identification of 593 SNPs in 161 loci and 294 indels in 122 loci within breeding lines, and 1,113 SNPs in 274 loci and 206 indels in 96 loci between cultivated varieties and *S. pimpinellifolium* (Van Deynze, van der Knaap and Francis, unpublished).

A promising method to identify SNPs is through oligo-based microarray hybridization of fluorescently labeled target sequences. Both cDNA and genomic DNA can be used as labeled targets during hybridization (Cui et al. 2005). For tomato, cDNA made from RNA extracted from various tissues of *S. lycopersicum* cv. Ohio7814 and *S. pimpinellifolium* LA1589 was hybridized to a NimbleGen array consisting of 15,270 tomato unigenes (Sim et al. 2007). In the NimbleGen design, each gene was represented by 12 perfect match and 12 mismatch oligonucleotide probes of 24 bp in length. Probes identified as outliers are potential single feature polymorphisms (SFPs). For the Ohio7814 and LA1589 hybridizations, 1,296 putative SFPs were identified. The verification of these SFPs was conducted by sequencing alleles for 216 putative SFPs, and 52% of the features identified as outliers were confirmed as polymorphic. SNPs accounted for 86% of the confirmed polymorphisms while indels accounted for 14% (Francis et al. unpublished).

Development of Markers with a priori Knowledge of Map Position

Once a PCR-based molecular linkage map is constructed, the need arises to develop markers in regions lacking sufficient coverage. Once these regions are identified, information from high-density molecular maps can be leveraged to develop markers that are targeted to specific regions of interest. The SGN database offers information about various tomato linkage maps covering the entire genome with more than 2,400 markers including RFLP, SSR and COS (Table 5). The sequences of RFLP markers can be retrieved, primers designed, alleles amplified, and sequencing performed to identify the nature of available polymorphism. In particular, RFLP markers with the TG prefix correspond to single or low copy genomic fragments which are particularly suitable for marker development of closely related species (e.g., Bai et al. 2004a; Frary et al. 2005). Primer information is available for the SSR and COS markers hosted at SGN, thus these markers can be tested directly for polymorphisms between the parents of a mapping population. Furthermore, BAC-clones from a *S. lycopersicum* cv. Heinz1706 genomic library are anchored to the genetic map and BAC-end or entire BAC-sequences are available at SGN. This sequence information allows users to design PCR primers and sequence the amplified products derived from parents of interest in order to develop markers targeted to specific regions. An added feature of the BAC-end sequence resources is that these segments are often derived from intergenic regions offering higher chances for polymorphism detection. Although the approach of marker development for targeted regions of the genome works well for interspecific crosses, this approach is only marginally

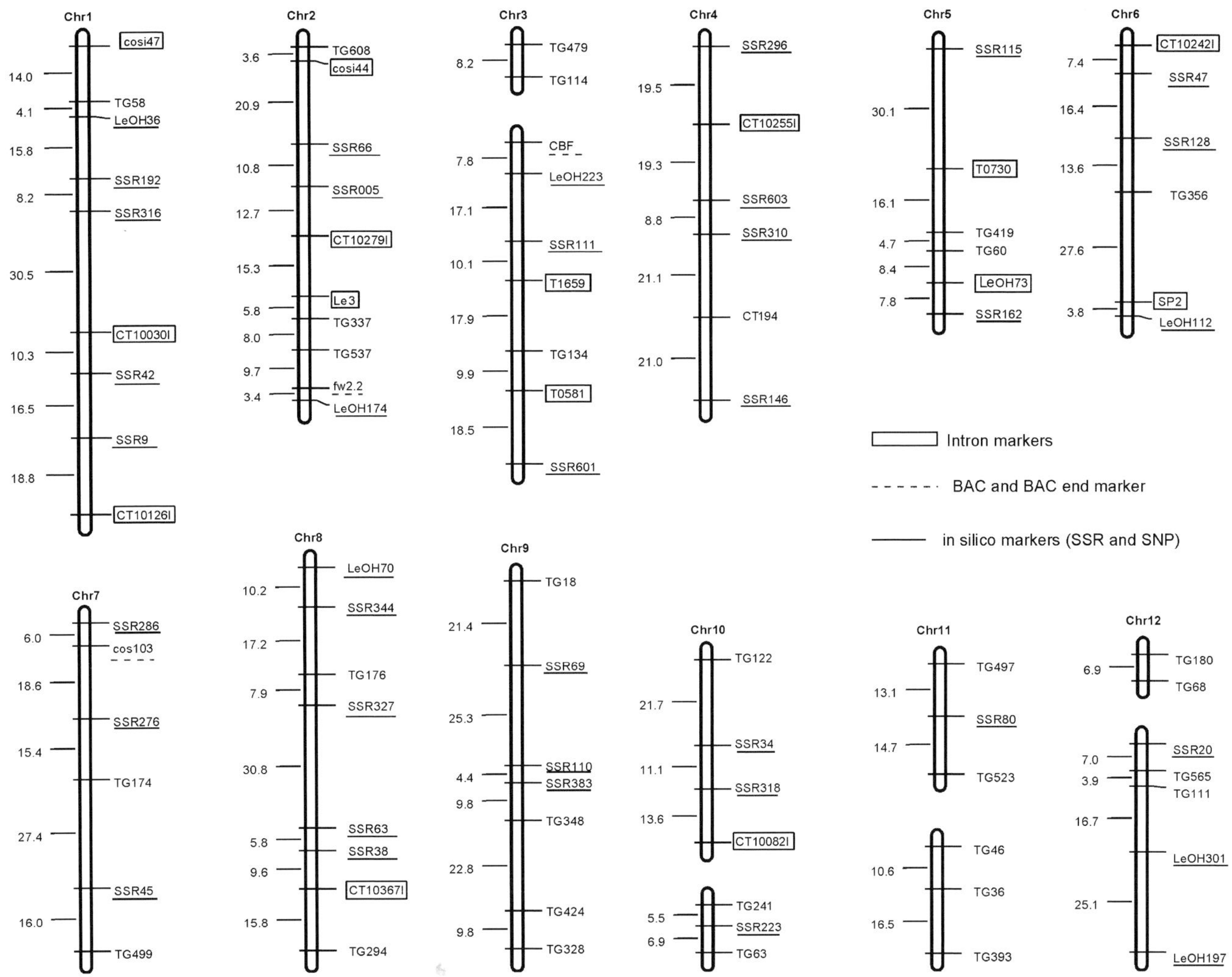

Fig. 2. PCR-based molecular linkage map of *S. pimpinellifolium* LA1589 × *S. lycopersicum* cv. Rio Grande F_2 population (Gonzalo and van der Knaap, unpublished). The map was constructed using Mapmaker v3.0 and the Kosambi mapping function. The markers are located on the right whereas map distances in cM are indicated on the left of the linkage groups. Marker information (primer sequences and polymorphisms) are available at SGN and www.tomatomap.net (Table 5). Markers obtained by in silico analysis are *underlined with a solid line*. Markers developed from BAC-ends or the entire BACs are *underlined with a dash line*. Markers obtained by mining intron sequences are *boxed*. Unlabeled markers are PCR-converted RFLP markers

Table 5. Web sites used for marker development in tomato (*S. lycopersicum*)

URL	Purpose	Features	Reference
http://www.sgn.cornell.edu	Solanaceae resource	Physical and genetic maps; genome and marker sequence information; bioinformatic tools	Mueller et al. 2005a
http://www.kazusa.or.jp/jsol/microtom/index.html	Tomato resource	Tomato full-length cDNA; gene expression; polymorphic markers	Tsugane et al. 2005
http://www.tomatomap.net	Tomato information source	Molecular map; sequence information; polymorphism within cultivated germplasm; genetic maps	Van Deynze, van der Knaap and Francis, unpublished
http://www.ncbi.nlm.nih.gov	Molecular biology information	Sequence information; publications; bioinformatics tools	Wheeler et al. 2006
http://www.tigr.org	Analysis of genome sequences	Genome, SSR and EST databases; bioinformatic tools	
http://www.dpw.wau.nl/pv/CAPStomato/	Tomato marker information	Tomato CAPS markers	Bai et al. 2004a
http://www.potatogenome.net/	Potato genome sequence portal	Potato genome sequence	
http://www.arabidopsis.org	*Arabidopsis* resource	Annotated genome sequence; gene expression; seed stocks; biochemical pathways; bioinformatics tools	Rhee et al. 2003
http://cgpdb.ucdavis.edu/sitemap.html	Composite database, COS marker discovery	COS markers for tomato; bioinformatics tools	
http://frodo.wi.mit.edu	Primer design	Primer3 program	Rozen and Skaletsky 2000
http://helix.wustl.edu/dcaps/dcaps	Primer design (dCAPS, CAPS)	dCAPS Finder2.0	Neff et al. 1998

effective within the cultivated germplasm pool due to the low rate of polymorphism (Yang and Francis, unpublished).

1.4.3 Application and Mapping of Markers

Multiple platforms for detecting polymorphisms exist, and the appropriate system depends on the goals of the research and the type of markers to be assayed. The appropriate genotyping platform is largely determined by the type of marker, desired throughput, cost efficiency, and available equipment. The following details are not intended to be comprehensive, but simply reflect some widely used genotyping platforms.

SSRs and Indels

For SSRs and indels, large polymorphisms between 20 bp to 1 kb can be detected using agarose gel electrophoresis. Fragment size differences from two to 20 bp are typically detected on polyacrylamide gels. The fragment separation can be performed by electrophoresis on non-denaturing polyacrylamide gels and stained with ethidium bromide or silver. One apparatus to perform this task is the vertical sequencer DASG-400 to DASG-600 (CBS SCIENTIFIC, California, USA). The throughput of this system is approximately 600 to 900 data points per day requiring substantial manual labor, although supply costs are low. Higher throughput can be achieved by size separation using denaturing polyacrylamide and a fluorescent dye-based automated detection system. Fluorochromes are covalently attached to one of the primers used in the amplification and the products are detected via laser excitation of the label. The LI-COR IR²®slab gel-based system (Li-COR, Nebraska, USA) uses two fluorochromes thereby increasing marker throughput to approximately 2,000 data points per day. This system requires manual labor in preparing the gel and loading of samples whereas automated analysis of band fragment sizes is performed using the genotyping software. Other fluorescent-based fragment separation systems utilize capillary electrophoresis technology identical to that used for dideoxy DNA sequencing. Manual labor is minimal when preparing the capillary, loading the samples, and detection, sizing and scoring the fragments. The Beckman CEQ8800® can detect fluorescently labeled fragments for eight or more markers simultaneously (http://www.beckmancoulter.com) with a throughput of approximately 1,536 data points per day. Although the throughput is not very high on the Beckman CEQ8800, the system is versatile and able to handle diverse sets of samples from users. The ABI system is another frequently used genotyping platform (Applied Biosystems, Foster City, CA, USA). From 16 (ABI 3100) to 96 (ABI3730*xl*) capillaries are run simultaneously, which greatly improves throughput. The four fluorochromes used in ABI chemistry and detection increase the number of multiplexed markers to at least 16 per capillary, resulting in daily throughput of 3,072 to 18,432 data points.

Base Substitutions

Detection of nucleotide changes that don't result in size differences requires other assays for detection. Cleaved amplified polymorphic sequence (CAPS) offer a convenient method of detection when the SNP alters a recognition site for a restriction enzyme (Konieczny and Ausubel 1993). After PCR amplification, the fragments are cut with the restriction enzyme that specifically digests one of the alleles. The digestion fragments are then separated on an agarose gel and scored. However, CAPS are limited to mutations that create or disrupt a restriction enzyme recognition site. When no recognition site is detected an alternative to CAPS is derived cleaved amplified polymorphic sequence (dCAPS), where PCR primer mismatches are used to create a polymorphism based on the target mutation (Neff et al. 1998). The program dCAPS Finder 2.0 (Table 5) facilitates the design of these PCR primers. Both CAPS and dCAPS markers are reliable and easy to apply. However, throughput is limited due to agarose gel electrophoresis. Furthermore, the analysis requires the use of endonucleases which are often expensive, thus limiting the wide-spread use of CAPS and dCAPS in molecular mapping.

Other SNP detection platforms discriminate between alleles using allele-specific primer extension (ASPE) assays. Allele discrimination is accomplished through a variety of detection systems and platforms (Kwok 2001). One of these platforms is offered by Luminex (http://www.luminexcorp.com), and uses the principles of flow cytometry to detect polymorphisms using ASPE. Two allele-specific primers are designed for each SNP, with the 3' base specific to the polymorphism and a tag attached to the 5' end. Successful extension with a biotin labeled base of the perfect matching primer allows products to be enriched and sorted using beads. Laser detection identifies the micro-bead and quantifies the signal. With 100 distinct sets of micro-beads, up to 50 bi-allelic SNPs can be assayed in a single reaction (Lee et al. 2004b) resulting in approximately 4,000 data points per day. Another procedure designed specifically for high-throughput SNP detection is offered by the Illumina Golden Gate Assay (http://www.illumina.com). In this procedure, hundreds of SNP-specific oligonucleotides are simultaneously hybridized directly to genomic DNA. These oligonucleotides contain target sequences for a set of universal primers and address sequences recognized by labeled beads in the array. Throughput is very high, although this system is only cost-effective when analyzing large datasets.

1.4.4 PCR-based Linkage Map of *S. lycopersicum* cv. Rio Grande × *S. pimpinellifolium* LA1589 F_2 Population

The development of a reference genetic map covering the entire genome with PCR-based markers demonstrates how public resources can be used to facilitate genetic investigations of tomato populations. Frary et al. (2005) reported on the first sequence-dependent PCR-based map for tomato for a population derived from a cross between *S. lycopersicum* and *S. pennellii*. Using many of the techniques reviewed above, a set of 87 PCR-based markers were mapped between two much more closely related red-fruited species, *S. lycopersicum* cv. Rio Grande and *S. pimpinellifolium* LA1589 (Gonzalo and van der Knaap, unpublished) (Fig. 2). Since markers that are polymorphic in subsets of species or within the cultivated germplasm pool have a higher chance of being polymorphic in other lines (Francis and Wang, unpublished), some of the previously developed PCR-based markers were directly applicable to this population. Fifteen markers were derived from the tomato intron sequence mining project (Van Deynze et al. 2006; Table 5), seven were derived from the in silico SNP project (Yang et al. 2004), 31 were SSRs available at SGN, whereas the remainder were converted from RFLP and BAC-ends or entire BACs using the sequence information available at SGN followed by sequencing of alleles. The map spans a total genetic distance of 974 cM with only a few gaps remaining (Fig. 2). All SSR markers and 21 indel markers were mapped on the Beckman CEQ8800 whereas nine SNP markers were mapped on the Luminex. The remaining markers were mapped as CAPS and dCAPS on agarose gels.

1.4.5 Future Approaches for Marker Development

Irrespective of the goal, genetic analyses of tomato continue to require development of markers. Thanks to the efforts of the International Tomato Sequencing Project (see Sect. 1.18) (Mueller et al. 2005b) and the Potato Genome Sequencing Consortium (Table 5), sequence information from these two species will substantially increase in the near future and significantly facilitate marker development. Additional sequence information from other tomato varieties and its wild relatives will also improve the efficiency of molecular marker development and implementation using some of the aforementioned techniques. Aside from the genotyping platforms described above, other systems are on the horizon for tomato. Chip-based technologies similar to the NimbleGen array have been exploited for other species using Genechips (Affymetrix; Gut 2004) and will likely soon be exploited for tomato populations.

1.5 Maps

No one is more keenly aware than the writer of the thanklessness of the task of attempting to prove [...] that certain genes [...] are not really located on the same chromosome ... (MacArthur 1934)

1.5.1 Map Types

Genetic Linkage Maps

The first Mendelian tomato linkage group was published by Hedrick and Booth (1907), where the *dwarf* and *ovate* loci were shown to be linked in an F_2 cross between *S. lycopersicum* varieties Livingston Stone and Dwarf Aristocrat (Tanksley 1993a). By 1934, 20 morphological markers had been mapped to ten linkage groups. This study scored over 48,000 plants derived from crosses among ten varieties that were homozygous for spontaneous mutations (MacArthur 1926, 1934). During the same decade researchers began to implement crosses with wild tomato species as a source of disease resistance. These disease resistance genes were also quickly located to linkage groups since the parents of the hybrid populations were more polymorphic compared to intraspecific *S. lycopersicum* crosses. For example, Langford (1937) generated data indicating *Cf* loci on chromosomes 6 and 11 using *S. pimpinellifolium*. By 1973, 257 morphological and disease resistance markers had been mapped (Linkage Committee 1973).

Molecular markers were adopted the following year when isozyme *Aps-1* was linked to the nematode resistance gene *Mi* on linkage group six (Rick and Fobes 1974). Seven loci of *Est*, *Prx*, and *Got* were added in 1977 using *S. pimpinellifolium* (Rick and Fobes 1977). In the mid-1980s, DNA-based RFLP markers were starting to be mapped, for example, in an interspecific cross between *S. lycopersicum* cv. VF36 and *S. pennellii* accession LA716 (Table 6, pop 8) (Bernatzky and Tanksley 1986; Vallejos et al. 1986). Chromosomal assignments of genomic clones by trisomic analysis were also confirmed using this population (Young et al. 1987). An IL population from a cross between *S. lycopersicum* cv. M82 and *S. pennellii* accession LA716 (Table 6, pop 9) also used RFLP markers (Eshed and Zamir 1995). PCR-based markers (RAPD, AFLP, SSR, CAPS) were added to tomato maps in the late 1990s (Table 6, pop 6, 8–11, 13–14, 17–19, 21) (Grandillo and Tanksley 1996a; Pillen et al. 1996a, b; Bonnema et al. 1997; Areshchenkova and Ganal 1999; Haanstra et al. 1999). The advent of the complete *Arabidopsis thaliana* genome sequence facilitated development of PCR-based COS markers using genes shared between distantly related plant taxa (Table 6, pop 10) (Fulton et al. 2002b). Additional recent reviews have listed tomato genetic link-

Table 6. Tomato linkage mapping populations. Crosses are listed in alphabetical order by species. Populations are numbered and maps of these populations are listed chronologically. Markers used for each map are listed, and whether the map is available in an online format

Linkage mapping population	Pop type[a]	Pop size	No. of markers	Marker type(s)	Online[b]	Reference
S. lycopersicum × *S. cheesmaniae*						
1. UC204B × LA483	F_2	350	71	RFLP		Paterson et al. 1991
S. lycopersicum × *S. chmielewskii*						
2. UC82B × LA1028	BC_1	237	70	RFLP, Isozyme		Paterson et al. 1988
S. lycopersicum × *S. habrochaites*						
3. E6203 × LA1777	BC_1	149	135	RFLP	SGN	Bernacchi and Tanksley 1997
4. NC84173 × PI126445	BC_1	145	171	RFLP, Resistance Gene Analog (RGA)		Zhang et al. 2002
S. lycopersicum × *S. lycopersicoides*						
5. VF36 × LA2951	BC_1	149	93	Isozyme, Morphological (Morph), RFLP	NCBI	Chetelat and Meglic 2000; Chetelat et al. 2000
S. lycopersicum × *S. lycopersicum* var. *cerasiforme*						
6. Cervil × Levovil	F_7 RIL	153	377	RFLP, RAPD, AFLP, Morph		Saliba-Colombani et al. 2000
S. lycopersicum × *S. neorickii*						
7. E6203 × LA2133	BC_2	170	133	RFLP		Fulton et al. 2000
S. lycopersicum × *S. pennellii*						
8. VF36 × LA716	F_2	46	84	RFLP	NCBI	Bernatzky and Tanksley 1986
	F_2	67	1,030	Isozyme, Morph, RFLP	SGN, NCBI	Tanksley et al. 1992
		42	368	RFLP, SSR		Broun and Tanksley 1996
		67	1,050	RFLP, EST		Pillen et al. 1996b
		42	909	AFLP		Haanstra et al. 1999
		67	19	SSR		Areshchenkova and Ganal 1999
		67	20	SSR		Areshchenkova and Ganal 2002
9. M82 × LA716	IL	50	375	RFLP	SGN	Eshed and Zamir 1995
		52	140	T-DNA		Gidoni et al. 2003
		50	20	CAPS		Yang et al. 2004

[a] IL introgression line; IBL inbred backcross line; RIL recombinant inbred line

[b] SGN http://www.sgn.cornell.edu/cview; NCBI http://www.ncbi.nlm.nih.gov/mapview; TMRD (Tomato Mapping Resources Database) http://www.tomatomap.net

Table 6. (continued)

Linkage mapping population	Pop type[a]	Pop size	No. of markers	Marker type(s)	Online[b]	Reference
10. LA925 × LA716	F_2	80	2,222	RFLP, CAPS, SSR, SNP, COS	SGN, NCBI	Fulton et al. 2002b
		83	1,579	CAPS, SSR		Frary et al. 2005
11. Allround × LA716	F_2	84	77	RFLP, SSR		Arens et al. 1995
		80	707	RFLP, AFLP		Haanstra et al. 1999
		84	38	RFLP, SSR		Bai et al. 2004a
S. lycopersicum × *S. peruvianum*						
12. E6203 × LA1706	BC_3	241	177	RFLP		Fulton et al. 1997
13. Solentos × LA2157	F_2	314	5	RFLP, CAPS		Bonnema et al. 1997
S. lycopersicum × *S. pimpinellifolium*						
14. M82 × LA1589	BC_1	257	120	RFLP, RAPD	SGN	Grandillo and Tanksley 1996a
15. NC84173 × LA722	BC_1	119	151	RFLP		Chen and Foolad 1999
16. E6203 × LA1589	BC_2F_6, IBL	196	127	RFLP	SGN	Doganlar et al. 2002c
17. Sun1642 × LA1589	F_2	46	101	RFLP, CAPS	TMRD	Yang et al. 2004
18. XF 98-7 × LA2184	F_2	142	112	SSR		Liu et al. 2005
19. Rio Grande LA 1589	F_2	94	87	SSR, CAPS, SNP		Gonzalo and van der Knaap, unpublished; see Sect. 1.4.4 and Fig. 2
S. peruvianum × *S. peruvianum*						
20. LA2157 × LA2172	BC_1	152	73	RFLP	NCBI	Van Ooijen et al. 1994
21. PI 128650 × PI 128657	BC_1	2,112	13	RFLP, RAPD		Pillen et al. 1996a

age maps (Foolad and Sharma 2005; Villalta et al. 2005).

Cytological and Cytogenetic Maps

Tomato aneuploid stocks have been used as tools for mapping loci for many decades (Lesley 1932). Populations developed from crosses between irradiated and mutant cultivars were extensively used to generate chromosome-based maps of tomato in the 1960s (Khush and Rick 1967b, 1968a, b) (Table 7, map 1). Radiation-induced chromosomal deficiencies allowed expression of recessive mutant alleles and localized mutations to chromosomal arms. In the 1990s, a resurgence of interest in cytology prompted the development of a physical recombination map used to quantify the distribution of crossovers along each chromosome (Sherman and Stack 1995) (Table 7, map 2). In the same decade a variety of in situ hybridization maps were also developed, linking fluorescent genetic markers to their physical chromosomal location (Fuchs et al. 1996; Zhong et al. 1996; Peterson et al. 1999; Wang et al. 2006), (Table 7, maps 3–7). These are termed cytogenetic maps, due to the connection made between genetic linkage and cytological position (Harper and Cande 2000). The linear order of markers is not always conserved between genetic and cytological positions (Peterson et al. 1999).

Physical Maps

Cloned genomic libraries have been used since late 1980s for identifying and isolating segments of the genome containing loci of interest (Ganal et al. 1989; Twell et al. 1989; Martin et al. 1993a; Alpert and Tanksley 1996). A recent 15x genomic clone BAC-library is being used as the physical map starting point for tomato genomic sequencing and positional cloning (Budiman et al. 2000) (Table 7, map 8).

Transposon Insert Map

Reverse genetic approaches involve altering gene function to study the resulting phenotypes. Tagged transposons have been used to generate such a population of 529 plants each containing one or two insertions. The locations of 140 single inserts have been anchored to a genetic map (Eshed and Zamir 1995) (Table 6, pop 9) using a combination of PCR and CAPS markers (Gidoni et al. 2003) (Table 7, map 9).

1.5.2 Maps Online

A number of the aforementioned maps are available on the internet at SGN (Mueller et al. 2005a), National Center for Biotechnology Information (NCBI) (Wheeler et al. 2004), and http://www.tomatomap.net (Van Deynze et al. 2006) (Tables 5–7).

The process of preparing genetic maps for display in the NCBI MapViewer includes populating the MapViewer database with ancillary information associated with the locus. This information includes the probe name, any GenBank accession(s) known, the UniSTS accession(s) for primers and the Entrez Gene id. This information is used to effect traversal of the databases within NCBI. Text searches of the data associated with loci can be performed. In addition to the search capability offered at the top of the homepage for MapViewer, http://www.ncbi.nlm.nih.gov/mapview, (encompassed by the large rectangle in upper right quadrant of Fig. 3) there are two indirect entry points – the link to PlantCentral http://www.ncbi.nlm.nih.gov/genomes/PLANTS/PlantList.html (encompassed by the rectangle in upper left quadrant of Fig. 3) or the link at the top of the plant list on the MapViewer home page (encompassed by the small rectangle in upper right quadrant of Fig. 3).

Proceeding via the indirect entry points (depicted by following the dashed lines) generates the query page – shown in the lower left quadrant of Fig. 3. Within the query page the query text ("TG128" in this example) is entered in the space outlined by a rectangle. Clicking the button encompassed by an ellipse generates the page in the lower right quadrant. Using the search capability on MapViewer is accomplished by adjusting the organisms at the arrow within the large rectangle, and entering the sought text in the box followed by clicking the "Go" button. The result is the lower right quadrant. Regardless of how one arrives at the lower right quadrant, clicking the "Display" button outlined by a rectangle generates Fig. 4.

The GenBank accession(s) associated with a locus are available for BLAST search. There are two indirect entry points – the link in PlantCentral http://www.ncbi.nlm.nih.gov/genomes/PLANTS/PlantList.html or the link at the top of the plant list on the MapViewer home page. Both of these are also included within the outlined areas in the top two quadrants of Fig. 3. As of this writing the nucleotide sequence for TG128

Table 7. "Other" maps listed by type, cultivar if known, number of plants, lines, or genome coverage. The number of components in the resource are indicated, and correspondence to genetic maps from populations in Table 6, if known

Map type	*S. lycopersicum* Cultivar	Lines or (coverage ×)	Items	Correspondence to genetic maps	Linkage group	Online	Reference
Cytogenetic	1. Irradiated Red Cherry	na[a]	74 deficiencies	35 Morph, Resist	all		Khush and Rick 1968a, 1968b
Cytological	2. "Cherry"	1	9,058 Recombination Nodules	na	all		Sherman and Stack 1995
Cytogenetic	3. "Cherry"	1	3 multicopy genes	rDNA, TGR1, THG2	all		Zhong et al. 1996
Cytogenetic	4. Moneymaker	1	2 YACs	CT277 (pop 8)	2		Fuchs et al. 1996
Cytogenetic	5. "Cherry"	1	3 clones	TG46, TG400, TG523 (pop 8)	11		Peterson et al. 1999
Cytogenetic	6. VFNT Cherry	1	11 BACs	CAPS, RFLP (pop 10)	2, 4, 7, 8, 9, 10, 11	SGN	Wang et al. 2006
Cytogenetic	7. na	na	63 BACs	COS, overgo (pop 10)	all	SGN	"Tomato FISH map" SGN, 2007
Physical	8. Heinz 1706	(15×)	129,024 clones	BAC-end sequence, RFLP	all	SGN	Budiman et al. 2000
T-DNA insert	9. Moneymaker	529	140 inserts	PCR/CAPS (pop 9)	all		Gidoni et al. 2003

[a] not available

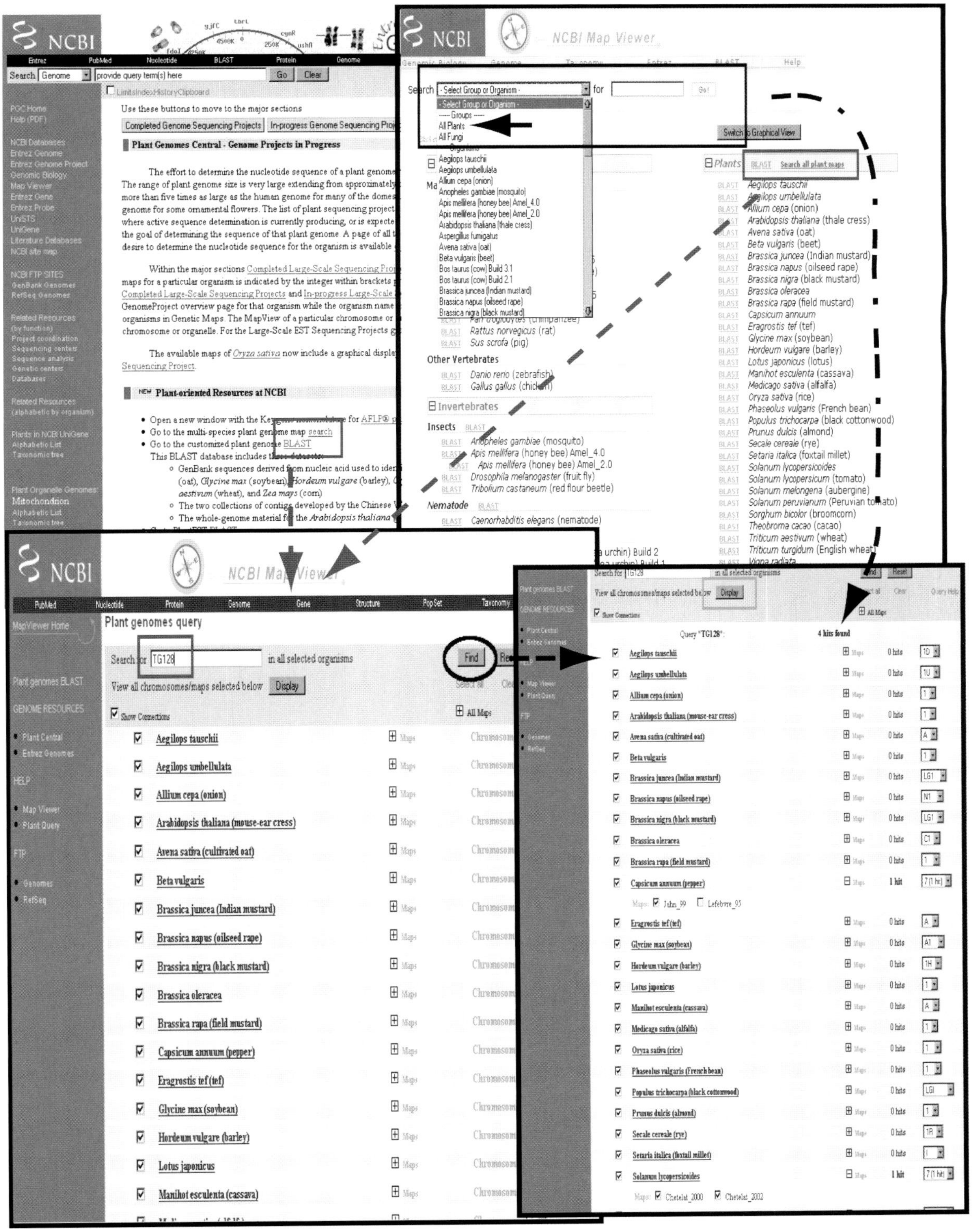

Fig. 3. Navigating to Genetic Maps at NCBI. This illustrates the link-mediated flow through NCBI web pages to perform text searches of plant genetic map information. Following the "search" link in either *rectangle-outlined areas* in the top two quadrants generates the lower left quadrant. The query text (in this example "TG128") is entered in the space *outlined by a rectangle* in the lower left quadrant. The lower right quadrant is generated by both *black arrows*. Clicking the "Display" button in the lower right quadrant generates Fig. 4

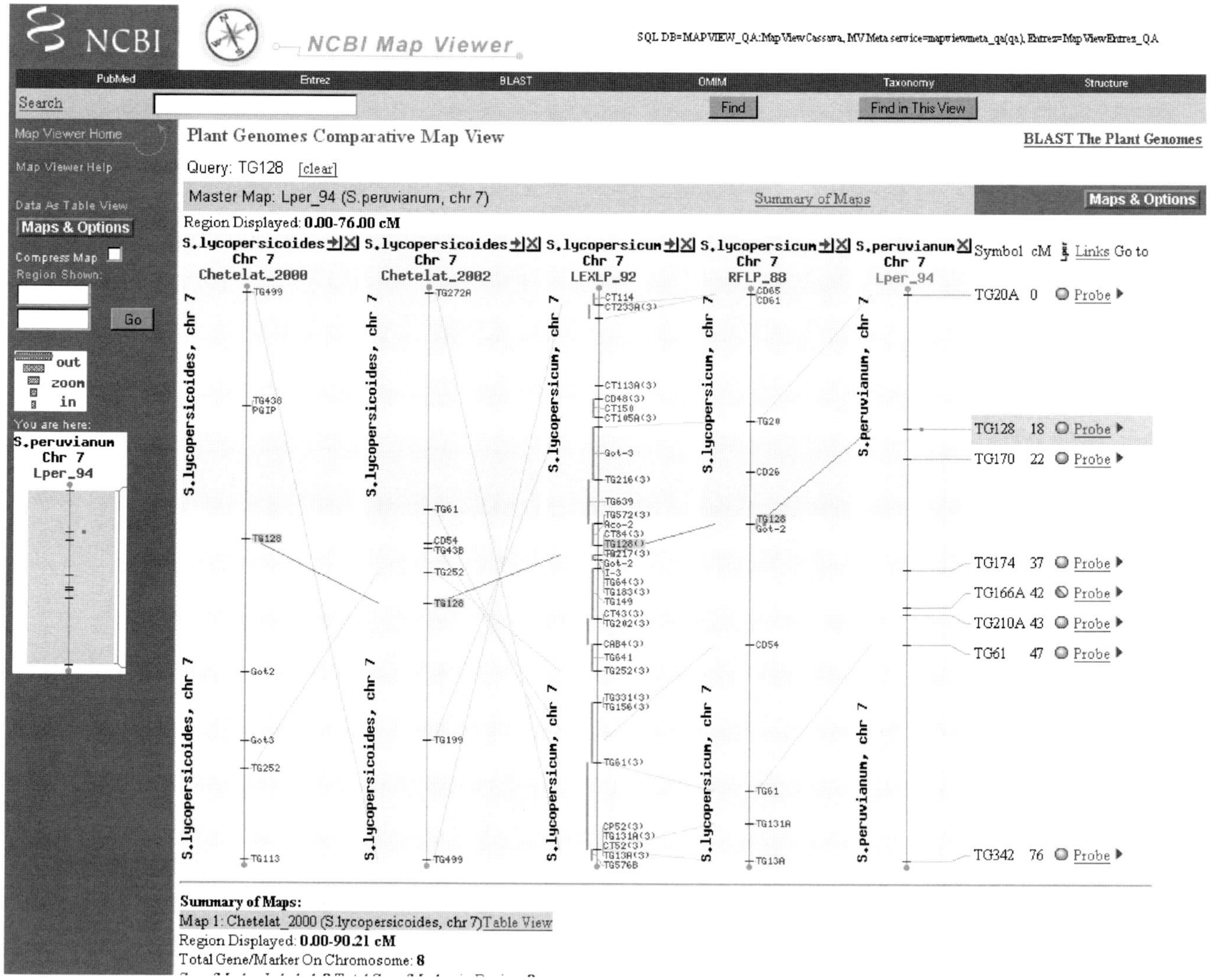

Fig. 4. NCBI comparative maps of (*from left to right*) *S. lycopersicum* × *S. lycopersicoides* (Chetelat and Meglic 2000) (Table 6, pop 5), *S. lycopersicoides* × *S. sitiens* (Pertuze et al. 2002), *S. lycopersicum* × *S. pennellii* (Tanksley et al. 1992) (Table 6, pop 8), *S. lycopersicum* × *S. pennellii* (Bernatzky and Tanksley 1986) (Table 6, pop 8), *S. peruvianum* × *S. peruvianum* (Van Ooijen et al. 1994) (Table 6, pop 20)

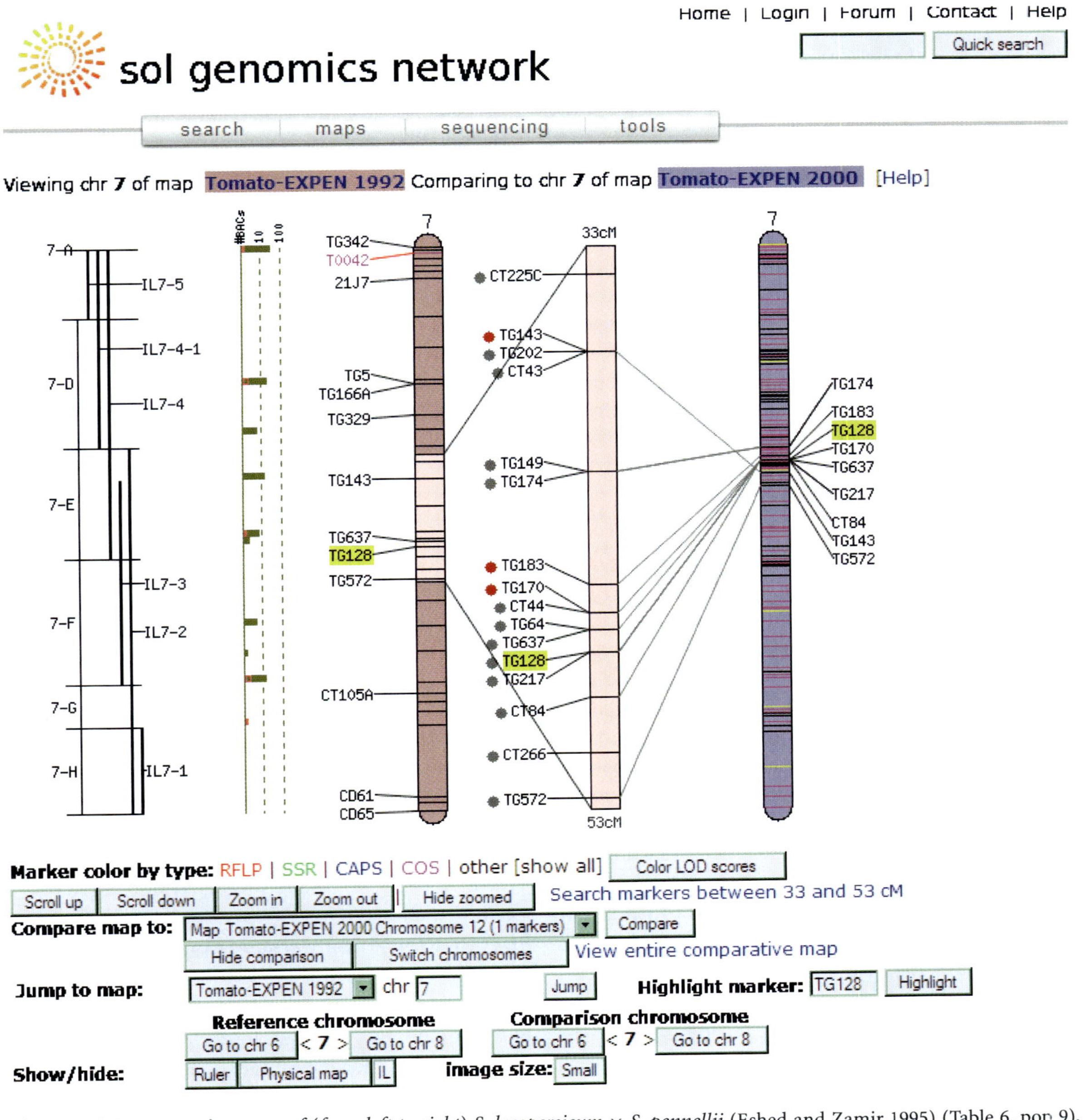

Fig. 5. SGN comparative maps of (*from left to right*) *S. lycopersicum* × *S. pennellii* (Eshed and Zamir 1995) (Table 6, pop 9), *S. lycopersicum* (Budiman et al. 2000) (Table 7, map 8), *S. lycopersicum* × *S. pennellii* (Tanksley et al. 1992) (Table 6, pop 8), *S. lycopersicum* × *S. pennellii* (Fulton et al. 2002b) (Table 6, pop 10)

has not been submitted to NCBI. If it were, the same display would be accessible using BLAST.

Figure 4 shows all of the maps that are currently available through the NCBI MapViewer genomic display resource that have the TG128 probe associated with a locus. A line connects the TG128 probe-associated loci.

SGN features a number of tomato and other Solanaceous crop genetic maps and associated data. Figure 5 shows the same region on chromosome 7 as displayed in Fig. 4, with marker TG128 highlighted. From left to right the maps shown are: an IL map; a representation of BAC coverage from a physical map; a genetic linkage map, the same linkage map zoomed in, and a second genetic linkage map.

Access to the maps of SGN is provided on the SGN homepage (http://sgn.cornell.edu) through the maps menu. The menu lists a number of frequently used maps; all available maps can be viewed by clicking on the "show all maps" menu position. Each map has a top-level overview page that shows all linkage groups and their relative sizes, an abstract with background information, and statistics about the map including marker counts and types of markers. On the map overview markers can be located using a text search box. Clicking on one of the linkage groups in the overview displays an enlarged view of the linkage group in the comparative browser as the reference chromosome. In this mode, clicking on the reference chromosome will zoom into the region clicked, and selecting a comparison linkage group will display the comparison between the two linkage groups. For the reference chromosome, additional data tracks can be displayed, if available, such as associated physical map information and available ILs. Markers and maps can be clicked to access related data.

In addition to the map data, SGN stores detailed information about molecular markers. This includes sequences for RFLP markers and primers, sequence source (GenBank accession or SGN EST or unigene sequence id) and PCR reaction conditions for CAPS, SSR and other PCR based markers. SGN also houses the master database for the COS (Fulton et al. 2002b) and COSII markers (see Sect. 1.4.2, 1.6.3). Currently, there are over 2,800 COSII markers available that have been developed to work on a maximum number of Asterid species (Wu et al. 2006). For the COSII, the database stores the alignment of the sequences that were used to develop the markers, the experimental PCR conditions for each accession that the markers were applied to, and sequence reads from the PCR products. The data for the different accessions is submitted by users who have used the primers in the species and accessions that are documented.

Using the SGN marker search, markers can be searched with a large number of search criteria, such as constraining the search to a species, chromosome, chromosome interval, marker type, LOD scores, association to a physical map, and more. The markers are also grouped into types such as COSII and COS.

1.6 Comparative Mapping in the Solanaceae

Comparative mapping is the side-by-side comparison of gene order and other genomic features in related individuals. It is a very useful tool in the analysis of genomes. Comparative mapping can aid in the (i) identification of gene function, and (ii) choice of additional markers for linkage mapping, marker-assisted selection, and positional cloning; it also provides insights into genome evolution. Our knowledge of the location and function of a morphological gene in tomato, for example, may provide insights into the function of a putative gene in the homologous region in pepper (*Capsicum* spp.). Knowledge of comparative maps can be of great assistance in positional cloning experiments, where additional genetic and physical markers from a dense map of a related species can be selected for application in the species of interest, provided an area of conserved synteny is under investigation. Questions about genome evolution have become increasingly interesting since the realization that the genomes of many crops have been shuffled (through inversions and translocations) over evolutionary time, but often retain large syntenic blocks when compared to related genera.

The ultimate comparative map between two genomes is a complete DNA sequence. Other types of maps are available, and plant geneticists and breeders work with the maps available for the given species of interest. In the Solanaceae, tomato genomics was initially the main force and is currently the driving force behind much of the progress in the study of genomics for the entire family (see Sect. 1.18).

1.6.1 Brief History of Comparative Mapping

Comparative mapping in plant genomes originated in the Solanaceae, through the early work of Tanksley and colleagues (Bernatzky and Tanksley 1986; Bonierbale et al. 1988; Tanksley et al. 1988). In these studies, isozyme and RFLP markers were placed onto maps of tomato and its close relatives within the Solanaceae, pepper, and potato (*Solanum tuberosum*). Orthology of loci, or identity by descent from a common ancestral sequence, was determined based on map position. For example, if a tomato marker was within a block of markers, all showing conserved synteny between tomato and pepper, then that marker was orthologous to the corresponding marker in pepper. This criterion is still used today.

These early studies were modest in terms of numbers of molecular markers, but they provided the first evidence for the existence of highly conserved syntenic blocks of loci in higher plant

Table 8. Chromosomal rearrangements in the Solanaceae as determined by comparative mapping. Data from Bonierbale et al. (1988), Tanksley et al. (1992), Prince et al. (1993), Livingstone et al. (1999), and Doganlar et al. (2002a)

Taxon	No. of chromosomes (2*n*)	No. of chromosomal rearrangements in comparison to tomato
Tomato (*Solanum lycopersicum*)	24	–
Potato (*Solanum tuberosum*)	48	5 paracentric inversions
Pepper (*Capsicum* spp.)	24	30 chromosome rearrangements
Eggplant (*Solanum melongena*)	24	23 inversions, 5 translocations
Tobacco (*Nicotiana tabacum*)	48	na[a]
Petunia (*Petunia hybrida*)	14 or 18	na

[a] Not available, molecular linkage maps have been published for tobacco and petunia (e.g., Strommer et al. 2002; Bindler et al. 2006; Julio et al. 2006), but the markers used do not allow for comparison with any tomato maps

genomes. Some of this information is summarized in Table 8. Comparative mapping is still proceeding in the Solanaceae (see Sect. 1.6.2). In most of these works tomato is central, and other plants' genomes are referenced against it. A variety of chromosomal rearrangements are postulated to have occurred during divergence from a common ancestor; this is based on breaks in conserved blocks of synteny when comparing maps of different genera within the family (Table 8). The effort that has gone into tomato mapping and sequencing has outpaced other members of the family, but tools and markers developed for tomato often work very well across many members of the family.

1.6.2 Mapping in Solanaceous Plants

Tomato (*S. lycopersicum*)

A large number of laboratories have produced molecular genetic maps of tomato. In 1992, a high-density map of the tomato genome was constructed (Tanksley et al. 1992) that continues to be used for comparative mapping (Table 6, pop 8). Many other maps have been developed for various purposes based on other populations (see Sect. 1.5) (Tables 6, 7).

Potato (*S. tuberosum*)

The first RFLP map of potato was published by Bonierbale et al. (1988), which first delineated the five inversions that characterize the differences between locus order in the tomato and potato genomes. Other groups subsequently published potato maps and compared them to tomato (e.g., Gebhardt et al. 1991). The high-density linkage map developed by Tanksley et al. (1992) for tomato remains an excellent resource for potato genetics, because a large number of orthologous markers were placed on both maps at the time it was published. Many additional maps in potato have been constructed, including a comparative map between potato and *Arabidopsis* (Gebhardt et al. 2003) showing 90 putative syntenic blocks. Recently an ultra-dense map of more than 10,000 AFLP markers was published (van Os et al. 2006).

Pepper (*Capsicum* spp.)

Mapping in pepper has revealed extensive chromosomal rearrangements between pepper and tomato (Tanksley et al. 1988; Prince et al. 1993). Several substantial linkage maps were subsequently released for pepper (Livingstone et al. 1999; Kang et al. 2001; Lee et al. 2004a; Minamiyama et al. 2006; Yi et al. 2006), including a multi-lab integrated map with 2,262 markers (Paran et al. 2004), which is currently the standard for pepper mapping.

Eggplant (*S. melongena*)

Using 233 markers derived from tomato cDNA, genomic DNA and ESTs, a genetic linkage map of eggplant was created (Doganlar et al. 2002a). Twenty-three paracentric inversions and five translocations explain the differences in marker order between the eggplant and tomato genomes.

Petunia (*Petunia hybrida*)

In petunia, a small initial linkage map containing 27 RFLP markers and 11 morphological or biochemical markers was published (Strommer et al. 2000). Subsequently, a much larger series of maps with a total of 800 AFLP markers was constructed (Strommer et al. 2002). Due to the type of marker, syntenic comparisons with tomato are not possible. The unexpected discovery in this paper was petunia's very low level of recombination, about ten-fold less in comparison with tomato.

Tobacco (*Nicotiana tabacum*)

The first genetic linkage map for the cultivated tetraploid tobacco was recently published (Bindler et al. 2007). For this work, SSR primers were identified through sequences provided by the Tobacco Genome Initiative (http://www.tobaccogenome.org), and 293 loci were successfully mapped. In addition to the Tobacco Genome Initiative, there is a European effort, the European Sequencing of Tobacco project (http://www.estobacco.info). Previous mapping had been done in members of the genus other than *Nicotiana tabacum*. For example, comparisons of the linkage maps of *N. sylvestris* and *N. plumbaginifolia* revealed significant intrageneric genome reorganization (Lin et al. 2001).

1.6.3 Markers and Orthology

Six major types of DNA markers have historically been used for genetic linkage mapping in higher plants including Solanaceae: RFLPs, RAPDs, AFLPs, SSRs, sequence characterized amplified regions (SCARs), and SNPs. In addition, COS markers have been specifically designed for comparative mapping. By comparing tomato EST sequences to the *Arabidopsis* genome sequence, Fulton et al. (2002b) identified a set of 1,025 genes, which they referred to as COS markers. COS markers are single or low-copy in both genera, thus probably orthologous, and potentially very useful for comparative mapping. Both computationally and through Southern hybridizations, Fulton et al. (2002b) found the COS markers to be present in many plants, including dicots and monocots. A potential disadvantage of the COS markers is that, because they are well conserved, they may show low polymorphism. Wu et al. (2006) followed up on COS marker development by designing a set of COSII markers, which are similar to the COS markers but have additional phylogenetic evidence to support orthology of sequences. In addition, a set of "universal PCR primers" has been designed which may enable amplification of these markers from a broad range of plant species. Attributes of DNA marker types used in comparative mapping are provided in Table 9.

The utility of a molecular marker for comparative mapping depends on level of polymorphism and ease of establishment of orthology when examining different crosses. Two loci in different individuals are orthologous if they have descended from a common ancestor. Orthology can be established fairly easily if the DNA marker detects only one locus. For multilocus markers, orthology is determined based on map position. Inferences about marker synteny in comparative genomics cannot be made unless the orthology of loci in different genomes can be determined with high confidence. Orthology is most commonly determined based on a combination of two factors: (i) presence of a single DNA fragment upon hybridiza-

Table 9. DNA markers used for comparative mapping. General properties are listed, but exceptions occur

Marker type	Polymorphism	Dominant or codominant	Number of loci detected
RFLP	Low	Codominant	Usually 1 to 2
RAPD	Low	Dominant	5 or more
AFLP	High	Dominant	50 to 100
SSR	High	Codominant	Usually 1
SCAR	Low	Variable	1
SNP	High	Codominant	1
COS	Low	Variable	Usually 1 to 2

Table 10. Comparative mapping of some genes and QTLs in the Solanaceae. Chromosomal locations, when known, are given by "c." followed by chromosome number

Category of QTLs	Tomato	Potato	Pepper	Eggplant	Petunia	Tobacco	Reference
Physiology/ Morphology	*invGE*, *Lin5*, fruit sugar, **c.9**	Invertases, cold sweetening, chip quality, **c.9**					Fridman and Zamir 2003; Li et al. 2005a
		Tuber skin color QTL, fruit shape QTL, **c.10**	Anthocyanin loci, fruit shape QTL *fs10.1*, **c.10**				Ben-Chaim et al. 2003
	ovate, **c.2**		Fruit size and shape *fw2.1*, *fs2.1*, **c.2**				Zygier et al. 2005; Ben-Chaim et al. 2006
	QTL for fruit weight *fw2.2*, *fw 9.2*, *fw11.1*			QTL for fruit weight *fw2.1*, *fw 9.1*, *fw11.1*			Doganlar et al. 2002b
	Ovate gene for fruit shape			Gene for fruit length *fl2.1*			Doganlar et al. 2002b
	Fruit shape QTL *fs10*		Fruit shape QTL *fs10*	Fruit shape QTL *ovs4.1*			Doganlar et al. 2002b
	Leaf C mutant for leaf shape			Lobing QTL *11ob6.1*			Frary et al. 2003a
	Various color genes/mutants	Various color genes/mutants			Various color genes/mutants		De Jong et al. 2004
			A	Multiple loci	*an2*		De Jong et al. 2004
Disease resistance	Resistance to true fungi, **multiple locations, overlapping with potato and pepper**	*Phytophthora* resistance, **multiple locations**	*Phytophthora* resistance, **multiple locations**				Grube et al. 2000b; Thabius et al. 2003; Ogundiwin et al. 2005
	Verticillium resistance, **c.9**	*Verticillium* resistance, **c. 2,6,9, and 12**					Simko et al. 2004a, b
	Nematode resistance PCR-NR_{14}	PCR-NR_{14} and *Globodera* resistance, **c.9**	*Meloidogyne incognita* resistance loci *Me3* and *Me4*, **c.12**				Bradshaw et al. 1998; Dijian-Caporalino et al. 2001; Caromel et al. 2003; Bryan et al. 2004
		Potato leafroll virus resistance, **c.11**				*N* gene for TMV resistance	Marczewski et al. 2001

Table 10. (continued)

Category of QTLs	Tomato	Potato	Eggplant	Pepper	Petunia	Tobacco	Reference
	Markers linked to pepper *Bs3* mapped to **c.2**			*Bs3* for *Xanthomonas* resistance, **c.3**			Pierre et al. 2000
	Resistance gene cluster, **c.11**	Resistance gene cluster, **c.11**		QTL for resistance to *cucumber mosaic virus*, **c.4, 6,** and **11**			Ben-Chaim et al. 2001
	Potyvirus resistance *Lv* gene, **c.9**, *Oidium* resistance, **c.6**			*Potyvirus* resistance Powdery mildew resistance, multiple QTL including **c.6, 9**			Grube et al. 2000a Lefebvre et al. 2003; Bai et al. 2004b

tion or PCR in both genomes being compared, and (ii) position on the linkage map relative to other markers of known (or presumed) orthology.

The detection of a single fragment upon PCR or Southern hybridization in two or more plant genomes is taken as evidence that the fragments are orthologous, barring any evidence to the contrary. Contrary evidence could include a firm map position that is not linked to any other locus that is linked to the putative ortholog in the other genome (although this could be evidence for fine-scale rearrangements) or DNA sequence that does not match.

Map position can allow one to distinguish between orthologous and paralogous loci corresponding to markers that detect multiple fragments. Figure 6 shows a simple hypothetical example of determining orthology vs. paralogy for an RFLP marker that has one copy in tomato and two in pepper. Because CD422b is in approximately the same map position in pepper as is CD422 in tomato, while CD422a is not, the most parsimonious interpretation is that CD422b is the ortholog to CD422 in tomato, while CD422a is the paralog.

Comparative QTL

There are several examples of successful comparative QTL mapping in the Solanaceae. Much of the comparative QTL mapping has been done with fruit

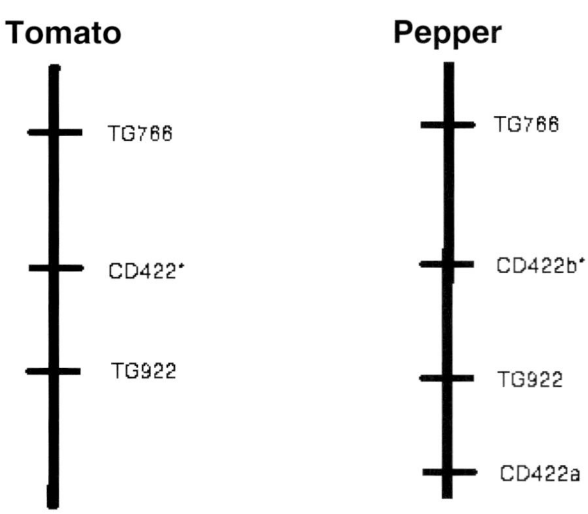

Fig. 6. Determination of marker orthology by map position. In this example, CD422 is located between TG766 and TG922 in the tomato genome. CD422 hybridizes to two fragments in the pepper genome that map to different loci. CD422b in pepper is designated orthologous to CD422 in tomato due to its similar map position. CD422a is paralogous

weight and shape characters (Doganlar et al. 2002b; Ben-Chaim et al. 2003; Frary et al. 2003a). Putative orthologous QTLs/genes controlling those phenotypes have been found in tomato, potato, pepper, and eggplant (Table 10). Most of the rest of the comparative QTL/gene mapping has been done on disease resistance loci. Orthologous QTLs have been found for oomycete, fungal, viral, and nematode disease resistances. Indeed, it seems that resistance gene clusters might be the rule rather than the exception (Table 10).

1.6.4 Solanaceae Comparative Mapping: Future Prospects

Comparative mapping proceeds at a rapid pace within the Solanaceae, because any mapping project that uses markers previously used in other mapping projects can be linked to those previous projects. The ultimate form of comparative mapping, comparative sequencing, is increasingly becoming feasible, with significant sequencing efforts being devoted to many members of the Solanaceae. As this book goes to press, approximately 19% of the tomato nuclear genome has been sequenced by the members of the International Tomato Genome Sequencing Project (see Sect. 1.18). In addition, both the tomato mitochondrial and chloroplast genomes are being sequenced by LAT-SOL and EU-SOL, two consortia of countries from Latin America and Europe. Sequencing efforts are also taking place in pepper (Lee et al. 2004a; Yi et al. 2006), potato (http://www.tigr.org/tdb/potato) and other members of the family. These efforts are greatly enhancing the resolution that can be achieved in comparative mapping in Solanaceae.

1.7 Principles of QTL Mapping and Characterization

In tomato, many important traits are under monogenic control (e.g., disease resistance, fruit color, shelf life, determinate growth). Molecular markers have been used to map and characterize these loci for decades and are frequently used by breeders for rapid screening of disease resistance genes. The availability of saturated molecular genetic maps of tomato has facilitated QTL mapping for traits related to fruit characteristics and yield, and for adaptation to abiotic stresses. Tomato is among the first plant species for which QTLs were detected (Paterson et al. 1988; Tanksley 1993b).

1.7.1 Principles of QTL Mapping

Molecular markers allow the analysis of the genetic control of quantitative traits by mapping QTL. Mapping QTL is based on a systematic search for association between the genotype at a marker locus and the average value of a phenotypic trait. It requires a segregating population, e.g., derived from the cross between two individuals with different values of the trait of interest. For each individual within the population, the genotype of marker loci distributed over the entire genome is determined and a saturated genetic map is constructed. Simultaneously, the value of the trait of interest is measured for each individual. Statistical methods are then used to find marker loci whose genotype is correlated with the trait and to estimate the genetic parameters of the QTL detected. Several biometric techniques have been proposed to find QTL, from the most simple, based on analysis of variance (ANOVA), applied marker by marker, to methods that simultaneously take into account two or more markers (Lynch and Walsh 1998). As a QTL can only be detected if the corresponding gene is polymorphic, the sample of QTL detected is specific to each progeny. The choice of parental lines is thus very important. For distantly related parents QTL detection is easiest, but major effect QTL may hamper the detection of lower effect QTL that would be detected if the major QTL were fixed.

Populations showing the highest efficiency in mapping QTL are those derived from crosses between two homozygous lines, such as F_2, RILs and BC. F_2s are the only populations allowing the dominance effect to be estimated, while a mixture of additivity and dominance is estimated with BC. Tanksley and Nelson (1996) proposed to search for QTL in advanced backcross (AB) BC_2, BC_3, and BC_4 populations. Although the power of QTL detection is reduced in comparison to F_2 or RIL, this strategy is attractive when screening positive alleles from a wild species, as it allows the identification of additive effects, it reduces linkage with unfavorable alleles around the QTL, and simultaneously advances the production of enhanced lines (see Sect. 1.8). The efficiency of detecting a particular

QTL in a segregating population is low partly because other QTLs are segregating and major QTLs are masking the minor ones. For this reason, Eshed and Zamir (1995) proposed the use of ILs in which each line possesses a unique segment from a wild progenitor introgressed into the same genetic background (see Sect. 1.8).

Wild species produce low quality fruit but can carry alleles at QTL which may improve agronomic traits. Such transgressive QTLs have been discovered frequently. Even when highly contrasted individuals have been chosen as parents of a population, it is not rare to find a QTL showing an effect opposite of that expected from the value of the parents. Results from advance backcross experiments in tomato have shown, for example, unexpected positive transgressions from wild relatives for various fruit traits (Bernacchi et al. 1998b).

Epistasis between QTLs has rarely been detected with classical populations (Tanksley 1993b), but this is mostly due to statistical limitations for the populations studied. One way of increasing the reliability of epistasis analysis is to eliminate the "background noise" due to other QTLs, using crosses between NILs (see Sect. 1.8.2) differing only by a chromosome fragment carrying a QTL (Eshed and Zamir 1996).

1.7.2 Stability of QTL Over Environment

The environment may have a significant impact on the effect of QTL: a QTL detected in one environment may no longer be detected in another, or its effect may vary in magnitude (see Sect. 1.8.2). This has been frequently observed, although environmental influence differs according to the characters and the range of environments studied (Paterson et al. 1991). Certain QTLs are detected in all or almost all environments tested, while others are specific to a single environment. Several statistical methods for the estimation of QTL × environment interactions have been proposed (Zeng 1993).

1.7.3 QTL Characterization: Fine-mapping, Positional Cloning and Candidate Genes

When a QTL is mapped within a genomic region, this implies that there is at least one gene responsible for a fraction of the phenotypic variation. Before trying to identify the gene responsible for the QTL, it first must be precisely characterized at the genetic level. It is critical to determine the number of QTLs segregating within the region. Indeed, several fine-mapping experiments identified several QTLs in regions where a single QTL was detected or where it was assumed that a single gene controlled two related traits (Paterson et al. 1990; Monforte et al. 2000b; Lecomte et al. 2004b). Several Mendelian mutations in tomato have been characterized by positional cloning, but very few QTLs have been definitively characterized at the molecular level (Paran and Zamir 2003; Salvi and Tuberosa 2005; Price 2006). Direct cloning of a QTL is more difficult than cloning of a major gene because the QTL only partially influences trait variation and its effect can only be appreciated by statistical methods. For this reason, the resources required are more considerable. The first QTL cloned by map-based cloning usually corresponds to QTL with strong effects that are independent of the environment. If nothing is known about the physiological and molecular determinism of a trait, positional cloning is the most straightforward method to characterize a QTL. Alternatively, if genes involved in the expression of the character, or related to developmental processes responsible for a trait (candidate genes) are known, it is possible to test whether polymorphism in one of them can explain variation in the trait. In both cases it is necessary to reduce the interval around the QTL through fine-mapping. The increasing number of molecular tools (Table 5, 16) and online maps (Table 5–7) in tomato will expedite this molecular characterization. Comparison of lines using metabolomic, transcriptomic, or proteomic techniques can help identify genes that can contribute to the phenotype (see Sect. 1.15.3).

1.8 Advanced-backcross and Introgression-line Breeding Strategies

The continued genetic improvement of crop species depends on the introgression of new alleles. Thus, it is imperative that new tools and concepts be developed that would allow the more efficient use of the genetic potential stored in gene banks and in exotic germplasm. Here we review two related molecular breeding strategies: the Advanced Backcross QTL

(AB-QTL) method and the Introgression Line (IL) strategy. These have been developed and tested in tomato as well as several other crops with the goal of increasing the efficiency with which natural biodiversity can be exploited to improve yield, adaptation and quality traits in cultivated crop species, while at the same time creating long-term resources for plant breeders.

1.8.1 Advanced Backcross QTL Analysis

AB-QTL was proposed by Tanksley and Nelson (1996) as a novel plant breeding scheme designed to integrate the processes of QTL discovery and variety development. Useful QTL alleles from unadapted germplasm (e.g., landraces, wild species) are identified while simultaneously being transferred to elite lines, thus streamlining the QTL identification and utilization pipeline while concurrently broadening the genetic diversity of the cultivated germplasm.

In tomato, the first AB-QTL mapping project started in 1995 as a molecular marker-assisted breeding experiment applied to processing tomatoes, in collaboration with several international processing companies. The main objectives of the project were to: (i) examine whether wild germplasm could be used as a source of new, agronomically beneficial QTLs, (ii) test whether the proposed marker-assisted breeding scheme would efficiently maximize QTL discovery, and (iii) develop new lines that would outperform elite commercial varieties for soluble solids content while continuing to maintain or improve other important characteristics for the processing industry, including yield, viscosity, firmness, color and fruit size. For the recurrent parent a commercially acceptable publicly available open-pollinated processing variety (cv. E6203) was agreed upon by all partners, and only wild tomato species were used as donor parents.

Five AB-QTL studies were conducted in tomato involving crosses with five wild *Solanum* tomato species: *S. pimpinellifolium* (LA1589) (Grandillo and Tanksley 1996b; Tanksley et al. 1996), *S. peruvianum* (LA1706) (Fulton et al. 1997), *S. habrochaites* (LA1777) (Bernacchi et al. 1998a, b), *S. neorickii* (LA2133) (Fulton et al. 2000), and *S. pennellii* (LA1657) (Frary et al. 2004).

These five species were chosen on the basis of genetic diversity and uniqueness, representing the broadest possible spectrum of wild species maintained in gene banks, with the hope that this would increase the frequency of new, previously undiscovered alleles. The use of a common recurrent parent allowed more direct cross-species comparisons. Of the five wild species tested, *S. pimpinellifolium* was the only red-fruited one and the most closely related to the cultivated tomato. At the other extreme was *S. peruvianum*, one of the most distantly related and also containing the highest amount of novel DNA and variation when compared to *S. lycopersicum* (Rick 1986; Miller and Tanksley 1990). All of these wild species have been the source of many major disease resistance genes, but no effort had previously been made to exploit the variation they contain for the improvement of quantitative traits.

For four of the studies marker analysis was conducted on BC_2 populations, and BC_3 or BC_2F_1 families or both were used for phenotypic analysis. In the *S. peruvianum* study genotypic analysis was postponed until the BC_3 generation and phenotypes were evaluated on the derived BC_4 families. The number of markers used for the molecular analyses ranged from a minimum of 121 in the *S. pimpinellifolium* study to a maximum of 174 markers in the *S. peruvianum* study. However, in the *S. peruvianum* and *S. pennellii* populations approximately 30% of the scored markers were fixed for the *S. lycopersicum* allele (SL) and could not be used for QTL mapping. Many of the markers fixed for the SL alleles corresponded to the chromosomal regions for which marker-assisted selection was applied to remove the wild parent allele in the BC_1 population (e.g., for the top of chromosome 1 as a result of selection at the self-incompatibility locus, *S*, to increase the fertility of the plant and, for the bottom of chromosome 6, as a result of selection at the self-pruning locus, *Sp*, to ensure that the plants would have a determinate growth habit which is essential for mechanical harvesting of processing tomatoes). For other regions of the genome the fixation for the SL allele may be the result either of sterility in early crosses or the result of genetic drift caused by the small size of the BC_1 populations used to develop the correspondent BC_2 generations (in the case of the *S. peruvianum* and *S. pennellii* studies). For most of the AB-QTL populations analyzed, however, the percentage of markers fixed for the SL allele was lower than 10%.

To reiterate, the main objective of this project was to focus on improving soluble solids content while maintaining or improving other traits important to the processing industry, including yield, viscosity, firmness, color, and fruit size. Therefore, each

AB-QTL population was evaluated for a wide array of traits, ranging from a minimum of 19 in the *S. habrochaites* study to 35 in the *S. peruvianum* AB-population. In all cases total yield, red yield and major fruit quality characteristics, including soluble solids content or Brix°, fruit color, viscosity, firmness and fruit pH, were measured. Due to frequent negative correlations between Brix° and yield, the derived parameter Brix° × yield is considered to be a more comprehensive biological and agricultural estimate for the productivity of processing tomatoes than yield alone (Eshed and Zamir 1995; Tanksley et al. 1996).

Agronomically favorable QTL alleles were identified in all five of these interspecific AB-QTL populations, for nearly half of the evaluated traits. For certain traits, such as pH and acidity, changes were not categorized as positive or negative, but rather were required to be kept within an acceptable range for processing purposes. But of the QTLs identified for which allelic effects could be deemed as favorable (+) or unfavorable (−), 20 of the 78 identified QTLs (26%), corresponding to 11 out of 23 traits (48%) had trait-improving alleles contributed from *S. pennellii* (Frary et al. 2004). Approximately the same percentage of traits with favorable wild-alleles were obtained with *S. habrochaites* (47%, Bernacchi et al. 1998b), while even higher percentages were observed for *L. pimpinellifolium* (88%, Tanksley et al. 1996), *S. peruvianum* (73%, Fulton et al. 1997) and *S. neorickii* (69%, Fulton et al. 2000).

It was possible for ten of the traits for which the effects could be defined either favorable or unfavorable (Brix°, viscosity, fruit color, stem retention, puffiness, firmness, total yield, fruit weight, red yield and Brix° × red yield) to be measured in all five AB-QTL studies; two other traits, cover, and maturity were measured in four of the five studies. Desirable alleles contributed from the wild species were identified not only for traits for which the wild parent showed a superior phenotype (e.g., Brix°, puffiness, cover) but also for those traits for which the wild phenotype was agronomically inferior (e.g., total yield, fruit weight, fruit color). The average percentage of favorable wild QTL alleles estimated across the five wild species ranged between a minimum of 3% for red yield to a maximum of 88% for Brix° (Grandillo and Tanksley 2005). Over the ten traits common to all five studies, the highest percentage of positive QTLs was identified in the *S. pimpinellifolium* study (44%), followed by the *S. peruvianum* (41%), *S. neorickii* (28%), *S. pennellii* (27%) and *S. habrochaites* (15%) studies (Grandillo and Tanksley 2005). Interestingly, those QTLs which were the most useful in terms of (i) having the most positive agronomic benefits, (ii) confirmation in NILs, and (iii) successful transfer into commercial varieties, are those which came from the most distantly related wild species (SD Tanksley, pers. comm.).

The AB-QTL strategy was also used in the first four AB populations to identify QTLs for biochemical properties that may contribute to the flavor of processed tomatoes, such as sugars and organic acids (Fulton et al. 2002a) (see Sect. 1.15.1). Flavor, a difficult characteristic to define and measure but very important to the industry, was assessed by a taste panel, as well as a derived trait, sugars/glutamic acid, which has been shown to be highly correlated with improved flavor (Bucheli et al. 1999). A total of 222 significant QTLs were identified for 15 evaluated traits. Studies, such as this, are important not only for the obvious purpose of developing new varieties with improved flavor but also for further analyses aimed at improving our knowledge of the biochemical pathways of fruit development.

Overall, these results show that in tomato, on average, for approximately 30% of QTLs for any given trait, an allele superior (from an agricultural viewpoint) to the cultivated parental allele can be identified in the wild species. Furthermore, after having sampled several wild species genomes the rate of discovery of "new" QTL alleles is still approximately 50% (Fulton et al. 2000; Frary et al. 2004), suggesting that continued sampling of exotic germplasm has not yet reached a steady state and there are still new and useful QTL alleles yet to be discovered.

1.8.2 ILs and "Exotic Libraries" for the Analysis of Quantitative Traits

These favorable wild QTL alleles are a valuable resource for breeding programs after they have been fixed in NILs or ILs and after the superior performance of the line has been validated in comparison to the cultivated recurrent parent in replicated field experiments.

From the tomato AB-QTL populations, MAS was applied to develop NILs that contain specific QTL alleles derived from the wild donors *S. habrochaites*, *S. pimpinellifolium*, and *S. peruvianum*, and that are able to significantly improve the performance of the elite variety (Tanksley et al. 1996; Bernacchi et al.

1998a, Monforte and Tanksley 2000a, b; Monforte et al. 2001; Yates et al. 2004). Bernacchi et al. (1998a) evaluated the agronomic performance of 23 NILs containing either *S. habrochaites* or *S. pimpinellifolium* introgressions, in five locations worldwide. This study revealed that 22 out of 25 (88%) of the quantitative factors exhibited, in at least one location, the phenotypic improvement that had been predicted in the previous QTL analysis of the BC_3 populations (Bernacchi et al. 1998a). This indicated that most of the QTLs detected in the BC_3 were not spurious and can be manipulated via MAS. However, the significance at which QTLs were detected in the BC_3 families as well as the degree of conservation of QTL across locations, as detected in the initial QTL analysis, seemed to be only modest predictors for QTL significance and QTL × environment interactions observed in the derived NILs. This study also highlighted the high frequency with which unexpected phenotypes (both beneficial and detrimental), not predicted by the original QTL analysis, can be observed in the derived NILs. Possible explanations for these findings include linkage drag, pleiotropic effects or epistatic interactions that could not be detected in the BC_3 mapping population perhaps because of the higher overall level of variance still present in these AB populations as compared to NILs.

What is particularly interesting is the magnitude of the improvements observed in the NILs compared to the control elite variety for many of the target traits (Bernacchi et al. 1998a). For example, per-location gains over the elite control (cv. E6203) ranged from 6 to 22% for soluble solids content; from 14 to 33% for fruit color; from 20 to 28% for total yield, from 15 to 48% for red yield, and from 9 to 59% for Brix° × red yield. The results obtained for soluble solids and yield can be compared to the historical genetic gains achieved for the same traits in processing tomatoes using conventional breeding strategies. For example, the per-location gains achieved for the trait Brix° × yield, after 4 years required for the first cycle of AB-QTL breeding using *S. habrochaites* as donor parent, ranged from 9 to 59%, resulting in yearly gains between 2.3 to 14.8%. For the same trait, the estimates of annual genetic gains obtained with conventional breeding were only of 0.9% and 1.5% for the Israeli and California data, respectively (Grandillo et al. 1999).

Given the amenable properties of introgression lines and the potential of exotic germplasm as a source of valuable genetic variation for the improvement of complex traits, Zamir (2001) proposed investing in the development of a genetic infrastructure of "exotic libraries" in order to enhance the rate of progress of introgression breeding based on wild species resources. An exotic library consists of a set of ILs, each of which carries a single, possibly homozygous, marker-defined chromosomal segment that originates from a donor exotic parent, in an otherwise homogeneous elite genetic background. The set of ILs would together represent the entire donor genome. Initially, one disadvantage of this type of population was the long time required for their development; however, the availability of numerous marker-screening technologies has now made the construction of such libraries a more efficient process that can be completed after ten generations of crossing and marker analysis (Young 1999).

In tomato, the first exotic library with genome wide coverage was developed by Eshed and Zamir (1995) from a cross between the wild green-fruited species *S. pennellii* (LA716) and the cultivated tomato *S. lycopersicum* (cv. M82). The library consisted of 50 RFLP-defined ILs and allowed the identification of yield-associated QTLs (Eshed and Zamir 1995). The results of that study highlighted the higher QTL mapping power of IL populations compared with conventional segregating populations such as F_2, BC_1, or RILs (Zamir and Eshed 1998). For example, while Eshed and Zamir (1995) detected a minimum of 18 and 23 QTLs for fruit size and Brix°, respectively, only a maximum of seven and four QTLs, respectively, were detected for the same traits when using standard mapping populations.

The efficiency of these libraries of ILs in detecting and mapping QTLs underlying traits of agronomic importance is due to the near-isogenic nature of the lines, such that any phenotypic difference between the cultivated recurrent parent and an IL, or the hybrid of the recurrent parent with an IL (ILH) is attributable only to the exotic parent genomic segment. This increases the ability to statistically identify smaller effect QTLs, as the "overshadowing effect" of independent major QTL is eliminated. Since the lines are identical to the recurrent parent, except for the single introgressed segment, their phenotypes generally resemble that of the cultivated parent. This reduces sterility and other problems that occur in breeding-population structures characterized by a high frequency of the exotic parent genome, and allows the lines to be evaluated for yield-associated traits. Another advantage of using IL libraries for QTL analysis is the simplicity of the statistical procedure used to detect QTL, which is based on the comparison of each IL with the recurrent parent for the trait of interest, and is therefore less affected by the issue of experiment-wide error. In addition, the epistatic effects that are mediated by other regions of the donor genome, with the exception of the loci contained in the same introgression line, are eliminated. Finally, the permanent nature of the IL libraries allows testing the phenotypic value of each introgression on multiple replicates. Replicated trials of the same line can be analyzed in different years and/or environments, which enable the estimation of the extent of QTL by environment interactions (Monforte et al. 2001; Liu et al. 2003; Gur and Zamir 2004). This provides more accurate estimates of the mean phenotypic values and increases the power of QTL detection. The permanent nature of these lines also allows several laboratories to collect data for different traits on the same lines, thereby facilitating the integration of data from independent studies and the creation of a comprehensive phenotypic database for general access (Zamir 2001).

The map resolution of a library of ILs is defined by the overlap between contiguous segments (bins) to which genes or QTL can be assigned by juxtaposing the lines (Pan et al. 2000; Liu et al. 2003). Bin lengths vary across the genome, depending on the number, length, and overlap of adjacent segments. Generally the map resolution of the initial ILs contributing to an IL library is relatively low; however, the ILs represent the starting point to develop appropriate sub-ILs that will allow the phenotypic effects of QTL to be fine-mapped to smaller intervals (Paterson et al. 1990).

In order to increase the mapping resolution of the *S. pennellii* exotic library, an additional 26 sub-ILs have been added, resulting in 76 lines which partition the entire genetic map into 107 bins defined by singular or overlapping segments (Pan et al. 2000; Liu et al. 2003). Over the past decade the 76 ILs and their hybrids have been evaluated for 20 different yield-associated, fruit morphology and biochemical traits. The resulting data are presented, in silico, in a search engine called "Real Time QTL" that displays a range of statistical and graphical outputs that describe in a user-friendly way the components of the genetic variation (http://zamir.sgn.cornell.edu/; Gur et al. 2004).

The *S. pennellii* IL library was successfully used to identify QTLs for several other traits including disease resistance (Astua-Monge et al. 2000), leaf dissection (Holtan and Hake 2003), fruit nutritional and antioxidant contents (Rousseaux et al. 2005), and tomato aroma (Tadmor et al. 2002). In the latter study, the analysis of the ripe fruits' volatiles of chromosome 8 ILs allowed the identification of *malodorous*, a wild species allele negatively affecting tomato aroma that was selected against during domestication.

Compared to populations segregating for an entire genome, ILs are a more powerful genetic tool to study the phenotypic effects of QTL interactions to better understand the nature of epistasis (Tanksley 1993). To address this issue, Eshed and Zamir (1996) crossed ten different homozygous *S. pennelli* ILs in a half-diallel scheme. The phenotypic values of the 45 double heterozygotes were compared to their respective single heterozygotes and the cultivated control for yield-associated traits (Eshed and Zamir 1996). A high proportion (28%) of the total tested interactions were epistatic ($P < 0.05$), and the detected epistasis was predominantly less-than-additive, i.e., the effect of the double heterozygotes was smaller than the sum of the effects of the corresponding single heterozygotes. The authors suggested the less-than-additive mode of interaction as an underlying genetic model to explain canalized characters, where the phenotype is kept within narrow boundaries despite genetic and environmental disturbances (Eshed and Zamir 1996). This implies that for quantitative traits affected by a large number of QTLs, the less-than-additive interaction ensures that a "loss" of an allele influencing a fitness trait will have a minimal effect on the phenotype.

ILs can also be used to obtain more precise estimates of the magnitude of QTL × genetic background interaction (G) (Eshed and Zamir 1995, 1996; Monforte et al. 2001; Gur and Zamir 2004; Lecomte et al. 2004b; Chaïb et al. 2006). The magnitude of QTL × G interaction is an important issue whenever QTL alleles identified in one genetic background need to be transferred via MAS into other genetic backgrounds. Monforte et al. (2001) found that QTL alleles on the *S. peruvianum* chromosome 4 IL (TA1160) maintained their effects across the four processing-type lines used as testers. Similarly, Gur and Zamir (2004) found that the effects on Brix° × yield of three *S. pennelli* introgressions carried in the line IL789 were consistent in all hybrid combinations obtained with four inbred tester lines. On the other hand, Lecomte et al. (2004a) found significant differences among three genetic backgrounds in the improvement of quality traits.

More recently, Semel et al. (2006) used the 76 *S. pennellii* ILs to assess the contribution of overdominant (ODO) effects on heterosis in the absence of epistasis. For this purpose, the *S. pennellii* ILs have been evaluated in the field, as homozygous and heterozygous lines, for 35 different yield and fitness traits. QTL analysis identified a total of 841 QTLs, and ODO QTLs were detected only for the reproductive traits. These results suggested that the true ODO model involving a single functional Mendelian locus is a more likely explanation for the heterosis observed in the ILs than the pseudoODO model.

To date, the *S. pennellii* exotic library is the most studied whole-genome IL population in tomato. Nevertheless, efforts are being invested in the public as well as in the private sectors to generate new library resources starting from interspecific crosses with different wild species of tomato.

For example, from the *S. habrochaites* (LA1777) AB-QTL population (Bernacchi et al. 1998a, b), Monforte and Tanksley (2000a) developed a set of 99 ILs and backcross recombinant inbred lines (BCRILs). Some of the lines contain multiple introgressions, and together they cover approximately 85% of the wild genome.

Similarly to *S. pennellii*, *S. habrochaites* is a green-fruited wild species; however, while *S. pennelli* is a drought tolerant species, *S. habrochaites* (which originates from high altitudes) is a cold tolerant species (Stevens and Rick 1986) (Table 1). *S. habrochaites* possesses several other characteristics that make it an attractive target as a source of useful QTLs for breeding including dense trichomes, insect resistance, and the synthesis of novel complex secondary metabolites (Stevens and Rick 1986; Van der Hoeven et al. 2000). The AB-QTL studies conducted by Bernacchi et al. (1998a, b) have provided evidence that this wild species also represents a good source of QTL alleles to improve yield and quality. Preliminary QTL studies conducted on some of the *S. habrochaites* ILs have shown that the lines differ in many traits including yield, leaf morphology, trichome density, as well as fruit characteristics such as shape, size and color. Favorable wild QTL alleles were detected for several of these traits (Monforte and Tanksley 2000b; Monforte et al. 2001; Yates et al. 2004). In addition, the lines differ in biochemical composition including anthocyanin content (Oyanedel 1999), soluble solids content of the fruit (Monforte and Tanksley 2000b; Monforte et al. 2001) and sesquiterpenes (Van der Hoeven et al. 2000). In order to improve the efficiency of introgression breeding based on this wild species, a new set of *S. habrochaites* ILs, each containing single homozygous wild introgressions, is currently being developed. This set of ILs should ensure whole genome coverage and increase the mapping resolution of the population (S Grandillo and SD Tanksley, pers. comm.).

Starting from the *S. pimpinellifolium* (LA1589) BC_2 population developed as part of the AB-QTL strategy, Doganlar et al. (2002c) derived a set of 196 inbred backcross lines (IBLs). On average the frequency of the cultivated parent allele in the IBL population was 88%, a value consistent with the expected value of 87%. These results confirmed that the recurrent parent's genome was progressively recovered through two backcrosses and four generations of selfing. The 196 lines were evaluated for 22 quantitative traits and a total of 71 significant QTLs were identified, for 48% of which the wild allele improved agronomic performance. Although this population cannot be defined as an "exotic library" yet, as the lines still contain multiple wild introgressions, they represent good starting material for the development of a whole genome population of ILs.

Another exotic library has been developed using the wild tomato-like nightshade *Solanum lycopersicoides* (LA2951) as donor parent, in the genetic background of cultivated tomato (*S. lycopersicum* cv. VF36) (Chetelat and Meglic 2000; Canady et al. 2005). *S. lycopersicoides* is native to high elevations (up to 3,600 m) in the Andes, where it survives exposure to freezing temperatures, and is resistant to several insect pests and pathogens that negatively impact tomato production (Chetelat and Meglic 2000). The population of *S. lycopersicoides* ILs consists of a primary subset of 56 lines which ensure coverage of approximately 96% of the *S. lycopersicoides* genome. Lines are homozygous whenever possible with a minimum number of introgressed segments per line. A secondary subset of 34 lines provides increased map resolution for specific regions. For this secondary subset, homozygotes were not recovered for certain introgressed segments, and several lines are maintained as heterozygous.

Recently, Finkers et al. (2007) reported the development of an IL population (BC5S2) of *S. habrochaites* LYC4 in the genetic background of *S. lycopersicum* cv. Moneymaker, an indeterminate growing tomato, by using AFLP as the platform for MAS. The population consists of 30 ILs (15 of which contain a single introgression), and together the lines provide an estimated coverage of 95% of the wild species genome. The population was used to identify QTL for resistance to *Botrytis cinerea*.

In addition to the aforementioned IL populations developed by the public sector, there are several other resources of exotic libraries deriving from crosses with different wild tomato species that have been developed by private companies (Peleman and van der Voort 2003; JD Peleman, pers. comm.). Moreover, several sets of ILs have been developed in *Arabidopsis thaliana* (Koumproglou et al. 2002), as well as in many other crop plant species including rice (Li et al. 2005b; Tian et al. 2006), wheat (Pestsova et al. 2006), barley (von Korff et al. 2004), lettuce (Jeunken and Lindhout 2004) and melon (Eduardo et al. 2005), and similar population structures have been developed and used in mouse genetics (Singer et al. 2004). These studies reflect an increasing interest within the scientific community to develop permanent congenic populations as efficient tools to understand the genetic and molecular bases of complex traits.

1.8.3 QTL Fine-mapping and Cloning Using Introgression Lines

Each IL generally contains only a small percentage of the donor genome (less than 5%), however these introgressions often influence several traits, including some undesirable ones. Therefore, it is often necessary to fine map QTL within an IL to reduce the length of the donor introgression. This allows not only the assessment of whether the effect on the phenotype is due to a single QTL or to several linked QTLs affecting the same trait, but also the verification of whether possible undesirable effects are caused by linkage drag of other genes or by pleiotropic effects of the selected QTLs (Eshed and Zamir 1995; Monforte and Tanksley 2000b; Monforte et al. 2001; Fridman et al. 2002; Frary et al. 2003b; Yates et al. 2004). Besides reducing linkage drag, the development of lines with smaller introgressions (sub-ILs) allows the identification of molecular markers more tightly linked to the QTL of interest, which can be used for MAS.

By substitution mapping it was possible, for example, to break linkages between poor yield, low fruit weight and high solids in a *S. habrochaites* IL mapping to the bottom of chromosome 1 (Monforte and Tanksley 2000b) or between orange fruit color and high sugars in a *S. chmielewskii* introgression (Frary et al. 2003b), and to distinguish between linkage drag and pleiotropy for several *S. habrochaites* QTL alleles mapped to the bottom of tomato chromosome 4 (Monforte et al. 2001). This same region of chromosome 4 has been more finely mapped using a series of lines containing small overlapping introgressions from *S. peruvianum* and *S. habrochaites* (Yates et al. 2004). The results show that QTLs for soluble solids content, fruit weight and stem scar are not allelic between the two wild species, which suggests that it may be possible to combine the *S. habrochaites* and the *S. peruvianum* alleles in a single line with the potential of obtaining improved lines characterized by extremely high soluble solids content.

ILs have proven to be invaluable starting material for the positional cloning of key genes underlying quantitative traits (Frary et al. 2000; Fridman et al. 2000, 2004; Yano et al. 2000; El-Din-El-Assal et al. 2001; Takahashi et al. 2001; Kojima et al. 2002). In tomato, one such QTL is *fw2.2*, a major fruit-size QTL that is believed to have played a major role during the domestication of this crop (Alpert and Tanksley 1995; Frary et al. 2000; Tanksley 2004). Natural genetic variation at this locus alone can change the size of fruit up to 30%, with the cultivated tomato allele contributing to this increase in fruit size (Frary et al. 2000). Cloning of *fw2.2* has shown that the locus codes for a repressor of cell division mainly during the cell division phase of fruit development (Frary et al. 2000; Cong et al. 2002). Large-fruit alleles of *fw2.2* are associated with a higher mitotic index during the cell division stage just after anthesis (Cong et al. 2002). FW2.2 is homologous to other plant proteins, but none of them has a known biological function. Interestingly, comparative sequencing of *fw2.2* locus in the genus *Solanum* showed that the fruit-weight phenotype was associated with variation in a few nucleotides in the promoter region rather than in the coding region (Nesbitt and Tanksley 2002), and the natural variants at the promoter of *fw2.2* were correlated with subtle changes in transcript levels as well as in the timing of gene expression (heterochronic allelic variation) (Cong et al. 2002).

Another example is given by *Brix9-2-5*, a *S. pennelli* QTL that increases sugar yield of tomato, that was mapped to a SNP in a gene encoding a flower- and fruit-specific apoplastic invertase (LIN5), which operates in sugar transport to the developing fruit (Fridman et al. 2000). QTL analysis representing five different tomato species delimited the functional polymorphism of *Brix9-2-5* to a single amino acid near the catalytic site of the invertase crystal, a point mutation that can alter enzyme kinetics and fruit sink strength (Fridman et al. 2004). Therefore, this study has also highlighted the power of using multi-species

IL resources for high-resolution analysis of complex phenotypes.

The use of ILs, which isolates a single QTL region, transformed the task of QTL cloning into one similar to that performed for simple Mendelian traits, with the exception that phenotyping requires more detailed measurements. Although the strategy of delimiting a QTL to a single gene using genetic approaches is extremely powerful and unbiased, it is still a time-consuming and technically demanding process (Fridman et al. 2000, 2004). Any additional information that could be associated with the observed traits in the ILs would therefore be useful in identifying the allele(s) responsible for a particular phenotype. Genomic technologies and methodologies that enable integration of the genetic components of QTL variation in genomic databases can help to accelerate the rate of QTL discovery. Furthermore, integrated strategies can reduce the list of candidate genes for target QTL (Wayne and McIntyre 2002).

Along these lines, the *S. pennellii* IL population has been used to test the potential of the candidate gene approach to identify candidate genes for QTL influencing the intensity of tomato fruit color (Liu et al. 2003), tomato fruit size and composition (Causse et al. 2004), and ascorbic acid content in fruit (Stevens et al. 2007). In all studies QTLs were mapped for the quantitative traits of interest along with the mapping analysis of genes encoding, respectively, enzymes of the carotenoid biosynthetic pathway, enzymes involved in the fruit primary carbon metabolism, and enzymes for the ascorbic acid synthesis and turnover pathways. While Causse et al. (2004) found a number of clear links between the presence of *S. pennellii* alleles of these genes and the observed trait, in the study conducted by Liu et al. (2003) on fruit color the number of QTLs that co-segregated within the same bins that contained the candidate gene was close to the number that was expected by chance alone. In a study conducted by Stevens et al. (2007), QTLs for ascorbic acid content were mapped not only on the *S. pennellii* IL population but also on two additional mapping populations. Among the candidate genes mapped, co-locations with mapped QTLs for ascorbic acid content were found between a monodehydroascorbate reductase (MDHAR) gene and a QTL mapped on bin 9-D, and a GDP-mannose epimerase (GME) gene and a QTL mapped on bin 9-J.

In order to further define the biochemical traits that are altered in each line, Overy et al. (2005) conducted an initial metabolomic profiling of the fruit pericarp of the two parents of the *S. pennellii* IL population and of six selected ILs. Principal Component Analysis (PCA) of the metabolite profiles revealed subtle differences in metabolism of the ILs when compared to their parents. A more comprehensive metabolic profiling of the *S. pennellii* IL population was pursued by Schauer et al. (2006), using a high-throughput gas chromatography-mass spectrometry (GC-MS) metabolite protocol in parallel with whole-plant phenotypic characterization (see Sect. 1.16.2). This approach allowed the identification of 889 single-metabolite QTLs, in addition to many other metabolic QTLs that influenced numerous compounds in a metabolic pathway, and 326 loci that modified yield-associated traits. The analysis indicated that at least 50% of the metabolic loci were associated with QTLs that influenced whole-plant yield-associated traits, and harvest index was identified as a regulator of the metabolite content of the mature fruit pericarp. The observation that plant morphology is a major factor affecting the metabolic composition of fruit at harvest time, suggests that this phenotype might regulate biological processes at various molecular levels.

A transcriptional profiling approach via cDNA microarray analysis was used by Baxter et al. (2005) on six non-overlapping *S. pennellii* ILs that share the common trait of increased ripe fruit soluble solids content and increased accumulation of fruit carbohydrate. This study provided evidence of genome-wide transcriptional changes and revealed links to mapped QTL and described traits (see Sect. 1.15.3).

Therefore, in the -omics era, ILs provide a new paradigm to improve the efficiency in discovery, candidate gene identification and cloning of target QTL. This can be achieved by combining the results derived from QTL position, DNA sequences, expression profiling data, and functional and molecular diversity analyses of candidate genes (Li et al. 2005b). However, in order to take advantage of the large amount of data that will be generated it is necessary to develop user-friendly bioinformatics management systems that will allow the integration of the entire range of statistical outputs derived from QTL analysis with genome information including gene content, expression and function (Gur et al. 2004).

1.8.4 QTL Pyramiding

The results obtained from the tomato QTL mapping studies along with those obtained for other crops suggest that substantial improvements in yield-associated phenotypes as well as other agriculturally important traits will unlikely be achieved with the introgression of a single QTL (Lawson et al. 1997; Ashikari and Matsuoka 2006). On the other hand, MAS pyramiding of newly discovered favorable wild QTL alleles from the same or from different wild donor species to obtain multiple-introgression lines could greatly improve crop performance.

From the tomato AB-QTL project, four favorable wild QTLs were identified that had positive effects on Brix° and Brix° × yield; these QTLs were pyramided into a single multi-QTL NIL whose performance in Brix° and Brix° × yield was far beyond any other commercial cultivar and nearly doubled the Brix° × yield over the original starting material (cv. E6203) (SD Tanksley, pers. comm.). A MAS pyramiding approach was pursued by Gur and Zamir (2004) to develop a multiple-introgression line (IL789) carrying three independent yield-promoting genomic regions derived from the drought-tolerant green-fruited wild species *S. pennellii* (Table 1) into the genetic background of the cultivated recipient genotype M82. The IL789 was crossed with four inbred tester lines, whose hybrids with M82 exhibit the highest Brix° × yield values, which permitted the assessment of the potential of the wild QTL in the context of high-yield genetic backgrounds – those close to the "yield barrier"– (Gur and Zamir 2004). The hybrids between IL789 and the four testers yielded 50% more than the leading commercial tomato hybrid (BOS3155) that was used as a control. This higher performance was observed under both wet and dry (10% irrigation) field conditions. The observation that the wild introgressions exerted the expected effects in diverse genetic backgrounds suggested that alleles similar to those of the wild species are not present in the cultivated tomato gene pool.

Overall these results indicate that QTL pyramiding represents a successful approach for producing new varieties; it also provides concrete evidence for the value of exotic germplasm as a rich source of new valuable QTL alleles. As more wild accessions are screened by means of the AB and "exotic libraries" breeding strategies, it will be these new combinations of QTLs, from various accessions, that will greatly improve the performance of elite germplasm.

1.8.5 Advanced-backcross and Introgression-line Breeding: Future Prospects

The processes of plant domestication and plant breeding have produced crops which exhibit characteristics that humans appreciate: high yield, good flavor, and large edible parts. However, these same processes have led to the diminished genetic variability of many cultivated species, leaving them vulnerable to new diseases and pests and making further improvements impossible without the input of new variation. The results obtained from the AB-QTL and IL strategies applied in tomato as well as in other crops have shown that the vast genetic variation available in the wild relatives of these crops, many of which are available in gene banks, is not only exploitable but can have great positive effects on all agronomic traits, including yield and other quantitative traits that have been intractable to MAS techniques. The AB-QTL and IL strategies not only make the exploitation of new genetic variation more efficient and practical, but create permanent resources from which plant breeders can draw for future needs.

1.9 Viral Resistance

Finding useful markers for MAS of disease resistance requires several conditions. These include selection of a durable resistance source, use of characterized pathogen strains, a robust phenotyping method, sufficient polymorphism among parents, and simple genetic control of the resistance trait. Many of the resistance genes or QTLs are pathogen strain or race-specific. Therefore, it is essential to characterize the target pathogen population in order to select proper resistance sources and specific resistance markers for MAS.

Since the pioneering study of the *Tm-2a* resistance gene (Young et al. 1988), at least 11 virus resistance genes and one QTL locus, all originating from wild relatives of tomato, have been mapped and associated with DNA-based molecular markers which allow for MAS (Table 11). Three of these 12 genes/QTLs (*Sw-5*, *Tm-2*, *Tm-2*^{2}) have been cloned as a result of the molecular marker work (Folkertsma et al. 1999; Brommonschenkel et al. 2000; Lanfermeijer et al. 2003, 2005).

Table 11. Resistance genes and markers for the most common tomato viral diseases

Agent and disease	Resistance gene(s)	Molecular marker	Chromosome	Source	Reference
Alfamovirus					
Alfalfa mosaic virus (AMV)	*Am*	AFLP	6	*S. habrochaites*	Parrella et al. 2004
Begomoviruses					
Tomato chlorotic mottle virus (TCMV)	*tcm-1*	na[a]	na	*S. peruvianum*/ *S. lycopersicum*[b]	Giordano et al. 2005; Ji et al. 2007
Tomato mottle virus (ToMoV)	*Ty-3*	RAPD, SCAR, CAPS	6	*S. chilense*	Griffiths 1998; Griffiths and Scott 2001;
	QTL	RAPD	6	*S. chilense*	Ji and Scott 2005; Ji and Scott 2006; Ji et al. 2007
Tomato yellow leaf curl virus (TYLCV)	*Ty-1*	RFLP	6	*S. chilense*	Zamir et al. 1994; Chagué et al. 1997;
	Ty-2	RFLP	11	*S. habrochaites*	Ji and Scott 2005; Agrama and Scott 2006;
	Ty-3	RAPD, SCAR, CAPS	6	*S. chilense*	Hanson et al. 2006; Ji and Scott 2006;
	QTL	RAPD	6	*S. pimpinellifolium*	Ji et al. 2007
	na	na	6	*S. cheesmaniae*	
	na	na	6	*S. peruvianum*	
Cucumovirus					
Cucumber mosaic virus (CMV)	*Cmr*	RFLP	12	*S. chilense*	Stamova and Chetelat 2000
Curtovirus					
Curly top virus (CTV)	Multigenic	na	na	*S. peruvianum*	Martin and Thomas 1986; Thomas and Mink 1998
aka[c] *Beet curly top virus* (BCTV)				*S. habrochaites* *S. pimpinellifolium*	
Poleroviruses					
Potato leaf roll virus (PLRV)	na	na	na	*S. peruvianum*	Hassan and Thomas 1988; Thomas and Mink 1998
Tomato yellow top virus (TYTV)	na	na	na	*S. peruvianum*	Hassan and Thomas 1988; Thomas and Mink 1998
Potexvirus					
Pepino mosaic virus (PepMV)	na	na	na	*S. chilense* *S. habrochaites*	Soler and Nuez 2000; Picó et al. 2002
Potato virus X (PVX)	na	na	na	*S. lycopersicum*	Rashid et al. 1989

[a] Not available
[b] The germplasm source of *tcm-1* is not definitively known

Table 11. (continued)

Agent and disease	Resistance gene(s)	Molecular marker	Chromosome	Source	Reference
Potyvirus					
Potato virus Y (PVY)	na	na	na	*S. pimpinellifolium*	Légnani et al. 1997; Parrella et al. 2002
and *Tobacco etch virus* (TEV)					
	pot-1	AFLP	3	*S. habrochaites*	
Tobamovirus					
Tomato mosaic virus (ToMV)	*Tm-1*	SCAR	2	*S. habrochaites*	Ohmori et al. 1996;
aka *Tobacco mosaic virus* (TMV)	*Tm-2*	Cloned	9	*S. peruvianum*	Lanfermeijer et al. 2003, 2005
	Tm-2[a] (aka *Tm-2*^{2})	Cloned	9	*S. peruvianum*	
Tospoviruses					
Groundnut bud necrosis virus (GBNV)	*Sw-5*	PCR, Cloned	9	*S. peruvianum*	Folkertsma et al. 1999
Groundnut ringspot virus (GRSV)	*Sw-5*	PCR, Cloned	9	*S. peruvianum*	Boiteux and Giordano 1993
Tomato chlorotic spot virus (TCSV)	*Sw-5*	PCR, Cloned	9	*S. peruvianum*	Boiteux and Giordano 1993
Tomato spotted wilt virus (TSWV)	*Sw-1a*	na	na	*S. peruvianum*	Finlay 1953; Stevens et al. 1992;
	Sw-1b	na	na	*S. lycopersicum*	Stevens et al. 1996b; Roselló et al. 1998;
	sw-2	na	na	*S. lycopersicum*	Folkertsma et al. 1999;
	sw-3	na	na	*S. lycopersicum*	Brommonschenkel et al. 2000;
	sw-4	na	na	*S. lycopersicum*	Roselló et al. 2001; Scott et al. 2005a;
	Sw-5	PCR, Cloned	9	*S. peruvianum*	Gordillo et al. 2007
	Sw-6	na	9	*S. peruvianum*	
	"Sw-7"[d]	na	na	*S. chilense*	

[c] Also known as

[d] *Sw-7* is not recognized in the literature as the gene responsible for this resistance at the time of this publication. Statistical evidence has been insufficient to report a single gene, nevertheless, all available evidence suggests resistance is conferred by a single dominant gene (Stevens unpublished data)

The 12 genes/QTLs represent resistance to six virus genera, *Alfamovirus*, *Begomovirus*, *Cucumovirus*, *Potyvirus*, *Tobamovirus*, and *Tospovirus* (Table 11), which are pathogenic to tomato. Interestingly, resistance to all the major viral diseases of tomato has been identified from wild tomato species. Furthermore, several new viral resistance genes have been successfully introgressed from wild relatives into commercially released lines during the last decade and MAS has played a major role in this process. MAS is of major importance within a wide variety of commercial tomato breeding companies, especially when breeding for disease resistance (MR Stevens, pers. comm.).

1.9.1 *Alfamovirus*

The Alfalfa mosaic virus (AMV; genus *Alfamovirus*; family *Bromoviridae*) infects over 600 species from 250 genera belonging to 70 plant families (Parrella et al. 2004). Although this disease is generally not considered a problem in tomato, it is distributed worldwide and has caused infection in tomato fields grown close to alfalfa (Zitter et al. 1991). The disease can be spread from 22 species of aphid vectors and some evidence suggests that the virus can be seed transmitted. These observations, combined with the discovery of necrotic strains of the virus have led to research on resistance in tomato (Finetti-Sialer et al. 1997; Parrella et al. 2000, 2004).

Accessions from *S. lycopersicum* var. *cerasiforme*, *S. pimpinellifolium*, *S. habrochaites*, *S. peruvianum*, and *S. pennellii* were screened for resistance to AMV (Parrella et al. 1997) and resistance was identified in three accessions of *S. habrochaites*. Additional work identified a single dominant gene, *Am* (Parrella et al. 1998), which confers resistance to a necrotic strain of the virus. Initially, *Am* was tagged using AFLP markers in conjunction with bulked segregant analysis (BSA) of resistant and susceptible segregates of 120 plants from a BC population derived from *S. habrochaites* parents (Parrella et al. 2000). Parrella et al. (2004) identified five AFLP markers linked to *Am* from 109 AFLP primer combinations. Subsequently, *S. habrochaites* ILs developed by Monforte and Tanksley (2000a) were employed to map *Am* to the short arm of chromosome 6 near the centromere. Curiously, *Am* is found in a disease resistance hotspot where *Mi* (nematode resistance; Vos et al. 1998), *Cf-2*/*Cf-5* (resistance to *Cladosporium fulvum*; Jones et al. 1993), *Ty-1* (resistance to TYLCV *Tomato yellow leaf curl virus;* Zamir et al. 1994), *Ty-3* (resistance to both TYLCV and ToMoV *Tomato mottle virus*; Ji and Scott 2006), *Ol-1* (resistance to *Oidium lycopersicum;* Van der Beek et al. 1994), and *Bw-5* (resistance to *Ralstonia solanacearum*; Thoquet et al. 1996a) are located.

1.9.2 *Begomoviruses*

Tomato chlorotic mottle virus (TCMV), ToMoV, and TYLCV are members of the *Begomovirus* genus (family *Geminiviridae*). These three viruses, for which resistance genes in tomato have been described, are among at least 35 viruses in this genus (Fauquet et al. 2003). These whitefly vectored viruses can be divided into two groups, those with a monopartite or bipartite genome (Ji et al. 2007). The cultivated tomato has provided little resistance to these viruses (Picó et al. 1998; Ji et al. 2007). The cultivar "Tyking" is the only reported *S. lycopersicum* source of resistance (believed to have been derived from a landrace); however, there is some evidence that the resistance was derived from a wild tomato species (Giordano et al. 2005; Ji et al. 2007). To date, there are five wild relatives of tomato (*S. habrochaites*, *S. chilense*, *S. pimpinellifolium*, *S. cheesmaniae*, *S. peruvianum*) that are purported to have resistance to various *Begomoviruses* (Zamir et al. 1994; Chagué et al. 1997; Giordano et al. 2005; Ji and Scott 2005, 2006; Agrama and Scott 2006; Hanson et al. 2006; Ji et al. 2007). Molecular markers have been identified that are linked to each of the four single genes/QTLs (*Ty-1*, *Ty-2*, *Ty-3*, *tcm-1*, Table 11) that reportedly provide resistance to *Begomoviruses*.

Ty-1 was derived from *S. chilense* and positioned on the short arm of chromosome 6, tightly linked to the RFLP marker TG97 (Zamir et al. 1994) and is part of the *Am*, *Mi*, *Cf-2*/*Cf-5*, *Ty-3*, *Ol-1* and *Bw-5* complex. The *Ty-2* gene, derived from *S. habrochaites*, was originally mapped between the RFLP markers TG36 and TG393 on the long arm of chromosome 11 (Hanson et al. 2006). Additional research refined the position of *Ty-2* to the telomeric end of the long arm of chromosome 11 and tightly linked to the CAPS marker TG105A (Ji et al. 2007). *Ty-3* is also part of the aforementioned resistance gene complex on chromosome 6, and is the most recent of *Begomovirus* resistance genes described (Ji and Scott 2006). This gene is flanked by the CAPS markers cLET-31-P16 and C2_At5g41480, which are separated by 0.5 centiMorgans (cM) (Ji and Scott 2006; Ji et al. 2007). Initial work to identify the location of *Ty-3* was accomplished with RAPD markers, which were later converted to SCAR markers.

The only recessive single gene (*tcm-1*) conferring resistance to a *Begomovirus* identified to date, was found using an F_6 line derived from "Tyking" with resistance to TCMV (Giordano et al. 2005). However, neither the genetic map location nor molecular markers linked to this gene was reported. "Tyking" has shown resistance to *Begomoviruses* in several studies (Fargette et al. 1996; Lapidot et al. 1997; Picó et al. 1998; Ji et al. 2007). Interestingly, a genomic region in the vicinity of *Ty-1* and *Ty-3* possibly originated from wild tomato species (Ji et al. 2007).

Resistance to ToMoV and TYLCV has been reported from several QTLs (Chagué et al. 1997; Griffiths 1998; Griffiths and Scott 2001). In each of these studies, RAPD markers were linked to some of the QTLs, all of which were located on chromosome 6. It is not likely that these QTLs are identical between each report since the studies were done on germplasm derived from two different species (*S. chilense* and *S. pimpinellifolium*), although they could be allelic.

1.9.3 *Cucumovirus*

Cucumber mosaic virus (*CMV*; genus *Cucumovirus*; family *Bromoviridae*) is aphid vectored and has a wide host range of over 800 plant species (Palukaitis et al. 1992). CMV is considered one of the more important tomato viruses in Europe and the Mediterranean (Stamova and Chetelat 2000). There are no reports indicating that the cultivated tomato has any natural resistance to CMV. However, resistance to CMV has been reported in several wild relatives of tomato (Gebre-Selassie et al. 1990; Stamova et al. 1990; Nitzany 1992; Phills et al. 1977; Stamova and Chetelat 2000). Nevertheless, introgression of that resistance into tomato has only been reported once (Stamova and Chetelat 2000). The dominant *Cmr* resistance gene, derived from *S. chilense*, was identified between the RFLP markers TG68 and CT79 on the short arm of chromosome 12.

1.9.4 *Curtovirus*

The *Curly top virus* (CTV), also known as *Beet curly top virus* (BCTV), is vectored by leafhoppers and belongs to the genus *Curtovirus*, and is similar to *Begomoviruses*. It is a member of the *Geminiviridae* family. Sources of resistance to this disease have been identified in *S. peruvianum*, *S. habrochaites*, and *S. pimpinellifolium* (Martin et al. 1971). After several decades of BCTV resistance research, Martin and Thomas (1986) concluded that the best method to prevent tomatoes from becoming infected with BCTV involved avoidance of the disease. This conclusion arose from repeated failures to introgress resistance into cultivated tomato from wild relatives. Nevertheless, a genetic solution may be possible. Thomas and Mink (1998) reported F_5 lines derived from (*S. lycopersicum* × *S. peruvianum*) interspecific hybrids that demonstrated an immunity type response after several different experiments. However, no additional studies of the progress of introgression of those "immunity" genetics into tomato have been reported.

1.9.5 *Poleroviruses*

Potato leaf roll virus (PLRV) and Tomato yellow top virus (TYTV) are both aphid-transmitted viruses which belong to the genus *Polerovirus* in the family *Lufeoviridae*. The only resistance reported for these viruses was identified in interspecific hybrids between *S. lycopersicum* and *S. peruvianum*. However, no additional research has been reported on either linked molecular markers or the inheritance of this resistance (Hassan and Thomas 1988; Thomas and Mink 1998).

1.9.6 *Potexvirus*

Both Pepino mosaic virus (PepMV) and Potato virus X (PVX) are in the *Potexvirus* genus in the *Flexiviridae* family. PepMV was first reported in greenhouse tomatoes in Europe in 1999 (van der Vlugt et al. 2000) and has since been found in tomatoes of North and South America (French et al. 2001;

Soler et al. 2002). Picó et al. (2002) reported that *S. chilense* and *S. habrochaites* both indicated resistance to PepMV. At this time, there is no published report as to the incorporation of this resistance into cultivated tomato. However, it is anticipated that in the near future there will be tomato lines resistant to this disease. More than one tomato breeding program is working on introgression into tomato of PepMV resistance from relatives of tomato (MR Stevens, pers. comm.).

Very little has been published on tomato resistance to PVX. A probable reason behind the lack of studies is that PVX has limited impact on tomato except in areas where the crop is grown near potato production. This situation has been especially noted in Pakistan and regions surrounding that country where the single published study was conducted (Rashid et al. 1989). Three tomato cultivars with moderate resistance were reported (Rashid et al. 1989).

1.9.7 *Potyvirus*

Potato virus Y (PVY) and Tobacco etch virus (TEV) are in the *Potyvirus* genus and belong to the family *Potyviridae*. PVY has had strain specific resistance reported making it difficult to widely use resistant germplasm (Nagai et al. 1992; Green and Hanson 1996; Légnani et al. 1997). Resistance has been reported in *S. habrochaites* and *S. pimpinellifolium* (Nagai et al. 1992; Stobbs et al. 1994; Green and Hanson 1996; Légnani et al. 1997; Thomas and McGrath 1988). From these studies only one gene was mapped using molecular markers, *pot-1* on chromosome 3 from *S. habrochaites* (Parrella et al. 2002). AFLPs were used to find markers linked to *pot-1*. Once identified as linked, fragments were cloned and mapped using the *S. habrochaites* (Monforte and Tanksley 2000a) ILs. The position of *pot-1* was located between TG135 and CT31 on chromosome 3.

1.9.8 *Tobamovirus*

Tobacco mosaic virus (*TMV*) is the type species of the *Tobamoviruses*, which have not yet been assigned to a virus family. TMV is closely related to Tomato mosaic virus (ToMV), and although the two species are distinct in protein composition, they have historically been considered synonymously. However, the species that predominates in tomato production worldwide is ToMV (Brunt 1986).

Resistance to these viruses is widely used for disease control and was bred into cultivated tomato from *S. habrochaites* and *S. peruvianum*. The source from *S. habrochaites* is a single dominant gene, *Tm-1*, which was mapped close to the centromere of chromosome 2 with several co-segregating RFLP markers (Tanksley et al. 1992). Ohmori et al. (1995b) identified six RAPD markers linked to *Tm-1*, which were subsequently converted to two codominant and four dominant SCAR markers (Ohmori et al. 1996). The *Tm-1* gene, however, has been overcome by several naturally occurring ToMV strains, which has reduced its importance in breeding programs (Lanfermeijer et al. 2003).

S. peruvianum was utilized in several cases to breed for resistance to TMV and ToMV. A single dominant gene, *Tm-2*, was identified from several sources including PI 126926 (Laterrot and Pecaut 1969), while another single dominant gene was identified from PI 128650 (Alexander 1963). The gene from PI 128650 was determined to be allelic to *Tm-2*, and was named *Tm-2*^{2} (Pecaut 1965). Because *Tm-2*^{2} provides resistance to a broader range of pathogenic strains than *Tm-2*, Cirulli and Alexander (1969) hypothesized that *Tm-2*^{2} may be a closely linked gene, rather than an allele of *Tm-2*. Based on this observation, Cirulli and Alexander (1969) suggested that *Tm-2*^{2} be called *Tm-2*^{a}, and both names have been used in the literature.

The *Tm-2*^{a} gene was the first virus resistance gene in tomato to be tagged by molecular markers. Young et al. (1988) utilized NIL to map *Tm-2*^{a} near the centromere on the long arm of chromosome 9, tightly linked (0.4 cM) to two RFLP markers, TG79 and TG101. Subsequently, Ohmori et al. (1995a) identified a total of 13 RAPD markers linked to the *Tm-2* locus, three of which were linked in coupling with the *Tm-2*^{a} allele and ten that were linked in coupling with the *Tm-2* allele. Further cloning and characterization of these markers produced three dominant SCAR markers linked to both *Tm-2* and *Tm-2*^{a} (Sobir et al. 2000). The *Tm-2*^{2} allele was subsequently cloned using a transposon tagging approach where only individuals with a disrupted *Tm-2*^{2} gene survived (Lanfermeijer et al. 2003). Lanfermeijer et al. (2003) determined that the *Tm-2* locus contained a single gene for which gene-specific PCR primers were designed. These primers were subsequently utilized to isolate the *Tm-2* allele (Lanfermeijer et al. 2005), which verified that *Tm-2* and *Tm-2*^{2} are allelic. From the limited sequence polymorphism at the locus, Lanfermeijer et al. (2005) developed CAPS markers to differentiate the *Tm-2*, *Tm-2*^{2}, and *tm-2* (susceptible) alleles. The molecular characterization of this locus has provided valuable insight into resistance to ToMV along with a PCR-based marker to distinguish alleles of resistant material in breeding programs. The determination of allelic composition is critical since the *Tm-2* allele has been overcome by some naturally occurring ToMV strains while the *Tm-2*^{2} allele has provided solid resistance.

1.9.9 *Tospoviruses*

Tomato spotted wilt virus (*TSWV*), which is vectored by at least ten species of thrips (Whitfield et al. 2005), is the type species of the genus *Tospovirus* (family *Bunyaviridae*) and was the only tospovirus name recognized until the early 1990s. Related tospo species that infect tomato include *Groundnut bud necrosis virus* (GBNV; Folkertsma et al. 1999), *Groundnut ringspot virus* (GRSV; Boiteux and Giordano 1993), and *Tomato chlorotic spot virus* (TCSV; Boiteux and Giordano 1993). Tospoviruses infect over 1,000 species of plants in both monocots and dicots (Parrella et al. 2003) and have been identified in all major tomato growing areas of the world.

Resistance to TSWV was first reported from wild tomato relatives in 1930 (Samuel et al. 1930). Early studies of the inheritance of resistance in tomato identified five resistant genes/alleles summarized by Stevens et al. (1992; Table 11). All five of these genes were strain-specific and had little resistance application beyond the area in which they were found. None are broadly used in present day tomato breeding programs for international markets (Paterson et al. 1989; Stevens et al. 1992; Gordillo et al. 2007). However, a single dominant TSWV resistance gene (*Sw-5*) was reported in a line from South Africa (van Zijl et al. 1986; Stevens et al. 1992). *Sw-5* is derived from *S. peruvianum*, which has since been shown to be resistant to a number of species of tospoviruses (Table 11). Since the identification of *Sw-5*, an additional allele of *Sw-5* and two more genes (*Sw-6* and *Sw-7*) have been reported (Roselló et al. 1998; Gordillo et al. 2007). To date only *Sw-5* has had molecular markers associated with it, although work is underway to find markers linked to *Sw-7* (MR Stevens, unpublished data).

Sw-5 is on the long arm of chromosome 9, very close to the telomere, and very tightly linked to the RFLP marker CT220. Several RAPD markers have been linked to this gene (Stevens 1993; Stevens et al. 1995, 1996a, b; Brommonschenkel 1996; Chagué et al. 1996). In addition, CAPS and SCAR-like markers linked to *Sw-5* have been developed (Stevens et al. 1996a; Folkertsma et al. 1999). *Sw-5* has been cloned by two separate teams using map-based cloning techniques (Brommonschenkel and Tanksley 1997; Folkertsma et al. 1999; Brommonschenkel et al. 2000). Since the cloning of *Sw-5*, a gene-specific PCR marker was developed based on its DNA sequence (Garland et al. 2005). According to representatives from private companies (MR Stevens, pers. comm.) and reports in the literature (Śmiech et al. 2000; Langella et al. 2004; Garland et al. 2005; Anfoka et al. 2006), *Sw-5* associated molecular markers are actively used for MAS.

1.10 Bacterial Resistance

1.10.1 Bacterial Canker

Few studies have been conducted to evaluate the degree of variation of bacterial canker caused by *Clavibacter michiganensis* subsp. *michiganensis*. Strains of *C. michiganensis* subsp. *michganensis* were highly genetically similar (>80%) based on repetitive-sequence-based PCR (rep-PCR) fingerprinting (Louws et al. 1998). Nonetheless, the pathogen could be grouped into four distinct genetic types designated A through D. Studies on mapping for resistance have used strains that were either highly aggressive or belonged to one of the predominant genetic groups. Sources of resistance to *C. michiganensis* subsp. *michiganensis* have been identified in several wild tomatoes. Durability of resistance from LA407 against several predominant strains in the USA has been proven (Francis et al. 2001). However, progress on breeding for resistance to the disease has been limited. This may be due to the complicated genetic control of resistance traits. Resistance loci from LA2157 (*S. peruvianum*) and LA407 (*S. hirsutum*) have been the most studied in terms of molecular genetic mapping; results are summarized in Table 12. At least two or three QTLs were identified with additive gene action and various gene interactions. *Rcm2.0* and *Rcm5.1* have been located in short chromosome intervals of 4.4 and 2.2 cM, respectively. Molecular markers within the intervals could be used for MAS.

1.10.2
Bacterial Speck

Strains of race 0 of *Pseudomonas syringae* pv. *tomato* causing bacterial speck are predominant in the world. This makes the single semi-dominant resistance gene, *Pto* highly durable. The original source of *Pto* is *S. pimpinellifolium*, possibly PI 370093 (Pitblado and Kerr 1980; Kerr and Cook 1983; Pitblado et al. 1984; Martin et al. 1991); the gene was named by Pitblado and MacNeill (1983). Due to its semi-dominant nature, disease is occasionally observed on commercial hybrids carrying only one copy of *Pto* (Buonaurio et al. 1996). *Pto* is tightly linked to *Fen*, a gene controlling sensitivity to fenthion. Thus, sensitivity to fenthion can be used to select for *Pto*. To confer resistance to race 0 of *P. syringae* pv. *tomato*, both *Pto* and *Prf* genes are required. Both genes have been cloned (Loh and Martin 1995; Salmeron et al. 1996), and thus, gene-based markers can be designed and used in MAS (Table 12).

1.10.3
Bacterial Spot

Molecular breeding for tomato bacterial spot caused by *Xanthomonas* spp. has been constrained by several factors. First, four species and five races of the pathogens have been described. The causal agents include *X. euvesicatoria* (race T1), *X. vesicatoria* (race T2), *X. perforans* (race T3, T4, T5), and *X. gardneri* (race T2) (Jones et al. 2004, 2005). Secondly, resistance traits in several known sources are quantitatively inherited. Thirdly, the three resistance sources Hawaii 7998, Hawaii 7981, and PI 114490 are either *S. lycopersicum* or *S. lycopersicum* var. *cerasiforme*, which are genetically highly similar to cultivars. Mapping results for resistance are summarized in Table 12. Hypersensitive resistance to race T1 was found in Hawaii 7998, which is associated with at least three QTLs (Yu et al. 1995). However, the hypersensitive response of Hawaii 7998 only contributed partially to field resistance. Therefore, evaluation of resistance reactions was later based on symptom severity on seedlings or in the field (Yang et al. 2005b). Molecular marker Rx3-L1 has been used in MAS for selecting *Rx3* and its efficacy has been demonstrated (Yang et al. 2005a). In locations where race T3 and T4 are predominant, *Xv4* would be a useful resistance gene. Markers TG599 and TG134 were located 8.4 or 11.9 cM from *Xv4* (Astua-Monge et al. 2000). Additional efforts in developing more closely linked markers would be necessary for applying MAS. This should be tractable because the origin of *Xv4* is *S. pennellii* (Table 12). The *Bs4* gene identified in cv. Money Maker (Ballvora et al. 2001) would not be useful in controlling bacterial spot on tomato, because it interacts with *avrBs4*, which is present in *Xanthomonas* strains that do not infect tomato (Bonas et al. 1993). Resistance to multiple races was found in PI 114490 and resistance to all races was not controlled by the same genes. However, a region on chromosome 11 was held in common for resistance to races T2, T3 and T4. The challenge to find useful resistance markers from PI 114490 has been the low polymorphism among *S. lycopersicum* lines.

1.10.4
BacterialW ilt

Ralstonia solanacarum is a species complex that causes bacterial wilt in tomato. The pathogen has been classified into five races, six biovars, and four phylotypes (Denny 2006). Strains of *R. solanacearum* are highly variable in their aggressiveness and cause variable severity of disease on tomatoes (Jaunet and Wang 1999). Therefore, it is important to use a well-characterized strain for resistance mapping studies. Sources of resistance to bacterial wilt in tomato have been identified primarily in cultivated tomatoes (Scott et al. 2005b). Classical genetic studies suggested a few major genes together with several minor genes conditioned resistance to bacterial wilt in tomato. Among the resistance sources, mapping has focused on L285 and Hawaii 7996, particularly the latter (Table 12). This is because of its durability at multiple locations and against diverse strains (Scott et al. 2005b). Mapping loci from Hawaii 7996 focused on populations derived from a cross with WVa700. A region on chromosome 6 linked to TG180 and TG240, which possibly contained two QTLs (Mangin et al. 1999), was associated with resistance against different pathogen strains and in different environments

Table 12. Molecular markers linked with resistance to bacterial canker, bacterial speck, bacterial spot, and bacterial wilt in tomatoes

Resistance source	Strain or Race	Mapping population[a]	QTL and markers in flanking region (shown in parentheses)[b]	Reference
Bacterial canker caused by *C. michiganensis* subsp. *michiganensis*				
LA2157 *S. peruvianum*	Cm542	F_2 of Solentos × LA2157	Three QTLs on chromosome 5 (TG363), 7 (TG61), 9 (TG254) with additive and co-dominant effects; SCAR markers based on TG61 on chr 7 were developed	van Heusden et al. 1999
LA407 *S. habrochaites*	A300, A226 and C290 belonged to predominant genetic groups	64 IBLs; F_2 of selected IBLs carrying Rcm2.0 or Rcm5.1 crossed with Ohio 86120 for map positioning	***Rcm2.0*** on chromosome 2 (TG537 to TG091; 4.4 cM); ***Rcm5.1*** on chromosome 5 (CT202 to TG358; 2.2 cM) with additive effects and additive-by-additive epistatic gene interactions	Kabelka et al. 2002; Coaker and Francis 2004
Bacterial speck caused by *Pseudomonas syringae* pv. *tomato*				
Rio Grande-PtoR *S. lycopersicum*	Strain PT11 (race 0)	F_2 of Rio Grande-PtoR and Rio Grande	***Pto*** on chromosome 5 co-segregated with TG538	Martin et al. 1993b
Rio Grande-76R	Strain T1 (race 0)	F_2 of a cross between a *prf* mutant (*prfPto*/*prfPto*) and Rio Grande (*Prfpto*/*Prfpto*)	***Prf*** is 0.12 cM from *Pto*; physically *Prf* lies between *Pto* (24 Kb from *Prf*) and *Fen* (500 bp from *Prf*)	Salmeron et al. 1996
Bacterial spot caused by *Xanthomonas* species				
Hawaii 7998 (H7998) *S. lycopersicum*	Race T1	H7998 x (H7998 × LA 716)	Three non-dominant and independent QTLs, i.e. ***rx-1*** (TG236) and ***rx-2*** (TG157) on chromosome 1 and ***rx-3*** (TG351) on chromosome 3	Yu et al. 1995
	Race T1	F_2 of Ohio 88119 × H7998; ABLs and IBLs for linkage confirmation	***Rx3*** (Rx3-L1, CosOH73, TOM49) on chromosome 3 contributed to reduced bacterial population (25%) and field resistance (41%)	Yang et al. 2005b
LA716 *S. pennellii*	Race T3	F_2 of H 7998 x LA716	A single dominant gene, ***Xv4*** (TG599 to TG134) on chromosome 3 conferred HR	Astua-Monge et al. 2000
Money Maker (MM) *S. lycopersicum*	Strains contained *avrBs4* that could induce hypersensitive responses on cultivated tomatoes	F_2 of MM × LA716	***Bs4*** co-segregated with TG432 on chromosome 5	Ballvora et al. 2001

[a] $F_{2:3}$: F2 derived F3 families
[b] Names of QTL or resistance genes are indicated in bold

Table 12. (continued)

Resistance source	Strain or Race	Mapping population[a]	QTL and markers in flanking region (shown in parentheses)[b]	Reference
PI 114490 *S. lycopersicum*	Race T2, T3, T4	166 IBL derived from OH 9242 × (FL 7600 × PI 114490)	A common region on chromosome 11 (SSR637, TOM196, TOM144) conferred resistance to races T2, T3, and T4	Yang et al. 2005a
Bacterial wilt caused by *Ralstonia solanacearum*				
L285 *S. lycopersicum*	UW364 (race 1, biovar 4)	71 F_2 of a cross between L285 and CLN286BC1F2-25-14-7	Three QTLs on chromosome 6 (CT184), 7 (TG51b), 10 (CT225b)	Danesh et al. 1994
Hawaii 7996	GMI8217 (race 1, bv1)	F_2 of H7996 x WVa700	Three QTL on chromosomes 4 (TG268), 6 (TG118; a major QTL), and 11 (GP162)	Thoquet et al. 1996a
S. lycopersicum		F_2 and $F_{2:3}$ of H7996 × WVa700	Six QTL on chromosomes 3 (GP226), 4 (K12 to GP165), 6 (TG178 to TG118 to TG73), 8 (CD40), 10 (CP105), and 11 (D6b to O10); Temporal analysis supports the presence of two QTL on chromosome 6 linked to TG180 and TG240	Thoquet et al. 1996b; Mangin et al. 1999
	Pss4 (race 1, bv3)	$F_{2:3}$ of H7996 × WVa700	Five QTL on chromosomes 2 (GP504), 6 (TG73; *Cf2*), 8 (CT135) and 12 (TG564); QTL linked to TG564 were associated with 70% of total variation	Wang et al. 2000
	JT519 (race 1)	$F_{2:3}$ and RIL of H7996 × WVa700	Three QTL on chromosomes 6 (***Bwr-6***; TG73), 8 (***Bwr-8***; CD40 to CT135), and 12 (***Bwr-12***; TG564)	Carmeille et al. 2006
	JT516 (race 3)	$F_{2:3}$ and RIL of H7996 × WVa700	Three QTLs on chromosomes 3 (***Bwr-3***; TG515), 4 (***Bwr-4***; CD73) and 6 (***Bwr-6***; TG73)	Carmeille et al. 2006

(Thoquet et al. 1996a, b; Wang et al. 2000; Carmeille et al. 2006). QTLs found within other chromosomal locations were not consistent over experiments, and terms such as strain-specific (Wang et al. 2000) or phylotype-specific (Carmeille et al. 2006) have been proposed. With the availability of a permanent mapping population, i.e., RIL population of Hawaii 7996 × WVa700, specificity of QTL can be verified. No known markers are suitable for MAS to transfer resistance from Hawaii 7996 to susceptible cultivars. This is mainly due to the low polymorphism between Hawaii 7996 and WVa700, which could possibly be solved by developing SNP markers.

1.11 Fungal Resistance

Conventional breeding is solely responsible for the integration of fungal disease resistance into commercial varieties (Scott and Gardener 2007). Tomatoes are affected by fungal pathogens at all stages of their life cycle and the majority of fungal resistances are due to single dominant genes (Laterrot 1997; Scott 2005). Table 13 and the following discussion reviews resistance genes and QTLs of important fungal diseases that have been mapped or are used in tomato breeding.

1.11.1 Alternaria Stem Canker

Alternaria stem canker caused by *Alternaria alternata* f. sp. *lycopersici* was first reported in California, but is now found in many tomato production areas (Grogan et al. 1975). Dark brown to black cankers form on infected plants near the soil line or just above the ground. This disease can girdle and destroy the plant. Foliar and fruit symptoms may also occur. Alternaria stem canker resistance is controlled by a single dominant gene designated *Asc*, which is incompletely dominant to a host-specific pathotoxin (Grogan et al. 1975; Clouse and Gilchrist 1987). Witsenboer et al. (1989) originally reported *Asc* to be located on chromosome 3 and linked to the *solanifolium* gene. RFLP markers TG134 and TG442 were reported to be tightly linked to *Asc* (van der Biezen et al. 1995; Mesbah et al. 1999).

1.11.2 Anthracnose Ripe Rot

Anthracnose ripe rot is a common disease caused by *Collelotrichum coccodes* and other *Colletotrichum* species (Barksdale 1972). Fruit exhibiting anthracnose symptoms have sunken, dark, circular lesions on ripe fruit. This disease is primarily found on processing tomatoes grown in humid regions. Resistance was first reported in 1964 (Robbins and Angell 1970) as partially dominant and polygenic. However an F_2 population studied by Stommel and Zhang (1998, 2001) identified three regions associated with resistance. The transfer of resistance into adapted breeding lines from unadapted, small-fruited *S. pimpinellifolium* germplasm was impeded, and the phenotypic variation explained by any single QTL was small.

1.11.3 Blackmold

A mold of ripe fruit, blackmold, is caused by the fungus *Alternaria alternata* and can lead to severe harvest losses in processing tomatoes (Pearson and Hall 1975). The disease progression begins with dry sunken lesions on the fruit which may extend into fruit cavities. RAPD genotyping of *A. alternata* identified two major phenetic groups and genetic diversity of the pathogen (Morris et al. 2000). QTLs for blackmold resistance were mapped by Robert et al. (2001) and found on chromosomes 2, 3, 9, and 12.

1.11.4 Corky Rot

Corky root rot is caused by *Pyrenochaeta lycopersici* Schn. & Ger. and causes root lesions that form a corky texture progressively destroying the root system (Gerlach and Schneider 1964). Resistance in the form of a recessive gene, *py-1*, has been integrated into cultivars from *S. peruvianum* (Laterrot 1983) and is found on the short arm of chromosome 3 between markers TG40 and CT31 (Doganlar et al. 1998). A new source of resistance controlled by a single dominant gene was reported by Stamova (2004) from a tomato line called Pirelly 38. Field tests with *py-1* and the new

Table 13. Resistance genes and markers for the most common tomato fungal diseases

Disease (pathogen)	Resistance gene(s) and/or QTLs	Molecular marker(s)	Chromosome(s)	Source	Reference
Alternaria Stem Canker (*Alternaria alternata* f. sp. *lycopersici*)	*Asc*	RFLP	3	*S. pennellii*	van der Biezen et al. 1995; Stommel and Zhang 1998; Mesbah et al. 1999
Anthracnose ripe rot (*Colletotrichum coccodes*)	Several QTLs	RAPD	Various	*S. lycopersicum*	Stommel and Zhang 1998, 2001
Blackmold (*Alternaria alternata*)	Several QTLs	na[a]	2, 3, 9, 12	*S. cheesmaniae*	Robert et al. 2001
Corky Root (*Pyrenochaeta lycopersici*)	*py-1*	RFLP	3	*S. peruvianum*	Doganlar et al. 1998
Early Blight (*Alternari solani*)	Several QTLs	na	Various	*S. lycopersicum*, *S. habrochaites*, *S. pimpinellifolium*	Foolad et al. 2002; Zhang et al. 2003; Foolad and Sharma 2005
Fusarium Crown and Root Rot (*Fusarium oxysporum* f. sp. *radidicis-lycopersici*)	*Frl*	linked to $Tm2^2$ gene	9	*S. peruvianum*	Vakalounakis et al. 1997
Fusarium Wilt (*Fusarium oxysporum* f. sp. *lycopersici*)	*I*, *I-1*, *I-2*, *I-2C*, *I-3*	RFLP	7, 11	*S. pimpinellifolium*, *S. pennellii*	Bournival et al. 1990; Sarfatti et al. 1991; Tanksley and Costello 1991; Ori et al. 1997; Simons et al. 1998
Gray Leafspot (*Stemphyllium* spp.)	*Sm*	RFLP	11	*S. pimpinellifolium*	Behare et al. 1991
Late Blight (*Phytophthora infestans*)	*Ph-1*, *Ph-2*, *Ph-3*, *Ph-4* and several QTLs	RFLP	7, 9, 10 and all chromosomes	*S. pimpinellifolium*, *S. habrochaites*	Pierce 1971; Frary et al. 1998; Moreau et al. 1998; Chunwongse et al. 2002; Brouwer and St. Clair 2003; Brouwer et al. 2004; Kole et al. 2006
Leaf Mold (*Cladosporium fulvum*)	*Cf-1*, *Cf-2*, *Cf-4*, *Cf-5*, *Cf-9*	RFLP	1, 6	*S. habrochaites*, *S. pimpinellifolium*	Jones et al 1993; Balint-Kurti et al. 1994; Lauge et al. 1998

[a] Not available

Table 13. (continued)

Disease (pathogen)	Resistance gene(s) and/or QTLs	Molecular marker(s)	Chromosome(s)	Source	Reference
Powdery Mildew (*Leveillula taurina* and *Oidium neolycopersici*)	*Lv, Ol-1, ol-2, Ol-3, Ol-4, Ol-5*	RFLP, PCR, RFLP, SCAR, CAPS, AFLP	4, 6, 12	*S. chilense, S. lycopersicum, S. habrochaites, S. neorickii*	Chunwongse et al. 1994; van der Beek et al. 1994; Huang et al. 2000; Bai et al. 2003; De Giovanni et al. 2004
Verticillium Wilt (*Verticillium dahliae* and *V. albo-atrum*)	*Ve*	RFLP, SCAR	9	*S. lycopersicum*	Schaible et al. 1951; Kawchuk et al. 1998; Diwan et al. 1999

resistance showed *py-1* did not resist the pathogen from the infested soil. However, more work needs be to be done on the inheritance of the gene and effectiveness of this new resistance against other isolates.

1.11.5 Early Blight

The fungus *Alternaria solani* causes the disease early blight and is found worldwide in locales with frequent rainfall or heavy dew. Early blight lesions are characterized by concentric rings of brown or black giving a classic target appearance. In addition, a yellow area surrounds the spots which coalesce to engulf the entire leaf, stem, or fruit. Initial efforts focused on incorporating stem resistance, particularly collar rot resistance from *S. lycopersicum* (Andrus et al. 1942), but also conferred a moderate level of leaf and fruit resistance. Along with additional sources of resistance (Foolad et al. 2000), early blight resistant cultivars have been developed which reduce the frequency of fungicide applications necessary for control, but do not eliminate the need for chemical control (Shoemaker and Gardner 1986; Gardner and Shoemaker 1999). Zhang et al. (2003) identified QTLs associated with early blight resistance in wild tomato species, but Graham et al. (2005) were not successful in identifying resistance from the *S. habrochaites* ILs (Monforte and Tanksley 2000).

1.11.6 Fusarium Crown and Root Rot

Fusarium crown and root rot (*Frl*) caused by the pathogen *Fusarium oxysporum* f. sp. *radidicis-lycopersici*, is often found in European greenhouses, but can also be found in locations with cool weather. Infected plants wilt similarly to those infected with fusarium wilt, but vascular browning is found only in the crown of the plant. Resistance is controlled by a single dominant gene, *Frl* (Berry and Oakes 1987; Vakalounakis 1988) derived from *S. peruvianum* and mapped to chromosome 9 (Vakalounakis 1997). Researchers selecting for the $Tm2^2$ gene found linkage to fusarium crown rot resistance (Laterrot and Couteadier 1989; Laterrot and Moretti 1992, 1996).

1.11.7
Fusarium Wilt

Fusarium wilt causes major crop losses worldwide. Currently, there are three races (races 1, 2, and 3) of the causal organism *Fusarium oxysporum* f. sp. *lycopersici*. Yellow older leaves are the first symptoms of the disease, this is followed by entire branches turning yellow, but often only one side or part of a leaf or branch will turn yellow. When the outer part of the vascular tissue is examined a characteristic red brown discoloration can be found which extends up into the plant. This disease can also be seed-borne and survive in the soil for long periods of time. Thus, care must be made to reduce contaminated machinery, infected plant debris and water as the fungus can be easily transported. Bohn and Tucker (1939) reported the first vertical resistance gene, *I*, found on chromosome 11 near TG523 (Scott and Gardner 2007). A gene for resistance to the second race, $I-2$, has also been mapped to chromosome 11 near TG105 (Laterrot 1976; Sarfatti et al. 1989). The $I-2$ gene has been cloned and is a complex locus (Ori et al. 1997; Simons et al. 1998). A gene for resistance to the third race was linked to the isozyme *Got-2* on chromosome 7 (Bournival et al. 1989). A fourth gene, $I-1$, has also been placed on chromosome 7, but is not an allele of $I-3$ (Sarfatti et al. 1991; Scott et al. 2004). Of these genes, two (*I* and $I-2$) have been used most often in breeding cultivars (Scott and Gardner 2007), however the $I-3$ gene is now being used in developing breeding lines and cultivars (Scott and Jones 1995).

1.11.8
Gray Leafspot

A complex of four species of the fungus *Stemphylium*; *S. solani*, *S. floridanum*, *S. botryosum*, and *S. vesicarum*, are responsible for gray leafspot. These fungi are found worldwide on tomatoes and exhibit similar symptoms. Shiny, smooth and gray lesions are found on the underside of leaves and later the foliage has a shot hole appearance. In severe cases, leaves turn yellow, this is followed by leaf drop, and in some cases, defoliation of the entire plant. A single incompletely dominant gene, *Sm*, is responsible for the resistance used today (Hendrix and Frazier 1949). The *Sm* gene is found on chromosome 11 (Behare et al. 1991) linked to the *I* and $I-2$ genes. A study by Kedar et al. (1967) found gametes with *Sm* and *I* were transferred preferentially thereby increasing their use in resistant cultivars.

1.11.9
Late Blight

Phytophthora infestans, an oomycetes close to fungus, causes late blight worldwide on both tomato and potato. Conditions conducive to development of the disease include cool temperatures and cloudy weather, which if ideal, can destroy the crop within a few days. Wild species are the source for high levels of resistance to late blight in both Solanaceous crops. Currently three named genes are being used in tomato breeding: *Ph-1* (Bonde and Murphy 1952; Gallegly 1952), *Ph-2* (Turkensteen 1973; Laterrot 1975) and *Ph-3* (Black et al. 1996). *Ph-1* has been of limited use as strains rapidly overcame the resistance (Conover and Walter 1953). *Ph-2* shows good resistance early, but often fails late in the season (Turkensteen 1973; Laterrot 1975). *Ph-3* has a high level of resistance, but has been overcome (Brusca 2003). These three *Ph* genes have been mapped to chromosomes 7, 10, and 9, respectively (Pierce 1971; Moreau et al. 1998; Chunwongse et al. 2002). QTLs were identified in a *S. habrochaites* accession (Brouwer and St. Clair 2004; Brouwer et al. 2004), but they require markers to efficiently transfer the resistance into cultivated tomato. Recently, Kole et al. (2006) have reported mapping of a gene *Ph-4* on chromosome 2 conferring resistance to late blight using an RIL population derived from a cross between a susceptible cultivated tomato variety and a resistant accession of *S. pimpinellifolium.*

1.11.10
Leaf Mold

Leaf mold is a disease problem found generally in protected culture due to the high humidity requirements of the pathogen, *Cladosporium fulvum.* Older leaves show the yellowish light-green blotches first, followed by a purplish or olive green mold on the lower side of the leaves coupled with a yellowing on the upper leaf surface. Lower leaves of infected plants will eventually turn yellow and drop off. Spread of the fungus is generally via the air or transport by workers or irrigation water (Watterson 1986). Langford (1937) discovered the first resistance gene, *Cf-1*. Twenty-four additional genes have been proposed by Kerr and coworkers in

Canada (Kanwar et al. 1980a, b). Haanstra (2000) revised this number to 19 due to identical specificities of five of the genes, but added two new genes for a total of 21 *Cf* genes. Virulent races of the pathogen appear quickly, thus Lindhout et al. (1989) have suggested at least two genes not overcome by current isolates be used in developing varieties. To date, 18 races have been reported (Lukyanenko 1991), but the nomenclature system is not uniform (Scott and Gardner 2007). Chromosome 5 has *Cf-2* and *Cf-5* with both genes being either allelic or tightly linked (Jones et al. 1993). Balint-Kurti et al. (1994) reported the two genes on chromosome 1 *Cf-4* and *Cf-9* to be allelic. Haanstra et al. (1999) and Haanstra (2000) reported additional loci, *Cf-ECP2* and *Cf-ECP5* also on chromosome 1 and linked to *Cf-4*. Currently breeders are using *Cf-2*, *Cf-4*, *Cf-5* and an unknown gene (Lukyanenko 1991; Scott and Gardner 2007).

1.11.11 Powdery Mildew

Two organisms cause the diseases referred to as powdery mildew: *Leveillula taurina* and *Oidium neolycopersici*. These diseases are primarily found in greenhouse production sites, but can also be found in the field on a wide range of crops. Symptoms begin on the upper leaf surfaces as yellow lesions which later develop into bright yellow irregular patches. From the lesions, necrotic areas develop, engaging the entire leaf. The two fungal species can be distinguished by the way in which the mycelium grows on the leaf (Lindhout et al. 1994). A dominant resistance gene to *L. taurica*, *Lv*, was identified in *S. chilense* (Yordanov et al. 1975) and mapped to chromosome 12 between CT121 and CT129. Proprietary markers are being developed to assist incorporating this resistance gene into cultivars (Scott and Gardner 2007). Several resistance genes have been found to the *Oidium* species (Lindhout et al. 1994; Mieslerova et al. 2000). Two incompletely dominant genes, *Ol-1* and *Ol-3*, located on chromosome 6 near the *Mi* gene were introduced from two *S. habrochaites* accessions and SCAR markers are available (Huang et al. 2000). Chromosome 4 carries the recessive gene, *ol-2*, and CAPS and AFLP markers are available for MAS (De Giovanni et al. 2004). Two additional genes *Ol-4* and *Ol-5* were identified on chromosome 6 (Bai et al. 2003; Bai 2004b). Three QTLs linked to quantitative genes have been identified by Bai et al. (2003). One was located in the same region as *Ol-1* and *Ol-3* on chromosome 6 and the other two on chromosome 12 near *Lv*.

1.11.12 Verticillium Wilt

Caused by the soil fungus, *Verticillium dahliae* and *V. albo-atrum*, Verticillium wilt is geographically distributed across a broad range of hot climates. A single dominant gene, *Ve*, was discovered in 1932 with resistance to race 1 (Schaible et al. 1951). The first resistant cultivars were released in 1952 (Cannon and Waddoups 1952) and today's cultivars carry the same resistance. Tracing the mapping of the *Ve* gene through scientific literature is problematic as it has been linked to chromosomes 4, 7, 9, and 12. In 1999, Diwan et al. (1999) proposed that *Ve* was located on the short arm of chromosome 9 near RFLP marker GP39. A codominant SCAR marker has been developed that should be useful in breeding (Kawchuk et al. 1998).

1.12 Insect Resistance

Tomatoes (*S. lycopersicum*) are susceptible to a wide array of arthropod pests (Lange and Bronson 1981; Berlinger 1986; Kennedy 2003). Currently available cultivars do not have sufficiently high levels of pest resistance to allow for significant reductions in the amount of pesticides used in the crop. Consequently, developing cultivars with increased levels of pest resistance is a major focus of some breeding programs, along with their adoption into integrated pest management programs aimed at reducing pesticide sprays and environmental impact.

The wild tomato species *S. habrochaites, S. peruvianum*, and *S. pennellii* have been reported to be sources of resistance to many tomato arthropod pests (Gentile and Stoner 1968; Gentile et al. 1968, 1969; Rick 1973; Kennedy and Yamamoto 1979; Williams et al. 1980; Ecole et al. 1999; Kennedy 2007). However, introgression of these resistances into the cultivated tomato has been limited by the difficulties in maintaining the necessary uniform infestations to select for resistance (Stevens and Rick 1986). Indirect selection techniques based on correlated traits with high heritability can be used to expedite introgression compared to expensive and slow direct selection methods (Juvik et al. 1982; Mutschler 2006).

Several allelochemicals are present in the wild relatives of the cultivated tomato and are associated with pest resistance: methyl-ketones such as 2-tridecanone in *S. habrochaites* (Williams et al. 1980; Fery and Kennedy 1987; Weston et al. 1989; Eigenbrode and Trumble 1993a, b; Maluf et al. 1997; Gonçalves et al. 1998), sesquiterpenes in *S. habrochaites* (Snyder et al. 1987; Eigenbrode et al. 1994), and acylsugars in *S. pennellii* (Goffreda et al. 1989). These compounds are found in glandular trichomes present in the leaf surface (Williams et al. 1980; Carter and Snyder 1985, 1986; Snyder and Carter 1985; Carter et al. 1989), and are often associated with moderately high or high heritability values (Maluf et al. 1997). Efficient demonstrations of indirect selection methods for high allelochemical content have been reported for 2-tridecanone or sesquiterpene (zingiberene) content and resistance to the South American tomato pinworm (Maluf et al. 1997), and spider mite repellence (Gonçalves et al. 1998). Quick and inexpensive techniques to measure these allelochemicals (Weston and Snyder 1990) for spider mite (*Tetranychus* spp.) repellence could also be indicative of their level of resistance to other pests.

S. pennellii accessions (particularly LA716) have shown a very high level of resistance to the whitefly *Bemisia tabaci/Bemisia argentifolii* complex, as well as to aphids (*Macrosiphum euphorbiae*, *Myzus persicae*), mites and Lepidopteran pests (Gentile et al. 1968, 1969; Juvik et al. 1982; Goffreda et al. 1989), including the South American tomato pinworm *Tuta* (= *Scrobipalpuloides*) *absoluta* (França et al. 1989). The multiple pest resistance of *S. pennellii* is due to the presence of type IV glandular trichomes and the glucose and sucrose esters of fatty acids (acylsugars) that they secrete (Gentile et al. 1968; Goffreda et al. 1989; Shapiro et al. 1994). A reduction of insects on plants has been shown using purified acylsugars, where feeding is reduced by *Macrosiphum euphorbiae* and *Myzus persicae*, a reduction in feeding, larval development and survival of *Helicoverpa zea* and *Spodoptera exigua*, and by reducing oviposition and feeding of the leaf miner *Liriomyza trifolii* and of the silver leaf whitefly *Bemisia argentifolii* (Goffreda et al. 1988, 1989; Hawthorne et al. 1992; Rodriguez et al. 1993; Juvik et al. 1994; Liedl et al. 1995). The presence of type IV glandular trichomes is controlled by at most two unlinked genes in crosses of *S. pennellii* with *S. lycopersicum* (Lenke and Mutschler 1984). A study on the inheritance of type IV trichome density, acylsugar accumulation levels, percentage of acylsugars and leaf area in interspecific populations between *S. pennellii* LA716 and LA1912 identified three major loci and ten loci with modest associations (Blauth et al. 1998). A MAS breeding program was also attempted to transfer the acylsugar trait to cultivated tomato, however the accumulation in plants with the five target regions were lower than that of the interspecific F_1, suggesting another unidentified region is necessary for high accumulation levels (Lawson et al. 1997).

1.13 Drought or Water-Deficit Stress Resistance

1.13.1 Genetics and Drought Resistance Characteristics of Tomato Species

S. lycopersicum is mesophytic; members of the species are not significantly drought or salt resistant. So while there is variability for drought and/or osmotic stress resistance within *S. lycopersicum*, this variability is limited (Foolad and Lin 1999; Srinivasa Rao et al. 2000; Fellner and Sawhney 2001); the best sources of resistance for cultivated tomato are from other species in the genus.

S. pennellii and *S. chilense* are indigenous to arid and semi-arid environments in South America (Rick 1973) (Table 1). *S. pennellii* is adapted to the coastal cliffs of Peru. This location has very low precipitation, but lots of early morning dew. *S. chilense* is adapted to the Atacama desert in northern Chile (Maldonado et al. 2003). This is the most arid temperate desert on the planet. *S. chilense* plants are often found in dry arroyos, with no other vegetation in the area (Rick 1973).

S. pennellii and *S. chilense* produce small green fruit, and they both have an indeterminate growth habit. *S. pennellii* has small thick rounded leaves, light green in color. This species also has a very small root to shoot ratio. *S. chilense* has thin finely divided leaves and a well-developed root system. These morphological differences are well matched to the environments in which they grow. *S. pennellii* appears adapted to arid environments by virtue of high water-use efficiency (WUE) (Martin and Thorstenson 1988) and an ability of its leaves to take up dew (Rick 1973). *S. pennellii* increased WUE under water deficit conditions, 3.42 g/kg at 25% field capacity compared with

2.71 g/kg at 100% field capacity, while the WUE for *S. lycopersicum* did not change in plants grown at these two field capacities 2.22 g/kg (O'Connell et al. 2007). Three major QTL or genetic loci were linked to the water-use phenotype in *S. pennellii* (Martin et al. 1989).

A second feature of the drought response of *S. pennellii* is stomatal response (Kebede et al. 1994). Stomata close rapidly upon water deficit stress; the detached leaf wilt rate reflects this adaptive response. In cultivated tomato, water is lost the most quickly, 6.3% leaf fresh wt/h, in *S. chilense* the rate is slower at 4.2% and in *S. pennellii* the rate is the slowest, 1.2% leaf fresh wt/h (O'Connell et al. 2007). The leaves of *S. pennellii* perceive the reduction in water availability once the leaf is cut off the plant and immediately close all of the stoma, reducing evaporative water loss. The leaves of *L. pennellii* have 20 times the amount of epicuticular lipid as cultivated tomato (Fobes et al. 1985). The cuticle of *S. pennellii* was much thicker than cultivated tomato, 5.5 μm vs. 1.5 μm. This trait also appears to be controlled by multiple genes as most of the F_2 individuals had intermediate cuticle dimensions, ranging between the two parental values (Treviño 1997; O'Connell et al. 2007). Together, the stomatal response and the waxy leaves allow *S. pennellii* to conserve leaf water despite a drying soil.

S. chilense has a very well-developed root system (Sánchez Peña 1999), presumably to explore deep soil layers in arroyos following seasonal rains. Chen and Tabaeizadeh (1992b) observed root growth increase during drought treatments, while leaf area and seedling growth rates were reduced in *S. chilense*. O'Connell et al. (2007) observed enhanced root development in mature *S. chilense* plants under non-stress conditions. *S. chilense* had a longer primary root, and twice the number of secondary roots as cultivated tomato. No differences were observed in distance of the root tip to the youngest secondary root or distance from the root base to oldest secondary root.

Measurements of plant water status during drought episodes also indicated that these two species are more drought resistant than cultivated tomato (Kahn et al. 1993; Sánchez Peña et al. 1995; O'Connell et al. 2007). The time until leaf water potentials dropped to the levels reported at wilt was different for each species. As expected, cultivated tomato usually wilted within 2 to 3 days after water was withheld; equivalent sized *S. pennellii* plants wilted within 4 to 6 days while *S. chilense* plants often wilted only after 15 days without water. Inspection of the leaf, osmotic and turgor potentials of these plants during a drought cycle demonstrate that *S. chilense* increased osmotic potential to much higher values than those observed for cultivated tomato or *S. pennellii*, −2.37 MPa vs. −1.62 or −1.20 MPa. This increase in osmotic potential maintains the leaf turgor potential, such that the leaves do not appear to wilt as soil water decreases; rather the leaves curl and become slightly brittle (Sánchez Peña 1999).

Other species may provide sources of drought resistance or tolerance; *S. pimpinellifolium*, has been investigated as a source of drought resistance during seed germination (Foolad et al. 2003). This species is more commonly investigated as a source for salt stress resistance (see Sect. 1.13.3). In this study, polyethylene glycol was used to generate the water deficit conditions during seed germination. Four QTLs were identified to be associated with seed germination drought tolerance, of which two were contributed by the *S. pimpinellifolium* parent and two by the cultivated tomato parent.

1.13.2 Drought-induced Changes in Gene Expression

In response to drought, salt and other osmotic stresses gene expression profiles change. These changes have been documented at the RNA and protein levels by many research groups, including many cases for tomato and related species (Bray 1988; Cohen and Bray 1990; Ho and Mishkind 1991; Chen and Tabaeizadeh 1992a, b; Thompson and Corlett 1995; Jin et al. 2000). Many of the changes in gene expression appear to be regulated by the plant hormone abscisic acid (ABA). ABA-dependent and independent routes for transcriptional regulation of drought, salt, and cold temperature stresses have been well described for *Arabidopsis* and predicted for other plants (Nakashima and Yamaguchi-Shinozaki 2006). The first event is the perception of the water deficit state, followed by a signal transduction cascade that may be initiated by altered levels or sensitivities to ABA or through other non-ABA based paths. Altered patterns of transcription are then observed, presumably in response to increased levels of specific transcription factors. These new transcripts then result in the accumulation of proteins/enzymes that synthesize osmoticum, stabilize membranes, bind water, thicken wax layers, alter ion flux, close stomata, etc. Also, transcription factors or other types of chromatin proteins may repress transcription. Photosynthesis rates are reduced during drought stress. This physiological change is mediated at a number of levels, but in particular, transcription is reduced for the genes for light harvesting chlorophyll a-b binding protein (*cab*) and Rubisco small subunit (*rbcS*) during drought stress (Bartholomew et al. 1991).

1.13.3 Genetics and Salt Stress Resistance Characteristics in Tomato Species

Several species of tomato are sources of salt resistance: *S. pimpinellifolium, S. cheesmaniae, S. galapagense, S. pennellii* and *S. peruvianum* (reviewed in Flowers 2004; Foolad 2004; Cuartero et al. 2006). As with drought, salt stress may have more than one definition, and resistance or tolerance to salt depends on the developmental stage of the plant (Foolad and Lin 1997). There are several possible physiological processes or components of salt tolerance and mapping of QTL for salt tolerance in tomato has been productive, particularly in interspecific crosses between cultivated tomato and *S. pimpinellifolium* (Foolad 1999; Foolad and Chen 1999; Foolad et al. 1997, 1998, 2001). These analyses have focused on salt tolerance expressed during seed germination and vegetative growth. QTLs for salt tolerance have been mapped to chromosomes 3, 5 and 9. Markers for these loci are predicted to be useful in MAS breeding programs (Foolad 2004).

Salt stress tolerance during fruit development has been investigated primarily in crosses between cultivated tomato and *S. pimpinellifolium* or *S. galapagense* (Monforte et al. 1996, 1997a, b, 1999). In these studies, total fruit weight and its components, fruit number, fruit weight and earliness were estimated and analyzed with molecular markers located throughout the genome. For the *S. lycopersicum* × *S. pimpinellifolium* analysis, a number of QTLs for salt tolerance were detected at loci known to be associated with fruit weight (Monforte et al. 1997a). In addition, there were differences depending on the specific *S. pimpinellifolium* accession and whether the plants were grown under saline or control conditions (Monforte et al. 1997a, b). When *S. galapagense* was the source of salt tolerance, results were confounded by a pleiotropic effect on earliness. F_2 progeny were bimodal for fruit development with an early and late population. The salt tolerance phenotype of *S. galapagense* was not observed among the F_2 in the early population, but could be detected in the late population (Monforte et al. 1999).

1.13.4 Drought Responsive Genes

By screening samples using microarrays, approximately 130 drought responsive genes were identified in *Arabidopsis* (Reymond et al. 2000; Seki et al. 2001). These drought responsive genes were placed into functional classes using the predicted amino acid sequence of the gene to identify possible gene functions; drought responsive genes were found in eleven of the 13 possible functional classes (Bray 2002).

A search of GenBank (March 2007) for drought responsive genes returned the following numbers. For "*Lycopersicon*" in the organism field and ABA, drought, or salt stress, there were 18, 12, and 16 entries, respectively. This search excluded ESTs. When the organism field was "*Arabidopsis*", there were 160, 138, and 51 entries for the same terms, respectively. If we presume that the number of genes annotated in *Arabidopsis* is close to the true number of drought-responsive genes, this suggests that ~10% of the drought responsive genes in tomato have been identified. In potato, over 1,400 transcripts have been identified as unique to potato leaf or root tissue under abiotic stress (Rensink et al. 2005). This set of transcripts included genes expressed in response to cold, drought, salt, and heat stress. While there was good overlap with genes expressed as stress responsive in *Arabidopsis* within these potato transcripts, there were also a number of transcripts that did not match any entry in an in-house non-identical amino acid database (Rensink et al. 2005). A number of the stress responsive transcripts in the potato set matched *Arabidopsis* genes with no functional annotation, which may help improve the annotation of the *Arabidopsis* genome.

The current challenge is to identify genes that confer drought- or salt-resistant phenotypes; multiple genes are predicted to be key as both phenotypes in tomato are inherited as quantitative traits. Further complicating the challenge is the stage-specific nature of the adaptive stress responses, seed germination vs. vegetative growth vs. fruit growth. The assays or screens that compare gene expression within one species with and without the stress are likely to identify genes expressed in all cells in response to the stress (e.g., gene products involved in osmotic adjustment mechanisms of cells). Drought resistance strategies that involve adapted or unique developmental patterns of gene expression or pre-existing states (constitutive expression) are not likely to be identified by screening differential treatments. Identification of these genes will rely on identification of the gene products associated with the QTLs for the trait, or by gene expression studies that include comparisons between genotypes as well as comparisons between environmental conditions over time in the drought or abiotic stress cycle. Microarrays or other high-throughput technologies to assay gene expression, in conjunction with sophisticated modeling of genetic networks, will undoubtedly be required for successfully carrying out these types of studies.

1.13.5 Genetic Engineering of Stress Resistance

Many genes whose expression is increased in response to drought or salt stresses have been cloned and characterized (Plant et al. 1991; Torres-Schumann et al. 1992; Chen et al. 1993, 1994; Fray et al. 1994; Yu et al. 1996; Trevino and O'Connell 1998; Harrak et al. 2001; Tirajoh et al. 2005; Yesbergenova et al. 2005). Transgenic expression of these genes has been used for two different objectives, to determine the function of the protein or gene product in the stress response; or to attempt to confer stress tolerance on the transgenic plant (e.g., Imai et al. 1995; Cortina and Culiánez-Macià 2005; Deguchi et al. 2006). All cases of drought and salt tolerances in wild tomato species are inherited as quantitative traits and were never simply inherited; transgenic plants engineered to express a single gene do not seem likely to result in a robust stress tolerant phenotype. A thorough review of this topic has been presented (Flowers 2004). Genetic engineering of expression of stress responsive genes is still an important approach to determining the biochemical and physiological function of these genes in plant responses to the environment. One example of this type of analysis was on a unique class of chromatin proteins, H1 histones.

1.13.6 Drought-Responsive H1 Histone

The association of H1 histones with nucleosomal arrays is believed to be primarily responsible for the formation of higher orders of chromatin compaction, thus inactivating the transcription of genes in that chromatin region. The family of plant H1 histones can be divided into two groups based on size, amino acid sequence, and expression characteristics (Ascenzi and Gantt 1997). The larger group, equivalent to the somatic linker histones in animals, is comprised of the H1 histones associated with the bulk of the chromatin. The second group is populated with H1 variant or subtypes that have a slightly smaller carboxy terminal domain, and more importantly, the members of this group are drought responsive (Wei and O'Connell 1996; Ascenzi and Gantt 1997; Bray et al. 1999). These drought-responsive H1 histones formed a separate clade in a dendrogram of H1 histones (O'Connell et al. 2007) and accumulated in the nucleus of drought-stressed plants (Ascenzi and Gantt 1997; Scippa et al. 2000). The abundance of the protein, however, is not sufficient to account for a general replacement of the H1 histones in the chromatin (Ascenzi and Gantt 1999b). Rather the abundance and pattern of expression of these variant H1 histones are likely to remodel the chromatin of selected regions of the genome, presumably around genes whose transcription is altered.

As mentioned above, repression of transcription of specific genes occurs during drought stress, notably the transcription of *cab* and *ribulose-bisphosphate carboxylase small subunit* (*rbcS*) (Bartholomew et al. 1991). Transcripts for the drought responsive H1 histone accumulated prior to the repression of *rbcS* transcription in tomato leaves during a drought cycle (Wei and O'Connell 1996). These results are consistent with the hypothesis that the function of the drought induced H1 histone is to repress transcription of genes during drought stress. However there were no changes in transcription of selected drought responsive genes in transgenic *Arabidopsis* overexpressing the drought-induced H1 histone (Ascenzi and Gantt 1999a). These plants did not have any alterations in water content during drought stress or any other measurable phenotypic differences. Scippa et al. (2004) used anti-sense constructs of the drought-induced H1 histone in tomato to investigate the function of this protein. They observed leaf developmental changes and an increased stomatal conductance rate in the transgenic plants with reduced expression of the drought-induced H1 histone. The authors concluded that this gene plays multiple roles during plant development associated with controlling plant water status.

1.13.7 Drought or Water-Deficit Stress Resistance: Summary

The genetic basis for abiotic stress resistances and tolerances in wild tomato species are quantitatively inherited; consequently it is unlikely that a single transgene expressed in cultivated tomato will confer agronomically relevant drought or salt tolerance. A complete understanding of the biochemical and physiological basis of abiotic stress resistances is essential for the development of crops able to yield harvestable product, i.e., tomato fruit, when grown with reduced water or poor quality water. With competing demands for agricultural water by urban development, water availability will continue to be a limitation for crop production. To best utilize a molecular approach, a full understanding of the physiological basis for the drought resistant phenotype is critical.

The best genetic sources of drought resistance for cultivated tomato are wild tomato species. There are two species with very different mechanisms for drought resistance; *S. chilense* invests in root growth, while *S. pennellii* regulates stomatal aperture efficiently during drought stress. Drought-induced alterations in gene expression, most commonly assayed at the transcriptional level, have been observed in a number of systems. Identification of genes associated with drought responses in plants, particularly in tomato, has been achieved through a variety of differential screens and comparisons. Inter-specific comparisons with drought-resistant species have been especially informative. The diversity of well characterized phenotypes in wild tomato species coupled with molecular genetic approaches are likely to result in a full understanding of the variety of ways that plants continue to grow in the face of limited water.

The molecular analysis of drought and salt responses in tomato and related species has identified a number of interesting genes and identified regions of the genome associated with selected stress resistance phenotypes. A candidate gene analysis for those regions or a biochemical or physiological assignment for these loci has yet to be accomplished; the full power of a genomics analysis of these responses has yet to be seen. The comparative analyses possible between cultivated tomato and wild tomato species that have evolved to survive in arid and semi-arid environments continues to be a productive research strategy. If these analyses are conducted using global gene expression (e.g., microarray) technology, then the results may give a more conclusive understanding of genetic responses during conditions of drought stress. Comparison of the genetic network for gene expression in resistant and susceptible genotypes will provide the missing information to understand the different strategies tomato has evolved to grow in environments with reduced or poor quality water. Some of these adaptive responses may be useful to plant breeders developing tomato genotypes for fruit production in areas likely to experience drought or reduced water availability.

1.14 Fruit Color and Nutritional Value

There is a wealth of genetic variability within modern and heirloom tomato cultivars, landraces, and wild species for improvement of fruit nutritive value. Progress in breeding for improved tomato nutritional value is largely influenced by availability of sufficient genetic diversity and knowledge of gene action. Breeding strategies for improved nutritional composition are similar to those for other traits in a cultivar development program. Based on knowledge gained through investigations of the character's heritability, the mode of inheritance, and existing genetic variability, appropriate breeding strategies based on phenotypic selection of individuals or families have been implemented to realize improvements in crop nutritional quality. Development of gene-specific probes and identification of markers tightly linked to phytonutrient constituent loci enables implementation of MAS strategies for genotypic selection of well-studied traits. Recent studies demonstrate renewed interest in characterization of *Solanum* tomato germplasm collections for fruit phytonutrients and devising selection indices to maximize breeding efficiency (Hanson et al. 2004; Rousseaux et al. 2005; Willits et al. 2005). Recent applications of metabolic profiling to tomato ILs and wild species provide new opportunities to link QTL with specific phytonutrients (Overy et al. 2005; Schauer et al. 2005, 2006; Schauer and Fernie 2006) (see Sect. 1.16).

The evidence of accumulated studies indicates a positive link between fruit and vegetable consumption and improved health. By virtue of the volume of tomato products that are consumed, tomatoes make important contributions to the dietary intake of vitamin A (9.5%) and vitamin C (11.5%) (USDA 2002).

Lycopene imparts red fruit color and also acts as a dietary antioxidant. In addition to these well-known vitamins and antioxidants, other compounds in tomato fruit with antioxidant properties include chlorogenic acid, rutin, plastoquinones, tocopherol, and xanthophylls (Beecher 1998; Leonardi et al. 2000). Tomatoes also contribute carbohydrates, fiber, flavor compounds, minerals, protein, fats and glycoalkaloids to the diet (Davies and Hobson 1981; Gundersen et al. 2001). Whereas nutrition studies have often focused on a single dietary phytonutrient, research clearly indicates that there are many bioactive compounds in food products and that it may be the combination of compounds that confer the beneficial health effects described (Laquatra et al. 2005).

1.14.1 Micronutrients

Considerable genetic variation exists in tomato for micronutrients with antioxidant activity or other health conferring properties (Hanson et al. 2004; Schauer et al. 2005). A number of these micronutrients, particularly carotenoids, have long been major objectives of breeding programs because of their contribution to the quality of fresh and processed tomato products. Increased recognition of their health promoting properties has stimulated new research to identify loci that influence their concentration in tomato.

Carotenoids

Consumption of tomato products is associated with decreased risk of cardiovascular disease and certain cancers, including prostate and cervical cancer (Giovannucci et al. 1995; Gerster 1997; Clinton 1998; Giovannucci 1999; Paiva and Russell 1999). β-carotene is well recognized as a pro-vitamin A carotenoid. More recent discussions note the positive association between lycopene and health, but also emphasize that there are a family of beneficial compounds in tomato (Laquatra et al. 2005). Lack of efficacy upon ingestion of purified carotenoid supplements suggests that well-studied carotenoids such as lycopene may act synergistically with other compounds in protecting human health (Ellinger et al. 2006).

Carotene biosynthesis in higher plants has been reviewed extensively (Bartley and Scolnik 1994; Hirschberg 2001). The availability of cDNAs that code for nearly all of the enzymes required for carotenoid biosynthesis in plants has stimulated considerable interest in engineering plants with altered carotenoid content (Cunningham and Gantt 1998). Numerous candidate loci are available for marker-based studies (Table 14). Numerous successful and unsuccessful efforts to alter fruit carotenoid composition by manipulating expression of carotenogenesis transgenes have been reported (Fray et al. 1995; Romer et al. 2000; Rosati et al. 2000; Dharmapuri et al. 2002; Fraser et al. 2002; Mehta et al. 2002). Although transgenic plant development has generally been considered a more expedient route to develop superior cultivars, Zamir (2001) argues that approximately 10 years are required to create a transgenic cultivar for testing and that this time investment is similar to that needed for development and testing of new lines developed in traditional breeding programs using exotic germplasm.

Fruit Color Mutants A large body of genetic data exists for simply inherited genes that influence carotenoid content in tomato. More than 20 genes have been characterized in tomato that influence the type, amount, or distribution of fruit carotenoids. Many of the available color variants were first identified as spontaneous mutants in cultivars of *S. lycopersicum*, but also occur in wild tomato species. The *Beta* (*B*) allele located on chromosome 6 was first characterized in transgressive orange-fruited segregants descended from a cross between *S. lycopersicum* and the green-fruited species *S. habrochaites* (Lincoln et al. 1943; Kohler et al. 1947). Inheritance studies suggested that high concentrations of β-carotene conditioned by *B* were controlled by a single gene exhibiting incomplete dominance (Lincoln and Porter 1950). Subsequent studies by Tomes et al. (1954) suggested that *B* was dominant but subject to influence by a modifier gene, *mo-B*, which segregated independently of *B*. Expression of the dominant modifier reduces β-carotene:lycopene ratios resulting in red-orange fruit. Utilizing molecular markers linked to *B* and *mo-B*, genotypic evaluations discounted incomplete dominance to explain inheritance of fruit carotene content, but revealed that *B* and *mo-B* were linked on chromosome 6 and did not segregate as independent genes (Zhang and Stommel 2000). *B* encodes a novel lycopene β-cyclase that converts lycopene to β-carotene (Ronen et al. 2000). The introduction of *B* from accessions of *S. galapagense*, *S. pimpinellifolium*, *S. chilense*, and *S. chmielewskii* has also been described (Rick 1956; Manuelyan et al. 1975; Chalukova 1988; Stommel and Haynes 1994).

Table 14. Loci that influence tomato fruit carotenoid and anthocyanin content

Locus	Chromosome and arm[a]	Allele synonyms	Synonym	Description	Reference
Carotenoid					
Apricot (*at*)	5	–	–	Fruit flesh yellow with pinkish blush	Jenkins and McKinney 1955
Beta-carotene (*B*)	6 L	–	–	Fruit flesh orange; increased β-carotene, reduced lycopene; encodes a chromoplast-specific lycopene β-cyclase	Lincoln and Porter 1950; Tomes et al. 1954; Ronen et al. 2000
		crimson (*c*)	og^c, *Cm, Cr, cm-2, cr-2*	Enhanced red color; increased lycopene, reduced β-carotene; phenotype similar to B^{og}	Thompson et al. 1967; Ronen et al. 2000
		minutum (*m*)	–	High β-carotene, low lycopene in ripe fruit	Chmielewski and Berger 1966
		old gold (*og*)	–	Corolla tawny orange; increased fruit lycopene	Rick and Smith 1953
Delta (*Del*)	12 S	–	–	Orange-red flesh; enhanced δ-carotene and α-carotene Fractions; encodes lycopene ε-cyclase	Tomes 1963; Ronen et al. 1999
Diospyros (*dps*)		–	–	Fruit tissue is dusky orange	Rick 1967
Green flesh (*gf*)	8 L	–	–	Chlorophyll retained in ripe fruit, normal lycopene synthesis; fruit reddish-brown	Kerr 1958a
Ghost (*gh*)	11 S	–	*ab*	Phytoene synthesis normal, no colored carotenoids; encodes a plastid terminal oxidase	Rick et al. 1959; Scolnik et al. 1987
Green ripe (*Gr*)	1 L	–	*gr*	Resembles *gf*, except that center of fruit turns red	Kerr 1958b; Barry et al. 2005
High pigment-1 (*hp-1*)	2 L	–	*hp, hp1, hp2, bs, dr*	Immature fruit dark green; increased levels of carotenoids and ascorbic acid in mature fruit; encodes the uv-damaged DNA-binding protein1	Clayberg et al. 1960; van Tuinen et al. 1997; Liu et al. 2004
		WB3 (*w*)	–	Similar to *hp-1*, but more extreme phenotype	Peters et al. 1989; Kerckhoffs and Kendrick 1997

[a] L long; S short

Table 14. Loci that influence tomato fruit carotenoid and anthocyanin content

Locus	Chromosome and arm[a]	Allele synonyms	Synonym	Description	Reference
High pigment-2 (*hp-2*)	1 S	–	*hp*	Similar to *hp1*; encodes deetiolated1 protein	Yen et al. 1997; Mustilli et al. 1999
		dark green (*dg*)	–	Immature fruit dark green; increased levels of carotenoids in mature fruit	Konsler 1973; Levin et al. 2003
		Jones (*j*)	*hp*	Similar to *hp2*	van Tuinen et al. 1997
High pigment-3 (*hp-3*)		–	–	Increased accumulation of carotenoids	Chetelat 2005
Intensified pigmentation (*Ip*)		–	–	Enhanced lycopene synthesis	Rick 1974
Modifier of B (*mo-B*)	6 L	–	*mo(B), moB*, I^B, i^B	Modifier of *B*; increases content of β-carotene in presence of *B*	Tomes et al. 1954; Zhang and Stommel 2000
Yellow (*r*)	3 S	–	–	Reduced polyenes, very low levels of carotenes; fruit flesh yellow; encodes phytoene synthase	Rick and Butler 1956; Fray and Grierson 1993
		– (*1s*)	r^2	Fruit flesh yellow	Stubbe 1963
		– (*2s*)	r^3, *r-2, r2*	Yellow fruit flesh; lighter yellow flowers	Stubbe 1960
		provisional4 (*prov4*)	*r*	Fruit flesh yellow	Chetelat 2005
		provisional5 (*prov5*)	*r*	Fruit flesh yellow	Chetelat 2005
		reddish yellow (*y*)	*ry*	Modifier for red color in yellow fruit	Young 1956
Sherry (*sh*)	10	–	–	Fruit flesh yellow with reddish tinge	Zscheile and Lesley 1967
Tangerine (*t*)	10 L	–	–	Orange fruit and stamens; colored carotenoids principally prolycopene; encodes carotenoid isomerase	MacArthur 1934; Isaacson et al. 2002
		– (2)	$t1^2$	Yellowish growing point; light green foliage; resembles *t* in flower and fruit color	Stubbe 1965
		virescent (*v*)	–	Fruit and flower color typical of *t*; irregular yellowing near growing point	Lesley and Lesley 1956
Colorless fruit epidermis (*y*)	1 S	–	–	Unpigmented fruit epidermis; colorless skin over red flesh results in pink fruit	Rick and Butler 1956
Anthocyanin					
Anthocyanin fruit (*Aft*)		–	–	Variable purple pigmentation; anthocyanin in skin and outer pericarp	Giorgiev 1972; Jones et al. 2003
Atroviolacium (*atv*)	7	–		Excess anthocyanin on fruit, stems and leaves	Rick 1964; Clayberg 1972
Aubergine (*Abg*)	10	–	–	Fruit epidermis purple, particularly on shoulder and where exposed to direct light	Rick et al. 1994a

Breakage of the linkage between *B* and *sp*, the gene for indeterminate growth habit, has eliminated this limitation of *B* for use in cultivars intended for commercial production (Stommel 2001; Stommel et al. 2005a, b).

In contrast with *B*, the recessive crimson mutant (*c*) enhances lycopene content at the expense of β-carotene (Thompson et al. 1967; Lee and Robinson 1980). Ronen et al. (2000) demonstrated that *crimson* is an allele of *B* and that null mutations in the *B* gene are responsible for the crimson phenotype. A second allele of *B*, *og*, similarly enhances fruit lycopene content. Crimson cultivars have been developed for their desirable dark red pigmentation. Recognition of health benefits attributable to lycopene in the diet has superseded any negative consequences of the loss in nutrients from reduced levels of β-carotene.

The recessive *tangerine* (*t*) mutant also conditions orange fruit color due to the accumulation of poly-cis-lycopene, also referred to as prolycopene (MacArthur 1934; Tomes 1963). Trans-lycopene is the principal form of lycopene in red tomato fruit. Located on chromosome 10, a clone of the *tangerine* gene, designated *CRTISO*, was shown to encode a carotenoid isomerase required during carotenoid desaturation (Isaacson et al. 2002). Analysis of two alleles of *t* demonstrated that in one case, loss of function in CRTISO was attributable to a deletion mutation in *CRTISO*, and in the second, expression of this gene was impaired. *CRTISO* is normally expressed in all green tomato tissues but is up-regulated during fruit ripening and in flowers. Evidence that cis-lycopene is more bioavailable than trans-lycopene (Boileau et al. 1999; Unlu et al. 2003) has focused considerable interest on this mutant in human nutrition-related studies. Fruit of the *tangerine* mutant also exhibit elevated phytoene and phytofluene. The dominant *delta* (*Del*) allele conditions increased fruit δ-carotene and reduced lycopene content, resulting in reddish-orange colored fruit (Tomes 1963). Ronen et al. (1999) demonstrated cosegregation of the *Crtl-e* locus encoding ε-cyclase with the *Del* mutation located on chromosome 12. ε-cyclase converts lycopene to δ-carotene. Transcript for *Crtl-e* was shown to increase 30-fold in ripening fruit of the *Del* mutant. Additional orange color variants include the *diospyros* (*dps*) mutant with dusky orange fruit.

Introgression of the non-allelic *high pigment-1* (*hp-1*) and *high pigment-2* (*hp-2*) (Van Tuinen et al. 1997; Yen et al. 1997) mutant alleles enhances total fruit carotenoid content 30 to 50% without significantly altering the relative percentage of different carotenoid constituents (Cookson et al. 2003). The *hp-2* allele encodes the tomato homolog of the nuclear protein *DEETIOLATED1* (*DET1*) from *Arabidopsis* that is involved in light signal transduction (Mustilli et al. 1999). The light hypersensitive *dark green* (*dg*) mutant (Konsler 1973) is allelic to *hp-2* (Levin et al. 2003). Additional studies demonstrate that *hp-1* is a mutation in a tomato *UV-DAMAGED DNA-BINDING PROTEIN 1* (*DDB1*) homolog whose *Arabidopsis* counterpart interacts with *DET1* (Liu et al. 2004). Additional alleles of *hp-1* and *hp-2*, *w* and *j*, respectively, have been identified that exhibit varying photoresponsiveness (Kerckhoffs and Kendrick 1997). Studies of these high pigment mutants reveal that light signal transduction regulates the carotenoid pathway in a manner that affects total fruit carotenoid content and that genes encoding components of light signal transduction may provide new genetic tools for manipulating fruit nutritional value (Yen et al. 1997; Liu et al. 2004). Liu et al. (2004) demonstrated that two tomato light signal transduction genes, *LeHY5* and *LeCOP1LIKE*, are positive and negative regulators of fruit pigmentation, respectively. Further studies reported additional putative light responsive genes that modulated carotenoid profiles in fruit of these light hypersensitive tomato mutants (Levin et al. 2004). Mature green fruit of these mutants is characteristically darker green due to elevated chlorophyll content (Baker and Tomes 1964; Palmieri et al. 1978). These high pigment mutations also increase fruit firmness and ascorbic acid levels (Jarret et al. 1984). Plants expressing both *crimson* and high pigment alleles produce fruit with lycopene levels three to four times that of conventional red-fruited tomatoes. Unfortunately, undesirable pleiotropic effects associated with these mutants have thus far limited their practical use.

The gene *Ip*, which has effects similar to that of the high pigment mutants, was described in progeny descended from a *S. lycopersicum* × *S. chmielewskii* cross (Rick 1974). Fruit expressing *Ip* also exhibit dark green immature fruit and intensified carotenoid pigmentation in ripe fruit. Unlike *hp* mutants, *Ip* behaves as a dominant gene and appears to have reduced detrimental effects on seed germination and plant vigor.

A variety of additional fruit color mutants have been characterized. The recessive *r* gene located on chromosome 3 is responsible for yellow fruit flesh, resulting in greatly reduced levels of polyenes and very low levels of colored carotenoids (Rick and Butler 1956). The *r* locus is transcriptionally regulated and corresponds to a null mutation for a chromoplast-specific phytoene synthase, *Psy1* (Fray and Grierson 1993; Fraser et al. 1994). A second phytoene synthase gene, *Psy2*, has been identified which is also expressed in ripening tomato fruit (Bartley and Scolnik 1993). However, its transcripts are relatively more abundant in mature leaves. Lois et al. (2000) proposed that a second gene, *DXS*, encoding the first enzyme of isoprenoid synthesis in plastids, works in concert with *Psy1* to control fruit carotenoid synthesis. The variant *ry* locus is an allele of *r* eliciting red color in yellow fruit (Young 1956). Variations on the yellow *r* mutant include the recessive *apricot* (*at*) locus on chromosome 5 (Jenkins and Mackinney 1955) and *sherry* locus on chromosome 10 (*sh*; Zscheile and Lesley 1967) that result in fruit that are characteristically yellow but with a pinkish/red blush at maturity. Expression of the *ghost* (*gh*) allele on chromosome 11 results in fruit that contain only phytoene and no colored carotenoids due to a block in the desaturation of phytoene (Rick et al. 1959; Scolnik et al. 1987). Incorporation of the recessive *y* allele results in a colorless fruit epidermis lacking normal yellow pigmentation (Rick and Butler 1956). The combination of *y* plus *r* results in pale yellow or "white" fruit. Presence of *y* in red-fleshed genotypes results in a pink fruit phenotype.

The recessive *green flesh* (*gf*) allele prevents the breakdown of chlorophyll that normally occurs in maturing fruit (Kerr 1958a). Retention of green chlorophylls in combination with lycopene in ripe fruit results in reddish-brown colored fruit. The *Green ripe* (*Gr*) mutant elicits green-pigmented flesh in ripe fruit (Kerr 1958b). Berry et al. (2005) mapped *Gr* to the long arm of chromosome 1 and determined that *Gr* is an ethylene insensitive mutant that may encode a novel ethylene signaling component.

Fruit Color QTL The described monogenic mutants have a dramatic effect on fruit pigmentation. Nonetheless, they have not contributed widely to enhanced carotenoid pigmentation in commercial cultivars. Extensive genetic and molecular characterization of simply inherited tomato pigment mutants has not established a molecular genetic basis for quantitatively inherited variation in fruit pigmentation. Not surprisingly, Liu et al. (2003) concluded that there is more to tomato fruit color than candidate genes involved in carotenoid biosynthesis.

Analogous to fruit firmness, soluble solids, and other fruit quality traits (see Sect. 1.15), QTLs associated with variation in fruit pigmentation have been described that begin to explain dissimilarity in intensity of red pigmentation in modern tomato cultivars (Table 15). Numerous QTLs introgressed from *S. pimpinellifolium* (Tanksley and Nelson 1996; Chen et al. 1999), *S. habrochaites* (Bernacchi et al. 1998a, b; Monforte et al. 2001; Kabelka et al. 2004; Yates et al. 2004), *S. pennellii* (Monforte et al. 2001; Liu et al. 2003; Frary et al. 2004), *S. peruvianum* (Fulton et al. 1997; Monforte et al. 2001; Yates et al. 2004), and *S. neorickii* (Fulton et al. 2000) have been described that influence fruit color. Analysis of QTLs identified in a *S. lycopersicum* cross also revealed loci associated with enhanced fruit color (Saliba-Colombani et al. 2001; Causse et al. 2002). Not surprisingly, these QTLs may have negative or positive effects on ripe fruit color and epistasis as well as pleiotropy may occur. Whereas QTL studies often focus upon the positive effect of loci introgressed from wild relatives of tomato, wild species alleles often have a negative effect on fruit color (Kabelka et al. 2004). Encouragingly, all of these studies identify some QTLs associated with fruit quality attributes that have also been identified by others in different interspecific tomato crosses. Conserved major loci and minor loci with positive epistatic effects will be of great interest in marker-assisted breeding strategies to improve tomato quality and nutritive value.

Vitamins

Vitamin A and vitamin C are the principal vitamins in tomato fruit. Tomatoes also provide moderate levels of folate and potassium in the diet and lesser amounts of vitamin E and several water-soluble vitamins.

Vitamin A β-carotene is the principal provitamin A carotenoid and is an essential nutrient in the human diet because of its retinoid activity (Tee 1992; Omenn et al. 1994). Epidemiological evidence indicates that increased intake of high β-carotene containing fruit and vegetables may be associated with a reduced risk of heart disease and certain cancers (Ziegler 1989; Doll 1990; Block et al. 1992; Omenn et al. 1994). Vitamin A deficiency has been described as one of the most serious nutritional disorders of children in the world, especially in developing countries (Sommer 1997; World Health Organization 2005). Adapted germplasm with increased levels of this provitamin A carotenoid have been developed (Tomes and Quack-

Table 15. QTL that influence fruit nutritive value identified via analysis of segregating populations developed from crosses between *S. lycopersicum* and wild tomato species

Germplasm	Trait	Number of QTL and influence (+)	(−)	Chromosomal locations	Reference
S. lycopersicum	Color	8	0	2, 3, 4, 8, 9, 11	Saliba-Colombani et al. 2001; Causse et al. 2002
S. habrochaites	Color	18	15	1, 2, 3, 4, 6, 8, 9, 10, 11	Bernacchi et al. 1998a, b; Monforte et al. 2001; Kabelka et al. 2004; Yates et al. 2004
S. neorickii	Color	18	24	1, 2, 4, 5, 7, 8, 9, 10, 11, 12	Fulton et al. 2000
S. pennellii	Color	12	10	2, 3, 4, 6, 7, 8, 9, 10, 11, 12	Monforte et al. 2001; Liu et al. 2003; Frary et al. 2004
	Ascorbic acid	1	5	3, 5, 10, 12	Rousseaux et al. 2005
	Total phenolics	3	6	3, 5, 6, 7, 8, 9	Rousseaux et al. 2005
S. peruvianum	Color	5	12	1, 3, 4, 6, 7, 8, 9, 10, 12	Fulton et al. 1997; Monforte et al. 2001; Yates et al. 2004
S. pimpinellifolium	Color	10	3	1, 4, 5, 6, 7, 10, 12	Chen et al. 1999; Tanksley and Nelson 1996

enbush 1958; Tigchelaar and Tomes 1974; Stommel 2001; Stommel et al. 2005a).

Vitamin C Fruits and vegetables supply approximately 91% of the vitamin C in the USA food supply. Within *S. lycopersicum* and its wild relatives, ascorbic acid levels range from 10 to 120 mg/100 g fresh weight (Lambeth et al. 1966; Hobson and Davies 1971). Depending upon cultivar, environment, fruit maturity and post-harvest treatment, ascorbic acid comprises 40 to 90% of the organic acids (Bradley 1946; Carangal et al. 1954; McClendon et al. 1959; Davies 1965). Malic acid is the principal organic acid. Stevens (1972) determined that citrate and malate concentration were controlled by single linked genes for each compound with the dominant alleles conditioning high citrate and low malate concentrations. Causse et al. (2003) reported additive inheritance for ascorbic acid content in 45 hybrids from parental lines that included ten large-fruited *S. lycopersicon* and three cherry-fruited (including one *S. pimpinellifolium*) types.

S. lycopersicum x *S. peruvianum* crosses have produced genotypes high in vitamin C. As noted previously, fruit of *high pigment* genotypes also contain increased levels of vitamin C. Although a negative relationship exists between vitamin C content and fruit size, demonstration that newer cultivars contain approximately 25% more vitamin C than those developed 20 years earlier suggests that additional incremental gains in genetic improvement of tomato vitamin C content may be achieved (Matthews et al. 1973; Burge et al. 1975).

In contrast with a well-understood ascorbic acid biosynthetic pathway in animals (Burns 1967), a pathway in plants was not proposed until 1998 (Wheeler et al. 1998). Agius et al. (2003) recently isolated the gene *GalUR* which encodes an NADPH-dependent D-galacturonate reductase and demonstrated that biosynthesis of ascorbic acid in strawberry fruit occurs through galacturonic acid, a component of cell wall pectins. Overexpression of *GalUR* in *Arabidopsis* enhanced vitamin C levels two- to three-fold, thus demonstrating the potential to manipulate vitamin C levels. Analysis of *S. pennellii* introgression lines identified six QTLs for fruit ascorbic acid content, most of which had a negative effect on ascorbic acid concentration (Rousseaux et al. 2005). More recently, Zou et al. (2006) utilized *S. pennellii* introgression lines to map 15 genes involved in tomato ascorbic acid biosynthesis and metabolism.

Folate Plant sources, principally green leafy vegetables and legume seeds (Scott et al. 2000; Konings et al.

2001), are the main source of dietary folate. Folic acid dietary fortification is practiced to offset birth defects, anemia, and increased risk of vascular disease and certain cancers (Lucock 2000; Krishnaswamy and Nair 2001; Molloy and Scott 2001). The biosynthetic pathway of folate is well characterized (Hanson and Gregory 2002; Goyer et al. 2004), and thus provides good opportunity for genetic improvement of plant folate content (Zhang et al. 2003; Hossain et al. 2004). Overexpression of GTP cyclohydrolase I in fruit of tomato transformants resulted in a 3- to 140-fold increase in levels of the folate precursor, pteridine, and an average 2-fold increase in folate content (de la Garza et al. 2004). Exogenous supply of folate precursors resulted in additional 10-fold increases in folate content, suggesting that additional genetic modifications in the folate biosynthetic pathway may further boost fruit folate content.

Glycoalkaloids

Glycoalkaloids and their toxic effects are commonly associated with Solanaceous species. Tomato accumulates the glycoalkaloids α-tomatine and dehydrotomatine in a 10:1 ratio (Madhavi and Salunkhe 1998). Tomatine consumption is associated with reduced LDL cholesterol and triglyceride levels. In contrast with potato glycoalkaloids, tomato glycoalkaloids appear to be less toxic for human consumption, presumably because they are eliminated from the body as an insoluble tomatine-cholesterol complex (Kozukue and Friedman 2003). A survey of wild tomato species for genetic variation in tomatine concentration identified the highest levels in an *S. chmielewskii* accession (Courtney and Lambeth 1977). Rick et al. (1994b) identified an unusual bitter-fruited *S. lycopersicum* accession that retains high tomatine levels in ripe fruit. Tomatine retention was controlled by a single recessive gene and presumably encodes a defective tomatine-degrading enzyme that is normally active in ripening fruit.

Minerals

The role of minerals in plant foods that have a positive effect on human health are well established (Lachance 1998). Tomato mineral composition is greatly influenced by plant nutrition, and as a result, has been well characterized in the context of mineral deficiency and the effect of these conditions on plant health. There is significant genotypic variation for mineral content in tomato fruit. Potassium, together with nitrate and phosphorous, constitutes approximately 93% of the total inorganic fruit constituents (Hobson and Davies 1971). The concentration of other minerals in the fruit is low. Hobson and Davies (1971) summarize prior reviews of tomato mineral composition.

Phosphorous levels in 25 divergent tomato accessions ranged from 3.1 to 6.7 mM (Paulson and Stevens 1974). Stevens and Paulson (1973) reported a strong genotype-environment interaction for fruit phosphorous concentration that was not attributed to variation in available soil phosphorous. The study suggested that few genes are involved in genetic control of fruit phosphorous levels and that additive and dominance effects, in addition to epistatic interactions, contribute to observed phenotypes.

Stevens (1972) reported potassium concentrations that ranged from 45.2 to 86.7 meq/liter among 55 divergent tomato lines. Potassium deficiency may contribute to poor fruit color and reduced acid content (Bradley 1946; Carangal et al. 1954). Positive correlations between potassium content and titratable acidity have been reported (Hobson and Davies 1971).

Heritable differences for calcium utilization efficiency in tomato have been documented (Giordano et al. 1982). The potential to manipulate mineral composition in tomato using transgene approaches is well demonstrated by a recent report wherein calcium content was increased up to 50% in carrot via expression of an *Arabidopsis* H^+/Ca^{2+} transporter (Park et al. 2004).

Amino Acids

Variability for reported amino acid composition in tomato fruit is likely due to both genotypic differences and plant nutrition (Davies 1966a; Freeman and Woodbridge 1960). Interest in amino acid content of tomato has generally focused upon their influence on fruit flavor. Attempts to influence fruit amino acid composition have been limited to studies evaluating the effect of different fertilizer regimes (Carangal et al. 1954; Davies 1964). Glutamic, α-aminobutyric, glutamine, and aspartic account for approximately 80% of the total free amino acids (Freeman and Woodbridge 1960). Free lysine and methionine constitute 20 to 25% and 12 to 18% of total lysine and methionine, respectively (Friedman 2002). The essential amino acid content of tomato is considered to be of good quality, being similar to soy protein (Friedman and Brandon 2001). Amino acid levels in fruit of wild tomato species were generally lower in comparison to the cultivated form (Schauer et al. 2005).

Phenolic Compounds

Flavonoids, phenolic acids, and polyphenols are the main classes of dietary phenolics (King and Young 1999). Flavonoids, which include anthocyanins, are the largest group of plant phenols and have been the subject of considerable research since they impart color to many horticultural commodities. Vinson et al. (1998) determined that vegetables high in phenolic compounds have antioxidant quality superior to that of the antioxidant vitamins A, C and E. In a survey of *S. lycopersicum* and *S. pimpinellifolium* accessions, total phenolic content was most closely associated with measures of antioxidant activity (Hanson et al. 2004). Rousseaux et al. (2005) noted large environmental interactions for fruit antioxidants and identified nine QTLs for total phenolic concentration in fruit of *S. pennellii* ILs.

Flavonoids Flavonoids comprise a large group of secondary plant metabolites and include anthocyanins, flavonols, flavones, catechins, and flavonones (Harborne 1994; Harborne and Williams 2000). As food constituents, flavonoids are believed to have numerous health promoting properties because of their antioxidant and free radical scavenging activity (Shahidi and Wanasundara 1992; Ames et al. 1993; Cook and Samman 1996; Hertog et al. 1997; Knekt et al. 1997; Rice-Evans et al. 1997; Sawa et al. 1999; Commenges et al. 2000; Harborne and Williams 2000; Nijveldt et al. 2001; Pietta 2000).

In tomato, the dominant *anthocyanin fruit* (*Aft*) gene elicits elevated levels of anthocyanin in the fruit skin and outer pericarp tissues, predominantly petunidin, and lesser amounts of malvidin and delphinidin (Giorgiev 1972; Jones et al. 2003). The recessive *atroviolacium* (*atv*; Rick 1963) and dominant *Aubergine* (*Abg*; Rick et al. 1994a) loci also result in varying degrees of anthocyanin accumulation in fruit epidermal tissues.

Genetic diversity for flavonoids in tomato and a well-characterized biosynthetic pathway make tomato an attractive crop in which to develop cultivars with enhanced flavonoid levels. Numerous efforts have focused on manipulation of transgene expression to enhance fruit flavonoids (Muir et al. 2001; Bovy et al. 2002; Colliver et al. 2002). Whereas conventional selection schemes have relied upon color assays or quantification of flavonoid products to identify unique genotypes that may be useful in breeding programs, Willits et al. (2005) devised an alternative strategy and identified two *S. chilense* accessions and a single *S. pennellii* accession that expressed structural genes of the anthocyanin biosynthetic pathway in the fruit peel and fruit flesh. Introgression of the *S. pennellii* accession into tomato produced progeny that accumulated high levels of quercetin in fruit flesh and peel.

Phenolic Acids Phenolic acids form a diverse group that includes the widely distributed hydroxybenzoic and hydroxycinnamic acids. Hydroxycinnamic acid esters of caffeic acid predominate in Solanaceous species and chlorogenic acid is typically the most abundant (Molgaard and Ravn 1988). Chlorogenic and related caffeoyl esters are among the most potent free radical scavengers in plant tissues (Sawa et al. 1999; Nakatani et al. 2000) and act as antioxidants in human erythrocytes and for low-density lipoproteins in vitro (Nardini et al. 1995; Lekse et al. 2001).

Silencing of candidate genes demonstrated that hydroxycinnamoyl transferase (HQT) was the primary route for chlorogenic acid accumulation in Solanaceous species (Niggeweg et al. 2004). HQT overexpression caused plants to accumulate higher levels of chlorogenic acid, with no demonstrable negative effects on levels of other soluble phenolics or lignin.

1.14.2 Macronutrients: Carbohydrates, Insoluble Solids, and Lipids

Plants contribute to the availability of dietary macronutrients including protein, fats, and carbohydrates. A large amount of data exists for these macronutrients in agronomic crops wherein these well-characterized constituents are significant crop yield components. In tomato, these constituents have been studied principally for their effects on fruit quality. Little attention has been given to their contribution to fruit nutritive value.

Modification of plant carbohydrate composition is a research area receiving closer scrutiny in light of the positive effects that fiber and complex carbohydrates can have on aspects of human health such as lipid metabolism and diabetes (Anderson 1990). Dietary plant fibers are derived from plant structural components and include both water-soluble and water-insoluble nonstarch polysaccharides and lignin. Genetic variability for the type and amount of soluble carbohydrates, pectin, gum, lignin, and other fiber components exists within tomato.

1.14.3 Carbohydrates

Considerable research effort has been dedicated to improving tomato fruit solids content because of the influence of solids on fruit quality (Stevens 1986). Sugars comprise 55 to 65% of the total soluble solids fraction and approximately 50% of the total solids in tomatoes. Considerable variability for soluble solids concentration is present within the cultivated tomato and its wild relatives. Soluble solids concentration of commercial hybrid cultivars generally ranges from 4.5 to 6.0% and can approach 15% in fruit of wild tomato species (Hewitt and Garvey 1987). Transient starch accumulation which occurs prior to fruit maturation contributes to sink strength and solids accumulation in developing fruit by maintaining a concentration gradient for sucrose between the leaf and the fruit (Dinar and Stevens 1981). Schaffer et al. (2000) demonstrated a relationship between increased levels of starch in immature *S. habrochaites* fruit and ADPglucose pyrophosphorylase activity. Increased starch levels in fruit of plants descended from *S. lycopersicum* × *S. habrochaites* crosses were attributed to a *S. habrochaites* derived introgression coding for the large subunit of ADPglucose pyrophosphorylase (*AGPaseL1*).

Fruit of the cultivated tomato and those of red-fruited wild tomato species accumulate the reducing sugars glucose and fructose as the principal storage sugars during fruit development. Little or no sucrose is detectable in the mature fruit. In contrast, fruit of the green-fruited wild tomato species accumulate significant quantities of sucrose in addition to reducing sugars (Davies 1966b). Biochemical factors associated with sucrose accumulation in *S. chmielewskii* (Yelle et al. 1991), *S. habrochaites* (Miron and Schaffer 1991), and *S. peruvianum* (Stommel 1992) have been described. Inheritance studies demonstrated that sucrose accumulation is controlled by a single recessive gene, *sucr*, in the green-fruited species *S. habrochaites* (Stommel and Haynes 1993) and *S. chmielewskii* (Chetelat et al. 1993). This locus maps to the pericentromeric region of chromosome 3. Transgenic tomato plants expressing a constitutive antisense invertase transgene had increased sucrose and decreased hexose storage concentrations and reduced levels of acid invertase in ripe fruit (Klann et al. 1996). Accumulated evidence indicates that *sucr* encodes an inactive invertase allele.

In typical ripe fruit of *S. lycopersicum*, slightly higher amounts of fructose than glucose result in glucose:fructose (G:F) ratios of 0.8 to 1.0 (Davies 1966b). In hexose-accumulating fruit derived from interspecific crosses with *S. habrochaites*, glucose concentrations are commonly low relative to those of fructose and result in much lower G:F ratios than those typically noted in *S. lycopersicum*. Observed segregation in *S. lycopersicum* × *S. habrochaites* populations indicated that G:F ratios were controlled by at least two genes (Stommel and Haynes 1993). More recent investigations demonstrated that a major locus (*FGR*) located on chromosome 4 influences G:F ratios in tomato fruit and that an additional genetic factor may be involved in determining the ratio of hexose sugars (Schaffer et al. 1999; Levin et al. 2000). *FGR* increases levels of fructose, relative to glucose, and exhibits an allelic dosage effect. An additional locus (*FK2*) located on chromosome 6 is epistatic to *FGR* and may decrease G:F (Levin et al. 2000). *FK2* is subject to marked genotype × environment interaction.

Numerous studies have identified QTLs introgressed into tomato from wild species that influenced fruit soluble solids content. Many of these QTLs had a positive impact on solids content, but negatively influenced fruit yield (Eshed and Zamir 1994b; Tanksley et al. 1996; Chen et al. 1999; Yates et al. 2004). Additional studies identified chromosomal segments from *S. chmielewskii* and *S. galapagense* that had a positive influence on fruit soluble solids while maintaining acceptable fruit size, pH, and yield (Triano and St. Clair 1995; Yousef and Juvik 2001). Collectively, these and other QTL studies underscore the contribution that multiple, non-allelic loci have on controlling soluble solids in tomato. Recent efforts to characterize a QTL that increases the hexose sugar component of soluble solids revealed a likely regulatory level for an apoplastic invertase gene (*LIN5*) and highlight the importance of intragenic recombination in genetic variability in soluble solids and other quantitatively inherited traits (Fridman et al. 2000).

1.14.4 Insoluble Solids

Tomato fruit insoluble solids contribute to fruit viscosity and firmness. Insoluble solids are comprised of water insoluble solids (WIS) and alcohol insoluble solids (AIS), the former being slightly larger. A strong relationship exists between AIS and viscosity, with fruit pericarp accounting for high correlation coefficients (Janoria and Rhodes 1974). The inheritance of AIS in a cross of high and low AIS cultivars demonstrated high heritability (0.68 and 0.75), additive genetic variation, and that less than three genes influence AIS levels (Stevens 1976). Fractionation of AIS into polyuronide and polysaccharide fractions demonstrated that water soluble polyuronides and water insoluble polyuronides accounted for approximately 90% of the differences in viscosity between a high and a low viscosity cultivar (Stevens 1976). Water soluble polyuronides are comprised of short and intermediate length chains as found in the fruit serum fraction. Water insoluble polyuronides represent the protopectin fraction. An increase in the water insoluble polysaccharides had the greatest potential for increasing fruit viscosity.

Genetic variation for texture and factors that contribute to AIS in tomato results from the interaction of numerous interacting QTLs. QTLs associated with AIS and its constituents have been described (Fulton et al. 2000; Causse et al. 2002; Frary et al. 2003b; Yates et al. 2004). These studies have shown that a few chromosomal regions on chromosomes 2 and 4 have a large influence on fruit biochemical composition and organoleptic quality as determined by both physical and sensory measures (Causse et al. 2002; Yates et al. 2004) (See Sect. 1.15.1).

Early work on the molecular genetics of fruit ripening and corresponding changes in fruit softening focused on polygalacturonase and its effects on ripening fruit. Antisense suppression of polygalacturonase accumulation demonstrated that the enzyme has only a minor effect on fruit softening, but has substantial effects on increasing viscosity of processed products and the integrity of stored fruit (Schuch et al. 1991; Kramer et al. 1992; Langley et al. 1994). Related efforts directed towards suppression of pectin methylesterase activity likewise had little influence on fruit firmness, but increased soluble solids of juice, serum viscosity, paste viscosity, and serum separation of processed juice (Tieman et al. 1992; Thakur et al. 1996).

At least seven tomato β-galactosidase genes are expressed during tomato fruit development, six of which are expressed during ripening and may influence fruit textural properties (Smith and Gross 2000). Antisense suppression of the tomato β-galactosidase 3 gene did not improve fruit firmness but resulted in fruit that processed into pastes with an increased proportion of insoluble solids and slightly increased viscosity (de Silva and Verhoeyen 1998). Similar studies examining the tomato β-galactosidase 4 gene produced fruit from antisense lines that were 40% firmer than controls. Ongoing studies of a number of tomato ripening mutants (*rin, nor, Nr, Cnr*) that exhibit altered fruit textural properties offer promise for further elucidation of fruit AIS constituents (Seymour 2002).

1.14.5 Lipids

Phytosterols are important structural components of plant membranes and stabilize phospholipid bilayers in plant cell membranes. Moreau et al. (2002) and Piironen et al. (2000) have recently reviewed phytosterols in foods and their health promoting properties. Phytosterols have received increased attention in the last 10 years because of their cholesterol-lowering properties. The primary interest in tomato phytosterols has focused on ripening related events involving phospholipids and their catabolism mediated by phospholipase D (Whitaker et al. 2001; Pinhero et al. 2003). Transformation of tomato with an antisense phospholipase D cDNA construct reduced phospholipase D activity 30 to 40% and resulted in firmer fruit with enhanced lycopene content, vitamin C, and flavor (Oke et al. 2003). Juice and sauce prepared from the transgenic fruit exhibited increased viscosity and levels of major flavor volatiles. Increased membrane stability due to decreased phospholipase D activity may account for the observed enhancement in fruit constituents.

Related studies highlight the potential for genetic modification of the, albeit low, composition of fatty acids in tomato fruit and effect changes in fatty acid composition and flavor volatile compounds. Using metabolic profiling, Schauer et al. (2005) determined that fruit fatty acids 16:0, 18:0, and 18:2 were higher in all wild tomato species with the exception of *S. pimpinellifolium*. Wang et al. (2001) demonstrated that expression of the yeast Δ9-desaturase gene in tomato resulted in changes in leaf fatty acid profiles that were accompanied by changes in volatiles derived from fatty acids. Using a similar strategy, Cook et al. (2002) expressed a Δ6-desaturase transgene in tomato resulting in fruit that produced γ-linolenic acid, octadecatetraenoic acid, and a reduced linoleic: α-linolenic acid ratio. Additional studies suggest that a chloroplast-targeted lipoxygenase, TomloxC, can utilize both linoleic and linolenic acids as substrates to generate volatile flavor compounds (Chen et al. 2004).

1.15 Fruit Quality

For processing tomato, fruit soluble solids content, pH, and paste viscosity are the major quality traits. According to distributors and retailers, fruit quality is essentially defined by shelf-life and firmness. During recent years, these criteria have been taken into account for breeding but have led to flavorless fruit. For consumers, quality is defined by the traits governing fruit attractiveness and fruit flavor. Nutritional quality (e.g., antioxidants, vitamins) is also important (See Sect. 1.14) but cannot be directly evaluated by consumers. Fruit size, shape, color and firmness are the first quality traits which attract consumers by visual and tactile stimulation. When eaten, the fruit expresses its organoleptic quality which involves the senses of taste and smell. Organoleptic quality is complex as it results from a combination of aroma, taste and texture components. These traits are often difficult to measure by methods other than sensory analysis. Some of the quality components are associated with each other, which makes the analysis even more complex. However, some of the major components of tomato flavor can be assessed by physical or chemical measurements. For example, flavor depends on the ratio of sugars and acids, and aroma is related to certain volatile compounds (Baldwin et al. 1998).

The large diversity in fruit size and shape of tomato illustrates the genetic variability available in this species. Each cultivar is characterized by a particular aroma profile, and a specific range of taste and texture. Genetic variation is the major source of fruit quality variation (Stevens 1986; Causse et al. 2003), but fruit organoleptic quality is also influenced by external factors such as the environment (Dorais et al. 2001) and conditions during fruit storage (Stern et al. 1994).

Identifying "robust" quality QTLs is a prerequisite for molecular breeding or for their molecular characterization. This robustness can be assessed by experimental studies of their stability over years, over different external conditions and over genetic backgrounds. These "robust" QTLs can also be identified by comparing data from different experiments and/or obtained with different populations. Molecular characterization of QTLs to date has been performed by positional cloning, but new genomic tools are starting to be applied for identifying candidate genes related to QTL variation.

1.15.1 Gene/QTL Mapping of Fruit Quality

Most tomato fruit quality traits are quantitatively inherited. Many studies have been performed by the groups of S. D. Tanksley (Cornell University, USA) and D. Zamir (The Hebrew University of Jerusalem, Israel) to map QTLs controlling yield and fruit quality related traits (Paterson et al. 1988, 1990, 1991; Azanza et al. 1994; Eshed and Zamir 1995; Goldman et al. 1995; Grandillo and Tanksley 1996b; Tanksley et al. 1996; Fulton et al. 1997, 2000, 2002a; Bernacchi et al. 1998b; Chen et al. 1999; Doganlar et al. 2002c; Frary et al. 2004). These studies were all performed on progeny derived from interspecific crosses between wild tomato species and processing tomato inbreds. Some quality traits of interest for processing tomato are common to fresh market tomato (e.g., sugar content, soluble solids content, pH, acidity, firmness) and QTL locations can be compared across the progenies. In most of the studies QTLs were detected, sometimes with strong effects. A few QTLs explaining a large fraction (20 to 50%) of the phenotypic variation, acting in concert with minor QTLs, are usually detected. Most of the QTLs act in an additive manner, but dominant and overdominant QTLs were detected (Paterson et al. 1988, 1991; de Vicente et al. 1993; Semel et al. 2006). Epistasis (interaction among QTLs) is rarely detected unless a specific experimental design is used (Eshed and Zamir 1996).

Fruit Weight and Fruit Shape QTLs

Grandillo et al. (1999) summarized the results of QTL mapping for fruit weight obtained in 17 studies based on progeny of various types and involving seven wild species. According to the studies, three to more than 18 QTLs were detected. Six QTLs explained more than 20% of the phenotypic variation. A common set of 28 QTLs could be identified that frequently segregated in at least two populations. Nevertheless only QTL cloning and complementation permits determination of whether each consensus QTL location corresponds to a single gene. Lippman and Tanksley (2001) studied progeny based on a cross between two genotypes of extreme fruit size. None of the detected six QTLs mapped to a novel location, but this was the first time that these six QTLs were detected together.

For fruit shape, Grandillo et al. (1999) identified a common set of 11 QTLs from the six studies where the fruit length:diameter ratio was segregating. In another study, three major QTLs were identified, *ovate* on chromosome 2, *sun* on chromosome 7 and *fs8.1* on chromosome 8 (van der Knapp et al. 2002). Locule number is another major component of fruit size. Several QTLs have been mapped for this trait (Lippman and Tanksley 2001; van der Knaap and Tanksley 2003; Barrero and Tanksley 2004) the major two corresponding to the mutations *fasciated* on chromosome 11 and *lc* on chromosome 2. A strong epistatic interaction between these two genes was shown, with locule number considerably increasing when both loci were homozygous for the alleles increasing locule number (Lippman and Tanksley 2001).

Sugar and Acid Content QTLs

Chromosomal regions carrying QTLs for sugar content or related traits (Brix°, fructose, glucose, or sucrose content), based on 14 populations involving eight different species (Paterson et al. 1988, 1990, 1991; Azanza et al. 1994; Eshed and Zamir 1995; Goldman et al. 1995; Grandillo and Tanksley 1996b; Tanksley et al. 1996; Fulton et al. 1997, 2000, 2002a; Bernacchi et al. 1998b; Chen et al. 1999; Saliba-Colombani et al. 2001; Doganlar et al. 2002c; Causse et al. 2004; Frary et al. 2004) are summarized in Fig. 7. From three to 19 QTLs were detected per progeny, with a total of 95 QTLs concentrated in 56 chromosomal regions. For the majority of QTLs the wild species alleles increased the trait value. In 28 regions, QTLs were detected in more than one population, and may possibly correspond to the same QTL. But the large number of regions involved suggests that many mechanisms are responsible in increasing fruit sugar content. The same results were obtained for acid content (Causse et al. 2002, 2004; Fulton et al. 2002a), with only a few regions common to acid and sugar content. In contrast, frequent colocations between QTLs for sugar content and fruit weight (Grandillo et al. 1999) with opposite allelic effects could be detected, suggesting pleiotropic effects of some common QTLs.

Levin et al. (2000) described a locus, *Fgr,* that modulates the fructose-glucose ratio in mature fruit, with a *S. habrochaites* allele yielding a higher ratio. More recently, Levin et al. (2004) showed that alleles of *S. habrochaites* at two loci interacted to increase this ratio. These loci remain to be characterized.

Fruit Firmness QTLs

Firmness is an important trait for fruit quality. It is related to shelf-life and to fruit texture, and thus has been frequently measured in genetic studies. Several long shelf-life mutants have contributed to the understanding of fruit maturation (Giovannoni 2001, 2004; Seymour et al. 2002). The first to be described, *Never ripe* (*Nr*, Rick 1956) was the first cloned (Wilkinson et al. 1995). The *Nr* gene encodes a protein homologous to ETR1 in *Arabidopsis.* ETR1 is an ethylene receptor, explaining the phenotype of the *Nr* mutant, which is insensitive to ethylene. The mutants ripening-inhibitor (*rin*, Robinson and Tomes 1968) and non-ripening (*nor*, Tigchelaar et al. 1973) are defective in ethylene biosynthesis. The corresponding genes encode MADS-box transcription factors (Vrebalov et al. 2002). The *Colourless nonripening* (*Cnr*) mutation, modified in cell-to-cell adhesion, is an epigenetic mutation in a SQUAMOSA promoter binding protein box (SBP-box) transcription factor (Manning et al. 2006).

Firmness has also been used in quantitative genetic studies. Figure 8 presents a summary of QTLs controlling fruit firmness in nine populations (Tanksley et al. 1996; Fulton et al. 1997, 2000; Bernacchi et al. 1998b; Causse et al. 2002; Doganlar et al. 2002c; Frary et al. 2003b, 2004; Walley and Seymour 2006). Forty-six QTLs controlling firmness were mapped using seven different populations. Firmness was measured by touching (30 QTLs), by mechanical instrumentation (11 QTLs), or by taste (5 QTLs). In some cases the QTLs obtained by these three different methods co-localized. More than half of the QTLs were grouped in clusters of three to four QTLs. These clusters were localized on chromosomes 1, 2, 4, 5, and 9–11. On chromosomes 2, 5, and 10, the genes *rin*, *nor*, and *Cnr* co-localized with firmness QTLs. It would be interesting to discover whether these QTLs are controlled by *rin*, *nor* and *Cnr*.

Volatile Compounds QTLs

QTLs for volatile compounds have been mapped in two populations. Saliba-Colombani et al. (2001) detected QTLs for 12 volatile compounds among 18 that were quantified in the progeny of a cross between cherry tomato inbred line Cervil and larger-fruited inbred line Levovil. Tieman et al. (2006) identified QTLs for 23 volatiles in the population of ILs derived from *S. pennellii.* Twenty-five loci altered in content of one or more volatiles were identified. Although ten volatiles were analyzed in both studies, only three QTLs were detected in the same regions, for phenylacetaldehyde on chromosome 8 (confirming the effect of the QTL *Malodorous*, named by Tadmor et al. 2002), on chromosome 9 for 2-methylbutanol, and

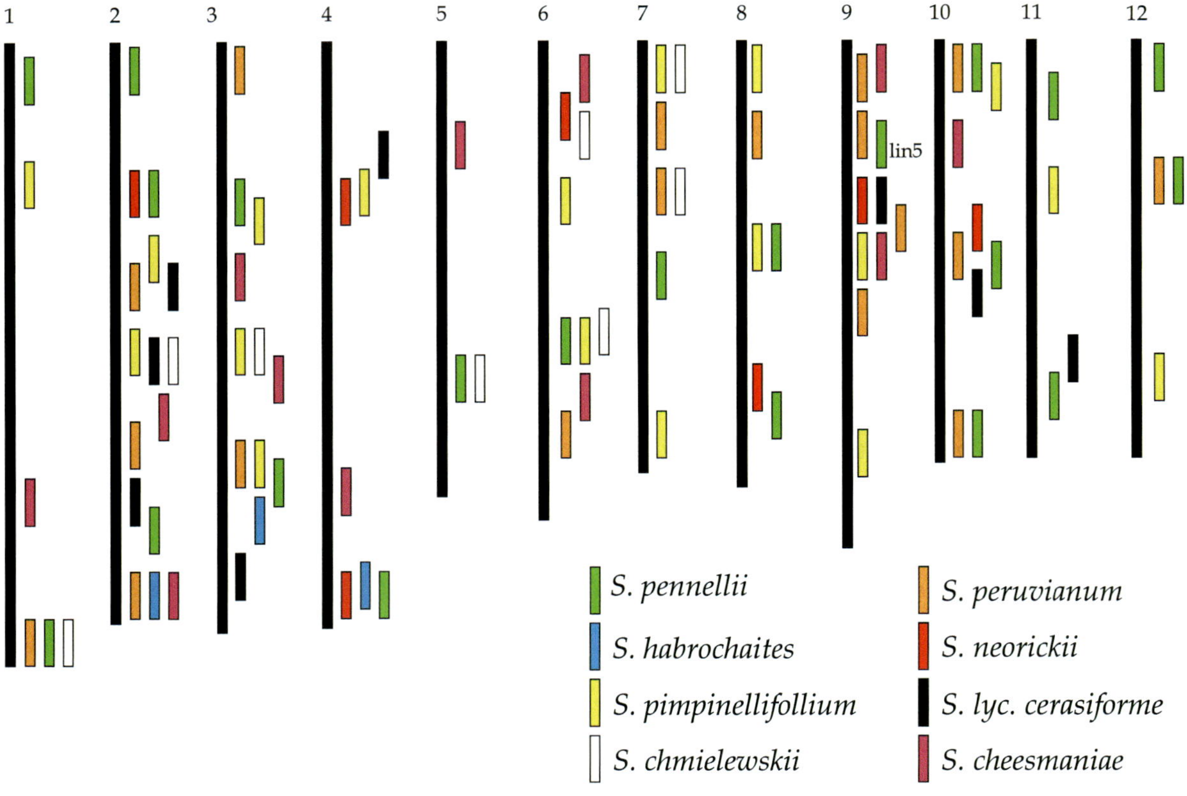

Fig. 7. Summary of QTL for sugar content or related traits (Brix° or hexose content) in one of the following progeny: *S. lycopersicum* × *S. cheesmaniae* F_2 population (Paterson et al. 1991); *S. lycopersicum* × *S. cheesmaniae* recombinant inbred population (Goldman et al. 1995); *S. lycopersicum* × *S. chmielewskii* F_2 and advanced backcross lines (Paterson et al. 1988, 1990; Azanza et al. 1994); *S. lycopersicum* × *S. habrochaites* advanced backcross population (Bernacchi et al. 1998b); *S. lycopersicum* × *S. neorickii* advanced backcross population (Fulton et al. 2000); *S. lycopersicum* × *S. pimpinellifolium* advanced backcross population (Tanksley et al. 1996; Doganlar et al. 2002c); *S. lycopersicum* × *S. pimpinellifolium* backcross populations (Grandillo and Tanksley 1996b; Chen et al. 1999); *S. lycopersicum* × *S. pennellii* introgression lines (Eshed and Zamir 1995; Causse et al. 2004); *S. lycopersicum* × *S. pennellii* advanced backcross population (Frary et al. 2004); *S. lycopersicum* × *S. peruvianum* advanced backcross population (Fulton et al. 1997); *S. lycopersicum cv cerasiforme* × *S. lycopersicum* recombinant inbred line population (Saliba-Colombani et al. 2001). The data concerning the advanced backcross involving *S. pimpinellifolium*, *S. peruvianum*, *S. neorickii* and *S. habrochaites* were summarized by Fulton et al. (2002a). The QTLs were positioned on the tomato reference map (Tanksley et al. 1992), based on the nearest marker

on chromosome 12 for pentanol. The content of some volatile compounds appeared strongly variable across years or environments (Tieman et al. 2006). This could partly explain the small number of QTLs common to the two studies. In both studies, QTLs for several volatiles were frequently in clusters. In a few cases these clusters corresponded to volatiles derived from the same metabolic pathway (related to fatty acid, carotenoid or amino acid degradation), suggesting the action of a gene within a single pathway. More frequently, co-localizations of QTLs for volatiles derived from various metabolic pathways were shown, suggesting a regulatory gene acting on several pathways.

Sensory Traits QTLs

Certain cherry tomato accessions have been shown to be most flavorful among tomato cultivars. In order to study organoleptic quality through all of its components, a QTL experiment was designed using the progeny of a cherry tomato line (Cervil, with high quality fruit) and a classical inbred line for fresh market (Levovil, producing large fruit with a common taste) (Causse et al. 2001, 2002; Saliba-Colombani et al. 2001). A population of 144 RILs was developed from crossing these two lines. Progeny were characterized for physical and chemical traits such as color, weight, firmness, pH and titratable acidity, sugars and soluble solids content, concentration of

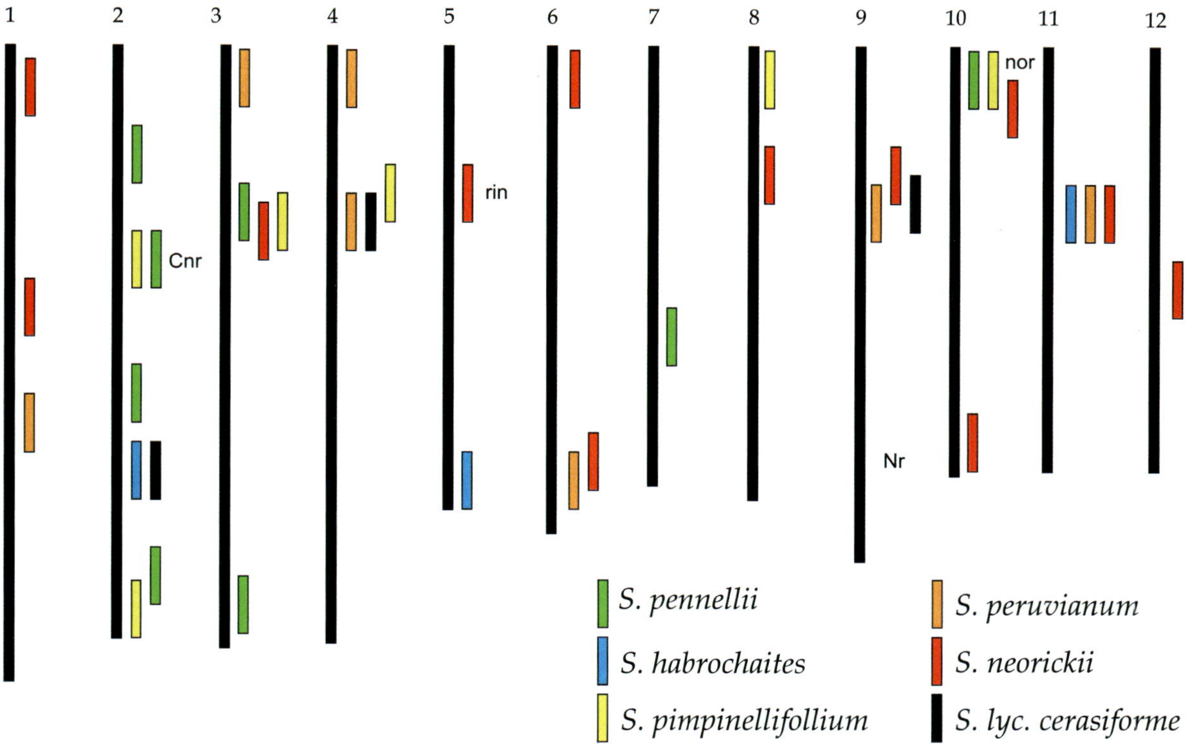

Fig. 8. Summary of QTL for firmness in one of the following progeny: *S. lycopersicum* × *S. pennellii* introgression lines (Causse et al. 2004, unpublished data); *S. lycopersicum* × *S. pennellii* advanced backcross population (Frary et al. 2004); *S. lycopersicum* × *S. habrochaites* advanced backcross population (Bernacchi et al. 1998b); *S. lycopersicum* × *S. pimpinellifolium* advanced backcross population (Tanksley et al. 1996; Doganlar et al. 2002c); *S. lycopersicum* × *S. peruvianum* advanced backcross population (Fulton et al. 1997); *S. lycopersicum* × *S. neorickii* advanced backcross population (Fulton et al. 2000); *S. lycopersicum cv cerasiforme* × *S. lycopersicum* (Saliba-Colombani et al. 2001). The QTLs were positioned on the tomato reference map (Tanksley et al. 1992) based on the nearest marker

12 aroma volatiles, and for sensory traits based on a panel of 56 judges. Judges were trained to quantify their sensory perceptions of taste (sourness, sweetness), aroma (lemon, candy, citrus, pharmaceutical aroma) and texture (meltiness, firmness, mealiness, juiciness). A genetic map was obtained from the RIL population and was used to map QTLs for the measured traits. For each trait, one to five QTLs were detected. These QTLs explained 9 to 45% of the phenotypic variation with some QTLs having major effects (e.g., those for six aroma volatiles) (Saliba-Colombani et al. 2001). Surprisingly, most QTLs were grouped within a few clusters (Causse et al. 2002). Alleles improving all fruit quality traits originated from the cherry tomato line (Causse et al. 2001, 2002). Despite the low heritability of sensory traits, QTLs could be detected for these traits and were often consistent with the location of QTL using mechanical measurements (Fig. 9).

1.15.2 Fruit Quality MAS

A marker-assisted backcross scheme was set up for introducing favorable alleles of five major QTL regions into three tomato lines with common taste (Lecomte et al. 2004a). The results demonstrated the usefulness of such an approach for improving fruit quality. In all three genetic backgrounds, the introduced regions had a favorable effect on the traits controlled by QTLs from the cherry tomato (donor line), with the exception of fruit weight. Consumer tests revealed that the prototypes were significantly preferred over their corresponding recurrent parents. However, some of the QTL effects were specific to a single genetic background. The genetic characterization of the QTLs for organoleptic quality traits was completed by studying the stability of the QTLs (Chaïb et al. 2006). This study confirmed that the QTLs were mainly specific to the

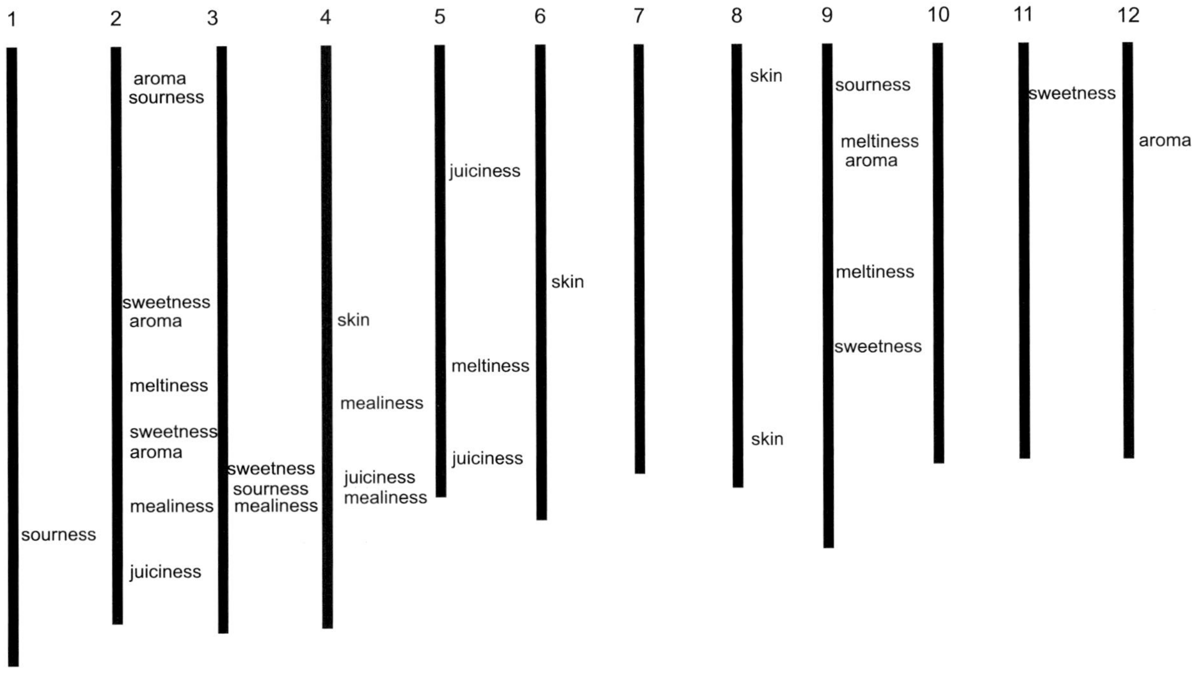

Fig. 9. QTLs detected for sensory traits in a recombinant inbred population derived from the cross between a cherry tomato and a large fruit line. Taste attributes were sweetness and sourness. Aroma attributes were overall aroma intensity, candy aroma, lemon aroma, citrus fruit (other than lemon) aroma, and pharmaceutical aroma. Texture attributes were flesh firmness, mealiness, meltiness, juiciness, and toughness of skin (adapted from Causse et al. 2001)

genetic background. Although approximately 50% of the QTLs were stable over years in the RIL population and in the genetic background used for QTL detection, a lower number of QTLs was detected in the two other genetic backgrounds.

Wild species, in spite of their unfavorable characteristics in comparison to cultivars, can carry alleles which may contribute to the improvement of most agronomic traits (de Vicente and Tanksley 1993; Bernacchi et al. 1998a) (See Sect. 1.8). Gur and Zamir (2004) made progress by pyramiding independent yield-promoting regions introduced from the wild species *S. pennellii*. Wild species may provide original aromas, either favorable to tomato quality, as found in an *S. peruvianum* accession (Kamal et al. 2001) or unfavorable as the *Malodorous* locus found in an *S. pennellii* accession (Tadmor et al. 2002).

1.15.3 Molecular Characterization of QTLs for Quality

Molecular characterization of QTLs consists of identifying the corresponding genes and the variation of either DNA sequence or gene expression explaining trait variation. High throughput genomic approaches can facilitate this by providing candidate genes for QTLs based on gene function. High throughput methods (transcriptomics, proteomics, metabolomics) allow the identification of changes in the content of mRNAs, proteins or metabolites by comparing two or more samples. First applied in *Arabidopsis thaliana*, these methods are now available in tomato. Initial results of their application to characterize QTLs controlling tomato fruit quality traits at the molecular level have been reported (see below). Traditionally, however, classical positional cloning has allowed the identification of genes corresponding to QTLs.

Isolation of Fruit Quality QTL

Fine mapping experiments that allow the precise mapping of QTL within a chromosomal region is the first step towards positional cloning (Paterson et al. 1990; Frary et al. 2003b; Lecomte et al. 2004b). Fine mapping may reveal the existence of several linked QTLs. For example, Lecomte et al. (2004b) identified two linked QTLs with moderate effects for fruit weight within a 20 cM region that were confounded as one major effect QTL.

Correspondence of map location of QTLs and genes related to carbon metabolism allowed the identification of several putative candidate genes (Causse et al. 2004), but the validation of their role in trait variation required fine-mapping and the identification of a causal polymorphism that was not obvious. Mutations of enzymes involved in carbon metabolism that alter sugar composition in fruit have been found in *S. chmielewskii* and in *S. habrochaites*. The *sucr* mutation in an invertase gene in *S. chmielewskii* provides fruit with sucrose instead of glucose and fructose (Chetelat et al. 1995). In *S. habrochaites*, an allele of the ADP glucose pyrophosphorylase enzyme was identified as being much more efficient than the allele of cultivated tomato, leading to an increase in the final sugar content of fruit (Schaffer et al. 2000).

Positional Cloning The first QTLs cloned in tomato influenced fruit quality traits (fruit weight, fruit shape, soluble solids) and were isolated through positional cloning. A fruit weight QTL (*fw2.2*) responsible for about 30% of the variation of this character was isolated using the classical strategy of high-resolution mapping by screening 3,472 F_2 plants, identifying 53 recombinants (between two markers 4.2 cM apart) and screening a YAC library. From a YAC likely to contain the responsible gene, a cosmid library was screened and three clones used to transform a tomato variety. The cosmid leading to differences in fruit size after transformation was sequenced and the two sequences corresponding to ORFs were used in a second round of transformation. This allowed the definitive identification of the clone corresponding to the QTL (Frary et al. 2000) (See Sect. 1.8.3). Nesbitt and Tanksley (2002) sequenced the *fw2.2* gene in a set of wild and cultivated accessions. They showed that the small-fruited accessions did not carry the same allele. No significant association could be detected between fruit size and *fw2.2* genotypes, highlighting the role of other QTLs. Their results suggested that the polymorphism responsible for the QTL variation was not in the coding region but probably in the promoter region. Recently, Cong and Tanksley (2006) showed that *fw2.2* regulates cell division by interacting with a kinase protein involved in the cell cycle.

The *ovate* gene, responsible for pear shape and elongated fruit shape was cloned by Liu et al. (2002). It corresponded to a nucleus-localized putative regulatory protein. A stop codon was responsible for the change in shape from round to elongated.

A major QTL for soluble solids content was mapped on chromosome 9 in the population of ILs derived from *S. pennellii* (Eshed and Zamir 1996). The line IL9-2-5 that contains a 9 cM genomic region of *S. pennellii* showed an increase in Brix° and thus restricted the QTL confidence interval to this region (See Sect. 1.8.3). The QTL was then cloned by a map-based cloning strategy (Fridman et al. 2000). The gene responsible for the QTL is an apoplastic invertase encoded by the *lin5* gene (Godt and Roitsch 1997). The mutation responsible for the trait variation was first delimited to a 484 bp region of the gene and then restricted to a single nucleotide mutation that leads to an amino acid change in the sequence of the invertase protein, positioned near the catalytic site of the enzyme (Fridman et al. 2004). A complex trait such as sugar content variation has thus been simplified into a SNP, introducing the concept of quantitative trait nucleotide (QTN, Fridman et al. 2004).

Transcriptomics for QTL Characterization Comparison of the transcriptomes of two samples leads to a list of differentially expressed genes. Several methods are available to compare transcriptomes including serial analysis of gene expression (SAGE), cDNA-AFLP, and differential display. Microarrays and DNA chips are currently the most accessible tools for transcriptomics.

Comparison of transcriptomes can be used to elucidate the modulation network of genes during a specific process or as a consequence of a specific mutation. For example, a transcriptomic analysis was performed by Alba et al. (2005) to identify genes implicated in fruit maturation. The transcript level of approximately 8,500 genes was followed during tomato fruit development in the pericarp. This study identified 869 genes differentially expressed during fruit development. Gene expression levels were strongly dependent on ethylene, which plays a central role in fruit ripening. In combination with the analysis of fruit development, transcript profiling of the *Nr* mutant (defective for an ethylene receptor, see above) was performed. From the 869 genes that were differentially expressed during fruit development, 37% were also affected by the *Nr* mutation. Together these data identified 72 candidate genes which could be responsible, in part, for the regulation of fruit development. It would be of great interest to map these genes to discover whether or not they co-localize with firmness QTL. This study further elucidated regulatory networks responsible for fruit ripening.

The comparison of transcriptome profiles of NILs can also identify differentially expressed genes, among which some may be located in genomic regions for which the lines differ. These genes then become candidates for QTLs.

Baxter et al. (2005) compared transcriptome profiles of six ILs derived from *S. pennellii* in the genetic background of the processing cultivar M82. All six ILs had a higher soluble solids content compared to M82. Each IL was characterized by a large set of genes differentially expressed (at 20 days post-anthesis) and 78% of significant changes were unique to a single line. Very few carbon related genes were altered in expression and very few genes differentially expressed were located in introgressed regions. This experiment gave clues as to which metabolic pathways were most perturbed by each introgression.

Proteomics for QTL Characterization Proteomic analysis reveals differentially expressed genes at the protein level (Rose et al. 2004). This approach is less frequently used than DNA chips but it reveals another level of genome expression, closer to the phenotype. Faurobert et al. (2007) have described proteome modifications during fruit development and ripening. The intensity of 1,791 spots which approximately corresponded to the same number of proteins was monitored. From these 1,791 spots, 148 were significantly differentially expressed. The corresponding proteins were identified for 90 of the 148 spots. Protein levels of genes involved in amino acid metabolism and protein synthesis were more abundant at early developmental stages than during ripening. Conversely, genes related to carbon metabolism were highly expressed in mature fruit.

Proteome analysis may also be used to analyze genetic variation (Pawlowskii et al. 2005; Rocco et al. 2006). Both quantitative (spot amount) and qualitative variation (protein shift due to allelic variation or spot absence due to a null allele) can be detected. The analysis of the two lines (Cervil and Levovil) used for mapping QTL for sensory quality described above (See Sect. 1.15.1) showed that more than 90% of the spot positions overlapped and that about 10% of the spots showed quantitative variation (Mihr et al. 2005). The comparison of proteome profiles of NILs derived from this cross allowed the identification of candidate proteins for fruit quality QTLs (Faurobert et al. 2006).

Metabolomics for QTL characterization New tools in biochemistry allow the simultaneous quantification of multiple metabolites present within a sample. Because organoleptic quality of a fruit is correlated to the amount of sugars, acids, and volatile compounds, these tools can be useful to characterize tomato quality (see Sect. 1.16). The evolution of metabolome profiling during fruit development has been described (Carrari et al. 2006), showing that the abundance of metabolites, from either primary or secondary metabolism, is very dynamic during fruit development. The variation of metabolite abundance is coordinated and strongly partitioned by metabolic pathways, illustrating the tight link between metabolites. Carrari et al. (2006) compared metabolite profiles to transcriptome profiles during tomato fruit development. Although some links were clearly evident between variation in metabolites and transcripts, gene transcript levels were less coordinated than metabolite levels. Metabolomic profiling combined with genetic studies may provide insights into the physiological bases of quantitative traits and yield clues as to which candidate genes to screen (Overy et al. 2005). Schauer et al. (2006) characterized metabolic profiles of fruit of the ILs derived from *S. pennellii.* They identified 889 QMLs (quantitative metabolic loci) among which half were associated with QTLs for yield-associated traits. This analysis highlights the value of combined genetic, physiological and biochemical profiling to identify the major components of fruit quality.

1.15.4 Fruit Quality: Conclusions

Tomato quality is a complex character due to its number of components and because it is dependent on conditions throughout the entire process of plant and fruit development. Genetic variation for fruit quality is extensive, especially if one considers the possibilities offered by wild tomato species. A few mutations have been shown to be involved in fruit quality, particularly in ripening, and QTL studies have revealed a number of genomic regions involved in the variation of quality traits. Several genomic hot-spots for fruit quality QTLs have been identified. These clusters of QTLs for several quality traits may permit simultaneous MAS for multiple traits. Very few QTLs have been identified at the molecular level, but one can expect a rapid increase in the number of genes identified in the near future, thanks to systems biology approaches combining transcriptomics, proteomics

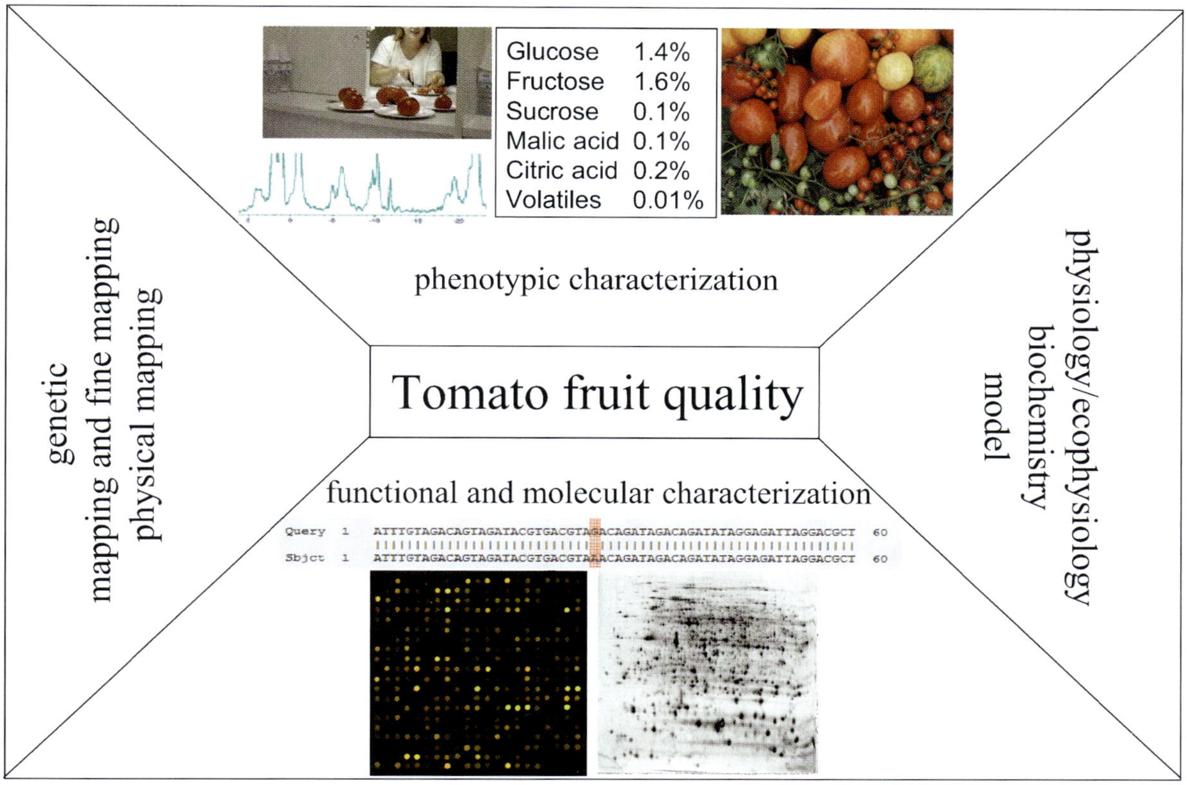

Fig. 10. Integrated approach for characterization of tomato fruit quality QTLs

and metabolomics studies and the information from the tomato genome sequence (Fig. 10). These discoveries will greatly facilitate breeding for improved fruit quality.

1.16 Metabolomics

A major goal of tomato metabolomics is to elucidate genes and biosynthetic pathways that can be exploited for crop improvement in productivity and quality. Tomato productivity involves net accumulation of photosynthetically synthesized compounds such as sugars and amino acids, and polymeric compounds derived from them, e.g., cellulose and starch. Fruit quality is determined by a complex blend of traits including taste, fragrance, and color. These traits are mostly attributed to metabolite composition. For example, accumulated sugars, amino acids, lipids, and various secondary metabolites affect fruit taste. Lipid, sugar, alcohol, aldehyde, and isoprenoid contents affect fruit flavor. Levels of carotenoids and flavonoids influence fruit color. Thus, "productivity" and "quality" can be defined in terms of metabolic profiles.

Recent development of chromatography-coupled MS technologies has facilitated the simultaneous detection of thousands of metabolite-derived mass peaks. Comprehensive acquisition of tomato metabolite data followed by high-throughput analysis has allowed an integrated approach (genomics, transcriptomics, metabolomics) to tomato research. This approach has the potential to identify priority target genes conferring tomato phenotypes (Carrari et al. 2006; Schauer et al. 2006). The application of MS technology has also revealed that tomato contains a large number of metabolites that have not yet been identified. Characterization of these unknown metabolites has become one of the major challenges in understanding the complexity of tomato metabolism.

1.16.1 Metabolomics Data Collection

Tomato contains a diverse range of metabolites with different molecular weights and polarities (Fig. 11). Thus, it is currently not possible to obtain an overview of the entire metabolome in a single analysis. Nevertheless, it has been shown that metabolite analyses with comprehensive coverage, if not complete, can provide novel insights into plant metabolic pathways (Urbanczyk-Wochniak et al. 2003; Hirai et al. 2004, 2005; Tohge et al. 2005). Comprehensive coverage is achieved through combinations of complementary extraction and detection methods.

Extraction

Optimum extraction conditions differ for different types of metabolites. To extract carbohydrates, amino acids, and fatty acids from tomato tissues using GC-MS, an extraction protocol originally developed for potato tuber (Roessner et al. 2000) has been widely used. To extract secondary metabolites for non-targeted analysis by liquid chromatography MS (LC-MS), a methanol extraction has been widely applied (Bino et al. 2005; Moco et al. 2006). A lipophilic solvent (e.g., chloroform) is required for extracting non-polar secondary metabolites, such as carotenoids. To extract volatile metabolites, solid phase micro-extraction (Tikunov et al. 2005) or solvent extraction (e.g., propanol, pentane) (Schmelz et al. 2003) have been reported.

Detection

MS is the primary detection method for plant metabolomic analysis. GC-MS is applicable to detect volatile metabolites (e.g., alcohols, monoterpenes, esters), and also non-volatile polar metabolites (primary metabolites, e.g., amino acids, sugars, lipids, organic acids). Metabolites can be identified by comparing fragment peak patterns and retention indices with those of standard compounds in databases. LC-MS is a versatile technology to analyze non-volatile secondary metabolites. However, in contrast to GC-MS, few mass spectral databases are currently available for LC-MS. The lack of a searchable LC-MS database is a major obstacle in identifying metabolites by LC-MS.

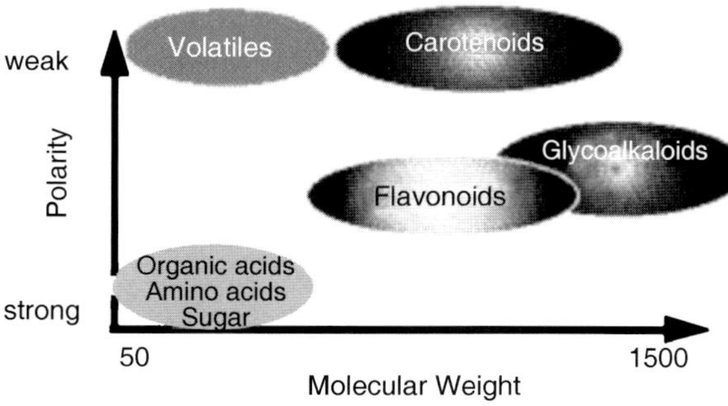

Fig. 11. Variation of tomato metabolites for molecular weights and polarities

Capillary electrophoresis MS (CE-MS) has been employed to analyze water-soluble metabolites (e.g., organic acids, nucleotides, amino acids, sugars, sugar phosphates). Although the application of CE-MS to the comprehensive profiling of plant metabolites has been limited to rice (Sato et al. 2004; Takahashi et al. 2006), CE-MS will likely be adopted more broadly due to its sensitivity, ease of sample preparation, and applicability to polar metabolites.

1.16.2 Metabolomics Data Analysis

Strategies for metabolomics data analysis can be classified into two categories, metabolic profiling and metabolite annotation (Fig. 12). Both strategies are being applied to tomato metabolomic data to fill in the gaps between metabolomics and genomics.

Metabolic Profiling

Metabolic profiling is defined as identification and quantification of metabolites (Hall 2006). Approximately 90 primary metabolites (Carrari et al. 2006) and 320 volatile metabolites (Tikunov et al. 2005) of tomato have been routinely profiled by GC-MS. A limited number of secondary metabolites have been successfully profiled using LC-MS due to their intrinsic complexity and the lack of a database. Metabolic profiling often includes estimating correlations between metabolites and between metabolites and other measures, such as gene expression and phenotypes. A correlation profile can provide critical clues to generate hypotheses concerning metabolic pathways, for example, assigning gene function or understanding gene interactions.

Cultivated varieties and wild species of tomato display a wide range of morphological and physiological phenotypes. Substantial variation of metabolite content is therefore expected. Metabolite profiles of *S. lycopersicum* and five wild species (*S. pimpinellifolium*, *S. neorickii*, *S. habrochaites*, *S. chmielewskii*, and *S. pennellii*) have been compared (Schauer et al. 2005). Extracts from leaves and fruits were analyzed using GC-MS, and levels of more than 80 metabolites (generally primary metabolites) were estimated. The leaf metabolite profile of *S. lycopersicum* was closest to *S. pimpinellifolium*, followed by *S. pennellii*,

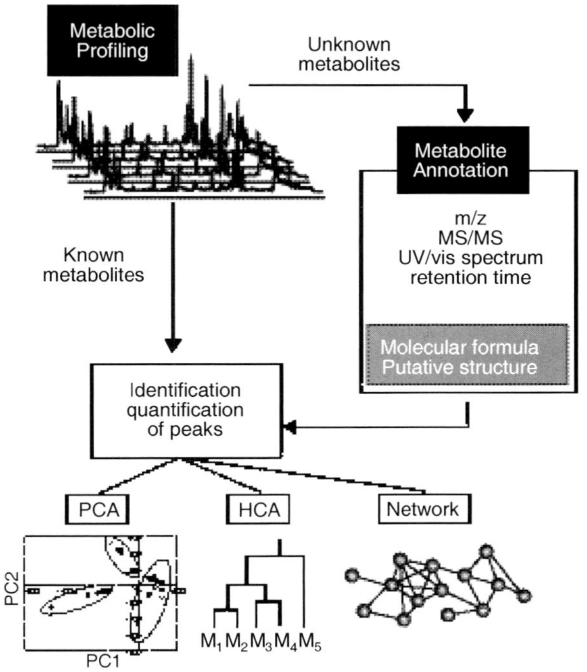

Fig. 12. Concepts of metabolic profiling and metabolite annotation. Metabolic profiling aims to identify and quantify metabolites. Metabolite annotation aims to provide chemical information concerning the peaks, especially unknown peaks, to allow consistent identification of peaks across multiple samples. Annotated peaks and known metabolites are then used in metabolic profiling analysis. Metabolic profiling data is statistically analyzed using PCA, hierarchical clustering analysis (HCA), and network analysis

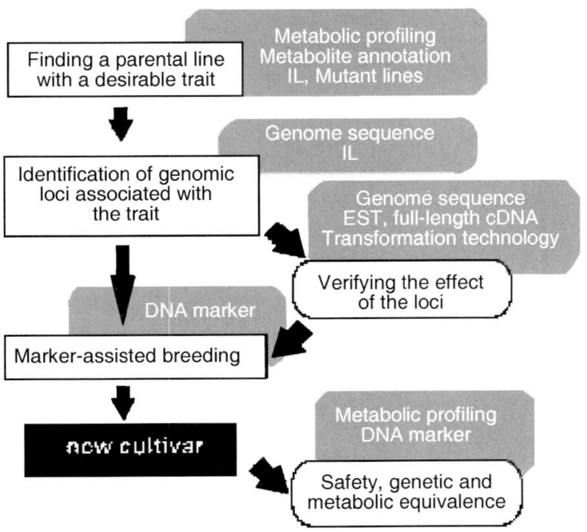

Fig. 13. Schematic relationships between molecular breeding and functional genomics. Functional genomics resources that potentially support each step of breeding are indicated (*gray boxes*)

S. chmielewskii, and *S. neorickii*. *S. habrochaites* was the most distinct from *S. lycopersicum*. In contrast to leaves, fruit metabolite profiles showed higher degree of variation. In particular, accumulation levels of sucrose, isocitrate, chlorogenate, and shikimate were higher in the wild species than in *S. lycopersicum*. Accumulation levels of glucose, fructose, α-tocopherol, and puturesine were lower in the wild species than in *S. lycopersicum*. The contents of amino acids and their derivatives were generally higher in *S. lycopersicum* compared to wild species, with the exception of higher γ-aminobutiric acid (GABA) in *S. pennellii*, and higher tryptophane in *S. habrochaites*.

Comprehensive metabolic profiling of *S. lycopersicum* × *S. pennellii* ILs using GC-MS was reported by Schauer et al. (2006). In this study 74 metabolites were quantified in ripe fruit pericarps from all 76 ILs. Metabolite contents tended to show increases in the ILs relative to *S. lycopersicum*. These increases were not limited to metabolites that were inherently more abundant in the wild species. For example, trichloroacetic acid cycle intermediates that were invariant across several species (Schauer et al. 2005) apparently increased in the ILs.

Comprehensive profiling of secondary metabolites using LC-Q time of flight MS (LC-Q-TOF-MS) was applied to compare *high pigment-2*^{dg} (*hp-2*^{dg}) mutant and wild-type (*wt*) *S. lycopersicum* alleles (Bino et al. 2005). The *hp-2*^{dg} allele is associated with increased levels of carotenoids, flavonoids, and ascorbic acid. The authors found 383 mass signals that were significantly higher in *hp-2*^{dg} compared to *wt*, and 62 mass signals that were significantly lower in *hp-2*^{dg} than *wt* in red ripe fruit. Only eight metabolites were successfully identified out of 445 mass signals.

To link metabolic profiling data to functional genomics, metabolites must be assigned to biological pathways and functions of genes involved in syntheses and degradation of metabolites must be elucidated. These goals can be achieved by combining metabolic profiling data with genetic, genomics, and phenotypic data. A correlation-based analysis is a powerful approach in analyzing putative relationships between metabolites and genes. In the absence of gene expression or genotypic data, a metabolite-to-metabolite correlation pattern can provide insights into a metabolic pathway. Tikunov et al. (2005) calculated pair-wise correlations between 322 volatiles detected by GC-MS. From the correlation matrix, the authors found that volatiles within a cluster were derived from a common biochemical precursor. Volatiles in five major clusters were derived from phenylalanine, leucine, isoleucine, lipid, and isoprenoid, respectively. This result suggested that the volatiles within the same metabolic pathway tended to accumulate in a coordinated manner.

Integration of metabolic profiling data with phenotypic traits was performed on 76 *S. lycopersicum* × *S. pennellii* ILs (Schauer et al. 2006). Profiles of 74 metabolites obtained by GC-MS were combined with data from nine yield-associated traits, and correlation coefficients between all possible pairs of metabolic and phenotypic traits were estimated. The authors illustrated the correlation patterns in the form of a network that was divided into three large modules. Module 1 comprised mainly phenotypic traits and sugar phosphates; module 2 comprised amino acids; and module 3 included sugars and organic acids. In the phenotypic traits module, harvest index and Brix° had relatively high numbers of total connections both within the module and to nodes in the two external modules, i.e., they served as network hubs. Brix° had high numbers of connections, probably because it is an integrator trait whose variance depends on the variance in many other traits. Interestingly, all other phenotypic traits were highly connected within module 1 and showed sparse connections to external modules. Most metabolites associated with phenotypic traits belonged to pathways of central metabolism. However, whether observed correlations directly reflected causal relationships cannot be concluded by a correlation-based study alone. Nevertheless, integrated analysis of metabolic profiling and phenotypic traits is potentially a powerful tool in building the foundation for molecular breeding.

Metabolite profiles are sometimes studied in conjunction with transcript profiles (Alba et al. 2005; Carrari et al. 2006). Carrari et al. (2006) performed a comprehensive parallel analysis of transcript and metabolites during tomato fruit development. Their study included the profiling of 92 tomato fruit metabolites by combined GC-MS and photodiode array high-performance liquid chromatography (PDA-HPLC) analyses, transcriptome analysis using a TOM1 microarray (Alba et al. 2004), and estimating metabolite-to-metabolite, transcript-to-transcript, and metabolite-to-transcript correlations. Correlations showed that metabolites belonging to the same class were regulated in a highly coordinated fashion, while transcripts belonging to the same functional group were relatively less coordinated. Moreover, metabolite-to-transcript correlations demonstrated that levels of metabolites sometimes displayed low correlations with transcript levels of genes within the pathway. For example, cell wall polysaccharides and genes associated with cell wall metabolism showed little correlation. These results suggested that metabolism was largely regulated at the post-transcriptional level. However, several high correlations between metabolites and transcripts were identified. Sugar phosphates, organic acids, and pigments (carotenoids, xanthophylls, chlorophylls) were highly correlated with ripening-related transcripts including ACC oxidase, ethylene receptor, ripening-inducible transcription factors, and ethylene response genes. The authors assessed the correlation between primary metabolism and pigment metabolism. Although the metabolite levels of TCA cycle intermediates (organic acids) showed few correlations to pigment levels, TCA cycle intermediates and pigments had significant correlations to the same ripening-associated transcripts.

The integration of metabolomic data with genomic data offers a promising approach to prioritize candidate genes for molecular breeding. To aid in understanding correlations between transcripts and metabolites by visualization of transcript changes within the context of metabolic pathways, the software MapMan, originally developed for *Arabidopsis* (Thimm et al. 2004), has been revised for Solanaceous plants as Solanaceous MapMan (Urbanczyk-Wochniak et al. 2006).

Metabolic Annotation

Metabolite annotation aims to collect and integrate as much information as possible from a given metabolite peak to provide information as to its identity and structure. Although there is no consensus as to the total number of tomato metabolites, 73 and 555 metabolites are registered in the categories *Lycopersicon* and *Solanum*, respectively, in the metabolite database Dictionary of Natural Products (http://www.chemnetbase.com/scripts/dnpweb.exe?welcome-main) as of March 2007. In two separate studies, 92 and 110 metabolites were found in tomato fruit by GC-MS and LC-Q-TOF-MS, respectively (Carrari et al. 2006; Moco et al. 2006). It has been reported that approximately 11,500 ions were detected from red tomato fruit extract by LC-Q-TOF-MS (Bino et al. 2005), which raises the possibility that the number of published tomato metabolites has not come close to approaching the total number. Many unknown metabolites await identification; this has been a major constraint in plant metabolomics due to the lack of standard chemicals corresponding to myriads of secondary metabolites. One solution is to develop a methodology to discriminate each unknown peak by providing a series of "annotations". An operational definition of metabolite annotation is to cluster detected peaks into peak groups that represent individual metabolites, and to provide chemical information concerning the peak groups, including *m/z* value, MS/MS fragmentation pattern, UV/visible absorption spectrum, retention time, putative molecular formula, putative structure predicted from MS/MS fragmentation patterns, and search results of databases and scientific literature. An annotation is useful for two aspects of biological analyses. First, it allows consistent identification of unknown peaks across a number of samples by using a combination of chemical data obtained for the peak. Second, a combination of *m/z* value and MS/MS fragmentation pattern provides critical information for predicting the molecular structure of the metabolite. MS/MS fragmentation patterns allow prediction of metabolite component moieties without elucidating metabolite stereo-structure. An annotation obtained by this procedure may serve as the primary, or mass spectral, annotation. A secondary, or biological annotation of a metabolite describes the tissues and environmental conditions under which a given metabolite was found. Concepts of metabolite annotation have been reviewed by Kind and Fiehn (2006) and Fiehn et al. (2005).

Metabolomics Databases

The most reliable class of metabolite annotation is provided when chemical information of an observed metabolite matches known compounds in natural product databases or descriptions from the literature. Moco et al. (2006) reported an analysis of fruit extracts pooled from 96 cultivars representing three major fruit types (cherry, Dutch beef, normal round) using LC-Q-TOF-MS. They annotated metabolites using *m/z*, retention time, absorption spectra, and MS/MS data and performed database searches against the Dictionary of Natural Products (http://www.chemnetbase.com/scripts/dnpweb.exe?welcome-main) and SciFinder (http://www.cas.org/SCIFINDER/). In addition, they compared MS data of the observed metabolites with the 60 tomato metabolites that had been reported in the literature. Consequently, they identified 26 metabolites based on literature reports, and 14 novel metabolites that were not previously reported in tomato fruit extracts. A large fraction of the identified metabolites belonged to the phenylpropanoid pathway. The authors created a tomato metabolite database MoToDB (http://appliedbioinformatics.wur.nl) based on MS data of metabolites that have been reported in literature (compiled data of 110 metabolites as of February 2007).

Iijima et al. (unpublished data) annotated metabolites from fruit of the cultivar Micro-Tom based on LC Fourier transform ion cyclotron resonance MS (LC-FTICR-MS) data. From peels and flesh of four maturation stages (immature green, mature green, orange, red ripe), approximately 700 peak groups that had both representative ion and isotopic ions in a series of consecutive scans were subjected to molecular formula prediction and database searches. Based on metabolite annotation, putative metabolic pathways between annotated metabolites were predicted. All MS data and annotations will be compiled into a database (Iijima et al. in prep).

1.16.3 Metabolomics and Molecular Breeding

Metabolomics facilitates the deduction of biologically relevant hypotheses as to putative metabolic pathways and genes responsible for metabolic reactions. The assignment of candidate genes for crop improvement will be accelerated as the body of tomato genomic data increases. High-throughput metabolic profiling holds vast potential

for screening promising tomato genes, varieties, and species for metabolite-oriented molecular breeding (Fig. 13). Firstly, profiling of complex secondary metabolites will support the breeding of novel tomato varieties with enhanced health benefits in addition to quality and productivity. Towards this end, metabolic profiling of a wide range of genotypes is required, along with correlations among metabolic traits and agronomic traits, and the development of associated molecular markers. Secondly, tomato is known to contain potentially toxic alkaloids (see Sect. 1.14.1). To check whether a novel overexpressing variety is free of potentially hazardous levels of metabolites will be important to consumers.

1.17 Bioinformatics

A comprehensive bioinformatics infrastructure is a critical resource for tomato researchers. Databases, websites, and associated tools allow fast and easy access to data such as sequences, annotation, mapping and expression information. This enables rapid progress and the generation of new hypotheses (Rhee et al. 2006). As one of the central genetic model systems of the Asterid clade, a large number of websites and informatics tools are available for tomato. The ongoing tomato genome sequencing project (Mueller et al. 2005b) (see Sect. 1.18) will continue to spawn many new bioinformatics efforts that will considerably expand the available tools for tomato and other members of the Solanaceae family. In 2003 the SOL project, an umbrella project for Solanaceae research around the world, published a whitepaper defining a vision for the next 10 years. This vision included the development of a strong bioinformatics infrastructure. The SOL project proposed to integrate resources into a network with a single entry point, the Solanaceae Genomics Network (SGN, http://www.sgn.cornell.edu, see http://sgn.cornell.edu/solanaceae-project). A selection of the major databases and resources that are currently available, including SGN, are described in Table 16. Several of these are described in more detail below.

1.17.1 Tomato Analyzer (http://www.oardc.ohio-state.edu/vanderknaap/tomato_analyzer.htm)

Most tools and databases are web-based. One tool, the Tomato Analyzer (Brewer et al. 2006) is standalone software. The Tomato Analyzer performs semi-automated phenotyping of fruit shape. It was originally designed for tomato but is easily applicable to other fruits or shape problems. It performs calculations based on scanned images to characterize shape using defined parameters and is also used to create a controlled vocabulary describing fruit shape.

1.17.2 TGRC (http://tgrc.ucdavis.edu)

The Tomato Genetics Resource Center (TGRC), founded by the late Charles M. Rick, is the central gene bank for tomato mutant stocks and wild relatives (Table 4). TGRC activities include the acquisition of germplasm, stock maintenance, seed distribution and a database. The database contains detailed information on collection information for the wild species, genetic loci, and photographic images of all mutants.

1.17.3 GRIN (http://www.ars-grin.gov/)

The Germplasm Resources Information Network (GRIN) web server provides germplasm information about plants, animals, microbes and invertebrates. This program is within the USDA-ARS. Passport data for over 5,000 accessions of cultivated tomato and several hundred accessions of wild tomato species conserved at PGRU (Table 4) are available through GRIN.

1.17.4 TIGR (http://www.tigr.org)

The Institute for Genomics Research (TIGR) maintains a database of plant sequences and transcript assemblies called Plant TA (http://plantta.tigr.org). This database contains sequences for several tomato species and provides transcript assemblies for other Solanaceous plants (Childs et al. 2007). In the past TIGR maintained the Plant Gene Indices database, but this database has since moved to the Computational Biology and Functional Genomics Laboratory at Harvard (http://compbio.dfci.harvard.edu/).

1.17.5 MiBASE (http://www.kazusa.or.jp/jsol/microtom/indexj.html)

MiBASE is specifically designed around the Micro-Tom tomato variety (Dan et al. 2006) and is housed at the Kazusa Institute in the Chiba Prefecture in Japan. The database offers comprehensive annotation for the MicroTom ESTs sequenced at the Kazusa Institute (Ya-

Table 16. Tomato data and resources available through major bioinformatics databases

Database	URL	Maps	Sequence data	Clone requests	Gene families	Pathways	Germplasm requests	Phenotypes	Gene expression
SGN	http://sgn.cornell.edu	X	X	X	X	X		X	
MiBASE	http://www.kazusa.or.jp/jsol/microtom/index.html		X	X	X	X			
TGRC	http://tgrc.ucdavis.edu						X	X	
GRIN	http://www.ars-grin.gov/npgs						X	X	
Genes That Make Tomatoes	http://zamir.sgn.cornell.edu/mutants						X	X	
TED	http://ted.bti.cornell.edu								X
TIGR	http://plantta.tigr.org		X						
TMRD	http://www.tomatomap.net	X							
GenBank	http://www.ncbi.nlm.nih.gov	X	X						X

mamoto et al. 2005), which were assembled into unigenes with other tomato EST sequences. MicroTom clones sequenced at the Kazusa Institute can be ordered at this site, and extensive biochemical pathway annotation information is available (Yano et al. 2006).

1.17.6 TED (http://ted.bti.cornell.edu)

The Tomato Expression Database is the primary database for tomato microarray data (Fei et al. 2006). The database consists of three parts: (1) the tomato Microarray Data Warehouse, serving as the central repository for raw gene expression data and experimental descriptions in a "minimal information about a microarray experiment" (MIAME) compliant format; (2) the Tomato Microarray Expression Database, storing the processed experimental results from a number of fruit development time course experiments; and (3) the Tomato Digital Expression Database, which contains expression analysis derived from EST abundance in non-normalized libraries. The database stores the array definitions for the clone-based TOM1 microarray as well as the oligo-based TOM2 microarray.

1.17.7 SGN (http://sgn.cornell.edu)

The SOL Genomics Network (SGN) (Mueller et al. 2005a), the front end website of the SOL project, is a comprehensive resource based on the clade-oriented database (COD) principle. Essentially, data from an entire clade of organisms is integrated into the database rather than data from a single model species. This allows more meaningful integration of diverse datasets for different organisms in a comparative and phylogenetic context. In this vein, SGN contains extensive comparative mapping data, sequence data from EST sequencing projects for tomato species, potato (*S. tuberosum*), pepper (*C. annuum*), eggplant (*S. melongena*), petunia (*P. hybrida*), *N. tabacum*, and species from closely related families such as coffee (*Coffea canephora* var. *robusta*) and snapdragon (*Antirrhinum majus*). All data are related to *Arabidopsis* and the emerging tomato genome sequence. The EST sequences are assembled into unigene sets that are annotated extensively based on annotation from sequence matches to *Arabidopsis* and GenBank sequences, Interpro domains, and detected features such as signal peptides, as well as other analyses. The unigene sequences form the basis of pre-calculated gene family "tribes" using the TribeMCL program (Enright et al. 2002). Member sequences of a multi-gene family are aligned and gene trees are calculated and stored in the database. These can be browsed online using the alignment viewer and tree browser tools. SGN recently introduced a locus database that contains annotated genetic loci from the literature for all species in the database. The locus database can be updated by the SGN user community. Each locus has an associated editor with privileges to edit the locus name, symbol, descriptions, and chromosomal location by logging on to the SGN website. Similarly, a phenotype database was introduced that allows users to submit mutant descriptions, images and other data about mutants for any of the species in the database.

SGN is one of the bioinformatics nodes of the tomato sequencing project (Mueller et al. 2005b), providing tools to the ten sequencing partners such as a BAC registry, project statistics, sequence repository, and viewers for the annotated sequence. Data in SGN is being mapped comprehensively to the emerging tomato reference sequence.

Tools available on SGN include BLAST (Altschul et al. 1990), the Intron Finder for Solanaceae ESTs, the CAPS Designer, and bulk downloads. A comprehensive FTP site with complete data sets is also available.

1.18 International Tomato Genome Sequencing Project

The International Tomato Genome Sequencing Project is a consortium of groups from ten different countries, including the USA, whose goal is to sequence the euchromatic regions of each of the 12 tomato chromosomes. The USA group is sequencing three of the chromosomes (1, 10, 11) and the remaining nine chromosomes are each being sequenced by a group from a different country. The breakdown on the countries sequencing the remaining nine chromosomes is as follows: Korea – chromosome 2, China – chromosome 3, UK – chromosome 4, India – chromosome 5, The Netherlands – chromosome 6, France – chromosome 7, Japan – chromosome 8, Spain – chromosome 9, and Italy – chromosome 12. Each group has acquired funding from agencies within their respective countries. The repository for all information and data related to the project can be found on SGN (http://www.sgn.cornell.edu) (see Sect. 1.17).

The tomato genome contains 950 Mb of DNA which is organized into 12 chromosomes ($n = x = 12$) (Arumuganathan and Earle 1991). Unlike the chromosomes of maize or rice, in which heterochromatin and euchromatin are interspersed, the heterochromatin in tomato is concentrated around the centromeres. This pericentric heterochromatin is largely devoid of genes and heavily populated by retrotransposons, but constitutes approximately 77% of the tomato DNA (Peterson et al. 1996; Van der Hoeven et al. 2002; Wang et al. 2006). In contrast, the distal portions of each tomato chromosome are comprised of largely contiguous stretches of gene-rich euchromatin, which corresponds to approximately 23% of the DNA (Peterson et al. 1996; Wang et al. 2006). Rather than sequencing the entire tomato genome (950 Mb), the international consortium has committed to sequencing the approximately 220 Mb of euchromatin which is estimated to contain approximately 85% of the predicted 35,000 genes that comprise the tomato genome (Van der Hoeven et al. 2002; Wang et al. 2006). The consortium has agreed that there are several advantages to sequencing a tiling path through the euchromatin rather than the entire tomato genome.

Sequencing the tomato genome is the cornerstone of a larger international effort: "The International Solanaceae Genome Initiative" or "SOL Initiative". The goal of this effort is to establish a network of information, resources and scientists to tackle two universal biological questions that the Solanaceae genomes are particularly suited to address: *(i) How can a common set of genes/proteins give rise to a wide range of morphologically and ecologically distinct organisms that occupy our planet? (ii) How can a deeper understanding of the genetic basis of plant diversity be harnessed to better meet the needs of society in an environmentally friendly and sustainable manner?*

The tomato reference sequence gathered through a BAC minimal tiling path approach will provide gene content and order, which is well documented to be similar to other Solanaceae genomes, thus making the phenotypic and evolutionary diversity in this family accessible for exploration at the sequence level. The tomato genome is connected to the other important members of the family by detailed comparative genetic maps (see Sect. 1.6) and the level of microsynteny is known to be well conserved with respect to gene content and order. In addition, because the Solanaceae represents a distinct and divergent clade of flowering plants, distant from *Arabidopsis*, *Medicago*, soybean, maize and rice, the tomato genome sequence will provide a rich resource for investigating the forces of gene and genome evolution over long periods of evolutionary time.

1.19 Reverse Genetics and TILLING[2]

1.19.1 Reverse Genetic Strategies in Tomato

The increasing abundance of gene sequence information in tomato and other plants is giving rise to new opportunities for studies of gene function. In a traditional forward genetic approach, a phenotype is investigated to determine the underlying genetic basis of a trait. In a reverse genetic approach, the aim is to alter a gene or its expression level in order to study the resulting phenotype. Ideally, a reverse genetic approach allows a genome wide interrogation of genes, so that the function of every gene can be studied. In tomato, a number of reverse genetic approaches are being developed to leverage the expanding genomic resources in this crop. Among these approaches are insertional mutagenesis, RNA interference (RNAi) and Targeting Induced Local Lesions IN Genomes (TILLING).

In insertional mutagenesis, a collection of lines is generated in which a transposon or modified element has been introduced and has randomly inserted into the genome in order to knock-out or tag genes. This type of approach has been extremely effective in *Arabidopsis* yielding access to a huge portion of that genome for functional analysis (Alonso et al. 2003). Because *Arabidopsis* is easily transformed, has a short generation time and has a small genome, large collections of insertion lines are available for this model dicot. In addition, access to the fully sequenced *Arabidopsis* genome allows mapping of the insertion sites to provide an in silico resource for investigators (Krysan et al. 2002; Sessions et al. 2002). Tomato is less amenable to the development of large populations of insertion lines because transformation methods are less efficient and the tomato genome is much larger than that of *Arabidopsis* (see Sect. 1.20). Despite these restrictions, a limited number of insertion lines have been produced and used for reverse genetic studies of gene function in tomato (Meissner et al. 2000; Gidoni et al. 2003). In some cases, the insertion element can be remobilized into nearby sequences in order to expand the utility of these collections (Gidoni et al. 2003).

Another useful tool for reverse genetics studies in plants is gene silencing through RNAi (Kusaba 2004; Watson et al. 2005). In this method, a transgene designed to produce a self-complementary RNA that forms a double stranded (ds) structure is introduced into a plant by transformation. The dsRNA is processed to small 21 to 24 nucleotide fragments that are then used to target sequence-specific degradation of endogenous RNA. In this way, expression of a target gene of interest can be suppressed and the resulting phenotype can be evaluated. However, the extent of gene silencing through RNAi can vary, so a number of transformants must be screened to ensure that suppression is achieved. In tomato, RNAi has been effectively applied to reduce the expression of a fruit allergen using transformation (Le et al. 2006). A major advantage of RNAi is that it can be used to simultaneously suppress multiple related genes as well as genes in tightly linked tandem arrays (Kusaba et al. 2003; Lawrence and Pikaard 2003). Another advantage of RNAi is that it can be used for tissue-specific suppression of genes when it would be detrimental to eliminate the expression of a gene throughout the plant. For example, fruit specific suppression of tomato DET1 by RNAi resulted in increased carotenoids and flavonoids in fruit and left the remainder of the plant unaffected (Davuluri et al. 2005). An alternate RNAi approach called virus-induced gene silencing (VIGS) utilizes a target sequence inserted into a viral genome to transiently infect plants and suppress target gene expression. In tomato, VIGS has been used to suppress genes expressed during fruit development (Fu et al. 2005). In addition, a high-throughput system for VIGS called Agrodrench has been developed to suppress expression of target genes in roots of tomato and other Solanaceae (Ryu et al. 2004).

TILLING is a powerful reverse genetic technique that allows a researcher to identify genetic variation in a gene of interest (Colbert et al. 2001; McCallum et al. 2000a). TILLING, much like a forward genetic screen, begins with a population or library of individuals with increased levels of genetic variation induced through mutagenesis. However, unlike forward genetic screens, TILLING ends in the identification of genetic lesions at the molecular level, which can then

[2] TILLING is a registered trademark of Arcadia Biosciences

be studied at the phenotypic level. Specifically, a mutagen is used to induce genetic variation in a population of thousands of individuals, from which genomic DNA and seed libraries are generated. These libraries serve as long-lasting, renewable repositories for the selection of plants whose genomes contain mutations of interest. The DNA library is pooled to increase efficiency and screened over targeted regions of the genome to discover novel genetic variation. Mutations at the level of SNPs can be identified rapidly and economically through enzymatic cleavage of mismatched DNA using CEL I endonuclease followed by gel electrophoresis of the fragments. The number of mutations that are discovered depends upon the size of the library, its mutation frequency, and the size of the target gene to be screened. These new mutations form an allelic series with a range of potential phenotypic effects that can help assess gene function (Hirschi 2003).

Reverse genetic approaches have proven extremely useful to study basic gene function. They also offer the potential to speed the development of novel commercial cultivars since functional genomics findings associated with traits of interest in one crop may also be relevant to another. TILLING, in particular, provides a means to access almost any gene for functional studies by increasing the genetic diversity of domestic tomato lines and is also suitable as a non-transgenic means of crop improvement.

1.19.2 Creating a TILLING Population

Mutagenesis

A TILLING population with a high mutation frequency will maximize efficient mutation detection and minimize the number of individuals that need to be screened to find mutations. For example, screening an 1 Kb gene in a population of 5,000 individuals with an average mutation frequency of 1 mutation per 250 Kb yields about 20 mutations in *Arabidopsis* (Greene et al. 2003). There is an approximately 60% chance that one of these 20 mutations will be a knockout (Henikoff et al. 2004). Increasing the population size to 12,000 individuals would raise the probability of finding a knockout mutation to 95%. But, if the mutation frequency were much lower, such as 1 mutation per 1,000 Kb, then approximately 20,000 individuals would need to be screened to find 20 mutations. In any case, a high mutation frequency must be balanced against detrimental effects such as lethality and sterility that are inherent in mutagenic treatments.

Chemical mutagenesis has been used for decades to increase available genetic diversity in plants and animals for forward screens and it has also been used successfully for generating TILLING populations. There are a number of chemicals available for the induction of large quantities of single base changes throughout the genome. Among the most commonly used are alkylating agents such as EMS, which induces primarily G/C to A/T transitions, and *N*-ethyl-*N*-nitrosourea (ENU), which induces mainly A/T to G/C transitions and A/T to T/A transversions (Kodym and Afza 2003). Mutagenic treatments that may work well in one species are not usually transferable to another, and even different varieties of the same crop may react differently to the same mutagenic treatment (Wu et al. 2005). As a result, the most effective chemical mutagen and treatment regime for a TILLING population must be determined empirically.

Seeds are the preferred starting material for chemical mutagenesis in tomato. The seeds are incubated in aqueous solutions of EMS or other mutagen and the optimal treatment parameters, including concentration of the mutagen and duration of the treatment, are determined. After treatment, seeds are washed extensively and sown. Lethality and delays in germination are often used as early indicators of the mutagenic effectiveness of the treatment. There is some correlation between the degree of lethality of treated seeds and the mutation frequency of the population although this correlation may vary considerably between plant species and even between varieties of the same species (Caldwell et al. 2004; Wu et al. 2005). For example, a high mutation frequency of 1 per 24 Kb can be achieved with little lethality in wheat (Slade et al. 2005). Conversely, a high degree of lethality was seen in a rice population with a mutation frequency of only of 1 per 1,000 Kb (Wu et al. 2005). Therefore, factors other than the extent of DNA damage may affect viability of M_1 seeds. For instance, it has been noted that different batches of mutagen can have different cytotoxic effects due to impurities (Loppes 1968) that could also contribute to lethality.

The most straightforward way to evaluate the effectiveness of different mutagenic treatments is to measure the mutation frequency directly by TILLING. For this purpose, small pilot populations of M_1 plants (800 to 1,000 plants per treatment) can be grown from treated seeds. M_1 plants are genetically chimeric since mutagenesis of the seed results in mutations in many different cells, but only the mutations in germ cells will be heritable and recovered in subsequent generations. Before committing to bring these pilot M_1 populations to maturity, one can measure the mutation frequencies in the growing M_1 seedlings by TILLING the pilot population over a small region of the genome. Although it has not been demonstrated that the mutation frequency in the somatic tissue of the M_1 population can be used to predict the precise mutation frequency of the M_2 population, it has been observed that a low mutation frequency at the M_1 stage indicates a low mutation frequency at the M_2 stage. Promising M_1 pilot populations can be expanded to the M_2 stage to both directly measure the mutation frequency in the M_2 population by TILLING and to assess the effects of the treatment on fertility in order to determine the optimal treatment.

Growing a TILLING Population

After determining the mutagenic treatment that gives a high mutation frequency suitable for screening without widespread lethality or sterility, a large population (>10,000) of M_1 plants is typically cultivated to provide sufficient numbers of plants for efficient TILLING (McCallum et al. 2000b). In the specific case of tomato, growing the M_1 plants in the field at commercial densities, instead of in the greenhouse, is usually more economical. This is especially true because the M_1 generation is often prolonged by 1 or 2 months, apparently due to delays in germination and to developmental setbacks experienced by the mutagenized plants. To produce plants with stably inherited mutations, the M_1 tomato plants are grown to maturity and allowed to self-fertilize. A single fruit is harvested from each fertile M_1 plant and the collected M_2 seed is considered to be equivalent to a new line or family, reflecting the unique collection of heritable genetic alterations in each M_1 plant (Caldwell et al. 2004; Hohmann et al. 2005). A single M_2 seed is planted from each line and labeled with a unique plant identifier and sampled for DNA. Once the plants reach maturity, M_3 seed is harvested and catalogued such that M_2 plant identifier, genomic DNA, and M_3 seed packet are linked in a database information management system. The result is a seed library comprising thousands of unique lines with corresponding genomic DNA that can be continually accessed for TILLING and replenished over time without repeating mutagenesis.

1.19.3 High-throughput Screening and Mutation Discovery

DNA Library Creation

Appropriate tissue collection and subsequent purification of genomic DNA from each M_2 plant is essential for successful recovery of the mutations induced during mutagenesis. In many plant species, tissue collection must be done at a young developmental stage to optimize quantity, quality, and purity of DNA. Sampling of older or damaged tissue should be avoided as it can adversely affect the quality and long-term stability of the extracted DNA. In tomato, high quality DNA can be extracted from tissue samples of the first two or three true leaves at approximately 1 month after planting. The apical meristem is left intact so that the plant can continue to grow and produce fruit for M_3 seed collection.

A high-throughput DNA extraction method is preferable, since TILLING is practiced on large DNA populations. The use of commercial DNA purification kits in combination with a liquid handling robot and a 96-well format allows a single researcher to produce over 1,000 DNA preparations per day. Liquid handlers also help with the normalization of large numbers of DNA samples to a standard DNA concentration. This in turn allows for the efficient pooling of DNA prior to mutation screening, ensuring that each individual in a pool is equally represented. Collection of the M_2 DNA samples into bar-coded tubes allows the DNA library to be linked to M_3 seed progeny of the plant. The DNA library then becomes a stable asset that can be used for the TILLING of many gene targets over many years.

Tomato Sequence Resources

Because TILLING is a PCR-based technique, only genes for which the DNA sequence is known can be screened for mutations. In 2003, an International Tomato Genome Sequencing Project was launched to sequence the euchromatic regions of the tomato genome with the goal of fully annotating the estimated 35,000 tomato genes (Mueller et al. 2005b) (see Sect. 1.18). According to SGN, sequencing of the tomato genome is 17% complete as of February 2007. Designing PCR primers for TILLING any genomic region in tomato will be greatly simplified once the genome has been fully sequenced. Until then, some excellent sources for tomato genetic information include unigenes, and genomic and mRNA sequences at the NCBI (http://www.ncbi.nlm.nih.gov), and an EST collection at the SGN website (http://www.sgn.cornell.edu). NCBI contains over 270,000 nucleotide sequences from both domesticated and wild tomato species and 226,728 *S. lycopersicum* ESTs and mRNAs clustered into 16,978 unigenes. SGN currently has 239,593 ESTs from *S. lycopersicum* assembled into 34,829 unigenes.

In addition to helping investigators characterize known tomato genes, TILLING can also be used to assign function to putative tomato homologs of genes that are well characterized in other species. In this case, the investigator can clone novel tomato genes by homology and then use TILLING to select tomato plants containing mutations in the putative homologs for phenotypic study. In particular, the fully sequenced *Arabidopsis* genome proves useful for identifying tomato homologs of a variety of well-characterized plant genes. Van der Hoeven et al. (2002) compared a large subset of tomato and *Arabidopsis* ESTs and found identifiable homologs in 70% of tomato unigenes. In addition, the splice junctions in *Arabidopsis* genes are often conserved in tomato homologs. The online program, COnsensus-DEgenerate Hybrid Oligonucleotide Primers (CODEHOP) can be used to design degenerate primers to conserved regions of protein families to increase the likelihood of obtaining functional homologs (http://bioinformatics.weizmann.ac.il/blocks/codehop.html). The first step in this process is using the Block Maker program (http://bioinformatics.weizmann.ac.il/blocks/blockmkr/www/make_blocks.html) to create "blocks" of protein homology (Henikoff et al. 2000), which are then reverse translated by CODEHOP and used to design degenerate primers for amplifying genomic DNA (Rose et al. 2003).

Once the DNA sequence for a tomato gene target is known, another online program known as CODDLE (COdons Optimized to Discover Deleterious LEsions) can be employed to predict which conserved areas of the gene would be most severely affected by mutation and also to calculate the types of alterations expected depending on the particular mutagen used (Till et al. 2003). For instance, only the codons for the amino acids tryptophan (TGG), arginine (CGA) and glutamine (CAA, CAG) can be mutated to stop codons (TGA, TAG, TAA) using EMS as a mutagen. For ENU, only the codons for the amino acids leucine (TTA, TTG), tyrosine (TAT), cysteine (TGT), lysine (AAA, AAG) or arginine (AGA) can be mutated to stop codons. Regions of a gene with the highest abundance of codons susceptible to conversion to stops are better targets for primer design if a stop mutation is desired. CODDLE thus predicts which regions of a gene will have the most severe impact on gene function and designs TILLING primers using the software Primer3 (Rozen and Skaletsky 2000).

Alternatively, if the investigator is interested in recovering all mutations in a gene regardless of severity, Primer3 can be used directly to design primers to introns or untranslated regions flanking each exon for maximum coverage of the coding region (http://frodo.wi.mit.edu/cgi-bin/primer3/primer3_www.cgi). In the absence of genomic sequence, cDNA and EST sequences from NCBI and SGN can be used to design primers to obtain intron sequence for TILLING primer design.

Mutation Discovery via TILLING

To increase the efficiency of the high-throughput screening process, DNA samples in a 96-well format are pooled together, allowing as many as 576 to 768 individuals to be screened simultaneously with 6-fold or 8-fold pooling. After pooling, PCR is used to amplify regions of target genes for mutation screening. With few exceptions, gene-specific primers are required for efficient mutation discovery. Primers that amplify more than one gene typically lead to decreased assay efficiency due to amplicon size differences that interfere with clear identification of cleavage products and decreased sensitivity of mutation detection in what is effectively a doubled pool size. In the most frequently used detection process (Colbert et al. 2001), each PCR primer is labeled at the 5' end with an infrared fluorescent dye (IRD) that emits in either the 700 nm or 800 nm range. Since a different dye is used for each primer, both ends of the complementary strands of DNA are uniquely labeled after PCR.

After amplification, the PCR products are denatured and cooled to allow the DNA strands to reanneal. When a mutation is present in the pool, the other individuals in the pool serve as wt strands, allowing heteroduplexes to form at any mismatched nucleotide between the mutant and the wt DNA strands. In addition to increasing throughput, pooling provides the wt DNA strands necessary for detection of homozygous mutations. Mismatches are detected using CEL I, an endonuclease purified from celery and available commercially as SURVEYOR (TRANSGENOMIC, Inc., Omaha, NE). CEL I is a member of the S1 nuclease family that specifically cleaves heteroduplexed

DNA on the 3' end of the mismatch (Oleykowski et al. 1998; Till et al. 2004).

Denatured cleavage products of CEL I mismatch digestion can then be visualized after size separation by electrophoresis on acrylamide gels. The LI-COR DNA analyzer (LI-COR Biosciences, Lincoln, NE) is the platform used most frequently for TILLING because it is able to detect as little as an estimated 100 attomoles of dye-labeled PCR product (Colbert et al. 2001). This high level of sensitivity enables the efficient detection of mutations in pooled samples. In addition, the relatively low cost of the instrument and short run times help keep costs low and efficiency high compared to other mutation discovery methods. Some other platforms used for mutation discovery or surveying natural variation include resequencing (Wienholds et al. 2002), denaturing HPLC (McCallum et al. 2000a), and separation of CEL I cleaved heteroduplexes by capillary electrophoresis (Perry et al. 2003; Cordeiro et al. 2006).

The LI-COR analyzer produces two images of the gel, one for the IRD 700 labeled product and the other for the IRD 800 labeled product. An important feature of the dual-labeling process is that two differentially labeled strands of DNA are produced when the PCR products are cut at heteroduplexes. The sizes of the IRD 800 labeled cleavage product and the IRD 700 labeled cleavage product added together equals the size of the full-length product (Fig. 14). This differential double end labeling of PCR products allows for fast and accurate confirmation of mutants since the cleaved heteroduplexes can easily be distinguished from PCR artifacts that migrate at the same position, greatly reducing the number of false positives.

After mutations are discovered within a pool, the individual containing the mutation must be identified. Every member in the pool can be sequenced to assign the genetic alteration to an individual, but this can be cost-prohibitive, especially if a large number of mutations are identified in a TILLING screen. Instead, each pool member can be doped with wt DNA to ensure heteroduplex formation and then rescreened to identify the individual containing the mutation. Alternatively, a pooling strategy can be set up such that individuals are arrayed in two orientations and screened twice. In this way, when a mutation is discovered the individual containing the mutation is automatically identified (Comai and Henikoff 2006). Regardless of the method used to deconvolute the pools, each identified mutation is sequenced to determine the exact nucleotide change. Finding the mutation in the sequence trace is made easier because the approximate base pair location can be inferred from the size of the CEL I cut product on the gel. Sequencing also indicates if the mutation is homozygous or heterozygous in the M_2 plant.

With the thousands of plants, seeds, and DNA samples that need to be tracked in a TILLING population, a database is an essential part of the process. M_2 seed must be uniquely identified and linked to the subsequent plant, DNA, and M_3 seed samples. M_2 plants may have lived out their life cycle when mutations are discovered, so it is often necessary to plant M_3 seed to recover mutations and also to bulk up seed for genotypic and phenotypic analysis.

Analysis of Mutations

Once the exact nucleotide change of a mutation is known through sequencing, online programs can be used to predict the effect on the encoded protein. One such program is called PARSESNP (Project Aligned Related Sequences and Evaluate SNPs) (Taylor and Greene 2003). This program uses the supplied reference DNA and mutation sequence information to locate and determine if the nucleotide change alters the protein sequence. The PARSESNP program also compares the sequence of the gene of interest to already compiled blocks of homology in a database to score a position specific scoring matrix (PSSM) difference (Henikoff and Henikoff 1996). Another useful software program, SIFT (Sorting Intolerant From Tolerant), can be used in concert with PARSESNP. SIFT searches current databases for similar sequences and provides a score based on amino acid conservation that predicts whether or not a change will be tolerated (Ng and Henikoff 2003). If a mutation has a high PSSM and a low SIFT score on the PARSESNP program output, it is likely that an amino acid change at that position is rare and is therefore predicted to have a severe effect on protein function. These predictions can help prioritize mutations for phenotypic analysis, especially when an abundance of mutations are discovered for a single gene target.

Phenotypic Analysis

Depending on the goals of the investigator, the desired phenotype can be a disruption of function to aid in basic research of a largely uncharacterized gene, or it can be to achieve a specific goal, such as improved agronomic or fruit quality traits. In either case, background mutations are the largest hurdle for any investigator in acquiring a clean phenotype.

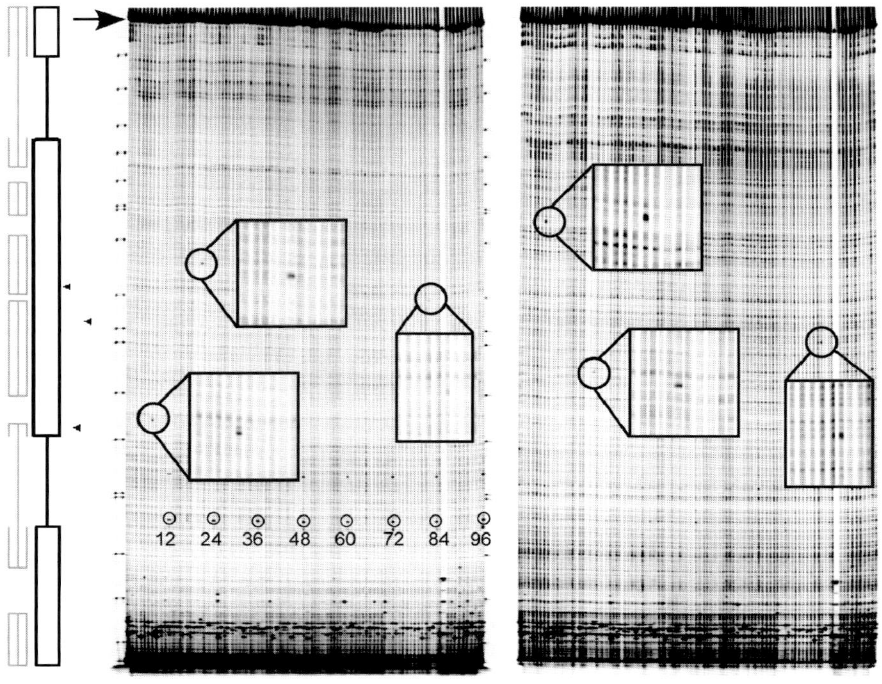

Fig. 14. Example of a 96-lane TILLING assay used for detection of mutations in 6-fold pools of tomato DNA. A gel image from the IRD 700 channel (*left*) and IRD 800 channel (*right*) with size ladders flanking the left image are shown. The *dark bands* at the top of the gels represent full-length (759 bp) uncut PCR products (*arrow*). CEL I cleaved heteroduplexes indicating the presence of a mutation are visible as dark bands (*circled*) shorter than the full-length product. The size of the cleavage products in the IRD 700 and IRD 800 channels corresponding to the same mutation add up to the full-length PCR product. *Insets*: enlargements of CEL I cut heteroduplexes. The PARSESNP output showing the gene model exons (*boxes*) and introns (*lines*) is shown to the left of the IRD 700 image. The *brackets* adjacent to the gene model indicate areas with homology to other proteins in the database. *Triangles* indicate the position of the mutations. A marker is added to every twelfth lane to improve the accuracy of the lane calls (*small circles*)

A population with a high mutation frequency is desired for efficient TILLING, but the resulting mutation loads in M_3 plants can greatly impede phenotypic analysis. Each M_3 plant can have hundreds or thousands of background mutations. Any of these background mutations can interfere with plant growth and reproduction, making it difficult to obtain tissue for analysis and in some cases, the background mutations can interfere specifically with the anticipated phenotype of the mutated gene. M_3 siblings that are wt for the mutation of interest but that contain a similar complement of background mutations as the homozygous plant are therefore essential controls in early phenotyping efforts. In cases where the M_2 plant is homozygous, no sibling controls will exist in the M_3 generation and the first controlled look at phenotype will take place after crossing to the unmutagenized parent line. The genotyping of progeny is greatly facilitated by the ability to use the identified mutation as its own molecular marker in a breeding program. Additional backcrosses will eliminate background mutations for advanced phenotypic analysis. For commercial crop development, mutations identified by TILLING can be introgressed into preferred germplasm or combined with other mutations to stack traits.

1.19.4 Reverse Genetics and TILLING: Conclusions

Sensitive mutation discovery by TILLING is opening up new prospects for functional genomics in a wide array of crops, especially those for which sequence information is limited or those in which transformation is not available or efficient. Some of the plants in which TILLING has been reported include *Arabidopsis*, lotus, wheat, maize, rice, barley, poplar, and soybean, with the list continuing to grow (Gilchrist and

Haughn 2005; Hohmann et al. 2005; Slade and Knauf 2005; Comai and Henikoff 2006). Arcadia Biosciences has also successfully applied TILLING in tomato, lettuce, castor, peanut, soybean, maize and rice (Slade et al. unpublished). One reason for its increasing popularity across diverse plant species is the general applicability of chemical mutagenesis and lack of requirements for special constructs or transformation protocols. However, it can take considerable time to develop a mutagenesis treatment optimized for each crop, especially one in which lethality is balanced with an adequate mutation frequency for efficient throughput. In addition to identifying novel induced diversity, "ecoTILLING" can be used to discover naturally occurring SNPs and small indels in different plant varieties and landraces (Comai et al. 2004; Gilchrist et al. 2006). This is an especially attractive way to uncover diversity in plants that are less amenable to the usual TILLING strategy, such as those with long generation times or in crops that are propagated vegetatively.

The genetic diversity that is created through chemical mutagenesis can range in effect from eliminating gene function entirely, to severely or slightly altering function, to having no effect at all. Such a range of potential phenotypes provides investigators with a full tool box for elucidating the function of unknown genes and can facilitate research of known genes where a reduction of function rather than a complete knock-out is desired. TILLING is also a non-transgenic method for introducing genetic diversity into domesticated varieties to provide novel alleles for commercially important traits in tomato and other Solanaceae and thus can be a powerful tool for crop improvement.

1.20 Genetic Engineering

Development of methods to transfer genes of interest into tomato has made it possible to use biotechnological approaches for the study of gene function. An essential prerequisite for a gene transfer system is an efficient and reliable method to regenerate plants from cells in various plant tissues. A significant amount of effort by many research groups has gone into development of an efficient regeneration system for tomato. There were several reports prior to the first reports on transformation (Behki and Leslley 1976; Kartha et al. 1976), and improvements on regeneration methods continued even after the first successful transformations were reported (Poonam et al. 2004). Regeneration from in vitro tomato cultures has been shown to be dependent upon the genotype (Stommel and Sinden 1991), explant type (Poonam et al. 2004), culture medium, namely the plant growth regulators (Uddin et al. 1988), and culture conditions (light, temperature) (Reynolds et al. 1982; Lercari et al. 2002; Poonam et al. 2004).

The earliest reports of *Agrobacterium tumefaciens*-mediated transformation of tomato were approximately 20 years ago (Horsch et al. 1985; McCormick et al. 1986). During the intervening years, there have been numerous reports of tomato transformation methods and projects that resulted in the recovery of transgenic lines that were successfully transferred to the greenhouse and field for further study. Plant material from sterile, in vitro-grown seedlings has been the primary source of explants for transformation with cotyledon segments being the most commonly used seedling part (Fillatti et al. 1987; Hamza and Chupeau 1993; van Roekel et al. 1993; Ellul et al. 2003; Frary and Van Eck 2004). However, hypocotyls segments from seedlings have also been used in some studies (Frary and Earle 1996; Park et al. 2003), although Frary and Earle (1996) reported that the time for recovery of transformants from hypocotyls took longer than recovery from cotyledon segments. There are also reports of using surface sterilized leaves from greenhouse-grown plants as a source of explants for transformations (Horsch et al. 1985; McCormick et al. 1986) and leaves from sterile, in vitro-grown plants (Sigareva et al. 2004). However, there are fewer reports where leaf segments were used for transformations as compared to the use of cotyledon segments.

When all the components necessary for an efficient regeneration and transformation system are in place, the plant material has the potential to regenerate a large number of shoots. To make recovery of transformants more efficient, a selectable marker gene is often included in the gene constructs used for transformations. These selectable marker genes can confer resistance to either antibiotics or herbicides. Therefore, when the appropriate concentration of either an antibiotic or herbicide component is incorporated into culture medium used for transformations, the non-transformed cells do not regenerate shoots, whereas, the cells containing the selectable marker and gene of interest can continue to grow and regenerate shoots. Various selectable markers have been used successfully for tomato transformation. The most commonly used selectable marker for tomato transformation has been the neomycin phosphotransferase gene *(nptII)*, which was isolated from the transposon Tn5 that was present in the bacterium strain *Escherichia coli* K12. This gene confers resistance to a range of aminoglycoside antibiotics including kanamycin, which is the most commonly used selection agent when *nptII* is used as the selectable marker gene. Another antibiotic resistance gene that has been used successfully for tomato transformation is the gene for hygromycin phosphotransferase (*hpt*), which is also from *E. coli* and confers resistance to the antibiotic hygromycin (Van Eck et al. 2006). Van Eck et al. (2006) also reported the use of the herbicide resistance gene, *bar*, from *Streptomyces hygmscopius* as a selectable marker for tomato transformation. The selection agent used for transformations with *bar* as the selectable marker was bialaphos. With interest in development of selection systems that would be more acceptable by the public, efforts are underway by a number of groups to find selection strategies that are not based on antibiotic or herbicide resistance. One such method was reported for tomato and is based on using a carbohydrate, mannose, as a selection agent. (Sigareva et al. 2004). For this selectable marker system, the phosphomannose isomerase gene (*pmi*) was used and mannose was incorporated into the culture medium to select for transformed cells. Non-transformed cells in the leaf explants could not grow on the mannose-containing medium, therefore only the transformed cells were able to grow and produce shoots.

For initial investigations of tomato transformation, genes were chosen for purposes of proof-of-concept and development of an efficient and reliable transformation method, rather than their influence on agronomically important traits. Following reports of efficient transformation methods for various genotypes, researchers have investigated traits of interest such as insect tolerance (Fischhoff et al. 1987), ripening (Vrebalov et al. 2002; Xiaong et al. 2005), bacterial disease resistance (Li and Steffens 2002), virus tolerance (Xu et al. 2004), and improved growth under environmental stresses (Hsieh et al. 2002; Jia et al. 2002; Mishra et al. 2002). Currently, there are no genetically engineered tomatoes on the market. However, there are several companies that have been involved or are currently developing genetically engineered tomatoes that will have a commercial application. For additional information on efforts for commercialization of genetically engineered tomatoes see http://www.geo-pie.cornell.edu/crops/tomato.html.

In addition to interest in traits of agronomic importance, tomato is also being investigated for production of plant-made pharmaceuticals, namely orally delivered vaccines (Van Eck et al. 2006). Characteristics that make tomatoes a good choice for production of orally delivered vaccines include high fruit yield, non-toxicity, reasonable biomass and protein content, and easy containment and production in greenhouses, even at commercial scales. Tomato has been investigated for production of vaccines for respiratory syncytial virus (Sandhu et al. 2000), cholera (Jiang et al. 2007), hepatitis E (Ma et al. 2003), and diarrheal illnesses caused by *E. coli* (Walmsley et al. 2003). Although there is some concern about consistency of amounts produced in each individual fruit, studies are in progress to produce freeze-dried material from batch harvests that can be packaged in capsules for delivery of controlled and effective amounts of the vaccine (Van Eck et al. 2006).

With the advent of genome sequencing for many plant species, functional genomics plays an important role in gene discovery, which requires methods for high-throughput generation of mutant and transgenic lines. Such is the case for tomato, where researchers are already investigating approaches for high-throughput transformation systems in anticipation of available tomato genome sequence. One important component of a high-throughput transformation system for functional genomics studies is the selection of a genotype that is amenable to transformation, has a fast generation cycle, and has a small size so that a large number of plants can be grown without the need for expansive growing areas. A tomato genotype that is being investigated for high-throughput transformation is Micro-Tom, a miniature-dwarf-determinate *S. lycopersicum* cultivar (Meissner et al. 1997). It can be grown at a high density, yields mature fruit within 70–90 days of sowing and is amenable to *Agrobacterium tumefaciens*-mediated transformation (Dan et al. 2006; Sun et al. 2006).

1.21 Tomato: Future Prospects

Since the early years when the first genetic markers were identified as being linked to disease resistance in tomato, there have been steady increases in developing molecular markers for genome mapping and MAS for numerous traits. Public and private sector breeders, as well as molecular marker service-providing companies, have placed increasing demands for new markers linked to additional genes influencing disease, pest, and abiotic stress resistances, fruit quality including flavor and nutritional value, and numerous other traits. More than 250,000 ESTs derived from more than 23 tomato cDNA libraries have been sequenced (Moore et al. 2002), and more than 30,000 unigenes have been defined (Van der Hooven et al. 2002; http://www.sgn.cornell.edu, http://www.ncbi.nlm.nih.gov). These unigenes represent a large set of candidate sequences for encoding critical developmental and physiological processes. With the sequencing of the tomato euchromatic portion of the genome (Mueller et al. 2005b), many additional candidate genes will be identified. High-throughput technologies of genomics, proteomics, and metabolomics enable the simultaneous quantification of the products of these genes during development (Alba et al. 2004; Fei et al. 2004) or in different plant tissues (Lemaire-Chamley et al. 2005). Techniques, such as comparative mapping, and tools, such as COS markers, facilitate synteny studies and expedite gene and QTL characterization (Fulton et al. 2002b). These studies will provide necessary data for hypothesis testing to rapidly elucidate the genes underlying favorable QTLs or mutations and subsequently incorporate them into cultivars.

Thousands of SNPs will be discovered that can be used for genome mapping, MAS, and positional cloning. Once a target gene has been characterized, it will be important to find new allelic variants within the large germplasm collections. An increased knowledge of gene function and regulation, as well as the development of more precise and efficient MAS, will help to avoid introgression of large segments and undesirable loci into elite lines. Mutational and transgenic tools, such as mutation libraries in a uniform genetic background (Menda et al. 2004), and techniques to screen for genetic lesions in specific genes (Comai and Henikoff 2006) will aid in the description of desirable alleles. Thanks to a wealth of novel tools and techniques, to its vast natural polymorphism at intra- and inter-specific levels, and to consumer demands for nutritious foods (Weimer 1999), tomato will retain its status as a valuable model crop for the foreseeable future.

Acknowledgement. This work was partly supported by the Research and Development Program for New Bio-Industry Initiatives, NIH, USDA, CSU Agricultural Research Initiative (CSU-ARI), CSU Program for Education and Research in Biotechnology (CSUPERB), California State University Fresno College of Science & Mathematics, California Pepper Commission, New Mexico Agricultural Experiment Station, and grants from the National Institutes of Health GM S06 GM08136 and USDA CSREES 2003-34287-13411. We thank Dr. D. Shibata (Kazusa DNA Research Institute) for helpful discussion on metabolomics.

References

Agius F, Gonzalez-Lamothe R, Caballero JL, Munoz-Blanco J, Botella MA, Valpuesta V (2003) Engineering increased vitamin C levels implants by overexpression of a D-galacturonic acid reductase. Nat Biotechnol 21:177–181

Agrama HA, Scott JW (2006) Quantitative trait loci for *Tomato yellow leaf curl virus* and *Tomato mottle virus* resistance in tomato. J Am Soc Hort Sci 131:267–272

Alba R, Fei ZJ, Payton P, Liu Y, Moore SL, Debbie P, Cohn J, D'Ascenzo M, Gordon JS, Rose JKC, Martin G, Tanksley SD, Bouzayen M, Jahn MM, Giovannoni J (2004) ESTs, cDNA microarrays, and gene expression profiling: tools for dissecting plant physiology and development. Plant J 39:697–714

Alba R, Payton P, Fei ZJ, McQuinn R, Debbie P, Martin GB, Tanksley SD, Giovannoni JJ (2005) Transcriptome and selected metabolite analyses reveal multiple points of ethylene control during tomato fruit development. Plant Cell 17:2954–2965

Alexander LJ (1963) Transfer of a dominant type of resistance to the four known Ohio pathogenic strains of *Tobacco mosaic virus* (TMV), from *Lycopersicon peruvianum* to *L. esculentum*. Phytopathology 53:869

Alonso JM, Stepanova AN, Leisse TJ, Kim CJ, Chen H, Shinn P, Stevenson DK, Zimmerman J, Barajas P, Cheuk R, Gadrinab C, Heller C, Jeske A, Koesema E, Meyers CC, Parker H, Prednis L, Ansari Y, Choy N, Deen H, Geralt M, Hazari N, Hom E, Karnes M, Mulholland C, Ndubaku R, Schmidt I, Guzman P, Aguilar-Henonin L, Schmid M, Weigel D, Carter DE, Marchand T, Risseeuw E, Brogden D, Zeko A, Crosby WL, Berry CC, Ecker JR (2003) Genome-wide insertional mutagenesis of *Arabidopsis thaliana*. Science 301:653–657

Alpert K, Grandillo S, Tanksley SD (1995) *fw2.2*: a major QTL controlling fruit weight is common to both red- and green-fruited tomato species. Theor Appl Genet 91:994–1000

Alpert K, Tanksley S (1996) High-resolution mapping and isolation of a yeast artificial chromosome contig containing *fw2.2*: A major fruit weight quantitative trait locus in tomato. Proc Natl Acad Sci USA 93:15503–15507

Altschul SF, Gish W, Miller W, Myers EW, Lipman DJ (1990) Basic local alignment search tool. J Mol Biol 215:403–410

Alvarez AE, van de Wiel CCM, Smulders MJM, Vosman B (2001) Use of microsatellites to evaluate genetic diversity and species relationships in the genus *Lycopersicon*. Theor Appl Genet 103:1283–1292

Ames BN, Shigenaga MK, Hagen TM (1993) Oxidants, antioxidants, and the degenerative diseases of aging. Proc Natl Acad Sci USA 90:7915–7922

Anderson JW (1990) Dietary fiber and human health. HortScience 25:1488–1495

Andrus CF, Reynard GB, Jorgensen H, Eades J (1942) Collar rot resistance in tomatoes. J Agric Res 65:339–346

Anfoka GH, Abhary M, Stevens MR (2006) Occurrence of *Tomato spotted wilt virus* (TSWV) in Jordan. EPPO Bull 36:517–522

Arens P, Odinot P, van Heusden AW, Lindhout P, Vosman B (1995) GATA-repeats and GACA-repeats are not evenly distributed throughout the tomato genome. Genome 38:84–90

Areshchenkova T, Ganal M (1999) Long tomato microsatellites are predominantly associated with centromeric regions. Genome 42:536–544

Areshchenkova T, Ganal MW (2002) Comparative analysis of polymorphism and chromosomal location of tomato microsatellite markers isolated from different sources. Theor Appl Genet 104:229–235

Arumuganathan K, Earle E (1991) Estimation of nuclear DNA content of plants by flow cytometry. Plant Mol Biol Rep 9:208–218

Ascenzi R, Gantt JS (1997) A drought-stress-inducible histone gene in *Arabidopsis thaliana* is a member of a distinct class of plant linker histone variants. Plant Mol Biol 34:629–641

Ascenzi R, Gantt JS (1999a) Molecular genetic analysis of the drought-inducible linker histone variant in *Arabidopsis thaliana*. Plant Mol Biol 41:159–169

Ascenzi R, Gantt JS (1999b) Subnuclear distribution of the entire complement of linker histone variants in *Arabidopsis thaliana*. Chromosoma 108:345–355

Ashikari M, Matsuoka M (2006) Identification, isolation and pyramiding of quantitative trait loci for rice breeding. Trends Plant Sci 11(7):344–350

Astua-Monge G, Minsavage GV, Stall RE, Vallejos E, Davis MJ, Jones JB (2000) *Xv4-vrxv4*: A new gene-for-gene interaction identified between *Xanthomonas campestris* pv. *vesicatoria* race T3 and the wild tomato relative *Lycopersicon pennellii*. Mol Plant-Micr Interact 13:1346–1355

Azanza F, Young TE, Kim D, Tanksley SD, Juvik JA (1994) Characterization of the effects of introgressed segments of chromosome 7 and 10 from *Lycopersicon chmielewskii* on tomato soluble solids, pH and yield. Theor Appl Genet 87:965–972

Bai Y, Feng X, van der Hulst R, Lindhout P (2004) A set of simple PCR markers converted from sequence specific RFLP markers on tomato chromosomes 9 to 12. Mol Breed 13:281–287

Bai Y, Huang CC, van der Hulst R, Meijer-Dekens F, Bonnema G, Lindhout P (2003) QTLS for tomato powdery mildew resistance (*Oidium lycopersici*) in *Lycopersicon parviflorum* G1.1601 co-localize with two qualitative powdery mildew resistance genes. Mol Plant-Micr Interact 16:169–176

Bai Y, van der Hulst R, Huang CC, Wei L, Stam P, and Lindhout P (2004) Mapping *Ol-4*, a gene conferring resistance to *Oidium neolycopersici* and originating from *Lycopersicon peruvianum* LA2172, requires multi-allelic, single-locus markers. Theor Appl Genet 109:1215–1223

Baker LR, Tomes ML (1964) Carotenoids and chlorophylls in two tomato mutants and their hybrids. Proc Am Soc Hort Sci 85:507–513

Balint-Kurti PJ, Dixon JS, Jones DA, Norcott KA, Jones JDG (1994) RFLP linkage analysis of the *Cf-4* and *CF-9* genes for resistance to *Cladosporium fulvum* in tomato. Theor Appl Genet 88:691–700

Ballvora A, Schornack S, Baker BJ (2001) Chromosome landing at the tomato *Bs4* locus. Mol Genet Genom 266:639–645

Barksdale TH (1972) Resistance in tomato to six anthracnose fungi. Phytopathology 62:660–663

Barrero LS, Tanksley SD (2004) Evaluating the genetic basis of multiple-locule fruit in a broad cross section of tomato cultivars. Theor Appl Genet 109:669–679

Barry CS, McQuinn RP, Thompson AJ, Seymour GB, Grierson D, Giovannoni J (2005) Ethylene insensitivity conferred by the *Green-ripe* and *Never-ripe 2* ripening mutants of tomato. Plant Physiol 138:267–275

Bartholomew DM, Bartley GE, Scolnik PA (1991) Abscisic acid control of *rbcS* and *cab* transcription in tomato leaves. Plant Physiol 96:291–296

Bartley GE, Scolnik PA (1993) cDNA cloning, expression during development, and genome mapping of *PSY2*, a second tomato gene encoding phytoene synthase. J Biol Chem 268:25718–25721

Bartley GE, Scolnik PA (1994) Molecular biology of carotenoid biosynthesis in plants. Annu Rev Plant Physiol Plant Mol Biol 45:287–301

Baudry E, Kerdelhue C, Innan H, Stephan W (2001) Species and recombination effects on DNA variability in the tomato genus. Genetics 158:1725–1735

Bauhin C (1623) Pinax theatri botanici. Ludovici Regis, pp 522

Baxter CJ, Sabar M, Quick WP, Sweetlove LJ (2005) Comparison of changes in fruit gene expression in tomato introgression lines provides evidence of genome-wide transcriptional changes and reveals links to mapped QTLs and described traits. J Exp Bot 56:1591–1604

Beecher GR (1998) Nutrient content of tomatoes and tomato products. Proc Soc Exp Biol Med 218:98–100

Behare J, Laterrot H, Sarfatti M, Zamir D (1991) Restriction fragment length polymorphism mapping of the *Stemphylium* resistance gene in tomato. Mol Plant-Micr Interact 4(5):489–492

Behki RM, Leslley SM (1976) In vitro plant regeneration from leaf explants of *Lycopersicon esculentum* (tomato). Can J Bot 54:2409–2414

Ben Chaim A, Grube RC, Lapidot M, Jahn M, Paran I (2001) Identification of quantitative trait loci associated with resistance to cucumber mosaic virus in *Capsicum annuum*. Theor Appl Genet 102:1213–1220

Ben Chaim A, Borovsky Y, De Jong W, Paran I (2003) Linkage of the *A* locus for the presence of anthocyanin and *fs10.1*, a major fruit-shape QTL in pepper. Theor Appl Genet 106:889–894

Ben Chaim A, Borovsky Y, Falise M, Mazourek M, Kang B-C, Paran I, Jahn M (2006) QTL analysis for capsaicinoid content in *Capsicum*. Theor Appl Genet 113:1481–1490

Berlinger MJ (1986) Pests. In: Atherton JG, Rudich J (eds) The tomato crop: a scientific basis for improvement. Kluwer Academic Press, London, pp 391–441

Bernacchi D, Beck-Bunn T, Emmatty D, Eshed Y, Inai S, Lopez J, Petiard V, Sayama H, Uhlig J, Zamir D, Tanksley S (1998a) Advanced backcross QTL analysis of tomato. II. Evaluation of near-isogenic lines carrying single-donor introgressions for desirable wild QTL-alleles derived from *Lycopersicon hirsutum* and *L. pimpinellifolium*. Theor Appl Genet 97:170–180, erratum 1191–1196

Bernacchi D, Beck-Bunn T, Eshed Y, Lopez J, Petiard V, Uhlig J, Zamir D, Tanksley S (1998b) Advanced backcross QTL analysis in tomato. I. Identification of QTLs for traits of agronomic importance from *Lycopersicon hirsutum*. Theor Appl Genet 97:381–397

Bernacchi D, Tanksley SD (1997) An interspecific backcross of *Lycopersicon esculentum* × *L. hirsutum* linkage analysis and a QTL study of sexual compatibility factors and floral traits. Genetics 147:861–877

Bernatzky R, Tanksley SD (1986) Towards a saturated linkage map in tomato based on isozymes and random cDNA sequences. Genetics 112:887–898

Berry CS, McQuinn RP, Thompson AJ, Seymour GB, Grierson D, Giovannoni J (2005) Ethylene insensitivity conferred by the *Green-ripe* and *Never-ripe 2* ripening mutants of tomato. Plant Physiol 138:267–275

Berry SZ, Oaks GL (1987) Inheritance of resistance to fusarium crown and root rot in tomato. HortScience 22:110–111

Bindler G, van der Hoeven R, Gunduz I, Plieske J, Ganal M, Rossi L, Gadani F, Donini P (2007) A microsatellite marker based linkage map of tobacco. Theor Appl Genet 114:341–349

Bino RJ, Ric de Vos CH, Lieberman M, Hall RD, Bovy A, Jonker HH, Tikunov Y, Lommen A, Moco S, Levin I (2005) The light-hyperresponsive *high pigment-2*^{dg} mutation of tomato: alterations in the fruit metabolome. New Phytol 166:427–438

Black LL, Wang TC, Hanson, PM, Chen JT (1996) Late blight resistance in four wild tomato accessions: effectiveness in diverse locations and inheritance of resistance. Phytopathology 86:S24

Blauth SL, Churchill GA, Mutschler MA (1998) Identification of quantitative trait loci associated with acylsugar accumulation using intraspecific populations of the wild tomato, *Lycopersicon pennellii*. Theor Appl Genet 96:458–467

Block G, Patterson B, Subar A (1992) Fruits, vegetable, and cancer prevention: A review of the epidemiological evidence. Nutr Cancer 18:1–29

Bohn GW, Tucker CM (1939) Immunity to fusarium wilt in the tomato. Science 89:603–604

Bohn GW, Tucker CM (1940) Studies on fusarium wilt of the tomato. I. Immunity in *Lycopersicon pimpinellifolium* Mill. and its inheritance in hybrids. Mo Agric Exp Stn Res Bull 311, pp 82

Bohs L (2005) Major clades in *Solanum* based on *ndhF* sequences. In: Keating RC, Hollowell VC, Croat T (eds) Festschrift for William G. Darcy: the legacy of a taxonomist (Monographs in Systematic Botany 104). Missouri: MBG Press, pp 27–49

Bohs L, Olmstead RG (1997) Phylogenetic relationships in *Solanum* (Solanaceae) based on *ndhF* sequences. Syst Bot 22:5–17

Boileau AC, Merchen NR, Wasson K, Atkinson CA, Erdman JW (1999) Cis-lycopene is more bioavailable than trans-lycopene in vitro and in vivo in lymph-cannulated ferrets. J Nutr 129:1176–1181

Boiteux LS, Giordano L de B (1993) Genetic basis of resistance against two *Tospovirus* species in tomato (*Lycopersicon esculentum*). Euphytica 71:151–154

Bonas U, Conrads-Strauch J, Balbo I (1993) Resistance in tomato to *Xanthomonas campestris* pv. *vesicatoria* is determined by alleles of the pepper-specific avirulence gene *avrBs3*. Mol Gen Genet 238:261–269

Bonde R, Murphy EF (1952) Resistance of certain tomato varieties and crosses to late blight. Maine Agric Exp Stn Bull 497:5–15

Bonierbale MW, Plaisted RL, Tanksley SD (1988) RFLP maps based on a common set of clones reveal modes of chromosomal evolution in potato and tomato. Genetics 120:1095–1103

Bonnema G, Schipper D, van Heusden S, Zabel P, Lindhout P (1997) Tomato chromosome 1: High-resolution genetic and physical mapping of the short arm in an interspecific *Lycopersicon esculentum* × *L. peruvianum* cross. Mol Gen Genet 253:455–462

Bournival BL, Scott JW, Vallejos CE (1989) An isozyme marker for resistance to race 3 of *Fusarium oxysporum* f. sp. *lycopersici* in tomato. Theor Appl Genet 78:489–494

Bournival BL, Vallejos CE, Scott JW (1990) Genetic analysis of resistances to races 1 and 2 of *Fursarium oxysporum* f.

sp. *lycopersici* from the wild tomato *Lycopersicon pennellii*. Theor Appl Genet 79:641–645

Bovy A, de Vos R, Kemper M, Schijlen E, Pertejo MA, Muir S, Collins G, Robinson S, Verhoeyen M, Hughes S, Santos-Buelga C (2002) High-flavonol tomatoes resulting from the heterologous expression of the maize transcription factor genes *LC* and *C*1. Plant Cell 14:2509–2526

Bradley DB (1946) Varietal and location influence on acid composition of tomato fruit. J Agric Food Chem 12:213–216

Bradshaw JE, Hackett CA, Meyer RC, Milbourne D, McNichol JW, Phillips MS, Waugh R (1998) Identification of AFLP and SSR markers associated with quantitative resistance to *Globodera pallida* (Stone) in tetraploid potato (*Solanum tuberosum* subsp. *tuberosum*) with a view to marker-assisted selection. Theor Appl Genet 97:202–210

Bray EA (1988) Drought- and ABA-induced changes in polypeptide and mRNA accumulation in tomato leaves. Plant Physiol 88:1210–1214

Bray EA (2002) Classification of genes differentially expressed during water-deficit stress in *Arabidopsis thaliana*: an analysis using microarray and differential expression data. Ann Bot 89:803–811

Bray EA, Shih T-Y, Moses MS, Cohen A, Imai R, Plant AL (1999) Water-deficit induction of a tomato H1 histone requires abscisic acid. Plant Growth Regul 29:35–46

Breto MP, Asins MJ, Carbonell EA (1993) Genetic variability in *Lycopersicon* species and their genetic relationships. Theor Appl Genet 86:113–120

Brewer MT, Lang L, Fujimura K, Dujmovic N, Gray S, van der Knaap E (2006) Development of a controlled vocabulary and software application to analyze fruit shape variation in tomato and other plant species. Plant Physiol 141:15–25

Brommonschenkel SH (1996) Genetic mapping and progress towards the positional cloning of the *Sw-5 Tomato spotted wilt virus* (TSWV) resistance gene in tomato (*Lycopersicon esculentum* Mill.). Cornell Univ, Ithaca, NY, USA, Ph.D. thesis, pp 64

Brommonschenkel SH, Frary A, Tanksley SD (2000) The broad-spectrum tospovirus resistance gene *Sw-5* of tomato is a homolog of the root-knot nematode resistance gene *Mi*. Mol Plant-Micr Interact 13:1130–1138

Brommonschenkel SH, Tanksley SD (1997) Map-based cloning of the tomato genomic region that spans the *Sw-5* tospovirus resistance gene in tomato. Mol Gen Genet 256:121–126

Broun P, Tanksley SD (1996) Characterization and genetic mapping of simple repeat sequences in the tomato genome. Mol Gen Genet 250:39–49

Brouwer DJ, Jones ES, St Clair DA (2004) QTL analysis of quantitative resistance to *Phytophthora infestans* (late blight) in tomato and comparisons with potato. Genome 47:475–492

Brouwer DJ, St Clair DA (2004) Fine mapping of three quantitative trait loci for late blight resistance in tomato using near isogenic lines (NILs) and sub-NILs. Theor Appl Genet 108:628–638

Brunt AA (1986) Tomato mosaic virus. In: Van Regenmortel MHV, Fraenkel-Conrat H (eds) The plant viruses. Plenum Press, New York, pp 181–204

Brusca JP (2003) Inheritance of tomato late blight resistance from 'Richter's Wild Tomato' and evaluation of late blight combinations in adapted fresh market tomato backgrounds. MS Thesis, North Carolina State Univ, Raleigh, USA

Bryan GJ, McLean K, Pande B, Purvis A, Hackett CA, Bradshaw JE, Waugh R (2004) Genetical dissection of H3-mediated polygenic PCN resistance in a heterozygous autotetraploid potato population. Mol Breed 14:105–116

Bucheli P, Voirol E, de la Torre R, Lopez J, Rytz A, Tanksley SD, Petiard V (1999) Definition of nonvolatile markers for flavor of tomato (*Lycopersicon esculentum* Mill.) as tools in selection and breeding. J Agric Food Chem 47:659–664

Budiman M, Mao L, Wood T, Wing R (2000) A deep-coverage tomato BAC library and prospects toward development of an STC framework for genome sequencing. Genome Res 10:129–136

Buonaurio R, Stravato VM, Cappelli C (1996) Occurrence of *Pseudomonas syringae* pv. tomato race 1 in Italy on *Pto* gene-bearing tomato plants. J Phytopathol 144:437–440

Burge J, Mickelsen O, Nicklow C, Marsh GL (1975) Vitamin C in tomatoes: comparison of tomatoes developed for mechanical or hand harvesting. Ecol Food Nutr 4:27–31

Burns JJ (1967) Ascorbic acid. In: Greenberg DM (ed) Metabolic pathways. 3rd edn, vol 1. Academic Press, New York, pp 394–411

Caldwell DG, McCallum N, Shaw P, Muehlbauer GJ, Marshall DF, Waugh R (2004) A structured mutant population for forward and reverse genetics in Barley (*Hordeum vulgare* L.). Plant J 40:143–150

Canady MA, Meglic V, Chetelat RT (2005) A library of *Solanum lycopersicoides* introgression lines in cultivated tomato. Genome 48:685–697

Cannon OS, Waddoups V (1952) Loran Blood and V.R. Moscow, two new *Verticillium* wilt resistance tomatoes for Utah. Utah Farm Home Sci 13:74

Carangal AR, Alban EK, Varner JE, Burrell RC (1954) The influence of mineral nutrition on the organic acids of the tomato, *Lycopersicon esculentum*. Plant Physiol 29:355–360

Carmeille A, Caranta EC, Dintinger EJ, Prior P, Luisetti EJ, Besse EP (2006) Identification of QTLs for *Ralstonia solanacearum race* 3-phylotype II resistance in tomato. Theor Appl Genet 114 (1):110–121

Caromel B, Mugniery D, Lefebvre V, Andrzejexski S, Ellisseche D, Kerlan MC, Rousselle P, Rousselle-Bourgeois F (2003) Mapping QTLs for resistance against *Globodera pallida* (Stone) Pa2/3 in a diploid potato progeny originating from *Solanum spegazzinii*. Theor Appl Genet 106:1517–1523

Carrari F, Baxter C, Usadel B, Urbanczyk-Wochniak E, Zanor MI, Nunes-Nesi A, Nikiforova V, Centero D,

Ratzka A, Pauly M, Sweetlove LJ, Fernie AR (2006) Integrated analysis of metabolite and transcript levels reveals the metabolic shifts that underlie tomato fruit development and highlight regulatory aspects of metabolic network behaviour. Plant Physiol 142:1380–1396

Carter CD, Sacalis JN, Gianfagna TJ (1989) Zingiberene and resistance to Colorado Potato Beetle in *Lycopersicon hirsutum* f. *hirsutum*. J Agric Food Chem 37:206–210

Carter CD, Snyder JC (1985) Mite responses in relation to trichomes of *Lycopersicon esculentum* × *L. hirsutum* F_2 hybrids. Euphytica 34:177–185

Carter CD, Snyder JC (1986) Mite responses and trichome characters in a full-sib F_2 family of *Lycopersicon esculentum* × *L. hirsutum*. J Am Soc Hort Sci 111:130–133

Cato SA, Gardner RC, Kent J, Richardson TE (2001) A rapid PCR-based method for genetically mapping ESTs. Theor Appl Genet 102:296–306

Causse M, Buret M, Robini K, Verschave P (2003) Inheritance of nutritional and sensory quality traits in fresh market tomato and relation to consumer preferences. J Food Sci 68:2342–2350

Causse M, Duffe P, Gomez MC, Buret M, Damidaux R, Zamir D, Gur A, Chevalier C, Lemaire-Chamley M, Rothan C (2004) A genetic map of candidate genes and QTLs involved in tomato fruit size and composition. J Exp Bot 55:1671–1685

Causse M, Saliba-Colombani V, Lecomte L, Duffe P, Rousselle P, Buret M (2002) QTL analysis of fruit quality in fresh market tomato: a few chromosome regions control the variation of sensory and instrumental traits. J Exp Bot 53:2089–2098

Causse M, Saliba-Colombani V, Lesschaeve I, Buret M (2001) Genetic analysis of organoleptic quality in fresh market tomato. 2. Mapping QTLs for sensory attributes. Theor Appl Genet 102:273–283

Chagué V, Mercier JC, Guénard M, de Courcel A, Vedel F (1996) Identification and mapping on chromosome 9 of RAPD markers linked to *Sw-5* in tomato by bulked segregant analysis. Theor Appl Genet 92:1045–1051

Chagué V, Mercier JC, Guénard M, de Courcel A, Vedel F (1997) Identification of RAPD markers linked to a locus involved in quantitative resistance to TYLCV in tomato by bulked segregant analysis. Theor Appl Genet 95:671–677

Chaïb J, Lecomte L, Buret M, Causse M (2006) Stability over genetic backgrounds, generations and years of quantitative trait locus (QTLs) for organoleptic quality in tomato. Theor Appl Genet 112:934–944

Chalukova M (1988) Carotenoid composition of the fruits of hybrids between *Lycopersicon esculentum* and some wild species of the genus *Lycopersicon*, IV. Progenies of lycopene and β-carotene BC_1P_1 hybrids of *L. chmielewskii*. Genet Breed 21:49–57

Charlesworth D (2003) Effects of inbreeding on the genetic diversity of populations. Phil Trans Roy Soc Lond Ser B Biol Sci 358:1051–1070

Chen FQ, Foolad MR, Hyman J, St Clair DA, Beelman RB (1999) Mapping of QTLs for lycopene and other fruit traits in a *Lycopersicon esculentum* × *L. pimpinellifolium* cross and comparison of QTLs across tomato species. Mol Breed 5:283–299

Chen G, Hackett R, Walker D, Taylor A, Lin Z, Grierson D (2004) Identification of a specific isoform of tomato lipoxygenase (TomloxC) involved in the generation of fatty acid-derived flavor compounds. Plant Physiol 136:2641–2651

Chen GQ, Foolad MR (1999) A molecular linkage map of tomato based on a cross between *Lycopersicon esculentum* and *L. pimpinellifolium* and its comparison with other molecular maps of tomato. Genome 42:94–103

Chen R-D, Campeau N, Greer AF, Bellemare G, Tabaeizadeh Z (1993) Sequence of a novel abscisic acid- and drought-induced cDNA from wild tomato (*Lycopersicon chilense*). Plant Physiol 103:301

Chen R-D, Tabaeizadeh Z (1992a) Alteration of gene expression in tomato plants (*Lycopersicon esculentum*) by drought and salt stress. Genome 35:385–391

Chen R-D, Tabaeizadeh Z (1992b) Expression and molecular cloning of drought-induced genes in the wild tomato *Lycopersicon chilense*. Biochem Cell Biol 70:199–206

Chen R-D, Yu L-X, Greer AF, Cheriti H, Tabaeizadeh Z (1994) Isolation of an osmotic stress- and abscisic acid-induced gene encoding an acidic endochitinase from *Lycopersicon chilense*. Mol Gen Genet 245:195–202

Chetelat RT (2004) Revised list of wild species stocks. Tomato Genet Coop Rep 54:52–76

Chetelat RT (2005) Revised list of monogenic stocks. Tomato Genet Coop Rep 55:48–69

Chetelat RT (2006) Revised list of miscellaneous stocks. Tomato Genet Coop Rep 56:37–56

Chetelat RT, De Verna JW, Bennett AB (1995) Introgression into tomato (*Lycopersicon esculentum*) of the *L. chmielewskii* sucrose accumulator gene (*sucr*) controlling fruit sugar composition. Theor Appl Genet 91:327–333

Chetelat RT, Ji Y (2007) Cytogenetics and evolution. In: Razdan MK, Mattoo AK (eds) Genetic improvement of Solanaceous crops, vol 2: tomato. Science Publ, Enfield, New Hampshire, USA, pp 73–108

Chetelat RT, Klann E, DeVerna JW, Yelle S, Bennett AB (1993) Inheritance and genetic mapping of fruit sucrose accumulation in *Lycopersicon chmielewskii*. Plant J 4:643–650

Chetelat RT, Meglic V (2000) Molecular mapping of chromosome segments introgressed from *Solanum lycopersicoides* into cultivated tomato (*Lycopersicon esculentum*). Theor Appl Genet 100:232–241

Chetelat RT, Meglic V, Cisneros P (2000) A genetic map of tomato based on BC(1) *Lycopersicon esculentum* × *Solanum lycopersicoides* reveals overall synteny but suppressed recombination between these homeologous genomes. Genetics 154:857–867

Chetelat RT, Rick CM, Cisneros P, Alpert KB, De Verna JW (1998) Identification, transmission, and cytological behavior of *Solanum lycopersicoides* Dun. monosomic alien addition lines in tomato (*Lycopersicon esculentum* Mill.). Genome 41:40–50

Childs KL, Hamilton JP, Zhu W, Ly E, Cheung F, Wu H, Rabinowicz PD, Town CD, Buell CR, Chan AP (2007) The TIGR Plant Transcript Assemblies database. Nucl Acids Res 35:D846–851

Chmielewski T, Berger S (1966) Investigations on the inheritance of high concentrations of provitamin A in tomato. Hodowla Roslin Aklim Nasienn 10:385–400

Chunwongse J, Bunn TB, Crossman C, Jinag J, Tanksley SD (1994) Chromosomal localization and molecular-marker tagging of the powdery mildew resistance gene (*Lv*) in tomato. Theor Appl Genet 89:76–79

Chunwongse J, Chunwongse C, Black L, Hanson P (2002) Molecular mapping of the *Ph-3* gene for late blight resistance in tomato. J Hort Sci Biotechnol 77:281–286

Cirulli M, Alexander LJ (1969) Influence of temperature and strain of *Tobacco mosaic virus* on resistance in a tomato breeding line derived from *Lycopersicon peruvianum*. Phytopathology 59:1287–1297

Clayberg CD (1972) Preliminary mapping of three chromosome 7 genes. Tomato Genet Coop Rep 22:4

Clayberg CD, Butler L, Rick CM, Young PA (1960) Second list of known genes in the tomato. J Hered 51:167–174

Clinton SK (1998) Lycopene: Chemistry, biology, and implications for human health and disease. Nutr Rev 56:35–51

Clouse SD, Gilchrist DG (1987) Interaction of the *asc* locus in F_8 paired lines of tomato with *Alternaria alternata* f. sp. *lycopersici* and AAL-toxin. Phytopathology 70:80–82

Coaker GL, Francis DM (2004) Mapping, genetic effects, and epistatic interaction of two bacterial canker resistance QTLs from *Lycopersicon hirsutum*. Theor Appl Genet 108 (6):1047–1055

Cohen A, Bray EA (1990) Characterization of three mRNAs that accumulate in wilted tomato leaves in response to elevated levels of endogenous abscisic acid. Planta 182:27–33

Colbert T, Till BJ, Tompa R, Reynolds S, Steine MN, Yeung AT, McCallum CM, Comai L, Henikoff S (2001) High-throughput screening for induced point mutations. Plant Physiol 126:480–484

Colliver S, Bovy A, Collins G, Muir S, Robinson S, de Vos CHR, Verhoeyen ME (2002) Improving the nutritional content of tomatoes through reprogramming their biosynthetic pathway. Phytochem Rev 1:113–123

Comai L, Henikoff S (2006) TILLING: practical single-nucleotide mutation discovery. Plant J 45:684–694

Comai L, Young K, Till BJ, Reynolds SH, Greene EA, Codomo CA, Enns LC, Johnson JE, Burtner C, Odden AR, Henikoff S (2004) Efficient discovery of DNA polymorphisms in natural populations by Ecotilling. Plant J 37:778–786

Commenges D, Scotet V, Renaud S, Jacqmin-Gadda H, Barberger-Gateau P, Dartigues JF (2000) Intake of flavonoids and risk of dementia. Eur J Epidemiol 16:357–363

Cong B, Liu J, Tanksley SD (2002) Natural alleles at a tomato fruit size quantitative trait locus differ by heterochronic regulatory mutations. Proc Natl Acad Sci USA 99:13606–13611

Cong B, Tanksley SD (2006) *FW2.2* and cell cycle control in developing tomato fruit: a possible example of gene co-option in the evolution of a novel organ. Plant Mol Biol 62:867–880

Conover RA, Walter JA (1953) The occurrence of a virulent race of *Phytophthora infestans* on late blight resistant tomato stocks. Phytopathology 43:344–345

Cook D, Grierson D, Jones C, Wallace A, West G, Tucker G (2002) Modification of fatty acid composition in tomato (*Lycopersicon esculentum*) by expression of a borage $\Delta 6$-desaturase. Mol Biotechnol 21:123–128

Cook NC, Samman S (1996) Flavonoids – chemistry, metabolism, cardioprotective effects, and dietary sources. Nutr Biochem 7:66–76

Cookson PJ, Kiano JW, Shipton CA, Fraser PD, Romer S, Schuch W, Bramley PM, Pyke KA (2003) Increases in cell elongation, plastid compartment size and phytoene synthase activity underlie the phenotype of the *high pigment-1* mutant of tomato. Planta 217:896–903

Cordeiro G, Eliott FG, Henry RJ (2006) An optimized ecotilling protocol for polyploids or pooled samples using a capillary electrophoresis system. Anal Biochem 355:145–147

Cortina C, Culiánez-Macià FA (2005) Tomato abiotic stress enhanced tolerance by trehalose biosynthesis. Plant Sci 169:75–82

Courtney WHI, Lambeth VN (1977) Glycoalkaloid content of mature green fruit *Lycopersicon* species. HortScience 12:550–551

Cox S (2000) From discovery to modern commercialism: the complete story behind *Lycopersicon esculentum*. http://www.landscapeimagery.com/articles.html

Cuartero J, Bolarin MC, Assins MJ, Moreno V (2006) Increasing salt tolerance in the tomato. J Exp Bot 57:1045–1056

Cui X, Xu J, Asghar R, Condamine P, Svensson JT, Wanamaker S, Stein N, Roose M, Close TJ (2005) Detecting single-feature polymorphisms using oligonucleotide arrays and robustified projection pursuit. Bioinformatics 21:3852–3858

Cunningham FX, Gantt E (1998) Genes and enzymes of carotenoid biosynthesis in plants. Annu Rev Plant Physiol Plant Mol Biol 49:557–583

Cutler KD (1998) From wolf peach to outer space - tomato history and lore. In: Tantalizing tomatoes: smart tips and tasty picks for gardeners everywhere. 21st Century Gardening Series, Brooklyn Botanic Garden, New York

Dan Y, Yan H, Munyikwa T, Dong J, Zhang Y, Armstrong CL (2006) MicroTom - a high-throughput model transformation system for functional genomics. Plant Cell Rep 25:432–441

Danesh D, Aarons S, McGill GE, Young ND (1994) Genetic dissection of oligogenic resistance to bacterial wilt in tomato. Mol Plant-Micr Interact 7:464–471

Darwin SC, Knapp S, Peralta IE (2003) Taxonomy of tomatoes in the Galapagos Islands: native and introduced species of *Solanum* section *Lycopersicon* (Solanaceae). Syst Biodiver 1:29–53

Davies JN (1964) Effect of nitrogen, phosphorous and potassium fertilizers on the non-volatile organic acids of tomato fruit. J Sci Food Agric 15:665–673

Davies JN (1965) The effect of variety on the malic and citric acid content of tomato fruit. Rep Glasshouse Crops Res Inst, pp 139–141

Davies JN (1966a) Changes in the non-volatile organic acids of tomato fruit during ripening. J Sci Food Agric 17:396–400

Davies JN (1966b) Occurrence of sucrose in the fruit of some species of *Lycopersicon*. Nature 209:640–641

Davies JN, Hobson GE (1981) The constituents of tomato fruit - The influence of environment, nutrition and genotype. Crit Rev Food Sci Nutr 15:205–280

Davuluri GR, van Tuinen A, Fraser PD, Manfredonia A, Newman R, Burgess D, Brummell DA, King SR, Palys J, Uhlig J, Bramley PM, Pennings HM, Bowler C (2005) Fruit-specific RNAi-mediated suppression of DET1 enhances carotenoid and flavonoid content in tomatoes. Nat Biotechnol 23:890–895

De Candolle A (1886) Origin of cultivated plants. Hafner Publishing Company, New York, pp 468 (reprint 1959)

De Giovanni C, Dell'Orco P, Bruno A, Ciccarese F, Lotti C, Ricciardi L (2004) Identification of PCR-based markers (RAPD, AFLP) linked to a novel powdery mildew resistance gene (*ol-2*) in tomato. Plant Sci 166:41–48

De Jong WS, Eannetta NT, De Jong DM, Bodis M (2004) Candidate gene analysis of anthocyanin pigmentation loci in the Solanaceae. Theor Appl Genet 108:423–432

de la Garza RD, Quinlivan EP, Klaus SMJ, Basset GJC, Gregory JF, Hanson AD (2004) Folate biofortification in tomatoes by engineering the pteridine branch of folate synthesis. Proc Natl Acad Sci USA 101:13720–13725

de Silva J, Verhoeyen ME (1998) Production and characterization of antisense-exogalactanase tomatoes. In: Kuiper HA (ed) Report of the demonstration programme on food safety evaluation of genetically modified foods as a basis for market introduction. The Netherlands Ministry of Economic Affairs, The Hague, pp 99–106

de Vicente MC, Tanksley SD (1993) QTL analysis of transgressive segregation in an interspecific tomato cross. Genetics 134:585–596

Deguchi M, Bennett AB, Yamaki S, Yamada K, Kanahama K, Kanayama Y (2006) An engineered sorbitol cycle alters sugar composition, not growth, in transformed tobacco. Plant Cell Environ 29:1980–1988

Denny T (2006) Plant pathogenic *Ralstonia* species. Gnanamanickam SS (ed) Plant-associated bacteria. Springer, Berlin Heidelberg New York

Dharmapuri S, Rosati C, Pallara P, Aquilani R, Bouvier F, Camara B, Giuliano G (2002) Metabolic engineering of xanthophyll content in tomato fruits. FEBS Lett 519:30–34

Dijian-Caporalino C, Pijarowski L, Fazari A, Samson M, Gaveau L, O'Byrne C, Lefebvre V, Caranta C, Palloix A, Abad P (2001) High-resolution genetic mapping of the pepper (*Capsicum annuum* L.) resistance loci *Me*$_3$ and *Me*$_4$ conferring heat-stable resistance to root-knot nematodes (*Meloidogyne* spp.). Theor Appl Genet 103:592–600

Dinar M, Stevens MA (1981) The relationship between starch accumulation and soluble solids content of tomato fruits. J Am Soc Hort Sci 106:415–418

Diwan N, Fluhr R, EshedY, Zamir D, Tanksley SD (1999) Mapping of *Ve* in tomato: a gene conferring resistance to the broad-spectrum pathogen, *Verticillium dahliae* race 1. Theor Appl Genet 98:315–319

Doganlar S, Dodson J, Gabor B, Beck-Bunn T, Crossman C, Tanksley SD (1998) Molecular mapping of the *py-1* gene for resistance to corky root rot (*Pyrenochaeta lycopersici*) in tomato. Theor Appl Genet 97:784–788

Doganlar S, Frary A, Daunay M-C, Lester RN, Tanksley SD (2002a) Comparative genetic linkage map of eggplant (*Solanum melongena*) and its implications for genome evolution in the Solanaceae. Genetics 161:1697–1711

Doganlar S, Frary A, Daunay M-C, Lester RN, Tanksley SD (2002b) Conservation of gene function in the Solanaceae as revealed by comparative mapping of domestication traits in eggplant. Genetics 161:1713–1726

Doganlar S, Frary A, Ku H-K, Tanksley SD (2002c) Mapping quantitative trait loci in inbred backcross lines of *Lycopersicon pimpinellifolium* (LA1589). Genome 45:1189–1202

Doll R (1990) Symposium on diet and cancer. An overview of the epidemiological evidence linking diet and cancer. Proc Nutr Soc 49:119–131

Dorais M, Papadopoulos AP, Gosselin A (2001) Greenhouse tomato fruit quality. Hort Rev 26:239–319

Ecole CC, Picanço M, Jham GN, Guedes RNC (1999) Variability in *Lycopersicon hirsutum* f. *typicum* and possible compounds involved in its resistance to *Tuta absoluta*. Agric For Entomol 1:249–254

Eduardo I, Arús P, Monforte AJ (2005) Development of a genomic library of near isogenic lines (NILs) in melon (*Cucumis melo* L.) from the exotic accession PI161375. Theor Appl Genet 112:139–148

Egashira H, Ishihara H, Takashina T, Imanishi S (2000) Genetic diversity of the '*peruvianum*-complex' (*Lycopersicon peruvianum* (L.) Mill. and *L. chilense* Dun.) revealed by RAPD analysis. Euphytica 116:23–31

Eigenbrode SD, Trumble JT (1993a) Antibiosis to beet armyworm (*Spodoptera exigua*) in *Lycopersicon* accessions. HortScience 28:932–934

Eigenbrode SD, Trumble JT (1993b) Resistance to beet armyworm, hemipterans, and *Liriomyzia* spp. in *Lycopersicon* accessions. J Am Soc Hort Sci 118:525–530

Eigenbrode SD, Trumble JT, Millar JG, White KK (1994) Topical toxicity of tomato sesquiterpenes to the beet armyworm and the role of these compounds in resistance derived from an accession of *Lycopersicon hirsutum* f. *typicum*. J Agric Food Chem 42:807–810

El-Din El-Assal S, Alonso-Blanco C, Peeters AJM, Koornneef M (2001) A QTL for flowering time in *Arabidopsis* reveals a novel allele of *CRY2*. Nat Genet 29:435–440

Ellinger S, Ellinger J, Stehle P (2006) Tomatoes, tomato products and lycopene in the prevention of prostrate cancer: do we have the evidence from intervention studies? Curr Opin Clin Nutr Metab Care 6:722–727

Ellul P, Garcia-Sogo B, Pineda B, Ríos G, Roig LA, Moreno V (2003) The ploidy level of transgenic plants in *Agrobacterium*-mediated transformation of tomato cotyledons (*Lycopersicon esculentum* L. Mill.) is genotype and procedure dependent. Theor Appl Genet 106:231–238

Enright AJ, Van Dongen S, Ouzounis CA (2002) An efficient algorithm for large-scale detection of protein families. Nucl Acids Res 30:1575–1584

Eshed Y, Zamir D (1994a) A genomic library of *Lycopersicon pennellii* in *Lycopersicon esculentum* – a tool for fine mapping of genes. Euphytica 79:175–179

Eshed Y, Zamir D (1994b) Introgressions from *Lycopersicon pennellii* can improve the soluble-solids yield of tomato hybrids. Theor Appl Genet 88:891–897

Eshed Y, Zamir D (1995) An introgression line population of *Lycopersicon pennellii* in the cultivated tomato enables the identification and fine mapping of yield associated QTL. Genetics 141:1147–1162

Eshed Y, Zamir D (1996) Less-than-additive epistatic interactions of quantitative trait loci in tomato. Genetics 143:1807–1817

FAOSTAT (2004) Food and Agriculture Organization of the United Nations Statistical Databases. http://faostat.fao.org/

Fargette D, Leslie M, Harrison BD (1996) Serological studies on the accumulation and localisation of three *Tomato leaf curl* geminiviruses in resistant and susceptible *Lycopersicon* species and tomato cultivars. Ann Appl Biol 128:317–328

Fauquet CM, Bisaro DM, Briddon RW, Brown JK, Harrison BD, Rybicki EP, Stenger DC, Stanley J (2003) Revision of taxonomic criteria for species demarcation in the family *Geminiviridae*, and an updated list of *Begomovirus* species. Arch Virol 148:405–421

Faurobert M, Mihr C, Bertin N, Pawłowski T, Negroni L, Sommerer N, Causse M (2007) Major proteome variations associated with cherry tomato pericarp development and ripening. Plant Physiol *143:1327–1346*

Faurobert M, Barre M, Chaïb J, Causse M (2006) Effect of quality QTL introgression on tomato fruit proteome. PAA/Solanaceae Conference, July 23–27, Univ of Wisconsin, Madison, USA

Fei Z, Tang X, Alba R, Giovannoni J (2006) Tomato Expression Database (TED): a suite of data presentation and analysis tools. Nucl Acids Res 34:D766–770

Fei ZJ, Tang X, Alba RM, White JA, Ronning CM, Martin GB, Tanksley SD, Giovannoni JJ (2004) Comprehensive EST analysis of tomato and comparative genomics of fruit ripening. Plant J 40:47–59

Fellner M, Sawhney VK (2001) Seed germination in a tomato male-sterile mutant is resistant to osmotic, salt and low-temperature stresses. Theor Appl Genet 102:215–221

Fery RL, Kennedy GG (1987) Genetic analysis of 2-tridecanone concentration, leaf trichome characteristics and tobacco hornworm resistance in tomato. J Am Soc Hort Sci 112:886–891

Fiehn O, Wohlgemuth G, Scholz M (2005) Setup and annotation of metabolomic experiments by integrating biological and mass spectrometric metadata. In: Ludascher B, Raschid L (eds) Data Integration in the Life Sciences. Springer, Berlin Heidelberg New York, pp 224–239

Fillatti JJ, Kiser J, Rose R, Comai L (1987) Efficient transfer of a glyphosate tolerance gene into tomato using a binary *Agrobacterium tumefaciens* vector. Bio/Technology 5:726–730

Finetti-Sialer M, Di Franco A, Papanice MA, Gallitelli D (1997) Tomato necrotic yellows induced by a novel strain of *Alfalfa mosaic virus*. J Plant Pathol 79:115–120

Finkers R, van Heusden AW, Meijer-Dekens F, van Kan JAL, Maris P, Lindhout P (2007) The construction of a *Solanum habrochaites* LYC4 introgression line population and the identification of QTLs for resistance to *Botrytis cinerea*. Theor Appl Genet *114:1071–1080*

Finlay KW (1953) Inheritance of spotted wilt resistance in the tomato. II. Five genes controlling spotted wilt resistance in four tomato types. Aust J Biol Sci 6:153–163

Fischhoff DA, Bowdish KS, Perlak FJ, Marrone PG, McCormick SM, Niedermeyer JG, Dean DE, Kusano-Kretzmer K, Mayer EJ, Rochester DE, Rogers SG, Fraley RT (1987) Insect tolerant transgenic tomato plants. Bio/Technology 5:807–813

Flowers TJ (2004) Improving crop salt tolerance. J Exp Bot 55:307–319

Fobes JF (1980) Trisomic analysis of isozymic loci in tomato species – segregation and dosage effects. Biochem Genet 18:401–421

Fobes JF, Mudd JB, Marsden MPF (1985) Epicuticular lipid accumulation on the leaves of *Lycopersicon pennellii* (Corr.) D'Arcy and *Lycopersicon esculentum* Mill. Plant Physiol 77:567–570

Folkertsma RT, Spassova MI, Prins M, Stevens MR, Hille J, Goldbach RW (1999) Construction of a bacterial artificial chromosome (BAC) library of *Lycopersicon esculentum* cv. Stevens and its application to physically map the *Sw-5* locus. Mol Breed 5:197–207

Foolad MR (1999) Comparison of salt tolerance during seed germination and vegetative growth in tomato by QTL mapping. Genome 42:727–734

Foolad MR (2004) Recent advances in genetics of salt tolerance in tomato. Plant Cell Tiss Org Cult 76:101–119

Foolad MR, Chen FQ (1999) RFLP mapping of QTLs conferring salt tolerance during the vegetative stage in tomato. Theor Appl Genet 99:235–243

Foolad MR, Chen FQ, Lin GY (1998) RFLP mapping of QTLs conferring salt tolerance during germination in an interspecific cross of tomato. Theor Appl Genet 97:1133–1144

Foolad MR, Lin GY (1997) Absence of a genetic relationship between salt tolerance during seed germination and vegetative growth in tomato. Plant Breed 116:363–367

Foolad MR, Lin GY (1999) Relationships between cold- and salt-tolerance during seed germination in tomato: germplasm evaluation. Plant Breed 118:45–48

Foolad MR, Ntahimpera N, Christ BJ (2000) Comparison of field, greenhouse and detached-leaflet evaluations of tomato germplasm for early blight resistance. Plant Dis 84:967–972

Foolad MR, Sharma A (2005) Molecular markers as selection tools in tomato breeding. 1st Intl Symp on Tomato Diseases, Orlando, Florida, USA, pp 225–240

Foolad MR, Stoltz T, Dervinis C, Rodriguez RL, Jones RA (1997) Mapping QTLs conferring salt tolerance during germination in tomato by selective genotyping. Mol Breed 3:269–277

Foolad MR, Zhang L, Khan A, Nino-Liu D, Lin GY (2002) Identification of QTLs for early blight (*Alternaria solani*) resistance in tomato using backcross populations of a *Lycopersicon esculentum* × *L. hirsutum* cross. Theor Appl Genet 104:945–958

Foolad MR, Zhang LP, Kin GY (2001) Identification and validation of QTLs for salt tolerance during vegetative growth in tomato by selective genotyping. Genome 44:444–454

Foolad MR, Zhang LP, Subbiah P (2003) Genetics of drought tolerance during seed germination in tomato: inheritance and QTL mapping. Genome 46:536–545

França FH, Maluf WR, Ferreira-Rossi PE, Miranda JEC, Coelho MCF, Castelo-Branco M, Resende AM (1989) Breeding for resistance to *Scrobipalpula absoluta* (Meyrick) among *Lycopersicon* accessions in Brazil. In: Green SK (ed) Management Practices for Tomato and Pepper Production in the Tropics. Asian Veg Res Dev Center, Taiwan, pp 113–122

Francis DM, Kabelka E, Bell J, Franchino B, St Clair D (2001) Resistance to bacterial canker in tomato (*Lycopersicon hirsutum* LA407) and its progeny derived from crosses to *L. esculentum*. Plant Dis 85:1171–1176

Franklin-Tong VE, Franklin FCH (2003) The different mechanisms of gametophytic self-incompatibility. Phil Trans Roy Soc Lond Ser B Biol Sci 358:1025–1032

Frary A, Earle ED (1996) An examination of factors affecting the efficiency of *Agrobacterium*-mediated transformation of tomato. Plant Cell Rep 16:235–240

Frary A, Doganlar S, Daunay M-C, Tanksley SD (2003a) QTL analysis of morphological traits in eggplant and implications for conservation of gene function during evolution of Solanaceous species. Theor Appl Genet 107:359–370

Frary A, Doganlar S, Frampton A, Fulton T, Uhlig J, Yates H, Tanksley S (2003b) Fine mapping of quantitative trait loci for improved fruit characteristics from *Lycopersicon chmielewskii* chromosome 1. Genome 46:235–243

Frary A, Fulton TM, Zamir D, Tanksley SD (2004) Advance backcross QTL analysis of a *Lycopersicon esculentum* × *L. pennellii* cross and identification of possible orthologs in the Solanaceae. Theor Appl Genet 108:485–496

Frary A, Graham E, Jacobs J, Chetelat RT, Tanksley SD (1998) Identification of QTL for late blight resistance from *L. pimpinellifolium* L3708. Tomato Genet Coop Rep 48:19–21

Frary A, Nesbitt TC, Frary A, Grandillo S, van der Knaap E, Cong B, Liu J, Meller J, Elber R, Alpert KB, Tanksley SD (2000) *fw-2.2*: a quantitative trait locus key to the evolution of tomato fruit size. Science 289:85–88

Frary A, Presting GG, Tanksley SD (1996) Molecular mapping of the centromeres of tomato chromosomes 7 and 9. Mol Gen Genet 250:295–304

Frary A, Van Eck J (2004) Organogenesis from transformed tomato explants. In: Peña L (ed) Methods in molecular biology, vol 286. Humana Press, Totowa, NJ, USA, pp 213–221

Frary A, Xu Y, Liu J, Mitchell S, Tedeschi E, Tanksley SD (2005) Development of a set of PCR-based anchor markers encompassing the tomato genome and evaluation of their usefulness for genetics and breeding experiments. Theor Appl Genet 111:291–312

Fraser PD, Romer S, Shipton CA, Mills PB, Kiano JW, Misawa N, Drake RG, Schuch W, Bramley PM (2002) Evaluation of transgenic tomato plants expressing an additional phytoene synthase in a fruit-specific manner. Proc Natl Acad Sci USA 99:1092–1097

Fraser PD, Truesdale MR, Bird CR, Schuch W, Bramley PM (1994) Carotenoid biosynthesis during tomato fruit development. Evidence for tissue-specific gene expression. Plant Physiol 105:405–413

Fray RG, Grierson D (1993) Identification and genetic analysis of normal and mutant phytoene synthase genes of tomato by sequencing, complementation and co-suppression. Plant Mol Biol 22:589–602

Fray RG, Wallace A, Fraser PD, Valero D, Hedden P, Bramley PM, Grierson D (1995) Constitutive expression of a fruit phytoene synthase gene in transgenic tomatoes causes dwarfism by redirecting metabolites from the gibberellin pathway. Plant J 8:693–701

Fray RG, Wallace A, Grierson D, Lycett GW (1994) Nucleotide sequence and expression of a ripening and water stress-related cDNA from tomato with homology to the MIP class of membrane channel proteins. Plant Mol Biol 24:539–543

Freeman JA, Woodbridge CG (1960) Effect of maturation, ripening and truss position on the free amino acid content in tomato fruits. Proc Am Soc Hort Sci 76:515–523

French CJ, Bouthiller M, Bernardy M, Ferguson G, Sabourin M, Johnson RC, Masters C, Godkin S, Mumford R (2001) First report of *Pepino virus* in Canada and the United States. Plant Dis 85:1121

Fridman E, Carrari F, Liu YS, Fernie AR, Zamir D (2004) Zooming in on a quantitative trait for tomato yield using interspecific introgressions. Science 305:1786–1789

Fridman E, Liu YS, Carmel-Goren L, Gur A, Shoresh M, Pleban T, Eshed Y, Zamir D (2002) Two tightly linked QTLs modify tomato sugar content via different physiological pathways. Mol Genet Genom 266:821–826

Fridman E, Pleban T, Zamir D (2000) A recombination hotspot delimits a wild-species quantitative trait locus for tomato sugar content to 484 bp within an invertase gene. Proc Natl Acad Sci USA 97:4718–4723

Fridman E, Zamir D (2003) Functional divergence of a syntenic invertase gene family in tomato, potato, and *Arabidopsis*. Plant Physiol 131:603–609

Friedman M (2002) Tomato glycoalkaloids: Role in the plant and in the diet. J Agric Food Chem 50:5751–5780

Friedman M, Brandon DL (2001) Nutritional and health benefits of soy proteins. J Agric Food Chem 49:1069–1086

Fu DQ, Zhu BZ, Zhu HL, Jiang WB, Luo YB (2005) Virus-induced gene silencing in tomato fruit. Plant J 43:299–308

Fuchs J, Kloos D, Ganal M, Schubert I (1996) In situ localization of yeast artificial chromosome sequences on tomato and potato metaphase chromosomes. Chrom Res 4:277–281

Fulton TM, Beck-Bunn T, Emmatty D, Eshed Y, Lopez J, Petiard V, Uhlig J, Zamir D, Tanksley SD (1997) QTL analysis of an advanced backcross of *Lycopersicon peruvianum* to the cultivated tomato and comparisons with QTLs found in other wild species. Theor Appl Genet 95:881–894

Fulton TM, Bucheli P, Voirol E, Lopez J, Petiard V, Tanksley SD (2002a) Quantitative trait loci (QTL) affecting sugars, organic acids and other biochemical properties possibly contributing to flavor, identified in four advanced backcross populations of tomato. Euphytica 127:163–177

Fulton TM, Grandillo S, Beck-Bunn T, Fridman E, Frampton A, Lopez J, Petiard V, Uhlig J, Zamir D, Tanksley SD (2000) Advanced backcross QTL analysis of a *Lycopersicon esculentum* × *Lycopersicon parviflorum cross*. Theor Appl Genet 100:1025–1042

Fulton TM, van der Hoeven R, Eanetta NT, Tanksley SD (2002b) Identification, analysis, and utilization of conserved ortholog set markers for comparative genomics in higher plants. Plant Cell 14:1457–1467

Gallegly ME (1952) Sources of resistance to two races of the tomato late blight fungus. Phytopathology 42:466

Ganal M, Young N, Tanksley S (1989) Pulsed field gel-electrophoresis and physical mapping of large DNA fragments in the *Tm-2a* region of chromosome-9 in tomato. Mol Gen Genet 215:395–400

Gardner RG, Shoemaker PB (1999) 'Mountain Supreme' early blight-resistant hybrid tomato and its parents, NC EBR-3 and NC EBR-4 tomato breeding lines. HortScience 34:745–746

Garland S, Sharman M, Persley D, McGrath D (2005) The development of an improved PCR-based marker system for *Sw-5*, an important TSWV resistance gene of tomato. Aust J Agric Res 56:285–289

Gebhardt C, Ritter E, Barone A, Debener T, Walkemeier B, Schachtschabel U, Kaufman H, Thompson RD, Bonierbale MW, Ganal MW, Tanksley SD, Salamini F (1991) RFLP maps of potato and their alignment with the homoeologous tomato genome. Theor Appl Genet 83:49–57

Gebhardt C, Walkemeier B, Henselewski H, Barakat A, Delseny M, Stuber K (2003) Comparative mapping between potato (*Solanum tuberosum*) and *Arabidopsis thaliana* reveals structurally conserved domains and ancient duplications in the potato genome. Plant J 34:529–541

Gebre-Selassie K, Laterrot H, Marchoux G, Ragozzino A, Sacchardo F (1990) Resistance to *Potato virus Y* and *Cucumber mosaic virus* in *Lycopersicon hirsutum*. Tomato Genet Coop Rep 40:12–13

Gentile AG, Stoner AK (1968) Resistance in *Lycopersicon* species to the tobacco flea beetle. J Econ Entomol 61:1347–1349

Gentile AG, Webb R, Stoner AK (1968) Resistance in *Lycopersicon* and *Solanum* to greenhouse whiteflies. J Econ Entomol 61:1355–1357

Gentile AG, Webb R, Stoner AK (1969) *Lycopersicon* and *Solanum* resistant to the carmine and the two-spotted spider mite. J Econ Entomol 62:834–836

Gerlach W, Schneider R (1964) Nachweise in Pyrenochaeta-stadiums bei Stammen des Korkwurzelerregers der Tomate. Phytopathology Z 50:262–269

Gerster H (1997) The potential role of lycopene for human health. J Am Coll Nutr 16:109–126

Gidoni D, Fuss E, Burbidge A, Speckmann GJ, James S, Nijkamp D, Mett A, Feiler J, Smoker M, de Vroomen MJ, Leader D, Liharska T, Groenendijk J, Coppoolse E, Smit JJ, Levin I, de Both M, Schuch W, Jones JD, Taylor IB, Theres K, van Haaren MJ (2003) Multi-functional T-DNA/Ds tomato lines designed for gene cloning and molecular and physical dissection of the tomato genome. Plant Mol Biol 51:83–98

Gilchrist EJ, Haughn GW (2005) TILLING without a plough: a new method with applications for reverse genetics. Curr Opin Plant Biol 8:211–215

Gilchrist EJ, Haughn GW, Ying CC, Otto SP, Zhuang J, Cheung D, Hamberger B, Aboutorabi F, Kalynyak T, Johnson L, Bohlmann J, Ellis BE, Douglas CJ, Cronk QC (2006) Use of Ecotilling as an efficient SNP discovery tool to survey genetic variation in wild populations of *Populus trichocarpa*. Mol Ecol 15:1367–1378

Gill BS (1983) Tomato cytogenetics – a search for new frontiers. In: Swaminathan MS, Gupta PK, Sinha V (eds) Cytogenetics of crop plants. Macmillan India, New Delhi, India, pp 456–480

Giordano LB, Gabelman WH, Gerloff GC (1982) Inheritance of differences in calcium utilization by tomatoes under low-calcium stress. J Am Soc Hort Sci 107:664–669

Giordano LB, Silva-Lobo VL, Santana FM, Fonseca MEN, Boiteux LS (2005) Inheritance of resistance to the bipartite *Tomato chlorotic mottle begomovirus* derived from *Lycopersicon esculentum* cv. 'Tyking'. Euphytica 143:27–33

Giorgiev C (1972) Anthocyanin fruit tomato. Tomato Genet Coop Rep 22:10

Giovannoni JJ (2001) Molecular biology of fruit maturation and ripening. Annu Rev Plant Physiol Mol Biol 52:725–749

Giovannoni JJ (2004) Genetic regulation of fruit development and ripening. Plant Cell 16:S170–S180

Giovannucci E (1999) Tomatoes, tomato-based products, lycopene, and cancer: Review of the epidemiologic literature. J Natl Canc Inst 91:317–331

Giovannucci E, Ascherio A, Rimm EB, Stampfer MJ, Colditz GA, Willett WC (1995) Intake of carotenoids and retinol in relation to risk of prostrate cancer. J Natl Canc Inst 87:1767–1776

Godt DE, Roitsch T (1997) Regulation and tissue-specific distribution of mRNAs for three extracellular invertase isoenzymes of tomato suggests an important function in establishing and maintaining sink metabolism. Plant Physiol 115:273–282

Goffreda JC, Mutschler MA, Ave DA, Tingey WM, Steffens JC (1989) Aphid deterrence by glucose esters in glandular trichome exudate of wild tomato *Lycopersicon pennellii*. J Chem Ecol 15:2135–2147

Goffreda JC, Mutschler MA, Tingey WM (1988) Feeding behavior of potato aphid affected by glandular trichomes of wild tomato. Entomol Exp Appl 48:101–107

Goldman IL, Paran I, Zamir D (1995) Quantitative trait locus analysis of a recombinant inbred line population derived from a *Lycopersicon esculentum* × *L. cheesmanii* cross. Theor Appl Genet 90:925–932

Gonçalves MIF, Maluf WR, Gomes LAA, Barbosa L (1998) Variation of 2-tridecanone level in tomato plant leaflets and resistance to two mite species (*Tetranychus* sp.). Euphytica 104:33–38

Gordillo LF, Stevens MR, Millard MA, Geary B (2007) Screening two *Lycopersicon peruvianum* collections for resistance to *Tomato spotted wilt virus*. Plant Dis (in press)

Gould WA (1983) Tomato production, processing and quality, AVI Publ, Westport, CT, USA

Goyer A, Illarionova V, Roooje S, Fischer M, Bacher A, Hanson AD (2004) Folate biosynthesis in higher plants. cDNA cloning, heterologous expression, and characterization of dihydroneopterin aldolases. Plant Physiol 135:103–111

Graham E, Wang TC, Hanson P (2005) Preliminary evaluation of LA/777 introgression lines for early blight resistance. Tomato Genet Coop Rep 55:15–18

Grandillo S, Ku HM, Tanksley SD (1999) Identifying the loci responsible for natural variation in fruit size and shape in tomato. Theor Appl Genet 99:978–987

Grandillo S, Tanksley SD (1996a) Genetic analysis of RFLPs, GATA microsatellites and RAPDs in a cross between *L. esculentum* and *L. pimpinellifolium*. Theor Appl Genet 92:957–965

Grandillo S, Tanksley SD (1996b) QTL analysis of horticultural traits differentiating the cultivated tomato from the closely related species *Lycopersicon pimpinellifolium*. Theor Appl Genet 92:935–951

Grandillo S, Tanksley SD (2005) Advanced backcross QTL analysis: Results and perspectives. In: Tuberosa R, Phillips RL, Gale M (eds) The wake of the double helix: from the green revolution to the gene revolution. Edizioni Avenue Media, Bologna, Italy, pp 115–132

Grandillo S, Zamir D, Tanksley SD (1999) Genetic improvement of processing tomatoes: a 20-year perspective. Euphytica 110:85–97

Green SK, Hanson P (1996) A new source of PVY resistance. Tomato Genet Coop Rep 46:9

Greene EA, Codomo CA, Taylor NE, Henikoff JG, Till BJ, Reynolds SH, Enns LC, Burtner C, Johnson JE, Odden AR, Comai L, Henikoff S (2003) Spectrum of chemically induced mutations from a large-scale reverse-genetic screen in Arabidopsis. Genetics 164:731–740

Griffiths PD (1998) Inheritance and linkage of geminivirus resistance genes derived from *Lycopersicon chilense* (Dunal) in tomato (*Lycopersicon esculentum* Mill.). Univ of Florida, Gainesville, Florida, USA, pp 170

Griffiths PD, Scott JW (2001) Inheritance and linkage of *Tomato mottle virus* resistance genes derived from *Lycopersicon chilense* accession LA 1932. J Am Soc Hort Sci 126:462–467

Grogan RG, Kimble KA, Misaghi I (1975) A stem canker disease of tomato caused by *Alternaria alternata* f. sp. *lycopersici*. Phytopathology 65:880–886

Grube RC, Blauth JR, Arnedo AMS, Caranta C, Jahn MK (2000a) Identification and comparative mapping of a dominant potyvirus resistance gene cluster in *Capsicum*. Theor Appl Genet 101:852–859

Grube RC, Radwanski ER, Jahn M (2000b) Comparative genetics of disease resistance within the Solanaceae. Genetics 155:873–887

Gundersen V, McCall D, Bechmann IE (2001) Comparison of major and trace element concentrations in Danish greenhouse tomatoes (*Lycopersicon esculentum* cv. Aromata F1) cultivated in different substrates. J Agric Food Chem 49:3808–3815

Gur A, Semel Y, Cahaner A, Zamir D (2004) Real-time QTL of complex phenotypes in tomato interspecific introgression lines. Trends Plant Sci 9(3):107–109

Gur A, Zamir D (2004) Unused natural variation can lift yield barriers in plant breeding. PLOS Biol 2:1610–1615

Gut IG (2004) An overview of genotyping and single nucleotide polymorphisms (SNP). In: Rapley R, Harbron S (eds) Molecular analysis and genome discovery. Wiley, Chichester, UK, pp 43–64

Haanstra JP, Wye C, Verbakel H, Meijer-Dekens F, van den Berg P, Odinot P, van Heusden AW, Tanksley S, Lindhout P, Peleman J (1999) An integrated high-density RFLP-

AFLP map of tomato based on two *Lycopersicon esculentum* × *L. pennelli* F_2 populations. Theor Appl Genet 99:254–271

Haanstra JPW (2000) Characterization of resistance genes to *Cladosporium fulvum* on the short arm of chromosome 1 of tomato. Wageningen Univ dissertation no. 2735, Wageningen, The Netherlands, pp 119

Hall RD (2006) Plant metabolomics: from holistic hope, to hype, to hot topic. New Phytol 169:453–468

Hamza S, Chupeau Y (1993) Re-evaluation of conditions for plant regeneration and *Agrobacterium*-mediated transformation from tomato (*Lycopersicon esculentum*). J Exp Bot 44:1837–1845

Hancock JF (2004) Plant evolution and the origin of crop species, CABI Publ, Wallingford, Oxon, UK, pp 313

Hanson AD, Gregory JF (2002) Synthesis and turnover of folates in plants. Curr Opin Plant Biol 5:244–249

Hanson P, Green SK, Kuo G (2006) *Ty-2*, a gene on chromosome 11 conditioning geminivirus resistance in tomato. Tomato Genet Coop Rep 56:17–18

Hanson PM, Yang R, Wu J, Chen J, Ledesma D, Tsou SCS, Lee T (2004) Variation for antioxidant activity and antioxidants in tomato. J Am Soc Hort Sci 129:704–711

Harborne JB (1994) The flavonoids. Advances in research since 1986. 1st edn. Chapman Hall, London

Harborne JB, Williams CA (2000) Advances in flavonoid research since 1992. Phytochemistry 55:481–504

Harper L, Cande W (2000) Mapping a new frontier; development of integrated cytogenetic maps in plants. Funct Integr Genom 1:89–98

Harrak H, Azelmat S, Baker EN, Tabaeizadeh Z (2001) Isolation and characterization of a gene encoding a drought-induced cysteine protease in tomato (*Lycopersicon esculentum*). Genome 44:368–374

Harrison L (2004) The Shaker book of the garden: with a facsimile of the original gardener's manual of 1843. Barron's, Hauppauge, NY

Hassan S, Thomas PE (1988) Extreme resistance to *Tomato yellow top virus* and *Potato leaf roll virus* in *Lycopersicon peruvianum* and some of its tomato hybrids. Phytopathology 78:1164–1167

Hawthorne DJ, Shapiro JA, Tingey WM, Mutschler MA (1992) Trichome-borne and artificially applied acylsugars of wild tomato deter feeding and ovoposition of the leafminer *Liriomyza trifolii*. Entomol Exp Appl 65:65–73

He C, Poysa V, Yu K (2002) Development and characterization of simple sequence repeat (SSR) markers and their use in determining relationships among *Lycopersicon esculentum* cultivars. Theor Appl Genet 106:363–373

Hedrick UP, Booth NO (1907) Mendelian characters in tomatoes. Proc Am Soc Hort Sci 5:19–24

Helentjaris T, King G, Slocum M, Siedenstrang C, Wegman S (1985) Restriction fragment polymorphisms as probes for plant diversity and their development as tools for applied plant breeding. Plant Mol Biol 5:109–118

Hendrix J, Frazier WA (1949) Studies of the inheritance of *stemphylium* resistance in tomatoes. Hawaii Agric Exp Stn Tech Bull 8, pp 24

Henikoff JG, Henikoff S (1996) Using substitution probabilities to improve position-specific scoring matrices. Comput Appl Biosci 12:135–143

Henikoff JG, Pietrokovski S, McCallum CM, Henikoff S (2000) Blocks-based methods for detecting protein homology. Electrophoresis 21:1700–1706

Henikoff S, Till BJ, Comai L (2004) TILLING. Traditional mutagenesis meets functional genomics. Plant Physiol 135:630–636

Hernández F (1651) Nova plantarum animalium et mineralium Mexicanorum historia. Vitalis Mascardi, Rome, Italy

Hertog MGL, Feskens EJ, Kromhout D (1997) Antioxidant flavonols and coronary heart disease risk. Lancet 349:699

Hewitt JD, Garvey TC (1987) Wild sources of high soluble solids in tomato. In: Nevins DJ, Jones RA (ed) Plant biology, vol 4, Tomato biotechnology. AR Liss, New York, pp 45–54

Hirai MY, Klein M, Fujikawa Y, Yano M, Goodenowe DB, Yamazaki Y, Kanaya S, Nakamura Y, Kitayama M, Suzuki H, Sakurai N, Shibata D, Tokuhisa J, Reichelt M, Gershenzon J, Papenbrock J, Saito K (2005) Elucidation of gene-to-gene and metabolite-to-gene networks in Arabidopsis by integration of metabolomics and transcriptomics. J Biol Chem 280:25590–25595

Hirai MY, Yano M, Goodenowe DB, Kanaya S, Kimura T, Awazuhara M, Arita M, Fujiwara T, Saito K (2004) Integration of transcriptomics and metabolomics for understanding of global responses to nutritional stresses in *Arabidopsis thaliana*. Proc Natl Acad Sci USA 101:10205–10210

Hirschberg J (2001) Carotenoid biosynthesis in flowering plants. Curr Opin Plant Biol 4:210–218

Hirschi KD (2003) Insertional mutants: a foundation for assessing gene function. Trends Plant Sci 8:205–207

Ho T-Y, Mishkind ML (1991) The influence of water deficits on mRNA levels in tomato. Plant Cell Environ 14:67–75

Hobson GE, Davies JN (1971) The tomato. In: Holme AC (ed) The biochemistry of fruits and their products, vol 2. Academic Press, New York, pp 437–482

Hohmann U, Jacobs G, Jung C (2005) An EMS mutagenesis protocol for sugar beet and isolation of non-bolting mutants. Plant Breed 124:317–321

Holtan HE, Hake S (2003) Quantitative trait locus analysis of leaf dissection in tomato using *Lycopersicon pennellii* segmental introgression lines. Genetics 165:1541–1550

Horkheimer H (1973) Alimentación y obtención de alimentos en el Perú prehispánico. Universidad Nacional Mayor de San Marcos, Lima

Horsch RB, Fry JE, Hoffmann NL, Eichholtz D, Rogers SG, Fraley RT (1985) A simple and general method for transferring genes into plants. Science 227:1229–1231

Hossain T, Rosenberg I, Selhub J, Kishore G, Beachy R, Schubert K (2004) Enhancement of folates in plants through metabolic engineering. Proc Natl Acad Sci USA 101:5158–5163

Hsieh TH, Lee JT, Yang PT, Chiu LH, Charng YY, Wang YC, Chan MT (2002) Heterology expression of the Arabidopsis *C-repeat/dehydration response element binding factor 1* gene confers elevated tolerance to chilling and oxidative stresses in transgenic tomato. Plant Physiol 129:1086–1094

Huang CC, Cui YY, Weng CR, Zabel P, Lindhout P (2000) Development of diagnostic PCR markers closely linked to the tomato powdery mildew resistance gene *Ol-1* on chromosome 6 of tomato. Theor Appl Genet 101:918–924

Imai R, Moses MS, Bray EA (1995) Expression of an ABA-induced gene of tomato in transgenic tobacco during periods of water deficit. J Exp Bot 46:1077–1084

Isaacson T, Ronen G, Zamir D, Hirschberg J (2002) Cloning of *tangerine* from tomato reveals a carotenoid isomerase essential for the production of β-carotene and xanthophylls in plants. Plant Cell 14:333–342

Janoria MP, Rhodes AM (1974) Juice viscosity as related to various juice constituents and fruit characters in tomatoes. Euphytica 23:533–562

Jarret RL, Sayama H, Tigchelaar EC (1984) Pleiotropic effects associated with the chlorophyll intensifier mutations high pigment and dark green in tomato. J Am Soc Hort Sci 109:873–878

Jaunet T, Wang J-F (1999) Variation in genotype and aggressiveness of *Ralstonia solanacearum* race 1 isolated from tomato in Taiwan. Phytopathology 89:320–327

Jenkins JA (1948) The origin of the cultivated tomato. Econ Bot 2:379–392

Jenkins JA, Mackinney G (1955) Carotenoids of the apricot tomato and its hybrids with yellow and tangerine. Genetics 40:715–720

Jeunken MJW, Lindhout P (2004) The development of lettuce backcross inbred lines (BILs) for exploitation of the *Lactuca saligna* (wild lettuce) germplasm. Theor Appl Genet 109:394–401

Ji Y, Chetelat RT (2003) Homoeologous pairing and recombination in *Solanum lycopersicoides* monosomic addition and substitution lines of tomato. Theor Appl Genet 106:979–989

Ji Y, Scott JW (2005) Identification of RAPD markers linked to *Lycopersicon chilense* derived *Begomovirus* resistant genes on chromosome 6 of tomato. Acta Hort 695:407–416

Ji Y, Scott JW (2006) *Ty-3*, a *Begomovirus* resistance locus linked to *Ty-1* on chromosome 6 of tomato. Tomato Genet Coop Rep 56:22–25

Ji Y, Scott JW (2007) Tomato. In: Singh RJ (ed) Genetic resources, chromosome engineering, and crop improvement series IV: Vegetable Crops. CRC Press, Boca Raton, FL, pp 59–113

Ji Y, Scott JW, Hanson P, Graham E, Maxwell DP (2007) Sources of resistance, inheritance, and location of genetic loci conferring resistance to members of the tomato-infecting *Begomoviruses*. In: Czosnek H (ed) Tomato yellow leaf curl virus disease: management, molecular biology, breeding for resistance. Kluwer Academic Publishers, Dordrecht (in press)

Jia GX, Zhu ZQ, Chang FQ, Li YX (2002) Transformation of tomato with the *BADH* gene from Atriplex improves salt tolerance. Plant Cell Rep 21:141–146

Jiang XL, He ZM, Peng ZQ, Qi Y, Chen Q, Yu SY (2007) Cholera toxin B protein in transgenic tomato fruit induces systemic immune response in mice. Transgenic Res *16:169–175*

Jin S, Chen CCS, Plant AL (2000) Regulation by ABA of osmotic-stress-induced changes in protein synthesis in tomato roots. Plant Cell Environ 23:51–60

Jones CM, Mes P, Myers JR (2003) Characterization and inheritance of the anthocyanin fruit (*Aft*) tomato. J Hered 94:449–456

Jones DA, Dickinson MJ, Balint-Kurti PJ, Dixon MS, Jones JDG (1993) Two complex resistance loci revealed in tomato by classical and RFLP mapping of the *Cf-2*, *Cf-4*, *Cf-5*, and *Cf-9* genes for resistance to *Cladosporium fulvum*. Mol Plant-Micr Interact 6:348–357

Jones JB, Lacy GH, Bouzar H, Stall RE, Schaad NW (2004) Reclassification of the Xanthomonads associated with bacterial spot disease of tomato and pepper. Syst Appl Microbiol 27:755–762

Jones JB, Lacy GH, Bouzar H, Minsavage GE, Stall RE, Schaad NW(2005) Bacterial spot - Worldwide distribution, importance and review. Acta Hort 695:27–33

Juvik JA, Berlinger MJ, Ben-David T, Rudich J (1982) Resistance among accessions of genera *Lycopersicon* and *Solanum* to four of the main insect pests of tomato in Israel. Phytoparasitica 10:145–156

Juvik JA, Shapiro JA, Young TE, Mutschler MA (1994) Acylglucose from wild tomato alters behavior and reduces growth and survival of *Helicoverpa zea* and *Spodoptera exigua* (Lepidoptera: Noctuidae). J Econ Entomol 87:482–492

Kabelka E, Franchino B, Francis DM (2002) Two loci from *Lycopersicon hirsutum* LA407 confer resistance to strains of *Clavibacter michiganensis* subsp. *michiganensis*. Phytopathology 92 (5):504–510

Kabelka E, Yang W, Francis DM (2004) Improved tomato fruit color within an inbred backcross line derived from *Lycopersicon esculentum* and *L. hirsutum* involves the interaction of loci. J Am Soc Hort Sci 129:250–257

Kahn TL, Fender SE, Bray EA, O'Connell MA (1993) Characterization of expression of drought- and ABA-regulated tomato genes in the drought-resistant species *Lycopersicon pennellii*. Plant Physiol 103:597–605

Kamal AHM, Takashina T, Egashira H, Satoh H, Imanishi S (2001) Introduction of aromatic fragrance into cultivated tomato from the 'peruvianum complex'. Plant Breed 120:179–181

Kanwar JS, Kerr EA, Harney PM (1980a) Linkage of *Cf-1* to *Cf-11* genes for resistance to tomato leaf mold *Cladosporium fulvum* Cke. Rep Tomato Genet Coop 30:20–21

Kanwar JS, Kerr EA, Harney PM (1980b) Linkage of *Cf-12* to *Cf-24* genes for resistance to tomato leaf mold

Cladosporium fulvum Cke. Rep Tomato Genetics Coop 30:22–23

Kartha KK, Gamborg OL, Shyluk JP, Constabel F (1976) Morphogenetic investigations on in vitro leaf culture of tomato (*Lycopersicon esculentum* Mill. cv. Starfire) and high frequency plant regeneration. Z Pflazenphysiol 77:292–301

Kawchuk LM, Hachey J, Lynch DR (1998) Development of sequence characterized DNA markers linked to a dominant verticillium wilt resistance gene in tomato. Genome 41:91–95

Kebede H, Martin B, Nienhuis J, King G (1994) Leaf anatomy of two *Lycopersicon* species with contrasting gas exchange properties. Crop Sci 34:108–113

Kedar N, Retig N, Katan J (1967) Non-random segregation of gene I for fusarium resistance in tomato. Euphytica 16:258–266

Kennedy GG (2003) Tomato, pests, parasitoids, and predators: tritrophic interactions involving the genus *Lycopersicon*. Annu Rev Entomol 48:51–72

Kennedy GG (2007) Resistance in tomato and other *Lycopersicon* species to insect and mite pests. In: Razdan MK, Mattoo AK (eds) Genetic improvement of Solanaceous crops, vol 2: tomato. Science Publ, Enfield, NH, USA, pp 488–519

Kennedy GG, Yamamoto RT (1979) A toxic factor causing resistance in a wild tomato to the tobacco hornworm and some other insects. Entomol Exp Appl 26:121–126

Kerckhoffs LHJ, Kendrick RE (1997) Photocontrol of anthocyanin biosynthesis in tomato. J Plant Res 110:141–149

Kerr EA (1958a) Linkage relations of *gf*. Tomato Genet Coop Rep 8:21

Kerr EA (1958b) Mutations for chlorophyll retention in ripe fruit. Tomato Genet Coop Rep 8:22

Kerr EA, Cook FI (1983) Ontario 7710 - a tomato breeding line with resistance to bacterial speck, *Pseudomonas syringae* pv. *tomato* (Okabe). Can J Plant Sci 63:1107–1109

Khush GS, Rick CM (1967a) Novel compensating trisomics of the tomato: cytogenetics, monosomic analysis, and other applications. Genetics 56:297–307

Khush GS, Rick CM (1967b) Tomato tertiary trisomics: Origin, identification, morphology and use in determining position of centromeres and arm location of markers. Can J Genet Cytol 9:610–631

Khush GS, Rick CM (1968a) Cytogenetic analysis of the tomato genome by means of induced deficiencies. Chromosoma 23:452–484

Khush GS, Rick CM (1968b) Tomato telotrisomics: origin, identification and use in linkage mapping. Cytologia 33:137–148

Kind T, Fiehn O (2006) Metabolomic database annotations via query of elemental compositions: mass accuracy is insufficient even at less than 1 ppm. BMC Bioinformatics 7:234

King A, Young G (1999) Characteristics and occurrence of phenolic phytochemicals. J Am Dietetic Assoc 99:213–218

Klann EM, Hall B, Bennett AB (1996) Antisense acid invertase (*TIV1*) gene alters soluble sugar composition and size in transgenic tomato fruit. Plant Physiol 112:1321–1330

Knapp S, Darwin SC (2007) Proposal to conserve the name *Solanum cheesmaniae* (L. Riley) Fosberg against *S. cheesmanii* Geras. (Solanaceae). Taxon *55:806–807*

Knekt P, Jarvinen R, Seppaanen R, Heliovaara M, Teppi L, Pukkala E, Aromaa A (1997) Dietary flavonoids and the risk of lung cancer and other malignant neoplasms. Am J Epidemiol 146:223–230

Kodym A, Afza R (2003) Physical and chemical mutagenesis. Methods Mol Biol 236:189–204

Kohler GW, Lincoln RE, Porter JW, Zscheile FP, Caldwell RM, Harper RH, Silver W (1947) Selection and breeding for high beta-carotene content (provitamin A) in tomato. Bot Gaz 109:219–225

Kojima S, Takahashi Y, Kobayashi Y, Monna L, Sasaki T, Araki T, Yano M (2002) *Hd3a*, a rice ortholog of the Arabidopsis FT gene, promotes transition to flowering downstream of *Hd1* under short-day conditions. Plant Cell Physiol 43:1096–1105

Kole C, Ashrafi H, Lin G, Foolad M (2006) Identification and molecular mapping of a new R gene, *Ph-4*, conferring resistance to late blight in tomato. Solanaceae Conf, Univ of Wisconsin, Madison, Abstr 449

Konieczny A, Ausubel FM (1993) A procedure for mapping *Arabidopsis* mutations using co-dominant ecotype-specific PCR-based markers. Plant J 4:403–410

Konings EJ, Roomans HH, Dorant E, Goldbohm RA, Saris WH, van den Brandt PA (2001) Folate intake of the Dutch population according to newly established liquid chromatography data for foods. Am J Clin Nutr 73:765–776

Konsler TR (1973) Three mutants appearing in 'Manapal' tomato. HortScience 8:331–333

Koumproglou R, Wilkes TM, Towson P, Wang XY, Beyon J, Pooni HS, Newbury HJ, Kearsey MJ (2002). STAIRS: a new genetic resource for functional genomic studies of *Arabidopsis*. Plant J 31(3):355–364

Kozukue N, Friedman M (2003) Tomatine, chlorophyll, β-carotene and lycopene content in tomatoes during growth and maturation. J Sci Food Agric 83:195–200

Kramer M, Sanders R, Bolkan H, Waters C, Sheehy RE, Hiatt WR (1992) Postharvest evaluation of transgenic tomatoes with reduced levels of polygalacturonase: processing, firmness and disease resistance. Postharvest Biol Tec 1:241–255

Krishnaswamy K, Madhavan Nair K (2001) Importance of folate in human nutrition. Brit J Nutr 85:S115–S124

Krysan PJ, Young JC, Jester PJ, Monson S, Copenhaver G, Preuss D, Sussman MR (2002) Characterization of T-DNA insertion sites in *Arabidopsis thaliana* and the implications for saturation mutagenesis. Omics 6:163–174

Kusaba M (2004) RNA interference in crop plants. Curr Opin Biotechnol 15:139–143

Kusaba M, Miyahara K, Iida S, Fukuoka H, Takano T, Sassa H, Nishimura M, Nishio T (2003) *Low glutelin content1*: a dominant mutation that suppresses the *glutelin* multigene family via RNA silencing in rice. Plant Cell 15:1455–1467

Kwok P-Y (2001) Methods for genotyping single nucleotide polymorphisms. Annu Rev Genom Hum Genet 2:235–258

Labate JA, Baldo AM (2005) Tomato SNP discovery by EST mining and resequencing. Mol Breed 16:343–349

Lachance PA (1998) Overview of key nutrients: micronutrient aspects. Nutr Rev 56:S34–S39

Lambeth VN, Straten EF, Fields ML (1966) Fruit quality attributes of 250 foreign and domestic tomato accessions. Univ Missouri Agric Expt Stn Res Bull 908

Lanfermeijer FC, Dijkhuis J, Sturre MJG, de Haan P, Hille J (2003) Cloning and characterization of the durable *Tomato mosaic virus* resistance gene *Tm-2*^{2} from *Lycopersicon esculentum*. Plant Mol Biol 52:1037–1049

Lanfermeijer FC, Warmink J, Hille J (2005) The products of the broken *Tm-2* and the durable *Tm-2*^{2} resistance genes from tomato differ in four amino acids. J Exp Bot 56:2925–2933

Lange WH, Bronson L (1981) Insect pests of tomato. Annu Rev Entomol 26:345–371

Langella R, Ercolano MR, Monti LM, Frusciante L, Barone A (2004) Molecular marker assisted transfer of resistance to TSWV in tomato elite lines. J Hort Sci Biotechnol 79:806–810

Langford AN (1937) The parasitism of *Cladosporium fulvum* cooke and the genetics of resistance to it. Can J Res 15:108–128

Langley KR, Martin A, Stenning R, Murray AJ, Hobson GE, Schuch WW, Bird CR (1994) Mechanical and optical assessment of the ripening of tomato fruit with reduced polygalacturonase activity. J Sci Food Agric 66:547–554

Lapidot M, Friedmann M, Lachman O, Yehezkel A, Nahon S, Cohen S, Pilowsky M (1997) Comparison of resistance level to *Tomato yellow leaf curl virus* among commercial cultivars and breeding lines. Plant Dis 81:1425–1428

Laquatra I, Yeung DL, Storey M, Forshee R (2005) Health benefits of lycopene in tomatoes – conference summary. Nutr Today 40:29–36

Laterrot H (1975) Selection pour la resistance au mildiou, *Phytophthora infestans* Mont., de Bary chez la tomate. Ann Amel Plant 24:129–149

Laterrot H (1976) Localisation chromosomique de 12 chez la tomate controlant la resistance au pathotype 2 de *Fusarium oxysporum* f. *lycopersici*. Ann Amel Plant 26:485–491

Laterrot H (1983) La utte genetiquecontre la maadie des racines liegueses de la tomate. Rev Hort 238:143–150

Laterrot H (1997) Breeding strategies for disease resistance in tomatoes with emphasis on the tropics: current status and research challenges. In: Maciel GA, Lopes GMB, Hayward C, Mariano RRL, de A Maranhao EA (eds) 1st Intl Symp on Trop Tomato Dis, ASHS Press, Alexandria, VA, USA, pp 126–132

Laterrot H, Couteadier Y (1989) Linkage between TMV and FORL resistances. Tomato Genet Coop Rep 39:21

Laterrot H, Moretti A (1992) FORL resistance in TM2^2-ah lines. Tomato Genet Coop Rep 42:24–25

Laterrot H, Moretti A (1996) FORL resistance in TM2 and TM2^a lines from the TGRC stock. Tomato Genet Coop Rep 46:17

Laterrot H, Pecaut P (1969) Gene *Tm-2*: new source. Tomato Genet Coop Rep 19:13–14

Lauge R, Dmitriev AP, Joosten MHA, De Wit PJGM (1998) Additional resistance gene(s) against *Cladosporium fulvum* present on the *Cf-9* introgression segment are associated with strong PR protein accumulation. Mol Plant-Micr Interact 11:301–308

Lawrence RJ, Pikaard CS (2003) Transgene-induced RNA interference: a strategy for overcoming gene redundancy in polyploids to generate loss-of-function mutations. Plant J 36:114–121

Lawson DM, Lunde CF, Mutschler MA (1997) Marker-assisted transfer of acylsugar-mediated pest resistance from the wild tomato, *Lycopersicon pennellii*, to the cultivated tomato, *Lycopersicon esculentum*. Mol Breed 3:307–317

Le LQ, Lorenz Y, Scheurer S, Fotisch K, Enrique E, Bartra J, Biemelt S, Vieths S, Sonnewald U (2006) Design of tomato fruits with reduced allergenicity by dsRNAi-mediated inhibition of ns-LTP (Lyc e 3) expression. Plant Biotechnol J 4:231–242

Lecomte L, Duffé P, Buret M, Servin B, Hospital F, Causse M (2004a) Marker-assisted introgression of 5 QTLs controlling fruit quality traits into three tomato lines revealed interactions between QTLs and genetic backgrounds. Theor Appl Genet 109:658–668

Lecomte L, Saliba-Colombani V, Gautier A, Gomez-Jimenez MC, Duffé P, Buret M, Causse M (2004b) Fine mapping of QTLs of chromosome 2 affecting the fruit architecture and composition of tomato. Mol Breed 13:1–14

Lee CY, Robinson RW (1980) Influence of the crimson gene (og^c) on vitamin content in tomato. HortScience 15:260–261

Lee NKM (1832) The cook's own book: being a complete culinary encyclopedia by a Boston housekeeper. Arno Press, New York, 1972, facsimile reprint

Lee S, Kim S-Y, Chung E, Joung Y-H, Pai H-S, Hur C-G, Choi D (2004a) EST and microarray analyses of pathogen-responsive genes in hot pepper (*Capsicum annuum* L.) non-host resistance against soybean pustule pathogen (*Xanthomonas axonopodis pv. glycines*). Funct Integr Genom 4:196–205

Lee S-H, Walker DR, Cregan PB, Boerma HR (2004b) Comparison of four flow cytometric SNP detection assays and their use in plant improvement. Theor Appl Genet 110:167–174

Lefebvre V, Daubeze A-M, van der Voort JR, Peleman J, Bardin M, Palloix A (2003) QTLs for resistance to powdery mildew in pepper under natural and artificial infections. Theor Appl Genet 107:661–666

Légnani R, Gebre-Selassie K, Marchoux G, Laterrot H (1997) Interactions between PVY pathogens and tomato lines. Tomato Genet Coop Rep 47:13–15

Lekse JM, Xia L, Morrow JD, May JM (2001) Plant catechols prevent lipid peroxidation in human plasma erythrocytes. Mol Cell Biochem 226:89–95

Lemaire-Chamley M, Petit J, Garcia V, Just D, Baldet P, Germain V, Fagard M, Mouassite M, Cheniclet C, Rothan C (2005) Changes in transcriptional profiles are associated with early fruit tissue specialization in tomato. Plant Physiol 139:750–769

Lenke CA, Mutschler MA (1984) Inheritance of glandular trichomes in crosses between *Lycopersicon esculentum* and *Lycopersicon pennellii*. J Am Soc Hort Sci 109:592–596

Leonardi C, Ambrosino P, Esposito F, Fogliano V (2000) Antioxidative activity and carotenoid and tomatine contents in different typologies of fresh consumption tomatoes. J Agric Food Chem 48:4723–4727

Lercari B, Manetti A, Bertram L (2002) Temporal and spatial pattern of light-dependent acquisition of competence for shoot formation in tomato hypocotyl. I Light pulse conditions. Adv Hort Sci 16:17–24

Lesley JW (1932) Trisomic types of the tomato and their relation to the genes. Genetics 17:545–559

Lesley JW, Lesley MM (1956) A virescent tangerine mutant in R_2 from seed treatment with radioactive isotope P^{32}. Tomato Genet Coop Rep 6:17–18

Levin I, Frankel P, Gilboa N, Tanny S, Lalazar A (2003) The tomato *dark green* mutation is a novel allele of the tomato homolog of the *DEETIOLATED1* gene. Theor Appl Genet 106:454–460

Levin I, Gilboa N, Teselson E, Shen S, Schaffer AA (2000) *Fgr*, a major locus that modifies fructose to glucose ratio in mature tomato fruits. Theor Appl Genet 100:256–262

Levin I, Lalazar A, Bar M, Schaffer AA (2004) Non GMO fruit factories strategies for modulating metabolic pathways in the tomato fruit. Ind Crop Prod 20:29–36

Li L, Steffens JC (2002) Over expression of polyphenol oxidase in transgenic tomato plants results in enhanced bacterial disease resistance. Planta 215:239–247

Li L, Strahwald J, Hofferbert HR, Lubeck J, Tacke E, Junghans H, Wunder J, Gebhardt C (2005a) DNA variation at the invertase locus *invGE/GF* is associated with tuber quality traits in populations of potato breeding clones. Genetics 170:813–821

Li Z-K, Fu B-Y, Gao Y-M, Xu J-L, Ali J, Lafitte HR, Jiang Y-Z, Rey JD, Vijayakumar CHM, Maghirang R, Zheng T-Q, Zhu L-H (2005b) Genome-wide introgression lines and their use in genetic and molecular dissection of complex phenotypes in rice (*Oryza sativa* L.). Plant Mol Biol 59:33–52

Liedl BE, Lawson DM, White KK, Shapiro JA, Cohen DE, Carson WG, Mutschler MA (1995) Acylsugar of wild tomato *Lycopersicon pennelli* alters settling and reduces oviposition of *Bemisia argentifolii* (Homoptera: Aleyrodidae). J Econ Entomol 88:742–748

Lin TY, Kao YY, Lin S, Lin RF, Chen CM, Huang CH, Wang CK, Lin YZ, Chen CC (2001) A genetic linkage map of *Nicotiana plumbaginifolia/Nicotiana longiflora* based on RFLP and RAPD markers. Theor Appl Genet 103:905–911

Lincoln RE, Porter JW (1950) Inheritance of beta-carotene in tomatoes. Genetics 35:206–211

Lincoln RE, Zscheile FP, Porter JW, Kohler GW, Caldwell RM (1943) Provitamin A and vitamin C in the genus *Lycopersicon*. Bot Gaz 105:113–115

Lindhout P, Korta W, Cislik M, Vos I, Gerlagh T (1989) Further identification of races of *Cladosporum fulvum* (*Fulvia fulva*) on tomato originating from the Netherlands, France and Poland. Neth J Plant Pathol 85:143–148

Lindhout P, Pet G, Van der Beek H (1994) Screening wild *Lycopersicon* species for resistance to powdery mildew (*Oidium lycopersicum*). Euphytica 72:43–49

Linkage Committee (1973) Linkage summary. Tomato Genet Coop Rep 23:9–11

Linnaeus C (1753) Species Plantarum, 1st ed. Stockholm: L. Salvius

Lippman Z, Tanksley SD (2001) Dissecting the genetic pathway to extreme fruit size in tomato using a cross between the small-fruited wild species *Lycopersicon pimpinellifolium* and *L. esculentum* var. giant heirloom. Genetics 158:413–422

Liu JP, Van Eck J, Cong B, Tanksley SD (2002) A new class of regulatory genes underlying the cause of pear-shaped tomato fruit. Proc Natl Acad Sci USA 99:13302–13306

Liu Y, Chen H, Wei Y, Zhuang T (2005) Construction of a genetic map and localization of QTLs for yield traits in tomato by SSR markers. Prog Nat Sci 15:793–797

Liu Y, Roof S, Ye Z, Barry C, van Tuinen A, Vrebalov J, Bowler C, Giovannoni J (2004) Manipulation of light signal transduction as a means of modifying fruit nutritional quality in tomato. Proc Natl Acad Sci USA 101:9897–9902

Liu Y, Zamir D (1999) Second generation *L. pennellii* introgression lines and the concept of bin mapping. Tomato Genet Coop Rep 49:26–30

Liu Y-S, Gur A, Ronen G, Causse M, Damidaux R, Buret M, Hirschberg J, Zamir D (2003) There is more to tomato fruit colour than candidate carotenoid genes. Plant Biotechnol J 1:195–207

Livingstone KD, Lackney VK, Blauth J, van Wijk R, Jahn MK (1999) Genome mapping in *Capsicum* and the evolution of genome structure in the Solanaceae. Genetics 152:1183–1202

Loh Y-T, Martin G (1995) The disease-resistance gene *Pto* and the fenthion-sensitivity gene *Fen* encode closely related functional protein kinases. Proc Natl Acad Sci USA 92:4181–4184

Lois LM, Rodriguez-Concepcion M, Gallego F, Campos N, Boronat A (2000) Carotenoid biosynthesis during tomato fruit development: regulatory role of 1-deoxy-D-xylulose 5-phosphate synthase. Plant J 22:503–513

Loppes R (1968) Ethyl methanesulfonate: an effective mutagen in *Chlamydomonas reinhardi*. Mol Gen Genet 102:229–231

Louws FJ, Bell J, Medina-Mora CM, Smart CD, Opgenorth D, Ishimaru CA, Hausbeck MK, de Bruijin FJ, Fulbright DW (1998) Rep-PCR-mediated genomic fingerprinting: A rapid and effective method to identify *Clavibacter micheganensis*. Phytopathology 88:862–868

Luckwill LC (1943a) The evolution of the cultivated tomato. J R Hort Soc 68:19–25

Luckwill LC (1943b) The genus *Lycopersicon*: an historical, biological, and taxonomic survey of the wild and cultivated tomatoes. Aberdeen Univ Studies 120:1–44

Lucock M (2000) Folic acid: Nutritional biochemistry, molecular biology, and role in disease processes. Mol Genet Metab 71:121–138

Lukyanenko AN (1991) Disease resistance in tomato. In: Kalloo G (ed) Genetic improvement of tomato. Springer, Berlin Heidelberg New York, pp 99–119

Lynch M, Walsh B (eds) (1998) Genetics and analysis of quantitative traits. Sinauer Associates, Sunderland, MA, USA

Ma Y, Lin SQ, Gao Y, Li M, Luo WX, Zhang J, Xia NS (2003) Expression of *ORF2* partial gene of hepatitis E virus in tomatoes and immunoactivity of expression products. World J Gastroenterol 9:2211–2215

MacArthur J (1926) Linkage studies with the tomato. Genetics 11:387–405

MacArthur J (1934) Linkage groups in the tomato. J Genet 29:123–133

MacBride JF (1962) Solanaceae, in Flora of Peru. Field Mus Nat Hist Ser 13, pp 3–267

Madhavi DL, Salunkhe DK (1998) Tomato. In: Salunkhe DK, Kadam SS (eds) Handbook of vegetable science. Marcel Dekker, New York, pp 171–201

Maldonado C, Squeo FA, Ibacache E (2003) Phenotypic response of *Lycopersicon chilense* to water deficit. Rev Chil Hist Nat 76:129–137

Maluf WR, Barbosa LV, Santa-Cecília LVC (1997) 2-Tridecanone-mediated mechanisms of resistance to the South American tomato pinworm *Scrobipalpuloides absoluta* (Meyrick, 1917) (Lepidoptera-Gelechiidae) in *Lycopersicon* spp. Euphytica 93:189–194

Mangin B, Thoquet P, Olivier J, Grimsley NH (1999) Temporal and multiple quantitative trait loci analyses of resistance to bacterial wilt in tomato permit the resolution of linked loci. Genetics 151:1165–1172

Manning K, Tor M, Poole M, Hong Y, Thompson AJ, King GJ, Giovannoni JJ, Seymour GB (2006) A naturally occurring epigenetic mutation in a gene encoding an SBP-box transcription factor inhibits tomato fruit ripening. Nat Genet 38:948–952

Manuelyan H, Yordanov M, Yordanova Z, Ilieva Z (1975) Studies on *β*-carotene and lycopene content in the fruits of *Lycopersicon esculentum* Mill. × *L. chilense* Dun. hybrids. Qual Plant Foods Hum Nutr 25:205–210

Marczewski W, Flis B, Syller J, Schoefer PR, Gebhardt C (2001) A major quantitative trait locus for resistance to potato leafroll virus is located in a resistance hotspot on potato chromosome XI and is tightly linked to N-gene-like markers. Mol Plant-Micr Interact 14:1420–1425

Marshall JA, Knapp S, Davey MR, Power JB, Cocking EC, Bennett MD, Cox AV (2001) Molecular systematics of *Solanum* section *Lycopersicum* (*Lycopersicon*) using the nuclear ITS rDNA region. Theor Appl Genet 103:1216–1222

Martin B, Nienhuis J, King G, Schaefer A (1989) Restriction fragment length polymorphisms associated with water use efficiency in tomato. Science 243:1725–1728

Martin B, Thorstenson Y (1988) Stable carbon isotope composition ($d^{13}C$), water use efficiency and biomass productivity of *Lycopersicon esculentum, Lycopersicon pennellii* and the F_1 hybrid. Plant Physiol 88:213–217

Martin GB, Brommonschenkel SH, Chunwongse J, Frary A, Ganal MW, Spivey R, Wu T, Earle ED, Tanksley SD (1993a) Map-based cloning of a protein-kinase gene conferring disease resistance in tomato. Science 262:1432–1436

Martin G, Carmen de Vicente M, Tanksley SD (1993b) High-resolution linkage analysis and physical characterization of the *Pto* bacterial resistance locus in tomato. Mol Plant-Micr Interact 6:26–34

Martin GB, Williams JGK, Tanksley SD (1991) Rapid identification of markers linked to a *Pseudomonas* resistance gene in tomato by using random primers and near-isogenic lines. Proc Natl Acad Sci USA 88:2336–2340

Martin MW, Cannon OS, Dewey WG (1971) Pedigree history of curly-top-resistant tomato lines released by USDA. Tomato Genet Coop Rep 21:23–24

Martin MW, Thomas PE (1986) Levels, dependability, and usefulness of resistance to *Tomato curly top* disease. Plant Dis 70:136–141

Matthews RF, Crill P, Burgis D (1973) Ascorbic acid content of tomato varieties. Proc Florida State Hort Soc 86:242–250

Matthiolus PA (1544) Di Pedacio Dioscoride Anazarbeo libri cinque della historia, et materia medicinale trodotti in lingua volgare Italiana, Venice

McCallum CM, Comai L, Greene EA, Henikoff S (2000a) Targeted screening for induced mutations. Nat Biotechnol 18:455–457

McCallum CM, Comai L, Greene EA, Henikoff S (2000b) Targeting induced local lesions IN genomes (TILLING) for plant functional genomics. Plant Physiol 123:439–442

McClendon JH, Woodmansee CW, Somers GF (1959) On the occurrence of free galacturonic acid in apples and tomatoes. Plant Physiol 34:389

McCormick S, Niedermeyer J, Fry J, Barnason A, Horsch R, Fraley R (1986) Leaf disc transformation of cultivated tomato (*L. esculentum*) using *Agrobacterium tumefaciens*. Plant Cell Rep 5:81–84

McCue G (1952) The history and use of the tomato: an annotated bibliography. Ann Missouri Bot Gar 39:289–348

McMeekin D (1992) Representations of pre-Columbian spindle whorls of the floral and fruit structure of economic plants. Econ Bot 46:171–180

Mehta R, Cassol T, Li N, Ali N, Handa AK, Mattoo AK (2002) Engineered polyamine accumulation in tomato enhances phytonutrient content, juice quality, and vine life. Nat Biotechnol 20:613–618

Meissner R, Chague V, Zhu Q, Emmanuel E, Elkind Y, Levy AA (2000) Technical advance: a high throughput system for transposon tagging and promoter trapping in tomato. Plant J 22:265–274

Meissner R, Jacobson Y, Melamed S, Levyatuv S, Shalev G, Ashri A, Elkind Y, Levy A (1997) A new model system for tomato genetics. Plant J 12:1465–1472

Menda N, Semel Y, Peled D, Eshed Y, Zamir D (2004) In silico screening of a saturated mutation library of tomato. Plant J 38:861–872

Mesbah LA, Kneppers RJA, Takken FLW, Laurent P, Hille J, Nijkamp HJJ (1999) Genetic and physical analysis of a YAC contig spanning the fungal disease resistance locus *Asc* of tomato (*Lycopersicon esculentum*). Mol Gen Genet 261:50–57

Mieslerova B, Lebeda A, Chetelat RT (2000) Variation in response of wild *Lycopersicon* and *Solanum* spp. against tomato powdery mildew (*Oidium lycopersici*). J. Phytopathol 148:303–311

Mihr C, Faurobert M, Pawłowski T, Bouchet JP, Sommerer N, Rossignol M, Negroni L, Causse M (2005) Proteome analysis of organoleptic quality in tomato. Acta Hort 682:277–283

Miller JC, Tanksley SD (1990) RFLP analysis of phylogenetic relationships and genetic variation in the genus *Lycopersicon*. Theor Appl Genet 80:437–448

Miller P (1754) The gardener's dictionary, Abridged 4th ed. London: John and James Rivington

Miller P (1768) The gardener's dictionary, Abridged 8th ed. London

Miller P (1807) The gardener's and botanist's dictionary, Posthumous ed. edited by Thomas Martin, Cambridge, London

Minamiyama Y, Tsuro M, Hirai M (2006) An SSR-based linkage map of *Capsicum annuum*. Mol Breed 18:157–169

Miron D, Schaffer AA (1991) Sucrose phosphate synthase, sucrose synthase, and invertase activities in developing fruit of *Lycopersicon hirsutum* Humb. and Bonpl. Plant Physiol 95:623–627

Mishra SK, Tripp J, Winkelhaus S, Tschiersch B, Theres K, Nover L, Scharf KD (2002) In the complex family of heat stress transcription factors, HsfA1 has a unique role as master regulator of thermotolerance in tomato. Genes Dev 16:1555–1567

Moco S, Bino RJ, Vorst O, Verhoeven HA, de Groot J, van Beek TA, Vervoort J, de Vos CH (2006) A liquid chromatography-mass spectrometry-based metabolome database for tomato. Plant Physiol 141:1205–1218

Molgaard P, Ravn H (1988) Evolutionary aspects of caffeoyl ester distribution in dicotyledons. Phytochemistry 27:2411–2421

Molloy AM, Scott JM (2001) Folates and prevention of disease. Public Health Nutr 4:601–609

Monforte AJ, Asins MJ, Carbonell EA (1996) Salt tolerance in *Lycopersicon* species. IV. Efficiency of marker assisted selection for salt tolerance improvement. Theor Appl Genet 93:765–772

Monforte AJ, Asins MJ, Carbonell EA (1997a) Salt tolerance in *Lycopersicon* species. V. Does genetic variability at quantitative trait loci affect their analysis? Theor Appl Genet 95:284–293

Monforte AJ, Asins MJ, Carbonell EA (1997b) Salt tolerance in *Lycopersicon* species. VI. Genotype-by-salinity interaction in quantitative trait loci detection: constitutive and response QTLs. Theor Appl Genet 95:706–713

Monforte AJ, Asins MJ, Carbonell EA (1999) Salt tolerance in *Lycopersicon* species. VII. Pleiotropic action of genes controlling earliness on fruit yield. Theor Appl Genet 98:593–601

Monforte AJ, Friedman E, Zamir D, Tanksley SD (2001) Comparison of a set of allelic QTL-NILs for chromosome 4 of tomato: Deductions about natural variation and implications for germplasm utilization. Theor Appl Genet 102:572–590

Monforte AJ, Tanksley SD (2000a) Development of a set of near isogenic and backcross recombinant inbred lines containing most of the *Lycopersicon hirsutum* genome in a *L. esculentum* genetic background: A tool for gene mapping and gene discovery. Genome 43:803–813

Monforte AJ, Tanksley SD (2000b) Fine mapping of a quantitative trait locus (QTL) from *Lycopersicon hirsutum* chromosome 1 affecting fruit characteristics and agronomic traits: breaking linkage among QTLs affecting different traits and dissection of heterosis for yield. Theor Appl Genet 100:471–479

Moore JA (1935) The early history of the tomato or loveapple. Missouri Bot Gard Bull 23:134–138

Moore S, Vrebalov J, Paxton P, Giovannoni J (2002) Use of genomics tools to isolate key ripening genes and analyse fruit maturation in tomato. J Exp Bot 53:2023–2030

Moreau P, Thoquet P, Olivier J, Laterrot H, Grimsley N (1998) Genetic mapping of *Ph-2*, a single locus controlling partial resistance to *Phytophthora infestans* in tomato. Mol Plant-Micr Interact 11:259–269

Moreau RA, Whitaker BD, Hicks KB (2002) Phytosterols, phytostanols, and their conjugates in foods: structural diversity, quantitative analysis, and health-promoting uses. Prog Lipid Res 41:457–500

Morris PF, Connolly MS, St Clair DA (2000) Genetic diversity of *Alternaria alternata* isolated from tomato in California assessed using RAPDs. Mycol Res 104:286–292

Mueller LA, Solow T, Taylor N, Skwarecki B, Buels R, Binns J, Lin C, Wright M, Ahrens R, Wang Y, Herbst E, Keyder E, Menda N, Zamir D, Tanksley SD (2005a) The SOL Genomics Network. A comparative resource for Solanaceae biology and beyond. Plant Physiol 138:1310–1317

Mueller LA, Tanksley SD, Giovannoni JJ, Van Eck J, Stack S, Choi D, Kim BD, Chen M, Cheng Z, Li C, Ling H, Xue Y, Seymour G, Bishop G, Bryan G, Sharma R, Khurana J, Tyagi A, Chattopadhyay D, Singh NK, Stiekema W, Lindhout P, Jesse T, Lankhorst RK, Bouzayen M, Shibata D, Tabata S, Granell A, Botella MA, Giuliano G, Frusciante L, Causse M, Zamir D (2005b) The Tomato Sequencing Project, the first cornerstone of the International Solanaceae Project (SOL). Comp Funct Genom 6:153–158

Muir SR, Collins GJ, Robinson S, Hughes S, Bovy A, de Vos CHR, van Tunen AJ, Verhoeyen ME (2001) Overexpression of petunia chalcone isomerase in tomato results in fruit containing increased levels of flavonols. Nat Biotechnol 19:470–474

Müller CH (1940a) A revision of the genus *Lycopersicon*. USDA Mis Publ 382:1–28

Müller CH (1940b) The taxonomy and distribution of the genus *Lycopersicon*. Nat Hort Mag 19:157–160

Mustilli AC, Fenzi F, Ciliento R, Alfano F, Bowler C (1999) Phenotype of the tomato *high pigment-2* mutant is caused by a mutation in the tomato homolog of *DEETIOLATED1*. Plant Cell 11:145–157

Mutschler MA (2006) Combing field and laboratory methods in tomato breeding strategies. Acta Hort 724:23–27

Nagai H, Siqueira WJ, Lourenção AL (1992) Tomato breeding for resistance to diseases and pests in Brazil. Acta Hort 301:91–97

Nakashima K, Yamaguchi-Shinozaki K (2006) Regulons involved in osmotic stress responsive and cold stress-responsive gene expression in plants. Physiol Plant 126:62–71

Nakatani N, Kayano S, Kikuzaki H, Sumino K, Katagiri K, Mitani T (2000) Identification, quantitative determination, and antioxidative activities of chlorogenic acid isomers in prune (*Prunus domestica* L.). J Agric Food Chem 48:5512–5516

Nardini M, Aquino MD, Tomassi G, Gentili V, Di Felice M, Scaccini C (1995) Inhibition of human low-density lipoprotein oxidation by caffeic acid and other hydroxycinnamic acid derivatives. Free Radical Bio Med 19:541–552

Neff MM, Neff JD, Chory J, Pepper AE (1998) dCAPS, a simple technique for the genetic analysis of single nucleotide polymorphisms: experimental applications in *Arabidopsis thaliana* genetics. Plant J 14:387–392

Nesbitt TC, Tanksley SD (2002) Comparative sequencing in the genus *Lycopersicon*: Implications for the evolution of fruit size in the domestication of cultivated tomatoes. Genetics 162:365–379

Ng PC, Henikoff S (2003) SIFT: Predicting amino acid changes that affect protein function. Nucl Acids Res 31:3812–3814

Nienhuis J, Helentjaris T, Slocum M, Ruggero B, Schaefer A (1987) Restriction fragment length polymorphism analysis of loci associated with insect resistance in tomato. Crop Sci 27:797–803

Niggeweg R, Michael AJ, Martin C (2004) Engineering plants with increased levels of the antioxidant chlorogenic acid. Nat Biotechnol 22:746–754

Nijveldt RJ, van Nood E, van Hoorn DEC, Boelens PG, van Norren K, van Leeuwen PAM (2001) Flavonoids: A review of probable mechanisms of action and potential applications. Am J Clin Nutr 74:418–425

Nitzany FE (1992) *Cucumber mosaic virus* in Israel. Phytopathol Med 14:16–20

O'Connell MA, Medina AL, Sanchez Pena P, Trevino MB (2007) Molecular genetics of drought resistance response in tomato and related species. In: Razdan MK, Mattoo AK (eds) Genetic improvement of Solanaceous crops. vol. 2: tomato. Science Publ, Enfield, NH, USA, pp 261–283

Ogundiwin EA, Berke TF, Massoudi M, Black LL, Huestis G, Choi D, Lee S, Prince JP (2005) Construction of 2 intraspecific linkage maps and identification of resistance QTLs for *Phytophthora capsici* root-rot and foliar-blight diseases of pepper (*Capsicum annuum* L.). Genome 48:698–711

Ohmori T, Murata M, Motoyoshi F (1995a) Identification of RAPD markers linked to the *Tm-2* locus in tomato. Theor Appl Genet 90:307–311

Ohmori T, Murata M, Motoyoshi F (1995b) RAPD markers linked to the *Tomato mosaic virus* resistance gene, *Tm-1*, in tomato. Jpn J Genet 70:179–184

Ohmori T, Murata M, Motoyoshi F (1996) Molecular characterization of RAPD and SCAR markers linked to the *Tm-1* locus in tomato. Theor Appl Genet 92:151–156

Oke M, Pinhero RG, Paliyath G (2003) The effects of genetic transformation of tomato with antisense phospholipase D cDNA on the quality characteristics of fruit and their processed products. Food Biotechnol 17:163–182

Oleykowski CA, Bronson Mullins CR, Godwin AK, Yeung AT (1998) Mutation detection using a novel plant endonuclease. Nucl Acids Res 26:4597–4602

Olmstead RG, Palmer JD (1997) Implications for the phylogeny, classification, and biogeography of Solanum from cpDNA restriction site variation. Syst Bot 22:19–29

Olmstead RG, Sweere JA, Spangler RE, Bohs L, Palmer JD (1999) Phylogeny and provisional classification of the Solanaceae based on chloroplast DNA. In: Nee M, Symon DE, Lester RN, Jessop JP (eds) Solanaceae IV, advances in biology and utilization. Roy Bot Gard, Kew, UK, pp 111–137

Omenn GS, Goodman G, Thornquist M, Grizzle J, Rosenstock L, Barnhart S, Balmes J, Cherniack MG, Cullen MR, Glass A, Keogh J, Meyskens F, Valanis B, Williams J (1994) The β-carotene and retinol efficacy trail (CARET) for chemoprevention of lung cancer in high risk populations: smokers and asbestos-exposed workers. Canc Res (suppl) 54:2038s–2043s

Ori N, Eshe Y, Paran I, Presting G, Aviv D, Tanksley SD, Zamir D, Fluhr R (1997) The I2C family from the wilt disease resistance locus *I2* belongs to the nucleotide binding, leucine-rich repeat superfamily of plant resistance genes. Plant Cell 9:521–532

Overy SA, Walker HJ, Malone S, Howard TP, Baxter CJ, Sweetlove LJ, Hill SA, Quick WP (2005) Application of metabolite profiling to the identification of traits in a population of tomato introgression lines. J Exp Bot 56:287–296

Oyanedel EA (1999) Quantitative trait loci analysis of chilling tolerance in tomato. PhD Dis, Cornell Univ, Ithaca, NY, USA

Paiva SAR, Russell RM (1999) Beta-carotene and other carotenoids as antioxidants. J Am Coll Nutr 18:426–433

Palmer JD, Zamir D (1982) Chloroplast DNA evolution and phylogenetic relationships in *Lycopersicon*. Proc Natl Acad Sci USA 79:5006–5010

Palmieri S, Martiniello P, Sorressi GP (1978) Chlorophyll and carotene content in *high pigment* and *green flesh* fruits. Tomato Gen Coop Rep 28:10

Palukaitis P, Roossinck MJ, Dietzgen RG, Francki RIB (1992) *Cucumber mosaic virus*. Adv Virus Res 41:281–348

Pan Q, Liu YS, Budai-Hadrian O, Sela M, Carmel-Goren L, Zamir D, Fluhr R (2000) Comparative genetics of nucleotide binding site-leucine rich repeat resistance gene homologues in the genomes of two dicotyledons: tomato and Arabidopsis. Genetics 155:309–322

Paran I, van der Voort JR, Lefebvre V, Jahn M, Landry L, van Schriek M, Tanyolac B, Caranta C, Ben Chaim A, Livingstone K, Palloix A, Peleman J (2004) An integrated genetic linkage map of pepper. Mol Breed 13:251–261

Paran I, Zamir D (2003) Quantitative traits in plants: beyond the QTL. Trends Genet 19:303–306

Park S, Kim CK, Pike LM, Smith RH, Hirschi KD (2004) Increased calcium in carrots by expression of an *Arabidopsis* H^+/Ca^{2+} transporter. Mol Breed 14:275–282

Park SH, Morris JL, Park JE, Hirschi KD, Smith RH (2003) Efficient and genotype-independent *Agrobacterium*-mediated tomato transformation. J Plant Physiol 160:1253–1257

Parrella G, Gognalons P, Gebre-Selassiè K, Vovlas C, Marchoux G (2003) An update of the host range of *Tomato spotted wilt virus*. J Plant Pathol 85:227–264

Parrella G, Hochu I, Gebre-Selassiè K, Gognalons P, Moretti A, Marchoux G, Caranta C (2000) Molecular tagging of the *Am* gene from *Lycopersicon hirsutum* f. *glabratum* PI 134417 using AFLP markers. Acta Physiol Plant 22:291–293

Parrella G, Laterrot H, Gebre-Selassiè K, Marchoux G (1998) Inheritance of resistance to *Alfalfa mosaic virus* in *Lycopersicon hirsutum* f. *glabratum* PI 134417. J Plant Pathol 80:241–243

Parrella G, Laterrot H, Marchoux G, Gebre-Selassiè K (1997) Screening *Lycopersicon* accessions for resistance to *Alfalfa mosaic virus*. J Genet Breed 51:75–78

Parrella G, Moretti A, Gognalons P, Lesage M-L, Marchoux G, Gebre-Selassiè K, Caranta C (2004) The *Am* gene controlling resistance to *Alfalfa mosaic virus* in tomato is located in the cluster of dominant resistance genes on chromosome 6. Phytopathology 94:345–350

Parrella G, Ruffel S, Moretti A, Morel C, Palloix A, Caranta C (2002) Recessive resistance genes against *Potyviruses* are localized in colinear genomic regions of the tomato (*Lycopersicon* spp.) and pepper (*Capsicum* spp.) genomes. Theor Appl Genet 105:855–861

Paterson AH, Damon S, Hewitt JD, Zamir D, Rabinowitch HD, Lincoln SE, Lander ES, Tanksley SD (1991) Mendelian factors underlying quantitative traits in tomato: comparison across species, generations, and environments. Genetics 127:181–197

Paterson AH, de Verna JW, Lanini B, Tanksley SD (1990) Fine mapping of quantitative trait loci using selected overlapping recombinant chromosomes, in an interspecies cross of tomato. Genetics 124:735–742

Paterson AH, Lander ES, Hewitt JD, Peterson S, Lincoln SE, Tanksley SD (1988) Resolution of quantitative traits into Mendelian factors by using a complete linkage map of restriction fragment length polymorphisms. Nature 335:721–726

Paterson RG, Scott SJ, Gergerich RC (1989) Resistance in two *Lycopersicon* species to an Arkansas isolate of *Tomato spotted wilt virus*. Euphytica 43:173–178

Paulson KN, Stevens MA (1974) Relationships among titratable acidity, pH and buffer composition of tomato fruits. J Food Sci 39:354–357

Pawlowski T, Grandbastien MA, Faurobert M, Mihr C, Causse M (2005) Proteomics of genetic diversity in relation to tomato fruit size. In: Mencarelli F, Tonutti P (eds) V Intl Post-harvest Symp, vol 682. ISHS, Verona, Italy, pp 285–288

Pearson RC, Hall DH (1975) Factors affecting the occurrence and severity of blackmold on ripe tomato fruit caused by *Alternaria alternata*. Phytopathology 65:1352–1359

Pecaut P (1965) Tomate: Resistance au virus de la mosaique du tabac (T.M.V.). Report of Station de'Amelioration des Plantes Maraicheres INRA, France, pp 50–53

Peleman JD, van der Voort JR (2003) Breeding by design. Trends Plant Sci 8:330–334

Peralta IE, Knapp S, Spooner DM (2007) The taxonomy of tomatoes: a revision of wild tomatoes (*Solanum* section *Lycopersicon*) and their outgroup relatives (*Solanum* sections *Juglandifolia* and *Lycopersicoides*). Syst Bot Mngr (in review)

Peralta IE, Knapp SK, Spooner DM (2005) New species of wild tomatoes (*Solanum* section *Lycopersicon*: Solanaceae) from Northern Peru. Syst Bot 30:424–434

Peralta IE, Spooner DM (2001) Granule-bound starch synthase (GBSSI) gene phylogeny of wild tomatoes (*Solanum* L. section *Lycopersicon* [Mill.] Wettst. subsection *Lycopersicon*). Am J Bot 88:1888–1902

Peralta IE, Spooner DM (2005) Morphological characterization and relationships of wild tomatoes (*Solanum* L. Section *Lycopersicon*). In: Keating RC, Hollowell VC, Croat T (eds) Festschrift for William G. Darcy: the legacy of a taxonomist (Monograph in Systematic Botany 104). MBG Press, Missouri, USA, pp 227–257

Peralta IE, Spooner DM (2007) History, origin and cultivation of tomato. In: Razdan MK, Mattoo AK (eds) Genetic improvement of Solanaceous crops, Vol 2: tomato. Science Publ, Enfield, NH, USA, pp 1–24

Perry JA, Wang TL, Welham TJ, Gardner S, Pike JM, Yoshida S, Parniske M (2003) A TILLING reverse genetics tool and a web-accessible collection of mutants of the legume *Lotus japonicus*. Plant Physiol 131:866–871

Pertuze R, Ji Y, Chetelat R (2002) Comparative linkage map of the *Solanum lycopersicoides* and *S. sitiens* genomes and their differentiation from tomato. Genome 45:1003–1012

Pestsova EG, Börner A, Röder MS (2006) Development and QTL assessment of *Triticum aestivum-Aegilops tauschii* introgression lines. Theor Appl Genet 112:634–647

Peters JL, Van Tuinen A, Adamse P, Kendrick RE, Koornneef M (1989) High pigment mutants of tomato exhibit high sensitivity for phytochrome action. J Plant Physiol 134:661–666

Peterson D, Lapitan N, Stack S (1999) Localization of single- and low-copy sequences on tomato synaptonemal complex spreads using fluorescence in situ hybridization (FISH). Genetics 152:427–439

Peterson D, Price H, Johnston J, Stack S (1996) DNA content of heterochromatin and euchromatin in tomato (*Lycopersicon esculentum*) pachytene chromosomes. Genome 39:77–82

Phills BR, Provvidenti R, Robinson RW (1977) Reaction of *Solanum lycopericoides* to vital diseases of the tomato. Tomato Genet Coop Rep 27:18

Picó B, Díez M, Nuez F (1998) Evaluation of whitefly-mediated inoculation techniques to screen *Lycopersicon esculentum* and wild relatives for resistance to *Tomato yellow leaf curl virus*. Euphytica 101:259–271

Picó B, Herraiz J, Riuz JJ, Nuez F (2002) Widening the genetic basis of virus resistance in tomato. Sci Hort 94:73–89

Pierce LC (1971) Linkage tests with *Ph* conditioning resistance to race 0, *Phytophthora infestans*. Tomato Genet Coop Rep 21:30

Pierre M, Noel L, Lahaye T, Ballvora A, Veuskens J, Ganal M, Bonas U (2000) High-resolution genetic mapping of the pepper resistance locus *Bs3* governing recognition of the *Xanthamonas campestris* pv *vesicatora* AvrBs3 protein. Theor Appl Genet 101:255–263

Pietta PG (2000) Flavonoids as antioxidants. J Nat Prod 63:1035–1042

Piironen V, Lindsay DG, Miettinen TA, Toivo J, Lampi AM (2000) Plant sterols: biosynthesis, biological function and their importance to human nutrition. J Sci Food Agric 80:939–966

Pillen K, Ganal M, Tanksley S (1996a) Construction of a high-resolution genetic map and YAC-contigs in the tomato *Tm-2a* region. Theor Appl Genet 93:228–233

Pillen K, Pineda O, Lewis CB, Tanksley SD (1996b) Status of genome mapping tools in the taxon Solonaceae. In: Paterson AH (ed) Genome mapping in plants. Academic Press, College Station, Texas, USA, pp 281–308

Pinhero RG, Almquist KC, Novotna Z, Paliyath G (2003) Developmental regulation of phospholipase D in tomato fruits. Plant Physiol Biochem 41:223–240

Pitblado RE, Kerr EA (1980) Resistance to bacterial speck (*Pseudomonas syringae* pv. *tomato*) in tomato. Acta Hort 100:379–382

Pitblado RE, MacNeill BH (1983) Genetic basis of resistance to *Pseudomas syringae* pv. *tomato* in field tomatoes. Can J Plant Pathol 5:251–255

Pitblado RE, MacNeill BH, Kerr EA (1984) Chromosomal identity and linkage relationships of *Pto*, a gene for resistance to *Pseudomonas syringae* pv. *tomato*. Can J Plant Pathol 6:48–53

Plant AL, Cohen A, Moses MS, Bray EA (1991) Nucleotide sequence and spatial expression pattern of a drought- and ABA-induced gene of tomato. Plant Physiol 97:900–906

Poonam B, Nanjappa A, Senaratna T, Midmore D (2004) Tissue culture studies of tomato (*Lycopersicon esculentum*). Plant Cell Tiss Org Cult 78:1–21

Price AH (2006) Believe it or not, QTLs are accurate! Trends Plant Sci 11:213–216

Prince JP, Pochard E, Tanksley SD (1993) Construction of a molecular linkage map of pepper and a comparison of synteny with tomato. Genome 36:404–417

Rashid F, Khalid S, Ahmad I, Mughal SM (1989) *Potato virus-X* (PVX) resistance in tomato cultivars. Trop Pest Mng 35:357–358

Rensink W, Hart A, Liu J, Ouyang S, Zismann V, Buell CR (2005) Analyzing the potato abiotic stress transcriptome using expressed sequence tags. Genome 48:598–605

Reymond P, Weber H, Damond M, Farmer EE (2000) Differential gene expression in response to mechanical wounding and insect feeding in *Arabidopsis*. Plant Cell 12:707–720

Reynolds JF, Bieber NE, Sun EL (1982) Environmental, genotype and pretreatment influences on regeneration of tomato *in vitro*. In Vitro 18:318–322

Rhee SY, Beavis W, Bernardini TZ, Chen G, Dixon D, Doyle A, Garcia-Hernandez M, Huala E, Lander G, Montoya M, Miller N, Mueller LA, Mundodi S, Reiser L, Tacklind J, Weems DC, Wu Y, Xu I, Yoo D, Yoon J, Zhang P (2003) The *Arabidopsis* Information Resource (TAIR): a model organism database providing a centralized, curated gateway to Arabidopsis biology, research materials and community. Nucl Acids Res 31:224–228

Rhee SY, Dickerson J, Xu D (2006) Bioinformatics and its applications in plant biology. Annu Rev Plant Biol 57:335–360

Rice-Evans CA, Miller NJ, Paaganga G (1997) Antioxidant properties of phenolic compounds. Trends Plant Sci 2:152–159

Rick CM (1956) New mutants. Tomato Genet Coop Rep 6:22–23

Rick CM (1963) Biosystematic studies on Galapagos Island tomatoes. Occ Pap Calif Acad Sci 44:59–77

Rick CM (1969) Controlled introgression of chromosomes of *Solanum pennellii* into *Lycopersicon esculentum* - segregation and recombination. Genetics 62:753–768

Rick CM (1972) Further studies on segregation and recombination in backcross derivatives of a tomato species hybrid. Biol Zentralbl 91:209–220

Rick CM (1973) Potential genetic resources in tomato species: clues from observation in native habitats. In: Srb AM (ed) Genes, enzymes and populations. Plenum Press, New York, pp 255–269

Rick CM (1974) High soluble-solids content in large-fruited tomato lines derived from a wild green-fruited species. Hilgardia 42:493–510

Rick CM (1986) Reproductive isolation in the *Lycopersicon peruvianum* complex. In: D'Arcy WG (ed) Solanaceae: biology and systematics. Columbia Univ Press, New York

Rick CM (1995) Tomato: *Lycopersicon esculentum* (Solanaceae). In: Smartt J, Simmonds NW (eds) Evolution of crop plants. Longman, Harlow, Essex, England, UK, pp 452–457

Rick CM, Barton DW (1954) Cytological and genetical identification of the primary trisomics of the tomato. Genetics 39:640–666

Rick CM, Butler L (1956) Cytogenetics of the tomato. Adv Genet 8:267–382

Rick CM, Chetelat RT (1995) Utilization of related wild species for tomato improvement. Acta Hort 412:21–38

Rick CM, Cisneros P, Chetelat RT, DeVerna JW (1994a) *Abg* – A gene on chromosome 10 for purple fruit derived from *S. lycopersicoides*. Tomato Genet Coop Rep 44:29–30

Rick CM, Deverna JW, Chetelat RT, Stevens MA (1986) Meiosis in sesquidiploid hybrids of *Lycopersicon esculentum* and *Solanum lycopersicoides*. Proc Natl Acad Sci USA 83:3580–3583

Rick CM, Fobes JF (1974) Association of an allozyme with nematode resistance. Tomato Genet Coop Rep 24:25

Rick CM, Fobes JF (1975) Allozyme variation in the cultivated tomato and closely related species. Bull Torrey Bot Club 102:376–384

Rick CM, Fobes JF (1977) Linkage relations of some isozymic loci. Tomato Genet Coop Rep 27:22–24

Rick CM, Fobes JF, Holle M (1977) Genetic variation in *Lycopersicon pimpinellifolium*: evidence of evolutionary change in mating systems. Plant Syst Evol 127:139–170

Rick CM, Fobes JF, Tanksley SD (1979) Evolution of mating systems in *Lycopersicon hirsutum* as deduced from genetic variation in electrophoretic and morphological characters. Plant Syst Evol 132:279–298

Rick CM, Holle M (1990) Andean *Lycopersicon esculentum* var. *cerasiforme* - genetic-variation and its evolutionary significance. Econ Bot 44:69–78

Rick CM, Kesicki E, Fobes JF, Holle M (1976) Genetic and biosystematic studies on two new sibling speces of *Lycopersicon* from interandean Peru. Theor Appl Genet 47:55–68

Rick CM, Smith PG (1953) Novel variation in tomato species hybrids. Am Nat 87:359–373

Rick CM, Tanksley SD (1981) Genetic variation in *Solanum pennellii* - Comparisons with 2 other sympatric tomato species. Plant Syst Evol 139:11–45

Rick CM, Thompson AE, Brauer O (1959) Genetics and development of an unstable chlorophyll deficiency in *Lycopersicon esculentum*. Am J Bot 46:1–11

Rick CM, Uylig JW, Jones AD (1994b) High α-tomatine content in ripe fruit of Andean *Lycopersicon esculentum* var. *cerasiforme*: developmental and genetic aspects. Proc Natl Acad Sci USA 91:12877–12881

Robbins ML, Angell FF (1970) Tomato anthracnose: Inheritance of reaction to *Colletotrichum cocciodes* in *Lycopersicon* spp. J Am Soc Hort Sci 95:469–471

Robert VJM, West MAL, Inai S, Caines A, Arntzen, Smith KJ, St Clair, DA (2001) Marker-assisted introgression of blackmold resistance QTL alleles from wild *Lycopersicon cheesmanii* to cultivated tomato (*L. esculentum*) and evaluation of QTL phenotypic effects. Mol Breed 8:217–233

Robertson LD, Labate JA (2007) Genetic resources of tomato (*Lycopersicon esculentum* var. *esculentum*) and wild relatives. In: Razdan MK, Mattoo AK (eds) Genetic Improvement of Solanaceous crops. vol 2: tomato. Science Publ, Enfield, NH, USA, pp 25–75

Robinson R, Tomes M (1968) Ripening inhibitor: a gene with multiple effect on ripening. Tomato Genet Coop Rep 18:36–37

Rocco M, D'Ambrosio C, Arena S, Faurobert M, Scaloni A, Marra M (2006) Proteomic analysis of tomato fruits from two ecotypes during ripening. Proteomics 6:3781–3791

Rodriguez AE, Tingey WM, Mutschler MA (1993) Acylsugars of *Lycopersicon pennelli* deter settling and feeding of the green peach aphid (Homoptera: Aphididae). J Econ Entomol *86:34–49*

Rodriguez F, Wu F, Tanksley S, Spooner D (2006) A multiple single-copy gene phylogenetic analysis of wild tomatoes (*Solanum* L. sect. *Lycopersicon* (Mill.) Wettst.) and their outgroup relatives. PAA/Solanaceae Conf, July 23–27, Univ of Wisconsin, Madison, WI, USA

Roessner U, Wagner C, Kopka J, Trethewey RN, Willmitzer L (2000) Technical advance: simultaneous analysis of metabolites in potato tuber by gas chromatography-mass spectrometry. Plant J 23:131–142

Romer S, Fraser PD, Kiano JW, Shipton CA, Misawa N, Schuch W, Bramley PM (2000) Elevation of the provitamin A content of transgenic tomato plants. Nat Biotechnol 18:666–669

Ronen G, Carmel-Goren L, Zamir D, Hirschberg J (2000) An alternative pathway to β-carotene formation in plant chromoplasts discovered by map-based cloning of *Beta* and *old-gold* color mutations in tomato. Proc Natl Acad Sci USA 97:11102–11107

Ronen GL, Cohen M, Zamir D, Hirschberg J (1999) Regulation of carotenoid biosynthesis during tomato fruit development: expression of the gene for lycopene epsilon-cyclase is down-regulated during ripening and is elevated in the mutant *Delta*. Plant J 17:341–351

Rosati C, Aquilani R, Dharmapuri S, Pallara P, Marusic C, Tavazza R, Bouvier F, Camara B, Giuliano G (2000) Metabolic engineering of beta-carotene and lycopene content in tomato fruit. Plant J 24:413–419

Rose JKC, Bashir S, Giovannoni JJ, Jahn MM, Saravanan RS (2004) Tackling the plant proteome: practical approaches, hurdles and experimental tools. Plant J 39:715–733

Rose TM, Henikoff JG, Henikoff S (2003) CODEHOP (COnsensus-DEgenerate Hybrid Oligonucleotide Primer) PCR primer design. Nucl Acids Res 31:3763–3766

Roselius K, Stephan W, Stadler T (2005) The relationship of nucleotide polymorphism, recombination rate and selection in wild tomato species. Genetics 171:753–763

Roselló S, Díez MJ, Nuez F (1998) Genetics of *Tomato spotted wilt virus* resistance coming from *Lycopersicon peruvianum*. Eur J Plant Pathol 104:499–509

Roselló S, Ricarte B, Díez MJ, Nuez F (2001) Resistance to *Tomato spotted wilt virus* introgressed from *Lycopersicon peruvianum* in line UPV 1 may be allelic to *Sw-5* and can be used to enhance the resistance of hybrids cultivars. Euphytica 119:357–367

Rousseaux MC, Jones CM, Adams D, Chetelat R, Bennett A, Powel A (2005) QTL analysis of fruit antioxidants in tomato using *Lycopersicon pennellii* introgression lines. Theor Appl Genet 111:1396–1408

Roxburgh W (1832) Flora Indica, or descriptions of Indian plants, vol. 2. W Thacker and Co, Calcutta, pp 691

Rozen S, Skaletsky HJ (2000) Primer3 on the WWW for general user and for biologist programmers. In: Krawetz S, Misener S (eds) Bioinformatics methods and protocols: methods in molecular biology. Humana Press, Totona, NJ, USA, pp 365–386

Ruiz H, Pavon JA (1797) Flora peruvianae, et chilensis Prodromus, sive novorum generum plantarum peruvianarum, et chilensium descriptions, et icons. edn 2. De órden del Rey, Madrid, Spain, pp 152

Ryu CM, Anand A, Kang L, Mysore KS (2004) Agrodrench: a novel and effective agroinoculation method for virus-induced gene silencing in roots and diverse Solanaceous species. Plant J 40:322–331

Saliba-Colombani V, Causse M, Gervais L, Philouze J (2000) Efficiency of RFLP, RAPD, and AFLP markers for the construction of an intraspecific map of the tomato genome. Genome 43:29–40

Saliba-Colombani V, Causse M, Langlois D, Philouze J, Buret M (2001) Genetic analysis of organoleptic quality in fresh market tomato. 1. Mapping QTLs for physical and chemical traits. Theor Appl Genet 102:259–272

Salmeron J, Oldroyd G, Rommens C, Scofield S, Kim H-S, Lavelle D, Dahlbeck D, Staskawicz B (1996) Tomato *Prf* is a member of the leucine-rich repeat class of plant disease resistance genes and lies embedded within the *Pto* kinase gene cluster. Cell 86:123–133

Salvi S, Tuberosa R (2005) To clone or not to clone plant QTLs: present and future challenges. Trends Plant Sci 6:297–304

Samuel G, Bald JG, Pittman HA (1930) Investigations on spotted wilt of tomatoes. Aust CSIR Bull 44:1–64

Sánchez Peña P (1999) Leaf water potentials in tomato (*L. esculentum* Mill.) *L. chilense* Dun. and their interspecific F_1. MS Thesis, New Mexico State Univ, Las Cruces, NM, USA

Sánchez Peña P, Fender SE, O'Connell MA (1995) Leaf water relations of *Lycopersicon chilense* during a drought cycle. Tomato Genet Coop Rep 45:40–41

Sandhu JS, Krasnyanski SF, Domier LL, Korban SS, Osadjan MD, Buetow DE (2000) Oral immunization of mice with transgenic tomato fruit expressing respiratory syncytial virus-F protein induces a systemic immune response. Transgenic Res 9:127

Sarfatti M, Abu-Abied M, Katan J, Zamir D (1991) RFLP mapping of *I1*, a new locus in tomato conferring resistance against *Fusarium oxysporum* f. sp. *lycopersici* race 1. Theor Appl Genet 82:22–26

Sarfatti M, Katan J, Fluhr R, Zamir D (1989) An RFLP marker in tomato linked to the Fusarium oxysporum resistance gene I2. Theor Appl Genet 78:755–759

Sato S, Soga T, Nishioka T, Tomita M (2004) Simultaneous determination of the main metabolites in rice leaves using capillary electrophoresis mass spectrometry and capillary electrophoresis diode array detection. Plant J 40:151–163

Sawa T, Nakao M, Akaike T, Ono K, Maeda H (1999) Alkylperoxyl radical scavenging activity of various flavonoids and other phenolic compounds: Implications for the anti-tumor-promoter effect of vegetables. J Agric Food Chem 47:397–402

Schaffer AA, Levin I, Oguz I, Petreikov M, Cincarevsky F, Yeselson Y, Shen S, Gilboa N, Bar M (2000) ADPglucose pyrophosphorylase activity and starch accumulation in immature tomato fruit: the effect of a *Lycopersicon hirsutum* derived introgression encoding for the large subunit. Plant Sci 152:135–144

Schaffer AA, Petreikov M, Miron D, Fogelman M, Spiegelman M, Bnei-Moshe Z, Shen S, Granot D, Hadas R, Dai N, Levin I, Bar M, Friedman M, Pilowsky M, Gilboa N, Chen L (1999) Modification of carbohydrate content in developing tomato fruit. HortScience 34:1024–1027

Schaible L, Cannon OS, Waddoups V (1951) Inheritance of resistance to verticillium wilt in a tomato cross. Phytopathology 41:986–990

Schauer N, Fernie AR (2006) Plant metabolomics: towards biological function and mechanism. Trends Plant Sci 11:508–516

Schauer N, Semel Y, Roessner U, Gur A, Balbo I, Carrari F, Pleban T, Perez-Melis A, Bruedigam C, Kopka J, Willmitzer L, Zamir D, Fernie AR (2006) Comprehensive metabolic profiling and phenotyping of interspecific introgression lines for tomato improvement. Nat Biotechnol 24:447–454

Schauer N, Zamir D, Fernie AR (2005) Metabolic profiling of leaves and fruit of wild species tomato: a survey of the *Solanum lycopersicoides* complex. J Exp Bot 56:297–307

Schenck PA (1851) The gardener's text book. Boston

Schmelz EA, Engelberth J, Alborn HT, O'Donnell P, Sammons M, Toshima H, Tumlinson III JH (2003) Simultaneous analysis of phytohormones, phytotoxins, and volatile organic compounds in plants. Proc Natl Acad Sci USA 100:10552–10557

Schuh W, Kanczler J, Robertson D, Hobson G, Tucker G, Grierson D, Bright S, Bird C (1991) Fruit quality characteristics of transgenic tomato fruit with altered polygalacturonase activity. HortScience 26:1517–1520

Scippa GS, Di Michele M, Onelli E, Patrignani G, Chiatante D, Bray EA (2004) The histone-like protein H1-S and the response of tomato leaves to water deficit. J Exp Bot 55:99–109

Scippa GS, Griffiths A, Chiantante D, Bray EA (2000) The H1 histone variant of tomato, H1-S, is targeted to the nucleus and accumulates in chromatin in response to water deficit stress. Planta 211:173–181

Scolnik PA, Hinton P, Greenblatt IM, Giuliano G, Delanoy MR, Spector DL, Pollock D (1987) Somatic instability of carotenoid biosynthesis in the tomato *ghost* mutant and its effect on plastid development. Planta 171:1–18

Scott J, Rebeille R, Fletcher J (2000) Folic acid and folates: the feasibility for nutritional enhancement in plant foods. J Sci Food Agric 80:795–824

Scott JW (2005) Perspectives on tomato disease resistance breeding; past, present and future. Acta Hort 695:217–224

Scott JW, Agrama HA, Jones JP (2004) RFLP-based analysis of recombination among resistance genes to fusarium wilt races 1, 2 and 3 in tomato. J Am Soc Hort Sci 129:394–400

Scott JW, Angell FF (1998) Tomato. In: Banga SS, Banga SK (eds) Hybrid cultivar development. Narosa Publ House, New Delhi, India, pp 451–475

Scott JW, Gardner RG (2007) Breeding for resistance to fungal pathogens. In: Razdan MK, Mattoo AK (eds) Genetic improvement of Solanaceous crops, vol 2: tomato. Science Publ, Enfield, NH, USA, pp 421–456

Scott JW, Jones JP (1995) Fla. 7547 and Fla. 7481 tomato breeding lines resistant to *Fusarium oxysporum* f. sp. *lycopersici* races 1, 2, and 3. HortScience 30:645–646

Scott JW, Stevens MR, Olson SM (2005a) An alternative source of resistance to *Tomato spotted wilt virus*. Tomato Genet Coop Rep 55:40–41

Scott JW, Wang JF, Hanson PM (2005b) Breeding tomatoes for resistance to bacterial wilt, a global view. Acta Hort 695:161–172

Seki M, Narusaka M, Abe H, Kasuga M, Yamaguchi-Shinozaki K, Carninci P, Hayashizaki Y, Shinozaki K (2001) Monitoring the expression pattern of 1300 Arabidopsis genes under drought and cold stresses by using a full-length cDNA microarray. Plant Cell 13:61–72

Semel Y, Nissenbaum J, Menda N, Zinder M, Krieger U, Issman N, Pleban T, Lippman Z, Gur A, Zamir D (2006) Overdominant quantitative trait loci for yield and fitness in tomato. Proc Natl Acad Sci USA 103:12981–12986

Sessions A, Burke E, Presting G, Aux G, McElver J, Patton D, Dietrich B, Ho P, Bacwaden J, Ko C, Clarke JD, Cotton D, Bullis D, Snell J, Miguel T, Hutchison D, Kimmerly B, Mitzel T, Katagiri F, Glazebrook J, Law M, Goff SA (2002) A high-throughput Arabidopsis reverse genetics system. Plant Cell 14:2985–2994

Seymour GB, Manning K, Eriksson EM, Popovich AH, King GJ (2002) Genetic identification and genomic organization of factors affecting fruit texture. J Exp Bot 53:2065–2071

Shahhidi F, Wanasundara PK (1992) Phenolic antioxidants. Crit Rev Food Sci Nutr 32:67–103

Shapiro JA, Steffens JC, Mutschler MA (1994) Acylsugars of the wild tomato *Lycopersicon pennellii* in relation to geographic distribution of the species. Biochem Syst Ecol 22:545–561

Sherman J, Stack S (1995) 2-dimensional spreads of synaptonemal complexes from Solanaceous plants. VI. High-resolution recombination nodule map for tomato (*Lycopersicon esculentum*). Genetics 141:683–708

Sherman JD, Stack SM (1992) Two-dimensional spreads of synaptonemal complexes from Solanaceous plants. V. tomato (*Lycopersicon esculentum*) karyotype and idiogram. Genome 35:354–359

Shi MM (2001) Enabling large-scale pharmacogenetic studies by high-throughput mutation detection and genotyping technologies. Clin Chem 47:164–172

Shoemaker PB, Gardner RG (1986) Resistance and fungicide application interval for tomato foliar diseases, 1985. Biol Cult Tests 1:24

Sigareva M, Spivey R, Willits MG, Kramer CM, Chang YF (2004) An efficient mannose selection protocol for tomato that has no adverse effect on the ploidy level of transgenic plants. Plant Cell Rep 23:236–245

Sim S-C, Yang W, van der Knaap E, Hogenhout S, Xiao H, Francis D (2007) Microarray-based SNP discovery for tomato genetics and breeding. In: Plant and Animal Genome XV Conf, San Diego, CA, USA, p 173

Simko I, Costanzo S, Haynes KG, Christ BJ, Jones RW (2004a) Linkage disequilibrium mapping of a *Verticillium dahliae* resistance quantitative trait locus in tetraploid potato (*Solanum tuberosum*) through a candidate gene approach. Theor Appl Genet 108:217–224

Simko I, Haynes KG, Ewing EE, Costanzo S, Christ BJ, Jones RW (2004b) Mapping genes for resistance to *Verticillium albo-atrum* in tetraploid and diploid potato populations using haplotype association tests and genetic linkage analysis. Mol Genet Genom 271:522–531

Simons G, Groenenkijk J, Wijbrandi J, Reijans M, Groenen J, Diergaard P, Van der Lee T, Bleeker M, Onstenk J, de Both M, Hring M, Mes J, Cornelissen B, Zabeau M, Vos P (1998) Dissection of the fusarium *I2*gene cluster in tomato reveals six homlogs and one active gene copy. Plant Cell 10:1055–1068

Singer JB, Hill AE, Burrage LC, Olszens KR, Song J, Justice M, O'Brien WE, Conti DV, Witte JS, Lander ES, Nadeau JH (2004) Genetic dissection of complex traits with chromosome substitution strains of mice. Science 304:445–448

Slade AJ, Fuerstenberg SI, Loeffler D, Steine MN, Facciotti D (2005) A reverse genetic, nontransgenic approach to

wheat crop improvement by TILLING. Nat Biotechnol 23:75–81
Slade AJ, Knauf VC (2005) TILLING moves beyond functional genomics into crop improvement. Transgenic Res 14:109–115
Śmiech M, Rusinowski Z, Malepszy S, Niemirowicz-Szczytt K (2000) New RAPD markers of *Tomato spotted wilt virus* (TSWV) resistance in *Lycopersicon esculentum* Mill. Acta Physiol Plant 22:299–303
Smith AF (1994) The tomato in America: early history, culture, and cookery. The Univ of South Carolina Press, Columbia, SC, pp 224
Smith DL, Gross KC (2000) A family of at least seven β-galactosidase genes is expressed during tomato fruit development. Plant Physiol 123:1173–1183
Smulders MJM, Bredemeijer G, Rus-Kortekaas W, Arens P, Vosman B (1997) Use of short microsatellites from database sequences to generate polymorphisms among *Lycopersicon esculentum* cultivars and accessions of other *Lycopersicon* species. Theor Appl Genet 97:264–272
Snyder JC, Carter CD (1985) Trichomes on leaves of *Lycopersicon hirsutum* and *L. esculentum* and their hybrids. Euphytica 34:53–64
Snyder JC, Johnson DA, Good DE, Weston PA (1987) Type VI trichome exudates from chemotypes of *L. hirsutum* and *L. hirsutum* f. *glabratum*. Tomato Genet Coop Rep 37:67–68
Sobir T, Ohmori M, Murata F, Motoyoshi F (2000) Molecular characterization of the SCAR markers tightly linked to the *Tm-2* locus of the genus *Lycopersicon*. Theor Appl Genet 101:64–69
Soler S, Nuez F (2000) Identificación de fuentes de resistencia en tomate al virus del mosaico del pepino dulce (PepMV). X Congreso De la Sociedad Española de Fitopatología, Valencia, Spain, p 199
Soler S, Prohens J, Díez MJ, Nuez F (2002) Natural occurrence of *Pepino mosaic virus* in *Lycopersicon* species in central and southern Peru. J Phytopathol 150:49–53
Sommer A (1997) Vitamin A deficiency, child health, and survival. Nutrition 13:484–485
Spooner DM, Anderson GJ, Jansen RK (1993) Chloroplast DNA evidence for the interrelationships of tomatoes, potatoes, and pepinos (Solanaceae). Am J Bot 80:676–688
Spooner DM, Peralta IE, Knapp S (2005) Comparisons of AFLPs with other markers for phylogenetic inference in wild tomatoes [*Solanum* L. section *Lycopersicon* (Mill.) Wettst.]. Taxon 54:43–61
Srinivasa Rao NK, Bhatt RM, Sadashiva AT (2000) Tolerance to water stress in tomato cultivars. Photosynthesis 38:465–467
Stamova BS, Chetelat RT (2000) Inheritance and genetic mapping of *Cucumber mosaic virus* resistance introgressed from *Lycopersicon chilense* into tomato. Theor Appl Genet 101:527–537
Stamova L (2004) Indication of dominant resistance to corky root. Proc Tomato Breeders Round Table, Annapolis, MD, USA
Stamova L, Christova D, Yordanov M (1990) Resistance to *Cucumber mosaic virus* (CMV). Rep Tomato Genet Coop 40:33–34
Stern DJ, Buttery RG, Teranishi R, Ling L, Scott K, Cantwell M (1994) Effect of storage and ripening on fresh tomato quality. Food Chem 49:225–231
Stevens MA (1972) Citrate and malate concentrations in tomato fruits: Genetic control and maturational effects. J Am Soc Hort Sci 97:655–658
Stevens MA (1976) Inheritance of viscosity potential in tomato. J Am Soc Hort Sci 101:152–155
Stevens MA (1986) Inheritance of tomato fruit quality components. In: Janick J (ed) Plant breeding reviews, vol 4. AVI Publ, Westport, CT, USA, pp 273–312
Stevens MA, Paulson KN (1973) Phosphorous concentrations in tomato fruits: inheritance and maturity effects. J Am Soc Hort Sci 98:607–610
Stevens MA, Rick CM (1986) Genetics and breeding. In: Atherton JG, Rudich J (eds) The tomato crop: a scientific basis for improvement. Chapman and Hall, London, pp 35–109
Stevens MR (1993) Studies on inheritance of resistance to *Tomato spotted wilt virus* in the genus *Lycopersicon*. Univ of Arkansas, Fayetteville, AK, USA
Stevens MR, Heiny DK, Griffiths PD, Scott JW, Rhoads DD (1996a) Identification of co-dominant RAPD markers tightly linked to the *Tomato spotted wilt virus* (TSWV) resistance gene *Sw-5*. Tomato Genet Coop Rep 46:27–28
Stevens MR, Heiny DK, Rhoads DD, Griffiths PD, Scott JW (1996b) A linkage map of the *Tomato spotted wilt virus* resistance gene *Sw-5* using isogenic lines and an interspecific cross. Acta Hort 431:385–392
Stevens MR, Lamb EM, Rhoads DD (1995) Mapping the *Sw-5* locus for *Tomato spotted wilt virus* resistance in tomatoes using RAPD and RFLP analyses. Theor Appl Genet 90:451–456
Stevens MR, Robbins MD (2007) Molecular markers in selection of tomato germplasm. In: Razdan MK, Mattoo AK (eds) Genetic improvement of Solanaceous crops. vol 2: tomato. Science Publ, Enfield, NH, USA, pp 239–260
Stevens MR, Scott SJ, Gergerich RC (1992) Inheritance of a gene for resistance to *Tomato spotted wilt virus* (TSWV) from *Lycopersicon peruvianum* Mill. Euphytica 59:9–17
Stevens R, Buret M, Duffé P, Garchery C, Baldet P, Rothan C, Causse M (2007) Candidate genes and QTLs affecting fruit ascorbic acid content in three tomato populations. Plant Physiol 143:1943–1953
Stobbs LW, Poysa V, Van Schagen JG (1994) Susceptibility of cultivars of tomato and pepper to a necrotic strain of *Potato virus Y*. Can J Plant Pathol 16:43–48
Stommel JR (1992) Enzymatic components of sucrose accumulation in the wild tomato species *Lycopersicon peruvianum*. Plant Physiol 99:324–328
Stommel JR (2001) USDA 97L63, 97L66, and 97L97: Tomato breeding lines with high fruit beta-carotene content. HortScience 36:387–388

Stommel JR, Sinden SL (1991) Genotypic differences in shoot-forming capacity of cultured leaf explants of *Lycopersicon hirsutum*. HortScience 26:1317–1320

Stommel JR, Abbott JA, Saftner RA (2005a) USDA 02L1058 and 02L1059: Cherry tomato breeding lines with high fruit beta-carotene content. HortScience 40:1569–1570

Stommel JR, Abbott JA, Saftner RA, Camp M (2005b) Sensory and objective quality attributes of beta-carotene- and lycopene-rich tomato fruit. J Am Soc Hort Sci 130:244–251

Stommel JR, Haynes KG (1993) Genetic control of fruit sugar accumulation in a cross of *Lycopersicon esculentum* × *L. hirsutum*. J Am Soc Hort Sci 118:859–863

Stommel JR, Haynes KG (1994) Inheritance of beta carotene content in the wild tomato species *Lycopersicon cheesmanii*. J Hered 85:401–404

Stommel JR, Zhang Y (1998). Molecular markers linked to quantitative trait loci for anthracnose resistance in tomato. HortScience 33:514

Stommel JR, Zhang Y (2001) Inheritance and QTL analysis of anthracnose resistance in the cultivated tomato (*Lycopersicon esculentum*). Acta Hort 542:303–310

Strommer J, Gerats AGM, Sanago M, Molnar SJ, Gerats T (2000) A gene-based RFLP map of petunia. Theor Appl Genet 100:899–905

Strommer J, Peters J, Zethof J, de Keukeleiere P, Gerats T (2002) AFLP maps of *Petunia hybrida*: building maps when markers cluster. Theor Appl Genet 105:1000–1009

Stubbe H (1960) Mutanten der wildtomate *Lycopersicon pimpinellifolium* (Jusl.) Mill. I Die Kulturpflanze 8:110–137

Stubbe H (1963) Mutanten der kulturtomate *Lycopersicon esculentum* Miller. IV Die Kulturpflanze 11:603–644

Stubbe H (1965) Mutanten der wildtomate *Lycopersicon pimpinellifolium* (Jusl.) Mill. III. Die Kulturpflanze 13:517–544

Sun HJ, Uchii S, Watanabe S, Ezura H (2006) A highly efficient transformation protocol for Micro-Tom, a model cultivar for tomato functional genomics. Plant Cell Physiol 47:426–431

Tadmor Y, Fridman E, Gur A, Larkov O, Lastochkin E, Ravid U, Zamir D, Lewinsohn E (2002) Identification of malodorous, a wild species allele affecting tomato aroma that was selected against during domestication. J Agric Food Chem 50:2005–2009

Takahashi H, Hayashi M, Goto F, Sato S, Soga T, Nishioka T, Tomita M, Kawai-Yamada M, Uchimiya H (2006) Evaluation of metabolic alteration in transgenic rice overexpressing dihydroflavonol-4-reductase. Ann Bot (Lond) 98:819–825

Takahashi Y, Shomura A, Sasaki T, Yano M (2001) *Hd6*, a rice quantitative trait locus involved in photoperiod sensitivity, encodes the alpha-subunit of protein kinase CK2. Proc Natl Acad Sci USA 98:7922–7927

Tanksley SD (1983) Molecular markers in plant breeding. Plant Mol Biol Rep 1:3–8

Tanksley SD (1993) Linkage map of tomato (*Lycopersicon esculentum*) (2 N = 24). In: O'Brien S (ed) Genetic maps: locus maps of complex genomes. Cold Spring Harbor Lab Press, Cold Spring Harbor, USA, pp 6.39–36.60

Tanksley SD (1993) Mapping polygenes. Annu Rev Genet 27:205–233

Tanksley SD (2004) The genetic, developmental, and molecular bases of fruit size and shape variation in tomato. Plant Cell 16:S181–S189

Tanksley SD, Bernatzky R, Lapitan NL, Prince JP (1988) Conservation of gene repertoire but not gene order in pepper and tomato. Proc Natl Acad Sci USA 85:6419–6423

Tanksley SD, Costello W (1991) The size of the *L. pennellii* chromosome 7 segment containing the $I-3$ gene in tomato breeding lines measured by RFLP probing. Tomato Genet Coop Rep 41:60

Tanksley SD, Ganal MW, Prince JP, De Vicente MC, Bonierbale MW, Broun P, Fulton TM, Giovannoni JJ, Grandillo S, Martin GB, Messeguer R, Miller JC, Miller L, Paterson AH, Pineda O, Röder MS, Wing RA, Wu W, Young ND (1992) High-density molecular linkage maps of the tomato and potato genomes. Genetics 132:1141–1160

Tanksley SD, Grandillo S, Fulton TM, Zamir D, Eshed T, Pétiard V, Lopez J, Beck-Bunn T (1996) Advanced backcross QTL analysis in a cross between an elite processing line of tomato and its wild relative *L. pimpinnellifolium*. Theor Appl Genet 92:213–224

Tanksley SD, Nelson JC (1996) Advanced backcross QTL analysis: a method for the simultaneous discovery and transfer of valuable QTLs from unadapted germplasm into elite breeding lines. Theor Appl Genet 92:191–203

Tanksley SD, Rick CM (1980) Isozymic gene linkage map of the tomato: applications in genetics and breeding. Theor Appl Genet 57:161–170

Taylor IB (1986) Biosystematics of the tomato. In: Atherton JG, Rudich J (eds) the tomato crop: a scientific basis for improvement. Chapman and Hall, London, pp 1–34

Taylor NE, Greene EA (2003) PARSESNP: A tool for the analysis of nucleotide polymorphisms. Nucl Acids Res 31:3808–3811

Tee ES (1992) Carotenoids and retinoids in human nutrition. Crit Rev Food Sci Nutr 31:103–163

Thabius A, Palloix A, Pflieger S, Daubeze A-M, Caranta C, Lefebvre V (2003) Comparative mapping of *Phytophthora* resistance loci in pepper germplasm: evidence for conserved resistance loci across Solanaceae and for a large genetic diversity. Theor Appl Genet 106:1473–1485

Thakur BR, Singh RK, Handa AK (1996) Effect of an antisense pectin methylesterase gene on the chemistry of pectin tomato (*Lycopersicon esculentum*) juice. J Agric Food Chem 44:628–630

Thimm O, Blasing O, Gibon Y, Nagel A, Meyer S, Kruger P, Selbig J, Muller LA, Rhee SY, Stitt M (2004) MAPMAN: a user-driven tool to display genomics data sets onto diagrams of metabolic pathways and other biological processes. Plant J 37:914–939

Thomas JE, McGrath DJ (1988) Inheritance of resistance to *Potato virus Y* in tomato. Aust J Agric Res 39:475–479

Thomas PE, Mink GI (1998) Tomato hybrids with nonspecific immunity to viral and mycoplasma pathogens of potato and tomato. HortScience 33:764–765

Thompson AE, Tomes ML, Erickson HT, Wann EV, Armstrong RJ (1967) Inheritance of crimson fruit color in tomatoes. Proc Am Soc Hort Sci 91:495–504

Thompson AJ, Corlett JE (1995) mRNA levels of four tomato (*Lycopersicon esculentum* Mill. L.) genes are related to fluctuating plant and soil water status. Plant Cell Environ 18:773–780

Thoquet P, Olivier J, Sperisen C, Rogowsky P, Laterrot H, Nigel G (1996a) Quantitative trait loci determining resistance to bacterial wilt in tomato cultivar Hawaii 7996. Mol Plant-Micr Interact 9:826–836

Thoquet P, Olivier J, Sperisen C, Rogowsky P, Prior P, Anais G, Mangin B, Bazin B, Nazer R, Nigel G (1996b) Polygenic resistance of tomato plants to bacterial wilt in the French West Indies. Mol Plant-Micr Interact 9(9):837–842

Tian F, Li de J, Fu Q, Zhu ZF, Fu YC, Wang XK, Sun CQ (2006) Construction of introgression lines carrying wild rice (*Oryza rufipogon* Griff.) segments in cultivated rice (*Oryza sativa* L.) background and characterization of introgressed segments associated with yield-related traits. Theor Appl Genet 112:570–580

Tieman DM, Harriman RW, Ramamohan G, Handa AK (1992) An antisense pectin methylesterase gene alters pectin chemistry and soluble solids in tomato fruit. Plant Cell 4:667–679

Tieman DM, Zeigler M, Schmelz EA, Taylor MG, Bliss P, Kirst M, Klee HJ (2006) Identification of loci affecting flavour volatile emissions in tomato fruits. J Exp Bot 57:887–896

Tigchelaar EC, Tomes M, Kerr E, Barman R (1973) A new ripening mutant, non-ripening (nor). Tomato Genet Coop Rep 35:20

Tigchelaar EC, Tomes ML (1974) 'Caro-Rich' tomato. HortScience 9:82

Tikunov Y, Lommen A, de Vos CH, Verhoeven HA, Bino RJ, Hall RD, Bovy AG (2005) A novel approach for nontargeted data analysis for metabolomics. Large-scale profiling of tomato fruit volatiles. Plant Physiol 139:1125–1137

Till BJ, Burtner C, Comai L, Henikoff S (2004) Mismatch cleavage by single-strand specific nucleases. Nucl Acids Res 32:2632–2641

Till BJ, Reynolds SH, Greene EA, Codomo CA, Enns LC, Johnson JE, Burtner C, Odden AR, Young K, Taylor NE, Henikoff JG, Comai L, Henikoff S (2003) Large-scale discovery of induced point mutations with high-throughput TILLING. Genome Res 13:524–530

Timms L, Jimenez R, Chase M, Lavelle D, McHale L, Kozik A, Lai Z, Heesacker A, Knapp S, Reiseberg L, Michelmore R, Kesseli R (2006) Analyses of synteny between *Arabidopsis thaliana* and species in the Asteraceae reveal a complex network of small syntenic segments and major chromosomal rearrangements. Genetics 173:2227–2235

Tirajoh A, Aung TST, McKay AB, Plant AL (2005) Stress responsive a-dioxygenase expression in tomato roots. J Exp Bot 56:713–723

Tohge T, Nishiyama Y, Hirai MY, Yano M, Nakajima J, Awazuhara M, Inoue E, Takahashi H, Goodenowe DB, Kitayama M, Noji M, Yamazaki M, Saito K (2005) Functional genomics by integrated analysis of metabolome and transcriptome of Arabidopsis plants over-expressing an MYB transcription factor. Plant J 42:218–235

Tomes ML (1963) Temperature inhibition of carotene synthesis in tomato. Bot Gaz 124:180–185

Tomes ML, Quackenbush FW (1958) Caro-Red, a new provitamin A rich tomato. Econ Bot 12:256–260

Tomes ML, Quackenbush FW, McQuistan M (1954) Modification and dominance of the gene governing formation of high concentrations of beta-carotene in the tomato. Genetics 39:810–817

Torres-Schumann S, Godoy JA, Pintor-Toro JA (1992) A probable lipid transfer protein gene is induced by NaCl in stems of tomato plants. Plant Mol Biol 18:749–757

Tournefort JPD (1694) Elemens de Botanique. l'Imprimerie royale, Paris, France

Treviño MB (1997) Drought-induced gene expression in plants: analysis of the non-specific lipid transfer protein gene family in the drought-tolerant tomato species *Lycopersicon pennellii*. PhD Dis, New Mexico State Univ, USA

Treviño MB, O'Connell MA (1998) Three drought-responsive members of the nonspecific lipid-transfer protein gene family in *Lycopersicon pennellii* show different developmental patterns of expression. Plant Physiol 116:1461–1468

Triano SR, St Clair DA (1995) Processing tomato germplasm with improved fruit soluble solids content. HortScience 30:1477–1478

Tsugane T, Watanabe M, Yano K, Sakuari N, Suzuki H, Shibata D (2005) Expressed sequence tags of full-length cDNA clones from the miniature tomato (*Lycopersicon esculentum*) cultivar Micro-Tom. Plant Biotechnol J 22:161–165

Turkensteen LJ (1973) Partial resistance of tomato against *Phytophthora infestans*, the late blight fungus. Center Agric Publ Docum, Wageningen, The Netherlands

Twell D, Wing R, Yamaguchi J, McCormick S (1989) Isolation and expression of an anther-specific gene from tomato. Mol Gen Genet 217:240–245

Uddin MR, Berry SZ, Bisges AD (1988) An improved shoot regeneration system for somaclone production in tomatoes. HortScience 23:1062–1064

Unlu Z, Clinton SK, Schwartz SJ (2003) Absorption of lycopene isomers following single meals containing tomato sauces with varying isomer patterns. FASEB J 17: Abstr No 456.9

Urbanczyk-Wochniak E, Luedemann A, Kopka J, Selbig J, Roessner-Tunali U, Willmitzer L, Fernie AR (2003) Parallel analysis of transcript and metabolic profiles: a new approach in systems biology. EMBO Rep 4:989–993

Urbanczyk-Wochniak E, Usadel B, Thimm O, Nunes-Nesi A, Carrari F, Davy M, Blasing O, Kowalczyk M, Weicht D, Polinceusz A, Meyer S, Stitt M, Fernie AR (2006) Conversion of MapMan to allow the analysis of transcript data from Solanaceous species: effects of genetic and environmental alterations in energy metabolism in the leaf. Plant Mol Biol 60:773–792

USDA (2002) Nutritive value of foods. Home and Garden Bulletin 72

USDA-ERS (2003) U.S. Tomato Statistics, 1960–2002. USDA-Economic Research Service. http://usda.mannlib.cornell.edu/MannUsda/viewDocumentInfo.do?documentID=1210

USDA-ERS (2006) Tomatoes at a glance. USDA-Economic Research Service. http://www.ers.usda.gov/Briefing/Tomatoes/

USDA-FAS (2005) The U.S. and world situation: fresh and processed tomatoes. USDA Foreign Agricultural Service, Horticultural & Tropical Products Division. http://www.fas.usda.gov/htp

Vakalounakis DJ (1988) The genetic analysis of resistance to fusarium crown and root rot of tomato. Plant Pathol 37:71–73

Vakalounakis DJ, Laterrot H, Moretti A, Ligoxigakis K, Smardas K (1997) Linkage between Frl (*Fusarium oxysporum* f. sp *radicis-lycopersici* resistance) and Tm-2 (tobacco mosaic virus resistance-2) loci in tomato (*Lycopersicon esculentum*). Ann Appl Biol 130:319–323

Vallejos C, Tanksley S, Bernatzky R (1986) Localization in the tomato genome of DNA restriction fragments containing sequences homologous to the ribosomal-RNA (45-S), the major chlorophyll-A/B binding polypeptide and the ribulose bisphosphate carboxylase genes. Genetics 112:93–105

Van der Beek JG, Pet G, Lindhout P (1994) Resistance to powdery mildew (*Oidium lycopersicon*) in *Lycopersicon hirsutum* is controlled by an incompletely-dominant gene *Ol-1* on chromosome-6. Theor Appl Genet 89:467–473

van der Biezen EA, Glagotskaya T, Overduin B, Nijkamp HJJ, Hille J (1995) Inheritance and genetic mapping of resistance to *Alternaria alternata* f. sp. *lycopersici* from *Lycopersicon pennellii*. Mol Gen Genet 247:453–461

Van der Hoeven R, Ronning C, Giovannoni J, Martin G, Tanksley S (2002) Deductions about the number, organization and evolution of genes in the tomato genome based on analysis of a large expressed sequence tag collection and selective genomic sequencing. Plant Cell 14:1441–1456

Van der Hoeven RS, Monforte AJ, Breeden D, Tanksley SD, Steffens JC (2000) Genetic control and evolution of sesquiterpene biosynthesis in *Lycopersicon esculentum* and *L. hirsutum*. Plant Cell 12:2283–2294

van der Knaap E, Lippman ZB, Tanksley SD (2002) Extremely elongated tomato fruit controlled by four quantitative trait loci with epistatic interactions. Theor Appl Genet 104 (2–3):241–247

van der Knaap E, Tanksley SD (2003) The making of a bell pepper-shaped tomato fruit: identification of loci controlling fruit morphology in Yellow Stuffer tomato. Theor Appl Genet 107:139–147

van der Vlugt RAA, Stijger CCMM, Verhoeven JTJ, Lesemann D-E (2000) First report of *Pepino mosaic virus* on tomato. Plant Dis 84:103

Van Deynze A, van der Knaap E, Francis D (2006) Development and application of an informative set of anchored markers for tomato breeding. In: Plant and Animal Genome XIV Conf, San Diego, CA, USA, P188

Van Eck J, Kirk DD, Walmsley AM (2006) Tomato (*Lycopersicon esculentum*). In: Wang K (ed) Methods in molecular biology: *Agrobacterium* protocols, vol 343. Humana Press, Totowa, NJ, USA, pp 459–473

van Heusden AW, Koornneef M, Voorrips RE, Brüggemann W, Pet G, Vrielink-van Ginkel R, Chen X, Lindhout P (1999) Three QTLs from *Lycopersicon peruvianum* confer a high level of resistance to *Clavibacter michiganensis* ssp. michiganensis. Theor Appl Genet 99:1068–1074

Van Ooijen J, Sandbrink J, Vrielink M, Verkerk R, Zabel P, Lindhout P (1994) An RFLP linkage map of *Lycopersicon peruvianum*. Theor Appl Genet 89:1007–1013

van Os H, Andrzejewski S, Bakker E, Barrena I, Bryan GJ, Caromel B, Ghareeb B, Isidore E, de Jong W, van Koert P, Lefebvre V, Milbourne D, Ritter E, van der Voort JN, Rousselle-Bourgeois F, van Vliet J, Waugh R, Visser RG, Bakker J, van Eck HJ (2006) Construction of a 10,000-marker ultradense genetic recombination map of potato: providing a framework for accelerated gene isolation and a genome-wide physical map. Genetics 173:1075–1087

van Roekel JSC, Damm B, Melchers LS, Hoekma A (1993) Factors influencing transformation frequency of tomato (*Lycopersicon esculentum*). Plant Cell Rep 12:644–647

van Tuinen A, Cordonnier-Pratt M, Pratt LH, Verkerk R, Zabel P, Koornneef M (1997) The mapping of phytochrome genes and photomorphogenic mutants of tomato. Theor Appl Genet 94:115–122

van Zijl JJB, Bosch SE, Coetzee CPJ (1986) Breeding tomatoes for processing in South Africa. Acta Hort 194:69–75

Villalta I, Reina-Sanchez A, Cuartero J, Carbonell EA, Asins MJ (2005) Comparative microsatellite linkage analysis and genetic structure of two populations of F6 lines derived from *Lycopersicon pimpinellifolium* and *L. cheesmanii*. Theor Appl Genet 110:881–894

Vinson JA, Hao Y, Su X, Zubik L (1998) Phenol antioxidant quantity and quality in foods: Vegetables. J Agric Food Chem 46:3630–3634

von Korff M, Wang H, Léon J, Pillen K (2004) Development of candidate introgression lines using an exotic barley accession (*Hordeum vulgare* ssp. *spontaneum*) as donor. Theor Appl Genet 109:1736–1745

Vos P, Simons G, Jesse T, Wijbrandi J, Heinen L, Hogers R, Frijters A, Groenendijk J, Diergaarde P, Reijans M, Fierens-

Onstenk J, de Both M, Peleman J, Liharska T, Hontelez J, Zabeau M (1998) The tomato *Mi-1* gene confers resistance to both root-knot nematodes and potato aphids. Nat Biotechnol 16:1365–1369

Vrebalov J, Ruezinsky D, Padmanabhan V, White R, Medrano D, Drake R, Schuch W, Giovannoni J (2002) A MADS-box gene necessary for fruit ripening at the tomato ripening-inhibitor (*Rin*) locus. Science 296:343–346

Walley C, Seymour G (2006) Investigating the polygenic nature of texture traits in tomato fruit. VI Intl Solanaceae Conf, Univ of Wisconsin, Madison, WI, USA

Walmsley AM, Alvarez ML, Jin Y, Kirk DD, Lee SM, Pinkhasov J, Rigano MM, Arntzen CJ, Mason HS (2003) Expression of the B subunit of *Escherichia coli* heat-labile enterotoxin as a fusion protein in transgenic tomato. Plant Cell Rep 21:1020–1026

Wang CL, Xing JS, Chin CK, Ho CT, Martin CE (2001) Modification of fatty acids changes the flavor volatiles in tomato leaves. Phytochemistry 58:227–232

Wang J-F, Olivier J, Thoquet P, Mangin B, Sauviac L, Grimsley NH (2000) Resistance of tomato line Hawaii 7996 to *Ralstonia solanacearum* Pss4 in Taiwan is controlled mainly by a major strain-specific locus. Mol Plant-Micr Interact 13:6–13

Wang Y, Tang X, Cheng Z, Mueller L, Giovannoni J, Tanksley SD (2006) Euchromatin and pericentromeric heterochromatin: Comparative composition in the tomato genome. Genetics 172:2529–2540

Warnock SJ (1988) A review of taxonomy and phylogeny of the genus *Lycopersicon*. HortScience 23:669–673

Watson JM, Fusaro AF, Wang M, Waterhouse PM (2005) RNA silencing platforms in plants. FEBS Lett 579:5982–5987

Watterson JC (1986) Diseases. In: Atherton JG, Rudich J (eds) The tomato crop: a scientific basis for improvement. Chapman and Hall, London, UK, pp 35–109

Wayne ML, McIntyre LM (2002) Combining mapping and arraying: an approach to candidate gene identification. Proc Natl Acad Sci USA 99:14903–14906

Wei T, O'Connell MA (1996) Structure and characterization of a putative drought-inducible H1 histone gene. Plant Mol Biol 30:255–268

Weide R, van Wordragen MF, Lankhorst RK, Verkerk R, Hanhart C, Liharska T, Pap E, Stam P, Zabel P, Koornneef M (1993) Integration of the classical and molecular linkage maps of tomato chromosome 6. Genetics 135:1175–1186

Weimer J (1999) Chapter 19: Accelerating the trend toward healthy eating: Public and private efforts. In: Frazao E (ed) America's eating habits: changes and consequences. Agric Info Bull 750, USDA/ERS, pp 385–401

Weston PA, Johnson DA, Burton HT, Snyder JC (1989) Trichome secretion composition, trichome densities, and spider mite resistance of ten accessions of *Lycopersicon hirsutum*. J Am Soc Hort Sci 114:492–498

Weston PA, Snyder JC (1990) Thumbtack bioassay: a quick method of measuring plant resistance to two-spotted spider mites (Acari: Tetranychidae). J Econ Entomol 83:501–504

Wheeler D, Church D, Edgar R, Federhen S, Helmberg W, Madden T, Pontius J, Schuler G, Schriml L, Sequeira E, Suzek T, Tatusova T, Wagner L (2004) Database resources of the National Center for Biotechnology Information: update. Nucl Acids Res 32:D35–D40

Wheeler DL, Barrett T, Benson DA, Bryant SH, Canese K, Chetvernin V, Church DM, DiCuccio M, Edgar R, Federhen S, Geer LY, Helmberg W, Kapustin Y, Kenton DL, Khovayko O, Lipman DJ, Madden TL, Maglott DR, Ostell J, Pruitt KD, Schuler GD, Schriml LM, Sequeira E, Sherry ST, Sirotkin K, Souvorov A, Starchenko G, Suzek TO, Tatusov R, Tatusova TA, Wagner L, Yaschenko E (2006) Database resources of the National Center for Biotechnology Information. Nucl Acids Res 34:D173–D180

Wheeler GL, Jones MA, Smirnoff N (1998) The biosynthetic pathway of vitamin C in higher plants. Nature 393:365–369

Whitaker BD, Smith DL, Green KC (2001) Cloning, characterization and functional expression of a phospholipase Dα cDNA from tomato fruit. Physiol Plant 112:87–94

Whitfield AE, Ullman DE, German TL (2005) Tospovirus-thrips interactions. Annu Rev Phytopathol 43:459–489

Wienholds E, Schulte-Merker S, Walderich B, Plasterk RH (2002) Target-selected inactivation of the zebrafish rag1 gene. Science 297:99–102

Wikström N, Savolainen V, Chase MW (2001) Evolution of the angiosperms: calibrating the family tree. Proc Roy Soc Lond Ser B Biol Sci 268:2211–2220

Wilkinson JQ, Lanahan MB, Yen HC, Giovannoni JJ, Klee HJ (1995) An ethylene-inducible component of signal-transduction encoded by Never-Ripe. Science 270:1807–1809

Williams WG, Kennedy GG, Yamamoto ET, Thacker JD, Borner J (1980) 2-Tridecanone - a naturally occurring insecticide from the wild tomato *Lycopersicon hirsutum* f. *glabratum*. Science 207:888–889

Willits MG, Kramer CM, Prata RTN, De Luca V, Potter BG, Steffens JC, Graser G (2005) Utilization of the genetic resources of wild species to create a nontransgenic high flavonoid tomato. J Agric Food Chem 53:1231–1236

Witsenboer HMA, van de Griend EG, Tiersma JB, Nijkamp HJJ, Hille J (1989) Tomato resistance to alternaria stem canker: localization in host genotypes and functional expression compared to non-host resistance. Theor Appl Genet 78:457–462

World Health Organization (2005) Vitamin A. 6 May, 2005. http://www.who.int/vaccines/en/vitaminamain.shtml

Wu F, Mueller LA, Crouzillat D, Petiard V, Tanksley SD (2006) Combining bioinformatics and phylogenetics to identify large sets of single copy, orthologous genes (COSII) for comparative, evolutionnary and systematics studies: a test case in the Euasterid plant clade. Genetics 174:1407–1420

Wu JL, Wu C, Lei C, Baraoidan M, Bordeos A, Madamba MR, Ramos-Pamplona M, Mauleon R, Portugal A, Ulat VJ,

Bruskiewich R, Wang G, Leach J, Khush G, Leung H (2005) Chemical- and irradiation-induced mutants of indica rice IR64 for forward and reverse genetics. Plant Mol Biol 59:85–97

Xiaong AS, Yao QH, Peng R, Li X, Han PL, Fan HQ (2005) Different effects on ACC oxidase gene silencing triggered by RNA interference in transgenic tomato. Plant Cell Rep 23:639–646

Xu P, Rogers SJ, Roossinck MJ (2004) Expression of anti-apoptotic genes *bcl-xL* and *ced-9* in tomato enhances tolerance to viral-induced necrosis and abiotic stress. Proc Natl Acad Sci USA 101:15805–15810

Yakovleff E, Herrera FL (1935) El mundo vegetal de los antiguos peruanos. Revista del Museo Nacional 3. Botánica etnológica, tomo 3, n°3:55. Lima

Yamamoto N, Tsugane T, Watanabe M, Yano K, Maeda F, Kuwata C, Torki M, Ban Y, Nishimura S, Shibata D (2005) Expressed sequence tags from the laboratory-grown miniature tomato (Lycopersicon esculentum) cultivar Micro-Tom and mining for single nucleotide polymorphisms and insertions/deletions in tomato cultivars. Gene 356:127–134

Yang W, Bai X, Kabelka E, Eaton C, Kamoun S, van der Knaap E, Francis D (2004) Discovery of single nucleotide polymorphisms in *Lycopersicon esculentum* by computer aided analysis of expressed sequence tags. Mol Breed 14:21–34

Yang W, Miller SA, Francis DM, Scott JW, Jones JB (2005a) Mining tomato genome sequence databases for molecular markers: application to bacterial resistance and marker assisted selection. Acta Hort 695:241–249

Yang W, Sacks EJ, Lewis Ivey ML, Miller SA, Francis DM (2005b) Resistance in *Lycopersicon esculentum* intraspecific crosses to race T1 strains of *Xanthomonas campestris* pv. *vesicatoria* causing bacterial spot of tomato. Phytopathology 95:519–527

Yano K, Watanabe M, Yamamoto N, Tsugane T, Aoki K, Sakurai N, Shibata D (2006) MiBASE: a database of a miniature tomato cultivar Micro-Tom. Plant Biotechnol 23:195–198

Yano M, Katayose Y, Ashikari M, Yamanouchi U, Monna L, Fuse T, Baba T, Yamamoto K, Umehara Y, Nagamura Y, Sasaki T (2000). *Hdl*, a major photoperiod sensitivity quantitative trait locus in rice, is closely related to the *Arabidopsis* flowering time gene CONSTANS. Plant Cell 12:2473–2484

Yates HE, Frary A, Doganlar S, Frampton A, Eannetta NT, Uhlig J, Tanksley SD (2004) Comparative fine mapping of fruit quality QTLs on chromosome 4 introgressions derived from two wild species. Euphytica 135:283–296

Yelle S, Chetelat RT, Dorais M, DeVerna JW, Bennett AB (1991) Sink metabolism in tomato fruit. IV. Genetic and biochemical analysis of sucrose accumulation. Plant Physiol 95:1026–1035

Yen HC, Shelton BA, Howard LR, Lee S, Vrebalov J, Giovannoni J (1997) The tomato high-pigment (*hp*) locus maps to chromosome 2 and influences plastome copy number and fruit quality. Theor Appl Genet 95:1069–1079

Yesbergenova Z, Yang G, Oron E, Soffier D, Fluhr R, Sagi M (2005) The plant Mo-hydroxylases aldehyde oxidase and anthine dehydrogenase have distinct reactive oxygen species signatures and are induced by drought and abscisic acid. Plant J 42:862–876

Yi G, Lee IM, Lee S, Choi D, Kim B-D (2006) Exploitation of pepper EST–SSRs and an SSR-based linkage map. Theor Appl Genet 114:113–130

Yordanov M, Stamova L, Stoyanova Z (1975) *Leveillula taurica* resistance in the tomato. Rep Tomato Genet Coop 25:24

Young ND (1999) A cautiously optimistic vision for marker-assisted breeding. Mol Breed 5:505–510

Young ND, Miller JC, Tanksley SD (1987) Rapid chromosomal assignment of multiple genomic clones in tomato using primary trisomics. Nucl Acids Res 15:9339–9348

Young ND, Zamir D, Ganal MW, Tanksley SD (1988) Use of isogenic lines and simultaneous probing to identify DNA markers tightly linked to the *Tm-2^a* gene in tomato. Genetics 120:579–585

Young PA (1956) *ry*, A modifier gene for red color in yellow tomato fruit. Tomato Genet Coop Rep 6:33

Yousef GG, Juvik JA (2001) Evaluation of breeding utility of a chromosomal segment from *Lycopersicon chmielewskii* that enhances cultivated tomato soluble solids. Theor Appl Genet 103:1022–1027

Yu L-X, Chamberland H, Lafontaine JG, Tabaeizadeh Z (1996) Negative regulation of gene expression of a novel proline-, threonine-, and glycine-rich protein by water stress in *Lycopersicon chilense*. Genome 39:1185–1193

Yu ZH, Wang JF, Stallt RE, Vallejos CE (1995) Genomic localization of tomato genes that control a hypersensitive reaction to *Xanthomonas campestris* pv. *vesicatoria* (Doidge) Dye. Genetics 141:675–682

Zamir D (2001) Improving plant breeding with exotic genetic libraries. Nat Rev Genet 2:983–989

Zamir D, Eshed Y (1998) Tomato genetics and breeding using nearly isogenic introgression lines derived from wild species. In: Paterson AH (ed) Molecular dissection of complex traits. CRC Press, NY, USA, pp 207–217

Zamir D, Ekstein-Michelson I, Zakay Y, Navot N, Zeidan M, Sarfatti M, Eshed Y, Harel E, Pleban T, van-Oss H, Kedar N, Rabinowitch HD, Czosnek H (1994) Mapping and introgression of a *Tomato yellow leaf curl virus* tolerance gene, *Ty-1*. Theor Appl Genet 88:141–146

Zeng ZB (1993) Theoretical basis for separation of multiple linked gene effects in mapping quantitative trait loci. Proc Natl Acad Sci USA 90:10972–10976

Zhang GF, Maudens KE, Storozhenko S, Mortier KA, Van Det Straeten D, Lambert WE (2003) Determination of total folate in plant material by chemical conversion into *para*-aminobenzoic acid followed by high performance liquid chromatography combined with on-line postcolumn derivatization and fluorescence detection. J Agric Food Chem 51:7872–7878

Zhang L, Lin GY, Nino-Liu DO, Foolad MR (2003) Mapping QTLs conferring early blight (*Alternaria solani*) resistance in a *Lycopersicon esculentum* × *L. hirsutum* cross by selective genotyping. Mol Breed 12:3–19

Zhang LP, Khan A, Nino-Liu D, Foolad MR (2002) A molecular linkage map of tomato displaying chromosomal locations of resistance gene analogs based on a *Lycopersicon esculentum* × *Lycopersicon hirsutum* cross. Genome 45:133–146

Zhang Y, Stommel JR (2000) RAPD and AFLP tagging and mapping of *Beta* (*B*) and *Beta* modifier (Mo_B), two genes which influence β-carotene accumulation in fruit of tomato (*Lycopersicon esculentum* Mill.). Theor Appl Genet 100:368–375

Zhong X, Fransz P, van Wennekes E, Zabel P, van Kammen A, de Jong J (1996) High-resolution mapping on pachytene chromosomes and extended DNA fibres by fluorescence in-situ hybridisation. Plant Mol Biol Rep 14:232–242

Ziegler RG (1989) A review of epidemiological evidence that carotenoids reduce the risk of cancer. J Nutr 119:116–122

Zitter TA, Jones JB, Jones JP, Stall RE (1991) Diseases caused by viruses. APS Press, St. Paul, MN, USA

Zou L, Li H, Ouyang B, Zhang J, Ye Z (2006) Cloning and mapping of genes involved in tomato ascorbic acid biosynthesis and metabolism. Plant Sci 170:120–127

Zscheile FP, Lesley JW (1967) Pigment analysis of sherry: Flesh color mutation resembling yellow in the tomato. J Hered 58:193–194

Zygier S, Ben Chaim A, Efrati A, Kaluzky G, Borovsky Y, Paran I (2005) QTL mapping for fruit size and shape in chromosomes 2 and 4 in pepper and a comparison of the pepper QTL map with that of tomato. Theor Appl Genet 111:437–444

2 Lettuce

David W. Still

Department of Plant Sciences & Technology, California State Polytechnic University, 3801 West Temple Avenue, Pomona, CA 91768, USA
e-mail: dwstill@csupomona.edu

2.1 Introduction

Lettuce (*Lactuca sativa* L.) is a member of the family Asteraceae (Compositae), a successful and diverse group of plants with a global distribution (Funk et al. 2005). The Asteraceae are thought to be the largest family of plants comprising some 23,000 to 30,000 species (Bremer 1994; Bayer and Starr 1998; Funk et al. 2005). Although the phylogeny of the dandelion tribe Cichorieae (Lactuceae) has been problematic (Funk et al. 2005), *Lactuca* species have been placed in subfamily Cichorioideae, tribe Cichorieae (Lactuceae) by Bentham (1873), Bremer (1994), and Cronquist (1955) on the basis of morphometric characteristics. More recent molecular analyses have supported this position and have helped clarify tribal placements (Bayer and Starr 1998; Funk et al. 2004, 2005).

Cultivated lettuce is believed to have been domesticated in the Mediterranean region from the wild species *Lactuca serriola* L. (Durst 1930; Harlan 1992) with the center of origin apparently southwest Asia (De Vries 1997). Lettuce is known to have been cultivated at least 4,500 years BP as long-leaved Cos-type lettuce was depicted on the walls of Egyptian tombs (Lindqvist 1960). Formally, cultivated lettuce is grouped into seven different types: Cos (a.k.a. Romaine), Cutting (a.k.a. Leaf), Stalk (or Asparagus), Butterhead, Crisphead (a.k.a. Iceberg or Cabbage), Latin, and Oilseed (Trehane 1995; De Vries 1997). All groups, except the Oilseed group, are selections within *Lactuca sativa*, while the Oilseed group are derived from either *L. serriola* or *L. sativa* or may be a hybrid between these two taxa. The growth habit, leaf texture, shape and color differ significantly among groups (images typical of the groups may be viewed at: http://www.cgn.wageningen-ur.nl/pgr/research/lettuce/cgncollection.htm).

The Cos group, named after the Greek island, is typified by plants with oblong upright rigid leaves and a prominent midrib with dark green leaves (De Vries 1997). According to Ryder (1986) this is the most common type in the Mediterranean region. The Cutting group forms no firm heads but instead produces a dense mass of leaves in the center of the plant. The leaf may vary from smooth margins to deeply lobed, and may include frilled, curled, or fringed leaves. This group is usually referred to as either greenleaf or redleaf based on leaf color, which ranges from green to yellowish-green and includes various shades of red depending on anthocyanin content and light intensity during growth. The Stalk lettuce group does not form heads and has prominent thickened fleshy stems and upright ovate leaves; both the leaves and stem may be eaten. The Butterhead group forms a small head of nearly spherical (orbicular) leaves surrounded by wrapper leaves. The leaf texture is distinct; leaf color varies from green to yellowish-green and cultivars with anthocyanin have been developed. The Butterhead group is widely used in Europe and was second in commercial importance in the early twentieth century in the US (Durst 1930). The Crisphead group forms tight, dense heads which are comprised of spherical leaves folded upon each other. The leaf is crispy and the veins are prominent. Leaf color varies from deep green to light green, with some genotypes containing anthocyanin. Crisphead lettuce is the most popular lettuce type in the US, although the market share of Cos and Cutting types have markedly increased in recent years (USDA 2005a). The Latin group forms a poorly organized rosette that is similar in appearance to the Butterhead group. The orbicular leaves of the Latin group are thick, have entire margins, and are green. This lettuce type is cultivated in the Mediterranean region including North Africa and in South America (see De Vries 1997). The Oilseed group is typified by a multistemmed upright growth habit with green leaves that are oblong to oblanceo-

Genome Mapping and Molecular Breeding in Plants, Volume 5
Vegetables
C. Kole (Ed.)

late. This group is cultivated for the high oil content of the seed.

2.1.1 Economic Importance

Lettuce is the second most valuable vegetable produced in the United States with farm-market receipts of over \$1.98 billion in 2005 (USDA 2005a). About 75% of all lettuce produced in the US is grown in California, and combined with Arizona these two states account for 96% of the total lettuce production (USDA 2005a). China produces almost one-half of the world's lettuce and at more than 11 million Mt their production is more than double that of the US (Table 1). Since 1980 worldwide lettuce production has increased ~2.7-fold with 22.4 million Mt produced in 2005 (FAO 2006). The increased production was primarily due to an increase in acreage planted as opposed to an increase in yield. From 1980 to 2005, lettuce yield (Kg/ha) increased by 118% compared to an increase of 149%, 135%, 146% and 128% for maize, potato, paddy rice and tomato, respectively (data not presented, FAO 2006). This might reflect the greater genetic resources developed for these crops or alternatively, the greater yield gain of these other species might be an effect of selecting for the harvested organ (seed, fruit or tuber) whereas the vegetative growth in lettuce cannot be further increased. In terms of biomass harvested per hectare, of all major agronomic and vegetable crops, only the tomato has a greater yield than lettuce (FAO 2006). If vegetative growth has been maximized in lettuce, any increase in yield can only be derived from improving disease and insect resistance.

Table 1. Top ten lettuce producing countries and world lettuce production for 2005 (FAO 2006)

Country	Metric tons	Ha planted
China	11,005,000	500,250
USA	4,976,880	131,280
Spain	920,000	39,000
Italy	846,800	43,604
India	790,000	120,000
Japan	530,000	22,000
France	526,000	16,500
Turkey	375,000	19,700
Mexico	243,406	11,290
Germany	200,000	8,200
World	22,382,250	1,015,159

2.1.2 Nutritional Value

The nutritional content of lettuce varies considerably with type (Table 2; Mou 2005; USDA 2005b). In general, Crisphead types are depauperate of nutritional value whereas the Cos and Cutting types have appreciable amounts of ascorbic acid, vitamin A, vitamin K, folate, and the carotenoids β-carotene and lutein+zeaxanthin (Table 2). The Cutting types are especially enriched with vitamin A and β-carotene, having about 15-fold higher amounts than Crisphead types. Cos types are relatively rich in vitamin C and folate compared to the other lettuce types.

Whereas the role of vitamins and minerals in human health are clearly established, other natural compounds such as flavonoids, tocopherols, and carotenoids, etc., are linked to human health only through epidemiological studies. Establishing a clear link between naturally occurring compounds and human health is an active area of research and efforts are underway in lettuce to increase the concentrations of these complex phytochemicals. Although lettuce is not the single richest source for many nutrients, it has the advantage that it is almost universally eaten raw thus preserving nutritional value that would otherwise be lost upon cooking.

Significant genetic variation in nutritional and phytochemical content exists within lettuce types. Mou (2005) screened 52 genotypes of Crisphead, Cutting, Cos, Butterhead, Latin, Stalk and the closely related taxa of *L. serriola* L., *L. saligna* L. and *L. virosa* L. for β-carotene and lutein content. In Crisphead lettuce types β-carotene concentration varied over fivefold (μ = 328 μg/100 g FW) while Butterhead types varied over fourfold, Cos types by twofold and Cutting type (greenleaf) by approximately twofold. Similar ranges in concentration within lettuce types were observed in lutein concentrations and a high correlation between β-carotene and lutein content was noted (Mou 2005).

Mainly due to their antioxidant activity carotenoids have been linked to human health, thus providing the impetus for increasing the carotenoid content in lettuce and other food groups (for review see Cooper 2004). Because per capita consumption of lettuce is high, lettuce may be a significant source of dietary antioxidants. Carotenoids are tetraterpene molecules (C_{40}) synthesized in the nonmevalonate

Table 2. Nutritional content of Cos, Butterhead, Cutting (green and red), Crisphead and Stalk lettuce types based on 100 g fresh weight (USDA 2005b)

Nutrient and units	Lettuce type Butterhead	Cos	Crisphead	Cutting[a]	Cutting[b]	Stalk
Protein (g)	1.35	1.23	0.90	1.36	1.33	0.85
Sugars, total (g)	0.94	1.19	1.76	0.78	0.48	–
Lipids:						
Fatty acids, total saturated (mg)	29	39	18	20	–	–
Fatty acids, total monounsaturated (mg)	8	12	6	6	–	–
Fatty acids, total polyunsaturated (mg)	117	167	74	82	–	–
Phytosterols (mg)	–	–	10	38	–	11
Minerals:						
Calcium (mg)	35	33	18	36	33	39
Iron (mg)	1.24	0.97	0.41	0.86	1.20	0.55
Magnesium (mg)	13	14	7	13	12	28
Phosphorous (mg)	33	30	20	29	28	39
Potassium (mg)	238	247	141	194	187	330
Sodium (mg)	5	8	10	28	25	11
Zinc (mg)	0.20	0.23	0.15	0.18	0.20	0.27
Copper (mg)	0.016	0.048	0.025	0.029	0.028	0.040
Manganese (mg)	0.179	0.155	0.125	0.250	0.203	0.688
Selenium (mcg)	0.6	0.4	0.1	0.6	1.5	0.9
Vitamins:						
Ascorbic acid (mg)	3.7	24.0	2.8	18	3.7	19.5
Thiamin (mg)	0.057	0.072	0.041	0.070	0.064	0.055
Riboflavin (mg)	0.062	0.067	0.025	0.080	0.077	0.070
Niacin (mg)	0.357	0.313	0.123	0.375	0.321	0.550
Pantothenic acid (mg)	0.150	0.142	0.091	0.134	0.144	0.183
Vitamin B-6 (mg)	0.082	0.074	0.042	0.090	0.100	0.050
Folate, total (mcg)	73	136	29	38	36	46
Vitamin A (IU)	3312	5807	502	7405	7492	3500
Vitamin E (mg)	0.18	0.13	0.18	0.29	0.15	–
Tocopherol, gamma (mg)	0.27	0.36	0.09	0.37	0.24	–
Vitamin K (phylloquinone) (mg)	102.3	102.5	24.1	173.6	140.3	–
Carotenoids:						
Carotene, beta (mcg)	1987	3484	299	4443	4495	–
Lutein + zeaxanthin (mcg)	1223	2312	277	1730	1724	–

[a] Greenleaf
[b] Redleaf

pathway that serve as minor accessory light-harvesting pigments absorbing and transferring light energy to chlorophyll molecules (Malkin and Niyogi 2000). They also play a major role in protecting the photosynthetic apparatus from photo-oxidative damage by preventing oxidation by singlet oxygen. Although not experimentally validated, lettuce plants bred for higher carotenoid concentrations may have a higher tolerance for environmental stress, which could be particularly beneficial in the high light environments of the low desert production areas of the US. Because carotenoids function to ameliorate oxidative stress under high light conditions, their concentration might be expected to be higher in plants grown under high light conditions and this has been observed. Carotenoid concentration was higher in lettuce plants grown in fields versus glasshouses, and in those maturing in the summer versus fall (Kimura and Rodgriguez-Amaya 2003). Because carotenoid synthesis is light responsive, it may not be possible to increase carotenoid concentration in Crisphead lettuce types via breeding because the tight compact head must be retained; instead, the carotenoid pathway may have to be stimulated via a constitutive promoter.

Other antioxidants, such as flavonols, are found in appreciable amounts in lettuce (Hohl et al. 2001). In Butterhead types, although present in the inner leaves, the highest concentrations of phenolic compounds (i.e., quercetin, chlorogenic acid, and chicoric acid) are found in the outer leaves. Concentrations of phenolic compounds increased in inner leaves by 10- to 20-fold once illuminated (Hohl et al. 2001). Fourfold differences in quercetin concentrations were observed among cultivars within Butterhead types (Hohl et al. 2001) and between lettuce types (DuPont et al. 2000). Although other fruits and vegetables may have higher concentrations of phenolic compounds, the peroxyl radical scavenging activity of lettuce (especially redleaf types) is higher when expressed on the basis of total phenolic concentration (Caldwell 2003).

2.1.3 Breeding Objectives

Similar to most crops, much of the breeding effort in lettuce is directed toward disease resistance. In virtually every region in which lettuce is grown the obligate biotrophic fungus downy mildew (*Bremia lactucae* Regal) is present and is considered the most significant disease affecting production (Ryder 1999). The inheritance of resistance to *Bremia lactucae* was first reported by Jagger and Whitaker (1940), and Crute and Johnson (1976) proposed a gene-for-gene interaction between the resistance and avirulence genes. Resistance has repeatedly been shown to be conferred by a single dominant gene (Sequeira and Raffray 1971; Ventura et al. 1971; Zink 1973). There are at least 19 dominant single downy mildew (*Dm*) genes that match specific avirulence genes in *B. lactucae* (Jeuken and Lindhout 2002; Grube and Ochoa 2005).

Resistance genes (*R*) are often members of multigene families and are found clustered throughout plant genomes (Michelmore and Meyers 1998; Hulbert et al. 2001). Each member of this gene family confers resistance to a specific strain of pathogen (Chin et al. 2001). The most common domain encoded by R genes is the leucine-rich repeat (LRR) which contributes to specificity in pathogen recognition (Dangl and Jones 2001). Of the three R gene classes that encode the LRR domain, the largest class encodes intracellular nucleotide binding site proteins (NBS-LRR; Dangl and Jones 2001). In general, the nucleotide binding site, leucine-rich repeats, and a serine/threonine protein kinase are conserved in *R* genes. In lettuce, at least 10 *Dm* genes are linked to the major resistance cluster and of these, *Dm3* is a member of the *RGC2* gene family, which is a NBS-LRR type gene (Meyers et al. 1998; Kuang et al. 2004). Recombination within the *Dm3* region in lettuce is repressed by about 18-fold relative to the genome average supporting the supposition of increased stability of resistance genes in terms of recombination (Chin et al. 2001). However, when the *Dm3* region was homozygous a high rate of spontaneous mutation (10^{-3} to 10^{-4} mutations per generation) was observed (Chin et al. 2001). Duplication and physical size of the clusters may influence gene stability as a greater number of duplication events and large cluster size are likely to increase misalignment and unequal crossing over. Stability and loss of function would also be influenced by the nature of the haplotypes comprising the heterozygote.

The *RGC2* genes are genetically complex. Based on sequence similarity of the 3′ region, intron similarity, frequency of sequence exchange and prevalence in natural populations, two types of *RGC2* genes were recognized (Type I and Type II; Kuang et al. 2004). The copy number ranged from 12 to 32 among the seven species and the LRR number varied from 40 to 47. The Type I genes exhibited frequent sequence exchanges among paralogs, homogenous intron sequences and heterogeneous exons while these features were not observed in Type II genes. In lettuce *RGC2* genes contain eight exons and seven introns and are up to ~15 kbp in length, whereas the average gene in *Arabidopsis* is ~2,000 bp and consists of 5 exons and 4 introns (The Arabidopsis Research Initiative 2000; Kuang et al. 2004). Diversity in the *RGC2* genes has been shaped by recombination and indel events, point mutations, and diversifying selection within the LRR domain, but the effects and targets within these genes are different between the Type I and Type II subsets (Kuang et al. 2004). Thus, size and complexity of *RGC2* genes serve as sources of variation that generate recognition specificity, but conversely the same genetic events can result in loss of function of these resistance genes.

Disease resistance to any single race of *B. lactucae* is short-lived so breeders are seeking methods in which to breed for durable resistance. One method is to use field resistance to screen genetic material as this provides selection pressure from multiple races of pathogens and environmental conditions are similar or identical to commercial production. Field resistance to downy mildew appears to be a quantitative trait indicating that genes other than *Dm* are involved (Norwood et al. 1983; Grube and Ochoa 2005). Crossing the resistant species *L. saligna* with susceptible *L. sativa* and screening the F_2 population for resistance allowed the detection of a single resistance gene (R39) and three quantitative trait loci (QTL) that accounted for 51% of the observed resistance to *Bremia* race NL14 (Jeuken and Lindhout 2002).

Nondomesticated germplasm serves as a valuable reservoir for nonrace specific resistance but newly developed genomic approaches are providing data that are helping to elucidate the complexity of disease resistance in plants. Microarray experiments have been useful in identifying genes that may aid in breeding for disease resistance (Langlois-Meurinne et al. 2005; Shi et al. 2006). Many of the genes identified may only be indirectly involved in resistance and may reveal heretofore unknown mechanisms by which plants cope with pathogens. Demonstration of the function and contribution of these genes identified by microarrays will be necessary before durable resistance is realized. Likewise, identification of QTL associated with disease resistance and subsequent cloning of the genes underlying these loci may reveal genes whose role in disease resistance was previously unknown.

Lettuce dieback is a disease transmitted by *Tombusviridae* viruses, including *tomato bushy stunt virus* and *lettuce necrotic stunt virus* (Obermeier et al. 2001). Cos cultivars appear to be susceptible to this disease whereas all modern Crisphead cultivars tested were resistant (Grube et al. 2005). Resistance to dieback of heirloom and primitive cultivars was observed at about the same frequency as modern cultivars. Resistance appeared to be conferred by a single dominant gene that mapped to linkage group 2 which contains the *Dm1/Dm3* cluster (Grube et al. 2005).

2.2 Genetic Maps: Intraspecific

The first genetic map of lettuce using molecular markers was reported in 1987 (Landry et al. 1987) from an F_2 population of 66 plants derived from a cross between the Crisphead cultivar "Calmar" and the Butterhead cultivar "Kordaat". The linkage map consisted of 53 genetic markers which included 41 RFLP loci, 5 downy mildew resistance genes, 4 isozyme loci and 3 morphological markers. The linkage map (created with LINKAGE-1, Suiter et al. 1983) covered 404 cM. Seven *Dm* genes were mapped: *Dm5/8* on LG 2, flanked by RFLP markers CL250 and CL202b; *Dm10* on LG 2; *Dm4* on LG 8; *Dm3* and *Dm1* on LG 9. Several *Dm* genes (*Dm10, Dm7, Dm11, Dm2, Dm14, Dm15, Dm16*) were linked to segregating markers but could not be placed on the map, while *Dm13* was unlinked. A detailed genetic map was later published using the genetic population just described (Kesseli et al. 1994). This map was created using MAPMAKER v2.0 (Lander et al. 1987) with a LOD threshold of 3.5 used to assign linkage groups. The map was comprised of 319 loci, and included 152 RFLP, 130 RAPD, 7 isozyme, 19 disease resistance and 11 morphological markers. These markers were assigned to 13 major and 4 minor linkage groups and the total map distance covered was 1,950 cM.

An intraspecific map was created by Still and colleagues (personal communication, Hayashi 2006). The mapping population consists of 123 F_8 families derived from single-seed descent from *L. sativa* cv. Diplomat (Crisphead) × *L. sativa* cv. Margarita (Butterhead). The parents were chosen on the basis of their use in the low desert production areas in the US and their divergent germination phenotypes. The map was created using JoinMap v.3.0 (van Ooijen and Voorrips 2001). Linkage groups were detected at a LOD threshold of 3.0 and the map consists of 150 AFLP markers and a seed coat color locus with a total map distance of 1,010.7 cM and an average marker interval of 7.2 cM.

2.3 Genetic Maps: Interspecific

Interspecific maps will facilitate the introgression of genes from wild germplasm into cultivated lettuce, aid in the study of transgenic gene flow from genetically engineered plants into the wild, and allow the processes of domestication to be elucidated if the progenitor species is included. Three research groups have created interspecific maps of lettuce for these purposes. Two independent F_2 populations of *L. sativa* × *L. saligna* were used to create genetic maps using AFLP markers (Jeuken et al. 2001). The two maps were integrated using 124 common markers and con-

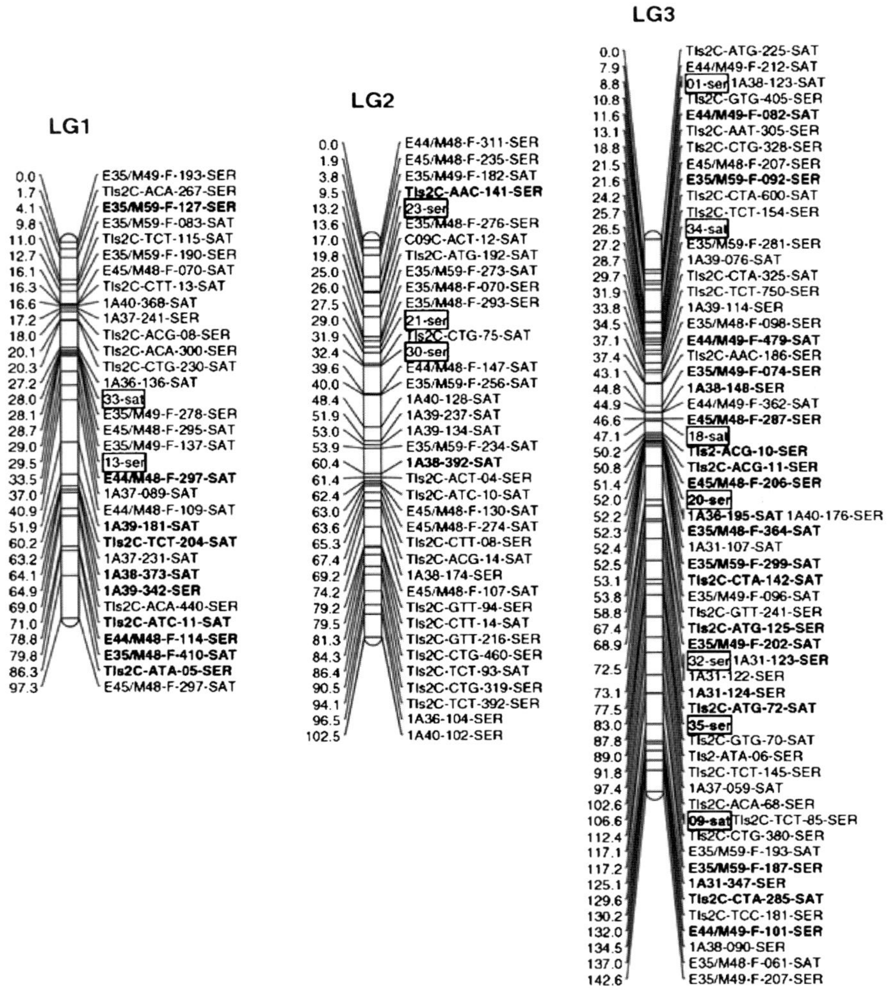

Fig. 1. Linkage map of an F_2 population derived from the cross between *L. sativa* × *L. serriola* (Syed et al. 2005). *Bold* markers show skewed segregation, NBS markers are *boxed*, SSAP markers are denoted with the prefix TIs2, AFLP markers by E or 1A; SER and SAT designate derivation of the marker from either the *L. serriola* or *L. sativa* parent, respectively (courtesy of Theoretical and Applied Genetics)

sisted of 476 AFLP markers and 12 SSRs on 9 linkage groups with a coverage of 854 cM with an average interval of 1.8 cM between markers. Population A consisted of 126 plants from a cross between *L. saligna* (CGN5271) × *L. sativa* cv. Olof (Butterhead) while population B consisted of 54 plants from a cross between *L. saligna* (CGN 11341) × *L. sativa* cv. Norden (Butterhead). Two linkage maps were created using JoinMap v.2.0 (Stam and van Ooijen 1995). Linkage groups were assigned at a LOD threshold of 6.0 and 4.0 for population A and B, respectively. Population A contained 412 markers on 10 linkage groups covering 895 cM whereas the linkage map of population B contained 294 AFLP markers and eight SSRs assigned to 10 linkage groups covering a total map length of 627 cM. The integrated map was created by assigning 488 (out of 533 total) markers to nine linkage groups with a LOD threshold of 6.0 (Jeuken et al. 2001).

More recently an F_2 population of 90 individuals was generated from an interspecific cross between *Lactuca serriola* (DH_M21) × *L. sativa* cv. Dynamite (Syed et al. 2005). The linkage map was created using JoinMap v.2.0 (Stam and van Ooijen 1995). AFLP markers were scored codominantly while SSAP and NBS were scored as dominant markers. A total of 458 (of 536 total) markers were assigned to 9 major and 1 minor linkage groups at a LOD threshold of 5.0 and the map covered 1,266 cM with the average distance between markers of 2.76 cM (Fig. 1). A total of 32% of the AFLP markers placed on this map were also found on the integrated linkage map of lettuce reported by Jeuken et al. (2001). The different marker types used by Syed et al. (2005) effectively distributed the markers more evenly across the genome than would otherwise be realized if utilizing a single marker type. Additionally, by using SSAP, a type of anchored AFLP,

Fig. 1. (continued)

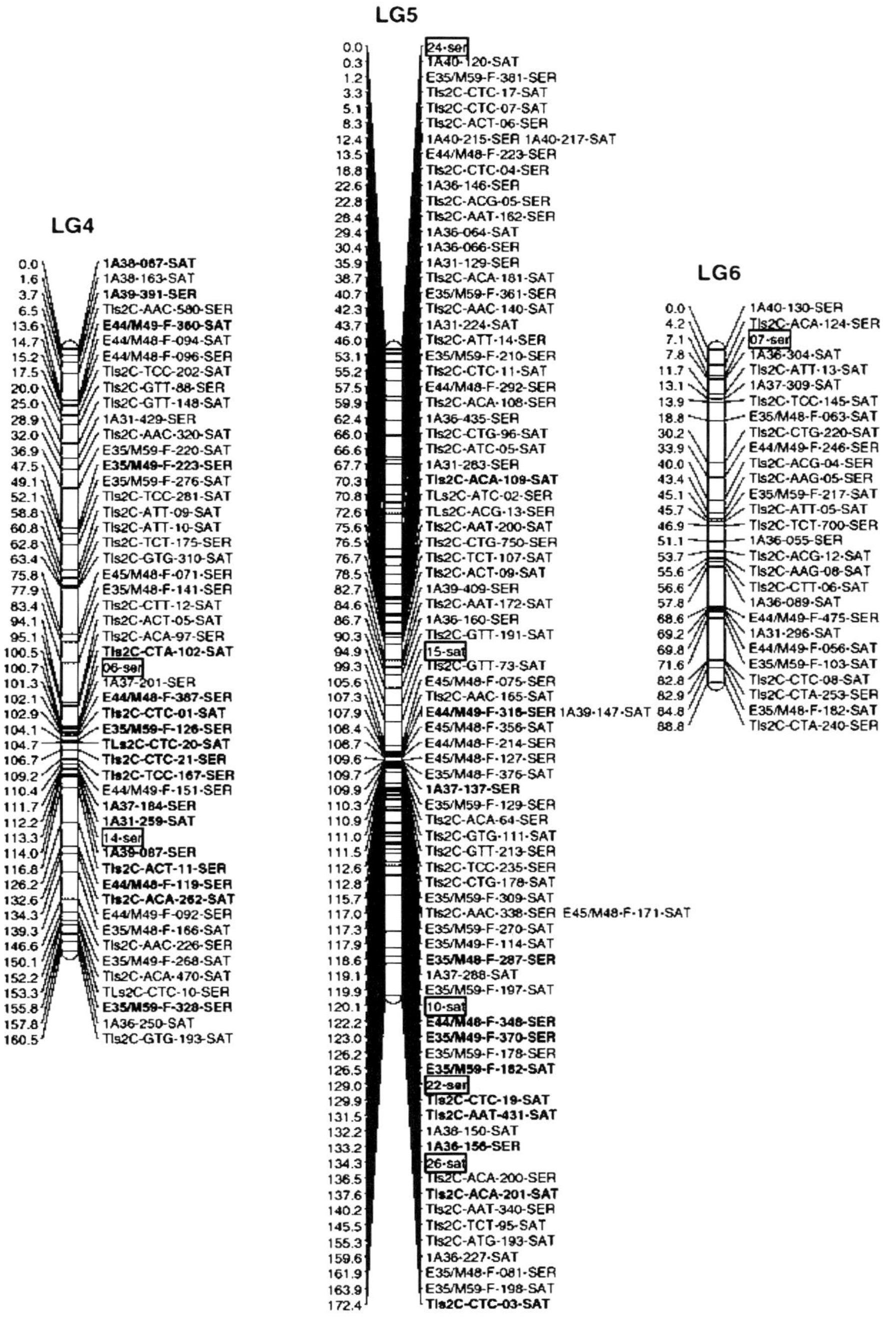

the detection of specific loci could be targeted (Waugh et al. 1997). In this case locus-specific markers were targeting long terminal repeat (LTR) sequences from *Ty1-copia* and *Ty3-gypsy* retrotransposons, which unlike DNA transposons, will not excise from their sites, so they serve as stable markers (Schulman et al. 2004). Targeting of the nucleotide binding site leucine-rich repeat R genes, called NBS profiling, enabled the detection of the map position of these genes. Targeting markers to potentially useful genes for which the physiological function has been determined is potentially a powerful method especially when combined with QTL mapping.

A third interspecific mapping population was generated by Michelmore and colleagues (http://cgpdb.ucdavis.edu/database/genome_viewer/viewer;

Fig. 1. (continued)

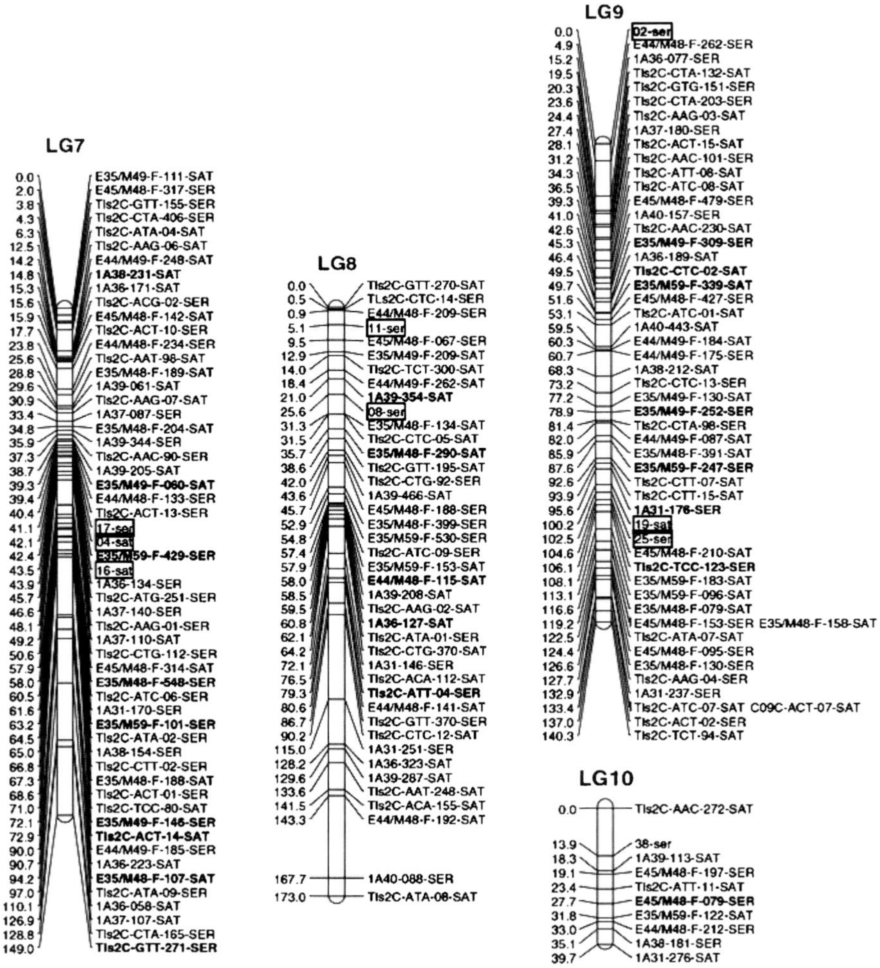

personal communication, Truco 2006). The mapping population consists of 103 F_8 families derived from single-seed descent from a *L. sativa* cv. Salinas (Crisphead) × *L. serriola* (UC96US23) F_2 population (Johnson et al. 2000). The map was created using JoinMap v.2.0 (Stam 1993; Stam and van Ooijen 1995) and consists of over 1700 AFLP, SSR and EST markers assigned to 9 linkage groups and spans 1,254 cM.

2.4 QTL Mapping

Genes and QTLs for disease resistance, plant growth and development, and seed germination traits have been mapped in lettuce. Using an $F_{2:3}$ *L. serriola* × *L. sativa* population, 13 QTLs with major effects on root architecture and water-use were mapped (Johnson et al. 2000). *Lactuca serriola* develops a deep tap root and extracts water and nutrients from deep soil zones whereas cultivated lettuce (*L. sativa*) has a short taproot and numerous lateral branches largely restricted to the upper soil layers (Jackson 1995). Because lettuce production requires relatively high inputs of fertilizers and water, the development of cultivars that are more deeply rooted would enhance agricultural sustainability and lower groundwater contamination by leaching. The QTLs in this study were detected with CIM using QTL Cartographer v1.112d (http://statgen.ncsu.edu/qtlcart/index.php) and markers were associated with the QTLs using a step-wise linear regression approach with the SRMAPQTL program. Significance probabilities ($P \leq 0.01$ or ≤ 0.05) of QTL were determined with the ZMAPQTL program with a 10-cM window and permutation analysis using 1000 iterations (Churchill

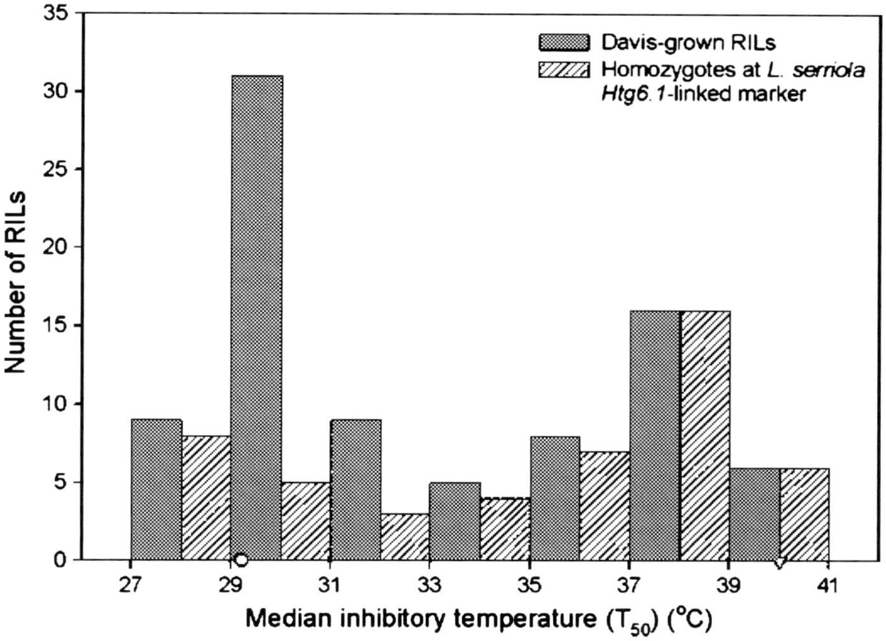

Fig. 2. Association of the *L. serriola*- derived QTL *Htg6.1* with the ability of lettuce seeds to germinate at high temperature (Argyris et al. 2005). A set of 103 RILs derived from an interspecific cross between *L. sativa* and *L. serriola* was grown in a field in Davis, CA and scored for their ability to germinate at a range of temperatures (Agryris et al. 2005) (courtesy of Theoretical and Applied Genetics)

and Doerge 1994). Each QTL accounted for 28 to 83% of the observed phenotypic variation. A QTL associated with the number of lateral roots along the bottom 5 cm of the taproot was located on linkage group 4 and accounted for 83% of the variation of that trait. With the exception of only one trait, the number of laterals in the bottom 5 cm of the taproot scaled for biomass in the greenhouse, only a single QTL was associated with each trait. This could be due to the relatively high stringency thresholds for QTL detection ($P < 0.05$ experiment-wise threshold level). It is likely that only major QTLs were detected while others with smaller effects probably exist but were not detected by the statistical methods employed in this study.

Seventeen QTLs for seed germination characteristics and seedling quality were mapped in the interspecific F_8 RILs derived from the cross between *L. sativa* cv. Salinas × *L. serriola* (UC96US23) (Argyris et al. 2005). The phenotypes scored included low (25 °C) and high (35 °C) temperature germination (LTG and HTG, respectively) and recovery of germination after far-red light exposure (FR-D). The plants producing the seed were grown in three environments: Davis, CA and Yuma, AZ in US and the Netherlands. Significant genotypic, environmental and G×E effects were observed for each germination trait with the exception of the environmental effect at high temperature. The QTL *Htg6.1*, derived from *L. serriola*, was present in all RIL families that had the ability to germinate at supraoptimal temperatures (Fig. 2). Heritabilites were estimated to be 0.30, 0.84 and 0.74 for LTG, HTG and FR-D, respectively. The genetic map was based on the F_7 RIL population and consisted of over 1700 AFLP, SSR and EST markers assigned to 9 linkage groups spanning 1,254 cM. The map was constructed using JoinMap v.2.0 (Stam 1993). The LOD threshold for the 17 QTLs ranged from 2.59 to 9.87 and corresponded to *Htg4.1* and *Htg6.1*, respectively.

Using the intraspecific F_8 RILs resulting from the *L. sativa* cross between cv. Diplomat and cv. Margarita, Still and colleagues mapped 43 QTLs associated with seed germination traits, including response to far-red light, high temperature germination, germination under osmotic stress, response to nitric oxides, and response to priming treatments (Hayashi 2006, personal communication; Plummer 2006, personal communication). The QTLs were mapped using MapQTL v4.0 (van Ooijen and Maliepaard 1996). The LOD levels for the various QTLs ranged from 2.52 to 7.95 for QTLs associated with germination at high temperatures under red light and the recovery of germination after exposure to far-red light, respectively. The amount of variation the QTLs accounted for in each of these traits ranged from 6.6% to 19.8%. Several of the germination loci collocated perhaps indicates a common mechanism such as integration of the signal transduction pathways and/or phytohormone synthesis or catabolism.

2.5 Development of Genomic Resources

Improvement of lettuce cultivars will be directed towards nutritional content, shelf life, and disease resistance through traditional and molecular-aided breeding methods. Some of these traits may be improved by genes from other sources far-removed from the *Lactuca* species. Gains in these characteristics will require the development of significant genomic resources as these traits are genetically complex or in some instances, genes coding for metabolic products are either missing or silent. To some extent wild germplasm may be used as a reservoir and efforts have been undertaken toward identifying and screening closely related species for resistance factors (see for example, Beharav et al. 2005). The development of virtually all horticultural traits will be facilitated by the development of genomic resources. To that end the Compositae Genome Project (CGP) has developed extensive EST data that have allowed lettuce to at least partially catch up with other more intensively studied species like tomato.

2.5.1 Sequence Information and EST Databases

The first phase of the CGP generated over 68,000 ESTs from *L. sativa* and *L. serriola*. The EST libraries were made from 10 pools of RNA from different tissues, developmental stages and environmental conditions of each of the two genotypes that had been used as parents for the core mapping populations for each species. These ESTs were assembled into ~19,523 lettuce unigenes and may be accessed at the Compositae Genome Project web site (http://cgpdb.ucdavis.edu/). The EST sequence information has been released to GenBank, and incorporated by TIGR into their gene indices and by MIPS into their SPUTNIK EST database. A second phase of EST generation by the CGP is in the process of generating over 560,000 ESTs from 17 species within the Compositae. This will include an additional 70,000 ESTs for *L. sativa* and *L. serriola* and for the first time provide ca. 90,000 reads for *L. saligna, L. virosa*, and *L. perennis* (Rieseberg and Michelmore 2006, personal communication). These reads will be available by the end of 2006.

With over 135,000 ESTs for *L. sativa* and *L. serriola*, gene discovery through EST sequencing has likely reached saturation (Michelmore 2006, personal communication). The gene space gap may be filled by sequencing of the methyl-filtrated (MF) and/or high C_0t fractions (single/low copy sequences) for lettuce and sunflower. The MF and high C_0t sequencing would supply the fraction of coding regions that is currently missing and provide valuable basic information such as sequences of introns, promoters, and 5′ and 3′ UTRs. Such information will be necessary to elucidate the mechanisms by which lettuce responds to its environment. Having data from the wild progenitor will also provide clues as to the evolutionary processes that occurred during the domestication of lettuce.

Whole genome microarray approaches have been shown in *Arabidopsis* and barley to be useful tools to simultaneously identify genes involved in stress responses and detect single feature polymorphisms (Borevitz et al. 2003; Gong et al. 2004; Cui et al. 2005) and single nucleotide polymorphisms in yeast (Gresham et al. 2006). An Affymetrix tiling chip is currently being developed for lettuce for massively parallel genetic mapping and sensitive expression analysis. This chip will have a total of 6.6 million features representing ~29,000 unigenes, each with ~200 oligonucleotides with a 2-bp stagger (Deynze and Michelmore 2006, personal communication).

2.5.2 Other Resources

To facilitate cloning, BAC libraries have been developed for lettuce. A 2× BAC library exists for lettuce with inserts ranging from 92.5 to 142 kb with an average insert size of ~111 kb (Frijters et al. 1997) although 10 to 15 times coverage is desirable. Additional libraries may exist in the private sector. As a foundation for sequencing, it would be useful to develop BAC fingerprints, generate a minimum tiling path, and sequence their ends. This would provide a framework for assemblies of random shotgun sequences and complement the genetically-defined bins of unigenes.

Development of mapping populations to address specific traits will facilitate mapping efforts and reveal basic information on the genetic mechanisms behind these traits. For example, several new RIL popula-

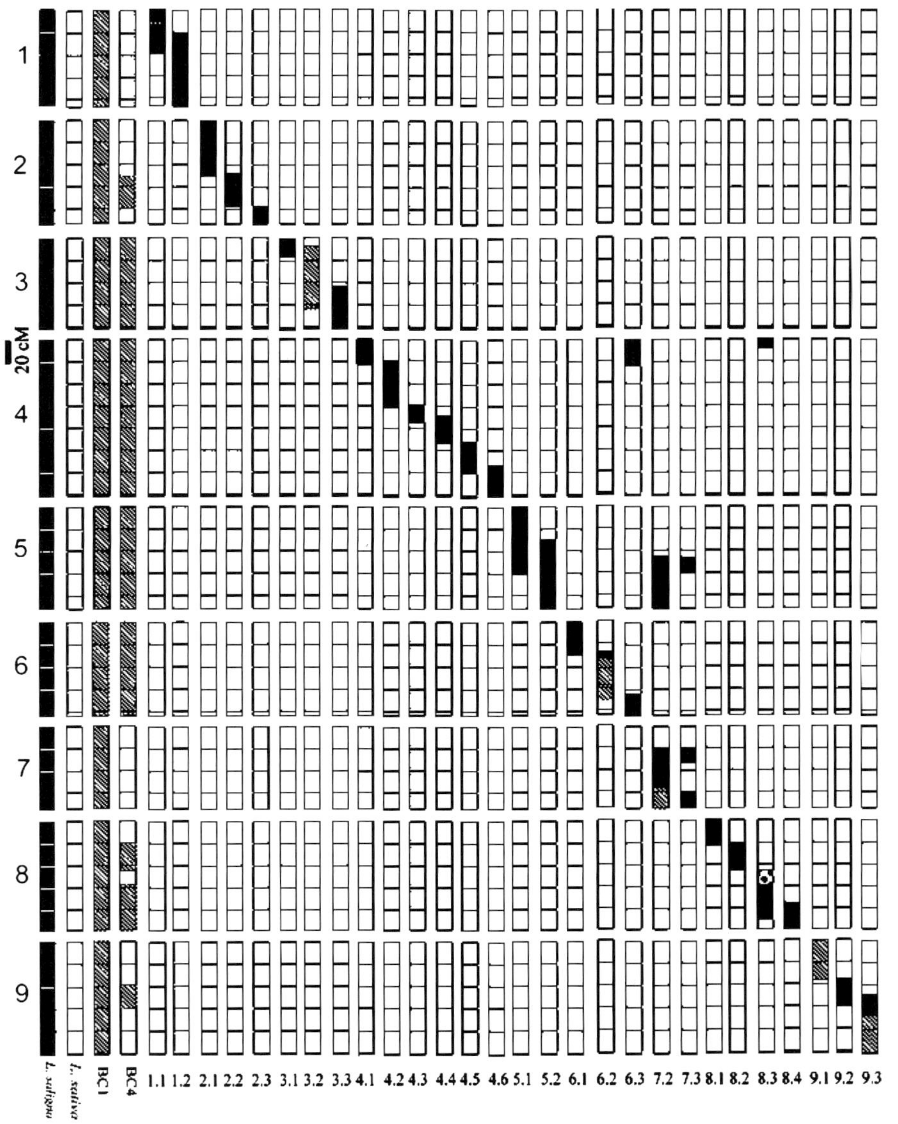

Fig. 3. A chromosomal view of genome coverage of the BC_1 and BC_4 generations and the genotypes of 28 backcross lines that cover >96% of the *L. saligna* genome. Each row represents one of the 9 chromosomes of *L. sativa*. *White*: homozygous *L. sativa*. *Black*: homozygous *L. saligna*. *Diagonal stripe*: heterozygous. The *dot* in chromosome 8, line 8.3 indicates uncertainty in the genotype in this region of the chromosome due to a lack of markers in this area (Jeuken and Lindhout 2004) (courtesy of Theoretical and Applied Genetics)

tions are being generated for lettuce; in particular, 300 RILs from an intraspecific *L. sativa* cv. Salinas × cv. Valmaine cross that will allow the dissection of the determinants of plant type. (Truco and Michelmore 2006, personal communication). The development of backcross inbred lines (BILs) will assist in marker-assisted selection, studying QTL×QTL interactions, more precisely estimate the contribution of the QTLs, and identify epistatic gene action. A set of 28 BILs has been developed in which *L. saligna* was introgressed into a *L. sativa* background (Jeuken and Lindhout 2004). Each of these lines contains a high percentage (90%) of the recurrent (*L. sativa*) parent and a low percentage of the donor parent (*L. saligna*) (Fig. 3).

2.5.3 Phenotypic Information

The genomic resources developed will never realize their full potential without the generation of accurate phenotypic data. The CGP research group have identified (or are identifying) QTLs for several vegetative and reproductive traits that include root architecture, bolting, branching, spines, leaf development, leaf succulence, shoot biomass, time to flowering, flower color, seed size, seed oil concentration and composition, seed germination under environmental stress, germination response to light spectra, mineral ion uptake traits, photosynthetic rate, water use efficiency, nutrient uptake, and disease resistance. Each of these traits is being mapped relative to candidate genes for nutrient ion transporters, enzymes associated with seed dormancy and germination, disease resistance and response, drought, salt, and high temperature stress tolerance, tocopherol and flavonoid synthesis, and floral, leaf and root development. More than 900 candidate genes have been identified in lettuce by the CGP research group and of those more than 55% have been mapped (Michelmore 2006, personal communication).

2.6 Future Prospects

As all genomic research is driven by technology, the goal of sequencing a mammalian-sized genome for $1000 recently announced by the National Human Genome Research Institute will have significant impacts on all aspects of the life sciences. Those in the plant field are likely to benefit the most from this because of the sheer number of different plant species and the relatively small amount of funding available for plant research. This technology will soon allow generation of whole-genome draft sequence data from different lettuce types, *Lactuca* species, and closely related Composites which will allow comparative analysis to be performed among these taxa. Such analyses will allow detection of key genes that are involved in the domestication process, disease resistance, metabolic pathways, and complex physiological traits such as seed germination, leaf morphology and reproduction. Identification of candidate genes will necessarily have to be followed up with functional analysis to confirm or refute their role and this will require both single-gene and global analysis approaches. In general, microarray and QTL mapping projects often identify several loci involved in the expression of a given trait. A significant effort must then be mounted to identify and clone the gene and demonstrate its function in vivo. This appears to be the remaining bottleneck in the discovery-to-application pipeline. In genetically complex traits, including diseases like cancer, the number of genes differentially regulated by experimental treatments and identified as candidate genes typically goes from an initial estimate of ~300 to somewhere around 15–30 key genes. Occasionally, a single gene may be key in changing a physiological process during the domestication process such as the *tb1* or *Sh4* in maize and rice, respectively (Doebley et al. 1995; Li et al. 2006). Only through the development of significant genetic and genomic resources will the roles of key genes be elucidated, leading to an increase in the efficiency of breeding.

References

Argyris J, Truco MJ, Ochoa O, Knapp SJ, Still DW, Lenssen GM, Schut JW, Michelmore RW, Bradford KJ (2005) Quantitative trait loci associated with seed and seedling traits in *Lactuca*. Theor Appl Genet 111:1365–1376

Bayer RJ, Starr JR (1998) Tribal phylogeny of the Asteraceae based on two noncoding chloroplast sequences, the *trnL* intron and *trnL*/*trnF* intergenic spacer. Ann Missouri Bot Gard 85:242–256

Beharav A, Lewinsohn D, Lebeda A, Nevo E (2005) New wild Lactuca genetic resources with resistance against Bremia lactucae. Genetic Resources Crop Evol 53:467–474

Bentham G (1873) Notes on the classification, history and geographical distribution of Compositae. J Linn Sco Bot 13:335–377

Borevitz JO, Liang D, Plouffe D, Chang H-S, Zhu T, Weigel D, Berry CC, Winzeler E, Chory J (2003) Large-scale identification of single-feature polymorphisms in complex genomes. Genome Res 13:513–523

Bremer K (1994) Asteraceae: Cladistics and Classification. Timber, Portland, OR

Caldwell CR (2003) Alkylperoxyl radical scavenging activity of redleaf lettuce (*Lactuca sativa* L.) phenolics. J Agri Food Chem 51:4589–4595

Chin DB, Arroyo-Garcia R, Ochoa OE, Kesseli RV, Lavelle DO, Michelmore RW (2001) Recombination and spontaneous mutation at the major cluster of resistance genes in lettuce (*Lactuca sativa*). Genetics 157:831–849

Churchill GA, Doerge R (1994) Empirical threshold values for quantitative trait mapping. Genetics 138:963–971

Cooper DA (2004) Carotenoids in health and disease: Recent scientific evaluations, research recommendations and the consumer. J Nutr 134:221S–224S

Cronquist A (1955) Phylogeny and taxonomy of the Compositae. Am Mid Naturalist 53:478–511

Crute IR, Johnson AG (1976) The genetic relationship between races of *Bremia lactucae* and cultivars of *Lactuca sativa*. Ann Appl Biol 83:125–137

Cui X, Xu J, Asghar R, Condamine P, Svensson JT, Wanamaker S, Stein N, Roose M, Close TJ (2005) Detecting single-feature polymorphisms using oligonucleotide arrays and robustified projection pursuit. Bioinformatics 21:3852–3858

Dangl JL, Jones JD (2001) Plant pathogens and integrated defence responses to infection. Nature 411:826–833

De Vries IM (1997) Origin and domestication of *Lactuca sativa* L. Genet Resour Crop Evol 44:165–174

Doebley J, Stec A, Gustus C (1995) *tesointe branched1* and the origin of maize: evidence for epistasis and evolution of dominance. Genetics 141:333–346

DuPont SM, Mondin Z, Williamson G, Price KR (2000) Effect of variety, processing, and storage on the flavonoid glycoside content and composition of lettuce and endive. J Agr Food Chem 48:3957–3964

Durst CE (1930) Inheritance in lettuce. Univ Illinois Agri Exp Sta Bull 356

FAO (2006) http://faostat.fao.org/faostat/sevlet/. Cited 20 April 2006

Frijters ACJ, Zhang Z, van Damme M, Wange G-L, Ronald PC, Michelmore RW (1997) Construction of a bacterial artificial chromosome library containing large *Eco*RI and *Hind*III genomic fragments of lettuce. Theor Appl Genet 94:390–399

Funk VA, Bayer RJ, Keeley S, Chan R, Watson L, Gemeinholzer B, Schilling E, Panero JL, Baldwin BG, Garcia-Jacas N, Susanna A, Jansen RK (2005) Everywhere but Antaractica: using a supertree to understand the diversity and distribution of the Compositae. Biol Skr 55:343–374

Funk VA, Chan R, Keeley SC (2004) Insights into the evolution of the tribe Arctoteae (Compositae: subfamily Cichorioideae s.s.) using *trnL-F, ndhF*, and ITS. Taxon 53:637–655

Gong J-M, Waner DA, Horie T, Li SL, Horie R, Abid KB, Schroeder JI (2004) Microarray-based rapid cloning of an ion accumulation deletion mutant in *Arabidopsis thaliana*. Proc Natl Acad Sci USA 101:15404–15409

Gresham D, Ruderfer DM, Pratt SC, Schacherer J, Dunham MJ, Botstein D, Kruglyak (2006) Genome-wide detection of polymorphisms at nucleotide resolution with a single DNA microarray. Science 311:1932–1935

Grube RC, Ochoa OE (2005) Comparative genetic analysis of field resistance to downy mildew in the lettuce cultivars "Grand Rapids" and "Iceberg". Euphytica 142:205–215

Grube RC, Wintermantel WM, Hand P, Aburomia R, Pink DAC, Ryder EJ (2005) Genetic analysis and mapping of resistance to lettuce dieback: a soilborne disease caused by tombusviruses. Theor Appl Genet 110:259–268

Harlan JR (1992) Crops and Man. Am Soc Agron, Madison, WI

Hohl U, Neubert B, Holger P, Schonhof I, Böhm H (2001) Flavonoid concentrations in the inner leaves of head lettuce genotypes. Eur Food Res Technol 213:205–211

Hulbert SH, Webb CA, Smith SM, Sun Q (2001) Resistance gene complexes: evolution and utilization. Annu Rev Phytopathol 39:285–312

Jackson LE (1995) Root architecture in cultivated and wild lettuce (*Lactuca* spp.). Plant Cell Environ 18:885–894

Jagger IC, Whitaker TW (1940) The inheritance of immunity from mildew (*Bremia lactucae*) in lettuce. Phytopathology 30:427–433

Jeuken M, Lindhout P (2002) *Lactuca saligna*, a nonhost for lettuce downy mildew (*Bremia lactucae*), harbors a new race-specific *Dm* gene and three QTLs for resistance. Theor Appl Genet 105:384–391

Jeuken M, van Wijk, R, Peleman J, Lindhout P (2001) An integrated interspecific AFLP map of lettuce (*Lactuca*) based on two *L. sativa* × *L. saligna* F_2 populations. Theor Appl Genet 103:638–647

Jeuken MJ, Lindhout P (2004) The development of lettuce backcross inbred lines (BILs) for exploitation of the *Lactuca saligna* (wild lettuce) germplasm. Theor Appl Genet 109:394–401

Johnson WC, Jackson LE, Ochoa O, van Wijk R, Peleman J, St. Clair DA, Michelmore RW (2000) Lettuce, a shallow-rooted crop, and *Lactuca serriola*, its wild progenitor, differ at QTL determining root architecture and deep soil water exploitation. Theor Appl Genet 101:1066–1073

Kesseli RV, Paran I, Michelmore RW (1994) Analysis of a detailed genetic linkage map of Lactuca sativa (Lettuce) constructed from RFLP and RAPD markers. Genetics 136:1435–1446

Kimura M, Rodriguez-Amaya D B (2003) Carotenoid composition of hydroponic leafy vegetables. J Agri Food Chem 51:2603–2607

Kuang H, Woo S-S, Meyers BC, Nevo E, Michelmore RW (2004) Multiple genetic processes result in heterogeneous rates of evolution within the major cluster disease resistance genes in lettuce. Plant Cell 16:2870–2894

Lander ES, Green P, Abrahamson J, Barlow A, Daly MJ, Lincoln SE, Newburg L (1987) MAPMAKER: An interactive computer package for constructing primary genetic linkage maps of experimental and natural populations. Genomics 1:174–181

Landry BS, Kesseli RV, Farrara B, Michelmore RW (1987) A genetic map of lettuce (*Lactuca sativa* L.) with restriction fragment length polymorphism, isozyme, disease resistance and morphological markers. Genetics 116:331–337

Langlois-Meurinne M, Gachon CMM, Saindrenan P (2005) Pathogen-responsive expression of glycosyltransferase genes UGT73B3 and UGT73B5 is necessary for resistance to *Pseudomonas syringae* pv *tomato* in Arabidopsis. Plant Physiol 139:1890–1901

Li C, Zhou A, Sang T (2006) Rice domestication by reducing shattering. Science 311:1936–1939

Lindqvist K (1960) On the origin of cultivated lettuce. Hereditas 46:319–350

Malkin R, Niyogi K (2000) Photosynthesis. In: Buchanan B, Gruissem W, Jones R (eds.) Biochemistry and Molecular Biology of Plants. American Society of Plant Physiology, Rockville, MD, pp 568–628

Meyers BC, Chin DB, Shen KA, Sivaramakrishnan S, Lavelle DO, Zhang Z, Michelmore RW (1998) The major resistance gene cluster in lettuce is highly duplicated and spans several megabases. Plant Cell 10:1817–1832

Michelmore RW, Meyers BC (1998) Clusters of resistance genes in plants evolve by divergent selection and a birth-and-death process. Genome Res 8:1113–1130

Mou B (2005) Genetic variation of Beta-carotene and lutein contents in lettuce. J Am Soc Hort Sci 130:870–876

Norwood JM, Crute IR, Johnson AG, Gordon P (1983) A demonstration of field resistance to lettuce downy mildew (*Bremia lactucae* Regel.) in progeny derived from cv. Grand Rapids. Euphytica 32:161–170

Obermeier C, Sears JL, Liu HY, Schlueter KO, Ryder EJ, Duffus JE, Koike ST, Wisler GC (2001) Characterization of distinct tombusviruses that cause disease of lettuce and tomato in the western United States. Phytopathology 91:797–806

Ryder EJ (1986) Lettuce breeding. In: Basset MJ (ed.) Breeding Vegetable Crops. AVI, Westport, CT

Ryder EJ (1999) Lettuce, Endive and Chicory. CABI, New York

Schulman AH, Flavell AJ, Ellis THN (2004) The application of LTR retrotransposons as genetic markers in plants. In: Miller W, Capy P (eds.) Mobile Genetic Elements: Protocols and Genomic Applications. Humana, Totowa, NJ

Sequeira L, Raffray JB (1971) Inheritance of downy mildew resistance in two plant introductions of *Lactuca sativa*. Phytopathology 61:578–579

Shi C, Thummler F, Melchinger AE, Wenzel G, Lubberstedt T (2006) Comparison of transcript profiles between near-isogenic maize lines in association with SCMV resistance based on unigene-microarrays. Plant Sci 170:159–169

Stam P (1993) Construction of integrated genetic linkage maps by means of a new computer package: JoinMap. Plant J 3:739–744

Stam P, van Ooijen JW (1995) JoinMap version 2.0: software for the calculation of genetic linkage maps. CPRO-DLO, Wageningen, The Netherlands

Suiter KA, Wendel JF, Case JS (1983) Linkage-1: a pascal computer program for the detection and analysis of genetic linkage. J Hered 74:203–204

Syed NH, Sorensen AP, Anonise R, van de Wiel C, van der Linden CG, van 't Westende W, Hooftman DAP, den Nijs HCM, Flavell AJ (2005) A detailed linkage map of lettuce based on SSAP, AFLP and NBS markers. Theor Appl Genet 112:517–527

The Arabidopsis Genome Initiative (2000) Analysis of the genome sequence of the flowering plant *Arabidopsis thaliana*. Nature 408:796–815

Trehane P (1995) International Code of Nomenclature for Cultivated Plants in 1995. Regnum Vegetable 133

USDA (2005a) Vegetables: Annual Summary 01.27.06. National Agricultural Statistics Service. http://usda.mannlib.cornell.edu/reports/nassr/fruit/pvg-bban/. Cited 2006

USDA (2005b) USDA National Nutrient Database for Standard Reference, Release 18. Nutrient Data Laboratory Home Page, http://www.nal.usda.gov/fnic/foodcomp. Cited 2006

van Ooijen JW, Maliepaard C (1996) MapQTL version 4.0: software for the calculation of QTL positions on genetic maps. CPRO-DLO, Wageningen, The Netherlands

van Ooijen JW, Voorrips RE (2001) JoinMap version 3.0: software for the calculation of genetic linkage maps. Plant Research International, Wageningen, The Netherlands

Ventura J, Netzer D, Globerson D (1971) Inheritance of resistance in lettuce to race 3 of downy mildew (*Bremia lactucae* Reg.). J Am Soc Hort Sci 96:103–104

Waugh R, McLean K, Flavell AJ, Pearce SR, Kumar A, Thomas BB, Powell W (1997) Genetic distribution of BARE-1-like retrotransposable elements in the barley genome revealed by sequence-specific amplification polymorphisms (S-SAP). Mol Gen Genet 253:687–694

Zink FW (1973) Inheritance of resistance to downy mildew (*Bremia lactucae* Reg.) in lettuce. J Am Soc Hort Sci 98:293–296

3 Radish

Yukio Kaneko[1], Chiaki Kimizuka-Takagi[2], Sang Woo Bang[1], and Yasuo Matsuzawa[1]

[1] Laboratory of Plant Breeding, Faculty of Agriculture, Utsunomiya University, 350 Minemachi, Utsunomiya 321-8505, Japan
e-mail: kaneko@cc.utsunomiya-u.ac.jp

[2] Mikado Seed Growers Co., Ltd., 2789-1 Shimo-Ohtaki, Ohtaki, Chiba Pref. 298-0202, Japan

3.1 Introduction

3.1.1 Origin and Distribution

Radish, *Raphanus sativus* L., is an annual vegetable of the family Cruciferae. The genus *Raphanus* originated in the regions from the Mediterranean to Black Sea and was classified under two sections of *Raphanis* DC. and *Hesperidopsis* Boiss. These two sections comprise six species including *Raphanus sativus*, *R. raphanistrum*, *R. microcarpus*, *R. rostras*, *R. landra* and *R. maritimus*, and one species *Raphanus aucheri*, respectively (Kitamura 1958). Kitamura (1958) also assumed that *R. sativus* could be developed around the Mediterranean region by natural or artificial crossing between *R. landra* and *R. maritimus*. Banga (1976) and Hida (1990) presumed that four wild species (*R. raphanistrum*, *R. maritimus*, *R. landra* and *R. rostras*) could be involved in the origin of radish.

From a study on the wild *R. sativus* and *R. raphanistrum* by Panetsos and Baker (1967), the differentiation of two species was confirmed. The wild *R. sativus* has distinguished features including a white flower or partially purple flower on a white background, and a tender and slightly thick pod made up of spongy parenchyma. On the other hand, *R. raphanistrum* has a yellow flower, and slender and hardy pod. The fully ripened pods break into pieces. In California, the above two species were growing with intermediate types that could originate from natural hybridization between them. The F_1 plants induced by artificial crossing between the two species showed intermediate features, chromosome configurations of 1IV+7II at first metaphase (MI) of pollen mother cells (PMCs) and moderate fertility of 50% in both pollen formation and seed setting. From these facts, it was suggested that the F_1 plants were carrying some reciprocal translocations in one pair of chromosomes. When the wild *R. sativus* type was naturally backcrossed with cultivated radish, some progenies did not form the quadrivalent chromosomes, and showed high seed setting. Therefore, it was suggested to induce gene flow (or introgression) from *R. raphanistrum* to radish cultivars. Eber et al. (1994, 1998) also confirmed these facts by the analysis of hybrids between the native *R. raphanistrum* and cultivated *R. sativus* in France. On the other hand, Kato and Fukuyama (1982) observed normal chromosome configuration of 9II in meiosis and high seed setting in F_1 hybrids between *R. sativus* and *R. raphanistrum* L. ssp. *landra*. From these results, it was assumed that the chromosomal reconstruction might happen even in *R. raphanistrum*.

Harberd (1976) proposed that the genus *Raphanus* could be included into a cytodeme from observations of chromosome number, chromosome configuration at MI in PMCs and fertility supported also by Prakash et al. (1999). Tsunoda (1979, 1991) assumed that the wild species of radish belonged to only *R. raphanistrum* and had their origin in the coastal area from the Mediterranean to Black Sea. *R. raphanistrum* was distributed widely in Russia and the New World, but not in Japan, China or India (Warwick 1993; Hinata 1995). Recently, the genus *Raphanus* was classified into two species, *R. sativus* and *R. raphanistrum*, where the latter comprises other wild species as the subspecies *R. raphanistrum* (Hinata 1995).

The wild radish, the so-called Hama-daikon, *R. sativus* var. *hortensis* f. *raphanistroides* Makino (Kitamura 1958), or *R. raphanistrum* ssp. *maritimus* (Hinata 1995), was naturally grown on the sea coast in East Asia. Another wild radish called Nora-daikon or No-daikon was naturally grown at the inland areas far from the coast. These wild radishes were thought to originate from the escape from the cultivated radish (Furusato and Miyazawa 1958; Kitamura 1958; Aoba 1981, 1988) or the migration

Genome Mapping and Molecular Breeding in Plants, Volume 5
Vegetables
C. Kole (Ed.)

of a weedy one becoming contaminated with cereals such as wheat and oat. However, the former view is supported by many researchers. Hama-daikon and Nora-daikon are regarded as the germplasm resources to investigate the origin of cultivated radish and to improve the radish crop. The molecular studies on the DNA and genomics, besides morphology, ecology and cytogenetics might provide valuable insight on origin, differentiation and domestication of radish.

3.1.2 Varietal Differentiation

Radish was used as food in ancient Egypt (2700 – 2200 BC) and was cultivated from the thirteenth century BC (Banga 1976). This crop was cultivated in Europe in the fifteenth or sixteenth century and was introduced to America in the twentieth century. Especially in East Asia, many varieties or cultivars were differentiated and grown since introduction to China 2,450 years ago and Japan 1,300 years ago (Kitamura 1958).

Radish was ecologically classified into five varieties, viz., var. *radicula* DC. (var. *sativus* Saz., hatsuka radish, north Chinese small radish), var. *niger* (Mill.) Pers. (var. *major* A. Voss., black radish, occidental small radish), var. *raphanistroides* Makino (var. *hortensis* Backer, north Chinese radish, south Chinese radish, and Japanese radish), var. *mougri* Helm (var. *caudatus* (L.) Hooker et Anderson, rat-tailed radish), and var. *olerfer* Netz. (var. *oleiformis* Pers. oil radish) (Kumazawa 1965; Banga 1976; Hida 1990). Hatsuka radish with its smaller root and early maturity is a popular garden vegetable, suitable for fresh vegetables all over the world. The north Chinese radish with its smaller, colored and starchy root is consumed for pickles, fresh vegetables and condiments. The south Chinese radish has characteristics such as pubescences on the leaf; long, round, low-starchy and juicy root; and early-bolting and blooming; and are used for salted vegetables and cooking. At first, Japanese radish was developed from south Chinese radish, introduced to Japan from China. The north Chinese radish was introduced later and grown in the restricted mountainous regions of central Japan such as Shinshu, San-in, Hokuriku and Tohoku provinces (Aoba 1979). The hatsuka-daikon was introduced from Europe at the beginning of the Meij period followed by the new Chinese varieties in the 1970s (Ashizawa 1984).

The cropping types of Japanese radish were classified into Autumn-Winter, Spring and Summer radish. Autumn-Winter radish includes Nerima, Miyashige, Shogoin, Shiroagari, Horyo, Moriguchi, Awa-bansei, Sakurajima, and local varieties called Ji-daikon that were adapted for the cultivation in the cold and warm regions. Spring and Summer radishes include Harufuku, Shigatsu, Ninengo, Kameido, and Shijyunichi and Minowase, respectively (Furusato and Miyazawa 1958; Kumazawa 1965). Radish is cultivated the entire year in Japan. In China, radish is classified into four ecotypes, i.e., south Chinese, middle Chinese, north Chinese and West Plateau types (Hirao 1982). Many varieties of south Chinese type have white root protruding from the culturing soil. A few ones of middle type are red-skinned and have white root, although they resemble the south Chinese type. These are also subdivided into Autumn-Winter, Spring and Summer types according to cropping practices. Many of the north Chinese types have a small and greenish root in addition to the red or white ones. West Plateau type needs earlier sowing period because of the low temperature at night during growth period at high elevation (10 – 15 °C). This type is late-bolting.

3.1.3 Karyotype

Radish, *Raphanus sativus* L., has RR genome with 18 somatic chromosome numbers. As for the karyotype of radish, Richharia (1937) described that all chromosomes were extremely small in size, about 0.9 – 2.5 μm in length at mitosis having median or submedian centromere, and that the genome comprised three tetrasomic (BB, DD, FF) and disomic chromosomes (A, C, E). In Indian radish, Mukherjee (1979) observed the median chromosomes of 1.1 to 4.0 μm in length and one or two chromosomes with secondary constriction. The detailed karyotypic analysis, however, did not advance only by the conventional techniques such as Heidenhain's iron hematoxylin staining and Feulgen squash method. Genome size was reported to be ca. 526 Mbp, or within the range of 468 – 516 Mbp for Chinese cabbage and 599 – 662 Mbp for cabbage (Arumuganathan and Earle 1991).

In comprehensive studies on the wide cross of Brassineae by Mizushima (1952, 1980), Harberd and McArthur (1980) and Prakash et al. (1999), it was confirmed that the F_1 hybrids between radish and the member of *Brassica* formed three to six of allosyndesis in addition to one autosyndesis of radish and suggested that radish was closely related to other *Brassica* crops in meiotic features. Spontaneous haploids were induced from chromosome addition lines of radish with a single chromosome of the related species. Haploid plants of radish exhibited the chromosome configuration of 3II+3I (Kaneko 1980; Kaneko et al. 2003) and 1III+2II +2I (Kaneko et al. 2003) at MI in PMCs suggesting that radish had autosyndesis in its genome as was reported to be the meiotic configuration of 1II+8I (Ramanujam 1941) and 1III+2II+3I for *B. rapa* (Armstrong and Keller 1981), 2II+5I (Thompson 1956) and 1III+1II+4I for *B. oleracea* (Armstrong and Keller 1982), and 2II+4I (Prakash 1973) for *B. nigra*. These facts support that *Brassica* members including radish evolved from a common ancestral form with $n = 6$ by chromosomal duplication.

3.1.4 Classical Genetic Analysis

By the progeny test for hybrid plants between radish cv. Shogoin and *R. sativus* ssp. Hama-daikon, Takebe (1944) suggested that the early-bolting trait of Shogoin was under the control of a single dominant gene in contrast to the late-bolting trait of Hama-daikon, whereas the traits of hardy and pubescent pods of Hama-daikon were controlled by a dominant gene in contrast to the trait of the thin and soft pod of Shogoin. A few plants with yellow cotyledon in the self-pollinated S_1 line of hatsuka-daikon cv. Cincinnati Market were known to terminate in course of time. The F_1 plants of Cincinnati Market × Golden Ball were normal, but the F_2 plants grown segregated into the ratio of 3:1 of normal and yellow cotyledon type. The chlorophyll abnormality expressed in Cincinnati Market was found to be controlled by a recessive gene that was designated as *xa* by Takebe (1938). Frost (1923) postulated the gene system ($R - r$) where two genes controlling root color and hardiness were linked with a crossing over value of $4.78 \pm 2.7\%$. Genetic analysis for root color was also studied by Hoshi et al. (1963), Ishikura et al. (1965) and Hoshi and Hosoda (1967). In a cross of red and white radish, all F_1 plants were purple, and the segregation of purple-red-white was observed in F_2. They found that purple pigmentation was due to a cyanidin 3-diglucosido-5-glucoside known as rubrobrassicin similar to one of *B. oleracea*, and that radish having only R gene expressed the red pigmentation by raphanusin (a pelargonidin 3-diglucosido-5-glucoside), and radish with R gene accompanied with the conditioning E gene showed purple by rubrobrassicin.

Tokumasu (1951) identified three plants exhibiting male-sterility (MS) controlled by a single nuclear recessive gene (*ms*). This MS system was due to the degeneration of male gametes by the enlargement of the tapetum layer after the microspore formation (Tokumasu 1957). Radish at large shows the physiological trait of seed vernalization for bolting and anthesis that is controlled by dominant genes against the trait of the green plant type (Kagawa 1971).

Hida (1990) summarized gene symbols or genetic modes as in Table 1.

In spite of the active studies on the classical genetic factors or genes controlling some traits, their mapping to radish chromosomes has not been reported thus far.

3.1.5 Economic Importance

The enlarged root of radish is commonly consumed as a fresh food such as salad and grated radish, salted, cooked with, pickles and the processed food such as dried radish. The other fresh food for salad known as kaiware, is a seedling with the elongated hypocotyl. Radish is also utilized as fodder. For instance, radish (*R. sativus* L. var. *oleiferous* Metzg.) is grown as a green manure crop for farmland in South America, and 5 ton per hectare is harvested for example in Paraguay (Florentin et al. 2001). In India and Southeast Asia, long and young pods of *R. caudatum* (*R. sativus* var. *caudatus*, rat-tailed radish) are eaten by boiling (Kitamura 1958; Banga 1976; Iwasa 1980; Yamaguchi 1983; Nomura et al. 1994).

3.1.6 Breeding Objectives

Breeding objectives have been focused on the ecological traits, for example, resistance to diseases, the adaptability for cultivating practices and the consumptive requirement. In Japan, the consumption trend of vegetable has changed in the last few decades. The large-sized vegetables such as radish, cabbage and Chinese cabbage are not in demand, but the compact and colored vegetables such as tomato, cucumber, let-

Table 1. A list of gene symbols in radish summarized by Hida (1990)

R^1 and R^2	Red color expressing to hypocotyl of seedling after germination (duplicated genes)
Pi	Pink color expressing to hypocotyl (this linked to *Ac1* of resistance for *Albugo candida* race 1 being 3.28 cM of genetic distance)
G	Aokubi meaning green protruding root is a single dominant against "white"
R	Red root
R^s	Root with horizontal red stripe
R^f	Red fresh (inside of enlarged root), these are multiple alleles having autarchic genes
B	Purple root coexisting with *R* gene
C	Fundamental substrate for color expression
Y^b	Black root
Y	Yellow root, multiple alleles related to $Y^b > Y$
R	Red root forming by pelargonidin 3-diglucosido-5-glucoside
E	Purple root (cyanidin 3-diglucosido-5-glucoside) coexisting with *R* gene
Apl and *Ar*	A shade of color in plant and root, respectively
Pu	Purple expressing remarkably in pod
gr	Slender root
gf1 and *gf2*	Yellowish green leaf with green spot
yg	Yellowish green leaf
yp	Yellow petal in mutant cultivated radish
a	Unanthosianidin, white petal and yellow seed
cp	Light yellow pollen
ms 1	Male-sterility of plant without pollen or with degenerated anther
C	Callus formation ability related to cytoplasmic gene
Rs and *R*1	Root regenerated ability from callus
Ac 1	Resistance for race 1 in *Albugo candida*
AR (R)	Resistance for *Aphanomyces raphani*
FOC 2 (R)	Resistance for race 2 in *Fusarium oxysporum* f. sp. *conglutinans*
RS (R)	Resistance for *Rhizoctonia solani*

tuce, and green pepper are preferred. However, this crop was cultivated in the arable lands of 41,500 ha occupying 8% of the cultivated acreages for vegetables, and producing 1,750,000 ton (42,200 kg per ha) in 2003. In Korea, the yield of 1.4 million tons was harvested from ca. 31,400 ha and the output in China increased to 38.8 million tons from ca. 1.2 million ha. Breeding for this crop is being practiced by the seed nursery companies both in Japan and Korea, and most of the varieties released from these companies are F_1 hybrids. Autumn radish is abundantly cultivated for the salted green, but the recent eating habits of the salt-decreased diets resulted in the consumptive reduction of not only the salted radishes but also the radish crop. Consumers now prefer radishes about 40 cm long, with a mild and smooth taste (less pungent) than the conventional lengthy ones. On the other hand, the farmers demand early and smooth production. The breeding goals are now to breed a variety of 35–38 cm in length, 7–8 cm in the diameter, 1.2–1.3 kg weight and 60–70 days of harvesting time. For these demands, a promising variety Taibyo-Miyashge Sobutori was released in 1974, and then Aokubi lines became popular in Japan. In the monograph *New Varieties of Vegetable Crops* published in 2003, the 30 varieties represented were all F_1 hybrids, although the domestic varieties such as Togakushi, Karami and Moriguchi were also commercial. Miyashige strain for Autumn-cropping, Minowase for Summer-cropping, and Tokinashi for Spring-cropping type were predominantly used as parental lines to breed the hybrids. Minowase has both virus resistance and heat tolerance and for that reason it has been used as an important germplasm in tropical and subtropical regions such as Thailand, India and Brazil.

The breeding objectives that are true for every cropping type are cold tolerance, late pore formation, virus resistance and flesh taste. In addition, introduction of the late-bolting trait and yellows disease resistances, the late-bolting trait and plumpness at low-temperature, and soft rot and yellows disease resistances were performed for the Autumn-, the Spring- and the Summer-radishes, respectively (Hida 1990).

Furthermore, the reduction of the floral differentiation ability by lower temperature, the unbolting trait at early Spring, resistance to virus and tolerance to heat are breeding goals especially in tropical and subtropical countries. As to the consumptive quality, breeding works are being focused on the moderate hardiness of root for Korean kimchi, some processing traits, scentless trait for the edible coloring, condimental traits for the soba noodle and so on.

3.1.7 Target Traits for Seed Production

Cultivation of radish largely depends upon adoption of hybrid varieties today because of their higher productivity and uniform traits. It is important to obtain abundantly superior F_1 seeds for sale. These have been accomplished by employing the self-incompatibility (SI) or the cytoplasmic male-sterility (CMS) system as in many *Brassica* crops. The self-incompatible system is controlled by an *S*-locus comprising multiple alleles showing a sporophytic type (Bateman 1952, 1955). In radish, about 20 *S*-alleles have been identified in commercial varieties (Bateman 1955; Okazaki and Hinata 1984). In addition, 25 – 34 *S*-alleles were detected to be present in wild populations of *R. raphanistrum* and 22 were identified in wild *R. sativus* (Sampson 1967; Karron et al. 1990).

Recently, Niikura and Matsuura (1999) and Niikura (2000) investigated the genetic variation in the *S*-alleles by analyzing the compatibility of 71 inbred lines and 223 landraces from 18 domestic and 5 foreign varieties maintained by sib-crossing among 3 to 5 plants. As a result, 37 alleles (designated as S^{201} to S^{237}) were identified for the genotypes of which S^{208}, S^{218} and S^{219} were prevalent in the varieties of Japanese radish. In addition, the dominant relationships were detected among *S*-alleles by the analysis of randomly selected lines with 16 homologous alleles of S^{201} to S^{217} except for S^{207}. Genotypes for the *S*-alleles were classified into 4 types as determined by Haruta (1962). Most of the varieties such as Awa bannsei, Moriguchi, Ninengo, Risyo, Shiro-agari, South Kyushu Ji-radish, Korean lines, Nepal lines were confirmed to carry the dominant *S*-alleles, whereas those of Hama-daikon, Miyashige, Nerima, Aki-tumari and Tropical lines were carrying the dominant *S*-alleles in lower frequency.

F_1 seed production by CMS system has been developed for onion, carrot, sugar beet, tomato, red pepper, and other crops (Namai 1991). In *Brassiceae*, the most promising CMS source was derived from the Japanese radish that was inspected in Kagoshima Prefecture of Japan in 1968 (Ogura 1968). The Ogura CMS type was introduced into the *Brassica* crops such as cabbage and oilseed rape at INRA in France (Bannerot et al. 1974, 1977; Dolourme et al. 1999). However, this type expressed chlorosis under low temperature and had low seed fertility under the conventional seed production. The recombination between the homoeologous domain of radish and oilseed rape (Paulmann and Robbelen 1988) and cell fusion (Pelletier et al. 1983; Sakai and Imamura 1994; Dolourme et al. 1999) were expected to overcome the above defects. Namai (1991) reported another two systems of the KS line derived from Japanese radish cv. Kozena and the CMS UK-1 line from F_1 plants between Chinese cv. Daibaika and an unknown cultivar. Sakai and Imamura (1994) developed a new CMS system by irradiation of 60 krad of X-rays to radish cv. Kozena with a CMS-associated gene, *orf125*, which had a sequence homologous to that of Ogura in the CMS-associated gene, *orf138*, except for 2 amino acid substitutions and a 39 bp deletion in the *orf138* coding region (Iwabuchi et al. 1999). Oilseed rape ($2n = 38$) with Kozena cytoplasm was derived from cybrids between CMS Kozena and oilseed rape cv. Wester. A potential method of F_1 seed production based on Kozena CMS and the Rf-line with restored pollen fertility discovered in Kozena has been developed (Sakai and Imamura 1994, 2003).

Sakai and Imamura (2003) also succeeded in the production of the rape-like plant ($2n = 40$), which had stable seed fertility without chlorosis by adding one pair of radish chromosome. Yamagishi and Terachi (1994) reported a new CMS radish cv. Utsugi-gensuke having Ogura cytoplasm.

3.2 Application of Molecular Markers

The molecular information on the radish genome has been accumulated based on the active studies on some members of the genus *Brassica*. Methodologies for isozymes, restriction fragment length polymorphism (RFLP) and random amplified polymorphic DNA (RAPD) have been devised and practically used.

3.2.1 Evolution and Phylogenetic Relationship

The evolutionary pathway in Cruciferae or the genetic relationship between some relatives was analyzed by RFLP and microsatellite (or SSR) markers derived from the nuclear DNA and a specific domain of DNA from a cytoplasmic organelle. As per the nuclear DNA analysis, radish belonged to the *nigra* lineage (Song et al. 1990; Thormann et al. 1994). Although, both an intergenic spacer of 5S rRNA gene and an internal transcribed region between 18S rRNA and 25S rRNA of radish were confirmed to be identical in these relatives (Yang et al. 1998, 1999). On the other hand, as to chloroplast and mitochondrial DNA, radish belonged to the *rapa/oleracea* lineage (Palmer et al. 1983; Palmer and Herbon 1988; Pradhan et al. 1992; Warwick and Black 1991, 1997). Moreover, it was suggested that radish is related more closely to the *rapa/oleracea* lineage than the *nigra* lineage by the analysis of the sequenced chloroplast noncoding region between *trn*T and *trn*F and between *trn*D and *trn*T. On the basis of these facts, Yang et al. (2002) postulated the hybrid origin of radish between the above two lineages.

Recently, microsatellites were applied to examine the *B. rapa* genome, and it led to the observation that 48.2% of the *B. rapa* genome is stranded with (GA)n and (CAA)n. More than 90% and 79% of primers designed from isolated microsatellite clones of *B. rapa* were applicable to other *Brassica* species and radish, respectively. It was suggested that the microsatellite markers developed in *B. rapa* were effective for genome or synteny analysis in Cruciferae including radish (Suwabe et al. 2002, 2004). The DNA analysis for nucleus and cytoplasm was confirmed to be effective to depict the genetic, genomic and phylogenetic relationships in Cruciferae.

3.2.2 Identification and Evaluation of Genotypes and their Purity

Molecular analysis has been effectively used as a technique to examine genetic relationships among radish varieties. By isozyme analysis, radish could be divided into two main clusters, namely European hatsuka-daikon and rat-tailed radish cv. Pakkihood and Japanese radishes. Japanese radishes also formed seven subclusters (Nomura et al. 1994). In RAPD analysis, radish comprised two clusters. One cluster included hatsuka-daikon originated from Europe and the other comprised Chinese radish and Japanese

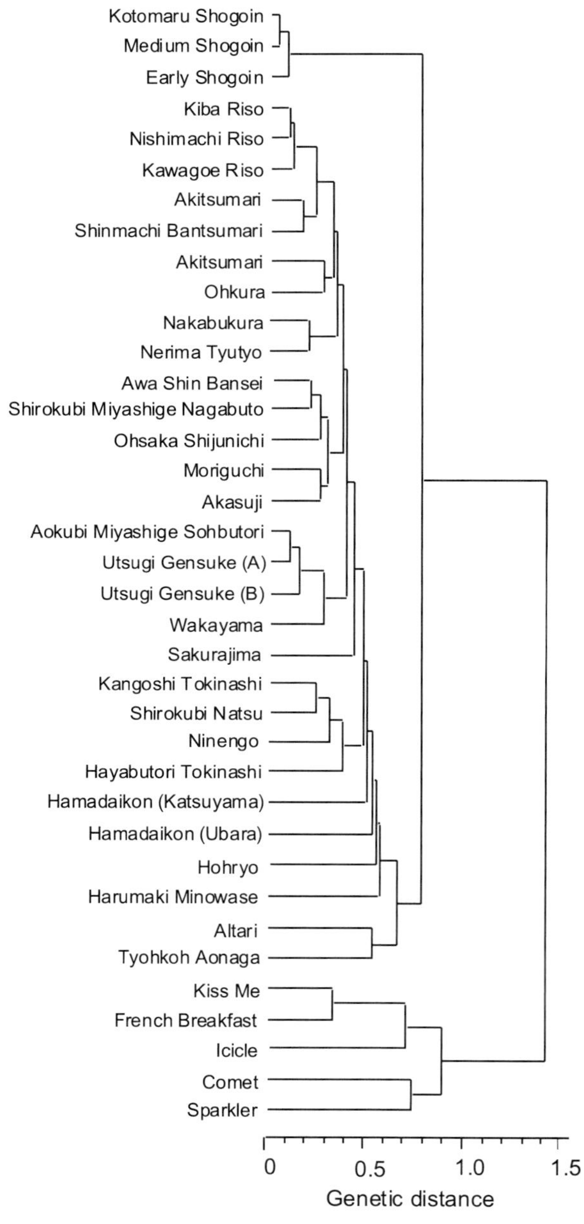

Fig. 1. Dendrogram showing genetic distance among radish varieties by UPGMA cluster analysis of Nei's distance calculated from RAPD data

group including wild radish Hama-daikon. The latter cluster also comprised eight subclusters (Takagi et al. 1994; Fig. 1). In addition, the collections of Hama-daikon from three districts were included in a cluster near one of Japanese and Chinese cultivated radishes. *R. raphanistrum*, on the other hand, was included in the most remote cluster suggesting the distant relationship in phylogeny (Yamagishi et al. 1998).

The isozyme and RAPD analyses have offered an effective method to clarify the genetic relationships of radish varieties where cultivars of Japanese radish were grouped into another type that was partly different from the one conventionally classified by Kumazawa (1965).

Information as to the differentiation of cultivated radishes was presented by analyzing the structure of the 3' region of the *Atp6* gene of mitochondria and a variation of base sequence (Yamagishi and Sasaki 2004). As a whole, 394 individuals from 27 Japanese, eight Chinese, and one Korean cultivar were examined by PCR analysis using three primers, which were produced from the base sequence inspected in the normal cytoplasm. As to the base sequence, 44% and 43% of individuals carried two distant varied types, and 75% of the cultivars belonged to one of four variants. A type except for two main types carried the cytoplasm resembling Ogura CMS, and another one to radish cv. black radish A. Two major types showed the structural rearrangements in mitochondrial genome at the site about 200 to 500 bp downstream from the *Atp6* gene. This resembled the *orfB* domain of mitochondria as clarified in a previous report (Terachi et al. 2001). By these facts, it could be assumed that most cultivars of Japanese cultivated radish were dual in origin as long as the cytoplasmic structure was concerned.

Yamagishi et al. (1998) evaluated the degree of fixation for each plant of a cultivar by RAPD analysis. They pointed out that the similarity index (SI value) among individuals in cultivated radishes was higher (0.70 for Chibetto-kei, 0.89 for Tokinashi) than those of the wild types of radish (0.505–0.727 for Hama-daikon and 0.669–0.696 for *R. raphanistrum*). Kaneko et al. (1999) studied the degree of fixation for the reverted-like plants ($2n = 18$) derived from the monosomic addition lines of radish ($2n = 19$) by RAPD analysis using 17 random primers. The SI values between generations (BC_1 to BC_4) are shown in Table 2. The values between Tokinashi and Shogoin and the one between Tokinashi and the original reverted radish-like plants ($2n = 18$) were 0.82 and 0.85, respectively. The value of Shogoin and the reverted radish-like plants was higher (0.95) than the two combinations mentioned above. By the exploration into the SI values between the reverted radish-like plants (BC_4) and the two distinctive cultivars, it could be assumed that the genetic background of the reverted radish-like plants ($2n = 18$) was converted from Shogoin to Tokinashi as the generations progressed.

Table 2. Similarity index (SI) among generations in the reverted radish-like plants ($2n = 18$) obtained from the radish cv. Tokinashi × original e-type MAL ($2n = 19$)

	Shogoin	Original[1]	Tokinashi
Shogoin			
Original[1]	0.95		
Tokinashi	0.82	0.85	
BC_4	0.84	0.87	0.98
BC_3	0.82	0.85	0.94
BC_2	0.84	0.87	0.92
BC_1	0.86	0.87	0.94
F_1	0.86	0.89	0.89

[1] Original: reverted radish-like plant possessing the genetic background of the radish cv. Shogoin, BC_4-F_1; reverted radish-like plants obtained by succesive backcrossing of the cv. Tokinashi × original e-type ($2n = 19$) with the cv. Tokinashi. Intracultivar SI values: in Shogoin = 0.94, in the cv. Tokinashi = 0.93, and in reverted radish-like plants ($2n = 18$) possessing the genetic background of the radish cv. Shogoin = 0.98

3.2.3 Confirmation of Hybridity

By isozyme analysis using esterase and peroxidase isozymes, the intergeneric hybridity of *Brassicoraphanus* between *B. rapa* ssp. *japonica* and *R. sativus* was identified (Kato and Tokumasu 1979). It was confirmed that the banding pattern of *Brassicoraphanus* comprised both bands of parental species in addition to the novel bands which might be produced by subunit reassociation. Hossain et al. (1988, 1990) reported that F_1 plants derived from *B. oleracea* × *R. sativus* were hybrids in the phosphoglucomutase isozyme. Marshall (1991) used phosphoglucose isomerase and leucine amino peptidase isozymes successfully to examine the selective fertilization among six parental lines in *R. raphanistrum*.

To introduce the trait of pod shattering resistance of radish into rape, Agnihorti et al. (1991) bred hybrids of *B. napus* × *Raphanobrassica* and tried to confirm the hybridity by RFLP analysis making use of both *Raphanus*-specific probe p337 and *Brassica*-specific probe pA7 encoding a 177-bp tandem repeat DNA fragment. Peter et al. (1995) performed RFLP analysis for hybrid plants between rape and radish using probes of pea r-DNA, *Arabidopsis thaliana* MS2 and the transposon-tagged cDNA constructed from a probe of *A. thaliana*.

RAPD analysis using total DNA as a template is widely examined for hybrid confirmation in various plants (Dubouzet et al. 1996; Kobayashi et al. 1996; Chrungu et al. 1999; Iizuka et al. 2001; Ishikawa et al. 2003; Tanaka et al. 2003). For example, Lefol et al. (1997) reported two hybrid plants from the crossing between *B. napus* and radish by conventional pollination and confirmed their hybridity by this analysis. Matsuzawa et al. (2000) also clarified the hybridity of *Raphanobrassica* (AARR, RRAA) derived from reciprocal crossings between radish and *B. rapa* by RAPD analysis with DNA-specific markers followed by the morphological and cytogenetical observations. These DNA-specific markers or protein markers such as isozyme, RFLP using a special domain and RAPD have been successfully used to confirm the hybrid nature.

3.2.4 Chromosome Identification

Chromosomal analysis has been carried out by examining the presence of a specific DNA marker such as RAPD, and on the other hand staining a specific site of chromosome or whole chromosome. The specific DNA markers were identified for each radish chromosome (Peterka et al. 2004) and chromosomes of other species in alien chromosome addition lines of radish (Kaneko et al. 2001; Bang et al. 2002). Hirai et al. (1995) visualized the presence of the repeated DNA sequence covering about 180 bp closely associated with the major C-heterochromatins at the proximal regions of the mitotic 18 chromosomes of radish. Five pairs of chromosomes of *Arabidopsis thaliana* ($2n = 10$), model plant for molecular biology, could not be discriminated by a conventional method because of their smaller size, but were successfully identified by the method of multicolored FISH using probes of two rRNA genes, cosmid clones and centromeric 180 bp repeated arrangement (Murata 1997). Radish has either four genes of 5S rRNA located near the centromere of a long arm in the longest chromosome, and four 25S rRNA located on a short arm of the same longest one, in addition to other genes for 5S or 25S rRNA located separately on the long arm of another two chromosomes (Schrader et al. 2000). More tiny chromosomes and their regions became identified by the combined application of DAPI and FISH.

Voss et al. (1999) tried to introduce the resistance to nematode from radish to rape through the radish chromosome addition line of rape by successive backcrossings. They could identify two radish chromosomes added to the rape genome and confirmed that one of them had a close relation to nematode resistance. Recently, Peterka et al. (2004) confirmed that a "d" chromosome of radish carried a nematode resistance gene (genes) by RAPD procedure. The chromosome inspected by Voss et al. (1999) might correspond to a "d" chromosome of radish reported by Peterka et al. (2004). Up till now, no one has succeeded in developing new chromosome addition lines with each individual radish chromosome that were evaluated for the genomic and chromosomal studies of radish in the near future.

3.2.5 Development of Markers for Self-Incompatibility and Cytoplasmic Male-Sterility Genes

By the PCR method, the male-sterility gene of Ogura CMS type was assumed to be the *orf138* of mitochondrial DNA that was confirmed to be coded only in the radish genome by a survey on 46 lines from 12 genera including 27 species of Cruciferae (Yamagishi and Terachi 1997). This CMS-specific gene was also found in the wild radish Hama-daikon and *R. raphanistrum*. The plants with the *orf138* gene of *R. raphanistrum* were collected from two lines of Turkey and one from California. Due to the fact that the genetic mode of *R. raphanistrum* was in accordance with the one of Hama-daikon, it was suggested that the gene of Hama-daikon would be derived from *R. raphanistrum* (Yamagishi 1999). Moreover, seven mutations, including six base substitutions and one deletion (39 bases), in *orf138* were detected when base sequencing was performed for this gene from 143 individuals of cultivated radish and wild radish. Two variations were specific to Hama-daikon and *R. raphanistrum* and a majority of Japanese radish had the genotype identical to Hama-daikon (Murayama et al. 1999).

A fertility restorer gene for Ogura CMS type was ascertained in both Hama-daikon and *R. raphanistrum*. This restorer gene was dimorphic where the one genotype was not found in Japanese radishes but in European and Chinese ones. As reported previously by RAPD (Takagi et al. 1994) and isozyme (Nomura et al. 1994) analysis, these facts might suggest that European and Japanese radishes were developed from independent ancestors (Yamagishi et al. 1998). It was confirmed that the fertility restorer gene carried in European radish, Hama-daikon and *R. raphanistrum* occupy a single locus being dominant. This gene is linked extremely close to the site of 1.2 cM with 410 bp band marker of primer OPH11. Seven molecular specific markers (RAPD markers) were also linked to the fertility restorer gene of fertile plants collected from European radish cv. Comet (Murayama et al. 1999).

Researches for genes related to self-incompatibility were carried out by the PCR-RFLP method (Niikura and Matsuura 1998). The SI gene in radish was inspected, referring to some information reported previously for *Brassica* species. Analysis of the PCR products with primers induced from the *S*-locus-specific glycoprotein (*SLG*) genes of *B. oleracea* and the total DNA extracted from radish as a template revealed that the gene S^{201} of radish conserved the homology with *SLG* of *B. oleracea* at the level of 88% in both DNA and amino acid arrangements.

3.3 Construction of Genetic Linkage Maps

In *Brassiceae*, genetic linkage maps were constructed in *B. napus* (Parkin et al. 1995; Sharpe et al. 1995), *B. rapa* (Song et al. 1991; Chyi et al. 1992; Teutonico and Osborn 1994; Truco et al. 1996; Kole et al. 1997; Ajisaka et al. 1999) and *B. oleracea* (Kianian and Quiros 1992; Kennard et al. 1994; Truco et al. 1996; Ramsay et al. 1996; Moriguchi et al. 1999). The RFLP and RAPD markers developed in the above studies are now being applied to develop the map of radish.

The genetic linkage map of radish was first reported by Bett and Lydiate (2003). In order to increase the number of polymorphic markers, they practiced the wide cross between *R. sativus* and *R. raphanistrum*. The map was constructed by using 171 *Brassica* RFLP probes in both 54 and 85 individuals of the BC_1 and F_2 population, respectively. A total of 144 probes (84%), i.e., 55 of pW probes, 40 of pN and six of cA derived from *B. napus*, 33 of pO from *B. oleracea* and 15 of pR from *B. rapa* were hybridized to radish DNA. As a result, 236 markers including 179 common markers in the BC_1 and F_2 populations were distributed over 915 cM and 844 cM, respectively.

Tsuro et al. (2005) reported a linkage map using 94 F_1 plants derived from a cross between open-pollinated cultivars, Huang-he hong-wan and Utsugi-Gensuke. Both the amplified fragment length polymorphism (AFLP) markers developed by Vos et al. (1995) and SSR markers of *B. rapa* previously reported by Suwabe et al. (2003) were assigned to the 11 linkage groups.

Using the inbred lines, HA2 and GSK3-1, which were selected from cultivated Japanese radish, Tokinashi exhibiting the late-bolting and Utsugi-Gensuke belonging to Miyashige-group, Kimizuka-Takagi and coworkers (personal communication, Kimizuka-Takagi 2006) in Mikado Seed Growers Co., Ltd. constructed the molecular linkage map of Japanese radish by RAPD and RFLP markers. To detect the polymorphism between the parents, European radish or Japanese radish cv. Shogoin belonging to the specific groups revealed by UPGMA cluster analysis from RAPD data (Fig. 1), they used plant materials derived from Tokinashi and Utsugi-Gensuke, because Japanese leading varieties known as Aokubi-radish have been bred from Miyashige and the late-bolting trait has mainly been introduced from Tokinashi for the development of Spring-cropping type.

The single F_1 plant was crossed to HA2 to develop a BC_1 population. One hundred twenty nine individuals were used for the marker screening and linkage analysis. The RFLP analysis was carried out using *Brassica* probes from Wester *Pst*I, Tobin *Eco*RI genomic and Wester cDNA libraries (Camargo 1994) and another Wester *Pst* I genomic library (Chyi et al. 1992). The RAPD by 10-mer and 12-mer oligonucleotides was also employed. The linkage analysis with computer software MAPL (Ukai et al. 1995) distributed the markers into nine linkage groups corresponding to the haploid chromosome number of radish and an additional short linkage block with 37 RFLP and 119 RAPD marker loci (Fig. 2). Total map length was 1,239.5 cM with 8.9 cM of average marker interval. Further mapping efforts using recombinant inbred lines (RIL) have been performed. We have

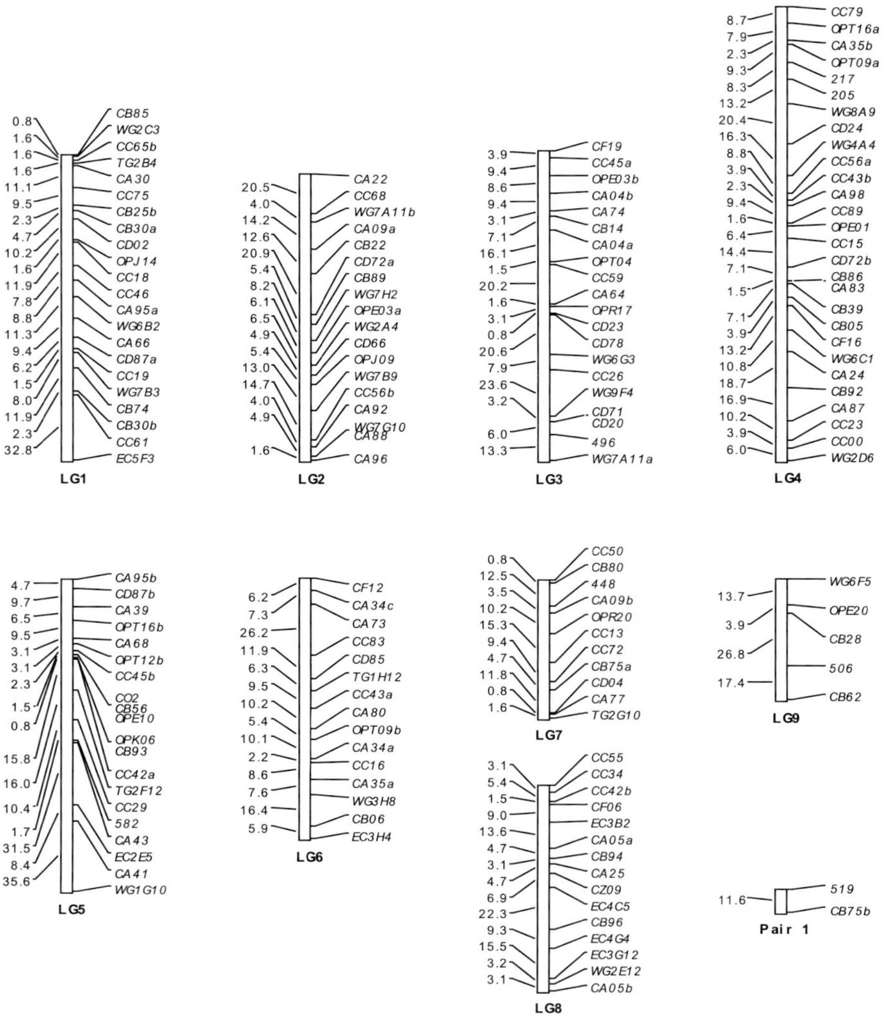

Fig. 2. BC_1 population-derived genetic linkage map of *Raphanus sativus* L. based on RFLP and RAPD markers

mapped two visible markers, *Ph* (purple hypocotyl) and *Pf* (purple flower), and disease resistance gene-related ACGM (Fourmann et al. (2002), chalcone isomerase, basic chitinase, cinnamyl alcohol dehydrogenase, EL13, cinnamate-4-hydroxylase and chitinase. A more precise linkage map could be developed based on our recent work.

Some mapping works have been focused on the fertility restorer gene, *Rfo*, of Ogura cytoplasmic male-sterility. Murayama et al. (1999) found the RAPD marker linked to *Rfo* and they converted it to a sequence characterized amplified region (SCAR; Paran and Michelmore 1993) marker, $OHP11_{410}$, from the cloned RAPD fragment. Moreover, AFLP and SCAR marker, STS190, converted from AFLP fragment was reported in the analysis of the same mapping population (Murayama et al. 2003). Recently, Giancola et al. (2003) reported *Rfo*-linked AFLP, ACGM (Brunel et al. 1999) and SCAR markers with no recombination by analysis of a total of 976 individuals.

3.4 Analysis of Quantitative Trait Loci

Quantitative Trait Loci (QTL) for the late-bolting trait were analyzed based on the map information shown in Fig. 2. The BC_1 plants were transplanted in the greenhouse in late autumn and the vernalized plants under the natural condition were examined for days-to-bolting (DTB). The phenotypic trait data for DTB exhibited continuous distribution suggesting polygenic control on the trait. The analysis using computer software QGENE (Nelson 1997) revealed the QTL on LG 5

Fig. 3. Radish cultivated in the open field infected by *Fusarium oxysporum* f. sp. *raphani*. The resistant variety (*left*) grows normally until the harvest, however, the susceptible one (*right*) dies

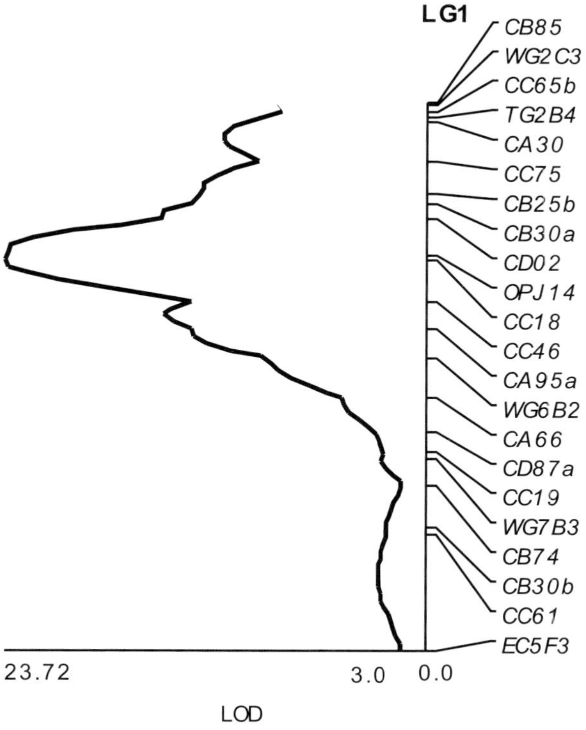

Fig. 4. LOD graph of LG 1 showing the location of a QTL associated with resistance to yellows caused by *F. oxysporum* f. sp. *raphani*

in the neighborhood of RFLP locus, TG2F12. This QTL explained 43.8% of the phenotypic variation. Subsequent analysis on RILs was performed under several conditions, i.e., vernalization requirement, different sowing dates with natural vernalization, seed vs. green plant vernalization, duration of vernalization and the long or short-day length after vernalization. No line could bolt under the nonvernalized condition. Besides on LG 5 from the result of BC_1, additional QTLs could be detected on some linkage groups, although the QTL on LG 5 could be activated under all conditions of the vernalized plants. Fine-mapping for DTB locus is attempted by using near-isogenic lines (NIL) that substituted the chromosomal region around the QTL on LG 5.

Yellows disease caused by *Fusarium oxysporum* f. sp. *raphani* has brought about tremendous damage to radish production in the continuous cropping field. In the infected field, susceptible varieties die completely (Fig. 3). It was reported that resistance to yellows is controlled by quantitative multiple genes anonymous in the dominant or recessive interaction (Peterson and Pound 1960; Ashizawa et al. 1979), although some major genes were also reported (Hida and Ashizawa 1985). In the artificial inoculation test for BC_1S_1 families, they became more susceptible than the resistant parent and the F_1 plant because the susceptible parent has been used for the backcrossing. The QTL for yellows resistance was detected in the neighborhood of RAPD loci, OPJ14 and CC18 on LG 1 (Fig. 4). Total phenotypic variation explained was 60.4% at this locus. Furthermore, in the analysis of RILs, two QTLs controlling yellows disease resistance were detected on LG 1. An allele controlling the shift from resistant to susceptible was also identified on LG 7. RILs could offer the potential plant materials to analyze QTLs for more complex traits of radish, e.g., root length, root enlargement, degree of neck-color (so-called Aokubi), disease resistance and some physiological disorders.

Recently, a quantitative description method of plant morphology was developed (Iwata et al. 1998, 2000, 2004a,b). It might be possible to map the QTLs for the features of the root and other traits in radish.

3.5 Marker-Assisted Breeding

The fragments detecting the RAPD markers mapped on radish linkage groups (Fig. 2) were excised from

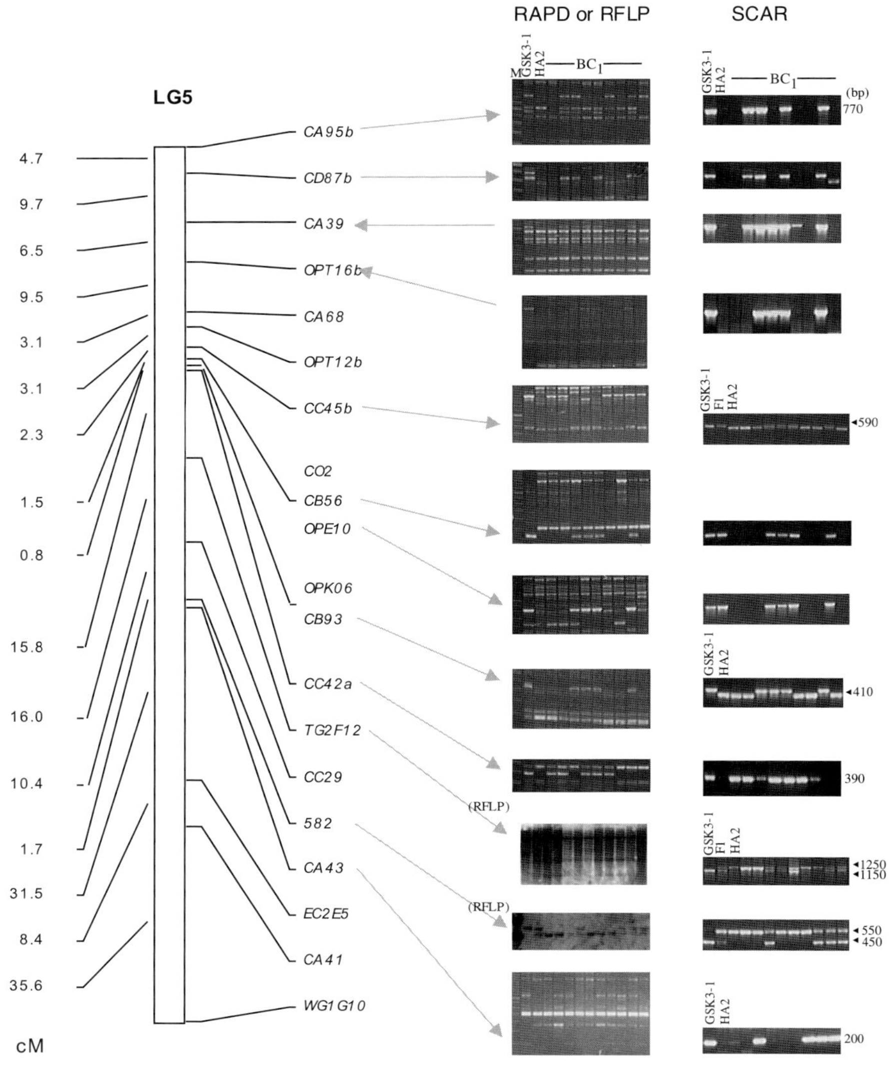

Fig. 5. Segregation patterns of the original RAPD or RFLP and the converted SCARs on radish LG 5 in a subset from BC_1 population

agarose gel and cloned into a plasmid vector. The cloned RAPD fragments and RFLP probes were sequenced to design PCR primers. For the confirmation of SCAR conversion, SCAR markers were compared to the original pattern of segregation in a subset of BC_1 mapping population. As a result, about 55% of mapped markers were successfully converted into SCAR. The LG 5 is shown in Fig. 5. Subsequently, a PCR-based RIL-map consisting of a total of 212 loci including SCAR, SSR and ACGM markers has been constructed (personal communication, Kimizuka-Takagi, 2006).

Marker-assisted backcross breeding (MABB) was attempted to introduce yellows resistance by using the BC_1 mapping population. The line possessing resistant allele and less remnant alleles for background loci was selected by the examination of graphical genotypes. After repeated MABB, we could select the NILs, which were introgressed from the resistant donor parent to susceptible one (Fig. 6). The NIL was proved to be resistant in the artificial inoculation practices, whereas all of the inoculated plants of the original line died from the pathogen (Fig. 7). Yellows resistance was also introduced into the elite inbred line that could be actually used as a parent of commercial F_1 hybrid varieties. In our recent study, the *S*-locus linked to yellows resistance was identified (personal communication, Kimizuka-Takagi, 2006). Therefore, the *S*-isogenic line could also be selected in addition to introgressing resistance by MABB. A BC_1 population was grown in the field, selected for the phenotype, and backcrossed again. In the BC_2 generation,

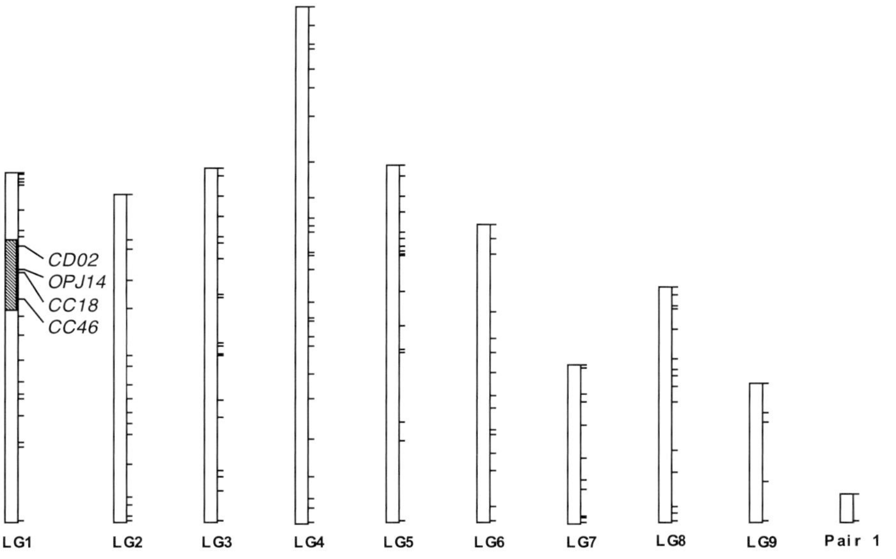

Fig. 6. Graphical representation of the near-isogenic line (NIL) that introgressed chromosomal segment involving a resistance to yellows. *Hatched box* indicates the introgressed region

Fig. 7. Artificially inoculated radish seedlings with *F. oxysporum* f. sp. *raphani*. The resistant donor parent, GSK3-1, and the near-isogenic line (NIL), HA2yr, survive, however, HA2 plants are completely dead

the plants selected by breeder's eye were genotyped by molecular markers. In this step, the BC_2 plant that possessed 52% of donor allele among mapped loci was selected and backcrossed again. Subsequent BC_3 generation was not conducted in the field. DNA was extracted from individual seedlings grown in the greenhouse. Marker-genotyping was performed at the remaining marker loci of nonrecurrent parental allele in BC_2 generation. It was assumed that some BC_3 plants had a chromosomal region containing yellows disease resistance on SCAR marker loci, SCC18 and SOPJ14, and the other background reverted to the recurrent parent completely. Moreover, BC_3 plants with a longer chromosomal segment involving *S*–locus

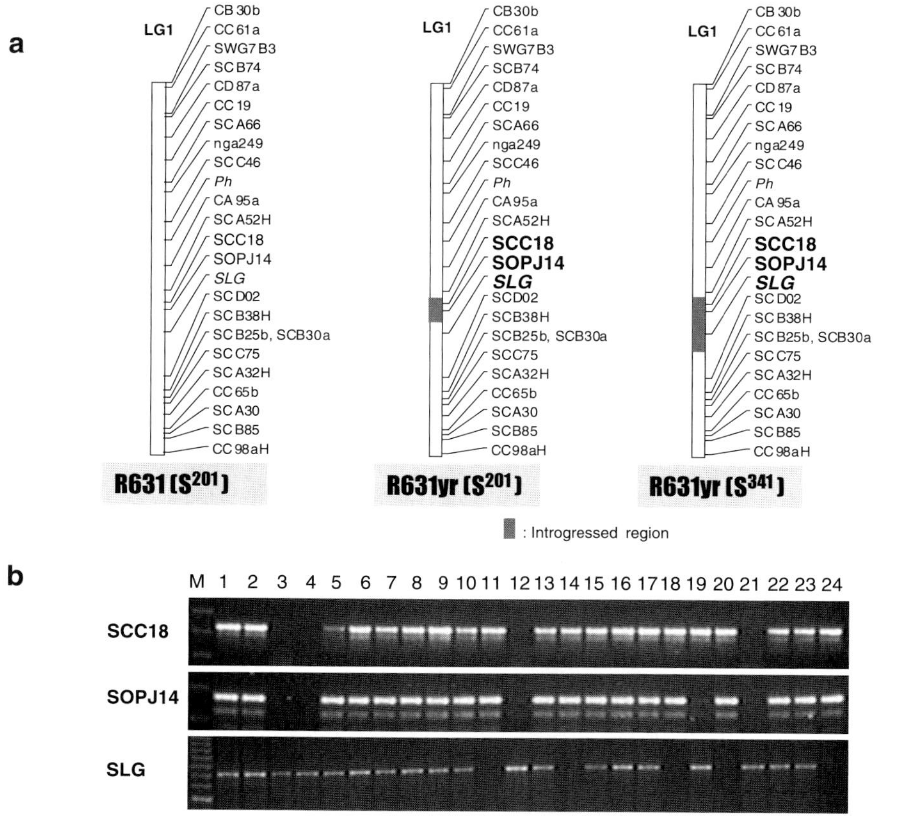

Fig. 8. Improvement of elite inbred line, R631, concerning with yellows resistance and self-incompatibility allele. Graphical representation of the NILs showing the introduced yellows resistance and cointrogressed *S*-allele (**a**), and the marker profiles in the segregant (**b**). Plants excepting of lanes, 3, 4, 12 and 21, possess resistance and the individuals of lanes, 11, 14, 18, 20 and 24, are cointrogressed an *S*-allele (S^{341}) from donor patent. Detection of *S*-allele was conducted by using primer pair for class II SLG (Lim et al. 2002). To confirm the amplified SLG fragment, PCR product was sequenced and aligned with DNA databases, and revealed significant homology to known *R. sativus* SLG. Linkage map derived from BIL are shown

(*SLG*) were selected (Fig. 8). It would be possible to convert the released commercial varieties to the yellows-resistance-version without changes of other agronomical traits. Development of an isogenic line for *S*-allele may lead to a double crossing system for F_1 seed production. MABB could also contribute to breed CMS lines. In fact, we could substitute the normal nucleus with Ogura cytoplasm (Ogura 1968) in only three years.

3.6 Map-Based Cloning

Recently, the fertility restorer gene (*Rf*) for cytoplasmic male-sterility has successively been cloned by the map-based cloning strategy. Desloire et al. (2003) reported on a combination of positional cloning and microsynteny analysis between *Arabidopsis* and radish. The tightly linked AFLP markers to the fertility restorer gene (*Rfo*) of Ogura cytoplasm were sequenced. It was identified that these markers were orthologous to *Arabidopsis* chromosome 1 (Giancola et al. 2003). The primers were newly designed from the sequence of the bacterial artificial chromosome (BAC) of *Arabidopsis*, and a high-resolution map was constructed by using 6907 radish segregant plants (Desloire et al. 2003). They tightly mapped *Rfo* to a genetic region of 0.042 cM where the markers M-F24D7.9 and M-F24D7.13 derived from *Arabidopsis* BAC clone, F24D7 were located. By screening a BAC library of radish, they finally delimited to a single clone, BAC64, and concluded that the *Rfo* was likely to correspond to pentatricopeptide-repeat (PPR) protein in its sequence from the sequence analysis. Similar results have been reported by Brown et al. (2003), although they accomplished the cloning of *Rfo* gene from the tightly linked *Brassica* RFLP marker as a starting-point for the walking. Some constructs of the variously predicted genes from genomic BAC or cosmid clones were introduced into *B. napus* having Ogura CMS by *Agrobacterium*-mediated transformation. They identified a construct, Bgl-5 that led to the recovery of pollen fertility stably inherited to T1 progeny. Since a single open-reading frame (ORF) named as g26 was predicted solely in the Bgl-5 radish DNA, it could be postulated that the g26 ORF corresponded to *Rfo*.

The polypeptide encoded by the g26 mRNA was 687 amino acids in length and contained 35 amino acid consensus sequence of PPR motif. Furthermore, much effort was also devoted to clone *Rfk1* which is the fertility restorer in Kozena CMS radish. Imai et al. (2003) detected four AFLP markers that cosegregated with the *Rfk1* locus and used to create a 163-kb lambda and cosmid contig. In the subsequent analysis, two cosmid clones covering the 43 kb region were identified, and then attempted to identify the fertility restorer by *Agrobacterium*-mediated transformation into the cybrid *B. napus* CMS, SW18, possessing Kozena mtDNA (Koizuka et al. 2003). As the plants with one of the lambda clones were pollen fertile, further transformation works were performed using ORFs within a lambda clone. They delimited *Rfk1* gene into 4.7-kb DNA fragments and identified *orf687* by RACE-PCR analysis. Brown et al. (2003) remarked that the sequences of the protein encoded by g26 for the fertility restoration of Ogura CMS and the *orf687* in Kozena CMS (Koizuka et al. 2003) were identical.

3.7 Future Works

In radish, a large amount of molecular biological information has been accumulating only in this decade. However, it has more unsettled aspects in comparison with rice and wheat. Useful molecular study has just started in radish. In this context, it is required to construct more detailed maps, to combine molecular markers and useful traits, e.g., disease resistances, bolting, and so on, to detect more QTLs on the chromosomes, and to match molecular maps and visualize chromosomes for identification of useful traits and introgression of desirable genes. The *Arabidopsis* genome information can be obtained via website (http://www.arabidopsis.org) and is a very useful resource to the *Raphanus* geneticists and breeders as well as other *Brassica* researchers. Although the genetic linkage map of *Raphanus sativus* will be more refined with further genetic markers, especially AFLPs, some gaps can also be filled by designing new markers from microsynteny analysis between *Raphanus* and *Arabidopsis*. Furthermore, the mapped loci including QTLs for agronomically important traits, e.g., disease resistance, late-bolting (flowering time) and morphological characters, will be related to known genes in *Arabidopsis*. Moreover, the development of cloning and discrimination of base sequence for useful genes are also important works.

References

Agnihorti A, Shivanna KR, Raina SN, Lakshmikumaran M, Prakash S, Jagannathan V (1991) Production of *Brassica napus* × *Raphanobrassica* hybrids by embryo rescue: an attempt to introduce shattering resistance into *B. napus*. Plant Breed 105:292–299

Ajisaka H, Kuginuki Y, Shiratori M, Ishiguro K, Enomoto S (1999) Mapping of loci affecting the cultural efficiency of microspore culture of *Brassica rapa* L. syn *campestris* L. using DNA polymorphism. Breed Sci 49:187–192

Aoba T (1979) Varieties and breeding in Japanese radish. The Iden 33:55–60 (in Japanese)

Aoba T (1981) Geographical distribution of domestic varieties in radish. In: Vegetables. Publication Department, Housei University, Tokyo, pp 232–251 (in Japanese)

Aoba (1988) Pedigree and variation of wild radish in Japan. Tech Agri 12:94–114 (in Japanese)

Armstrong KC, Keller WA (1981) Chromosome pairing in haploids of *Brassica campestris*. Theor Appl Genet 59:49–52

Armstrong KC, Keller WA (1982) Chromosome pairing in haploids of *Brassica oleracea*. Can J Genet Cytol 24:735–739

Arumuganathan K, Earle ED (1991) Nuclear DNA content of some important plant. Plant Mol Biol Rep 9:208–218

Ashizawa M, Hida K, Yoshikawa H (1979) Studies on the breeding *Fusarium* resistance in radish. I. Screening of radish varieties for *Fusarium* resistance. Bull Veg Ornam Crops Res Sta 6:39–70 (in Japanese)

Ashizawa M (1984) Transition of crop sorts and varieties in vegetables. Rep Veg Ornam Crops Res Stn 17:214–236 (In Japanese)

Bang SW, Kaneko Y, Matsuzawa Y, Bang KS (2002) Breeding of *Moricandia arvensis* monosomic chromosome addition lines (2n=19) of alloplasmic (*M. arvensis*) *Raphanus sativus*. Breed Sci 52:193–199

Banga O (1976) Radish, *Raphanus sativus* (Cruciferae). In: Simmonds NW (ed.) Evolution of Crop Plants. Longman, London, pp 60–62

Bannerot H, Boulidard L, Cauderon Y, Tempe J (1974) Transfer of cytoplasmic male sterility from *Raphanus sativus* to *Brassica oleracea*. Eucarpia Meet, Cruciferae. Dundee, Scotland, pp 52–54

Bannerot H, Boulidard L, Chupeau Y (1977) Unexpected difficulties met with the radish cytoplasm in *Brassica oleracea*. Cruciferae News 2:16

Bateman AJ (1952) Self-incompatibility systems in angiosperms. I. Theory. Heredity 6:285–310

Bateman AJ (1955) Self-incompatibility systems in angiosperms. III. Cruciferae. Heredity 9:53–68

Bett KE, Lydiate DJ (2003) Genetic analysis and genome mapping in *Raphanus*. Genome 46:423–430

Brown GG, Formanova N, Jin H, Wargachuk R, Dendy C, Patil P, Laforest M, Zhang J, Cheung WY, Landry BS (2003) The radish *Rfo* restorer gene of Ogura cytoplasmic male sterility encodes a protein with multiple pentatricopeptide repeats. Plant J 35:262–272

Brunel D, Froger N, Pelletier G (1999) Development of amplified consensus genetic markers (ACGM) in *Brassica napus* from *Arabidopsis thaliana* sequence on known biological function. Genome 42:387–402

Camargo LEA (1994) Mapping RFLP and quantitative trait loci in *Brassica oleracea*. Dissertation, University of Wisconsin, Madison, WI

Chrungu B, Verma N, Mohanty A, Pradhan A, Shivanna KR (1999) Production and characterization of interspecific hybrids between *Brassica maurorum* and crop brassicas. Theor Appl Genet 98:608–613

Chyi YS, Hoenecke ME, Sernyk JL (1992) A genetic linkage map of restriction fragment length polymorphism loci for *Brassica rapa* (syn. *campestris*). Genome 35:746–757

Desloire S, Gherbi H, Laloui W, Marhadour S, Clouet V, Cattolico L, Falentin C, Giancola S, Renard M, Budar F, Small I, Caboche M, Delourme R, Bendahmane A (2003) Identification of the fertility restoration locus, *Rfo*, in radish, as a member of the pentatricopeptide-repeat protein family. EMBO Rep 4:588–594

Dolourme R, Horvais R, Vallee P, Renard M (1999) Double low restored F_1 hybrids can be produced with the Ogu-INRA CMS in rapeseed. Proceedings of the 10th International Rapeseed Congress, Canberra, Australia, 1999

Dubouzet JG, Etoh T, Arisumi K, Yoshitake T (1996) A diagnostic test to confirm interspecific *Allium* hybrids using random amplified polymorphic DNA from crude leaf DNA extracts. J Jpn Soc Hort Sci 65:321–326

Eber F, Chevre AM, Baranger A, Vallee P, Tanguy X, Renard M (1994) Spontaneous hybridization between a male-sterile oilseed rape and two weeds. Theor Appl Genet 88:362–368

Eber F, Boucherie R, Broucqsault LM, Bouchet Y, Chevre AM (1998) Spontaneous hybridization between vegetable crops and weeds. 1. Garden radish (*Raphanus sativus* L.) and wild mustard (*Sinapis arvensis* L.). Agronomie 18:489–497

Florentin MA, Penalva M, Calegari A, Derpsch R (2001) Abonos verdes y rotacion de cultivos en siembra directa. Pequenas propiedades. In: Proyecto Conservacion de Suelos Magtz Deag San Lorenzo, Paraguay, pp 36–37 (in Spanish)

Fourmann M, Barret P, Froger N, Baron C, Charlot F, Delourme R, Brunel D (2002) From *Arabidopsis thaliana* to *Brassica napus*: development of amplified consensus genetic markers (ACGM) for construction of a gene map. Theor Appl Genet 105:1196–1206

Frost HB (1923) Heterosis and dominance of size factors in *Raphanus*. Genetics 8:116–153

Furusato K, Miyazawa A (1958) Varieties of Japanese radish viewed from horticulture. In: Nishiyama I (ed.) Japanese Radish. Jpn Soc From Sci, Tokyo, pp 138–161 (in Japanese)

Giancola S, Marhadour S, Desloire S, Clouet V, Falentin-Guyomarc'h H, Laloui W, Falentin C, Pelletier G, Renard M, Bendahmane A, Delourme R, Budar F (2003) Characterization of a radish introgression carrying the Ogura fertility restorer gene *Rfo* in rapeseed, using the *Arabidopsis* genome sequence and radish genetic mapping. Theor Appl Genet 107:1442–1451

Harberd DJ (1976) Cytotaxonomic studies of *Brassica* and related genera. In: Vaughan JG, MacLeod AJ, Jones BMG (eds.) The Biology and Chemistry of the Cruciferae. Academic, London, pp 47–68

Harberd DJ, McAuthur ED (1980) Meiotic analysis of some species and genus hybrids in the *Brassiceae*. In: Tsunoda S, Hinata K, Gomez-Campo C (eds.) *Brassica* Crops and Wild Allies. Biology and Breeding. Jpn Sci Soc Press, Tokyo, pp 65–87

Haruta T (1962) Studies on the genetics of self- and cross-incompatibility in cruciferous vegetables. Rep Agri Nagaoka Farm Takii Co 2:1–169

Hida K (1990) Root crops: Radish. In: Matsuo T (ed.) Collected Data of Plant Genetic Resources, vol. 2. Koudansha, Tokyo, pp 823–834 (in Japanese)

Hida K, Ashizawa M (1985) Breeding of radishes for *Fusarium* resistance. JARQ 19:133–138

Hinata K (1995) Radish Asahi Encyclopedia. World Plant 67:206–207 (in Japanese)

Hirai K, Irifune K, Tanaka R, Morikawa H (1995) Molecular and cytological characterization of a highly repeated DNA sequence in *Raphanus sativus*. Genome 38:1237–1243

Hirai M, Harada T, Kubo N, Tsukada M, Suwabe K, Matsumoto S (2004) A novel locus for clubroot resistance in *Brassica rapa* and its linkage markers. Theor Appl Genet 108:639–643

Hirao R (1982) Chinese vegetables (1), the varieties and cultivation. Agri Hort 57:653–660 (in Japanese)

Hoshi T, Takemura E, Hayashi K (1963) Genetic modification of hydroxylation pattern in radish anthocyanins. Studies in anthocyanins, XLII. Bot Mag Tokyo 76:431–439

Hoshi T, Hosoda T (1967) Genetical studies on anthocyanins in *Brassica* crops. Jpn J Breed 17:47–51

Hossain MM, Inden H, Asahira T (1988) Intergeneric and interspecific hybrids through in vitro ovule culture in the cruciferae. Plant Sci 58:121–128

Hossain MM, Inden H, Asahira T (1990) In vitro culture of intergeneric hybrids between *Brassica oleracea* and *Raphanus sativus*. Sci Hort 41:181–188

Iizuka M, Kudo N, Kimura Y, Ogiwara I (2001) Interspecific hybrids between *Spiraea thunbergii* Sieb. ex Blume, and *S. japonica* L. fil. via ovule culture. J Jpn Soc Hort Sci 70:767–773

Imai R, Koizuka N, Fujimoto H, Hayakawa T, Sakai T, Imamura J (2003) Delimitation of the fertility restorer locus *Rfk1* to a 43-kb contig in Kosena radish (*Raphanus sativus* L.). Mol Genet Genom 269:388–394

Ishikura N, Hoshi T, Hayashi K (1965) Crystallization and characterization of the basic triglucoside common to all components in purple pigment of hybrid radish. Studies on anthocyanins, XLV. Bot Mag Tokyo 78:8–13

Ishikawa S, Bang SW, Kaneko Y, Matsuzawa Y (2003) Production and characterization of intergeneric somatic hybrids between *Moricandia arvensis* and *Brassica oleracea*. Plant Breed 122:233–238

Iwabuchi M, Koizumi N, Fujimoto H, Sakai T, Imamura J (1999) Identification and expression of the kosena radish (*Raphanus sativus* cv. Kosena) homolog of the ogura radish CMS-associated gene, *orf138*. Plant Mol Biol 39:183–188

Iwasa S (1980) Root crops: Radish. In: Vegetables in Tropics. Youkendo, Tokyo, pp 293–300 (in Japanese)

Iwata H, Niikura S, Matsuura S, Takano Y, Ukai Y (1998) Evaluation of variation of root shape of Japanese radish (*Raphanus sativus* L.) based on image analysis using elliptic Fourier descriptors. Euphytica 102:143–149

Iwata H, Niikura S, Matsuura S, Takano Y, Ukai Y (2000) Diallel analysis of root shape of Japanese radish (*Raphanus sativus* L.) based on elliptic Fourier descriptors. Breed Sci 50:73–80

Iwata H, Niikura S, Matsuura S, Takano Y, Ukai Y (2004a) Genetic control of root shape at different growth stage in radish (*Raphanus sativus* L.). Breed Sci 54:117–124

Iwata H, Niikura S, Matsuura S, Takano Y, Ukai Y (2004b) Identification between genetic effects and soil type in diallel analysis of root shape and size of Japanese radish (*Raphanus sativus* L.). Breed Sci 54:313–318

Kagawa A (1971): Studies on the inheritance of flower inductive habits in *Brassica* crops. Bull Fac Agri Gifu Univ 31:41–62 (in Japanese)

Kaneko Y (1980) Haploid plant of *Raphanus sativus*. C. I. S. 28:21–23

Kaneko Y, Bang SW, Matsuzawa Y (1999) Early-bolting trait and RAPD markers in the specific monosomic addition line of radish carrying the e-chromosome of *Brassica oleracea*. Plant Breed 119:137–140

Kaneko Y, Yano H, Bang SW, Matsuzawa Y (2001) Production and characterization of *Raphanus sativus-Brassica rapa* monosomic chromosome addition lines. Plant Breed 120:163–168

Kaneko Y, Bang SW, Torii-Abe J, Eduardo RB, Matsuzawa Y (2003) Chromosome pairing in haploid plants of radish derived from alien monosomic addition lines. Plant Breed 122:450–452

Karron JD, Marshall DL, Oliveras DM (1990) Numbers of sporophytic self-incompatibility alleles in populations of wild radish. Theor Appl Genet 79:457–460

Kato M, Tokumasu S (1979) An electrophoretic study of esterase and peroxidase isozyme in *Brassicoraphanus*. Euphytica 28:339–349

Kato M, Fukuyama T (1982) Production of *Raphanus sativus* Aokubi-miyashige with *R. raphanistrum* Seiyo-nodaikon cytoplasm. I. Process nucleus substitution and selection in the fourth and sixth generations. Bull Exp Farm Coll Agri Ehime Univ 4:29–37

Kennard WC, Slocum MK, Figdore SS, Osborn TC (1994) Genetic analysis of morphological variation in *Brassica oleracea* using molecular markers. Theor Appl Genet 87:721–732

Kianian SF, Quiros CF (1992) Generation of a *Brassica oleracea* composite RFLP map: linkage arrangement among various populations and evolutionary implications. Theor Appl Genet 84:544–554

Kikuchi M, Ajisaka H, Kuginuki Y, Hirai M (1999) Conversion of RAPD markers for a clubroot resistance gene of *Brassica rapa* into sequence-tagged sites (STSs). Breed Sci 49:83–88

Kitamura S (1958) Varieties and transitions of radish. In: Nishiyama I (ed.) Japanese Radish. Jpn Soc From Sci Tokyo, Japan, pp 1–19 (in Japanese)

Kobayashi N, Akabane M, Handa T, Takayanagi K (1996) Inheritance of morphological characters and RAPD markers in intersubgeneric hybrids of Azalea, (*Rhododendron kiusianum* Makino × *R. indicum* (L.) Sweet) × *R. japonicum* (A. Gray) Suringer f. *flavum* Nakai. J Jpn Soc Hort Sci 65:145

Koizuka N, Imai R, Fujimoto H, Hayakawa T, Kimura Y, Kohno-Murase J, Sakai T, Kawasaki S, Imamura J (2003) Genetic characterization of a pentatricopeptide repeat protein gene, *orf687*, that restores fertility in the cytoplasmic male-sterile Kosena radish. Plant J 34:407–415

Kole C, Kole P, Vogelzang R, Osborn TC (1997) Genetic linkage map of a recombinant inbred population of *Brassica rapa*. J Hered 88:553–557

Kumazawa S, (1965) Radish. In: Vegetable Gardening. Youkendo, Tokyo, pp 295–322 (in Japanese)

Lefol E, Seguin-Swartz G, Downey RK (1997) Sexual hybridization in crosses of cultivated *Brassica* species with the crucifers *Erucastrum gallicum* and *Raphanus raphanistrum*: Potential for gene introgression. Euphytica 95:127–139

Lim S, Cho HJ, Lee SJ, Cho YH, Kim BD (2002) Identification and classification of *S*–haplotypes in *Raphanus sativus* by PCR-RFLP of the S locus glycoprotein (SLG) gene and the S locus receptor kinase (SRK) gene. Theor Appl Genet 104:1253–1262

Marshall DL (1991) Nonrandom mating in wild radish: Variation in pollen donor success and effects of multiple paternity among one- to six-donor pollinations. Am J Bot 78:1404–1418

Matsuzawa Y, Funayama T, Kamibayasi M, Konnai M, Bang SW, Kaneko Y (2000): Synthetic *Brassica rapa-Raphanus sativus* amphidiploid lines developed by reciprocal hybridization. Plant Breed 119:357–359

Mizushima U (1952) Karyogenetic studies on *Brassiceae*. Gihodo, Tokyo, Japan, pp 1–112 (in Japanese)

Mizushima U (1980) Genome analysis in *Brassica* and allied genera. In: Tsunoda S, Hinata K, Gomez-Campo C (ed.) *Brassica* Crops and Wild Allies. Biology and Breeding. Jpn Sci Soc Press, Tokyo, pp 89–106

Moriguchi K, Kimizuka-Takagi C, Ishii K, Nomura K (1999) A genetic map based on RAPD, RFLP, isozyme, morphological markers and QTL analysis for clubroot resistance in *Brassica oleracea*. Breed Sci 49:257–265

Mukherjee P (1979) Karyotypic variation in ten strains of Indian radish (*Raphanus sativus* L.). Cytologia 44:347–352

Murata M, Heslop-Harrison JS, Motoyoshi F (1997) Physical mapping of the 5S ribosomal RNA genes in *Arabidopsis thaliana* by multi-color fluorescence in situ hybridization with cosmid clones. Plant J 12:31–37

Murayama S, Yamagishi H, Terachi T (1999) Identification of RAPD and SCAR markers linked to a restorer gene for Ogura cytoplasmic male sterility in Radish (*Raphanus sativus* L.) by bulked segregant analysis. Breed Sci 49:115–121

Murayama S, Habuchi T, Yamagishi H, Terachi T (2003) Identification of a sequence-tagged site (STS) marker linked to a restorer gene for Ogura cytoplasmic male sterility in radish (*Raphanus sativus* L.) by non-radioactive AFLP analysis. Euphytica 129:61–68

Namai H (1991) Breeding using cytoplasmic male sterility (CMS) in radish. Jpn Soc Seed Prod Breed Engine in 1991, pp 165–181 (in Japanese)

Nelson JC (1997) QGENE: software for marker-based genomic analysis and breeding. Mol Breed 3:239–245

Niikura S, Matsuura S (1998) Identification of self-incompatibility alleles (S) by PCR-RFLP in radish (*Raphanus sativus* L.). Euphytica 102:379–384

Niikura S, Matsuura S (1999) Genetic variation of the *S*-alleles and level of self-incompatibility in Japanese cultivated radish (*Raphanus sativus* L.). Breed. Res 1:211–220

Niikura S (2000) Studies on the self-incompatibility related characteristics of radish (*Raphanus sativus* L.). Dissertation, University of Hokkaido, Japan, pp 1–137 (in Japanese)

Nomura K, Yoneda K, Uchiyama H, Koyama T (1994) Agricultural and botanical study of rat-tailed radish cv. Pakki food. 3. Relationships between Pakki food and Japanese cultivated radish. Trop Agri Res 38:25–26 (in Japanese)

Ogura H (1968) Studies on the new male sterility in Japanese radish, with special reference to the utilization of this sterility towards the practical raising of hybrid seeds. Mem Fac Agri Kagoshima Univ 6:39–78

Okazaki K, Hinata K (1984) Analysis of *S*-alleles and *S*-glycoproteins in F_1-hybrid varieties of Japanese radish (*Raphanus sativus* L.). Jpn J Breed 34:237–245

Palmer JD, Shields CR, Cohen DB, Orton TJ (1983) Chloroplast DNA evolution and the origin of amphidiploid *Brassica* species. Theor Appl Genet 65:181–189

Palmer JD, Herbon LA (1988) Plant mitochondrial DNA evolves rapidly in structure, but slowly in sequence. J Mol Evol 28:87–97

Panetsos CA, Baker HG (1967) The origin of variation in wild *Raphanus sativus* (Cruciferae) in California. Genetica 38:243–274

Parkin IAP, Sharpe AG, Keith DJ, Lydiate DJ (1995) Identification of the A and C genomes of amphidiploid *Brassica napus* (oilseed rape). Genome 38:1122–1131

Paran I, Michelmore RW (1993) Development of reliable PCR-based markers linked to downy mildew resistance genes in lettuce. Theor Appl Genet 85:985–993

Paulmann W, Robbelen G (1988) Effective transfers of cytoplasmic male sterility from radish (*Raphanus sativus* L.) to rape (*Brassica napus* L.). Plant Breed 100:299–309

Pelletier G, Primard C, Vedel F, Chetrit P, Remy R, Rousselle P, Renard M (1983) Intergeneric cytoplasmic hybridization in Cruciferae by protoplast fusion. Mol Gen Genet 191:244–250

Peter L, Metz J, Nap JP, Stiekema WJ (1995) Hybridization of radish (*Raphanus sativus* L.) and oilseed rape (*Brassica napus* L.) through a flower-culture method. Euphytica 83:159–168

Peterson JL, Pound GS (1960) Studies on resistance in radish to *Fusarium oxysporum* f. *congulatinans*. Phytopathology 50:807–812

Peterka H, Budahn H, Schrader O, Ahne R, Schutze W (2004) Transfer of resistance against the beet cyst nematode from radish (*Raphanus sativus*) to rape (*Brassica napus*) by monosomic chromosome addition. Theor Appl Genet 109:30–41

Piao ZY, Park YJ, Choi SR, Hong CP, Park JY, Choi YS, Lim YP (2002) Conversion of AFLP marker linked to clubroot resistance gene into SCAR markers. J Kor Soc Hort Sci 43:653–659

Piao ZY, Deng YQ, Choi SR, Park YJ, Lim YP (2004) SCAR and CAPS mapping of *CRb*, a gene conferring resistance to *Plasmodiophora brassicae* in Chinese cabbage (*Brassica rapa* ssp. *pekinensis*). Theor Appl Genet 108:1458–1465

Pradhan AK, Prakash S, Mukhopadhyay A, Pental D (1992) Phylogeny of *Brassica* and allied genera based on variation in chloroplast and mitochondrial DNA patterns: molecular and taxonomic classifications are incongruous. Theor Appl Genet 85:331–340

Prakash S (1973) Haploidy in *Brassica nigra* Koch. Euphytica 22:613–614

Prakash S, Takahata Y, Kirti PB, Chopra VL (1999) Cytogenetics. In: Gomez-Campo C (ed.) Biology of *Brassica* coenospecies. Elsevier, Amsterdam, pp 59–106

Ramanujam S (1941) A haploid plant in toria (*Brassica campestris* L.). Proc Ind Acad Sci 14:25–34

Ramsay LD, Jennings DE, Bohuon EJR, Arthur AE, Lydiate DJ, Kearsey MJ, Marshall DF (1996) The construction of a substitution library of recombinant backcross lines in *Brassica oleracea* for the precision mapping of quantitative trait loci. Genome 39:558–567

Richharia RH (1937) Cytological investigation of *Raphanus sativus*, *Brassica oleracea*, and their F_1 and F_2 hybrids. J Genet 34:19–44

Sakai T, Imamura J (1994) Somatic hybridization between radish (*Raphanus sativus*) and rapeseed (*Brassica napus*).

In: Bajaj YPS (ed.) Biotechnology in Agriculture and Forestry, vol 27. Somatic Hybridization in Crop Improvement. Springer, Berlin Heidelberg New York, pp 320–333

Sakai T, Imamura J (2003) Development and utilization of hybrid-seed production system in rapeseeds (*Brassica napus* L.). Breed Res 5:93–102

Sampson DR (1967) Frequency and distribution of self-incompatibility alleles in *Raphanus raphanistrum*. Genetics 56:241–251

Schrader O, Budahn H, Ahne R (2000) Detection of 5S and 25S rRNA genes in *Sinapis alba*, *Raphanus sativus* and *Brassica napus* by double fluorescence in situ hybridization. Theor Appl Genet 100:665–669

Sharpe AG, Parkin IAP, Keith DJ, Lydiate DJ (1995) Frequent nonreciprocal translocations in the amphidiploid genome of oilseed rape (*Brassica napus*). Genome 38:1112–1121

Song KM, Osborn TC, Williams PH (1990) *Brassica* taxonomy based on nuclear restriction fragment length polymorphisms (RFLPs). 3. Genome relationships in *Brassica* and related genera and the origin of *B. oleracea* and *B. rapa* (syn. *campestris*). Theor Appl Genet 79:497–506

Song KM, Suzuki JY, Slocum MK, Williams PH, Osborn TC (1991) A linkage map of *Brassica rapa* (syn. *campestis*) based on restriction fragment length polymorphism loci. Theor Appl Genet 82:296–304

Suwabe K, Iketani H, Nunome T, Kage T (2002) Isolation and characterization of microsatellites in *Brassica rapa* L. Theor Appl Genet 104:1092–1098

Suwabe K, Tsukazaki H, Iketani H, Hatakeyama K, Fujimura M, Nunome T, Fukuoka H, Matsumoto S, Hirai M (2003) Identification of two loci for resistance to clubroot (*Plasmodiophora brassicae* Woronin) in *Brassica rapa* L. Theor Appl Genet 107:997–1002

Suwabe K, Iketani H, Nunome T, Ohyama A, Hirai M, Fukuoka H (2004) Characteristics of microsatellites in *Brassica rapa* genome and their potential utilization for comparative genomics in Cruciferae. Breed Sci 54:85–90

Takagi C, Komooka H, Kimizuka Y (1994) Variety-classification by random amplified polymorphic DNA in radish (*Raphanus sativus* L.). J Jpn Soc Hort 62 (Suppl 2):42 (in Japanese)

Takebe T (1938) On inheritance of root color in *Raphanus sativus* LINN. Jpn J Genet 14:39–50 (in Japanese)

Takebe T (1944) Studies on the hybrids between radish and Hama-daikon. J Jpn Soc Hort 15:74–77 (in Japanese)

Tanaka J, Ota-Metoku S, Takeda Y (2003) Garden-variety camellia Robiraki derived from crossing between *Camellia japonica* as seed parent and *C. sinensis* as pollen parent. Application of RAPD and SSR marker analysis to tea breeding by interspecific hybridization. Breed Res 5:149–154

Terachi T, Yamaguchi K, Yamagishi H (2001) Sequence analysis on the mitochondrial *orfB* locus in normal and Ogura male-sterile cytoplasm from wild and cultivated radishes. Curr Genet 40:276–281

Thompson KF (1956) Production of haploid plants of Marrow-stem kale. Nature 178:748

Thormann CE, Ferreira ME, Camargo LEA, Tivang JG, Osborn TC (1994) Comparison of RFLP and RAPD markers to estimate genetic relationships within and among cruciferous species. Theor Appl Genet 88:973–980

Tokumasu S (1951) Male sterility in Japanese radish (*Raphanus sativus* L.). Sci Bull Fac Agri Kyushu Univ 13:83–89

Tokumasu S (1957) Histological studies on pollen degeneration in male-sterile Japanese radish. Jpn J Breed 6:249–254

Truco MJ, Hu J, Sadowski J, Quiros CF (1996) Inter- and intra-genomic homology of the *Brassica* genomes: implications for their origin and evolution. Theor Appl Genet 93:1225–1233

Tsunoda S (1979) Ecology of wild species of the tribe *Brassiceae* and origin of cultivation. In: Recent Advances in Breeding. Jpn Soc Breed 20:41–45 (in Japanese)

Tsunoda S (1991) The history of cultivated plants in the tribe *Brassiceae*. (1) Cultivated plants and wild allies in the tribe *Brassiceae*. Agri Hort 66:233–239 (in Japanese)

Tsuro M, Suwabe K, Kubo N, Matsumoto S, Hirai M (2005) Construction of a molecular linkage map of radish *(Raphanus sativus* L.) based on AFLP and *Brassica*-SSR markers. Breed Sci 55:107–111

Ukai Y, Ohsawa R, Saito A, Hayashi T (1995) MAPL: a package of computer programs for construction of DNA polymorphism linkage maps and analysis of QTL. Breed Sci 45:139–142

Vos P, Hogers R, Bleeker M, Reijans M, van de Lee T, Hornes M, Frijters A, Pot J, Peleman J, Kuiper M, Zabeau M (1995) AFLP: a new technique for DNA fingerprinting. Nucl Acids Res 23:4407–4414

Voss A, Luhs WW, Snowdon RJ, Friedt W (1999) Development and molecular characterization of rapeseed (*Brassica napus* L.) resistant against beet cyst nematodes. Proceedings of the 10th International Rapeseed Congress, Canberra, Australia, 1999

Warwick SI (1993) Guide to the wild germplasm of *Brassica* and allied crops. IV. Wild species in the tribe Brassiceae (Cruciferae) as sources of agronomic traits. Tech Bull 17E:1–19

Warwick SI, Black LD (1991) Molecular systematics of *Brassica* and allied genera (Subtribe *Brassicinae*, *Brassiceae*): chloroplast genome and cytodeme congruence. Theor Appl Genet 82:81–92

Warwick SI, Black LD (1997) Phylogenetic implications of chloroplast DNA restriction site variation in subtribes Raphaninae and Cakilinae (Brassicaceae, tribe Brassiceae). Can J Bot 75:960–973

Yamagishi H, Terachi T (1994) Molecular and biological studies on male-sterile cytoplasm in Cruciferae. II. The origin of Ogura male-sterile cytoplasm inferred from the segregation pattern of male sterility in the F_1 progeny of wild and cultivated radishes (*Raphanus sativus* L.). Euphytica 80:201–206

Yamagishi H, Terachi T (1997) Molecular and biological studies on male-sterile cytoplasm in Cruciferae. IV. Ogura-type cytoplasm found in the wild radish, *Raphanus raphanistrum*. Plant Breed 116:323–329

Yamagishi H, Tateishi M, Terachi T, Murayama S (1998) Genetic relationships among Japanese wild radishes (*Raphanus sativus* f. *raphanistroides* Makino), cultivated radishes and *R. raphanistrum* revealed by RAPD. J Jpn Soc Hort Sci 67:526–531

Yamagishi H (1999) Detection of the cytoplasmic male sterility genes, *orf138*, *orf263* and *orf224*, in cruciferous plants by polymerase chain reactions. Bull Inst Nat Land Utilization Devel Kyoto Sangyo Univ 20:145–155

Yamagishi H, Sasaki J (2004) Differentiation of the cytoplasm of Japanese cultivated radishes (*Raphanus sativus*) into four types in the 3' region of mitochondrial *Atp6*. Beed Sci 54:189–195

Yamaguchi M (1983) Winter radish, Lobok, Daikon: *Raphanus sativus* (Longipinnatus group). In: translated by Sakiyama R, World Vegetables. Youkendo, Tokyo, pp 224–227 (in Japanese)

Yang YW, Tseng PF, Tai PY, Chang CJ (1998) Phylogenetic position of *Raphanus* in relation to *Brassica* species based on 5S rRNA spacer sequence data. Bot Bull Acad Sin 39:153–160

Yang YW, Lai KN, Tai PY, Ma DP, Li WH (1999) Molecular phylogenetic studies of *Brassica*, *Rorippa*, *Arabidopsis* and allied genera based on the internal transcribed spacer region of 18S-25S rDNA. Mol Phylo Evol 13:455–462

Yang YW, Tai PY, Chen Y, Li WH (2002) A study of the phylogeny of *Brassica rapa*, *B. nigra*, *Raphanus sativus*, and their related genera using noncoding regions of chloroplast DNA. Mol Phylo Evol 23:268–275

4 Carrot

James M. Bradeen[1] and Philipp W. Simon[2]

[1] Department of Plant Pathology, University of Minnesota, 495 Borlaug Hall, 1991 Upper Buford Circle, St. Paul MN 55108, USA
e-mail: jbradeen@umn.edu

[2] USDA-ARS & Department of Horticulture, University of Wisconsin-Madison, 1575 Linden Drive, Madison, WI 53706, USA

4.1 Introduction

4.1.1 History of Cultivation

When and how humans first consumed carrot (*Daucus carota* L.) is debatable. Banga (1957a) cited archeological evidence of carrot seeds in close association with campfire remains and pottery chards in Switzerland and Southern Germany dating to 2000 – 3000 BC. He indicated, however, that while this is evidence ancient humans intentionally collected carrot, it is not proof of carrot cultivation. He further suggested that carrot may have been used first as a medicinal herb or that its seed was used for medicinal or culinary purposes. Although writings of the Roman scholar Pliny the Elder (23 – 79 BC) described several root crops grown at the time, Banga concluded, based on botanical descriptions and etymologies, that none of the crops is the carrot grown today. Heywood (1983) concurred, concluding that root vegetables referred to in ancient writings as "pastinaca" were most probably parsnips (*Pastinaca sativa)* and not carrot. The first widely accepted evidence of cultivation of modern day carrot comes from tenth-century Afghanistan and Iran. From there cultivation of the root vegetable spread throughout Asia and Europe (Banga 1957a).

Early carrots were commonly purple, yellow, or white. From these, modern Eastern carrots, with slightly dissected, pubescent, grey-green foliage and purple or yellow roots, and modern Western carrots, with deeply dissected, mostly glabrous, yellow-green foliage and orange, yellow, red, or white roots were developed (Small 1978). Banga (1957b) traced the development and European adaptation of the carrot by examining writings and paintings dating from the mid-sixteenth century. He concluded that purple carrots, apparently introduced from the Middle East, were widely grown throughout Western Europe beginning in the fourteenth century and extending into the nineteenth century. By the seventeenth century, yellow carrots, also introduced from the Middle East, were gaining popularity throughout Europe and gradually replaced purple cultivars. Yellow carrots were popular well into the nineteenth century, but in many locations these cultivars were gradually relegated to the status of livestock food (Banga 1957b). Orange (high carotene) carrots were first depicted in European paintings during first half of the seventeenth century. Several hypotheses have been proposed to explain the origin of orange carrots. Banga (1957b) concluded that orange carrots were developed from yellow types through gradual selection, with the orange cultivars "late horn" and "half-long horn" originating in the Netherlands during the seventeenth century. Small (1978) reviewed alternative hypotheses for the origin of the orange carrot, including the possibility that crosses between cultivated and wild germplasm may have played a role in the generation of enhanced pigment types. Heywood (1983) argued for an origin of orange carrot that involved both hybridization between cultivated and wild types and human selection for improved pigmentation.

Vavilov (1951) described the center of origin for Eastern carrot as the Central Asiatic Center, including northwest India, Afghanistan, Tadjikistan, Uzbekistan, and western Tian-Shan. Heywood (1983) described Afghanistan as a center of diversity and likely site of origin for Eastern carrot. Vavilov (1951) described the center of origin for the Western carrot as the Near-Eastern Center (Asia Minor, Transcaucasia, Iran, and the highlands of Turkmenistan), noting a multitude of cultivated types in Asia Minor. Banga (1957a) later proposed that present day Asia Minor is the primary gene center for the species.

Genome Mapping and Molecular Breeding in Plants, Volume 5
Vegetables
C. Kole (Ed.)

4.1.2 Carrot Biology

Carrot is an herbaceous species with alternate, compound leaves organized as a rosette. At the flowering stage, leaf internodes lengthen radically and the main stem terminates in an inflorescence. Individual flowers are small with five petals and five anthers. Flowers are collectively organized as a flattened umbrella-shaped umbel (Fig. 1). The umbel is a defining characteristic for carrot and related taxa. Secondary, tertiary, and higher order umbels developing from side branches are common. Umbels in cultivated carrot are usually white and may be tinged with yellow or green (Rubatzky et al. 1999). Although carrot flowers are usually hermaphroditic, male-sterility is an important characteristic for hybrid seed production and controlled crosses. Male-sterile flowers may be of the brown anther or petaloid type. Brown anther male-sterility (Welch and Grimball 1947) causes the anthers to abort and shrivel without dehiscence. In petaloid male-sterile flowers, the functional anther is replaced by petals, giving individual flowers a semidouble appearance. Carrot flowers are epigynous with an ovary composed of two locules, each containing a single ovule. Flowers are protandrous, facilitating outcrossing. The carrot fruit is a dry schizocarp composed of two winged or ribbed mericarps, each a single seed. Ribs are commonly spined. Carrot seed and other tissues have a distinctive odor due to the presence of volatile compounds.

Carrot is cultivated as an annual crop harvested in the vegetative phase. Flowering or bolting in cultivated carrots is initiated after a vernalization period of at least six weeks and the crop is typically bred on an annual cycle. Selection for lack of bolting is primary among breeding goals, as some wild carrots and unadapted cultivars begin to flower with little or no vernalization.

Carrot has a thick, fleshy taproot that develops from the hypocotyl. The root is the primary organ of agricultural importance. In cross section, the root consists of an inner xylem core surrounded by phloem tissue. In some cultivars the distinction between xylem and phloem can be quite evident. Roots range in length from 5 cm to more than 50 cm and are generally conical in shape (Fig. 1), although there is tremendous diversity in root shapes in the cultivated carrot. Rubatzky et al. (1999) illustrate 19 different classes of carrot root shapes ranging from short, nearly conical Paris Market types to long, distinctively tapered types such as Altringham and Imperial Long Scarlet. Root shape is commercially significant and may be a determining factor for whether a carrot cultivar is important for processing or fresh market sales. In the US, commercially important root shapes included the tapered Imperator types (fresh market, prepacked, and "baby" or "cut and peeled" carrots), and the blockier Chantenay and Danvers types (processed, including diced and sliced carrots) (Rubatzky et al. 1999). The conical Nantes is a popular fresh market root type in Europe, while Chantenay, Danvers, and Flakkee types predominate in the European processing market (Rubatzky et al. 1999). Although root shape is commonly used to define carrot cultivars, root shape classes are not well-defined genetically and crosses between two root shape classes may yield progeny with an entirely unique root shape (Rubatzky et al. 1999). Consistently, Bradeen et al. (2002) demonstrated poor correlation between 14 root shape classes and genetic similarity estimates calculated from AFLP and ISSR data. The authors conclude that root shape may be conditioned by relatively few genes. As described above, carrot roots vary considerably in color as well as shape. Modern cultivated carrot may be white, yellow, red, purple, or orange (Fig. 1). The xylem core may be pigmented differently from the phloem.

Carrot is a diploid species with nine chromosome pairs ($2n = 2x = 18$). Relatively little cytology has been reported for the species. Individual carrot chromosomes are relatively short and are uniform in length. Sharma and Bhattacharyya (1954) described secondary constrictions on three of the chromosomes and a satellite on one chromosome. Zenkteler (1962) reported an average of two chiasmata per bivalent. The haploid nucleus contains approximately 1 pg of DNA and the genome is estimated at 473 Mbp (Arumuganathan and Earle 1991; Bennett and Leitch 1995), four times larger than the *Arabidopsis* genome, similar in size to the rice genome, half the size of tomato genome, and one-fifth the size of the maize genome. Forty percent of the carrot genome is highly repetitive. The species is outcrossing with high levels of morphological (Small 1978; Heywood 1983) and molecular variation (St. Pierre et al. 1990; St. Pierre and Bayer 1991; Nakajima et al. 1998; Bradeen et al. 2002). Inbreeding depression can be severe, but no system of self-incompatibility has been described.

Carrot provides a tractable system for the study of somatic embryogenesis (reviewed by Feher et al. 2003), so the species has for many years attracted the

A

B

Fig. 1. Carrot morphology. In *Daucus carota* and related species, individual flowers are borne on undivided pedicels originating from a common node (A). The resulting umbrella shaped racemose inflorescence is known as an umbel and is a defining characteristic of the Apiaceae. Carrot roots vary widely in shape and color (B). Root shape is a significant production characteristic but root classes are not genetically distinct and relatively few genes may condition root shape (Rubatzky et al. 1999; Bradeen et al. 2002). Carrot root colors include white, yellow, orange, red, and purple (black). The xylem core may be pigmented differently than the phloem. Important carrot root pigments include carotenes (orange, yellow), xanthophylls (yellow), anthocyanin (purple/black), and lycopene (red)

attention of molecular biologists. However, although valued as a vegetable throughout most of the world, carrot is a crop of relatively minor economic importance. Consequently, until recently, few molecular resources relevant to carrot improvement have been developed. But important resources have been developed in recent years and there would appear to be substantial opportunity for carrot improvement through the application of molecular markers and maps, and the isolation of important genes.

4.1.3 Taxonomy and Classification

Cultivated carrot belongs to family Apiaceae (= Umbelliferae), subfamily Apioideae, genus *Daucus*, section *Daucus*, species *D. carota* L. The classification of this crop plant, however, is complicated at nearly every taxonomic level.

Due to morphological plasticity, relationships between the Apiaceae and related plant taxa are controversial. For example, some authors argue that the mostly herbaceous, temperate Apiaceae is actually a specialized group within the woody, tropical Araliaceae (Thorne 1968; Judd et al. 1994). Most modern classifications, however, support recognition of two separate but closely related families belonging to the order Apiales. (Dahlgren 1980; Takhtajan 1980; Cronquist 1981), a scheme that is also supported by DNA sequence analysis of the chloroplast genes *rbcL* (Plunkett et al. 1996b) and *matK* (Plunkett et al. 1997), and the nuclear ribosomal 26S large subunit (Chandler and Plunkett 2004). Molecular data support a close relationship between order Apiales and subclass Asteridae (Plunkett et al. 1996b), consistent with the classification of Dahlgren (1980).

Relationships within the Apiaceae are also convoluted. The family contains between 2500 and 3750 species classified into 300 to 462 genera (Cronquist 1981; Heywood 1983; Pimenov and Leonov 1993). Defining characteristics include a racemose inflorescence organized as an umbel and a dry, generally ridged fruit. Many members of this family give off a characteristic odor. Constance (1971) details historical classification schemes for the family. Among modern classification systems, three morphologically defined subfamilies and a series of tribes have been recognized (Drude 1898; Heywood 1983; Pimenov and Leonov 1993). Characteristics of the inflorescence and fruits have been particularly important in developing these classification systems. Plunkett et al. (1996a) used DNA sequence analysis of the chloroplast *matK* gene to examine relationships within the Apiaceae. They demonstrate generally poor molecular support for morphologically defined subfamilies and tribes and conclude that current classification systems may be artificial (Plunkett et al. 1996a). Similar conclusions were reached by Downie et al. (1998) via DNA sequence analysis of nuclear ribosomal internal transcribed spacer (ITS1 and ITS2) and the intron of chloroplast gene *rpoC1*. However, a monophyletic origin for the economically important subfamily Apioideae (carrot, celery, parsnip, fennel, parsley, cumin, coriander, dill, and anise) has been supported at the morphological (Judd et al. 1994) and molecular (Downie et al. 1998) levels. Downie et al. (1998) identified several well-supported clades within subfamily Apioideae, but these clades do not correspond to morphologically defined tribes.

The genus *Daucus* includes at least 20 species (Sáenz Laín 1981; Pimenov and Leonov 1993) mostly of European and Mediterranean origin. However, the genus is cosmopolitan and representatives are native to or have become widely distributed throughout much of the world, particularly in temperate areas. Based on morphological and anatomical features, Sáenz Laín (1981) organized *Daucus* species into five sections: *Daucus* L. (12 species), *Anisactis* DC. (three species), *Platyspermum* DC. (three species), *Chrysodaucus* Thell. (one species), and *Meoides* Lange (one species). Carrot (*D. carota)* is the only important cultivated form of the genus. The species, however, includes several wild or feral subspecies in addition to the cultivated carrot.

Nakajima et al. (1998) used RAPD and AFLP markers to characterize relationships between five *Daucus* spp., including *D. carota*. All *D. carota*, including seven carrot (*D. carota* ssp. *sativus)* cultivars or inbred lines and accessions of six wild *D. carota* ssp., could be distinguished from other *Daucus* spp. by both RAPD and AFLP. In a similar study, Vivek and Simon (1999b) examined relationships among eight wild *Daucus* spp., the wild *D. carota* ssp. *drepanensis*, and eight cultivated *D. carota* ssp. *sativus* cultivars or inbreds, representing four of the five sections defined by Sáenz Laín (1981). Restriction site data from the chloroplast genome strictly support the sections of Sáenz Laín (1981), although the authors caution that a single accession of section *Anisactis* was available for study and that additional accessions of all sections might provide more robust support (Vivek and Simon 1999b).

Several authors have specifically examined diversity and taxa relationships within *D. carota*. In his taxonomic treatment of the species, Heywood (1983) recognized 13 morphologically defined *D. carota* subspecies divided into two broad groups: the *carota* group [ssp. *carota, sativus* (Hoffm.) Arc., *maritimus* (Lam.) Batt., *major* (Vis.) Arc., *azoricus* Franco, and *parviflorus* (Desf.) Thell.] and the *gingidium* group [ssp. *gadecaei* (Rony & Camus) Heywood, *gummifer* Hooker fil., *commutatus* (Paol.) Thell., *hispanicus* (Gouan) Thell., *hispidus* (Arc.) Heywood, *drepanensis* (Arc.) Heywood, and *ruseus* (Onno) Heywood & Okeke]. However, the species has been found to be an extremely diverse genetic continuum using morphological (Small 1978) and molecular approaches (St. Pierre et al. 1990; St. Pierre and Bayer 1991; Nakajima et al. 1998; Bradeen et al. 2002). For example, using AFLP markers, Nakajima et al. (1998) demonstrated that two carrot cultivars of Japanese origin appear more closely related to wild *D. carota* ssp. than to carrot cultivars of European origin. Similarly, in a study of 124 *D. carota* populations that included 51 wild or feral accessions, AFLP and ISSR data revealed that certain accessions of wild *D. carota* from at least four different subspecies were imbedded within a cluster of carrot cultivars (Bradeen et al. 2002). These observations are consistent with suggestions that the species as presently recognized is actually a species complex, ranking among the most demanding classification issues in the Apiaceae (Small 1978; Heywood 1983). No modern study has found good morphological or molecular evidence of distinct Eastern and Western carrot groups (Small 1978; Nakajima et al. 1998; Bradeen et al. 2002). Further phylogenetic evaluation of the family Apiaceae, the subfamily Apoideae, and *Daucus carota* is clearly warranted.

4.1.4 Carrot Production and Consumer Quality

World carrot production has more than quadrupled during the past 45 years from 5.8 million Mt in 1961 to 23.6 million Mt in 2004 (FAOSTAT data 2004; Fig. 2). Prominent increases are evident for China, especially during the period from 1984 to 2004 (Fig. 2). However, increases in production are evident worldwide, especially in temperate regions, and are due primarily to increases in production area with only modest gains in average yield (Fig. 2). Simon (2000) reviewed recent trends in carrot production in the US, indicating that consumer popularity of baby or cut and peeled carrots has led to increases in US carrot production area over the past two decades. Modest enhancements

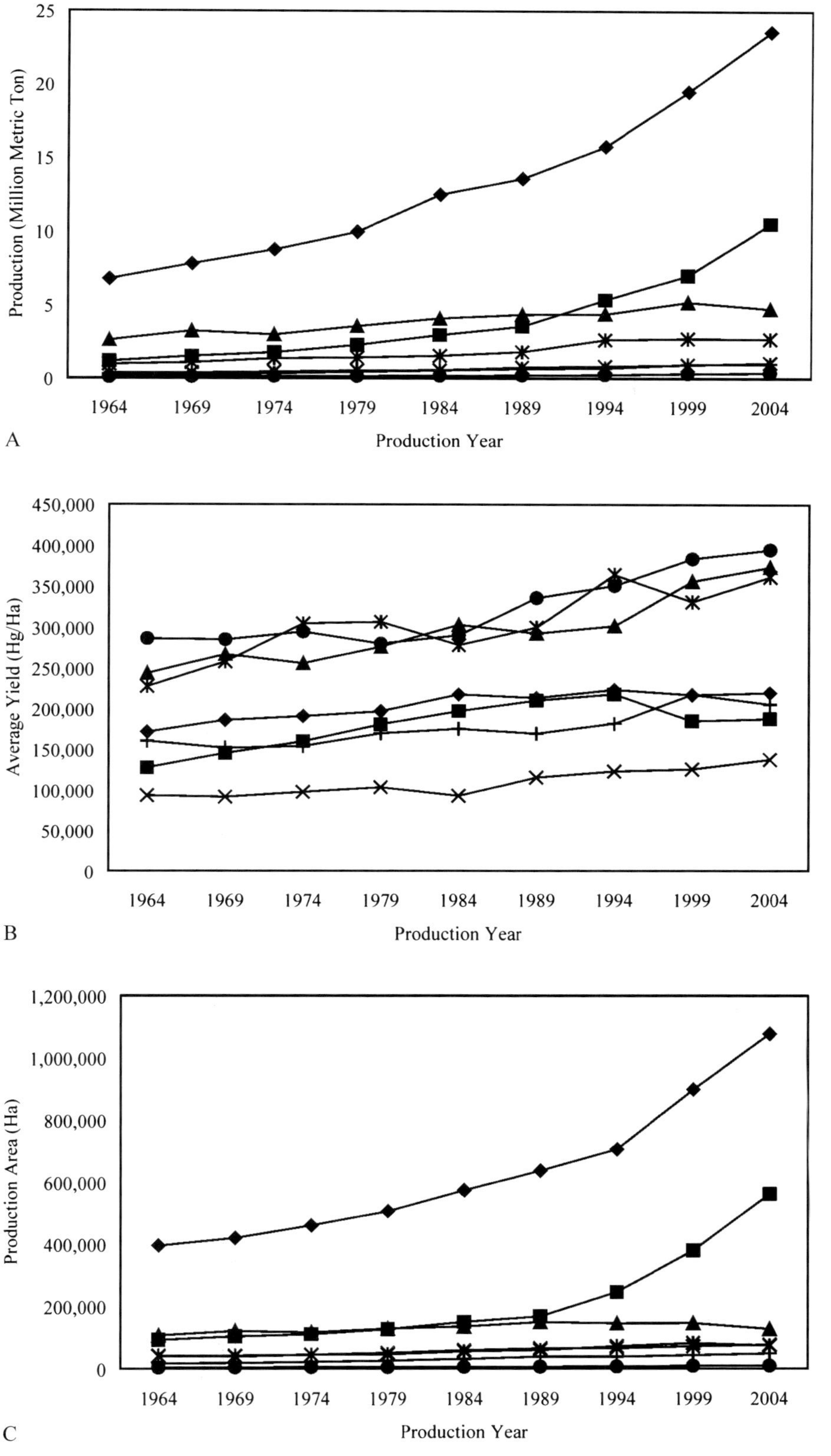

Fig. 2. Trends in world carrot production (1964 – 2004). Carrot production statistics for World (♦), Africa (■), Asia (▲), Europe (×), North and Central America (∗), Oceania: Australia and New Zealand (•), and South America (+) for the period 1964 – 2004. **A** Total annual carrot production in metric ton. **B** Average yield in hg/ha. **C** Production area in ha. During the past four decades, worldwide carrot production has increased dramatically. Regional increases in carrot production are particularly evident in Asia. Total production increases are mostly due to increases in production area with only modest gains in average yield (adapted from FAOSTAT data 2004)

Table 1. World carrot (*Daucus carota* L.) production in metric tons (1995–2004) (adapted from FAOSTAT data 2004)

Region	% World production[a]	World rank	Annual production (Mt) 1995	1996	1997	1998	1999	2000	2001	2002	2003	2004
World	100.0	–	16,585,471	17,892,910	18,913,509	18,732,727	19,474,059	20,816,401	21,096,175	21,627,184	23,584,128	23,607,214
China	28.3	1	3,656,179	4,376,379	4,477,698	4,514,679	4,846,009	5,740,164	6,111,984	7,126,460	8,093,079	8,292,500
United States	9.6	2	1,878,000	2,043,000	2,266,950	2,128,300	1,941,950	1,858,610	1,837,440	1,677,870	1,900,110	1,900,110
Russian Federation	7.2	3	1,182,070	1,292,010	1,424,150	1,160,240	1,372,200	1,597,090	1,575,070	1,474,100	1,735,260	1,825,000
Poland	4.2	4	814,311	794,128	799,428	991,955	906,477	946,736	921,911	692,073	834,621	800,000
Japan	3.4	5	724,700	736,200	716,100	648,100	676,700	681,700	691,300	644,500	653,700	660,000
United Kingdom	3.2	6	517,200	617,400	623,100	617,600	673,200	710,500	763,900	717,400	587,300	650,000
France	3.2	7	637,851	644,060	652,000	662,000	672,000	658,376	649,489	727,645	688,426	469,280
Italy	2.5	8	396,443	346,185	457,829	471,806	509,849	642,065	590,997	561,442	577,811	580,000
Ukraine	2.2	9	408,500	318,600	449,000	379,700	384,700	496,500	463,000	445,400	530,000	500,000
Netherlands	1.9	10	368,900	350,000	350,000	277,000	448,000	385,000	378,000	422,000	432,000	430,000
Germany	1.9	11	297,411	340,082	312,723	371,950	379,544	431,541	444,448	414,960	426,038	420,000
India	1.7	12	300,000	320,000	340,000	340,000	340,000	350,000	350,000	350,000	350,000	350,000
Spain	1.6	13	304,592	301,541	356,695	333,250	400,562	425,357	383,311	270,128	270,128	270,000
Mexico	1.6	14	199,588	219,500	306,753	319,926	363,368	376,847	355,903	378,517	378,517	378,517
Indonesia	1.5	15	247,179	269,837	227,322	332,846	286,536	326,693	300,648	282,248	355,802	355,802
Canada	1.5	16	288,710	320,400	290,510	324,842	294,183	261,284	279,050	286,496	301,818	293,400
Australia	1.4	17	238,539	249,926	257,405	266,531	256,608	283,304	320,908	331,129	305,669	305,669
Morocco	1.2	18	223,525	224,500	221,400	229,000	220,850	209,510	198,370	233,390	332,210	332,210
Turkey	1.2	19	250,000	270,000	240,000	232,000	239,000	235,000	230,000	235,000	240,000	240,000
Argentina	1.1	20	210,000	254,542	241,166	210,615	221,791	257,471	220,000	225,000	230,000	230,000

[a] % World production and World rank calculated based on average annual production for the period 1995–2004 (adapted from FAOSTAT data 2004)

in average carrot yield are due to adaptation of improved cultural practices, increased farm mechanization, and the development of improved cultivars, especially hybrids. Use of hybrid carrot cultivars by US growers accounted for more than 40% increase in average yield between 1955 and 1975 (Simon 2000). Little concerted breeding effort has been dedicated to developing improved carrot cultivars for tropical or subtropical areas (Simon 2000) and average carrot yield in warm regions such as Africa has remained consistently modest during the period from 1964 to 2004 (Fig. 2). During the past ten years (1995–2004), China, the US, and the Russian Federation have been the largest producers of carrot, together accounting for 45% of world production (FAOSTAT data 2004; Table 1).

Carrot is not a significant source of calories in the human diet. However, among fruits and vegetables, carrot ranks seventh in overall contribution to nutrition (Alasalvar et al. 2001). Carrots are a significant or potential source of dietary nutrients in the form of plant pigments, including carotenoids, anthocyanins, and other flavonoids. The health benefits of these compounds, including protection against certain forms of cancer, reduction of the risk of cardiovascular disease, and scavenging of free radicals (reviewed by van den Berg et al. 2000 and Stintzing and Carle 2004), have led to consumer interest in natural products rich in carotenoids and anthocyanins.

Approximately 40 carotenes, including the α- and β-carotenes present in carrot, are provitamin A precursors (Simpson 1983). Provitamin A compounds are enzymatically converted to vitamin A in the human body. In the US, carrots currently account for 30% of total dietary vitamin A (Simon 1992). Before synthetic sources of carotene became available in the 1940s, carotene was commercially extracted from carrots (Heywood 1983). In recent years, increased consumer interest in natural foods has led to commercial use of carrot carotenes as nutritional amendments and food coloring (Simon et al. 1997; Rubatzky et al. 1999). Worldwide, vitamin A deficiency remains a significant problem, particularly in developing countries, and is a leading cause of blindness among children (van den Berg et al. 2000). Carrot cultivars differ significantly in total root carotenoid content (Simon and Wolff 1987; Heinonen 1990), with α- and β-carotenes present at concentrations of 530 to 35,833 and 1161 to 64,350 μg/100 g root tissue, respectively (van den Berg et al. 2000). Selection for improved carotenoid content has been highly successful for carrot (Welsh 1997; Rubatzky et al. 1999).

Carrots are also a potential source for the antioxidant anthocyanins. Anthocyanins impart a purple or reddish color and occur at high concentrations in the purple Eastern carrot cultivars. Stintzing and Carle (2004) reviewed the role of anthocyanins in plants and concluded that in the carrot root, anthocyanins might play a role as monosaccharide transporters or osmotic adjusters that protect plant cells from drought and cold temperatures. Alasalvar et al. (2001) reported carotenoid levels and consumer preference for orange and purple carrots. In that study, purple carrots had more than twice the α- and β-carotene present in orange carrots and were perceived by consumers to be sweeter than orange carrots, although total sugars and calculated relative sweetness was lower for purple than for orange carrots. The authors concluded that purple carrots may be a viable alternative to orange rooted varieties (Alasalvar et al. 2001). Surles et al. (2004) extended the range of colors to include orange, purple, red, yellow, and white, and concluded that while all colors were acceptable, white and orange were preferred in sensory evaluations. Furthermore, they found the orange carrot had a similar carotene content to purple. The inconsistent conclusions drawn from these two studies confirm observations that flavor and the different color classes of carrot are usually independent of each other, biochemically and genetically. Linkage studies have only included a few of the genes for these quality traits thus far, but as those traits are better mapped, limitations in manipulating these traits in a breeding program will be revealed. Anthocyanin is commercially extracted from purple carrot for use as a natural food dye (Downham and Collins 2000), a trend that is likely to continue. Breeders have initiated development of purple carrot breeding lines (Simon et al. 1997).

4.1.5 Carrot Breeding

Although intentional or unintentional selection for improved carrot types has undoubtedly occurred since the dawn of carrot cultivation, concerted breeding began only in the mid-twentieth century. Carrot is an outcrossing species that suffers from inbreeding depression. Individual carrot flowers are small and flowers in the same umbel may open at different times, making hand emasculation on a large scale impractical. However, it is possible for breeders to achieve directed controlled crosses on a limited scale using hand emasculation followed by pollination via bluebottle flies in enclosed pollination chambers (Rubatzky et al. 1999). On a larger scale, controlled pollination can be achieved by allowing open pollination between desired parents. Although umbels of each parent will contain self-pollinated and cross-pollinated seeds, the resulting progeny can often be distinguished on the basis of plant vigor and root size: progeny resulting from self-pollinations are less vigorous due to inbreeding depression. The discovery of brown anther genic cytoplasmic male-sterility (CMS) in carrot by Welch and Grimball (1947), Cornell petaloid CMS by Munger (personal communication, 1953), and Wisconsin petaloid CMS by Morelock et al. (1996) enabled breeders to conduct controlled crosses on a large scale in enclosed mesh cages. CMS also made possible the commercial production of hybrid carrot seed.

Phenotypic recurrent selection is an important breeding approach for the development of open pollinated breeding lines and inbred populations for hybrid generation. Peterson and Simon (1986) detailed methods for carrot inbred development and hybrid testing. Because inbred lines frequently are poor seed yielders, three way crosses have been extensively utilized for carrot hybrid seed production. However, in recent years, there has been an emergence of F_1 hybrid development (Simon 2000).

Carrot breeders select for multiple agricultural and quality traits. Among important agricultural traits, breeders select for slow bolting habit, uniform and desired root length, shape, and surface texture, upright foliage, moderate canopy size, well-developed root color, and small, less conspicuous xylem core. Breeders might also screen for resistance to important diseases including *Alternaria* leaf blight (caused by the fungus *A. dauci*), aster yellows (caused by a mycoplasma), *Cercospora* leaf blight (caused by the fungus *C. carotae*), bacterial leaf spot (caused by *Xanthomonas campestris* pv. *carotae*), and cavity spot (caused by the fungal pathogen *Pythium* spp.). Carrot fly (*Psila rosae*), aphids (various species), leafhoppers (e.g., *Macrosteles quadrilineatus*) and nematodes, especially the root-knot nematode (*Meloidogyne incognita* and *M. javanica*) are important pests of carrot and breeders may also select for resistance to these agents. Important quality traits include carotene and anthocyanin content, as described above. Breeders also select for taste components including high sugars and low volatile terpenoids. Breeders of carrots intended for processing or culinary use might select for improved canning, freezing, or cooking characteristics. There is a current need for selection of improved carrot types adapted to tropical lowland environments. This list of agricultural and quality traits is by no means exhaustive and breeders in various regions of the world may have additional needs. Recent, detailed reviews of carrot breeding methods and goals have been presented by Rubatzky et al. (1999) and Simon (2000).

4.2 Map Construction

Molecular markers for map construction can be broadly divided into two categories: biochemical markers, such as isozymes, and DNA-based markers. Allelic variation at isozyme loci is assumed to be selectively neutral, allowing accumulation of gene mutations at isozyme loci that are revealed as modified protein mobilities through starch or polyacrylamide electrophoretic gels. Isozyme data are generally codominant, allowing differentiation between heterozygous individuals and homozygotes of either parental class. The utility of isozymes for linkage map construction in plants is substantially limited by the paucity of loci that can be unambiguously scored. In contrast, DNA markers are virtually unlimited in number and the application of DNA markers to linkage map construction is restricted primarily by a researcher's ability to identify polymorphisms. For outcrossing species, such as carrot, this has not been a significant difficulty. Several DNA-based marker systems have been utilized for map construction in carrot, including restriction fragment length polymorphism (RFLP), random amplified polymorphic DNA (RAPD) and amplified fragment length polymorphism (AFLP) marker systems.

The RFLP method, first described by Botstein et al. (1980), is based on DNA-DNA hybridization. Capitalizing upon the specificity of a restriction enzyme for its target DNA sequence, the method is highly accurate and reproducible. RFLP data are generally codominant in nature. However, the RFLP method is time-consuming, relatively expensive, and cumbersome, requiring isolation and manipulation of microgram quantities of genomic DNA. The PCR-based RAPD method was described by Williams et al. (1990) and is comparatively rapid, inexpensive, and technologically simple, requiring only nanogram quantities of genomic DNA. RAPD markers, however, are generally dominant in nature, sacrificing information about the genetic composition of heterozygotes. Several authors have also documented issues with the reliability and repeatability of RAPD markers for mapping and lineage assessment applications (Echt et al. 1992; Riedy et al. 1992; Heun and Helentjaris 1993; Thormann et al. 1994; Hallden et al. 1996). Nevertheless, RAPD markers have been successfully utilized to construct linkage maps in several plant species (Kennard et al. 1994; Mudge et al. 1996; Sondur et al. 1996; Bucci et al. 1997; Davis and Yu 1997; Jiang et al. 1997; King et al. 1998; Boiteux et al. 2000). AFLP markers were first reported by Vos et al. (1995). The method combines the specificity and repeatability of restriction digestion with the ease and rapidity of detection of PCR. The method has proven highly robust and has been widely used for linkage map construction in plants (Ballvora et al. 1995; Becker et al. 1995; Meksem et al. 1995; Mackill et al. 1996; Bendahmane et al. 1997; Brigneti et al. 1997; Li et al. 1998; Vuylsteke et al. 1999; Bradeen et al. 2001). AFLP data, like RAPD data, are generally dominant in nature.

4.2.1 Nuclear Linkage Maps

Linkage mapping in carrot is summarized in Table 2. As is true for many plant species, the first reported molecular linkage map in carrot was composed solely of isozyme markers (Westphal and Wricke 1991). Using BC_1 populations generated from self-pollinated S1 carrot lines, Westphal and Wricke (1991) reported band variation and linkage analysis for 10 different isozyme systems. For three systems (6-phosphogluconic dehydrogenase (6-PGD), phosphoglucomutase (PGM), and phosphoglucose isomerase (PGI)) the authors reported evidence of duplication of loci; duplication was confirmed for 6-PGD and PGM by segregation analysis. A total of 12 loci were assigned to four linkage groups. Given the difficulty of identifying appropriate isozyme markers, Westphal and Wricke (1991) concluded that DNA-based markers such as RFLP and RAPD markers might be more efficient for the construction of detailed linkage maps for carrot. Nevertheless, the isozyme linkages reported by these authors were the first linkage map for carrot and laid a foundation for subsequent mapping efforts.

Later, Schulz et al. (1993) combined RFLP, RAPD, and isozyme markers to construct a carrot linkage map. Four S1 populations generated from the carrot cultivars Finaê, Primatoê, and Gustoê (two populations), each with a minimum of 70 individuals, were analyzed. Segregation analysis was completed for 10 isozyme, 14 RFLP, and 28 RAPD markers. The authors also used six RAPD fragments that were monomorphic in one or more mapping population as RFLP probes, bringing the total number of mapped molecular markers to 58. Across mapping populations, these loci were organized into eight linkage groups, short of the nine linkage groups expected for a saturated map of carrot. Three markers remained unlinked. Linkage groups ranged in composition from two to nine markers and in length from 1.7 cM (two markers) to 78.6 cM (seven markers). Average distance between markers was estimated at 13.1 cM. Subsequently, Westphal and Wricke (1997), using the same mapping populations, expanded the map to approximately 200 markers including AFLP and microsatellite markers in addition to isozyme, RFLP, and RAPD markers. The final map consisted of nine main linkage groups with "several" markers remaining unmapped.

Vivek and Simon (1999a) reported a carrot linkage map composed of 106 molecular markers. Markers were mostly AFLPs (96 markers), but also included six RFLPs, two RAPDs, and two microsatellite-derived markers. The authors utilized a mapping population of 103 F_2 individuals generated by self-pollinating a single F_1 individual from the cross B9304 × YC7262. Using a conservative LOD of 4.0, they defined 11 linkage groups, each composed of five to 15 markers. Seventeen additional markers were assigned to linkages of only two to three markers; 12 markers were unlinked. The resulting molecular map was 524.1 cM in length with an average distance of 4.9 cM between mapped markers. The authors estimate the total size of the carrot genome to be 900 cM (Vivek and Simon 1999a).

The most recently reported and most complete linkage map for carrot was constructed by merging marker data from two independent F_2 populations (Santos and Simon 2004). Map merging is a procedure

Table 2. History of linkage map construction for carrot (*Daucus carota* L.)

Mapping population Type	Parent cross	No.individuals	Marker type (no.)	Total no. markers	No. linkage groups	Map length (cM)	Reference
S1	Various	Various	Isozyme (12)	12	4	114.0 (estimated)	Westphal and Wricke (1991)
S1	4 Populations: (a) cv. Finaê ⊗ (b) cv. Primatoê ⊗ (c) cv. Gustoê ⊗ (d) cv. Gustoê ⊗	≥70	RFLP (14) RAPD (12) Isozyme (10) RAPD Probes (6) Morphological (1)	58	5–8	151.3–283.5	Schulz et al. (1994)
S1	4 Populations: (a) cv. Finaê ⊗ (b) cv. Primatoê ⊗ (c) cv. Gustoê ⊗ (d) cv. Gustoê ⊗	≥70	RFLP (unspecified) RAPD (unspecified) Isozyme (unspecified) AFLP (unspecified) Microsatellite (unspecified) Morphological (1)	~200	9	Unspecified	Westphal and Wricke (1997)
F2	B9304 × YC7262	103	AFLP (96) RFLP (6) Morphological (3) RAPD (2) Microsatellite (2)	109	11	534.4	Vivek and Simon (1999)
F2	2 Populations: (a) B493 × QAL (b) Brasilia × HCM	 183 160	 AFLP (250) AFLP (287)	 250 287	 9 9	 1114.0 1188.0	Santos and Simon (2004)
Merged	Pops (a) + (b)	–	AFLP (138)	138	6 (+3 unmerged)	517.6	

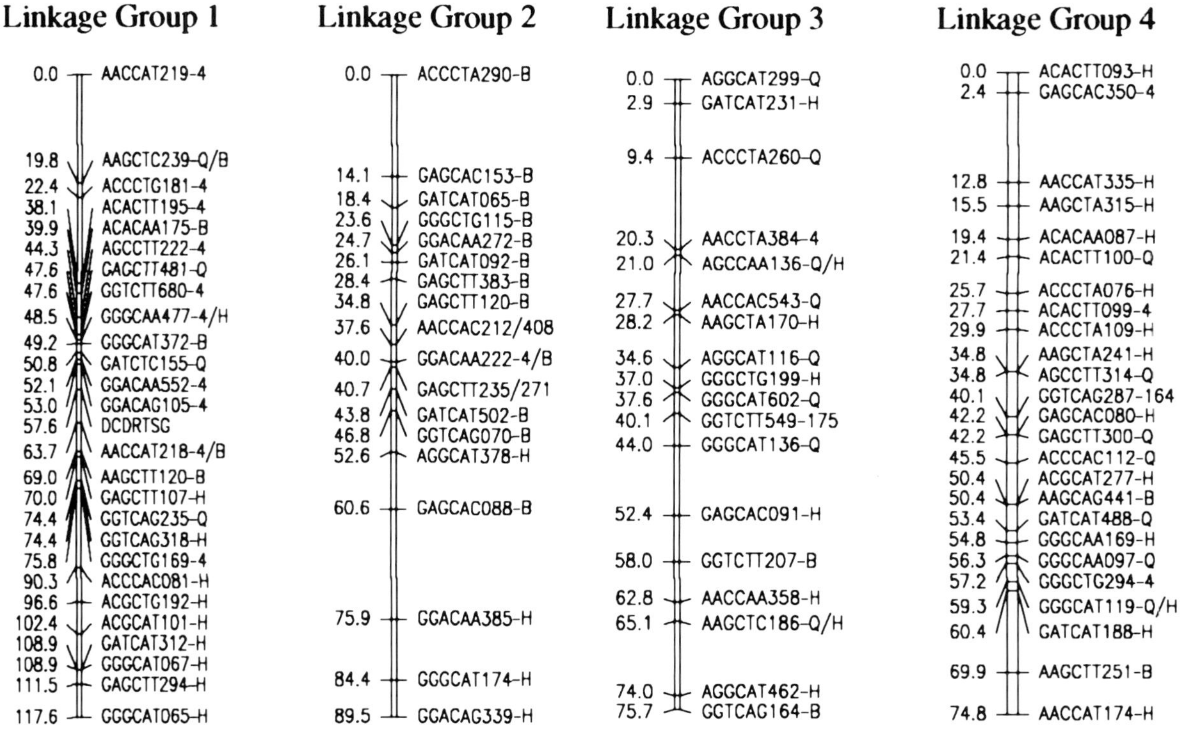

Fig. 3. Merged carrot linkage groups. Map merging is the statistical integration of two or more linkage maps by first building a scaffold map from markers common to more than one population and subsequently placing population-specific markers relative to scaffold markers. Santos and Simon (2004) generated AFLP and PCR marker data for carrot F_2 mapping populations originating from the crosses Brasilia × HCM and B493 × QAL. The populations shared in common 28 AFLP and two PCR markers which were used for scaffold map construction. A total of 108 population-specific markers were subsequently placed upon the scaffold map. The result is six merged linkage groups, each encompassing between 68.9 and 117.6 cM and composed of 18 to 30 markers. Marker names reflect AFLP primer combinations and fragment sizes. Codominant AFLP markers are indicated by the listing of two fragment sizes separated by a *slash*. Parental origin of each marker is indicated as: "B" for Brasilia, "H" for HCM, "4" for B493, and "Q" for QAL. Map merging scaffold markers shared between populations are indicated by the listing of two parental types separated by a *slash*. Map distances in cM are shown to the *left* of each linkage group. Three additional linkage groups from each population were not merged due to a lack of shared scaffold markers (figure adapted from Santos and Simon 2004)

allowing integration of marker data generated in different populations into a common, unified map (Stam 1993). Merging requires that a framework constructed from markers shared between populations be established. Markers unique to a single population are then statistically placed within the framework. This is completed for all markers from all populations, resulting in a single, unified map. Map merging has been successfully accomplished for several plant species (Sebastian et al. 2000; Bradeen et al. 2001; Paran et al. 2004). The F_2 carrot populations utilized by Santos and Simon (2004) were constructed from disparate germplasm. The first cross was between an inbred carrot line (B493) and a wild carrot (*D. carota* ssp. *carota;* QAL). The second cross was between an open pollinated Brazilian cultivar (Brasilia) and an elite, high carotene line (HCM). AFLP data were generated for each of these populations using two restriction enzyme combinations: *Eco*RI/*Mse*I and *Pst*I/*Mse*I. A limited number of PCR-based, codominant markers were also typed for both populations (Santos and Simon 2004). Key to successful map merging is identifying markers common to each mapping population. Due to the dominant nature of most AFLP markers and the complex banding patterns generated by this procedure, verifying that similarly sized AFLP markers from separate populations are identical by descent is of paramount importance. Previously, Santos and Simon (2002b) demonstrated that 84% of AFLP markers of common size generated from the same F_2 mapping

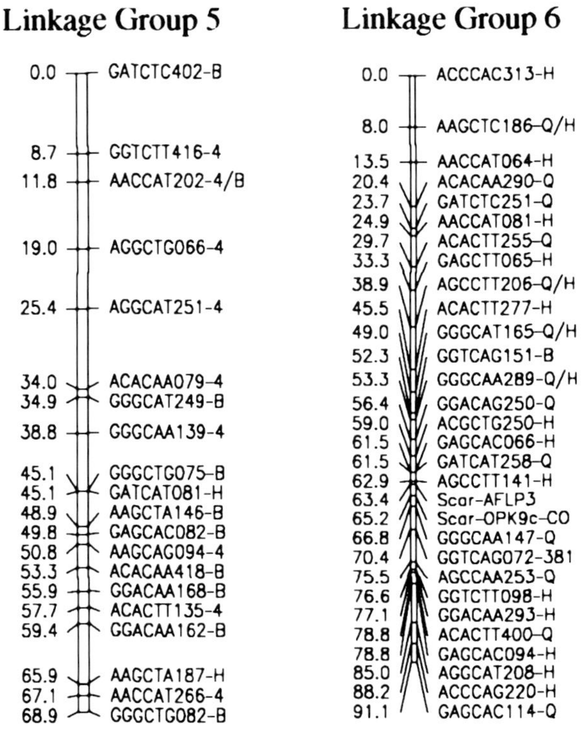

Fig. 3. (continued)

populations shared greater than 91% DNA sequence identity. For map merging, 28 AFLP markers and two PCR markers common to both mapping populations were used to construct a mapping framework upon which additional, population-specific markers were placed. In total, six linkage groups could be merged (Fig. 3). The merged linkage groups ranged in composition from 18 to 30 markers and in length from 68.9 cM (20 markers) to 117.6 cM (27 markers). The average distance been markers on the merged linkage groups was 3.75 cM. Three additional linkage groups from each population were not merged due a paucity of shared markers (Santos and Simon 2004).

To date, linkage maps have not been reported for *Daucus* species other than *D. carota*. In addition to carrot, the Apiaceae includes several cultivated species including celery, parsnip, fennel, parsley, cumin, coriander, dill, and anise. Among these species, linkage mapping has been reported only for celery (Huestis et al. 1993; Yang and Quiros 1995). Markers mapped in celery include RFLP, RAPD, isozyme, and morphological markers but a lack of shared markers between the celery and carrot maps precludes the use of existing maps for comparative mapping between these species.

4.2.2 Evaluation of Marker Systems for Linkage Mapping

Several marker systems have been utilized for linkage mapping in carrot. Each marker system has advantages and disadvantages. Vivek and Simon (1999a) summarized observed efficiencies for RFLP, RAPD, and AFLP markers for linkage map construction in carrot. The parental genotypes utilized to generate their F_2 mapping population were phenotypically distinct. One parent, B9304, was derived exclusively from Western carrot germplasm. The second parent, YC7262, included in its background Eastern carrot germplasm (YC7262 is a full-sib of carrot inbred B7262 (Simon et al. 1997)). Nevertheless, of 300 genomic DNA clones evaluated as RFLP probes, only 30 (10%) were both clearly scorable and polymorphic between the parental genotypes. The authors further reported a 20% (33 primers) polymorphism rate between parental genotypes in a survey of 164 RAPD primers, but only two of the RAPD markers could be scored unambiguously (Vivek and Simon 1999a). By comparison, Schulz et al. (1993) reported an observed polymorphism rate of 33% (20/60 RAPD primers) for four mapping populations generated by self-pollinating *D. carota* cultivars. Differences in genetic background could account for differences in observed polymorphism rates for RAPD markers used in these two studies. Finally, of 404 AFLP bands generated using seven primer combinations, Vivek and Simon (1999a) identified 164 polymorphic fragments for an observed polymorphism rate of 42%. Thus, the authors found that AFLP is more than two times as efficient as RFLP and RAPD methodologies at generating markers for carrot linkage map construction. Consistently, Nakajima et al. (1998) reported that while both AFLP and RAPD markers were useful for phylogenetic applications in *Daucus*, the AFLP system yielded more than four times as many useful markers per reaction. Protocols for RAPD (Boiteux et al. 1999) and AFLP (Briard et al. 2000) optimized for carrot have been published.

4.2.3 Organellar Physical Maps

CMS is a trait of major importance to carrot breeding and hybrid seed production. In carrot, male-sterility is conditioned by genes encoded in the mitochondrial genome. However, factors encoded in the nuclear genome may restore male-fertility. This system is correctly referred to as a genic CMS system since both nuclear and organellar factors interact to control the expression of sterility. Borner (1995) reviewed the genetics of CMS in carrot. The mitochondrial genome of higher plants is dynamic in arrangement and may consist of multiple subgenome circular DNA molecules rather than a single large master chromosome (Folkerts and Hanson 1991; Yamato et al. 1992; Janska and Mackenzie 1993; Kubo et al. 2000). Bowes and Wolyn (1998) examined restriction patterns of the mitochondrial genomes of three male-fertile maintainer carrot lines and six petaloid male-sterile lines originating from three geographical locations. They documented considerable mitochondrial diversity among male-fertile types, consistent with a dynamic mitochondrial genome arrangement, but relatively little diversity among petaloid male-sterile lines, even when lines originated from regions separated by more than 1,500 km. The authors concluded petaloid CMS lines probably arose from a common mitochondrial genome rearrangement event.

Robison and Wolyn (2002) constructed a physical map of the mitochondrial genome of petaloid male-sterile carrot line W259A. The map was constructed by ordering cosmid fragments based on restriction site analyses. The completed map is approximately 255 kb in length. The authors identified three large repeated regions of the genome ranging in size from 0.53 to 2.0 kb. Repeated mitochondrial genome regions are common in higher plant species. Recombination among repeat regions is thought to enable the reversible formation of subgenomic circular DNA molecules and may result in stoichiometric differences in genome regions. Consistently, Robison and Wolyn (2002) demonstrated via Southern hybridization that each of the three repeated regions identified in the CMS carrot mitochondrial genome undergo rearrangement in planta. Thus, they documented the organization of the CMS mitochondrial genome into subgenomic circular DNA molecules and concluded that the overall arrangement of the CMS carrot mitochondrial genome precludes its existence as a single large circular chromosome. The authors further demonstrate stoichiometric differences in mitochondrial genome composition in planta. Stoichiometric differences in specific mitochondrial genome regions have been implicated in CMS of several plant species. Although expression or editing of certain mitochondrial genes has been correlated with male-sterility in carrot (Szklarczyk et al. 2000; Nakajima et al. 2001; Rurek et al. 2001), no specific gene or region of the CMS carrot mitochondria has been conclusively associated with male-sterility.

4.3 Gene Mapping

In addition to RAPD, RFLP, and isozyme markers, Schulz et al. (1994) mapped a single carrot morphological marker, *phen1*. The *phen1/phen1* phenotype was described as complicated by the authors and apparently included small, dark green, curled leaves and a propensity for annual flowering without vernalization, although root development appeared normal. Although *phen1* appears to be of little agricultural significance, its incorporation into a linkage map is the first example of gene mapping by phenotype in carrot. Since then, several agriculturally important genes have been mapped in carrot including single genes conditioning root phenotypes, disease resistance, and male-sterility.

Carotene and anthocyanins are among the important pigments in carrot roots, imparting orange/yellow and purple coloration, respectively. The genetics of carotene synthesis in the carrot root is complex, involving several enhancer genes and inhibitor genes. Kust (1970) examined the inheritance of pigmentation in the carrot xylem core, describing several genes including *Y2*. The *Y2* locus is a dominant suppressor of carotene accumulation in the carrot core, *y2/y2* individuals possessing an α-, β-, and γ-carotene rich orange core and *Y2*– individuals possessing a xanthophyll-rich yellow or carotene depleted white core, depending on genetic background (Buishand and Gabelman 1979). Bradeen and Simon (1998) studied 103 F_2 individuals of the cross B9304 × YC7262, which segregated for core color. Using bulked segregant analysis (BSA; Michelmore et al. 1991) combined with F_2 mapping, the authors identified six AFLP markers linked to and flanking the *Y2* locus. Markers were located between 3.8 and 15.8 cM from the gene. Using the same F_2 mapping population, Vivek and Simon (1999a) subsequently identified a single AFLP marker located 2.2 cM from the *Y2* locus, assigning the locus to one end of linkage group B. Anthocyanin accumulation in the carrot phloem is conditioned by the *P1* locus, with purple (*P1*) dominant to nonpurple (*p1*) (Simon 1996). Simon (1996) studied the inheritance of *P1* and *Y2* in F_2 and BC populations originating from Eastern carrot germplasm and concluded that the two loci are unlinked. Consistent with this, Vivek and Simon (1999a) mapped *P1* to linkage group A, independent of *Y2*. *P1* is flanked by AFLP markers mapping 1.7 and 8.1 cM away from the gene.

Sucrose and the reducing sugars glucose and fructose account for more than 95% of free sugars and up to 60% of total carbohydrates in the carrot root. Sugars are a major component of carrot flavor making sugar content and composition important characteristics for carrot breeding efforts. Freeman and Simon (1983) demonstrated that while carrot inbred lines in that study did not vary substantially in total sugar content, they varied substantially in sugar composition. In some inbred lines, reducing sugars accounted for only 10 – 12% of total sugars while in other inbred lines reducing sugars accounted for 51 – 78% of total sugars (Freeman and Simon 1983). The authors further demonstrated that sugar composition is conditioned by a single locus they named *Rs*. *Rs*-genotypes store a higher percentage of reducing sugars (glucose and fructose) and *rs/rs* genotypes store a higher percentage of sucrose. Freeman and Simon (1983) concluded that the Rs protein may affect production or function of invertase, the enzyme responsible for the conversion of sucrose to glucose and fructose. This, in fact, turned out to be correct, as the *rs* allele has recently been found to be a naturally occurring knockout mutant of a carrot invertase isozyme which produces no functional enzyme (Yau and Simon 2003). Vivek and Simon (1999a) mapped *Rs* to one end of linkage group C, 8.1 cM away from an AFLP marker. Mapping results are consistent with inheritance data indicating that *Rs* is genetically unlinked to *Y2* and *P1* (Simon 1996).

The *Mj-1* locus conditions resistance to the root-knot nematode (*Meloidogyne javanica*). Derived from the Brazilian carrot cultivar Brasilia, *Mj-1* consists of a single dominant locus or two closely linked (10–20 cM) loci inherited as a unit (Simon et al. 2000). Boiteux et al. (2000) utilized BSA on an F_2 population of 442 individuals derived from the cross Brasilia-1252 (resistant parent) × B6274 (susceptible parent). Resistant and susceptible F_2 bulks were screened with 1000 RAPD primers, revealing 63 fragments putatively linked to *Mj-1*. Subsequent F_2 analysis allowed the authors to identify and map four highly reproducible RAPD fragments linked to and flanking *Mj-1* (Boiteux et al. 2000). Markers were mapped 0.8 to 5.7 cM away from resistance. All RAPD markers were in coupling phase with resistance (i.e., presence of the RAPD fragment was linked to the *Mj-1* resistance allele).

Robison and Wolyn (2002) recovered mitochondrial gene fragments from carrot genomic DNA using PCR primers developed from mitochondrial genes cloned in other plant species. They subsequently used the fragments to map more than 20 genes and pseudogenes on a 255 kb physical map of the CMS carrot mitochondrial genome. Three genes, *cox1*, *cox2*, and *atp8* were associated with repeated genome regions that reversibly exist as subgenomic circular DNA molecules and may be present at nonstochiometric ratios. The agricultural significance of these genes or the role they may play in conditioning male-sterility is not clear at present.

Several additional agriculturally important genes have been described and studied at the genetic level (reviewed in Simon 2000), including additional genes conditioning root color and disease resistance. The incorporation of these and additional genes into burgeoning molecular maps of carrot would appear to be a significant goal.

4.4 Quantitative Trait Loci (QTL) Detected

In addition to single genes conditioning agriculturally important traits, several QTL have been identified in carrot through segregation analysis. To date, QTL conditioning synthesis in carrot roots of provitamin A α- and β-carotenes, the carotene lycopene, and precursors in the carotene pathway have been mapped. Santos and Simon (2002a) generated an F_2 mapping population of 160 individuals originating from the cross Brasilia (medium orange) × HCM (a high carotene line; carotene content ranging from 460 to 499 ppm). This population was also used for the map merging experiment detailed above (Santos and Simon 2004). The authors measured α- and β-carotene, lycopene, and the precursors ζ-carotene and phytoene in F_2 progeny via high-performance liquid chromatography. A total of 287 AFLP markers were mapped. AFLP markers associated with QTL conditioning each of the traits measured were detected. QTL originated from both Brasilia and HCM and individually explained from 3.7 to 13.2% of observed phenotypic effect based on interval analyses. In total, eight QTL for α-carotene were detected. Collectively, these QTL account for 40% of observed phenotypic variation. Three QTL for β-carotene synthesis were detected, explaining 21% of observed variation. A single factor conditioning lycopene synthesis was detected. That factor originated from Brasilia and accounted for 7.2% of total phenotypic variation. Four QTL accounted for 16.3% of observed variation in ζ-carotene accumulation and five QTL accounted for 28% of observed variation in phytoene accumulation. Significantly, Santos and Simon (2002a) demonstrated a propensity for QTL involved in carotene synthesis and accumulation to cluster in common genome regions. The authors concluded that clustering of genes involved in a common pathway may be evolutionarily advantageous.

4.5 Marker-Assisted Germplasm Maintenance and Breeding

In addition to map construction and gene mapping, molecular markers have been utilized to improve carrot breeding approaches. Specifically, molecular markers have been used to optimize maintenance strategies for carrot germplasm, to identify carrot breeding lines and test hybrid seed purity, and for marker-aided selection (MAS).

It is the ideal of a genebank to maintain germplasm collections in such a way as to preserve both common and rare alleles. Yet financial and biological limitations can result in erosion of allelic diversity and loss of rare alleles. le Clerc et al. (2003) reviewed the challenges facing curators of crop germplasm collections and explored the effects of reproducing population size on allelic frequency. Working with a homogeneous population of the open pollinated carrot cultivar Géante de Milly Tilques, the authors established five populations composed of two, eight, 15, 39, and 70 individuals. Random leaf samples were collected from each population. Individuals within a population were allowed to intermate but populations were kept reproductively isolated from one another. Seeds were harvested and planted and leaf tissue was collected from 11 to 33 second generation individuals. le Clerc et al. (2003) generated AFLP data from populations produced in the first and second generations of seed increase. The authors reported that balanced seed sampling, in which each individual contributes equally to the next generation, resulted in less allelic drift, as measured by AFLP, than did bulk seed sampling, in which each individual contributes all of its seed to the next generation (i.e., highly prolific individuals would contribute disproportionately to the next generation). The authors concluded that a minimum of 80 carrot roots per generation should be utilized for germplasm maintenance.

Grzebelus et al. (2001) discussed the challenges of commercial F_1 hybrid seed production for carrot. Inbreeding depression in carrot can be severe, precluding the development of highly homozygous inbred lines. Consequently inbred lines and the resulting F_1 generation retain some level of genetic heterogeneity. Environmental conditions might sometimes lead to partial male-fertility in CMS lines used as females for F_1 seed production, potentially leading to self pollination. Additionally, as with any outcrossing species, there is potential for rogue pollen to contaminate F_1 seed production plots. Grzebelus et al. (2001) explored genetic heterogeneity within and between carrot inbred lines using AFLP. They further tested whether AFLP markers can be used routinely to test F_1 seed pedigree and purity. AFLP data were generated separately for 18 individuals from each of four inbred lines, including male-sterile lines and the corresponding maintainer lines. Using eight AFLP primer pairs, 86 polymorphic markers were identified, with one to six AFLP markers being unique to each inbred line. Polymorphic AFLP markers were subsequently used to characterize eight F_1 hybrids generated from the same inbred lines. AFLP markers correctly reflected the known pedigrees. The authors concluded that AFLP markers are useful to study genetic relationships between breeding lines and suggested this might ultimately predict combining ability. However, AFLP markers may be of limited use for routine testing of carrot F_1 seed pedigree and purity due to their dominant nature and observed genetic heterogeneity within parental inbred lines.

While PCR-based markers systems such as AFLP and RAPD are useful for linkage map construction, gene mapping, and documentation and exploration of allelic diversity, they are not well-adapted to high-throughput MAS. Markers for MAS should be closely associated with and, ideally, flanking the gene of interest. They must also be technologically simple, rapid, and inexpensive. Codominant markers are preferred for MAS since they allow differentiation between homozygous classes and the heterozygous class. AFLP and RAPD markers are mostly dominant. This and the need for polyacrylamide gel electrophoresis and

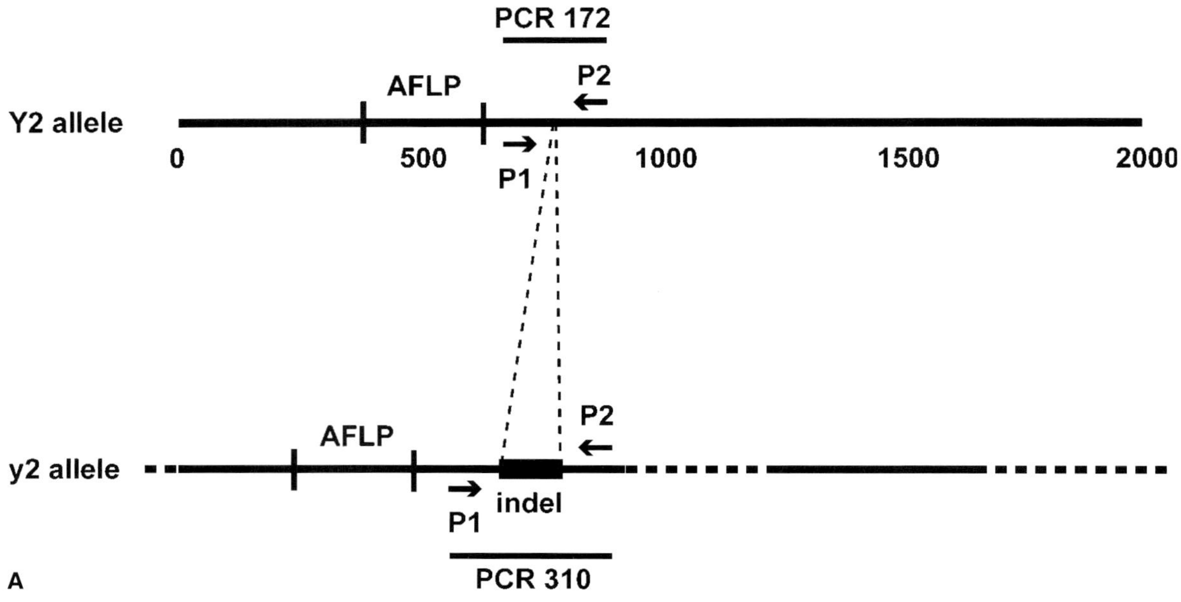

Fig. 4. Conversion of a dominant carrot AFLP marker to a simple PCR, codominant form. The carrot *Y2* locus conditions carotene accumulation in the carrot root xylem core. Bradeen and Simon (1998) mapped the *Y2* locus using AFLP markers. One dominant marker mapping 6.6 cM from *Y2* was selected for conversion to a simple PCR marker form. **a** The 264 bp AFLP marker was cloned and sequenced. Using primers developed from the AFLP fragment sequence, the authors generated more than 2 kb of genome sequence associated with the *Y2* (dominant) allele via inverse PCR (*solid horizontal line* labeled Y2 allele). Fragment size and positions in base pairs are shown below the *Y2* region. Generated sequence included the cloned AFLP fragment. Approximately 1.2 kb of sequence associated with the *y2* (recessive) allele was generated via standard PCR, cloning, and sequencing (*solid* and *dashed horizontal line* labeled y2 allele; *dashed regions* were not sequence characterized). The authors identified an indel of 138 bp, the presence of which is associated with the *y2* (recessive) allele, in regions downstream of the AFLP fragment. PCR primers P1 and P2, which flank the indel, were developed. PCR primers yield a fragment of 172 bp associated with the *Y2* (dominant) allele or a fragment of 310 bp associated with the *y2* (recessive) allele. **b** Photo of an agarose gel. First lane contains size marker. Lanes 2–13 are PCRs using primer pair P1/P2 and carrot genomic DNA of indicated genotypes. *Y2*– and *y2*–specific fragments are indicated with *arrows*. *Y2/Y2* and *y2/y2* homozygotes, and *Y2/y2* heterozygotes are readily distinguished. A heteroduplex composed of one DNA strand from the *Y2*–specific PCR fragment and one DNA strand from the *y2*–specific PCR fragment is evident in heterozygote lanes. The molecular nature of the heteroduplex was independently confirmed (Bradeen and Simon 1998)

radioactive or fluorescent labeling make AFLP markers impractical for large-scale MAS. The dominant nature and issues with repeatability and reliability make RAPD markers impractical for large-scale MAS. Nevertheless, markers associated with traits of interest can be particularly useful for selection of phenotypes that are time consuming, difficult, or expensive to identify. Because carrot is a biennial species, MAS also has the potential of streamlining the selection process by reducing the need for costly and time consuming large scale grow-outs of breeding material. As markers associated with agriculturally significant traits have been identified in carrot, researchers have attempted conversion of AFLP and RAPD markers to simple PCR forms that are better suited for MAS. To date, several nuclear and mitochondrial markers with characteristics suitable for MAS have been developed.

Bradeen and Simon (1998) identified linkage between the *Y2* locus, which conditions carotene accumulation in the carrot xylem core, and six AFLP markers. The authors cloned and sequenced an AFLP marker of 264 bp. The dominant marker, located 6.6 cM from the *Y2* locus was linked in repulsion with high carotene accumulation (i.e., presence of the AFLP fragment = *Y2* allele = low carotene accumulation). Using primers designed from the AFLP marker, the authors generated a genomic fragment incorporating the AFLP marker and greater than 2 kb in length via inverse PCR. Genomic DNA of the *Y2/Y2* mapping parent was used as template for inverse PCR. The sequence of the 2 kb fragment was used to design a series of standard PCR primers which were used to characterize the corresponding genomic region in the *y2/y2* mapping parent. The authors identified an indel

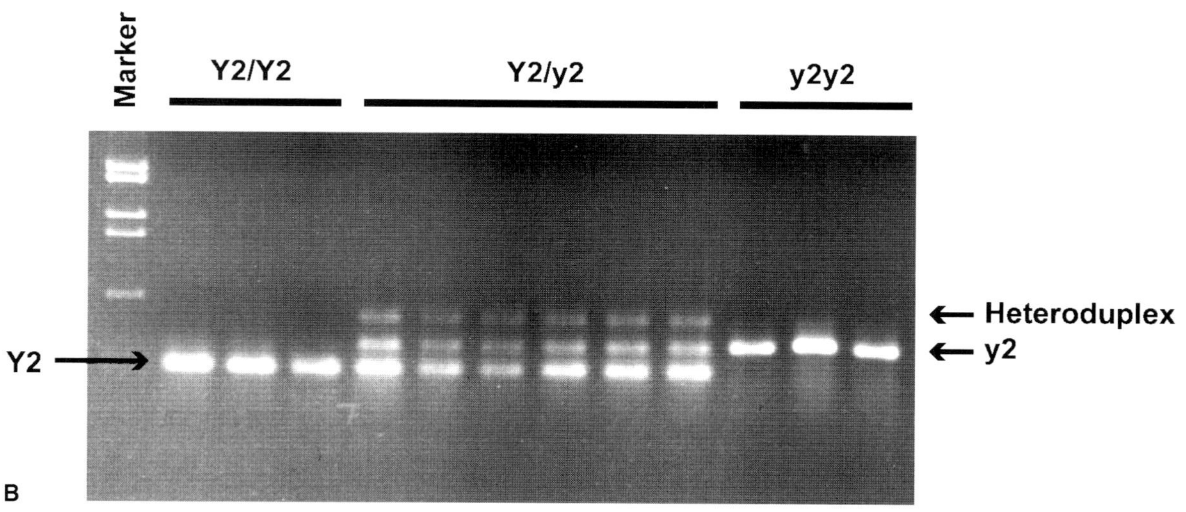

Fig. 4. (continued)

of 138 bp, the presence of which was associated with the *y2* allele (Fig. 4). PCR primers flanking the indel were developed. For the *Y2* allele, the PCR primers generated a fragment of 172 bp. For the *y2* allele, the PCR primers generated a fragment of 310 bp (172 bp + 138 bp indel = 310 bp; Fig. 4). Thus the dominant AFLP marker was converted to a codominant, simple PCR marker. While the need for a molecular marker for the *Y2* locus is minimal since high carotene (orange xylem core) and low carotene (yellow xylem core) genotypes could be visually distinguished, the authors noted that the conversion of a *Y2* AFLP marker to a form well-suited for MAS served as an important model for other markers and plant species (Bradeen and Simon 1998).

Boiteux et al. (2000) identified RAPD markers linked to the carrot *Mj-1* locus for resistance to the root-knot nematode. The RAPD markers flanked the *Mj-1* gene and recombination between the two extreme markers was 6.5 cM. The authors subsequently surveyed a collection of 121 carrot cultivars and breeding lines of diverse backgrounds for the presence of each of the markers, confirming a low level of recombination between the markers. The authors concluded that these markers could be useful for MAS, allowing not only the targeted incorporation of *Mj-1* but also reducing the potential for incorporating undesirable genes via linkage drag. Later Boiteux (2000) reported an expanded RAPD map of the *Mj-1* region and conversion of two RAPD markers tightly linked ($0.0-0.5 \pm 0.1$ cM) to *Mj-1* to codominant simple PCR markers. Conversion to codominant marker forms was made possible by the existence of $\sim$ 100 bp indels, the presence of which was associated with a specific *Mj-1* allele. The design of longer, site-specific PCR primers flanking the indels resulted in a PCR fragment size polymorphism between regions associated with different *Mj-1* alleles. Boiteux et al. (2004) examined *Mj-1* allelic dosage effect on resistance phenotype, utilizing the converted markers to distinguish *Mj-1/Mj-1* and *mj-1/mj-1* homozygotes and *Mj-1/mj-1* heterozygotes in an F_2 population of 396 individuals originating from the cross Brasilia-1252 (resistant; *Mj-1*—) × B6274 (susceptible; *mj-1/mj-1*). The authors concluded that additional genomic regions might interact with *Mj-1* to impart resistance to the root-lesion nematode. The authors also predicted that the converted markers they have developed may be useful for MAS, enabling efficient incorporation of the *Mj-1* locus (Boiteux et al. 2004).

Yau et al. (2005) confirmed that the *rs* allele is the result of an insertion sequence of 2.5 kb integrated into the first intron region of acid soluble invertase isozyme II. Since this isozyme plays a major role in converting sucrose transported in the developing storage root to reducing sugars, homozygous recessive plants store primarily sucrose rather than reducing sugars. Three primers were designed to differentiate *Rs/Rs*, *Rs/rs* and *rs/rs* plants with simple PCR amplification. Codominant, PCR-based markers allowed genotyping of the *Rs* locus in one-week-old carrot seedlings whereas mature carrot roots were needed to make this evaluation previously, and homozygous dominant plants could not be differentiated from heterozygotes without lengthy progeny testing. Marker-assisted selection of carrot root sugar type was exercised and complete accuracy in predicting sugar type was realized in subsequent generations. These PCR-based markers are being applied in carrot breeding programs screening for this trait in segregating populations, for studying the distribution and origins of this trait in domestic and wild carrots, and for identifying seed mixtures as low as 10% *Rs/-* or 1% *rs/rs*.

Mitochondrial markers distinguishing male-fertile and CMS carrot lines have also been targeted for conversion to simpler forms. Nakajima et al. (1999) identified six dominant RAPD markers that could distinguish male-fertile and CMS lines. Each marker was cloned and end sequenced and longer, site-specific PCR primers were developed. Using a combination of three converted markers, the authors could distinguish male-fertile from CMS lines and Cornell from Wisconsin petaloid cytoplasms. In a subsequent study, Bach et al. (2002) sought additional markers to distinguish male-fertile and male-sterile cytoplasms, arguing that the carrot mitochondrial genome is diverse and simultaneous application of multiple markers would expand the applicability of the assays to a broader range of germplasm. The authors first identified polymorphic RFLP markers associated with six conserved genes in the male-fertile carrot line K831B and the petaloid male-sterile line K826A. Subsequent sequence analysis revealed that polymorphisms were due to several factors including gene duplication, upstream and downstream rearrangements, duplication of flanking regions, and truncated and chimeric genes. Next the authors designed a series of PCR primers to capitalize upon observed genome differences between male-sterile and male-fertile lines. A diverse carrot germplasm collection of eight male-fertile and nine male-sterile lines were characterized using the converted PCR markers. In total, 17 markers that could distinguish male-fertile and male-sterile cytoplasms were developed. Three additional markers that could distinguish between three different mitochondrial configurations in male-sterile lines were also developed (Bach et al. 2002). With the recent discovery that the expression of MADS box genes homologous to GLOBOSA and DEFICIENS are associated with cytoplasmic carrot male-sterility (Linke et al. 2003), markers might be developed to better target genomic regions responsible for this trait.

4.6 Cloned Genes

Several genes have been cloned from carrot using a variety of methodologies. Examples include genes cloned via isolation and sequencing of functional proteins followed by PCR to recover the corresponding gene (Ojima et al. 1997; Worrall and Elias 1998; Hoffmann-Sommergruber et al. 1999; Meyer et al. 1999; Sano et al. 2004), screening of cDNA libraries using heterologous probes or PCR primers developed from previously characterized genes in other plant species (Sebkova et al. 1995; Sturm et al. 1999a; Hirner et al. 2001), and isolation of genes displaying differential expression in various tissues, during different physiological stages, or under varying environmental conditions (Nagata et al. 1993; Wurtele et al. 1993; Lin et al. 2003). To date, map-based cloning has not been a significant approach for the isolation of carrot genes. Due to carrot's status as a model organism for somatic embryogenesis, many of the cloned genes are involved in this process (e.g., Nagata et al. 1993; Wurtele et al. 1993; Schmidt et al. 1997; Sano et al. 2004). Additionally, genes conditioning anthocyanin accumulation (Hirner et al. 2001), cold tolerance (Worrall and Elias 1998; Meyer et al. 1999), and carrot allergens (Smith et al. 1993; Hoffmann-Sommergruber et al. 1999) have been identified.

Among cloned genes with obvious agricultural significance are genes conditioning carbohydrate accumulation in the root. Sucrose is a disaccharide molecule synthesized in the leaves and transported to the root. Sucrose is subsequently cleaved into simple hexose sugars and utilized as a carbon source by non-photosynthesizing organs. Two enzymes are responsible for the conversion of sucrose to simple sugars. Sucrose synthase catalyzes the reversible conversion of sucrose to UDP-glucose and fructose. The enzyme also plays a role in starch and cellulose biosynthesis and other cellular activities. The gene appears to play an important role in plant growth and is potentially a major determinant of carrot yield. Invertase catalyzes the irreversible cleavage of sucrose to glucose and fructose. Thus, invertase is a major determinant of carrot flavor. Genes encoding both enzymes have been isolated from carrot and subsequently characterized in planta.

Sebkova et al. (1995) used a sucrose synthase fragment derived from potato to identify the corresponding cDNA clone from a carrot root library. The authors dubbed the identified locus *Susy*1**Dc1*. The authors compared the deduced *Susy*1**Dc1* amino acid sequence with deduced amino acid sequences from previously isolated sucrose synthase genes from bean, tomato, potato, Arabidopsis, maize, rice, and barley. The carrot protein was 85% identical to those of tomato and potato and 68–73% identical to sucrose synthase proteins from monocot species. Based on Southern hybridization to genomic DNA, the authors concluded that the carrot genome encodes one or two sucrose synthase genes (Sebkova et al. 1995). Consistently, Sturm et al. (1999b) subsequently reported the isolation of a second sucrose synthase gene, *Susy*1**Dc2*. *Susy*1**Dc2* was identified by screening a carrot genomic library with a fragment derived from *Susy*1**DC1*. Both of these genes consist of at least 13 predicted exons of similar size and organization. At the nucleotide level, *Susy*1**Dc1* and *Susy*1**Dc2* are 85% identical in exon regions and 45% identical in intron regions. Deduced gene products are 87% identical. Interestingly, the authors further demonstrated that the two sucrose synthase genes differed markedly in their expression patterns. Northern analyses revealed that *Susy*1**Dc1* is expressed in leaves, roots, flowers, and developing seeds, but *Susy*1**Dc2* is expressed exclusively in carrot flowers (Sturm et al. 1999b), consistent with the apparent absence of *Susy*1**Dc2* clones in a carrot root cDNA library (Sebkova et al. 1995). Sturm et al. (1999b) documented strongest *Susy*1**Dc1* transcription and, by correlation (Sebkova et al. 1995), strongest sucrose synthase activity in actively growing tissues, including the base of the carrot stem and the tip of the carrot root. Tang and Sturm (1999) further explored the in planta function of sucrose synthase by disrupting its activity in carrot. Transgenic carrot lines containing an antisense copy of *Susy*1**Dc1* regulated by the constitutive CaMV35S promoter were generated. As expected, the authors reported significant reduction of sucrose synthase transcript in the root of antisense carrot lines. Sucrose levels in antisense lines were significantly elevated relative to nonantisense lines, with a corresponding reduction in glucose, UDP-glucose, fructose, starch, and cellulose. Antisense plants were significantly smaller than control plants but gross leaf and root morphologies were unaltered. Significantly, the authors reported that expression of antisense *Susy*1**Dc1* did not significantly alter leaf to root dry weight ratios in most lines. Based on these observations, they concluded that while sucrose synthase appears to play an important role in metabolic activities associated with plant growth, it is not significantly involved in sucrose partitioning in carrot (Tang and Sturm 1999).

There are at least three primary forms of plant invertases, distinguished largely by their pH optima and localization within the cell. Acid invertases may be extracellular (cell wall associated) or vacuole-localized. Extracellular acid invertases may play an important role in phloem unloading, thereby maintaining a sucrose gradient between source and sink plant organs. Vacuole or soluble acid invertases may play an important role in the regulation of simple sugar levels in mature plant tissues. A third type of invertase, the neutral invertase, is localized within the cytosol. A definitive role of neutral invertase has not been documented. Carrot genes encoding all three types of invertase have been cloned. Sturm and Chrispeels (1990) reported the isolation from carrot of a gene encoding extracellular acid invertase. The gene was isolated from a λ -gt11 expression library by immunodetection using antibodies against purified carrot extracellular acid invertase protein. Using a similar approach, Unger et al. (1994) subsequently reported the cloning of carrot genes encoding three different isoforms of soluble acid invertase. Similarity between deduced amino acid sequences of the carrot extracellular acid invertase and the soluble acid invertase isoforms ranged from 62 to 64%. Sturm et al. (1999a) used partial protein sequence from a purified carrot neutral invertase to identify a corresponding *Arabidopsis* expressed sequence tag in silico. PCR primers were generated and used to amplify an *Arabidopsis* fragment that was subsequently used to probe a carrot cDNA library. Identity of the isolated carrot cDNA clone was verified by expression in *E. coli*. The deduced amino acid sequence of the carrot neutral invertase gene showed low homology to acid invertases and the authors concluded that the carrot acid and neutral invertase genes are not related by evolutionary origin (Sturm et al. 1999a).

To explore the role of invertase in plant development, Tang et al. (1999) generated transgenic carrot lines carrying antisense copies of acid invertase genes regulated by the CaMV35S promoter. Antisense transgenic plants displayed reduced acid invertase transcript. Antisense transgenics were phenotypically abnormal at all growth stages, but in vitro plantlets grown on media enriched with the hexose simple sugars fructose and glucose appeared phenotypically normal. The authors could obtain mature antisense transgenic plants only if plantlets were first cultivated on hexose media and then transferred to soil. The resulting plants had two to three times more leaves than nontransgenic controls. Plants carrying an antisense copy of extracellular acid invertase failed to develop normal tap roots and had a leaf to root dry weight ratio more than fifty times that of control plants. In contrast, plants carrying an antisense copy of soluble acid invertase developed phenotypically normal but diminutive tap roots and had a leaf to root dry weight ratio of only 4.5 times that of control plants. The authors concluded that invertase plays a critical role in regulating source sink relations in carrot.

A direct role of soluble acid invertase in sucrose partitioning and flavor development in the carrot root was established by Yau and Simon (2003) using a candidate gene approach. The dominant *Rs* allele conditions accumulation in carrot roots of the simple hexose sugars, fructose and glucose. Homozygous *rs/rs* individuals accumulate sucrose in the roots. The *Rs* locus was genetically described by Freeman and Simon (1983), who documented the role of this locus as a key determinant of carrot root sugars and flavor. Towards characterization of the molecular basis of the *Rs* locus, Yau and Simon (2003) generated near isogenic *Rs/Rs* and *rs/rs* lines from the carrot inbred B4367. The authors used RT-PCR to detect transcripts of key enzymes in the sucrose/simple sugar pathway including sucrose synthase, extracellular acid invertase, and two isoforms (designated isoform I and II) of soluble acid invertase. Expression patterns and RT-PCR amplicon sizes were identical between B4367Rs and B4367rs lines for sucrose synthase, extracellular acid invertase, and soluble acid invertase isoform I. However, transcript of soluble acid invertase isoform II, while present in roots of B4367Rs, was completely absent in roots of B4367rs. In leaves, the authors noted the expected single, ~2.0 kb RT-PCR amplicon of soluble acid invertase isoform II in B4367Rs, but two amplicons of ~2.7 and 3.2 kb in B4367rs. The wildtype ~2.0 kb amplicon was not observed in leaves of B4367rs. Subsequent characterization of genomic DNA identified a ~2.5 kb insert in the largest intron of the *rs* allele. The authors demonstrated that the insert results in differential mRNA processing, leading to nonfunctional transcript of variable sizes in the leaves of B4367rs. Yau and Simon (2003) further reported that, in parallel mapping studies (Vivek and Simon 1999a), perfect linkage was observed between the genetically defined *Rs* locus and the observed mutant soluble acid invertase isoform II allele, providing strong evidence that the *Rs* locus encodes for soluble acid invertase isoform II.

4.7 Future of Molecular Breeding of Carrot

The future will most certainly bring continued development of linkage maps for carrot. The next generation of maps will be more highly saturated, representing a larger percentage of the carrot genome with markers more closely spaced. Agriculturally significant genes will continue to be localized. Genes conditioning disease resistance, root phenotypes including pigmentation and sugar and terpenoid content, and male-sterility, including the nuclear restorers of fertility in male-sterile cytoplasms (e.g., *Rf*), are candidates for gene mapping in the near future. Candidate gene analysis is underway in carrot in an attempt to associate structural genes in pigment biosynthetic pathways to major color genes and QTL (Just 2004). Improved resolution of the petaloid male-sterile carrot mitochondrial genome may identify candidate genes responsible for this trait. The mitochondrial genome of male-fertile carrot lines will almost certainly be mapped as well, allowing comparative mapping with male-sterile lines. MAS may be more widely integrated into carrot breeding programs as converted markers linked to agriculturally significant traits are generated. MAS has particular promise for traits that are costly, difficult, or time-consuming to measure. As linkage maps for celery and other Apiaceae species are developed, there may be an interest in comparative mapping with the carrot nuclear genome, leading to an improved understanding of the evolutionary origin of agriculturally important genes and overall genome arrangement in this family. Positional cloning in carrot cannot be far behind. Already large insert libraries have been constructed for the species. The root-knot nematode resistance gene, *Mj-1*, and genes conditioning carotenoid and anthocyanin biosynthesis are among likely candidates for positional cloning. Comparative transcriptome analyses are likely to continue in carrot and have the potential of speeding discovery of agriculturally important genes. Genomics resources such as expressed sequence tag libraries, microarrays, in situ hybridization methodologies, and transposon-tagging systems are likely to be developed.

References

Alasalvar C, Grigor JM, Zhang D, Quantick PC, Shahidi F (2001) Comparison of volatilves, phenolics, sugars, antioxidant vitamins, and sensory quality of different colored carrot varieties. J Agri Food Chem 49:1410–1416

Arumuganathan K, Earle ED (1991) Nuclear DNA conents of some important plant species. Plant Mol Biol Rep 9:208–218

Bach IC, Olesen A, Simon PW (2002) PCR-based markers to differentiate the mitochondrial genome of petaloid and male-fertile carrot (*Daucus carota* L.). Euphytica 127:353–365

Ballvora A, Hesselbach J, Niewohner J, Leister D, Salamini F, Gebhardt C (1995) Marker enrichment and high-resolution map of the segment of potato chromosome VII harbouring the nematode resistance gene Gro1. Mol Gen Genet 249:82–90

Banga O (1957a) The development of the original European carrot material. Euphytica 6:64–76

Banga O (1957b) Origin of the European cultivated carrot. Euphytica 6:54–63

Becker J, Vos P, Kuiper M, Salamini F, Heun M (1995) Combined mapping of AFLP and RFLP markers in barley. Mol Gen Genet 249:65–73

Bendahmane A, Kanyuka K, Baulcombe DC (1997) High-resolution genetical and physical mapping of the Rx gene for extreme resistance to potato virus × in tetraploid potato. Theor Appl Genet 95:153–162

Bennett MD, Leitch IJ (1995) Nuclear DNA amounts in angiosperms. Ann Bot 76:113–176

Boiteux LS (2000) Characterization of the *Meloidogyne javanica* resistance locus employing molecular markers and isolation of candidate disease resistance loci in the carrot (*Daucus carota* L.) genome. Dissertation, University of Wisconsin

Boiteux LS, Belter JG, Roberts PA, Simon PW (2000) RAPD linkage map of the genomic region encompassing the root-knot nematode (*Meloidogyne javanica)* resistance locus in carrot. Theor Appl Genet 100:439–446

Boiteux LS, Fonseca MEN, Simon PW (1999) Effects of plant tissue and DNA purification method on randomly amplified polymorphic DNA-based genetic fingerprinting analysis in carrot. J Am Soc Hort Sci 124:32–38

Boiteux LS, Hyman JR, Bach IC, Fonseca MEN, Matthews WC, Roberts PA, Simon PW (2004) Employment of flanking codominant STS markers to estimate allelic substitution effects of a nematode resistance locus in carrot. Euphytica 136:37–44

Borner T, Linke B, Nothnagel T, Scheike R, Schulz B, Steinborn R, Brennicke A, Stein M, Wricke G (1995) Inheritance of nuclear and cytoplasmic factors affecting male sterility in *Daucus carota*. Adv Plant Breed 18:111–122

Botstein D, White RL, Skolnick M, Davis RW (1980) Construction of a genetic linkage map in man using restriction fragment length polymorphisms. Amer J Hum Genet 32:314–331

Bowes CE, Wolyn DJ (1998) Phylogenetic relationships among fertile and petaloid male-sterile accessions of carrot, *Daucus carota* L. Theor Appl Genet 96:928–932

Bradeen JM, Bach IC, Briard M, le Clerc V, Grzebelus D, Senalik DA, Simon PW (2002) Molecular diversity analysis of cultivated carrot (*Daucus carota* L.) and wild *Daucus* populations reveals a genetically nonstructured composition. J Am Soc Hort Sci 127:383–391

Bradeen JM, Simon PW (1998) Conversion of an AFLP fragment linked to the carrot Y2 locus to a simple, codominant, PCR-based marker form. Theor Appl Genet 97:960–967

Bradeen JM, Staub JE, Wye C, Antonise R, Peleman J (2001) Towards an expanded and integrated linkage map of cucumber (*Cucumis sativus* L.). Genome 44:111–119

Briard M, le Clerc V, Grzebelus D, Senalik D, Simon PW (2000) Modified protocols for rapid carrot genomic DNA extraction and AFLP analysis using silver stain or radioisotopes. Plant Mol Biol Rep 18:235–241

Brigneti G, Garcia Mas J, Baulcombe DC (1997) Molecular mapping of the potato virus Y resistance gene Rysto in potato. Theor Appl Genet 94:198–203

Bucci G, Kubisiak TL, Nance WL, Menozzi P (1997) A population consensus, partial linkage map of Picea abies Karst. based on RAPD markers. Theor Appl Genet 95:643–654

Buishand JG, Gabelman WH (1979) Investigations on the inheritance of color and carotenoid content in phloem and xylem of carrot roots (*Daucus carota* L.). Euphytica 28:611–632

Chandler GT, Plunkett GM (2004) Evolution in Apiales: nuclear and chloroplast markers together in (almost) perfect harmony. Bot J Linn Soc 144:123–147

Constance L (1971) History of the classification of Umbelliferae (*Apiaceae)*. In: Heywood VH (ed.) The Biology and Chemistry of the Umbelliferae. Bot J Linn Soc, London, pp 1–8

Cronquist A (1981) An Integrated System of Classification of Flowering Plants. Columbia University Press, New York

Dahlgren RMT (1980) A revised system of classification of the angiosperms. Bot J Linn Soc 80:91–124

Davis TM, Yu H (1997) A linkage map of the diploid strawberry, Fragaria vesca. J Hered 88:215–221

Downham A, Collins P (2000) Coloring our foods in the last and next millennium. Intl J Food Sci Technol 35:5–22

Downie SR, Ramanath S, Katz-Downie DS, Llanas E (1998) Molecular systematics of Apiaceae subfamily Apioideae: phylogenetic analyses of nuclear ribosomal DNA internal transcribed spacer and plastid *rpoC1* intron sequences. Am J Bot 85:563–591

Drude CGO (1898) Umbelliferae. In: Engler A, Prantl K (eds.) Die natürlichen Pflanzenfamilien, vol 3(8). Wilhelm Engelmann, Leipzig, Germany, pp 63–250

Echt CS, Erdahl LA, McCoy TJ (1992) Genetic segregation of random amplified polymorphic DNA in diploid cultivated alfalfa. Genome 35:84–87

Feher A, Pasternak TP, Dudits D (2003) Transition of somatic plant cells to an embryogenic state. Plant Cell Tiss Org Cult 74:201–228

Folkerts O, Hanson MR (1991) The male sterility-associated *pcf* gene and the normal *atp9-1* gene in petunia are located on different mitochondrial DNA molecules. Genetics 129:885–895

Freeman RE, Simon PW (1983) Evidence for simple genetic control of sugar type in carrot (*Daucus carota* L.). J Am Soc Hort Sci 108:50–54

Grzebelus D, Senalik D, Jagosz B, Simon PW, Michalik B (2001) The use of AFLP markers for the identification of carrot breeding lines and F1 hybrids. Plant Breed 120:526–528

Hallden C, Hansen M, Nilsson NO, Hjerdin A (1996) Competition as a source of errors in RAPD analysis. Theor Appl Genet 93:1188–1192

Heinonen MI (1990) Carotenoids and provitamin A activity of carrot (*Daucus carota* L.) cultivars. J Agri Food Chem 38:609–612

Heun M, Helentjaris T (1993) Inheritance of RAPDs in F1 hybrids of corn. Theor Appl Genet 85:961–968

Heywood VH (1983) Relationships and evolution in the *Daucus carota* complex. Isr J Bot 32:51–65

Hirner AA, Veit S, Seitz HU (2001) Regulation of anthocyanin biosynthesis in UV-A-irradiated cell cultures of carrot and in organs of intact carrot plants. Plant Sci 161:315–322

Hoffmann-Sommergruber K, O'Riordain G, Ahorn H, Ebner C, Laimer da Camara Machado M, Puhringer H, Scheiner O, Breiteneder H (1999) Molecular characterization of Dau c 1, the Bet v 1 homologous protein from carrot and its cross-reactivity with Bet v 1 and Api g 1. Clin Exp Allergy 29:840–847

Huestis GM, McGrath JM, Quiros CF (1993) Development of genetic markers in celery based on restriction fragment length polymorphisms. Theor Appl Genet 85:889–896

Janska H, Mackenzie SA (1993) Unusual mitochondrial genome organization in cytoplasmic male-sterile common bean and the nature of cytoplasmic reversion to fertility. Genetics 135:869–879

Jiang C, Lewis ME, Sink KC (1997) Combined RAPD and RFLP molecular linkage map of asparagus. Genome 40:69–76

Judd WS, Sanders RW, Donoghue MJ (1994) Angiosperm family pairs: preliminary phylogenetic analyses. Harvard Papers Bot 5:1–51

Just B (2004) Genetic mapping of carotenoid pathway structural genes and major gene QTLs for carotenoid accumulation in wild and domesticated carrot (*Daucus carota* L.). Dissertation, University of Wisconsin, Madison

Kennard WC, Poetter K, Dijkhuizen A, Meglic V, Staub JE, Havey MJ (1994) Linkages among RFLP, RAPD, isozyme, disease-resistance, and morphological markers in narrow and wide crosses of cucumber. Theor Appl Genet 89:42–48

King JJ, Bradeen JM, Bark O, McCallum JA, Havey MJ (1998) A low-density genetic map of onion reveals a role for tandem duplication in the evolution of an extremely large diploid genome. Theor Appl Genet 96:52–62

Kubo T, Nishizawa S, Sugawara A, Itchoda N, Estiati A, Mikami T (2000) The complete nucleotide sequence of the mitochondrial genome of sugar beet (*Beta vulgaris* L.) reveals a novel gene for tRNACys(GCA). Nucl Acids Res 28:2571–2576

Kust AF (1970) Inheritance and differential formation of color and associated pigments in xymlem and phloem of carrot, *Daucus carota,* L. Dissertation, University of Wisconsin, Madison

le Clerc V, Briard M, Granger J, Delettre J (2003) Genebank biodiversity assessments regarding optimal sample size and seed harvesting techniques for the regeneration of carrot accessions. Biodiversity Conserv 12:2227–2236

Li X, van Eck HJ, Rouppe van der Voort JNAM, Huigen DJ, Stam P, Jacobsen E (1998) Autotetraploids and genetic mapping using common AFLP markers: the R2 allele conferring resistance to *Phytophthora infestans* mapped on potato chromosome 4. Theor Appl Genet 96:1121–1128

Lin H, Zhang L, Yang Z, Huang M, Wu N (2003) Molecular characterization of poly(A)-binding protein from *Daucus carota*. DNA Seq 14:147–153

Linke B, Nothnagel T, Borner T (2003) Flower development in carrot CMS plants: mitochondria affect the expression of MADS box genes homologous to GLOBOSA and DEFICIENS. Plant J 34:27–37

Mackill DJ, Zhang Z, Redona ED, Colowit PM (1996) Level of polymorphism and genetic mapping of AFLP markers in rice. Genome 39:969–977

Meksem K, Leister D, Peleman J, Zabeau M, Salamini F, Gebhardt C (1995) A high-resolution map of the vicinity of the R1 locus on chromosome V of potato based on RFLP and AFLP markers. Mol Gen Genet 249:74–81

Meyer K, Keil M, Naldrett MJ (1999) A leucine-rich repeat protein of carrot that exhibits antifreeze activity. FEBS Lett 447:171–178

Michelmore RW, Paran I, Kesseli RV (1991) Identification of markers linked to disease-resistance genes by bulked segregant analysis: a rapid method to detect markers in specific genomic regions by using segregating populations. Proc Natl Acad Sci USA 88:9828–9832

Morelock TE, Simon PW, Peterson CE (1996) Wisconsin wild: another petaloid male-sterile cytoplasm for carrot. HortScience 31:887–888

Mudge J, Andersen WR, Kehrer RL, Fairbanks DJ (1996) A RAPD genetic map of Saccharum officinarum. Crop Sci 36:1362–1366

Nagata R, Wada M, Satoh S, Kamada H, Harada H (1993) A cDNA differentially expressed between embryogenic and nonembryogenic carrot (*Daucus carota* L.) cells. Plant physiol 103:1021

Nakajima Y, Oeda K, Yamamoto T (1998) Characterization of genetic diversity of nuclear and mitochondrial genomes in *Daucus* varieties by RAPD and AFLP. Plant Cell Rep 17:848–853

Nakajima Y, Yamamoto T, Muranaka T, Oeda K (1999) Genetic variation of petaloid male-sterile cytoplasm of carrots revealed by sequence-tagged sites (STSs). Theor Appl Genet 99:837–843

Nakajima Y, Yamamoto T, Muranaka T, Oeda K (2001) A novel *orfB*-related gene of carrot mitochondrial genomes that is associated with homeotic cytoplasmic male sterility (CMS). Plant Mol Biol 46:99–107

Ojima A, Shiota H, Higashi K, Kamada H, Shimma Y, Wada M, Satoh S (1997) An extracellular insoluble inhibitor of cysteine proteinases in cell cultures and seeds of carrot. Plant Mol Biol 34:99–109

Paran I, Rouppe van der Voort J, Lefebvre V, Jahn M, Landry L, van Schriek M, Tanyolac B, Caranta C, Chaim AB, Livingstone K, Palloix A, Peleman J (2004) An integrated genetic map of pepper (*Capsicum* spp.). Mol Breed 13:251–261

Peterson CE, Simon PW (1986) Carrot breeding. In: Bassett MJ (ed.) Breeding Vegetable Crops. AVI, Westport, CT, pp 321–356

Pimenov MG, Leonov MV (1993) The Genera of the Umbelliferae: a Nomenclator. Royal Botanical Gardens, Kew, UK

Plunkett GM, Soltis DE, Soltis PS (1996a) Evolutionary patterns in Apiaceae: inferences based on *matK* sequence data. Syst Bot 21:477–495

Plunkett GM, Soltis DE, Soltis PS (1996b) Higher level relationships of Apiales (Apiaceae and Araliaceae) based on phylogenetic analysis of *rbcL* sequences. Am J Bot 83:499–515

Plunkett GM, Soltis DE, Soltis PS (1997) Clarification of the relationship between Apiaceae and Araliaceae based on *matK* and *rbcL* sequence data. Am J Bot 84:565–580

Riedy MF, Hamilton WJ, Aquadro CF (1992) Excess of non-parental bands in offspring from known primate pedigrees assayed using RAPD PCR. Nucl Acids Res 20:918

Robison MM, Wolyn DJ (2002) Complex organization of the mitochondrial genome of petaloid CMS carrot. Mol Genet Genom 268:232–239

Rubatzky VE, Quiros CF, Simon PW (1999) Carrots and Related Vegetable *Umbelliferae*. CABI, New York

Rurek M, Szklarczyk M, Adamczyk N, Michalik B, Augustyniak H (2001) Differences in editing of mitochondrial *nad3* transcripts from CMA and fertile carrots. Acta Biochim Polonica 48:711–717

Sáenz Laín C (1981) Research on *Daucus* L. (*Umbelliferae*). Anal Jardin Bot Madrid 37:481–533

Sano T, Nishimoto M, Saburi W, Kimura A, Yasuda H, Uchibatake M, Ohwada T, Masuda H (2004) Isolation and characterization of cDNA encoding P-19.5 protein accumulated preferentially at early stage of carrot somatic embryogenesis. Plant Sci 167:1211–1217

Santos CAF, Simon PW (2002a) QTL analyses reveal clustered loci for accumulation of major provitamin A carotenes and lycopene in carrot roots. Mol Genet Genom 268:122–129

Santos CAF, Simon PW (2002b) Some AFLP amplicons are highly conserved DNA sequences mapping to the same linkage groups in two F2 populations of carrot. Genet Mol Biol 25:195–201

Santos CAF, Simon PW (2004) Merging carrot linkage groups based on conserved dominant AFLP markers in F2 populations. J Am Soc Hort Sci 129:211–217

Schmidt ED, Guzzo F, Toonen MA, de Vries SC (1997) A leucine-rice repeat containing receptor-like kinase marks somatic plant cells competent to form embryos. Development 124:2049–2062

Schulz B, Westphal L, Wricke G (1993) Linkage groups of isozymes, RFLP and RAPD markers in carrot (*Daucus carota* L. *sativus*). Euphytica 74:67–76

Schulz B, Westphal L, Wricke G (1994) Linkage groups of isozymes, RFLP and RAPD markers in carrot (*Daucus carota* L. *sativus*). Euphytica 74:67–76

Sebastian RL, Howell EC, King GJ, Marshall DF, Kearsey MJ (2000) An integrated AFLP and RFLP Brassica oleracea linkage map from two morphologically distinct doubled-haploid mapping populations. Theor Appl Genet 100:75–81

Sebkova V, Unger C, Hardegger M, Sturm A (1995) Biochemical, physiological, and molecular characterization of sucrose synthase from *Daucus carota*. Plant Physiol 108:75–83

Sharma AK, Bhattacharyya NK (1954) Further investigations on several genera of Umbelliferae and their relationship. Gentica 30:1–68

Simon PW (1992) Genetic improvement of vegetable carotene content. Biotechnology and Nutrition Third International Symposium, London 1992. Butterworth-Heinemann, London, pp 291–300

Simon PW (1996) Inheritance and expression of purple and yellow storage root color in carrot. J Hered 87:63–66

Simon PW (2000) Domestication, historical development, and modern breeding of carrot. Plant Breed Rev 19:157–190

Simon PW, Matthews WC, Roberts PA (2000) Evidence for simply inherited dominant resistance to *Meloidogyne javanica* in carrot. Theor Appl Genet 100:735–742

Simon PW, Rubatzky VE, Bassett MJ, Strandberg JO, White JM (1997) B7262, purple carrot inbred. HortScience 32:146–147

Simon PW, Wolff XY (1987) Carotene in typical and dark orange carrots. J Agri Food Chem 35:1017–1022

Simpson KL (1983) Relative value of carotenoids as precursors of vitamin A. Am Proc Nutr Soc 42:7–17

Small E (1978) A numerical taxonomic analysis of the *Daucus carota* complex. Can J Bot 56:248–276

Smith MW, Ito M, Yamada T, Suzuki T, Komamine A (1993) Isolation and characterization of a cDNA clone for plant nuclear antigen 21D7 associated with cell division. Plant Physiol 101:809–817

Sondur SN, Manshardt RM, Stiles JI (1996) A genetic linkage map of papaya based on randomly amplified polymorphic DNA markers. Theor Appl Genet 93:547–553

St. Pierre MD, Bayer RJ (1991) The impact of domestication on the genetic variability in the orange carrot, cultivated *Daucus carot*a ssp. *sativus* and the genetic homogeneity of various cultivars. Theor Appl Genet 82:249–253

St. Pierre MD, Bayer RJ, Weis IM (1990) An isozyme-based assessment of the genetic variability within the *Dau-*

cus carota complex (Apiaceae: Caucalideae). Can J Bot 68:2449–2457

Stam P (1993) Construction of integrated genetic linkage maps by means of a new computer package: JoinMap. Plant J 3:739–744

Stintzing FC, Carle R (2004) Functional properties of anthocyanins and betalains in plants, food, and in human nutrition. Trends Food Sci Technol 15:19–38

Sturm A, Chrispeels MJ (1990) cDNA cloning of carrot extracellular beta-fructosidase and its expression in response to wounding and bacterial infection. Plant Cell 2:1107–1119

Sturm A, Hess D, Lee H, Lienhard S (1999a) Neutral invertase is a novel type of sucrose-cleaving enzyme. Physiol Plant 107:159–165

Sturm A, Lienhard S, Schatt S, Hardegger M (1999b) Tissue-specific expression of two genes for sucrose synthase in carrot (*Daucus carota* L.). Plant Mol Biol 39:349–360

Surles RL, Weng N, Simon PW, Tanumihardjo SA (2004) Carotenoid profiles and consumer sensory evaluation of specialty carrots (*Daucus carota* L.) of various colors. J Agri Food Chem 52:3417–3421

Szklarczyk M, Oczkowski M, Augustyniak H, Borner T, Linke B, Michalik B (2000) Organisation and expression of mitochondrial atp9 genes from CMS and fertile carrots. Theor Appl Genet 100:263–270

Takhtajan AL (1980) Outline of the classification of flowering plants (Magnoliophyta). Bot Rev 46:225–359

Tang G, Luscher M, Sturm A (1999) Antisense repression of vacuolar and cell wall invertase in transgenic carrot alters early plant development and sucrose partitioning. Plant Cell 11:177–189

Tang G, Sturm A (1999) Antisense repression of sucrose synthase in carrot (*Daucus carota* L.) affects growth rather than sucrose partitioning. Plant Mol Biol 41:465–479

Thormann CE, Ferreira ME, Camargo LEA, Tivang JG, Osborn TC (1994) Comparison of RFLP and RAPD markers for estimating genetic relationships within and among cruiferous species. Theor Appl Genet 88:973–980

Thorne RF (1968) Synopsis of a putatively phylogenetic classification of the flowering plants. Aliso 6:57–66

Unger C, Hardegger M, Lienhard S, Sturm A (1994) cDNA cloning of carrot (*Daucus carota)* soluble acid beta-fructofuranosidases and comparison with the cell wall isoenzyme. Plant Physiol 104:1351–1357

van den Berg H, Faulks R, Granado HF, Hirschberg J, Olmedilla B, Sandmann G, Southon S, Stahl W (2000) The potential for the improvement of carotenoid levels in foods and the likely systemic effects. J Sci Food Agri 80:880–912

Vavilov NI (1951) The origin, variation, immunity and breeding of cultivated plants. Chron Bot 13:1–366 (translated from Russian by Chester S)

Vivek BS, Simon PW (1999a) Linkage relationships among molecular markers and storage root traits of carrot (*Daucus carota* L. ssp. *sativus)*. Theor Appl Genet 99:58–64

Vivek BS, Simon PW (1999b) Phylogeny and relationships in *Daucus* based on restriction fragment length polymorphisms (RFLPs) of the chloroplast and mitochondrial genomes. Euphytica 105:183–189

Vos P, Hogers R, Bleeker M, Reijans M, Lee T, Hornes M, Frijters A, Pot J, Peleman J, Kuiper M, Zabeau M (1995) AFLP: a new technique for DNA fingerprinting. Nucl Acids Res 23:4407–4414

Vuylsteke M, Mank R, Antonise R, Bastiaans E, Senior ML, Stuber CW, Melchinger AE, Lubbersted T, Xia XC, Stam P, Zabeau M, Kuiper M (1999) Two high-density AFLP linkage maps of Zea mays L.: analysis of distribution of AFLP markers. Theor Appl Genet 99:921–935

Welch JE, Grimball EL (1947) Male sterility in carrot. Science 106:594

Welsh RM (1997) Agronomic problems related to provitamin A carotenoid-rich plants. Eur J Clin Nutr 51:S34–S38

Westphal L, Wricke G (1991) Genetic and linkage analysis of isozyme loci in *Daucus carota* L. Euphytica 56:259–267

Westphal L, Wricke G (1997) Construction of a linkage map of *Daucus carota* L. *sativus* and its application for the mapping of disease resistance and restorer genes. J Appl Genet 38A:13–19

Williams JGK, Kubelik AR, Livak KJ, Rafalski JA, Tingey SV (1990) DNA polymorphisms amplified by arbitrary primers are useful as genetic markers. Nucl Acids Res 18:6531–6535

Worrall D, Elias L (1998) A carrot leucine-rich-repeat protein that inhibits ice recrystallization. Science 282:115–117

Wurtele ES, Wang H, Durgerian S, Nikolau BJ, Ulrich TH (1993) Characterization of a gene that is expressed early in somatic embryogenesis of *Daucus carota*. Plant Physiol 102:303–312

Yamato K, Ogura Y, Kanegae T, Yamada Y, Ohyama K (1992) Mitochondrial genome structure of rice suspension culture from cytoplasmic male-sterile line (A-58CMS): reappraisal of the master circle. Theor Appl Genet 83:279–288

Yang XF, Quiros CF (1995) Construction of a genetic linkage map in celery using DNA-based markers. Genome 38:36–44

Yau Y, Simon PW (2003) A 2.5-kb insert eliminates acid soluble invertase isozyme II transcript in carrot (*Daucus carota* L.) roots, causing high sucrose accumulation. Plant Mol Biol 53:151–162

Yau Y, Santos K, Simon PW (2005) Molecular tagging and selection for sugar type in carrot roots with codominant, PCR-based markers. Mol Breed 16:1–10

Zenkteler M (1962) Microsporogenesis and tapetal development in normal and male-sterile carrots (*Daucus carota* L.). Am J Bot 49:341–348

5 *Brassica rapa*

Masashi Hirai[1] and Satoru Matsumoto[2]

[1] Kyoto Prefectural University, Kita-Inayazuma, Seika Kyoto 619-0244, Japan
e-mail: mhirai@kab.seika.kyoto.jp
[2] National Institute of Vegetable and Tea Science, National Agriculture and Bio-oriented Research Organization, Ano, Mie 514-2392, Japan

5.1 Introduction

5.1.1 Origin of *Brassica rapa*

Brassica rapa is believed to have originated in the Mediterranean areas and Central Asia (Prakash and Hinata 1980; Gómez-Campo and Prakash 1999), where this species is mainly recognized as the turnip. Tsunoda (1980) observed a wild population of *B. rapa* in Asia Minor. The crop was introduced to East Asia along the Silk Road in ancient times. In China, the turnip has been recorded in literature from the fifth century BC (Li 1981). Leafy vegetables in this species may develop from turnip-type plants. Pak-choi was recorded in China in the seventh century AD. Various leafy vegetables have since been differentiated in East Asia, which is believed to be a secondary center for the genetic diversity of *B. rapa*. This species shows as wide a morphological diversity as *Brassica oleracea*. On the contrary, oleiferous *B. rapa* has developed in the area of Central Asia, Afghanistan, and North-West India (Prakash and Hinata 1980; Gómez-Campo and Prakash 1999).

5.1.2 Vegetable Crops of *B. rapa*

Chinese Cabbage

Although nonheading leafy-type crops of *B. rapa* have been recorded in ancient China and Korea (Pyo 1981), tightly heading Chinese cabbage is a relatively new crop, first recorded in China in the eighteenth century. This crop may have originated from either a nonheading or a loosely heading type (Li 1981), and is now widely cultivated in China, Japan, and Korea. It is most important in Korea for making the traditional fermented pickles, kimchi. Chinese cabbage is also used in Japan for traditional fermented pickles and as a cooked vegetable. Originally an autumn crop, Chinese cabbage is now grown all year round. A number of cultivars have been bred for production in various growing seasons.

Turnip

This crop is thought to be an original type of *B. rapa* and is cultivated worldwide, where it has a wide genetic diversity in terms of root shape, size, and color. This crop is sometimes used as fodder in Europe.

Pak-choi and Nonheading Leafy Crops

Originally cultivated in China, Pak-choi is nonheading with enlarged white petioles. Numerous nonheading leafy vegetables with wide morphological variation are known in China, Korea, and Japan. Some of these varieties are only known locally.

Mizuna

Mizuna is sometimes classified as mustard, but it is not as hot. A remarkable characteristic of this crop is branching. The stem branches at the basal part without elongation, and then numerous leaves grow. The leaves and petioles are very narrow. Originally found as a local vegetable in Kyoto in central Japan, it is now widely grown throughout Japan, and has sometimes been found in China and Korea.

Flower Vegetables

Morphologically different types of flower buds with growing stems and leaves are used as cooked vegetables in some areas. For example, Hanana is widely cultivated in Japan as a cooked vegetable in early spring. Hongcaitai grows numerous flower shoots with anthocyanin pigment, and is cultivated in China. A vegetable known as broccoli-raab in the US originated in the Mediterranean area, and produces broccoli-type

Genome Mapping and Molecular Breeding in Plants, Volume 5
Vegetables
C. Kole (Ed.)

massive flower buds. Most of them bolt without vernalization.

5.1.3 Conventional Breeding

Major members of this species are self-incompatible and pure line selection is impossible. Therefore, most cultivars are mass-selected. A single cultivar has a wide genotypic variation, even if it shows a uniform morphological appearance. The first hybrid cultivar of Chinese cabbage was released by the Takii Seed Co., Japan, in 1950. After that, hybrid cultivars of Chinese cabbage have become common in Japan, Korea, and other countries. They show a more uniform appearance, and are suitable for simultaneous harvesting. As such, they are welcome by growers and have a big market. Self-incompatibility is used for hybrid seed production. A simple system for checking *S*-alleles has been developed (Nishio et al. 1994; Sakamoto and Nishio 2001). Anther culture or microspore culture is possible in this species. Kenshin-type Chinese cabbage, a high temperature-tolerant type, shows high efficiencies in anther culture and microspore culture (Sato et al. 1989a,b). This technique is sometimes used for genetic analysis and breeding.

As Chinese cabbage is cultivated all year round, cultivars with various types of growth characters are bred and used. Early-maturing types can be harvested within 50 days after sowing, while late-maturing Chinese cabbage types require prolonged cultivation for as long as 120 days. Resistances to high and low temperatures are needed for summer and winter cultivation, respectively. Bolting resistance is required in cultivars for spring harvest.

The main diseases of Chinese cabbage are clubroot, which is caused by *Plasmodiophora brassicae*; softrot, which is caused by *Erwinia carotovora*; downy mildew, which is caused by *Peronospora parasitica*; and various virus diseases caused by *turnip mosaic virus* and other viruses. White rust caused by the fungus *Albugo candida* affects various *Brassica* crops including turnip rape and vegetable crops of *B. rapa* in cool seasons. Browning of young leaf margins is called tip burn, and is caused by calcium deficiency. Black specks sometimes appear on white midribs, and this phenomenon is called Gomasho, and is thought to be a physiological disorder. However, the cause of this disorder is still unknown.

5.2 Construction of Molecular Maps

Several groups have reported linkage maps of *B. rapa*. Song et al. (1991) reported a map of *B. rapa* based on an F_2 population derived from a cross between Chinese cabbage and spring broccoli using restriction fragment length polymorphism (RFLP) markers. The map covered 1,850 cM. Chyi et al. (1992) reported an RFLP map covering 1,876 map units, and Teutonico et al. (1994) published an RFLP-based map covering 1,785 cM. Kole et al. (1997) published the first molecular map based on an immortal population in this species. They developed an RFLP map covering 890 cM using a recombinant inbred population. Recently, a Korean group reported a detailed map based on a doubled haploid population derived from a cross between a Kenshin-type Chinese cabbage and a standard type Chinese cabbage (Choi et al. 2004). This is recognized as a reference map for the numbering (R1 to R10) and orientation of linkage groups of this species by the Multinational *Brassica* Genome Project (http://www.brassica.info/). Suwabe et al. (2006) reported a linkage map that includes as many as 113 microsatellite markers. A comparison of linkage maps has been made using common RFLP probes, but the RFLP study was very laborious and time-consuming. Moreover, a single RFLP probe sometimes detects more than two loci in the *B. rapa* genome (Song et al. 1991; Chyi et al. 1992; Kole et al. 1997). In contrast, a polymerase chain reaction (PCR) study with microsatellite-specific primers is very simple and rapid, and most primer pairs detect only one locus. Suwabe's map may be used as a standard for this species. Correspondence of these two maps will be shown.

Microsatellite markers are now extensively developed and are used for the genetic analysis and comparative genomics of *Brassica* species (Saal et al. 2001; Suwabe et al. 2002, 2004; Lowe et al. 2004; http://www.brassica.info/ssr/SSRinfo.htm).

5.3 Comparison of Crucifer Genomes

Relationship of *Brassica* genomes has been clarified by U (1935). Three main diploid *Brassica* species, *B. rapa*, *B. nigra*, and *B. oleracea*, have A, B, and C genomes, respectively. They have similar genome sizes and are believed to have the triplicated genomes of the ancestral cruciferous plant, the genomic size of which is similar to that of *Arabidopsis thaliana* (Lagercrantz 1998; Axelsson et al. 2001; Rana et al. 2004). An RFLP study revealed extensive duplication in the *B. rapa* genome (Song et al. 1991). Truco et al. (1996) illustrated the phylogenetic relationships among the chromosomes of three *Brassica* genomes.

Although many studies have been conducted on comparisons between the *Arabidopsis* and *Brassica* genomes, they were mainly undertaken using *B. oleracea* and *B. nigra* (Lagercrantz 1998). Suwabe et al. (2006) compared *Arabidopsis* and *B. rapa* genomes and pointed out that small segments of the *Arabidopsis* genome are scattered throughout the *B. rapa* genome. An extensive rearrangement should have occurred during the evolution of the *B. rapa* genome. After arranging BAC clones of *Brassica* species and probing with *Arabidopsis* genes, Rana et al. (2004) clearly demonstrated the microstructure of *Brassica* genomes, i.e., triplication and a subsequent loss of duplicated genes during the evolution of *Brassica* genomes from an ancestral genome.

5.4 Genetic Analysis and Linkage Markers of Agricultural Traits

5.4.1 Clubroot

Clubroot, caused by the obligate biotroph *P. brassicae*, is a serious soil-borne disease in China, Korea, Japan, and other areas. Kuginuki et al. (1997) first reported linkage markers for a clubroot resistance locus. Five resistance loci, *CRa, Crr1, Crr2, Crr3*, and *CRb*, were found as major genes (Matsumoto et al. 1998; Suwabe et al. 2003; Hirai et al. 2004; Piao et al. 2004). Among them, *Crr1* and *Crr2* were mapped in linkage groups R8 and R1, respectively. In contrast, *Crr3* and *CRb* were mapped in R3. Because the genome regions for *Crr1*, *Crr2*, and *CRb* correspond to chromosome 4 of *A. thaliana*, they may differentiate from the same part of the ancestral genome (Suwabe et al. 2004, 2006). The triplicated positions are all active as clubroot resistance genes in the *B. rapa* genome. *Crr3* was found to correspond to chromosome 3 of *A. thaliana* (Table 1).

5.4.2 White Rust

A disease caused by *A. candida*, white rust is a problem in turnip rape and vegetable *B. rapa*. Kole et al. (1996) reported mapping of a gene *ACA1* on linkage group 4 controlling resistance to the race AC2 and found its close linkage to a trait locus for pubescence. Later on Kole et al. (2002b) found that resistance to two races, AC2 and AC7 of the pathogen was controlled by the same locus, *Aca1*, or two closely linked loci on linkage group 4 of *B. rapa*. In addition, they detected a QTL on linkage group 2 with small effect for the race AC2.

5.4.3 Flowering and Physiological Characters

Brassica rapa contains both annual (flowers without vernalization) and biennial (requires vernalization for flowering) types. Most studies on the genetic variation of flowering time were based on genetic differences between annual and biennial plants. Teutonico and Osborn (1995) studied the flowering time of *Brassica rapa* and found two QTLs in linkage groups 2 and 8. Osborn et al. (1997) found a QTL region *VFR2* on linkage group 8 with pronounced effect. This region was homologous to *Arabidopsis* chromosome 5, where *Flowering Locus C (FLC)* and *CONSTANS (CO)* were located. A QTL with a smaller effect, *VFR1*, was homologous to the top of chromosome 4 of *A. thaliana*, where *FRIGIDA (FRI)* and *LUMINIDEPENDES (LD)* were located. A further study, using a Mendelized population showing monogenic segregation for vernalization, evidenced for homology of *VFR2* to *FLC* of *A. thaliana* (Kole et al. 2001). Examining the variation in flowering time without vernalization, Axelsson et al. (2001) found that two QTLs in a *B. rapa* map corresponded to chromosome 5 of *A. thaliana*, where *CO* and *FLC* are located.

The aforementioned studied variation does not include late flowering traits after vernalization: in other words, bolting resistance. Chinese cabbage cultivars for spring harvest require this trait. A QTL study on a very late flowering line of *B. rapa* found a locus with an additive effect as large as 69 days, as well as a linkage marker (Ajisaka et al. 2001).

Cold tolerance is an important agronomic trait for vegetable *B. rapa* and turnip rape. A QTL analysis showed that several loci affected the winter sur-

Table 1. Clubroot resistance genes detected in *Brassica rapa*

CR locus	Map position[1]	Origin[2]	Gene source	Reference
Cra	–	–	T136-8	Matumoto et al. 1998
Crr1	R8	Chromosome 4	Siloga	Kuginuki et al. 1997; Suwabe et al. 2003
Crr2	R1	Chromosome 4	Siloga	Suwabe et al. 2003
Crr3	R3	Chromosome 3	Milan White	Hirai et al. 2003
CRb	R3	Chromosome 4	Gelria R[3]	Piao et al. 2004

[1]Linkage group named under international agreement
[2]Chromosome of *Arabidopsis* showing homology
[3]A CR locus derived from a turnip, Gelria R was mapped very near to CRb

vival and freezing tolerance of *B. rapa* (Kole et al. 2002a). They compared the genomic regions containing these QTLs to those of *B. napus* and found homology of some regions. Besides, they reported linkage of certain QTLs controlling vernalization, flowering time, winter survival and freezing tolerance in each of these two species. Schrantz et al. (2002) cloned four *Flowering Locus C* (FLC) homologs, *BrFLC1*, *BrFLC2*, *BrFLC3* and *BrFLC5* in *B. rapa*. Using populations derived from a crossing of biennial and annual plants, they find cosegregation of flowering time and three *FLC* genes, *BrFLC1*, *BrFLC2*, and *BrFLC5*. Since vernalization has outstanding effect on the flowering of *B. rapa*, it is interesting that which *FLC* gene has a main effect on vernalization in this species.

5.4.4 Morphological Traits

Song et al. (1995) reported the mapping of several morphological traits. These included pubescence, leaf lobes, petiole characters, and flowering. Among them, pubescence and leaf lobes were mapped as single dominant genes, although some effect of minor genes was recognized. Teutonico and Osborn (1994) reported the loci for pubescence (*Pub1*), seed color (*Yls1*) and presence of erucic acid (*Eru*). Some of the modern Chinese cabbage cultivars have yellow inner leaves in the head. Matsumoto et al. (1998) mapped a gene, *Oy*, for yellow color development. A gibberellic acid-insensitive dwarf gene, *dwf2*, was mapped to the bottom of linkage group R6 (Muangprom and Osborn 2004).

5.5 Marker-Assisted Selection

As mentioned in the previous paragraphs, linkage markers for some traits were developed in the vegetable *Brassica rapa*. Most of these are linkage markers for monogenic traits. However, many commercially important traits are polygenic, and the genetics of these traits are yet to be studied. Verticillium wilt and Erwinia soft rot are problem diseases in the Chinese cabbage industry, but the genetic basis of resistance to these diseases has not been studied because a reproducible scoring system for them has been difficult to establish. Heading is an indispensable trait of Chinese cabbages, and is thought to be controlled by a number of loci, but its genetics also remain to be studied. Some disease-resistance traits have been introduced into Chinese cabbages from nonheading *B. rapa* lines by backcross methods. Breeders' efforts have been predominantly directed toward the recovery of tight-heading traits by backcross breeding. The identification of suitable linkage markers for the heading trait should greatly assist these breeders. Similarly, early crops and crop productivity are important traits in the vegetable *B. rapa*. Because most cultivars are hybrids and these traits may be affected by hybrid vigor, special populations should be developed to provide genetic information useful in the breeding of commercial hybrid cultivars.

Because new cultivars are mainly bred by seed companies, their pedigrees and methods of selection are not disclosed. We infer from the limited information available on their breeding that some seed companies are already using molecular linkage markers. However, their use is limited to a few cases, such as in clubroot-resistance traits.

B. rapa is a highly polymorphic crop. A number of genotypes occur within a morphologically homogeneous line. This is particularly convenient in the search for linkage markers. However, it sometimes makes the application of DNA marker techniques to breeding difficult. The polymorphisms reported in the literature are not always reproducible, even when the same parental strains are used, because of the polymorphisms within the lines used as parents. Therefore, a single linkage marker for one locus is not always useful for breeders. A number of linkage markers should be developed for marker-assisted selection in *B. rapa*. A linkage map containing sufficient anchor markers, such as simple sequence repeats, may also be useful for breeders.

5.6 Genome Sequencing and Map-Based Cloning

The Multinational *Brassica* Genome Project (http://brassica.bbrc.ac.uk/ and http://www.niab.go.kr) has commenced a whole-genome sequencing study. The project uses bacterial artificial chromosome (BAC) libraries from a Chinese cabbage variety, Chiifu, and two reference populations, one with 86 doubled haploid (DH) lines and another with at least 200 recombinant inbred lines. These materials were established by Chungnam National University, Korea. The preliminary stage of the project involves BAC-end sequencing, and is now underway. When the 1000 seed BACs are anchored to the *Arabidopsis* genome sequence and a *B. rapa* genetic map drawn, the next stage will be whole-genome sequencing. The construction of a physical map and whole-genome sequencing of the *B. rapa* genome should provide useful information for breeders and geneticists working with *Brassica* crops.

B. rapa has a rather small genome that shows high homology to that of the model plant, *Arabidopsis thaliana*. It is thought to be good material for map-based cloning. A number of studies have investigated clubroot-resistance genes in *B. rapa*. The map-based cloning of clubroot-resistance genes is progressing in Korea and Japan.

References

Ajisaka H, Kuginuki Y, Yui S, Enomoto S, Hirai M (2001) Identification and mapping of a quantitative trait locus controlling extreme late-bolting in Chinese cabbage (*Brassica rapa* L. ssp. *pekinensis* syn. *campestris* L.) using bulked segregant analysis. Euphytica 118:75–81

Axelsson T, Shavorskaya O, Lagercrantz U (2001) Multiple flowering time QTLs within several *Brassica* species could be the result of duplicated copies of one ancestral gene. Genome 44:856–864

Chyi YS, Hoenecke ME, Sernyk JL (1992) A genetic linkage map of restriction fragment length polymorphism loci for *Brassica rapa* (syn. *campestris*). Genome 35:746–757

Choi SR, Kim JH, Han TH, Piao ZY, Plaha P, Koo DH, Bang JW, King GJ, Teakle GR, Allender CJ, Lydiate DJ, Beynon E, Osborn TC, Lim YP (2004) Construction of genetic linkage map based on molecular markers and analysis of QTLs controlling morphological traits in Chinese cabbage (*Brassica rapa* L. ssp. *Pekinensis*). Joint Meeting of the 14th Crucifer Genetics Workshop and the 4th ISHS Symposium on *Brassicas*, 24–28 October 2004, Abstract Book, p 151

Gómez-Campo C, Prakash S (1999) Origin and Domestication. In: Gómez-Campo C (ed.) Biology of *Brassica* Coenospecies. Elsevier, Amsterdam, pp 33–58

Hirai M, Harada T, Kubo N, Tsukada M, Suwabe K, Matsumoto S (2004) A novel locus for clubroot resistance in *Brassica rapa* and its linkage markers. Theor Appl Genet 108:639–643

Kole C, Kole P, Vogelzang R, Osborn TC (1997) Genetic linkage map of a *Brassica rapa* recombinant inbred population. J Hered 88:553–557

Kole C, Thormann CE, Karlsson BH, Palta JP, Gaffney P, Yandell B, Osborn TC (2002a) Comparative mapping of loci controlling winter survival and related traits in oilseed *Brassica rapa* and *B. napus*. Mol Breed 9:201–210

Kole C, Williams PH, Rimmer SR, Osborn TC (2002b) Linkage mapping of genes controlling resistance to white rust (*Albugo candida*) in *Brassica rapa* (syn. *campestris*) and comparative mapping to *Brassica napus* and *Arabidopsis thaliana*. Genome 45:22–27

Kole C, Quijada, P, Michaels SD, Amasino RM, Osborn TC (2001) Evidence for homology of flowering-time genes *VFR2* from *Brassica rapa* and *FLC* from *Arabidopsis thaliana*. Theor Appl Genet 102:425–430

Kole C, Teutonico R, Mengistu A, Williams PH, Osborn TC (1996) Molecular mapping of a locus controlling resistance to *Albugo candida* in *Brassica rapa*. Phytopathology 86:367–369

Kuginuki Y, Ajisaka H, Yui M, Yoshikawa H, Hida K, Hirai M (1997) RAPD markers linked to a clubroot-resistance locus in *Brassica rapa* L. Euphytica 98:149–154

Lagercrantz (1998) Comparative mapping between *Arabidopsis thaliana* and *Brassica nigra* indicates that *Brassica* genomes have evolved through extensive genome replication accompanied by chromosome fusions and frequent rearrangements. Genetics 150:1217–1228

Li CW (1981) The origin, evolution, taxonomy and hybridization of Chinese cabbage. In: Talekar, NS, Griggs TD (eds.) Chinese Cabbage. AVRDC, Tainan, pp 3–10

Lowe AJ, Moule C, Trick M, Edwards KJ (2004) Efficient large-scale development of microsatellites for marker and mapping applications in *Brassica* crop species. Theor Appl Genet 108:1103–1112

Matsumoto E, Yasui C, Ohi M, Tsukada M (1998) Linkage analysis of RFLP markers for clubroot resistance and pigmentation in Chinese cabbage (*Brassica rapa* ssp. *pekinensis*). Euphytica 104:79–86

Muangprom A, Osborn TC (2004) Characterization of a dwarf gene in *Brassica rapa*, including the identification of a candidate gene. Theor Appl Genet 108:1378–1384

Nishio T, Sakamoto K, Yamaguchi J (1994) PCR-RFLP of *S* locus for identification of breeding lines in cruciferous vegetables. Plant Cell Rep 13:546–550

Osborn TC, Kole C, Parkin IAP, Sharpe AG, Kuiper M, Lydiate DJ, Trick M (1997) Comparison of flowering time genes in *Brassica rapa*, *B. napus* and *Arabidopsis thaliana*. Genetics 146:1123–1129

Piao, ZY, Deng YQ, Choi SR, Park YJ, Lim YP (2004) SCAR and CAPS mapping of CRb, a gene conferring resistance to *Plasmodiophora brassicae* in Chinese cabbage (*Brassica rapa* ssp. *pekinensis*). Theor Appl Genet 108:1458–1465

Prakash S, Hinata K (1980) Taxonomy, cytogenetics and origin of crop brassicas, a review. Opera Bot 55:3–57

Pyo HK (1981) Historical observations on the cultivar development of Chinese cabbage in Korea. In: Talekar NS, Griggs TD (eds.) Chinese Cabbage. AVRDC, Tainan, pp 41–47

Rana D, van den Boogaart T, O'Neill CM, Hynes L, Bent E, Macpherson L, Park JY, Yong Pyo Lim YP, Bancroft I (2004) Conservation of the microstructure of genome segments in *Brassica napus* and its diploid relatives. Plant J 40:725–771

Saal B, Plieske J, Hu J, Quiros CF, Struss D (2001) Microsatellite markers for genome analysis in *Brassica*. II. Assignment of rapeseed microsatellites to the A and C genomes and genetic mapping in *Brassica oleracea* L. Theor Appl Genet 102:695–699

Sakamoto K, Nishio T (2001) Distribution of *S*–haplotypes in commercial cultivars of *Brassica rapa*. Plant Breed 120:155–161

Sato T, Nishio T, Hirai M (1989a) Varietal differences in embryogenic ability in anther culture of Chinese cabbage (*Brassica campestris* ssp. *pekinensis*). Jpn J Breed 39:149–158

Sato T, Nishio T, Hirai M (1989b) Plant regeneration from isolated microspore culture of Chinese cabbage (*Brassica campestris* ssp. *Pekinesis*). Plant Cell Rep 8:486–488

Schrantz ME, Quijida P, Sung S-B, Lukrnd, L, Amisano R, Osborn TC (2002) Characterization and effects of the replicated flowering time gene *FLC* in *Brassica rapa*. Genetics 162:1457–1468

Song KM, Suzuki JY, Slocum MK, Williams PH, Osborn TC (1991) A linkage map of *Brassica rapa* (syn. *campestris*) based on restriction fragment length polymorphism loci. Theor Appl Genet 82:296–304

Song K, Slocum MK, Osborn TC (1995) Molecular marker analysis of genes encoding morphological variation in *Brassica rapa* (syn. *campestris*). Theor Appl Genet 90:1–10

Suwabe K, Iketani H, Nunome T, Kage T, Hirai M (2002) Isolation and characterization of microsatellite in *Brassica rapa* L. Theor Appl Genet 104:1092–1098

Suwabe K, Iketani H, Nunome T, Ohyama A, Hirai M, Fukuoka H (2004) Characteristics of microsatellites in *Brassica rapa* genome and their potential utilization for comparative genomics in *Cruciferae*. Breed Sci 54:85–90

Suwabe K, Tsukazaki H, Iketani H, Hatakeyama K, Fujimura M, Nunome T, Fukuoka H, Matsumoto S, Hirai M (2003) Identification of two loci for resistance to clubroot (*Plasmodiophora brassicae* Woronin) in *Brassica rapa* L. Theor Appl Genet 107:997–1002

Suwabe K, Tsukazaki H, Iketani H, Hatakeyama K, Kondo M, Fujimura M, Nunome T, Fukuoka H, Hirai M, Matsumoto S (2006) SSR-based comparative genomics between *Brassica rapa* and *Arabidopsis thaliana*: the genetic origin of clubroot resistance. Genetics, in press

Teutonico RA, Osborn TC (1995) Mapping loci controlling vernalization requirement in *Brassica rapa*. Theor Appl Genet 91:1279–1283

Teutonico RA, Osborn TC (1994) Mapping of RFLP and qualitative trait loci in *Brassica rapa* and comparison to the linkage maps of *B. napus*, *B. oleracea*, and *Arabidopsis thaliana*. Theor Appl Genet 89:885–894

Truco MJ, Hu J, Sadowski J, Quros CF (1996) Inter- and intragenomic homology of the *Brassica* genomes: implications for their origin and evolution. Theor Appl Genet 93:1225–1233

Tsunoda S (1980) Ecophysiology of wild and cultivated forms in *Brassica* and allied genera. In: Tsunoda S, Hinata K, Gomez-Campo C (eds.) *Brassica* Crops and Wild Allies. Japan Scientific Societies, Tokyo, Japan, pp 109–120

U N (1935) Genome-analysis in *Brassica* with special reference to the experimental formation of *B. napus* and peculiar mode of fertilization. Jpn J Bot 7:389–452

6 Beet

J. Mitchell McGrath[1], Massimo Saccomani[2], Piergiorgio Stevanato[2], and Enrico Biancardi[3]

[1] USDA-ARS Sugarbeet and Bean Research, 494 PSSB Michigan State University East Lansing, MI 48824-1325, USA
e-mail: mitchmcg@msu.edu
[2] Dipartimento di Biotecnologie Agrarie, Università di Padova, viale dell'Università 16, 35020 Legnaro (Pd) Italy
[3] CRA – Istituto Sperimentale per le Colture Industriali, Sezione di Rovigo, viale Amendola 82, 45100 Rovigo, Italy

6.1 Introduction

6.1.1 Brief History of the Crop

The earliest use of beets, likely prehistorical, was leaves harvested from wild plants and used for food (Coons 1936; Ford-Lloyd et al. 1975; de Bock 1986; Lange et al. 1999). Selection likely first transformed the annual habit into the biennial habit characteristic of all current crop types, conserving and propagating germplasm for their leaf quality, the only part utilized at that time (Biancardi 1999). Sweet, swollen roots were probably selected from leafy beets, likely bearing resemblance to the chard of today, cultivated in Assyrian, Greek and Roman gardens (Ford-Lloyd et al. 1975). Traits such as the swollen red root were desired since the middle ages in Europe (Pink 1993), originally selected for its use as a leaf vegetable in the Mediterranean region and then later for use as a fresh or stored root vegetable (Campbell 1976) and as one of the first sources of dietary sweeteners available during the winter months. Later in the Middle Ages, the use of beet root was expanded to include animal feed, and the fodder beet became an important component of European agriculture by the nineteenth century. Beets grown exclusively for sucrose are of relatively recent origin (von Lippmann 1925), economic production was begun in Germany and some years later by edict in Napoleonic France under British blockade of sucrose from tropically grown sugar cane (Winner 1993). Beets with higher levels of sucrose were selected from a white fodder beet variety. The White Silesian variety is still considered to be the primary source of sugar beet germplasm grown today (Fischer 1989). In the following century, sugar beet cultivation expanded to other temperate climates of the world and more recently into warmer climates such as in Northern Africa (Winner 1993). Sea beet (*Beta vulgaris* ssp. *maritima*), the presumed ancestor of the cultivated types, is now common along the Mediterranean coastline and the central and northern Atlantic coasts of Europe and to a lesser extent inland (Ford-Lloyd et al. 1975). Dissemination of wild seed may often be by ocean currents since the fruit is buoyant and most extant wild populations are found within 10 meters of mean sea level (Doney et al. 1990).

Cultivated varieties include leaf beet (e.g., chard), garden beet (e.g., table or red), fodder beet and sugar beet. Molecular marker evidence suggests greater diversity is present in wild *Beta vulgaris* ssp. *maritima* relative to cultivated germplasm. All types are freely crossable and give fertile offspring, although distorted segregation in advanced generations may be observed presumably due to high genetic load in this outcrossing species. It is clear that molecular markers are useful for characterizing sugar beet germplasm (Mita et al. 1991). Ninety-five percent of molecular markers tested discriminated between *Beta* species in sections *Beta* and *Patellares* and 43% of the genomic clones detected variation between tested *Beta vulgaris* cultivars. A relatively large number of *Beta* accessions were examined at ribosomal RNA encoding genes (Santoni and Bervillé 1992). Most *Beta vulgaris* germplasm, including sugar beets, was monomorphic for a particular type. Interestingly, variants were found in Swiss chard. *Beta vulgaris* ssp. *maritima* contained the greatest diversity among all accessions analyzed. Another study scored 111 polymorphic fragments across 41 diverse accessions of cultivated and wild beet (Jung et al. 1993). Genetic diversity in cultivated sugar beet germplasm was found to be low compared with other beet types and wild species. Diversity within crop types has been investigated most intensively within sugar beet (Jung et al. 1993; Hjerdin et al. 1994; Kraft et al. 1997; McGrath et al. 1999; Wang and Goldman 1999), currently the most important economic crop

Genome Mapping and Molecular Breeding in Plants, Volume 5
Vegetables
C. Kole (Ed.)

of the group, and while diversity is reduced in sugar beet relative to sea beets, evidence is consistent with all crop types having been selected from within the wild sea beet germplasm pool. Wild species diversity is viewed as a potential source of novel agronomic alleles (Frese et al. 2001).

6.1.2 Botanical Description

Cultivated beets are herbaceous, dicotyledonous plants in the genus *Beta*, in the *Chenopodiaceae* family. The genus is divided into four sections (*Beta*, *Corollinae*, *Nanae*, and *Patellares*) and includes, as well as all cultivated beets in section *Beta*, 11 other species with little or no commercial value but useful as sources of genetic traits (Ford-Lloyd et al. 1975; Lewellen 1992; Letschert et al. 1994). Species of sections *Patellares*, *Corollinae*, and *Nanae* have more limited geographic distribution than section *Beta*, and are found on various European islands of the Atlantic Ocean and coastal and inland locations from Greece to Iran (Ford-Lloyd et al. 1975; de Bock 1986). Each section is described as being progressively more difficult to hybridize with *Beta vulgaris* (Coons 1954; Letschert et al. 1994) and showing less affinity with *B. vulgaris* in chromosome pairing behavior (Nakamura et al. 1991) and repetitive DNA sequences (Schmidt and Heslop-Harrison 1993, 1998).

The cultivated crop, generally harvested in the first year after sowing, is the nonreproductive tissues, either petioles or leaves in the case of the chard and leafy types, or roots in the remaining crop types where end use in suggested in the common name. Leaves differentiate to form a rosette; their size can vary in relationship to genotype, plant stage, climatic conditions and the presence of leaf diseases (Klotz 2005). The first pairs of leaves are horizontally oriented to maximize light interception and subsequent leaves have a more erect position. In root types, a conical and lengthened taproot forms early during development and continues to enlarge during the growing season. Sucrose and pigments accumulate in vacuoles of parenchyma cells, located in between concentric cortical rings that are a unique and distinguishing feature of beets (Artschwager 1926; Hayward 1938; Doney et al. 1981; Elliott and Weston 1993).

Following a period of near freezing temperatures with long nights (vernalization) (Smit 1983; Elliott and Weston 1993), the rosette forms into a flower stalk with indeterminate inflorescences. Flowers are perfect and wind pollinated, and insect pollination occurs at a low frequency (Artschwager 1927). Inflorescences are green and sessile, and their number varies from two to many. The calyx is composed of five parts, which are adherent at the base of the ovary. There are five stamens inserted in a ring at the base, which secrete uncharacterized aromatic substances. The anthers are separated into two loggias, each of these made up of two pollen sacks. A pistil and a tricarpelate ovary, positioned on the structure that includes the ovule, form the gynaecium. The style is very short and terminates with a three- or four-lobed stigma that persists in the mature fruit. Beet pollen is spherical from 15 to 20 μm in diameter; each flower can produce up to 85,000 pollen grains (Knapp 1958). Self-fertilization rarely occurs, partially because the male and female organs of the same flower become active at different times but also due to a complex system of self-incompatibility (Lundqvist et al. 1973). The duration of the flowering period may be 40 or more days. After fertilization, flowers in the same cluster borne in axils gradually bond at the base to form the seed cluster (seedball), a corky and round structure of about 4–6 mm in diameter botanically classified as a glomerule or utricle. The true seed has a thin, pigmented seed coat that is easily separated from the seed, and contains a maternally-derived perisperm that serves as a carbohydrate reserve, the vestigial endosperm and the embryo consisting of two lipid-rich cotyledons and the axis. On germination, the seed imbibes water through the vascular architecture left by the peduncle, and the axis elongates and forces open the operculum (seedcap) at the sites of lowest tissue resistance (Taylor et al. 2003). If the seed has been placed at the correct depth in the soil (about 2 cm), emergence occurs in one to three weeks depending on temperature. The relatively small true seed is part of the problem in obtaining uniform emergence.

The normal two year breeding cycle can be hastened by seed or seedling vernalization, or by the use of a single dominant gene *B* that confers an annual habit (Abegg 1936; Bosemark 1993). The outcrossing behavior of beet can be circumvented by a dominant gene for self-fertility, S^f (Owen 1942). Inbreeding depression is one consequence of using S^f in a breeding program. A number of inbred lines have been developed using S^f in conjunction with the Mendelian recessive for male-sterility (*aa*), and this has allowed alternative breeding schemes beyond traditional population improvement methods (i.e., mass selection, recurrent selection), which dominates most sugar beet and other crop-use breeding (Hecker and Helmerick 1985; Bosemark 1993). Red beet breeding has relied on S^f for many years (Goldman and Navazio 2003).

DNA content (C-value) of *Beta vulgaris* is reported to be 714 to 758 million base pairs per haploid genome ($n = x = 9$), with variation reported among subspecies (Bennett and Smith 1976; Arumuganathan and Earle 1991). The nine chromosomes of sugar beet are morphologically similar at mitotic metaphase, with the exception of centromeres either metacentric or submetacentric and the presence of a terminal constriction, or satellite, on chromosome 1 (Bosemark and Bormotov 1971; Nakamura et al. 1991). The terminal constriction on chromosome 1 carries the major cluster of 18S-5.8S-25S ribosomal RNA genes and ca. 20 copies at an unlinked locus (Schmidt et al. 1994). The 5S ribosomal RNA genes of *Beta vulgaris* have been cytologically and genetically located to an interstitial site near the centromere on chromosome IV (Schmidt et al. 1994; Schondelmaier et al. 1997). Most crops are diploid ($2n = 2x = 18$), although triploid hybrids are common in sugar beet (Bosemark 1993), and species in other sections have been described from diploid to pentaploid, all based on $x = 9$ chromosomes (Smith 1980). Monosomic and nullisomic plants have not been recovered, indicating the true diploid nature of the crop, and cytogenetic results are supported by linkage analyses of molecular markers where a lack of extensive chromosome duplication is documented (Schondelmaier et al. 1996; Halldén et al. 1998). However, duplicated genes may be more common. In preliminary experiments using 17 ESTs as probes against a BAC (bacterial artificial chromosome) library with six sugar beet genome equivalents, an average of 13 BAC clones per probe was identified (2.6 genes per probe per genome equivalent), with a range of 1 to 39 BAC clones identified per probe (McGrath et al. 2004).

Trisomic series have been obtained (Butterfass 1964; Romagosa et al. 1987), all but one are morphologically distinguishable. Chromosome nomenclature, defined in genetic linkage maps, has only recently been standardized, based on work by Schondelmaier and Jung (1997) integrating previous cytogenetic information based on the Butterfass trisomic series. Thus, many published maps are not concordant. It is presumed that individual chromosomes are homoeologous within the genus *Beta*, and between the crop types, but this remains to be demonstrated.

Highly repetitive DNA sequences constitute 60% or more of the beet genome (Flavell et al. 1974). Excluding ribosomal RNA repeats, the highly repetitive fraction of the genome consists of many families of short (140 to 160 nt) repeating units each with high copy number ($>10^5$ copies per genome) (Schmidt and Heslop-Harrison 1996) and transposable element-like sequences (Schmidt et al. 1995; Staginnus et al. 2001). Each chromosome in sugar beet has a characteristic pattern of repeat-sequence distribution, further supporting the true diploid nature of beet with little or no duplication of the primary chromosome set (Schondelmaier et al. 1996; Halldén et al. 1998). Highly repetitive sequence diversity is high among *Beta* genomes, especially between sections, and has proven an advantage in characterizing interspecific hybrids in *Beta* (Desel et al. 2002).

The sugar beet mitochondrial genome is 368,799 bp and has a 43.9% G+C content (Kubo et al. 2000). The beet mitochondria genome, as represented by a single male-fertile genotype, is over twice as large as the chloroplast genome (368 kb), and encodes 59 recognizable genes. Duplicated sequences, introns, unidentified open reading frames, and foreign sequences imported from the chloroplast and nucleus comprise much of the mitochondrial genome. Twenty-three mitochondrial cytotypes have been described in beets, and nonrandom associations between chloroplast and mitochondrial cytotypes may indicate common cytoplasmic ancestry in some populations of sea beets (Desplanque et al. 2000). Only two cytotypes have been economically important in the deployment of cytoplasmic male-sterility for hybrid seed production; sterile and fertile or normal (Owen 1945). These two mitochondrial genomes differ by at least 15 structural rearrangements including 35 kb of DNA inserted in five regions of the sterile cytoplasm (Kubo et al. 1999). Ivanov et al. (2004) identified specific markers associated with male-sterility in sugar beet that correspond to the transcribed genes of the mitochondrial genome. The interactions between nuclear-cytoplasmic genome are fundamental in determining the expression of cytoplasmic male-sterility used in producing commercial sugar beet hybrids (Bosemark 1993).

6.1.3 Economic Importance

Sugar beet crop supplies about a quarter of the world's consumption of sugar; sugar cane produces the remainder. While cane grows in tropical climates, sugar beet finds its best conditions in continental climates that are characterized by moderate temperatures and uniformly distributed rains. The cultivation of sugar beet is distributed among 40 countries over a total surface area of approximately 7 Mha, from which about 37 Mt of sugar are produced. The largest sugar beet cultivated areas are situated in European countries (Ukraine, Russia, Germany, France, UK, Italy, Spain, etc.). Large cultivations also exist in Asia (China, Turkey, Iran, Japan, Moldova, etc.), in the Americas (USA, Canada and Chile) and in northern Africa (Morocco, Egypt and Tunisia). Production statistics of other crop types are not widely available. Table beet and chard are grown nearly worldwide, especially important in Eastern Europe, but generally for local markets. Commercial production of red table beet for canning in the US rarely exceeds 6,000 ha per year. Fodder beet is important in Europe and Canada.

According to Alexander (1971), sugar beet taproot constituents can be divided in water (75%) and dry matter (25%). Soluble solids (20%) and insoluble solids (5%) are the main components of the dry matter. Sucrose is about 16% of the soluble solids and the remainder (4%) are so-called nonsugars or impurities, which can be eliminated or reduced due to their negative effects on sucrose crystallization. Nitrogenous compounds that are particularly noxious to sugar processing, compose 1.8% of nonsugars. Among the nitrogen-free organic compounds (1.4%) can be cited glucose and fructose, which are monosaccharides derived mainly from sucrose. The remainder is composed of soluble mineral matter (0.8%). Table beets have similar nutrient profiles, with perhaps a slightly reduced proportion of sucrose. Table beet products are good dietary sources of potassium and folic acid, low in protein, and the betalain pigments have potential as antioxidants.

6.1.4 Breeding Objectives

Selection objectives in each of the cultivated types were, and are, quite different. An impressive modification of the plant morphology is evident, not only among the cultivated types themselves, but also among extant wild beet populations. In garden, fodder, and sugar beets, the shape and the composition of the root became completely different from wildtypes, whereas in the leaf beet, only the foliar apparatus has been remarkably modified (Biancardi 1999). Common pathogens do not discriminate between crop types, so breeding for resistance is a common feature of all beet improvement programs. Often, resistances will have been identified in sugar beet and then transferred to other crop types (Goldman and Navazio 2003).

Specific breeding objectives of table beets are root shape and color, and for chard are leaf and petiole characters and color. The primary pigments in beet are the betalains, a unique class of alkaloid pigments found primarily in the Caryophyllales and some fungi (Stafford 1994). Betalain pigments are comprised of the red-violet betacyanins and the yellow betaxanthins. Both are derived from betalamic acid following the cleavage of L-DOPA between the 4- and 5-positions, and differ from one another by conjugation of a substituted aromatic nucleus in the 1,7-diazaheptamethinium chromophore (Fischer and Dreiding 1972; Clement et al. 1994). The cleavage of L-DOPA results in two intermediates, 4,5-secodopa and cyclodopa glucoside. The former intermediate is converted into betalamic acid, which in turn condenses with cyclodopa glucoside to form both betacyanin and betaxanthin. Glycosylation occurs both before and after the condensation reaction, and both pigment molecules contain glucose residues.

Alleles at two linked loci (*R* and *Y*) condition production of betalain pigment in the beet plant (Keller 1936). Color patterning in the beet plant is affected by these *R* locus alleles as well as alleles at the *Y* locus. Red roots are observed only in the presence of dominant alleles at the *R* and *Y* loci, while white roots are conditioned by recessive alleles at both loci. A *yy* condition coupled with *rr*, which is characteristic of most sugar beet cultivars, produces no betacyanin and produces betaxanthin only in the hypocotyls. Betalain pigments extracted from red beet roots provide a natural alternative to synthetic red dyes. Betalains have been successfully used in commercial food coloring operations for a number of years (von Elbe et al. 1974). Red beet dye use is increasing in a number of products, and breeding for increased dye concentrations has the potential to be substituted for other dyes while simultaneously providing antioxidants to the diet. Several investigations suggest additional loci play a role in the quantity of betalain synthesized in the beet root (Watson and Gabelman 1984). Betalain pigment concentration responds to selection in a quantitative fashion. Pigment levels increased an average of 45% in three cycles of selection (Wolyn and Gabelman 1990), and additional gains have been possible (Goldman et al. 1996).

For fodder beet and sugar beet, the primary breeding objective is yield (Frandsen 1958; Knapp 1958; Barocka 1985). For sugar beet, the prime concern is, of course, the yield of white sugar, which at its most fundamental level is a product of a beet's sucrose percentage and its weight. Early breeding resulted in quick improvement in percent sucrose. By 1900, sucrose levels had risen from the 3–6% level reported in the earliest materials to the 12–18% level commonly seen in modern varieties. The genetics of sucrose percentage were detailed by Savitsky (1940). That data suggests two to four major genes control sucrose percent in crosses among divergent types such as sugar (15–20% sucrose), fodder and red beets (3–12% sucrose each) or chard types (12–15% sucrose). Further experiments have confirmed that sucrose percentage is a quantitatively controlled trait with high heritability (Culbertson 1942; Powers 1957; Powers et al. 1963; Zhao et al. 1997). Processing quality is an important sugar beet breeding objective, and is affected by the proportion of sucrose to total soluble solids. By-products of factory processing are: (1) molasses used for the production of ethanol, glutamate, glycine-betaine, and as a nitrogen source in bioreactors; (2) pulp for animal feed; and (3) lime ($CaCO_3$) used for improving acid soils (McGinnis 1971). Processing quality includes several characters that affect the quantity of sugar extractable from the processed roots. Many of the traits that influence quality are under genetic control, but the effect of environment, cultural practices and storage conditions frequently prevails and confound the genetic differences. Among the impurity components (called also nonsugars) sodium, potassium, and amino-nitrogen received the main attention (Campbell 2002). In many cases, their concentration in the roots can be reduced with few mass selection cycles (Powers et al. 1963), suggesting that additive genetic variance is prevailing in determining the single factors of processing quality.

The main target of sugar beet breeding is the development of varieties with the maximum sugar yield at the lowest economical and environmental costs (Knapp 1958). Therefore, the cultivated varieties must be adapted to specific agro-climatic conditions occurring in the different production environments (Barocka 1985). Sugar beet breeders are also involved in the traits related to seed multiplication of commercial varieties. From this point of view, production of seed with high germination ability is important. This trait influences the uniformity of field populations and has a significant effect on sugar yield. The plantlets need to develop quickly, so that the leaves can cover the interrows as soon as possible for the best light interception.

The improvement of sugar yield may be achieved by increasing not only the photosynthesis efficiency and the sucrose accumulation in the roots, but also the traits related to make easier the mechanical harvesting and to reduce the postharvest storage losses. It is also possible to improve the physical and chemical traits of the root tissues to enhance the efficiency of the extraction processes, such as decreasing through selection the concentration of components, including nitrogen compounds, monosaccharides, sodium, potassium, etc. (Campbell 2002).

In the southern areas of cultivation characterized by mild winters, autumn sowing to avoid summer drought is practised. In this case, the varieties must be carefully selected for bolting resistance. In fact, the winter conditions of low temperature and photoperiod favor the beginning of the reproductive phase which are the origins of the development of stalks, flowers and seed. The presence of bolted or annual beets lowers the sugar yield and the seed developed by these beets give rise to infestations of weed beets that become very difficult to control (Smit 1983).

6.1.5 Classical Breeding Achievements

Sugar beet, as the other types of cultivated beet, was initially diploid ($2n = 2x = 18$). The first tetraploid sugar beet families, having twice ($2n = 4x = 36$) the normal chromosomes, were obtained around 1940 with the employment of the mutagenic properties of colchicine (Schwanitz 1938). Plants from the two ploidy levels were crossed, producing triploid ($2n = 3x = 27$) hybrids (Bosemark 1993). Triploid hybrids, manifesting morphological characteristics intermediate to the parental ploidy levels, were commercially of interest especially in Europe, where these hybrids display a slight productive superiority, at times yielding 10% more than the diploid average. In some cases, better disease resistance, as against Cercospora leaf spot, was observed (Skaracis and Smith 1987).

Two exceptional advances in sugar beet breeding during the twentieth century have had a tremendous impact on the economic viability of sugar beet production in the US and the world, cytoplasmic male-sterility (CMS) and monogerm seed. These developments have some, lesser impact on other crop types. Restoration of male-fertility in a sterile cytoplasm is conditioned by alleles at two unlinked loci, *X* and *Z* (Owen 1945). Both must be recessive in the seed parent for expression of male-sterility. Maintainer lines, known as O-types in sugar beet and B-lines in table beet, are also doubly recessive but fertile because of a normal cytoplasm. The process of selecting for O-type requires that each individual in the population under development be tested against a CMS-tester line and the progeny evaluated for male-sterility. Although some phenotypic differences in the restorative abilities of *X* and *Z* have been reported (e.g., male-fertility is higher when an *X* allele is dominant versus *Z*), in practice it is difficult to discriminate between their effects. In addition, the O-type must carry as many useful traits as possible, including high sucrose concentration, high general combining ability, tolerance to a wide range of biotic and abiotic stresses and good seed yield potential, as well as monogerm seed.

The second exceptional advance in sugar beet breeding has been the development of genetically monogerm seed (*m*), first found as a variant in a commercial seed production field (Savitsky 1950). Current agronomic practices require that O-types also be monogerm where each seed ball has only one seed embedded within it. In the wild condition, beet seedballs contain an average of three to four seed each. Many seeds germinate and compete with one another, and it was necessary for growers to spend >100 hours/hectare for singling the crop to a stand of about 100,000 plants per hectare. The selection of genetic monogerm seed and the use of precision seed drills permitted the required stand without hand singling.

All commercial sugar beets grown in the developed world are hybrids. Most sugar beet breeding programs have at least two components; a seed parent development program geared towards developing O-type populations and a pollinator program geared towards breeding for problems associated with unique and diverse growing regions. New characters are difficult to introgress into open-pollinated seed parents. Pollen parents are generally multigerm, open-pollinated and mass selected for disease resistance, and thus it is easier to fix resistance alleles in pollinators than in seed parent lines. Locally adapted hybrid combinations are made using pollinators selected for performance under region-specific growing conditions. Hybrids are generally heterozygous for disease resistance alleles, and this can reduce efficacy of resistance relative to breeding lines. In some areas of the world, multigerm open-pollinated, as well as hybrid, varieties are popularly used because either stand establishment is problematic or labor costs to singulate seedlings are low.

In several cultivation areas, the growers require varieties endowed with resistance to specific diseases, because the cultivation of resistant varieties reduces costs, toxicity and environmental damage associated with chemical protection. In some cases, the genetic resistance is the only means for avoiding losses, for instance resistance to the diseases rhizomania and curly top (Coons 1936; Hecker and Helmerick 1985). Several types of genetic resistance against beet diseases (e.g., rhizomania, Cercospora leaf spot, cyst nematodes) have been transferred from wild beet species to the cultivated varieties (Lewellen 1992; Biancardi et al. 2002). Severe reductions in sugar yield are frequently caused by the cyst nematodes (*Heterodera schachtii*). Resistance from *Beta procumbens*, was transferred to sugar beet by Savitsky (1975), but the yield penalty of this resistance has been difficult to overcome. Programs on hybridization with the species of section *Patellares* were initiated later in Europe (Speckmann and de Bock 1982). Various nematode-resistant monosomic additions in diploid sugar beets were established, each carrying a chromosome segment from *Beta procumbens* (Yu 1982).

Genetic resistances for other pathogens were identified and commercially exploited, e.g., against Rhizoctonia root rot, downy mildew, powdery mildew, etc. (Whitney and Duffus 1986; Biancardi et al. 2005), but each of these pathogens remains a problem today. Many efforts for developing resistances to abiotic stresses were made in different countries. An extensive list of germplasm releases by the USDA over the last 70 years underscores the importance of breeding and germplasm improvement to industry (Doney 1995). Systematic screening of the 2500+ *Beta* accessions in the National Plant Germplasm System identified additional sources of resistance within the primary gene pool (Panella 1996).

Cercospora leaf spot caused by the fungus *Cercospora beticola* is perhaps the most problematic disease of humid and temperate zones. The fungus develops typical necrotic lesions on the leaves. Only one source of partial, quantitative, genetic resistance is available for breeding (Skaracis and Biancardi 2000). A second qualitative type of resistance named C2 has been reported when plants are infected with Cercospora strains from California (Lewellen and Whitney 1976). The first mentioned type of Cercospora leaf spot resistance is controlled by at least four gene pairs with variable effects depending on the severity of infection (Smith and Gaskill 1970). Chemical treatments, together with genetic resistance, provide quite effective protection.

Rhizomania is caused by the *beet necrotic yellow vein virus* (*BNYVV*) carried and transferred to sugar beet roots by the fungus *Polymyxa betae*. The virus is common in most cultivation areas and causes losses of up to 80% of the potential sugar yield. Resistant varieties, which currently provide the unique protection against the disease, are the result of three decades of breeding efforts. The first source of resistance to rhizomania was discovered in Cercospora leaf spot resistant germplasm derived from the multigerm variety Alba P (Biancardi et al. 2002). Based upon segregating F_2 populations, this resistance was classified as quantitative. A more resistant variety Rizor was released in 1985 and this resistance was recognized as monogenic and dominant as hybrids produced segregated in a pattern typical of a single dominant gene, *Rz1*. An additional monogenic resistance (*Rz2*) was identified in a sea beet population coded WB42 (Scholten and Lange 2000).

6.1.6
Classical Mapping Efforts

Inheritance of two color genes, *R* conferring a red root phenotype and *Y* conferring a yellow root phenotypes was the first demonstration of linkage (ca. 8 cM) in beets (Keller 1936), followed closely with demonstration of this group's linkage with the bolting gene, *B* (ca. 15 cM from *R*) (Abegg 1936). This now famous $Y - R - B$ linkage group was further extended (summarized in Smith 1980), and until 1980, only one other linkage group was found based on segregation of morphological markers alone. Goldman and Austin (2000) proposed a gene for blotchy color distribution (*bl*) in the root linked to the $Y - R - B$ group, however this should not be confused with an earlier gene conferring black root (*bl*) whose linkage has not yet been assigned (Smith 1980). Relatively few morphological phenotypes have been examined for Mendelian segregation and linkage beyond the 42 summarized by Smith (1980), except for the (linkage not determined) fasciated flower stalk (*ffs*) characterized in red beet by Goldman (1998). A stem fasciation character (*fas*) was also found in sugar beet to be loosely linked with monogerm seed (*m*) at ca. 27 cM (Wagner et al. 1992).

The relatively few morphological markers described in beet arises as a consequence of its outcrossing nature, where it is difficult to obtain inbred lines needed to uncover recessive alleles in populations. The ease by which pollen disseminates by air currents also hinders calculating precise linkage relationships since pollen contamination (as well as putative lethal alleles) disturb segregation ratios, although this is changing with wider deployment of the dominant S^f allele that renders controlled-cross F_1 hybrids self-fertile. Sensitivity of morphological marker phenotype expression due to environmental variability is also a concern, and isozyme characterization has been useful in developing further genetic loci in beet. Most of this work has concentrated on sugar beet, however results should hold throughout the species, and gene nomenclature should be standardized across the species without regard to crop type.

A number of isozyme marker systems have been deployed to examine linkage and genetic diversity in beets, with better results than using morphological markers due to their general independence from environmental conditions. Various authors have investigated seed storage proteins (van Geyt and Smed 1984) and as many as 13 isozyme systems (Wagner et al. 1992), with polymorphism evident in most cases, and greater diversity found among nonsugar and wild allies as compared to sugar beet where examined (Abe and Shimamoto 1990; Aicher and Saunders 1990). The locus nomenclature has not been standardized, and distorted segregation is common. Linkage was found between various isozyme markers (Abe and Tsuda 1987; van Geyt et al. 1990; Abe et al. 1992, 1993), but more importantly linkage of isozyme loci with morphological traits was characterized. The association between the color locus, *R*, and isocitrate dehydrogenase (*Icd2*) was uncovered and extended multiple times (Smed et al. 1989; Wagner et al. 1992; Abe et al. 1993; Pillen et al. 1993). Stem fasciation, monogerm seed, and four isozyme loci have been linked (Abe et al. 1992; Wagner et al. 1992), and other tentative linkage groups could be defined via cosegregation of CMS restorer loci with isozyme loci (Abe et al. 1992; Wagner et al. 1992; Pillen et al. 1993). Two groups have suggested linkage between a phosphoglucose isomerase (PGI) locus and the S^f self-fertility locus segregating in self-fertile by self-sterile derived genetic populations based on highly distorted segregation ratios (Aicher and Saunders 1990; Abe et al. 1993). Trisomic analyses, based on the Butterfass (1964) series, located five isozyme loci to four chromosomes (Lange et al. 1993; Oleo et al. 1993). More recently, a malate dehydrogenase isozyme tightly linked to root-knot nematode (*Meloidogyne* spp.) resistance has been described (Yu et al. 2001), and further developments have generated a cleaved amplified polymorphic sequence (CAPS) marker near this locus (Weiland and Yu 2003).

6.2 Construction of Genetic Maps

Isozyme and morphological markers were integrated with restriction fragment length polymorphic markers (RFLPs) using both anonymous and named DNA probes (Pillen et al. 1993). A number of genetic maps have been constructed with molecular markers (Barzen et al. 1992, 1995; Pillen et al. 1992, 1993; Uphoff and Wricke 1995; Halldén et al. 1996; Schondelmaier et al. 1996; Nilsson et al. 1997; Schumacher et al. 1997; Hansen et al. 1999; Rae et al. 2000). All but two maps (Yu 2004; Trebbi 2005) have been constructed from sugar beet and other crop types are not yet represented, although the fundamental genetic basis is unlikely to be much different, but allele frequencies likely vary and fixation of "crop use specific" alleles might be expected. Maps have been constructed using anonymous genomic restriction fragment length polymorphic (RFLP), random amplified polymorphic DNA (RAPD), amplified fragment length polymorphic (AFLP) markers, and simple sequence repeats (SSR), and where possible maps have included morphological and isozyme markers. Most SSR markers to date were developed in the private sector and their availability is restricted. Single nucleotide polymorphisms (SNP) are becoming available for mapping in sugar beet (Schneider et al. 2001; Möhring et al. 2004).

The number of markers ranged from 85 to 413 markers and the genetic distance summed across all nine linkage groups (corresponding to the basic chromosome number of nine in *Beta*) for each map ranged from 621 to 1,057 cM. The reason for a large difference in genetic map length is not clear, but it is not related to the number of markers mapped. Most maps have shown a strong clustering of markers in one or two regions of each linkage group, suggesting restricted

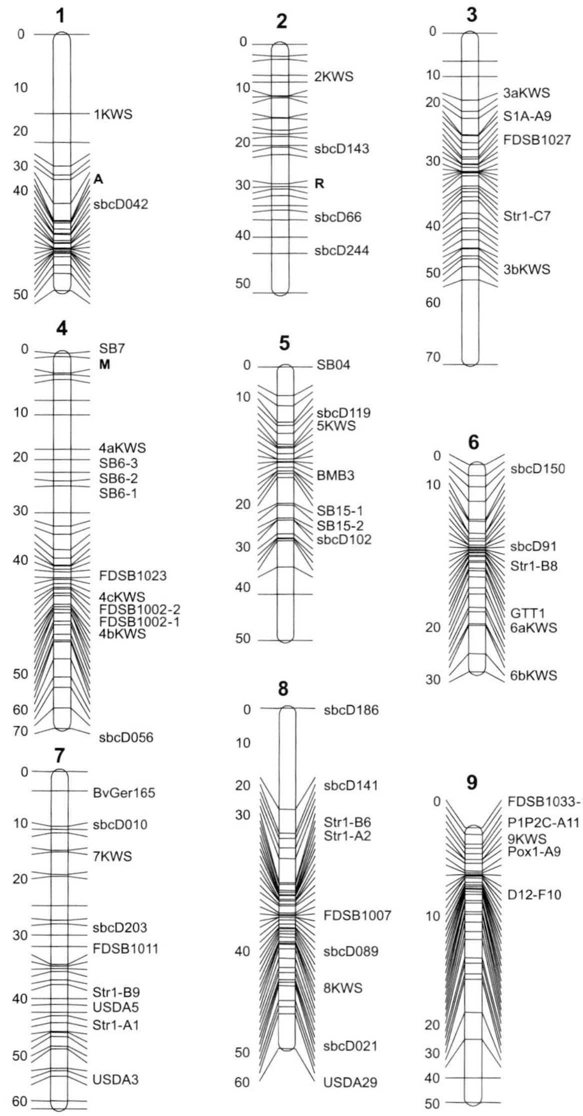

Fig. 1. Genetic map of *Beta vulgaris* (after Trebbi 2005). Chromosomes are numbered according to Schondelmaier and Jung (1997), using SSR markers previously assigned to linkage groups by KWS SAAT AG, Einbeck, Germany (*suffix KWS*). All other named markers except those with *single letter* designations are fragment length polymorphisms associated with cDNA clones. Single letter designations are morphological loci conferring nuclear male-sterility (*A*), red betalain pigment production (*R*), and multigerm seed (*M*). Not shown, but indicated with a *line* on each linkage group, are positions of scored AFLP markers. Total map length is 512 cM

genetic recombination but also arising from the type of marker used (Nilsson et al. 1997). Each of the genetic maps shows variation in the number of detected duplicated loci, ranging from less than 2% (Pillen et al. 1993) to over 38% (Barzen et al. 1995). Up to three bivalents have been detected in meiotic chromosome analyses of haploid sugar beets (Yu 1980; Cistue et al. 1985), suggesting some duplicated chromosome regions exist in beet. The map of Trebbi (2005) involved a cross between sugar beet and table beet using predominantly AFLP markers (Fig. 1), and Yu (2004) using similar markers mapped a sugar × wild beet population. Comparative genome analyses using *Arabidopsis* conserved genes were mapped in a number of species including beets with a conclusion that conserved synteny blocks extend among unrelated dicot plant families (Dominguez et al. 2003).

Particularly noteworthy is the work of Schondelmaier and Jung (1997) who defined molecular, isozyme, and morphological linkage groups based on the Butterfass (1964) trisomic series, thus establishing a common nomenclature for beet linkage groups. Inconsistencies persist in the literature regarding chromosome assignments, although many maps contain a few markers in common, such as the red color locus (*R*) or monogermity (*m*).

6.3 Gene Mapping and Marker-Assisted Selection

Genes for annual vs. biennial habit (*B*), restoration of cytoplasmic male-sterility (*X* and *Z*), nematode resistance, sugar yield, Cercospora leaf spot resistance, and rhizomania resistance have received considerable attention as a result of their considerable economic importance to the beet sugar industry.

In addition to linkage with and isozyme marker (Abe et al. 1993), the annual habit gene *B* was tightly flanked (ca. 5 cM) with RFLP markers (Boudry et al. 1994). AFLP markers were used to saturate this region, resulting in recovery of four markers within 1 cM or less, including two that showed no recombination with the bolting locus (El-Mezawy et al. 2002). A dense physical map has been constructed around this locus in preparation for map-based cloning (Hohmann et al. 2003).

Loci involved in restoration of male-fertility in a sterile cytoplasm, *X* and *Z*, have been located on Butterfass chromosomes 3 and 4, respectively (Schondelmaier and Jung 1997). Locus *X* was located terminally on chromosome 3 (Pillen et al. 1993; Uphhoff and Wricke 1995). A quantitative trait loci (QTL) approach to mapping restorer genes was taken by Hjerdin-Panagopoulos et al. (2002). Two QTLs, 15 cM apart and explaining 79% of the variability, were detected on chromosome 4, and in a different population, another QTL on chromosome 3 explained 72% of the phenotypic variance, and these QTLs likely correspond to *Z* and *X*, respectively, although which specific QTL on chromosome 4 relates to *Z* is uncertain.

Desplanque et al. (2000) characterized a number of mitochondrial types within wild beet populations, in addition to the commonly used Owen (1945) S- and N-cytotypes. Using bulked segregant analysis and AFLP markers, Touzet et al. (2004) described a novel restorer locus for the G-cytotype on chromosome 8. Previously, Laporte et al. (1998) demonstrated linkage of RAPD markers with monogermity (*m*) and a restorer for the H-cytotype, and suggested that this may not be novel since Owens gene *Z* and *m* are on Chromosome 4. Touzet and Budar (2004) describe some of the potential gene functions that could be implicated in CMS fertility restoration.

Marker analyses for corrected sugar yield and sugar content, as well as amino-nitrogen, sodium, and potassium, the primary solutes involved in loss of sucrose to molasses during processing, were performed in two segregating populations tested in a number of environments in order to identify QTLs associated with these characters (Weber et al. 2000). QTLs were discovered but they generally mapped in different chromosomal locations in the two populations and only a few were stably expressed in the same population across environments. Expressed sequence tags (ESTs) predicted to function in carbohydrate and nitrogen metabolism (e.g., candidate genes; Schneider et al. 1999), were used to map QTL for seven traits related to sugar production (Schneider et al. 2002), including corrected sugar yield, root yield, ion balance, sugar content, amino nitrogen, potassium, and sodium. Phenotypic evaluation was of test cross progeny grown in six locations. Twenty-one QTLs were detected for these traits, and four of these were found across different environments (root and corrected sugar yield located on chromosome 4, sucrose content on chromosome 9, and potassium level on chromosome 2). Trebbi and McGrath (2003) examined QTLs for sucrose content in a sugar beet by table beet cross (Trebbi 2005), with particular focus on sucrose content as a proportion of dry matter content, and found 13 QTLs distributed throughout the beet genome.

Sugar beet cyst nematode resistance, morphological, and isozyme markers were placed on the maps of Wagner et al. (1992) and Uphoff and Wricke (1995). Resistance to the beet cyst nematode (*Heterodera schachtii*) is found among wild species of the section *Patellares*, and introgression into beets has been possible via chromosome translocation. At least three different resistance genes have been defined, i.e., *Hs1* in the homoeologous chromosomes 1 of the three species in this section, *Hs2* in the homoeologous groups to chromosome 7 of *B. procumbens* and *B. webbiana*, and *Hs3* in chromosome 8 of *B. webbiana* (Kleine et al. 1998). Species-specific DNA probes were used to identify wild chromosome segments containing *Hs1* in segregating progenies of monosomic alien addition lines (Schmidt et al. 1990; Jung and Hermann 1991; Jung et al. 1992; Salentjin et al. 1992). RAPD markers were also linked to resistant genes Hs^{pro-1} (Halldén et al. 1997) and Hs^{pat-1}, where two out of six markers were closely linked to the Hs^{pat-1} and one of them was converted to a sequence tagged site, usable for gene cloning (Salentjin et al. 1995). With the use of genome-specific satellite markers and chromosomal break-point analyses, the $Hs1^{pro-1}$ locus was positionally cloned (Cai et al. 1997) and redeployed by genetic transformation (e.g., Koenig and Buttner 2004) in the hopes the yield penalty shown by translocation stocks could be averted.

QTLs for Cercospora leaf spot resistance have been identified (Nilsson et al. 1999; Schafer-Pregl 1999; Setiawan et al. 2000; Weber et al. 2000), where five genes previously were implicated in its expression (Smith and Gaskill 1970). Five QTLs accounting for 7 to 18% of phenotypic variation each (based on 221 AFLPs and 46 RFLPs), located on linkage groups 1, 2, 9 and two on linkage group 3, were detected by composite interval mapping (Nilsson et al. 1999). Schafer-Pregl et al. (1999) analyzed QTLs under natural and artificial inoculation and repeated at different plant stages, where three major QTLs were detected on chromosomes 2, 6 and 9 in all conditions, and suggested three additional QTLs on chromosomes 4 and 5 in an F_2 population only. In artificial epiphytotics an additional QTL on chromosome 3 was seen. Setiawan et al. (2000) characterized four QTLs located to chromosomes 3, 4, 7 and 9.

Resistance to powdery mildew (*Erysiphe betae*) has been found in beet. Several QTLs for oligogenic resistance, explaining 27% of the phenotypic variance, have been identified. Monogenic resistance has also been described (Janssen et al. 2003).

Rhizomania is perhaps the most important recent disease of sugar beet, and marker-assisted selection has been instrumental in deploying resistance (Biancardi et al. 2002). Two resistances are known, derived from different sources, named *Rz1* and *Rz2* (Scholten and Lange 2000). Barzen et al. (1992) identified a RFLP marker linked to *Rz1* on chromosome 4, and subsequently sequenced RAPD markers to develop sequence characterized amplified region (SCAR) markers within 2 cM of the *Rz1* locus (Barzen et al. 1997). This locus was also described by Pelsy and Merdinoglu (1996) using bulked segregant analysis (BSA) to first identify linked RAPDs, mapping those in an F_2 population, and performing a QTL analysis that demonstrated ca. 67% of the phenotypic variance associated with a single region, and likely found by Giorio et al. (1997) in a different sugar beet accession. Discovery of *Rz2* and its linkage with *Rz1* was done with RAPD markers and subsequently converted to sequence tagged site (STS) markers, separated by a distance of ca. 20 cM on chromosome 4 (Scholten et al. 1997, 1999).

Disease resistance genes often show common nucleotide sequence motifs, and the general class of resistance genes (R genes) and their analogs (RGA) are of considerable interest. Hunger et al. (2003) used degenerate primers designed using R gene sequences to recover 47 genomic and cDNA RGAs showing expected motifs and domains for 21 nucleotide binding sites (NBS): leucine-rich repeat (LRR) R gene domains and 19 for serine (threonine) protein kinase domain R genes. RGAs were mapped to all chromosomes, identifying alleles associated with rhizomania and Cercospora resistance, within a cluster of nine RGAs on chromosome 7. Interestingly, neither Hunger et al. (2003) nor Tian et al. (2004) could recover any clones corresponding to the TIR-type R gene class in sugar beet, suggesting that beets have lost this particular type of R gene during its evolution.

6.4 Advanced Works

Over 22,000 *B. vulgaris* expressed sequence tags from beet are available and significant information is available on their putative functions. For

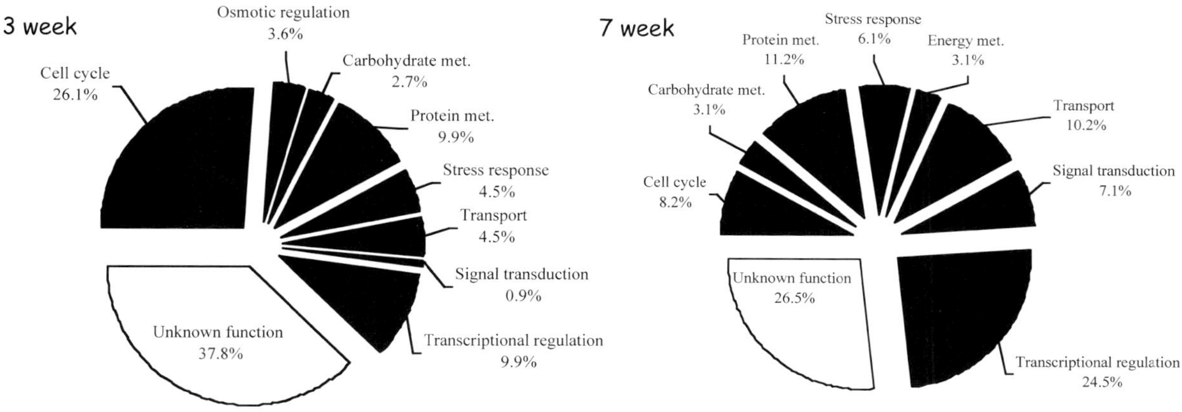

Fig. 2. Functional class distribution of sequences recovered from 3 and 7-week-old beet taproots. Note that 3 WAE (weeks after emergence) has a richer representation of cell cycle and osmotic regulation putative gene products, and 7 WAE functional categories are more highly represented in transport, signal transduction, and transcriptional regulation functional categories (Trebbi 2005)

instance, ESTs collapse into 12,950 tentative consensus (TC) sequences, representing 11,677 peptides (http://sputnik.btk.fi/). The database openSputnik is well-populated with physical characteristics (length, GC content, etc.), annotative attributes (developmental stage, tissue, clone library, etc.) and functional attributes (functional distribution, rankings, gene ontologies, etc.). A Michigan State University (MSU) website (http://genomics.msu.edu/sugarbeet/) contains static clustering and BLAST results against *Arabidopsis* and nonredundant nucleotide sequence space. Both databases are incomplete. The Sputnik database is geared towards comparative evolutionary genomics (Rudd 2005). All TCs have been assigned putative functions, where similarities are identified via BLAST, by comparing against the *Arabidopsis thaliana* genome sequence and/or the database of nonredundant sequences, and results were posted on public websites http://genomics.msu.edu/sugarbeet/, http://sputnik.btk.fi/ (accessed 04/04/2006). The majority of sequenced clones (ESTs) were preselected prior to sequencing to remove a large proportion of redundant transcripts, and thus represents a unigene set of over 10,000 uniquely expressed gene sequences covering four (young and mature roots, leaf, and flower) important developmental stages of beet growth (Herwig et al. 2002). Recently, The Institute for Genomic Research (TIGR) compiled a gene index of beets, resulting in a comprehensive functionally annotated resource of the available nucleotide sequence data (http://www.tigr.org/tdb/tgi/plant.shtml, accessed 04/04/2006).

Taproot specific ESTs were recovered by Kloos et al. (2002), and root EST macroarrays confirmed and extended these results (Bellin et al. 2002). Enzymes of the glyoxylate pathway, deduced through EST approaches, were expressed at high levels in stress germination environments (de los Reyes et al. 2003) as was a germin-like protein gene thought to be an oxalate oxidase providing germination-promoting hydrogen peroxide from seed reserves of oxalic acid (de los Reyes and McGrath 2003). Trebbi and McGrath (2003) examined differential gene expression analyses performed on root tissues sampled from the first to the ninth week after emergence (WAE) via 134 cDNA-AFLP primer and showed 1121 differences, suggesting a transition from juvenile to adult plant growth occurs at by 6 WEA. Analyses of 442 EST sequences obtained by subtractive hybridization supported this transition as some transcript classes were underrepresented at one of the two time points examined (Fig. 2).

6.5 Future Scope of Works

Beyond marker analyses of traditionally important agronomic and disease traits, which needs to continue and expand, further insight into the growth and development of the beet including differences between crop types as well as the responses beets exhibit towards the environment, will need increased attention. Early season growth (e.g., the first 10 weeks) is a critical phase of the beet's life, not only to have good field stands but also to acquire metabolic capacity for agronomic productivity. Early season development includes acquisition of disease resistance (from acute symptoms with devastating effects to chronic symptoms that reduce yield potential), and development of the taproot. This change from seedling to adult vegetative growth coincides, in the field, with warming temperatures (and greater seedling disease), increased growth rate, and increased light interception. Yield of sucrose is directly proportional to the interception of solar irradiation, and maximal interception of sunlight does not occur until the crop canopy is fully developed usually past the summer solstice. Critical insight into gene expression during early growth will help increase biomass production and reduce seedling mortality. Most (if not all) constructive agronomic processes are in place by the tenth week after emergence. Disease losses are a constant concern through the growing season and during postharvest storage, but are caused by a relatively small number of pathogens for which genetic resistance is generally available.

Transition to adult growth is a new concept, for beets, where there are at least two developmental phenotypes associated with a transition; development of supernumerary cortical rings by at least three weeks postgermination, and the sharp increase in sucrose content occurring roughly between four and seven weeks after germination. Whether these disparate phenomena constitute a developmental phase shift between juvenile and adult plant growth is speculative, but clearly a number of developmental events must happen that allow the beet taproot to act as a mature storage organ.

Integrating gene discovery with sugar beet breeding will result in a set of annotated genes useful for recognition and prediction of beet development and response to environment, including challenges associated with biotic and abiotic stresses. Such knowledge ultimately could be used to manipulate biochemical pathways to maximize beet agronomic and horticultural performance. Basic knowledge about the range of biochemical functions as well as their tissue localizations and genetic contributions will allow development of rational molecular breeding objectives.

References

Abe J, Guan GP, Shimamoto Y (1992) Sugarbeet linkage group II constructed by isozyme and marker genes. Proc Jpn Soc Sugar Beet Technol 34:44–50

Abe J, Guan GP, Shimamoto Y (1993) Linkage maps for nine isozyme and four marker loci in sugarbeet (*Beta vulgaris* L.). Euphytica 66:117–126

Abe J, Shimamoto Y (1990) Isozyme variation in sugarbeet and its wild relatives. Proc Jpn Soc Sugar Beet Technol 32:107–112

Abe J, Tsuda C (1987) Genetic analysis for isozyme variation in the section *Vulgares*, genus *Beta*. Jpn J Breed 37:253–261

Abegg FA (1936) A genetic factor for the annual habit in beets and linkage relationship. J Agri Res 53:493–511

Aicher LD, Saunders JW (1990) Inheritance studies and clonal fingerprinting with isoenzymes in sugarbeet. Crop Sci 30:1064–1072

Alexander JT (1971) Factors affecting quality. In: Johnson RT, Alexander JT, Bush GE, Hawkes GR (eds.) Advances in Sugar Beet Production. Iowa State University Press, Ames, IA, pp 371–380

Artschwager E (1926) Anatomy of the vegetative organs of the sugar beet. J Agri Res 33:143–176

Artschwager E (1927) Development of flowers and seed in the sugar beet. J Agri Res 34:1–25

Arumuganathan K, Earle ED (1991) Nuclear DNA content of some important plant species. Plant Mol Biol Rep 9:208–218

Barocka KH (1985) Zucker und Futterrüben (*Beta vulgaris* L.). In: Hoffmann W, Mudra A, Plarre W (eds.) Lehrbuch der Züchtung landwirtschaftlicher Kulturpflanzen, vol 2. Paul Parey, Berlin, pp 245–287

Barzen E, Mechelke W, Ritter E, Seitzer JF, Salamini F (1992) RFLP markers for sugar-beet breeding: chromosomal linkage maps and location of major genes for rhizomania resistance, monogermy and hypocotyl color. Plant J 2:601–611

Barzen E, Melchelke W, Ritter E, Schulte-Kappert E, Salamini F (1995) An extended map of the sugarbeet genome containing RFLP and RAPD loci. Theor Appl Genet 90:189–193

Barzen E, Stahl R, Fuchs E, Borchardt DC, Salamini F (1997) Development of coupling-repulsion-phase SCAR markers diagnostic for the sugar beet Rr1 I allele conferring resistance to rhizomania. Mol Breed 3:231–238

Bellin D, Werber M, Theis T, Schulz B, Weisshaar B, Schneider K (2002) EST sequencing, annotation and macroarray transcriptome analysis identify preferentially root-expressed genes in sugar beet. Plant Biol 4:700–710

Bennett MD, Smith JB (1976) Nuclear DNA amounts in angiosperms. Phil Trans R Soc London Series B 274:227–274

Biancardi E, Lewellen RT, De Biaggi M, Erichsen AW, Stevanato P (2002) The origin of rhizomania resistance in sugar beet. Euphytica 127:383–397

Biancardi E (1999) Miglioramento genetico. In: Casarini B, Biancardi E, Ranalli P (eds.) La Barbabietola da Zucchero negli Ambienti Mediterranei. Edagricole, Bologna, Italy, pp 57–145

Biancardi E, Campbell L, Skaracis GN, De Biaggi M (2005) Genetics and Breeding of Sugar Beet. Science Publisher, Enfield, NH

Bosemark NO (1993) Genetics and breeding. In: Cooke DA, Scott RK (eds.) The Sugar Beet Crop: Science into Practice. Chapman and Hall, London, pp 67–119

Bosemark NO, Bormotov VE (1971) Chromosome morphology in a homozygous line of sugar beet. Hereditas 69:205–211

Boudry P, Wieber R, Saumitou LP, Pillen K, Dijk Hv, Jung C, Van DH (1994) Identification of RFLP markers closely linked to the bolting gene *B* and their significance for the study of the annual habit in beets (*Beta vulgaris* L.). Theor Appl Genet 88:852–858

Butterfass T (1964) Die Chloroplastenzahlen in verschiedenartigen Zellen trisomer Zuckerrüben (*Beta vulgaris* L.). Z Bot 52:46–77

Cai DG, Kleine M, Kifle S, Harloff HJ, Sandal NN, Marcker KA, Klein-Lankhorst RM, Salentijn EMJ, Lange W, Stiekema WJ, Wyss U, Grundler FMW, Jung C (1997) Positional cloning of a gene for nematode resistance in sugar beet. Science 275:832–834

Campbell GKG (1976) Sugar beet. In: Simmonds NW (ed.) Evolution of Crop Plants. Longman, London, UK, p 25

Campbell LG (2002) Sugar beet quality improvement. In: Basra AS, Randhawa LS (eds.) Quality Improvement in Field Crops. Food Products Press, Binghamton, NY, pp 395–413

Cistue L, Romagosa I, Tsuchiya T, Lasa JM (1985) Karyotype analysis in haploid sugar beet. Bot Gaz 146:259–263

Clement JS, Mabry TJ, Wyler H, Dreiding AS (1994) Chemical review and evolutionary significance of the betalains. In: Behnke HD, Mabry TJ (eds.) Caryophyllales: Evolution and Systematics. Springer, Berlin Heidelberg New York, pp 247–261

Coons GH (1936) Improvement of the sugar beet. USDA Yearbook of Agriculture, pp 625–656

Coons GH (1954) The wild species of *Beta*. Proc Am Soc Sugar Beet Technol 8:142–147

Culbertson JO (1942) Inheritance of factors influencing sucrose percentage in *Beta vulgaris*. J Agri Res 64:153–172

de Bock TSM (1986) The genus *Beta*: domestication, taxonomy, and interspecific hybridization for plant breeding. Acta Hort 182:335–343

de los Reyes BG, McGrath JM (2003) Cultivar-specific seedling vigor and expression of a putative oxalate oxidase germin-like protein in sugar beet (*Beta vulgaris* L.). Theor Appl Genet 107:54–61

de los Reyes BG, Myers SJ, McGrath JM (2003) Differential stress-induction of glyoxylate cycle enzymes as a marker for seedling vigor in sugar beet (*Beta vulgaris*). Mol Gen Genom 269:692–698

Desel C, Jansen R, Gue DD, Schmidt T (2002) Painting of parental chromatin in *Beta* hybrids by multicolor fluorescent *in situ* hybridization. Ann Bot London 89:171–181

Desplanque B, Viard F, Bernard J, Forcioli J, Saumitou-Laprade P, Cuguen J, van Dijk H (2000) The linkage disequilibrium between chloroplast DNA and mitochondrial DNA haplotypes in *Beta vulgaris* ssp. *maritima* (L.): the usefulness of both genomes for population genetic studies. Mol Ecol 9:141–154

Dominguez I, Graziano E, Gebhardt C, Barakat A, Berry S, Arus P, Delseny M, Barnes S (2003) Plant genome archaeology: evidence for conserved ancestral chromosome segments in dicotyledonous plant species. Plant Biotechnol J 1:91–99

Doney DL, (1995) USDA-ARS sugarbeet releases. J Sugar Beet Res 32:229–257

Doney DL, Whitney ED, Terry J, Frese L, Fitzgerad P (1990) The distribution and dispersal of *Beta vulgaris* L. ssp. *maritima* germplasm in England, Wales and Ireland. J Sugar Beet Res 27:29–37

Doney DL, Wyse RE, Theurer JC (1981) The relationship between cell size, yield, and sucrose concentration of the sugarbeet root. Can J Plant Sci 61:447–453

Elliott MC, Weston GD (1993) Biology and physiology of the sugar-beet plant. In: Cooke DA, Scott RK (eds.) The Sugar Beet Crop: Science into Practice. Chapman and Hall, London, UK, pp 37–66

El-Mezawy A, Dreyer F, Jacobs G, Jung C (2002) High-resolution mapping of the bolting gene *B* of sugar beet. Theor Appl Genet 105:100–105

Fischer HE (1989) Origin of the Weisse schlesische Rübe (white Silesian beet) and resynthesis of sugar beet. Euphytica 41:75–80

Fischer N, Dreiding AS (1972) Biosynthesis of betalaines. On the cleavage of the aromatic ring during the enzymatic transformation of dopa into betalamic acid. Helv Chim Acta 55:649–658

Flavell RB, Bennet MD, Smith JB (1974) Genome size and the proportion of repeated nucleotide sequence DNA in plants. Biochem Genet 12:257–269

Ford-Lloyd BV, Williams ALS, Williams JT (1975) A revision of *Beta* section *Vulgares* (*Chenopodiaceae*), with new light on the origin of cultivated beets. Bot J Linn Soc 71:89–102

Frandsen KJ (1958) Breeding of forage beets. In: Roemer T, Rudorf W (eds.) Handbuch der Pflanzenzüchtung, vol 3. Paul Parey, Berlin, pp 284–311

Frese L, Desprez B, Ziegler D (2001) Potential of genetic resources and breeding strategies for base-broadening in *Beta*. In: Cooper HD, Spillane C, Hogdkin T (eds.) Broadening the Genetic Base of Crop Production. IPGRI/FAO, Rome, pp 295–309

Giorio G, Gallitelli M, Carriero F (1997) Molecular markers linked to rhizomania resistance in sugar beet, *Beta vulgaris*, from two different sources map to the same linkage group. Plant Breed 116:401–408

Goldman IL (1998) Inheritance of *ffs*, a gene conditioning fasciated flower stem in red beet. J Am Soc Hort Sci 123:632–634

Goldman IL, Eagen KA, Breitbach DN, Gabelman WH (1996) Simultaneous selection is effective in increasing betalain pigment concentration but not total dissolved solids in red beet (*Beta vulgaris* L.). J Am Soc Hort Sci 121:23–26

Goldman IL, Austin D (2000) Linkage among the *R*, *Y* and *Bl* loci in table beet. Theor Appl Genet 100:337–343

Goldman IL, Navazio JP (2003) History and breeding of table beet in the United States. Plant Breed Rev 22:357–388

Halldén C, Sall T, Olsson K, Nilsson NO, Hjerdin A (1997) The use of bulked segregant analysis to accumulate RAPD markers near a locus for beet cyst nematode resistance in *Beta vulgaris*. Plant Breed 116:18–22

Halldén C, Ahrén D, Hjerdin A, Säll T, Nilsson NO (1998) No conserved homoeologous regions found in the sugar beet genome. J Sugar Beet Res 35:1–13

Halldén C, Hjerdin A, Rading IM, Säll T Fridlundh B, Johannisdottir G, Tuvesson S, Äkesson C, Nilsson NO (1996) A high-density RFLP linkage map of sugarbeet. Genome 39:634–645

Hansen M, Kraft T, Christiansson M, Nilsson NO (1999) Evaluation of AFLP in *Beta*. Theor Appl Genet 98:845–852

Hayward HE (1938) The Structure of Economic Plants: Chapter IX: *Chenopodiaceae*. MacMillan, New York, pp 247–282

Hecker RJ, Helmerick RH (1985) Sugarbeet breeding in the United States. In: Russell GE (ed.) Progress in Plant Breeding: 1. Butterworths, London, pp 37–61

Herwig R, Schulz B, Weisshaar B, Hennig S, Steinfath M, Drungowski M, Stahl D, Wruck W, Menze A, O'Brien J, Lehrach H, Radelof U (2002) Construction of a unigene cDNA clone set by oligonucleotide fingerprinting allows access to 25,000 potential sugar beet genes. Plant J 32:845–857

Hjerdin A, Säll T, Nilsson NO, Bornmann CH, Hallden C (1994) Genetic variation among wild and cultivated beet of the Section *Beta* as revealed by RFLP analysis. J Sugar Beet Res 1:59–67

Hjerdin-Panagopoulos A, Kraft T, Rading IM, Tuvesson S, Nilsson NO (2002) Three QTL regions for restoration of Owen CMS in sugar beet. Crop Sci 42:540–544

Hohmann U, Jacobs G, Telgmann A, Gaafar RM, Alam S, Jung C (2003) A bacterial artificial chromosome (BAC) library of sugar beet and a physical map of the region encompassing the bolting gene *B*. Mol Gene Genom 269:126–136

Hunger S, Di Gaspero G, Möhring S, Bellin Schafer-Pregl R, Borchardt DC, Durel CE, Werber M, Weisshaar B, Salamini F, Schneider K (2003) Isolation and linkage analysis of expressed disease-resistance gene analogs of sugar beet (*Beta vulgaris* L.). Genome 46:70–82

Ivanov MK, Revenko AS, Dymshits DM (2004) Cytoplasmic male sterility-associated structural variation of the mitochondrial genome regions containing *rps3* and *orf215* in sugar beet (*Beta vulgaris* L.). Mol Biol 38:345–350

Janssen GJW, Nihlgard M, Kraft T (2003) Mapping of resistance genes to powdery mildew (*Erysiphe betae*) in sugar beet. Intl Sugar J 105:448–451

Jung C, Herrmann RG (1991) A DNA probe for rapid screening of sugar beet (*Beta vulgaris* L.) carrying extra chromosomes from wild beets of the *Procumbentes* section. Plant Breed 107:275–279

Jung C, Koch R, Fischer F, Brandes A, Wricke G, Herrmann RG (1992) DNA markers closely linked to nematode resistance genes in sugar beet (*Beta vulgaris* L.) mapped using chromosome additions and translocations originating from wild beets of the *Procumbentes* section. Mol Gen Genet 232:271–278

Jung C, Pillen K, Frese L, Fahr S, Melchinger AE (1993) Phylogenetic relationships between cultivated and wild species of the genus *Beta* revealed by DNA fingerprinting. Theor Appl Genet 86:449–457

Keller W (1936) Inheritance of some major color types in beets. J Agri Res 52:27–38

Kleine M, Voss H, Cai D, Jung C (1998) Evaluation of nematode-resistant sugar beet (*Beta vulgaris* L.) lines by molecular analysis. Theor Appl Genet 97:896–904

Kloos DU, Oltmanns H, Dock C, Stahl D, Hehl R (2002) Isolation and molecular analysis of six taproot expressed genes from sugar beet. J Exp Bot 53:1533–1534

Klotz K (2005) Anatomy and Physiology. In: Biancardi E, Campbell L, Skaracis GN, De Biaggi M (eds.) Genetics and Breeding of Sugar Beet. Science Publisher, Enfield, NH, pp 9–18

Knapp E (1958) *Beta* Rüben. In: Roemer T, Rudorf W (eds.) Handbuch der Pflanzenzüchtung, vol. 3 Paul Parey, Berlin, pp 196–284

Koenig R, Buttner G (2004) Strategies for the detection of potential Beet necrotic yellow vein virus genome recombinations which might arise as a result of growing A type coat protein gene-expressing sugarbeets in soil containing B type virus. Transgenic Res 13:21–28

Kraft T, Fridlund B, Hjerdin A, Sall T, Tuvesson S, Hallden C (1997) Estimating genetic variation in sugar beets and wild beets using pools of individuals. Genome 40:527–533

Kubo T, Nishizawa S, Mikami T (1999) Alterations in organization and transcription of the mitochondrial genome of cytoplasmic male-sterile sugar beet (*Beta vulgaris* L.). Mol Gen Genet 262:283–290

Kubo T, Nishizawa S, Sugawara A, Itchoda N, Estiati A, Mikami T (2000) The complete nucleotide sequence of the mitochondrial genome of sugar beet (*Beta vulgaris* L.) reveals a novel gene for tRNACys(GCA). Nucl Acids Res 28:2571–2576

Lange W, Oleo M, de Bock TSM, D'Haeseleer M, Jacobs, M (1993) Chromosomal assignment of three enzyme coding loci (*Icd1*, *Nad-Mdh1* and *Aco1*) using primary trisomics in beet (*Beta vulgaris* L.). Plant Breed 111:177–184

Lange W, Brandenburg WA, de Bock TSM (1999) Taxonomy and cultonomy of beet (*Beta vulgaris* L.). Bot J Linn Soc 130:81–96

Laporte V, Merdinoglu D, Saumitou Laprade P, Butterlin G, Vernet P, Cuguen J (1998) Identification and mapping of RAPD and RFLP markers linked to a fertility restorer gene for a new source of cytoplasmic male sterility in *Beta vulgaris* ssp. *maritima*. Theor Appl Genet 96:989–996

Letschert JPW, Lange W, Frese L, Van Den Berg RG (1994) Taxonomy of *Beta* section *Beta*. J Sugar Beet Res 31:69–85

Lewellen RT (1992) Use of plant introductions to improve populations and hybrids of sugarbeet. In: Use of Plant Inductions in Cultivar Development, Part 2. CSSA Spl Pub No 20, Crop Sci Soc Am, Madison, WI, pp 117–136

Lewellen RT, Whitney ED (1976) Inheritance of resistance to race C2 of *Cercospora beticola* in sugar beet. Crop Sci 16:558–561

Lundqvist A, Østerbye U, Larsen K, Linde-Laursen I (1973) Complex self-incompatibility systems in *Ranunculus acris* L. and *Beta vulgaris* L. Hereditas 74:161–168

McGinnis RA (1971) Chemistry of the beet and processing materials. In: McGinnis RA (ed.) Beet Sugar Technology, 3rd edn. Beet Sugar Development Foundation, Fort Collins, CO, pp 25–63

McGrath JM, Derrico CA, Yu Y (1999) Genetic diversity in selected, historical USDA sugarbeet germplasm releases and *Beta vulgaris* ssp. *maritima*. Theor Appl Genet 98:968–976

McGrath JM, Shaw RS, de los Reyes BG, Weiland JJ (2004) Construction of a sugar beet BAC library from a hybrid that combines diverse traits. Plant Mol Biol Rep 22:23–28

Mita G, Dani M, Casciari P, Pasquali A, Selva E, Minganti C, Piccardi P (1991) Assessment of the degree of genetic variation in beet based on RFLP analysis and the taxonomy of *Beta*. Euphytica 55:1–6

Möhring S, Salamini F, Schneider K (2004) Multiplexed, linkage group-specific SNP marker sets for rapid genetic mapping and fingerprinting of sugar beet (*Beta vulgaris* L.). Mol Breed 14:475–488

Nakamura C, Skaracis GN, Romagosa I (1991) Cytogenetics and breeding in sugarbeet. In: Tsuchiya T, Gupta PK (eds.) Chromosome Engineering in Plants: Genetics, Breeding and Evolution. Elsevier, New York, pp 295–314

Nilsson NO, Halldén C, Hansen M, Hjerdin A, Säll T (1997) Comparing the distribution of RAPD and RFLP markers in a high-density linkage map of sugar beet. Genome 40:644–651

Nilsson NO, Hansen M, Panagopoulos AH, Tuvesson S, Ehlde M, Christiansson M, Rading IM, Rissler M, Kraft T (1999) QTL analysis of Cercospora leaf spot resistance in sugar beet. Plant Breed 118:327–334

Oleo M, Lange W, D'Haeseleer M, Debock TSM, Jacobs M (1993) Isozyme analysis of primary trisomics in beet (*Beta vulgaris* L.). Genetical characterization and techniques for chromosomal assignment of two enzyme-coding loci: leucine aminopeptidase and glutamate oxaloacetate transaminase. Theor Appl Genet 86:761–768

Owen FV (1942) Inheritance of cross and self-sterility and self-fertility in *Beta vulgaris*. J Agri Res 64:679–698

Owen FV (1945) Cytoplasmically inherited male-sterility in sugar beets. J Agri Res 71:423–440

Panella L (ed.) (1996) Report on the status of *Beta* germplasm in the United States. Compiled by the Sugarbeet Crop Germplasm Committee, http://www.ars-grin.gov/npgs/cgc_reports/beta/ (accessed October 29, 2005)

Pelsy F, Merdinoglu D (1996) Identification and mapping of random amplified polymorphic DNA markers linked to a rhizomania resistance gene in sugar beet (*Beta vulgaris* L) by bulked segregant analysis. Plant Breed 115:371–377

Pillen K, Steinrücken G, Herrmann RG, Jung C (1993) An extended linkage map of sugar beet (*Beta vulgaris* L.) including nine putative lethal genes and the restorer gene *X*. Plant Breed 111:265–272

Pillen K, Steinrücken G, Wricke G, Herrmann RG, Jung C (1992) A linkage map of sugar beet (*Beta vulgaris* L). Theor Appl Genet 84:129–135

Pink DAC (1993) Beetroot. In: Kalloo G, Bergh BO (eds.) Breeding Vegetable Crops. Pergamon, Oxford, pp 473–477

Powers L, Schmehl WR, Federer WT, Payne MG (1963) Chemical genetic and soil studies involving thirteen characters in sugar beets. J Am Soc Sugar Beet Technol 12:393–448

Powers L (1957) Identification of genetically-superior individuals and the prediction of genetic gains in sugar beet breeding programs. J Am Soc Sugar Beet Technol 9:408–432

Rae SJ, Aldam C, Dominguez I, Hoebrechts M, Barnes SR, Edwards KJ (2000) Development and incorporation of microsatellite markers into the linkage map of sugar beet (*Beta vulgaris* spp.). Theor Appl Genet 100:1240–1248

Rudd S (2005) openSputnik: a database to ESTablish comparative plant genomics using unsaturated sequence collections. Nucl Acids Res 33:D622–D627; doi:10.1093/nar/gki040

Romagosa I, Cistue L, Tsuchya T, Lasa JM, Hecker RJ (1987) Primary trisomics in sugar beet. II. Cytological identification. Crop Sci 27:435–439

Salentjin EMJ, Sandal NN, Lange W, de Bock TSM, Krens FA, Marcker KA, Stiekema WJ (1992) Isolation of DNA markers linked to a beet cyst nematode resistance locus in *Beta patellaris* and *Beta procumbens*. Mol Gen Genet 235:432–440

Salentjin EMJ, Arensdereuver MJB, Lange W, Stiekema WJ, Klein-Lankhorst RM (1995) Isolation and characterization of RAPD-based markers linked to the beet cyst-nematode resistance locus *HS1(Pat-1)* on chromosome 1 of *B. patellaris*. Theor Appl Genet 90:885–891

Santoni S, Bervillé A (1992) Characterization of the nuclear ribosomal DNA units and phylogeny of *Beta* L. wild forms and cultivated beets. Theor Appl Genet 83:533–542

Savitsky VF (1940) Genetics of sugar beets. USDA-National Agricultural Library Translation 5825. Translated from Russian, p 135

Savitsky VF (1950) Monogerm sugar beets in the United States. Proc Am Soc Sugar Beet Technol 6:156–159

Savitsky H (1975) Hybridization between *Beta vulgaris* and *B. procumbens* and transmission of nematode (*Heterodera schactii*) resistance to sugar beet. Can J Genet Cytol 17:197–209

Schafer-Pregl R, Borchardt DC, Barzen E, Glass C, Mechelke W, Seitzer JF, Salamini F (1999) Localization of QTLs for tolerance to *Cercospora beticola* on sugar beet linkage groups. Theor Appl Genet 99:829–836

Schmidt T, Junghans H, Metzlaff M (1990) Construction of *Beta procumbens*-specific DNA probes and their application for the screening of *B. vulgaris - B. procumbens* (2n = 19) addition lines. Theor Appl Genet 79:177–181

Schmidt T, Heslop-Harrison JS (1993) Variability and evolution of highly repeated DNA sequences in the genus *Beta*. Genome 36:1074–1079

Schmidt T, Kubis S, Heslop-Harrison JS (1995) Analysis and chromosomal localization of retrotransposons in sugar beet (*Beta vulgaris* L.): LINES and Ty1-*copia*-like elements as major components of the genome. Chrom Res 3:335–345

Schmidt T, Heslop-Harrison JS (1996) The physical and genomic organization of microsatellites in sugar beet. Proc Natl Acad Sci USA 93:8761–8765

Schmidt T, Schwarzacher T, Heslop-Harrison JS (1994) Physical mapping of ribosomal RNA genes by fluorescent in situ hybridization and structural analysis of 5S ribosomal RNA genes and intergenic spacer sequences in sugar beet (*Beta vulgaris*). Theor Appl Genet 88:629–636

Schmidt T, Heslop-Harrison JS (1998) Genomes, genes and junk: the large-scale organization of plant chromosomes. Trends Plant Sci 3:195–199

Schneider K, Borchardt DC, Schafer-Pregl R, Nagl N, Glass C, Jeppsson A, Gebhardt C, Salamini F (1999) PCR-based cloning and segregation analysis of functional gene homologs in *Beta vulgaris*. Mol Gen Genet 262:515–524

Schneider K, Weisshaar B, Borchardt DC, Salamini F (2001) SNP frequency and allelic haplotype structure of *Beta vulgaris* expressed genes. Mol Breed 8:63–74

Schneider K, Schafer-Pregl R, Borchardt DC, Salamini F (2002) Mapping QTLs for sucrose content, yield and quality in a sugar beet population fingerprinted by EST-related markers. Theor Appl Genet 104:1107–1113

Scholten OE, Klein-Lankhorst RM, Esselink DG, DeBock TSM, Lange W (1997) Identification and mapping of random amplified polymorphic DNA (RAPD) markers linked to resistance against beet necrotic yellow vein virus (BNYVV) in *Beta* accessions. Theor Appl Genet 94:123–130

Scholten OE, Lange W (2000) Breeding for resistance to rhizomania in sugar beet: A review. Euphytica 112:219–231

Scholten OE, de Bock TSM, Klein-Lankhorst RM, Lange W (1999) Inheritance of resistance to BNYVV in *Beta vulgaris* conferred by a second gene for resistance. Theor App Genet 99:740–746

Schondelmaier J, Steinrücken G, Jung C (1996) Integration of AFLP markers into a linkage map of sugar beet (*Beta vulgaris* L). Plant Breed 115:231–237

Schondelmaier J, Jung C (1997) Chromosomal assignment of the nine linkage groups of sugar beet (*Beta vulgaris* L.) using primary trisomics. Theor Appl Genet 95:590–596

Schondelmaier J, Schmidt T, Heslop-Harrison JS, Jung C (1997) Genetic and chromosomal localization of the 5S rDNA locus in sugar beet (*Beta vulgaris* L.). Genome 40:171–175

Schumacher K, Schondelmaier J, Barzen E, Steinrücken G, Borchardt D, Webber WE, Salamini F (1997) Combining different linkage maps in sugar beet (*Beta vulgaris* L.) to make one map. Plant Breed 116:23–38

Schwanitz F (1938) Die Herstellung polyploider Rassen bei *Beta* Rüben und Gemüsearten durch Behandlung mit Colchicin. Züchter 10:278–279

Setiawan A, Koch G, Barnes SR, Jung C (2000) Mapping quantitative trait loci (QTLs) for resistance to Cercospora leaf spot disease (*Cercospora beticola* Sacc.) in sugar beet (*Beta vulgaris* L.). Theor Appl Genet 100:1176–1182

Skaracis GN, Biancardi E (2000) Breeding for Cercospora resistance in sugar beet. In: *Cercospora beticola* Sacc. Biology, Agronomic Influence and Control Measures in Sugar Beet, vol 2. Advances in Sugar Beet Research. IIRB, Brussels, pp 177–195

Skaracis GN, Smith GA (1987) Ploidy level effect on resistance to Cercospora leaf spot in sugar beet. In: Proceedings of the 24th Meeting, Am Soc Sugar Beet Technol, p 26

Smed E, van Geyt JPC, Oleo M (1989) Genetical control and linkage relationships of isozyme markers in sugar beet (*B. vulgaris* L.). 1. Isocitrate dehydrogenase, adenylate kinase, phosphoglucomutase, glucose phosphate isomerase and cathodal peroxidase. Theor Appl Genet 78:97–104

Smit AL (1983) Influence of external factors on growth and development of sugar beet (*Beta vulgaris* L.). Agri Res Rep 914, Wageningen, The Netherlands

Smith GA (1980) Sugarbeet. In: Fehr WR, Hadley HH (eds.) Hybridization of Crop Plants. American Society of Agronomy, Madison, WI, pp 601–616

Smith GA, Gaskill JO (1970) Inheritance of resistance to Cercospora leaf spot in sugarbeet. J Am Soc Sugar Beet Technol 16:172–180

Speckmann GJ, de Bock TSM (1982) The production of alien monosomic additions in *Beta vulgaris* as a source for the introgression of resistance to beet root nematode (*Heterodera schachtii*) from *Beta* species of the section *Patellares*. Euphytica 31:313–323

Stafford HA (1994) Anthocyanins and betalains: evolution of the mutually exclusive pathways. Plant Sci 101:91–98

Staginnus C, Huettel B, Desel C, Schmidt T, Kahl G (2001) A PCR-based assay to detect En/Spm-like transposon sequences in plants. Chrom Res 9:591–605

Taylor AG, Goffinet MC, Pikuz SA, Shelkovenko TA, Mitchell MD, Chandler KM, Hammer DA (2003) Physicochemical factors influence beet (*Beta vulgaris* L.) seed germination. In: Nicolás G, Bradford KJ, Côme D, Pritchard HW (eds.) The Biology of Seeds: Recent Research Advances. CABI, Wallingford, UK, pp 433–440

Tian Y, Fan LJ, Thurau T, Jung C, Cai D (2004) The absence of TIR-type resistance gene analogs in the sugar beet (*Beta vulgaris* L.) genome. J Mol Evol 58:40–53

Touzet P, Budar F (2004) Unveiling the molecular arms race between two conflicting genomes in cytoplasmic male sterility? Trends Plant Sci 9:568–570

Touzet P, Hueber N, Burkholz A, Barnes S, Cuguen J (2004) Genetic analysis of male fertility restoration in wild cytoplasmic male sterility G of beet. Theor Appl Genet 109:240–247

Trebbi D, McGrath JM (2003) Sucrose accumulation during early sugar beet development. In: Proceedings of the American Society of Beet Sugar Technologists. Joint Meeting of the International Institute for Beet Research and the American Society of Beet Sugar Technolgists, San Antonio, TX, 26 February–6 March 2003, pp 267–271

Trebbi D (2005) Genetic analysis of sucrose accumulation in sugar beet (*Beta vulgaris* L.). Dissertation, Michigan State University

Uphoff H, Wricke G (1995) A genetic map of sugar beet (*Beta vulgaris*) based on RAPD markers. Plant Breed 114:355–357

van Geyt JPC, Smed E (1984) Polymorphism of some marker enzymes of the sugarbeet (*Beta vulgaris* L.) investigated by polyacrylamide gel electrophoresis and starch gel electrophoresis. Zeit Pflanzenzüchtung 92:295–308

van Geyt JPC, Smed E, Oleo M (1990) Genetic control and linkage relationships of isozyme markers in sugar beet (*B. vulgaris* L.). 2. NADP- and NAD-specific malate dehydrogenases, 6-P-gluconate dehydrogenase, shikimate dehydrogenase, diaphorase and aconitase. Theor Appl Genet 80:595–601

von Elbe JH, Pasch JH, Adams JP (1974) Betalains as food colorants. Proc IV Intl Congr Food Sci and Technol 1:485–492

von Lippmann EO (1925) Geschichte der Rübe (*Beta*) als Kulturpflanze. Springer, Berlin Heidelberg New York

Wagner H, Weber WE, Wricke G (1992) Estimating linkage relationship of isozyme markers and morphological markers in sugar beet (*Beta vulgaris* L.) including families with distorted segregation. Plant Breed 108:89–96

Wang M, Goldman IL (1999) Genetic distance and diversity in table beet and sugar beet accessions measured by randomly amplified polymorphic DNA. J Am Soc Hort Sci 124:630–635

Watson JF, Gabelman WH (1984) Genetic analysis of betacyanin, betaxanthin, and sucrose concentrations in roots of table beet. J Am Soc Hort Sci 109:386–391

Weber WE, Borchardt DC, Koch G (2000) Marker analysis for quantitative traits in sugar beet. Plant Breed 119:97–106

Weiland JJ, Yu MH (2003) A cleaved amplified polymorphic sequence (CAPS) marker associated with root-knot nematode resistance in sugarbeet. Crop Sci 43:1814–1818

Whitney ED, Duffus JE (1986) Compendium of Beet Diseases and Insects. APS, St. Paul, MN

Winner C (1993) History of the crop. In: Cooke DA, Scott R (eds.) The Sugar Beet Crop: Science into Practice. Chapman and Hall, London, pp 1–35

Wolyn DJ, Gabelman WH (1990) Selection for betalain pigment concentrations and total dissolved solids in red table beets. J Am Soc Hort Sci 115:165–169

Yu MH, Pakish LM, Zhou H (2001) An isozyme marker for resistance to root-knot nematode in sugarbeet. Crop Sci 41:1051–1053

Yu MH (1980) Meiotic chromosome behavior in monoploid sugarbeet. Can J Genet Cytol 22:375–380

Yu MH (1982) Interpretation of mechanism for nematode resistance in sugar beet. J Am Soc Sugar Beet Technol 21:351–361

Yu Y (2004) Genetics of *Aphanomyces* disease resistance in sugarbeet (*Beta vulgaris*), AFLP mapping and QTL analyses. Dissertation. Michigan State University, p 100

Zhao B, Mackay IJ, Caligari PDS, Mead R (1997) The efficiency of between and within full-sib family selection in a recurrent selection programme in sugar beet (*Beta vulgaris* L.). Euphytica 95:355–359

7 Capsicums

Ilan Paran[1], Arnon Ben-Chaim[2], Byoung-Cheorl Kang[2], and Molly Jahn[2]

[1] Department of Plant Genetics and Breeding, Agricultural Research Organization, The Volcani Center, P.O. Box 6, Bet Dagan 50250, Israel
e-mail: iparan@volcani.agri.gov.il
[2] Department of Plant Breeding, Cornell University, Ithaca, NY 14853, USA

7.1 Introduction

7.1.1 Brief History of the Crop

The genus *Capsicum* (vegetable pepper) originated in South America with two centers of domestication, one in Central America and the other in the Andean region of South America. Five species, among approximately 30 in the genus, were independently domesticated and have been cultivated primarily for use as a spice and vegetable for thousands of years (Andrews 1995). *Capsicum* returned to Europe with Columbus' expedition in the fifteenth century, subsequently moving rapidly around the world. Following its arrival in Western Europe, both pungent and nonpungent forms of the species, *C. annuum* came into wide cultivation, assuming a particularly important role in the cuisines of some parts of Europe, e.g., Hungary, West Africa and many regions of Asia.

In addition to its importance as a food and spice, the capsaicinoid compounds, responsible for pungent sensation when the fruits are consumed, have been and continue to be used widely for diverse medicinal applications (Bosland and Votava 2000). Because capsaicinoids interact specifically with the mammalian pain receptor, VR1, they have been widely used as topical analgesics (Caterina et al. 2000). While pain control is the most familiar application, hundreds of publications each year report various pharmacological applications of these molecules from weight loss to cancer.

In the eighteenth and early nineteenth century, relatively large, nonpungent *C. annuum* types with fleshier pericarp or fruit walls (bell peppers) became dominant in some parts of the world, especially in Western Europe and North America. In Latin America, Asia and Africa, the pungent types were still favored, still part of every meal eaten in many regions on these continents. Today *Capsicum* varieties are utilized for a diverse range of food products as vegetable and spice, fresh, dehydrated or processed, in medicine, in pest and animal control and even in law enforcement, which make this crop of immense cultural and economic importance (Bosland and Votava 2000).

7.1.2 Taxonomy

Capsicum, a genus of about 30 species, is found in the tribe, Solanae, in the family, *Solanaceae* (Hunziker 2001). *C. annuum* still predominates in agriculture worldwide and comprises a species complex with *C. chinense* and *C. frutescens,* most familiar as the habanero and tabasco peppers, respectively. Cultivated *C. annuum* types include bell pepper, jalapeno (known as chipotle pepper when roasted), New Mexico chile, ancho, Anaheim, and banana pepper, to name just a few. Many local varieties and landraces are also grown widely mostly in Latin America. In addition to this complex, two other domesticated species, *C. pubescens* and *C. baccatum*, are grown primarily in Latin America. *C. pubescens* has distinctive purple flowers and black seeds and was domesticated in the Peruvian and Bolivian Andes, therefore it appears to be adapted to relatively cool conditions. Wild forms of all domesticated species except *C. pubescens* are known (Pickersgill 1997).

7.1.3 Morphology

Because the fruit or pod, technically a berry, is the commodity of the pepper plant, fruit morphology,

Genome Mapping and Molecular Breeding in Plants, Volume 5
Vegetables
C. Kole (Ed.)

flavor and pungency are the characteristics of most economic importance within the genus. A tremendous wealth of genetic variation is known with respect to fruit traits such as size, shape, color, and flavor, resulting in more than 50 commercially recognized pod types. The major pod types are described by Bosland (1992) and by Andrews (1995).

7.1.4 Karyotype, Ploidy Level and Genome Size

Most *Capsicum* species are diploid with 12 pairs of chromosomes, however the taxon, *C. ciliatum*, basal to the genus has 13 pairs of chromosomes and likely consists of multiple species (personal communication, Bohs, Nee, 2006). Polyploidy is not frequent in *Capsicum* although reports of tetraploids exist (summarized by Lippert et al. 1966). Haploids occur frequently as twin seedlings but are of little relevance for crop improvement. The production of doubled haploids by tissue culture is a common procedure, used extensively in crop breeding to maximize recovery of homozygous recessive genotypes of interest and used experimentally to create populations for inheritance studies and the construction of genetic maps (Pochard et al. 1983; Lefebvre et al. 1995). A set of trisomic lines were produced by Pochard and used for the chromosomal assignment of several mutations (Pochard 1977). Various types of chromosomal rearrangements are prevalent in the genus both within and between species (Pickersgill 1997; Onus and Pickersgill 2004). The best characterized is a reciprocal translocation between chromosomes 1 and 8 in *C. annuum* and *C. chinense* (Livingstone et al. 1999). The genome size of pepper was estimated by flow cytometry at 7.65 pg/nucleus for *C. annuum* and 9.72 pg/nucleus for *C. pubescens* (Belletti et al. 1998). Genome size in nucleotides is estimated to be about 3,000 Mbp (Arumuganathan and Earle 1991).

7.1.5 Economic Importance

Pepper is ranked third or fourth among vegetable crops worldwide, grown in most countries in the world with production acreages estimated at more than 7.5 million (www.chilepepperinstitute.org/Statistics.htm). The world's largest producer is China (more than 8 million tons in 2001), followed by Mexico and Turkey, which produced 1,900,000 and 1,400,000 tons, respectively, in 2001. In addition to its role as an internationally traded commodity, pepper is an important component of household gardens for home consumption in most parts of the world.

7.1.6 Chemical Composition and Nutritional Value

Capsicum may be best known for its biosynthesis of capsaicinoids, a family of up to 25 related alkaloid analogs produced in epidermal cells of the placenta or dissepiment of the fruit that account for the pungent or "hot" sensation when consumed. It has been hypothesized that this trait evolved to deter mammalian herbivory, where crushing molars and acidic digestion are detrimental to seed survival. The favored agents of dispersal are various species of birds. Avian species do not perceive pain in response to capsaicin and are attracted by the brightly colored fruit (Tewksbury and Nabhan 2001). Furthermore, birds roost in trees, under which optimal light environments may be found for wild peppers, and capsaicin may act to retard movement of the seed through the bird's digestive tract, effectively expanding the seed shadow from a source plant (personal communication, Tewksbury, 2006).

A major classification of pepper genotypes is based on the presence or absence of pungency in the fruit. Pungent peppers are called chiles, chillis or chilis; while nonpungent varieties may be called sweet peppers, although sugar content may vary dramatically. Hot peppers also vary dramatically in their pungency level and other flavor characteristics, e.g., habanero peppers are extremely hot while jalapeno peppers are pungent, but quite mild.

Carotenoid and anthocyanin pigments are responsible for fruit color and for nutritional value of *Capsicum* fruit. The most common colors of pepper fruits are green, red, yellow and orange and chocolate or purple. Green peppers are usually not mature; the mature colors ranging from lemon or yellow through orange, peach and red result from the accumulation of different carotenoids in fruit chromoplasts during fruit ripening. The predominant red pigments are capsanthin and capsorubin, the yellow and orange pigments are lutein, β-carotene (provitamin A), zeaxanthin, violaxanthin and antheraxanthin. Peppers are ranked first in antioxidant content among vegetables with very high levels of vitamin C

(Palevitch and Craker 1995). Consumption of a single pepper fruit is enough to meet an adult RDA for this vitamin. A list of nutrients of different peppers is provided at www.chilepepperinstitute.org.

7.1.7 Breeding Objectives

There are many types of peppers that are being utilized for different purposes, each with different quality requirements and traits required for successful production. The major traits under selection are: yield, fruit color and color intensity, size, shape, degree of pungency, pericarp thickness, flowering time, fruit set at extreme temperature, concentration of fruit set, growth habit (adaptation to open field or greenhouse production, mechanical or hand harvesting) and disease resistance. The major market types and their important fruit quality traits were summarized by (Poulos 1994). These include: fresh consumption; fresh for processing into sauce, paste, canned or pickled product; dried for spice (whole fruits and powder); oleoresin extraction, and ornamental types.

7.1.8 Classical Genetic Mapping

Probably the earliest report of genetic linkage in *Capsicum* was reported by Deshpande (1933); however, the four cases of putative linkage were not confirmed in subsequent studies. The first study in which firm evidence for genetic linkage was obtained was among three genes that control fruit shape (*O*), purple immature fruit color (*A*) and yellow immature fruit color (G_1) reported by Peterson (1959). The most comprehensive linkage study using morphological markers was conducted by Pochard (1977) using a set of 11 trisomic lines. By trisomic analyses, eight genes were localized to specific chromosomes. Three genes, *C*, now known as *Pun1* (presence of capsaicin in the fruit), *xantha 3* and *xantha 8* were found as linked to each other in the trisomic line JA. Two genes, *L* (resistance to TMV) and *MoA* (modifier of anthocyanin accumulation) were found as linked to each other in the trisomic line BR. The genes *up* (erect fruit), *A* (anthocyanin accumulation) and *y* (now *Ccs*) (yellow fruit color) were found to be unlinked but were assigned to the trisomic lines, NO, RO and IN, respectively. Linkage between partial resistance to CMV and susceptibility to TMV was reported by Pochard et al. (1983); however, genetic distance between the two loci was not provided.

The first study in which isozymes were used for linkage mapping in *Capsicum* was reported by Tanksley (1984) who mapped 14 isozymes in an interspecific cross of *C. annuum* and *C. chinense*, of which nine were arranged in four linkage groups. A reciprocal translocation that differentiates the two parental species was identified as linked to the isozyme loci *Skdh1* and *Pgm2*.

7.1.9 Classical Breeding Achievements

The literature of pepper breeding was summarized in two comprehensive reviews by Greenleaf (1986) and by Poulos (1994). Much progress has been achieved in increasing yield and quality traits in the last century. A major contribution to yield and quality enhancement was the introduction of disease resistance genes, often from wild germplasm, into elite backgrounds and the introduction of commercially produced hybrid seed. Resistance to TMV was identified in *C. chinense* (L^3) and *C. chacoense* (L^4) and both genes were quickly introduced to numerous cultivars (Boukema 1980; Berzal-Herranz et al. 1995; de la Cruz et al. 1997). More recently, resistance to TSWV was identified in *C. chinense* (Boiteux and De-Avila 1994), resulting in commercially available resistant varieties in the past few years. Resistance to bacterial leaf spot disease (*Bs2*) was identified in *C. chacoense* (Cook and Guevara 1984) and subsequently introduced to numerous cultivars. Since this time, a number of other genes that control various strains of *Xanthomonas campestris* pv. *vesicatoria* have been identified that are both strain-specific and genera (Kim and Hartmann 1985; Hibberd et al. 1987). Additional genes for resistance to *potyviruses*, CMV, nematodes, *Phytophthora capsici* and powdery mildew have been identified in several *Capsicum* species and are being utilized in breeding programs worldwide.

A major objective in the past 20 years has been the development of varieties and equipment to facilitate the mechanical harvest of pepper. This has proven a complex and challenging objective and work continues in both public and private sector settings.

7.1.10 Limitations of Classical Endeavors and Utility of Molecular Mapping

Modern varieties of crop plants usually represent a small fraction of the variation that exists in their gene pool. Wild *Capsicum* species were used mainly for the introgression of disease resistance genes, rarely considered as a potential source for valuable genes for quantitatively inherited traits related to yield because of their inferior performance and the possibility of unfavorable linkages. Quantitative trait loci (QTL) mapping for many traits in diverse species have identified transgressive alleles reviewed by Paran (2003). At these loci, allelic effects opposite to those predicted by the parental values were detected, indicating that alleles with favorable effects on yield related traits exist in the wild parents. The use of wide crosses, combined with molecular mapping to identify the favorable alleles and manage linkage drag will allow deployment of a wider range of alleles in elite *Capsicum* germplasm.

7.2 Construction of Genetic Maps

7.2.1 Brief History of Mapping Efforts

The first genetic map of pepper consisting of molecular markers was reported by Tanksley et al. (1988) who used common RFLP markers to genotype a pair of segregating populations derived from crosses between the tomato, *Solanum lycopersicum* and a wild species, and pepper, *C. annuum* crossed to a wild species. From these two genetic maps populated with common markers, the syntenic relationship between the pepper and tomato genomes was assessed. At this level of resolution, it was clear that, in general, gene repertoire was conserved. Gene order, however, showed considerable rearrangement. Several maps with more comprehensive coverage of the genome, also based primarily on tomato RFLP markers, followed shortly thereafter (Prince et al. 1993; Lefebvre et al. 1995). When attempts were made to convert tomato-based RFLP markers to PCR-based markers for use in further genetic studies and breeding of pepper, primers often failed to amplify the pepper genome.

The most comprehensive comparative map of pepper was developed by Livingstone et al. (1999). This map included more than a thousand loci including many of pepper origin (RFLP, RAPD and mostly AFLP markers). In addition, several hundred loci were detected by RFLP with tomato probes, allowing considerably improved resolution in assessment of syntenic relationships between tomato and pepper. This study revealed that macrosynteny has generally been retained since divergence of these genera, and that 98% of the tomato genome and 95% of the pepper genome can be accounted by about 20 relatively large linkage blocks. Since the publication of this study, additional departures from synteny have been identified, often as a consequence of small internal rearrangements within an otherwise conserved genome segment. An RFLP map based on markers that are mostly pepper-originated was developed by Kang et al. (2001). This map was subsequently expanded to include 46 pepper SSR markers (Lee et al. 2004). The most saturated map was constructed using pooled data from six individual inter- and intraspecific maps (Paran et al. 2004). Despite the use of numerous markers, however, a few small unlinked linkage groups remained, indicating that the pepper map is still incomplete and/or parents involved in mapping crosses differ by undefined chromosomal rearrangements.

7.2.2 First Generation Maps

Detailed characteristics of the genetic maps in *Capsicum* are presented in Table 1. Several pepper maps were constructed using F_2 or BC_1 populations from interspecific crosses of *C. annuum* × *C. chinense*. These include the map of Tanksley et al. (1988), Prince et al. (1993), Livingstone et al. (1999) and Kang et al. (2001). Because of at least one reciprocal translocation between *C. annuum* and *C. chinense*, chromosomes 1 and 8, involved in the translocation, are merged into a single linkage group in the latter populations. To date, two maps based on crosses of *C. annuum* × *C. frutescens* were constructed. The first is a BC_2 population from a cross of *C. annuum* Maor × *C. frutescens* BG 2816 (*C. frutescens*) constructed using RFLP markers (Rao et al. 2003). The second is an F_2 population from the cross of *C. annuum* NuMex RNaky × *C. frutescens* 14-6 constructed from public and proprietary SSR marker sets described below anchored

Table 1. Characteristics of pepper maps

Cross	Population type	Population size	Marker types	No. of markers	Number of linkage groups	Length (cM)	Reference
Doux des landes (*C. annuum*) × PI 159234 (*C. chinense*)	BC1	46	RFLP	85	14	nd[1]	Tanksley et al. 1988
NuMex R Naky (*C. annuum*) × PI 159234 (*C. chinense*)	F2	46	RFLP	192	19	720	Prince et al. 1993
NuMex R Naky (*C. annuum*) × PI 159234 (*C. chinense*)	F2	75	AFLP, RFLP, RAPD	1007	13	1246	Livingstone et al. 1999
TF68 (*C. annuum*) × Habanero (*C. chinense*)	F2	107	RFLP, AFLP	580	16	1320	Kang et al. 2001
TF68 (*C. annuum*) × Habanero (*C. chinense*)	F2	107	RFLP, SSR	333	15	1762	Lee et al. 2003
Maor (*C. annuum*) × Perennial (*C. annuum*)	F2	180	RFLP, RAPD, AFLP	177	12	1740	Ben Chaim et al. 2001
Maor (*C. annuum*) × BG 2816 (*C. frutescens*)	BC2	248	RFLP	92	12	1100	Rao et al. 2003
H3 (*C. annuum*) × Vania (*C. annuum*)	DH	101	AFLP, RFLP, RAPD	543	20	1513	Lefebvre et al. 2002
Perennial (*C. annuum*) × Yolo Wonder (*C. annuum*)	DH	114	AFLP,RAPD, RFLP	630	26	1668	Lefebvre et al. 2002
Yolo Wonder (*C. annuum*) × Criollo de Morelos 334 (*C. annuum*)	F2	151	RFLP, RAPD, AFLP	208	18	685	Lefebvre et al. 2002
NuMex RNaky (*C. annuum*) × 14-6 (*C. frutescens*)	F2	234	AFLP, RFLP, SSR	728	16	1358	Ben Chaim (personal commincation, 2006)
NuMex RNaky (*C. annuum*) × PI 159234 (*C. chinense*)	F2	100	RFLP, SSR	426	15	1304	Ben Chaim (personal commincation, 2006)

[1] Not determined

by RFLP and cloned genes (personal communication, Ben Chaim 2006) (Table 1). Unlike the sterility observed in F_1 and F_2 progenies from *C. annuum* and *C. chinense* crosses, progenies from crosses between *C. annuum* and *C. frutescens* typically do not exhibit large decreases in fertility; suggestion major chromosomal rearrangements do not differentiate these two species.

In addition to the interspecific maps, several intraspecific maps from crosses within *C. annuum* have been constructed and are represented in the integrated map. A map containing 85 markers derived from three partial maps constructed with double haploids (DH) lines was reported by Lefebvre et al. (1995), updated with additional markers including RFLP, RAPD, AFLP and cloned genes (Lefebvre et al. 2002). An intraspecific *C. annuum* map from a different cross (Perennial × Maor), consisted of 177 AFLP and RFLP markers (Ben-Chaim et al. 2001a). Currently the pepper linkage groups are not assigned to chromosomes; however, chromosomes can be differentiated by C-banding (Moscone et al. 1993) and FISH with mapped DNA fragments (Sadder and Weber 2002), and will eventually enable assignment of genetic linkage groups to pepper chromosomes. The putative centromeric regions of the linkage groups are marked by clustering of markers as a result of reduced recombination in the these regions.

7.2.3
Second Generation Maps

An integrated genetic map of pepper was constructed by merging segregation data from six mapping populations (Paran et al. 2004; www.plbr.cornell.edu/psi). These populations included maps built from the interspecific F_2 cross reported by Livingstone et al. (1999), an interspecific BC_1 cross of *C. annum* × *C. chinense* (personal communication, Tanyolac, Paran 2006), and the four intraspecific *C. annuum* maps described above (Ben-Chaim et al. 2001b; Lefebvre et al. 2002). The integrated map included a total of 2262 loci covering 1,832 cM distributed in 13 linkage groups. Map integration improved the average marker density throughout the genome to 1 marker per 0.8 cM; however, because of uneven marker distribution, 15 gaps of at least 10 cM between adjacent markers still remain in the map.

7.2.4
Integration of Molecular Markers Based on Microsatellites (SSR)

Two interspecific SSR-based maps based on F_2 populations derived from a *C. frutescens* × *C. annuum* (FA03) and *C. annuum* × *C. chinense* (AC99) were created via collaboration between several universities, research institutes and private companies holding large sets of proprietary SSR markers. All available marker sets were mapped on the same population creating relatively dense maps (personal communication, Jahn 2006). All loci will be reported, together with the originator of the marker at http://www.sgn.cornell.edu/. To date, the FA03 map contains a total of approximately 728 markers covering 1,358 cM grouped to 12 linkage groups and associated with the 12 chromosomes of pepper. Five smaller linkage groups could be associated with four of the main linkage groups (4, 7, 11, 12), but did not group with them at statistically significant scores. The AC99 map currently contains a total of about 450 markers covering 1,304 cM. Again, 12 linkage groups could be associated with the 12 chromosomes of pepper with three smaller groups, which could be associated with linkage groups 4, 8, 11, but did not group with them at a statistically significant level.

Among the markers used to construct these maps, 150 SSR loci are common to both maps and can be used to describe colinearity. Selected tomato RFLP markers with known locations on the pepper genome were also used to anchor these maps to previous linkage maps of both pepper and tomato/potato. The agreement between the two maps and the integrated map (Paran et al. 2004) is twofold. First, the grouping of the two populations was similar. Second, no major translocation was observed. In both maps the markers were unevenly distributed throughout the genome, and marker classes tended to cluster.

7.3
Gene Mapping

7.3.1
Disease Resistance

Monogenic resistance genes for major diseases caused by viruses (*potyviruses*, TSWV, TMV, CMV), nematodes and bacteria have been incorporated into commercial cultivars and used as targets for mapping and developing linked markers for use in marker-assisted selection (MAS) (Table 2).

Resistance to *Potyvirus*

Several genes for *potyvirus* resistance are known in *Capsicum* (Kyle and Palloix 1997). The *pvr1* locus, first described in the 1950s by Cook is now known to be identical to the *pvr2* locus (Kang et al. 2005) and was mapped to chromosome 4 (Murphy et al. 1998). This locus is now known to encode the eukaryotic translation initiation factor, eIF4E (Ruffel et al. 2002; Kang et al. 2005). Based on chronological priority the *pvr2* locus has been redesignated, *pvr1*, and alleles formerly assigned to *pvr2* have been redesignated to indicate the correct locus identity as follows: $pvr2^1$ is now $pvr1^1$ and $pvr2^2$ is now $pvr1^2$. It is unclear whether *pvr5* is distinct from *pvr1* because genetic data are not available and this gene has also been mapped to the same region in chromosome 4 (Caranta et al. 1997a; Murphy et al. 1998). The corresponding region in tomato (the short arm of chromosome 3) was shown to contain *pot-1*, a resistance gene against similar *potyviruses* (Parrella et al. 2002). This correspondence of pepper and tomato *potyvirus* resistance genes is one of relatively few exceptions to the general lack of colinearity of genes that confer resistance to identical pathogen taxa in the *Solanaceae* family (Grube et al. 2000a; Jahn et al. 2000), the others all involving resistance to *Phytophthora*. This picture is changing as more and more disease resistance genes are mapped in the *Solanaceae*. Increasingly, it is clear that cross-generic groupings of resistance genes may contain overlapping sets of specificities (personal communication, Jahn 2006).

Another locus at which recessive *potyvirus* resistance occurs is *pvr6*, mapped in the vicinity of the RFLP marker TG57 in chromosome 3 (Caranta et al. 1996), likely to be a variant allele at the gene eIF(iso)4E (Kang et al. 2005). The resistant allele originated from *C. annuum* Perennial and was detected in a double haploid line that was resistant to pepper veinal mottle virus (PVMV). Resistance to PVMV was only observed when *pvr6/pvr6* occurred in a $pvr1^2/pvr1^2$ background. A cluster of at least two, perhaps more dominant *potyvirus* resistance genes is located in chromosome 10 and it contains the dominant *Pvr4* and *Pvr7* genes (Caranta et al. 1999; Grube et al. 2000b; Arnedo-Andres et al. 2002).

Resistance to *Tomato Spotted Wilt Virus* (*TSWV*)

In a ddition to the potyvirus resistance genes, the genomic region in chromosome 10 contains a dominant gene, *Tsw*, that confers resistance to TSWV that belongs to the tospoviruses group (Jahn et al. 2000; Moury et al. 2000). This region, therefore, comprises the first cluster of dominant resistance genes in pepper.

Resistance to Root-Knot Nematode

Several genes conferring resistance to root-knot nematodes have been reported in pepper. Me_3 and Me_4 were shown to be linked at approximately 10 cM and mapped to one of two chromosomes 7 or 12 (Djian-Caporalino et al. 2001). An important nematode resistance gene in pepper that is commonly used in commercial cultivars is *N* (Thies and Fery 2000). The relationship of this allele to other nematode resistance in pepper has not been established, so it is possible that *N* is an allele at a locus with another designation or it may be an independent locus. Efforts to map resistance conferred by *N* are currently underway (personal communication, Thies 2006).

Resistance to Bacterial Spot

Bacterial spot caused by *Xanthomonas campestris* is an important disease of pepper spread worldwide. At least four dominant genes, *Bs1*, *Bs2*, *Bs3* and *Bs4* have been identified and used in breeding. Additionally, a nonhypersensitive resistance controlled by two recessive genes, *bs5* and *bs6*, was identified by Jones et al. (2002). Phenotypically similar nonhypersensitive resistance termed general defense reaction controlled by a single recessive gene, *gdr*, was reported by Csillery et al. (2004). The map location of these recessive genes has not been determined yet. *Bs2* is the first resistance gene cloned in pepper by a map-based cloning approach (Tai et al. 1999b); however, its genomic position has not been reported yet. Bulked segregant analysis was used to identify AFLP markers linked to *Bs3* (Pierre et al. 2000). Although, the gene and the markers were not mapped in pepper, the AFLP fragments were cloned and mapped in tomato chromosome 2 within an interval of 1.6 cM between TG33 and TG31.

Resistance to *Tobacco Mosaic Virus* (*TMV*)

Resistance to TMV is conferred by the *L* gene for which a series of alleles have been identified (Boukema 1980). *L* was first assigned to chromosome BR by tri-

Table 2. Gene tagging with molecular markers

Trait	Gene symbol	Population	Marker type	Chromosome	Distance (cM)	Reference
Resistance to *potyviruses*	pvr1	NuMex Rnaky (*C. annuum*) × PI 159234 (*C. chinense*	RFLP	4	5.4	Murphy et al. 1998
Resistance to *potyviruses*	pvr2	Vania (*C. annuum*) × H3 (*C. annuum*)	RAPD, RFLP	4	2.4	Caranta et al. 1997a; Parrella et al. 2002
Resistance to *potyvirus*	Pvr4	Yolo Wonder (*C. annuum*) × Criollo de Morelos 334 (*C. annuum*)	AFLP, CAPS	10	2.1	Caranta et al. 1999
Resistance to *potyvirus*	pvr6	Yolo Wonder (*C. annuum*) × Perennial (*C. annuum*)	RFLP	3	9.8	Caranta et al. 1996
Resistance to TSWV	Tsw	NuMex Rnaky (*C. annuum*) × PI 159234 (*C. chinense*)	RAPD	10	3.4	Jahn et al. 2000
Resistance to TSWV	Tsw	PI 195301 (*C. frutescens*) × PI 152225 (*C. chinense*)	RAPD, CAPS	10	0.9	Moury et al. 2000
Resistance to nematodes	Me_3 Me_4	Yolo Wonder (*C. annuum*) × PM687 (*C. annuum*)	AFLP	Tomato 12	0.5	Djian-Caporalino et al. 2001
Resistance to TMV	L	Yolo Wonder (*C. annuum*) × Perennial (*C. annuum*)	RFLP	11	6	Lefebvre et al. 1995
Resistance to bacterial spot	Bs3	ECW-30R (*C. annuum*) × PI 197409 (*C. annuum*)	AFLP	Tomato 2	1	Pierre et al. 2000
Presence of pungency	C	Maor (*C. annuum*) × BG 2816 (*C. frutescens*)	RFLP	2	0	Blum et al. 2002
Yellow fruit color	y	Lamuro (*C. annuum*) × Lamuga (*C. annuum*)	RFLP	6	0	Lefebvre et al. 1998
Mature fruit color	C2	4751(*C. annuum*) × PI 152225 (*C. chinense*)	RFLP	4	0	Thorup et al. 2000
Brown fruit color	cl	4590 (*C. annuum*) × PI 159234 (*C. chinense*)	RFLP	1	3.8	Efrati et al. 2005
Anthocyanin accumulation	A	5226 (*C. annuum*) × PI 159234 (*C. chinense*)	RFLP	10	0	Borovsky et al. 2004
Soft flesh and deciduous	S	Maor (*C. annuum*) × BG 2816 (*C. frutescens*)	RFLP	6	0	Rao and Paran, 2003
Upright fruit	up	Yolo Wonder (*C. annuum*) × Perennial (*C. annuum*)	RFLP	12	16	Lefebvre et al. 2002
Fasciculate	fa	5219 (*C. annuum*) × BG 2816 (*C. frutescens*)	RFLP	6	0	Elitzur and Paran (personal commincation, 2006)

somic analysis (Pochard 1977). Subsequently, *L1* was mapped in *C. annuum* to chromosome 11 in the vicinity of the tomato RFLP marker TG36 (Lefebvre et al. 1995; Ben-Chaim et al. 2001a). The *L4* allele originated from *C. chacoense* was tagged by a RAPD marker which was also converted to a SCAR marker 1.5 cM away from *L* (Matsunaga et al. 2003).

7.3.2 Fruit Traits

Pungency

Pungency, caused by synthesis of capsaicinoids in the placenta of hot pepper fruit is a characteristic unique to the genus, *Capsicum*. The ability to produce capsaicinoids is determined by a single dominant gene *C*, also known as *Pun1*. Pepper breeders most commonly determine the presence or absence of pungency by tasting or high-pressure liquid chromatography. Neither of these methods, however, are suitable for large-scale screening of segregating populations, therefore, the mapping of *Pun1* and development of efficient system for MAS application should be very beneficial for almost any pepper breeding program. The *C* locus was found as loosely linked to RFLP markers from chromosome 2 in several mapping populations (Tanksley et al. 1988; Lefebvre et al. 1995; Ben-Chaim et al. 2001b). Further mapping experiments identified the tomato RFLP marker TG205 as cosegregating with *C* and in addition a PCR-based CAPS marker tightly linked to *C* was developed (Blum et al. 2002).

Systematic screens of candidate structural genes from the predicted model of the capsaicinoid biosynthetic pathway identified an acyltransferase gene obtained in a suppressive subtractive hybridization (SSH) library of pungent pepper fruit perfectly colocalized with the *Pun1* locus. The nonpungent allele at this locus has a 2.5 kb deletion that disrupts the promoter and first predicted exon. Based on this deletion PCR-based markers were developed and used for marker-assisted breeding (Stewart et al. 2005).

Fruit Color

Fruit color is produced by several biochemical pathways that occur in the pericarp. Carotenoids determine mature color, while anthocyanins and chlorophyll determine fruit color when immature. The inheritance of mature fruit color was originally reported to be controlled by three independent genes *Y*, *C1* and *C2* (Hurtado-Hernandez and Smith 1985). Because carotenoid biosynthesis is one of the best-characterized biochemical pathways in plants, the enzymes and genes in the pathway are well known, and were evaluated as candidates for association with fruit color loci. This approach suggested that the gene coding for capsanthin-capsorubin synthase (CCS) is a candidate for *Y* because complete cosegregation of *Y* and CCS was observed in several independent populations (Lefebvre et al. 1998; Popovsky and Paran 2000). RFLP analysis of the *Y* locus indicated that the recessive allele results from a deletion in the coding region of the gene. A similar approach identified complete cosegregation between *C2* and the gene coding for phytoene synthase (PSY) (Thorup et al. 2000; Huh et al. 2001).

Brown color in mature pepper fruit results from the simultaneous accumulation of red carotenoids and chlorophyll pigments as a consequence of impaired chlorophyll catabolism during ripening. This mutation is controlled by a single recessive gene, *chlorophyll retainer* (*cl*) (Smith 1950). A double recessive mutant at *y* and *cl* results in a green mature color. The *cl* locus was mapped recently to chromosome 1 (Efrati et al. 2005); however, no known genes from the chlorophyll catabolism pathway cosegregated with this mutation.

Purple immature fruit color results from the accumulation of anthocyanin pigments after anthesis, controlled by the single dominant gene *A*. The *A* locus was mapped to chromosome 10 (Ben Chaim et al. 2003a) and subsequently shown to cosegregate with the Petunia gene *Anthocyanin2* (*An2*), an R2R3 MYB transcription factor that regulates anthocyanin biosynthesis (Borovsky et al. 2004).

Fruit Texture

Ripe fruits of wild peppers are small, held erect, undergo rapid texture deterioration (termed soft flesh) and are deciduous. These traits make wild fruits adapted to seed dispersal by birds but were eliminated from most cultivated peppers during domestication. The soft flesh and deciduous nature is controlled by a single dominant gene *S* mapped to chromosome 10. The soft flesh trait was found to cosegregate with the polygalacturonase (PG) gene from tomato (Rao and Paran 2003). The complete linkage of PG and *S*, its increased expression in soft flesh fruits compared to bell peppers and the resemblance of the soft flesh and deciduous phenotype to PG-associated phenotypes in other plants, all strongly indicate that PG is a candidate gene for *S*.

Plant Architecture

One of the most popular phenotypes utilized for ornamental pepper breeding is the *fasciculate* (*fa*) mutation. The *fa* mutation is controlled by a single recessive gene and it is characterized by the formation of clusters of flowers and fruits caused by reduced internodes length after initiation of flowering and by reduced flowering time (personal communication, Elitzur, Paran 2006). The *fa* locus was mapped to chromosome 6 and cosegregated with the *SELF PRUNING* (*SP*) gene that controls inflorescence architecture in tomato (personal communication, Elitzur, Paran 2006).

7.4 QTLs Detected

A list of pepper QTL studies and the major findings is provided in Table 3.

7.4.1 Disease Resistance

Cucumber Mosaic Virus (*CMV*)

CMV is one of the most important viruses infecting pepper worldwide. Resistance to CMV was discovered in the Indian *C. annuum* accession Perennial (Singh 1992). Double haploid progenies from the cross of Perennial and the bell variety Yolo Wonder were evaluated for a component of CMV resistance or tolerance, cell-to-cell movement (Caranta et al. 1997b). Two QTL regions were detected on linkage groups 6 and 12 with additive effects, explaining 24% and 19% of the phenotypic variation, respectively, and one digenic interaction between the main effect QTL in chromosome 12 and TG66 in chromosome 3 that contributed 33% of the phenotypic variation. For all QTLs, the alleles originated from Perennial contributed to an increased level of the resistance.

A similar cross between Perennial and another bell-type variety, Maor, was analyzed for whole plant resistance to CMV by Ben-Chaim et al. (2001a). Four main effect QTLs were detected in chromosomes 4, 6, 11 and 13, for which all the alleles that originated from Perennial contributed to an increased level of the resistance. The QTL with the largest effect was *cmv11.1* linked to the RFLP marker TG36 explaining between 16% to 33% of the phenotypic variation for the trait. Additional two digenic interactions were detected, the first involved *cmv11.1* and TG191 in chromosome 2 and the second, two markers in chromosome 3, neither of which were associated with CMV resistance. In both interactions, increased resistance involved one allele from Perennial and a second allele from Maor. Interestingly *cmv11.1* is tightly linked in repulsion to *L* that confers resistance to TMV as originally observed by (Pochard et al. 1983). When this region is saturated with molecular marker loci, it should be possible to identify recombinants between *L* and *cmv11.1*, bringing the resistance alleles into the coupling phase.

Partial resistance to CMV in the cultivar Vania originated from *C. baccatum*, and was studied in double haploid progenies from the cross of Vania and H3 (Caranta et al. 2002). Four main effect QTLs were detected, of which the QTL, *cmv12.1* from Vania, had the largest effect on the resistance and explained between 45 to 63% of the phenotypic variation. Two additional digenic interactions in which the alleles from H3 increased the level of resistance were detected. Overall, there was a little conservation in QTL positions reported in the three CMV resistance studies. This may be a consequence of different virus testing methods, different resistance sources or mechanisms, and/or to the lack of common markers among the three maps.

Phytophthora capsici

A comparative QTL study was performed in three mapping populations segregating for resistance originating from distinct sources, Vania, Perennial and Criollo de Morelos (CM) 334 (Thabuis et al. 2003). A total of 18 chromosomal regions were detected with additive effect QTL; 10 digenic interactions were also detected in the three crosses. The QTL with the largest effect (>50% of the explained variation for most resistance components) was detected in chromosome 5 in all three populations by two inoculation methods (root and stem tests) and four resistance components. A subsequent study also distinguished this QTL consistently in germplasm exhibiting the highest level of resistance to *P. capsici* (Quirin et al. 2005). In addition to this locus on chromosome 5, a QTL was detected

Table 3. QTL studies in pepper

Trait	Population	No. of QTLs	Major effect QTLs[1]	Transgressive QTL[2]	Reference
Resistance to CMV	Yolo Wonder (*C. annuum*) × Perennial (*C. annuum*)	4	*AG03_2.1*	–	Caranta et al. 1997b
Resistance to CMV	Maor (*C. annuum*) × Perennial (*C. annuum*)	7	*cmv11.1*	+	Ben Chaim et al. 2001a
Resistance to CMV	Vania (*C. annuum*) × H3 (*C. annuum*)	7	*cmv12.1*	+	Caranta et al. 2002
Resistance to *Phytophthora*	Vania X H3; Yolo Wonder × Perennial; Yolo Wonder × Criollo de Morelos 334	18	*rec5.1, sta5.1*	+	Thabuis et al. 2003
Resistance to *potyviruses*	Yolo Wonder (*C. annuum*) × Perennial (*C. annuum*)	11	*AC10_0.3, TG56 CT135, TG124*	+	Caranta et al. 1997c
Resistance to powdery mildew	Vania (*C. annuum*) × H3 (*C. annuum*)	7	*Lt_6.1*	+	Lefebvre et al. 2003
Resistance to Anthracnose	Jatilaba (*C. annuum*) × PRI95030 (*C. chinense*)	4	*B1*	+	Voorrips et al. 2004
Capsaicinoid content	Maor (*C. annuum*) × BG 2816 (*C. frutescens*)	1	*cap*	–	Blum et al. 2003
Capsaicinoid content	NuMex RNaky (*C. annuum*) × 14-6 (*C. frutescens*)	3	*total 7.1*	–	Ben Chaim (personal communication, 2006)
Fruit size	Maor (*C. annuum*) × Perennial (*C. annuum*)	5	–	–	Ben Chaim et al. 2001b
Fruit size	Maor (*C. annuum*) × BG 2816 (*C. frutescens*)	8	–	–	Rao et al. 2003
Fruit shape	Maor (*C. annuum*) × Perennial (*C. annuum*)	3	*fs3.1*	–	Ben Chaim et al. 2001b
Fruit shape	Maor (*C. annuum*) × BG 2816 (*C. frutescens*)	5	*fs3.1*	+	Rao et al. 2003
Fruit shape	5226 (*C. annuum*) × PI 159234 (*C. chinense*)	1	*fs10.1*	–	Ben Chaim et al. 2003a
Fertility restoration	{Yolo Wonder (*C. annuum*) × Perennial (*C. annuum*)} × 77013A (*C. annuum*)	5	*E39/M48-Dp*	+	Wang et al. 2004

[1] QTL with effect larger than 20% explained phenotypic variation

[2] QTL for which the allele with an increased effect is originated from the lower value parent

on chromosome 10 in two of the three crosses; all other QTL were cross-specific. For most QTLs except for the locus on chromosome 5, the alleles originated from the resistant parents contributed to increased resistance.

Potyviruses

The doubled haploid progenies from the cross of the resistant accession Perennial and the susceptible cultivar Yolo Wonder were analyzed for resistance to two isolates of PVY and to *potyvirus* E (Caranta et al. 1997a). A total of 11 chromosomal regions had significant effect on the resistance. Only one QTL in the region containing *pvr1* (formerly *pvr2*) on chromosome 4 was detected for all three viruses. Two additional QTLs were detected for the two PVY strains, while all other QTLs were strain-specific. For the major effect QTL, likely to be *pvr1*, the alleles originated from Perennial increased the level of resistance as expected; however, for several minor effect QTLs, the susceptible parent Yolo Wonder contributed to an increase in the level of resistance. Similarly the two digenic interactions that were detected involved one allele from Perennial and a second allele from Yolo Wonder.

Powdery Mildew

Powdery mildew is caused by the fungal pathogen *Leveillula taurica*. Resistance to the disease was found in the Ethiopian *C. annuum* accession H3. QTL analysis in the double haploid progenies from the cross of H3 and the susceptible cultivar Vania revealed the presence of seven genomic regions involved with the resistance (Lefebvre et al. 2003). The most significant QTL was *Lt_6.1* detected in chromosome 6. Resistant alleles at the QTLs originated from H3 except in one case. One digenic interaction for resistance was detected that involved alleles from Vania. Several of the QTLs mapped to positions that could correspond to loci that control powdery mildew resistance in tomato caused by *Oidium lycopersici*.

Anthracnose

QTL analysis for resistance to two *Colletotrichum* races that causes the anthracnose disease was done in an F_2 population derived from an interspecific cross of *C. annuum* and *C. chinense* (Voorrips et al. 2004). One major QTL (B1) present on linkage group B that is involved in resistance against both *Colletotrichum* strains was detected by all three resistance tests. Additional three minor QTLs were detected, for which one (B2), the susceptible allele was inherited from the resistant *C. chinense* parent. Because the molecular map consisted mainly AFLP markers it was not possible to integrate it with RFLP-based maps; therefore, the location of the QTLs can not be compared at present to other disease resistance loci.

The mapping of resistance gene analogs derived from conserved regions in known genes conferring qualitative resistance and defense response genes in pepper revealed possible colocalization with some QTLs for several diseases reported above (Pfieleger et al. 1999, 2001). This suggests that the genes that control quantitative resistance may be partly conserved with major genes controlling qualitative resistance and defense resistance genes; however, due to the relatively low resolution of these analyses, more stringent evaluation of this hypothesis is necessary before definitive conclusions can be advanced.

7.4.2 Fruit Traits

Fruit Size

Two QTL studies were performed in crosses of the large blocky cultivar Maor with the small-fruited accessions Perennial (*C. annuum*) and BG2816 (*C. frutescens*) (Ben-Chaim et al. 2001b; Rao et al. 2003). Five and eight QTLs were detected in the cross of Maor × Perennial and Maor × BG 2816, respectively. None of the QTLs had a major effect larger than 20% of the explained phenotypic variation. For all QTLs, the alleles originated from the large blocky parent increased fruit size.

Fruit Shape

The same populations analyzed for fruit size were also evaluated for fruit shape (index of fruit length to fruit width). Three and five QTLs for fruit shape were detected in the crosses of Maor × Perennial and Maor × BG 2816, respectively. In contrast to fruit size QTLs which had relatively small effects, a major QTL, *fs3.1*, that accounted for up to 67% of the phenotypic variation in Maor × Perennial and 24% of the phenotypic variation in Maor × BG 2816 was detected. The effect of *fs3.1* was verified in advanced backcross inbred lines and in a third cross involving a *C. annuum* blocky cultivar and *C. chinense* PI 152225 (Ben Chaim et al. 2003b). This indicates the conservation of *fs3.1* as a major QTL that differentiates elongated from blocky fruits in *Capsicum*. A second major fruit shape QTL, *fs10.1*, was detected on chromosome 10 in a cross of 5226, a *C. annuum* line with round fruit and PI 159234, *C. chinense* accession with elongated fruit (Ben-Chaim et al. 2001b). This QTL which explained up to 44% of phenotypic variation for fruit shape was found linked to the *A* locus, confirming an early report describing linkage between fruit shape and fruit color genes (Peterson 1959). This linkage relationship between *fs10.1* and *A* resembles a similar linkage relationship in potato in which anthocyanin and tuber shape genes are linked, suggesting the possibility that homologous genes control the shape pattern of different organs such as fruit and tuber.

Pungency

Quantitative variation in capsaicinoids (capsaicin and dihydrocapsaicin) content measured by HPLC analysis was examined in a cross of the sweet blocky parent Maor and the pungent wild accession BG 2816 (*C. frutescens*). Bulked segregant analysis (Michelmore et al. 1991) was employed with RAPD markers screened on bulks constructed from the two extremes of the HPLC distribution (Blum et al. 2003). This screen identified three RAPD markers that were mapped in addition to RFLP markers in chromosome 7 and identified a major QTL termed *cap* at which the allele from BG 2816 contributed to the increased level of pungency and it explained up to 38% of the phenotypic variation for this trait. To determine whether known structural genes from the capsaicinoids biosynthesis pathway could define candidates for *cap* or other QTL for capsaicinoid content, 12 genes from the pathway were mapped; however, none had a significant effect on this trait.

The content of three capsaicinoid analogs was measured on the FA03 population, an interspecific F_3 population involving *C. frutescens* described above, grown in three environments (personal communication, Paran, Jahn, 2006). In contrast to a previous study that suggested high variability in capsaicinoid content due to environmental factors (Harvell and Bosland 1997), the genetic correlations between the different environments were high ($R^2 > 0.7$). The genetic model for the control of capsaicinoids level in this cross suggests three QTLs and one digenic interaction, and an additional OTL that controls the ratio between the main capsaicinoids (personal communication, Ben Chaim, 2006). This genetic model explains up to 50% of the variation of pungency level in this population.

Fertility Restoration

Male-sterility in *Capsicum* can define an important system for the production of commercial hybrid seeds, therefore analysis of male-sterility and fertility restoration has been the focus of study. QTL analysis for fertility restoration of nuclear-cytoplasmic male-sterility was performed in a population of double haploid progenies of Yolo Wonder and Perennial, fertility restorer lines crossed to a cytoplasmic-genic male-sterile line (Wang et al. 2004). One major QTL for fertility restoration was detected in chromosome 6, accounting for 20% to 69% of the phenotypic variation. Four additional QTLs with minor effect were also detected at which alleles from both parents contributed to fertility restoration. This illustrates that the genetic background of both hybrid parents can stabilize fertility restoration, which complicates the utilization of nuclear-cytoplasmic male-sterility for hybrids production.

7.5 Marker-Assisted Breeding

The only marker-assisted selection program documented in pepper is for resistance to *P. capsici*, although allele-specific molecular markers and markers linked to a number of useful traits have been reported. Four QTLs for resistance were identified in the small-fruited *C. annuum* accession Perennial and were transferred by backcrossing into a bell pepper background (Thabuis et al. 2004a). Validation of the additive QTL effects and one interaction between two QTLs was done in the BC_1 S_1 generation. All the transferred chromosomal regions had an effect on the resistance although the magnitudes of the effects of most QTLs were reduced compared to the original mapping population. In a second study, resistance to *Phytophthora* identified in *C. annuum* Criollo de Morelos 334 was transferred to a bell pepper background by phenotypic recurrent selection (Thabuis et al. 2004b). The resistance level was evaluated during the selection cycles by tests with different severity levels and alleles frequencies at resistance QTLs were monitored in parallel. While loss of resistance level and loss of QTL allele for resistance was frequently observed in the low severity test, more stringent selection resulted in higher resistance levels and higher frequencies of favorable resistant alleles, as might be expected. While a side-by-side comparison of phenotypic and marker-assisted selection schemes is yet to be determined, both studies provided validation of QTLs in breeding populations and show the potential for use of markers to improve complex traits, either as means to identify favorable haplotypes subsequently selected based on phenotype, or perhaps in recurrent cycles where markers might serve as means by which resistance would be indirectly selected.

A SCAR marker was developed as a breeding tool to assist in selection of the major resistance QTL *Phyto.5.2* by screening diverse resistant and susceptible germplasm with RAPD primers (Quirin et al. 2005). A RAPD band present only in genotypes showing the highest level of resistance was cloned and converted to a SCAR marker that mapped back to an inferred position for *Phyto.5.2* on chromosome 5. The wide range differentiation of resistant and susceptible genotypes by this marker may provide an excellent tool for selecting peppers highly resistant to *P. capsici*.

7.6 Map-Based Cloning

The *Bs2* gene that confers resistance to the bacterial spot pathogen *Xanthomonas campestris* was isolated by a positional cloning approach (Tai et al. 1999a,b). Molecular markers linked to *Bs2* were identified by screening near-isogenic lines that differ for the resistance gene. Tightly linked AFLP markers that flank *Bs2* were found and allowed physical mapping of the resistance gene region by analyzing high molecular weight DNA in gel electrophoresis. These markers were subsequently used to screen a yeast artificial chromosome (YAC) library of pepper and identified positive clones that contained the appropriate region (Tai and Staskawicz 2000). This library was constructed from the cultivar Early Calwonder-123R provided by Robert Stall, which contained three resistance genes against *Xanthomonas campestris* and was estimated to have a 3× coverage of the pepper genome. The *Bs2* locus was fine-mapped in a large F_2 population of 1800 individuals, isolated and confirmed. Transgenic expression of *Bs2* in tomato and tobacco conferred resistance to *X. campestris*, indicating that all downstream elements necessary for resistance may be broadly present in solanaceous plants.

Other genes that account for well known phenotypes in pepper that have been isolated include *POLYGALCTURONASE* (Rao and Paran 2003), *CAPSANTHIN CAPSORUBIN SYNTHASE* (Lefebvre et al. 1998), *PHYTOENE SYNTHASE* (Thorup et al. 2000; Huh et al. 2001), *Pun1* (Stewart et al. 2005) and *pvr1* (Ruffel et al. 2002; Kang et al. 2005).

In addition to the YAC library used to clone *Bs2*, at least three additional large-insert libraries have been constructed in bacterial artificial chromosome (BAC) vectors. The first library had 10× genome coverage of a *C. annuum* doubled haploid line homozygous for the *pvr1* (Ruffel et al. 2004). The second library contained 12 genome equivalents from *C. annuum* Criollo de Morelos 334 and to date has been used in studies of several target regions including the *L*, *Y* and *C2* loci (Yoo et al. 2003). The third library of 8.5 genome equivalents was constructed from the Mexican *C. frutescens* BG 2816 accession (personal communication, Giovannoni, Jahn 2006) and was used for the isolation of the *FASCICULATE* gene (personal communication, Elitzur, Paran 2006). Other large insert libraries have been constructed including one for *C. chinense* that are proprietary, and therefore largely unavailable to researchers. These libraries originate from different backgrounds and provide a rich resource for *Capsicum* genomics that can be utilized for the isolation of genes and regulatory elements that control agricultural important traits and for comparative sequence analyses of specific genomic regions with other *Solanaceous* plants.

7.7 Future Scope of Work

Although considerable progress has been achieved in pepper mapping, the pepper map is not yet complete. Future mapping in pepper will rely on integration of sequence information from other *Solanaceae* and from other plant species. The international *Solanaceae* genome project (SOL) put forward as its goal to determine the sequence of the tomato genome as a public reference genome for other solanaceous taxa. Genome sequencing of tomato and of other *Solanaceae* (potato, tobacco) will enhance significantly future mapping experiments in pepper as the availability of markers will be limited only by detection of polymorphism between parents. Identification of genes that control pepper development and production will take advantage of more efficient positional cloning and candidate gene mapping approaches. High-resolution mapping will be limited mainly by population size and by phenotyping ability and not by the lack of markers. Functional genomics experiments such as determination of expression profiles by microarrays, metabolic profiling and screening mutant populations are currently underway in pepper. These complementary approaches will allow the identification of many new genes and their function in *Capsicum*.

Acknowledgement. This work was supported in part by The United States-Israel Binational Agricultural Research and Development Fund Grant No. IS- 3225-01C, USDA IFAFS Plant Genome Award No. 2001-52100-11347, NSF Plant Genome Award No. 0218166, NSF Metabolic Biochemistry Award No. 0412056, the California Pepper Improvement Foundation/California Pepper Commission and a generous gift from Kalsec, Inc.

References

Andrews J (1995) The domesticated *Capsicums*, new edn. University of Texas Press, Austin, TX

Arnedo-Andres S, Gil-Ortega R, Luis-Arteaga M, Hormaza I (2002) Development of RAPD and SCAR markers linked to the *Pvr4* locus for resistance to PVY in pepper (*Capsicum annuum* L.). Theor Appl Genet 105:1067–1074

Arumuganathan K, Earle ED (1991) Nuclear DNA content of some important plant species. Plant Mol Biol Rep 9:208–219

Belletti P, Marzachi C, Lanteri S (1998) Flow cytometric measurement of nuclear DNA content in *Capsicum* (*Solanaceae*). Plant Syst Evol 209:85–91

Ben Chaim A, Grube R, Lapidot M, Jahn M, Paran I (2001a) Identification of quantitative trait loci associated with resistance to cucumber mosaic virus in *Capsicum annuum*. Theor Appl Genet 102:1213–1220

Ben Chaim A, Paran I, Grube R, Jahn M, van Wijk R, Peleman J (2001b) QTL mapping of fruit related traits in pepper (*Capsicum annuum*). Theor Appl Genet 102:1016–1028

Ben Chaim A, Borovsky E, De Jong W, Paran I (2003a) Linkage of the *A* locus for the presence of anthocyanin and *fs10.1*, a major fruit-shape QTL in pepper. Theor Appl Genet 106:889–894

Ben Chaim A, Borovsky E, Rao GU, Tanyolac B, Paran I (2003b) *fs3.1*: a major fruit shape QTL conserved in *Capsicum*. Genome 46:1–9

Berzal-Herranz A, de la Cruz A, Tenllado F, Diaz-Ruiz JR, Lopez L, Sanz AI, Vaquero C, Serra MT, Garcia-Luque I (1995) The *Capsicum L3* gene-mediated resistance against the tobamoviruses is elicited by the coat protein. Virology 209:498–505

Blum E, Liu K, Mazourek M, Yoo EY, Jahn M, Paran I (2002) Molecular mapping of the *C* locus for presence of pungency in *Capsicum*. Genome 45:702–705

Blum E, Mazourek M, O'Connell M, Curry J, Thorup T, Liu K, Jahn M, Paran I (2003) Molecular mapping of capsaicinoid biosynthesis genes and QTL analysis for capsaicinoid content in *Capsicum*. Theor Appl Genet 108:79–86

Boiteux LS, De-Avila AC (1994) Inheritance of a resistance specific to tomato spotted wilt Tospovirus in *Capsicum chinense* 'PI 159236'. Euphytica 75:139–142

Borovsky Y, Oren-Shamir M, Ovadia R, De Jong W, Paran I (2004) The *A* locus that controls anthocyanin accumulation in pepper encodes a MYB transcription factor homologous to *Anthocyanin2* of Petunia. Theor Appl Genet 109:23–29

Bosland PW (1992) Chiles: a diverse crop. Hort Technol 2:7–10

Bosland PW, Votava EJ (2000) Peppers: Vegetable and Spice Capsicums. CABI, New York

Boukema IW (1980) Allelism of genes controlling resistance to TMV in *Capsicum* L. Euphytica 29:433–439

Caranta C, Palloix A, Gebre-Selassie K, Lefebvre V, Moury B, Daubeze AM (1996) A complementation of two genes originating from susceptible *Capsicum annuum* lines confers a new and complete resistance to pepper vienal mottle virus. Phytopathology 86:739–743

Caranta CA, Palloix A, Lefebvre V (1997a) QTLs for a component of partial resistance to cucumber mosaic virus in pepper: restriction of virus installation in host-cell. Theor Appl Genet 94:431–438

Caranta C, Lefebvre V, Palloix A (1997b) Polygenic resistance of pepper to potyviruses consists of a combination of isolate specific and broad spectrum quantitative trait loci. Mol Plant Micr Interact 10:872–878

Caranta C, Thabuis A, Palloix A (1999) Development of a CAPS marker for the *Pvr4* locus: a tool for pyramiding potyvirus resistance genes in pepper. Genome 42:1111–1116

Caranta C, Pflieger S, Lefebvre V, Daubeze AM, Thabuis A, Palloix A (2002) QTLs involved in the restriction of cucumber mosaic virus (CMV) long-distance movement in pepper. Theor Appl Genet 104:586–591

Caterina MJ, Leffler A, Malmberg AB, Martin WJ, Trafton J, Petersen-Zeitz KR, Koltzenburg M, Basbaum AI, Julius D (2000) Impaired nociception and pain sensation in mice lacking the capsaicin receptor. Science 288:306–313

Cook AA, Guevara YG (1984) Hypersensitivity in *Capsicum chacoense* to race 1 of the bacterial spot pathogen of pepper. Plant Dis 68:329–330

de la Cruz A, Lopez L, Tenllado F, Diaz-Ruiz JR, Sanz AI, Vaquero C, Serra MT, Garcia-Luque I (1997) The coat protein is required for the elicitation of the *Capsicum* L_2 gene-mediated resistance against the tobamoviruses.Mol Plant Micr Interact 10:107–113

Csillery G, Szarka E, Sardi E, Mityko J, Kapitany J, Nagy B, Szarka J (2004) The unity of plant defense: genetics, breeding and physiology. In: Proceedings of the 12th Meeting on Genetics and Breeding of *Capsicum* and Eggplant, Noordwijkerhout, The Netherlands, 2004, pp 147–153 de la Cruz A, Lopez L, Tenllado F, Diaz-Ruiz JR, Sanz AI, Vaquero C, Serra MT, Garcia-Luque I (1997) The coat protein is required for the elicitation of the *Capsicum* L_2 gene-mediated resistance against the tobamoviruses. Mol Plant Micr Interact 10:107–113

Deshpande RB (1933) Studies in Indian chillies. 3. Inheritance of some characters in *Capsicum* L. Ind J Agri Sci 3

Djian-Caporalino C, Pijarowski L, Fazari A, Samson M, Gaveau L, O'Byrne C, Lefebvre V, Caranta C, Palloix A, Abad P (2001) High-resolution genetic mapping of the pepper (*Capsicum annuum* L.) resistance loci Me_3 and Me_4 conferring heat-stable resistance to root-knot nematodes (*Meloiogyna* spp.). Theor Appl Genet 103:592–600

Efrati A, Eyal Y, Paran I (2005) Molecular mapping of the *chlorophyll retainer* (*cl*) mutation in pepper (*Capsicum* spp.) and screening for candidate genes using tomato ESTs homologous to structural genes of the chlorophyll catabolism pathway. Genome 48:347–351

Greenleaf WH (1986) Pepper breeding. In: Bassett MJ (ed.) Breeding Vegetable Crops. AVI, Westpost, CT, pp 67–134

Grube RC, Radwanski ER, Jahn M (2000a) Comparative genetics of disease resistance within the solanaceae. Genetics 155:873–887

Grube RC, Blauth JR, Arnedo-Andres MS, Caranta C, Jahn M (2000b) Identification and comparative mapping of a dominant potyvirus gene cluster in *Capsicum*. Theor Appl Genet 101:852–859

Harvell KP, Bosland PW (1997) The environment produces a significant effect on pungency of chiles (*Capsicum annuum* L.). HortScience 32: 1992

Hibberd AM, Bassett MJ, Stall RE (1987) Allelism tests of three dominant genes for hypersensitive resistance to bacterial spot of pepper. Phytopathology 77:1304–1307

Huh JH, Kang BC, Nahm SH, Kim S, Ha KS, Lee MH, Kim BD (2001) A candidate gene approach identified phytoene synthase as the locus for mature fruit color in red pepper (*Capsicum* spp.). Theor Appl Genet 102:524–530

Hunziker AT (2001) Genera Solanacearum: The genera of *Solanaceae* ilustrated, arranged according to a new system. ARG Grantner, Ruggell

Hurtado-Hernandez H, Smith PG (1985) Inheritance of mature fruit color in *Capsicum annuum* L. J Hered 76:211–213

Jahn M, Paran I, Hoffmann K, Radwansky E, Livingstone K, Grube R, Aftergoot E, Lapidot M, Moyer J (2000) Genetic mapping of the *Tsw* locus for resistance to the *Tospovirus* tomato spotted wilt virus in *Capsicum* spp. and its relationship to the *Sw-5* gene for resistance to the same pathogen in tomato. Mol Plant Micr Interact 13:673–682

Jones JB, Minsavage GV, Roberts PD, Johnson RR, Kousik CS, Subramanian S, Stall RE (2002) A nonhypersensitive resistance in pepper to the bacterial spot pathogen is associated with two recessive genes. Phytopathology 92:273–277

Kang B, Nahm SH, Huh JH, Yoo HS, Yu JW, Lee MH, Kim BD (2001) An interspecific (*Capsicum annuum* × *C. chinense*) F_2 linkage map in pepper using RFLP and AFLP markers. Theor Appl Genet 102:531–539

Kang BC, Yeam IH, Frantz DJ, Murphy JF, Jahn MM (2005) The *pvr1* locus in pepper encodes a translation initiation factor eIF4E that interacts with *Tobacco etch virus* VPg. Plant J 42:392–405

Kim BS, Hartmann RW (1985) Inheritance of a gene (*Bs3*) conferring hypersensitive resistance to *Xanthomonas campestris* pv *vesicatoria* in pepper (*Capsicum annuum*). Plant Dis 69:233–235

Kyle MM, Palloix A (1997) Proposed revision of nomenclature for potyvirus resistance genes in *Capsicum*. Euphytica 97:183–188

Lee JM, Nahm SH, Kim YM, Kim BD (2004) Characterization and molecular genetic mapping of microsatellite loci in pepper. Theor Appl Genet 108:619–627

Lefebvre V, Palloix A, Caranta C, Pochard E (1995) Construction of an intraspecific integrated linkage map of pepper using molecular markers and double haploid progenies. Genome 38:112–121

Lefebvre V, Kuntz M, Camara B, Palloix A (1998) Capsanthin-capsorubin synthase gene: a candidate gene for the *y* locus controlling the red fruit color in pepper. Plant Mol Biol 36:785–789

Lefebvre V, Pflieger S, Thabuis A, Caranta C, Blattes A, Chauvet JC, Daubeze AM, Palloix A (2002) Towards the saturation of the pepper linkage map by alignment of three intraspecific maps including known-function genes. Genome 45:839–854

Lefebvre V, Daubeze AM, Rouppe van der Voort J, Peleman J, Bardin M, Palloix A (2003) QTLs for resistance to powdery mildew in pepper under natural and artificial infections. Theor Appl Genet 107:661–666

Lippert LF, Smith PG, Bergh BO (1966) Cytogentics of vegetable crops. Garden pepper, *Capsicum* spp. Bot Rev 32:24–55

Livingstone KD, Lackney VK, Blauth J, Wijk VR, Jahn MK (1999) Genome mapping in *Capsicum* and the evolution of genome structure in the *Solanaceae*. Genetics 152:1183–1202

Matsunaga H, Saito T, Hirai M, Nunome T, Yoshida T (2003) DNA markers linked to pepper mild mottle virus (PMMoV) resistant locus (L^4) in *Capsicum*. J Jpn Soc Hort Sci 72:218–220

Michelmore RW, Paran I, Kesseli RV (1991) Identification of markers linked to disease-resistance genes by bulked segregant analysis: a rapid method to detect markers in specific genomic regions by using segregating populations. Proc Natl Acad Sci USA 88:9828–9832

Moscone EA, Lambrou M, Hunziker AT, Ehrendorfer F (1993) Giemsa C-banded karyotyes in *Capsicum* (*Solanaceae*). Plant Syst Evol 186:213–229

Moury B, Pflieger S, Blattes A, Lefebvre V, Palloix A (2000) A CAPS marker to assist selection of tomato spotted wilt virus (TSWV) resistance in pepper. Genome 43:137–142

Murphy JF, Blauth JR, Livingstone KD, Lackney VK, Jahn M (1998) Genetic mapping of the *pvr1* locus in *Capsicum* spp. and evidence that distinct potyvirus resistance loci control responses that differ at the whole plant and cellular levels. Mol Plant Micr Interact 11:943–951

Onus AN, Pickersgill B (2004) Unilateral incompatibility in *Capsicum* (*Solanaceae*): occurrence and taxonomic distribution. Ann Bot (Lond) 94:289–295

Palevitch D, Craker LE (1995) Nutritional and medical importance of red pepper (*Capsicum* spp.). J Herbs Spices Med Plants 3:55–83

Paran I (2003) Marker assisted utilization of exotic germplasm. In: Nguyen HT, Blum A (eds.) Physiology and Biotechnology Integration for Plant Breeding, Marcel Dekker, New York

Paran I, Rouppe van der Voort J, Lefebvre V, Jahn M, Landry L, van Schriek M, Tanyolac B, Caranta C, Ben Chaim A, Livingstone K, Palloix A, Peleman J (2004) An integrated genetic linkage map of pepper (*Capsicum* spp.). Mol Breed 13:251–261

Parrella G, Ruffel S, Moretti A, Morel C, Palloix A, Caranta C (2002) Recessive resistance genes against potyviruses are localized in colinear genomic regions of the tomato (*Lycopersicon* spp.) and pepper (*Capsicum* spp.) genomes. Theor Appl Genet 105:855–861

Peterson PA (1959) Linkage of fruit shape and color genes in *Capsicum*. Genetics 44:407–419

Pfieleger S, Lefebvre V, Caranta C, Blattes A, Goffinet B, Palloix A (1999) Disease resistance gene analogs as candidates for QTLs involved in pepper-pathogen interactions. Genome 42:1100–1110

Pfieleger S, Palloix A, Caranta C, Blattes A, Lefebvre V (2001) Defense response genes colocalize with quantitative disease resistance loci in pepper. Theor Appl Genet 103:920–929

Pickersgill B (1997) Genetic resources and breeding of *Capsicum* spp. Euphytica 96:129–133

Pierre M, Noel L, Lahaye T, Ballvora A, Veuskens J, Ganal M, Bonas U (2000) High-resolution genetic mapping of the pepper resistance locus *Bs3* governing recognition of the *Xanthomonas campestris* pv *vesicatora* AvrBs3 protein. Theor Appl Genet 101:255–263

Pochard E, Dumas de Valuix R, Florent A (1983) Linkage between partial resistance to CMV and susceptibility to TMV in the line Perennial: analysis on androgenetic homozygous lines. *Capsicum* Eggplant News 2:34–35

Pochard E (1977) Localization of genes in *Capsicum annuum* L. by trisomic analysis. Ann Amelior Plantes 27:255–266

Popovsky S, Paran I (2000) Molecular analysis of the *Y* locus in pepper: its relation to capsanthin-capsorubin synthase and to fruit color. Theor Appl Genet 101:86–89

Poulos JM (1994) Pepper breeding (*Capsicum* spp.): achievements, challenges and possibilities. Plant Breed Abstr 64:143–155

Prince JP, Pochard E, Tanksley SD (1993) Construction of molecular linkage map of pepper and a comparison of synteny with tomato. Genome 36:404–417

Quirin EA, Ogundiwin EA, Prince JP, Mazourek M, Briggs MO, Chlanda TS, Kim KT, Falise M, Kang BC, Jahn MM (2005) Development of sequence characterized amplified region (SCAR) primers for the detection of *Phyto.5.2*, a major QTL for resistance to *Phytophthora capsici* Leon. in pepper. Theor Appl Genet 110:605–612

Rao GU, Ben Chaim A, Borovsky E, Paran I (2003) Mapping of yield related QTLs in pepper in an interspecific cross of *Capsicum annuum* and *C. frutescens*. Theor Appl Genet 106:1457–1466

Rao GU, Paran I (2003) Polygalacturonase: a candidate gene for the soft flesh and deciduous fruit mutation in *Capsicum*. Plant Mol Biol 51:135–141

Ruffel S, Dussault MH, Palloix A, Moury B, Bendahmane A, Robaglia C, Caranta C (2002) A natural recessive gene against *potato virus Y* in pepper corresponds to the eukaryotic initiation factor 4E (eIF4E). Plant J 32:1067–1075

Ruffel S, Caranta C, Palloix A, Lefebvre V, Caboche M, Bendahmane A (2004) Structural analysis of the eukaryotic initiation factor 4E gene controlling potyvirus resistance in pepper: exploitation of a BAC library. Gene 338:209–216

Sadder M, Weber G (2002) Comparison between genetic and physical maps in *Zea mays* L. of molecular markers linked to resistance against *Diatraea* spp. Theor Appl Genet 104:908–915

Singh J (1992) Breeding multiple resistant lines in Chilli Pepper at PAU, Ludhiana. In: Proceedings of Convention of Genetics and Breeding on Capsicum and Eggplant, Rome, 7–10 September 1992, pp 127–131

Smith PG (1950) Inheritance of brown and green mature color in peppers. J Hered 41:138–140

Stewart CS, Kang BC, Liu K, Mazourek M, Yoo EY, Moore SL, Kim BD, Paran I, Jahn MM (2005) The *Pun1* gene in pepper encodes a putative acyltransferase. Plant J 42:675–688

Tai TH, Dahlbeck D, Stall RE, Peleman J, Staskawicz BJ (1999a) High-resolution genetic physical mapping of the region containing the *Bs2* resistance gene of pepper. Theor Appl Genet 99:1201–1206

Tai TH, Dahlbeck D, Clark ET, Gajiwala P, Pasion R, Whalen M, Stall RE, Staskawicz BJ (1999b) Expression of the *Bs2* pepper gene confers resistance to bacterial spot disease in tomato. Proc Natl Acad Sci USA 96:14153–14158

Tai TH, Staskawicz BJ (2000) Construction of a yeast artificial chromosome library of pepper (*Capsicum annuum* L.) and identification of clones from the *Bs2* resistance locus. Theor Appl Genet 100:112–117

Tanksley SD (1984) Linkage relationships and chromosomal locations of enzyme-coding genes in pepper (*Capsicum annuum* L.). Chromosoma 8:352–360

Tanksley SD, Bernatzky R, Lapitan NL, Prince JP (1988) Conservation of gene repertoire but not gene order in pepper and tomato. Proc Natl Acad Sci USA 85:6419–6423

Tewksbury JJ, Nabhan GP (2001) Directed deterrence by capsaicin in chilies. Nature 412:403–404

Thabuis A, Palloix A, Pflieger S, Daubeze AM, Caranta C, Lefebvre V (2003) Comparative mapping of *Phytophthora* resistance loci in pepper germplasm: evidence for conserved resistance loci across solanaceae and for a large genetic diversity. Theor Appl Genet 106:1473–1485

Thabuis A, Palloix A, Servin B, Daubeze AM, Signoret P, Hospital F, Lefebvre V (2004a) Marker-assisted introgression of *phytophthora capsici* resistance QTL alleles into a bell pepper line: validation of additive and epistatic effects. Mol Breed 14:9–20

Thabuis A, Lefebvre V, Bernard G, Daubeze AM, Phaly E, Pochard E, Palloix A (2004b) Phenotypic and molecular evaluation of a recurrent selection program for a ploygenic resistance to *phytophthora capsici* in pepper. Theor Appl Genet 109:342–351

Thies JA, Fery RL (2000) Characterization of resistance conferred by the *N* gene to *Meloidogyne arenaria* races 1 and 2, *M. hapla*, and *M. javanica* in two sets of isogenic lines of *Capsicum annuum* L. J Am Soc Hort Sci 125:71–75

Thorup TA, Tanyolac B, Livingstone KD, Popovsky S, Paran I, Jahn M (2000) Candidate gene analysis of organ pigmentation loci in the *Solanaceae*. Proc Natl Acad Sci USA 97:11192–11197

Voorrips RE, Finkers R, Sanjaya L, Groenwold R (2004) QTL mapping of anthracnose (*Colletotrichum* spp.) resistance in a cross between *Capsicum annuum* and *C. chinense*. Theor Appl Genet 109:1275–1282

Wang LH, Zhang BX, Lefebvre V, Huang SW, Daubeze AM, Palloix A (2004) QTL analysis of fertility restoration in cytoplasmic male-sterile pepper. Theor Appl Genet 109:1058–1063

Yoo Y, Kim S, Kim YH, Lee CJ, Kim BD (2003) Construction of a deep coverage BAC library from *Capsicum annuum*, CM334. Theor Appl Genet 107:540–543

8 *Brassica oleracea*

Danuta Babula[1], Małgorzata Kaczmarek[1], Piotr A. Ziółkowski[2], and Jan Sadowski[1,2]

[1] Institute of Plant Genetics, Polish Academy of Sciences, Strzeszyńska 34, 60-479 Poznań, Poland
e-mail: dbab@igr.poznan.pl
[2] Department of Biotechnology, Adam Mickiewicz University, Międzychodzka 5, 60-371 Poznań, Poland

8.1 Introduction

8.1.1 Brief History of the Crop

The *Brassica oleracea* species is a classic and excellent model of morphological diversity forms within a single species. Hence, it represents one of the most spectacular illustrations of structural evolution in plants under domestication. *B. oleracea* is a perennial herb found largely in Europe, especially in selected costal regions (Thompson 1976; Tsunoda et al. 1980; Song et al. 1988a; Gray 1989; Kalloo and Bergh 1993; Fig. 1) and it includes at least six cultivated and one wild subspecies.

The most primitive *B. oleracea*, designated subspecies *oleracea* (= *B. oleracea* var. *sylvestris* L.; a low, shrubby plant with woody lower parts of stems and herbaceous upper parts) is considered to be the ancestral progenitor of all horticultural types. It inhabits the coastal rocky cliffs of the Mediterranean, northern Spain, western France, southern and southwestern Britain (Helm 1963; Tsunoda et al. 1980; Figs. 1 and 2e). In contrast, however, Snogerup (1980) suggested that the various horticultural types were derived from different wild species. Selective breeding for special characteristics of the plant followed domestication at least 2,500 years ago. As a result, within *B. oleracea* six distinct vegetables, collectively known as cole crops have been produced from this single species. They include cauliflower (*B. oleracea* ssp. *botrytis* L.) and broccoli (*B. oleracea* ssp. *italica* L.) with the enlarged inflorescence, kohlrabi (*B. oleracea* ssp. *gongylodes* L.) with the enlarged stem, kale (*B. oleracea* ssp. *medullosa* Thell.) with the marrowstem, cabbage (*B. oleracea* ssp. *capitata* L.) with the enlarged, twisted leaves and apical buds, and Brussels sprouts (*B. oleracea* ssp. *gemmifera* DC) with the enlarged lateral buds (Kalloo and Bergh 1993; Fig. 2). These cultivar forms of *B. oleracea* reveal considerable differences for many traits, such as for example biennial vs. annual habit, temperature requirements for flower induction, leaf dimensions, heading habit and flower development. The traits responsible for morphological divergence have been intensively studied at the genetic and molecular levels. For example, a specific cauliflower phenotype with a large dense structure of inflorescence is controlled by *APETALA1* (*AP1*) and *CAULIFLOWER* (*CAL*) genes (Mandel et al. 1992; Kempin et al. 1995). Interestingly, alleles at these loci determine the development of cauliflower curd or broccoli head phenotypes (Kop et al. 2003). Other vegetables within *B. oleracea* species are: collard grown in the southern US (*B. oleracea* ssp. *acephala* DC), curly kale/feathered cabbage (*B. oleracea* ssp. *sabellica* L.), and *B. oleracea* ssp. *alboglabra* (L. H. Bailey) Musil (Chinese kale), which has been domesticated in China (Prakash and Hinata 1980; Quiros et al. 1994). The status of *B. oleracea* ssp. *alboglabra* is questionable and sometimes it is considered as a distinct species (Snogerup 1980). *B. oleracea* ssp. *alboglabra* (A12DH genotype) due to a short generation time and self-compatibility (Bohuon et al. 1996) was selected in the 1990s as a model for analysis of genome C (the Multinational *Brassica* Genome Project; http://brassica.bbsrc.ac.uk/). Other plants within the *B. oleracea* species are the bizarre tree cabbage (*B. oleracea* var. *exaltata*) and Jersey kale, which may be up to 3 meters tall (Parker and Cox 1974). In the recent years morphologically simple "rapid-cycling" genotypes have been derived with a life cycle of 6 to 8 weeks (Williams and Hill 1986). Moreover, ornamental kale has recently become very popular.

Many subspecies of *B. oleracea* exhibit a high degree of genomic similarity and hence they are intercrossable to generate fully fertile progenies. It confirms the fact that *B. oleracea* is characterized by genome plasticity and it is ideal for genetic analysis

Genome Mapping and Molecular Breeding in Plants, Volume 5
Vegetables
C. Kole (Ed.)

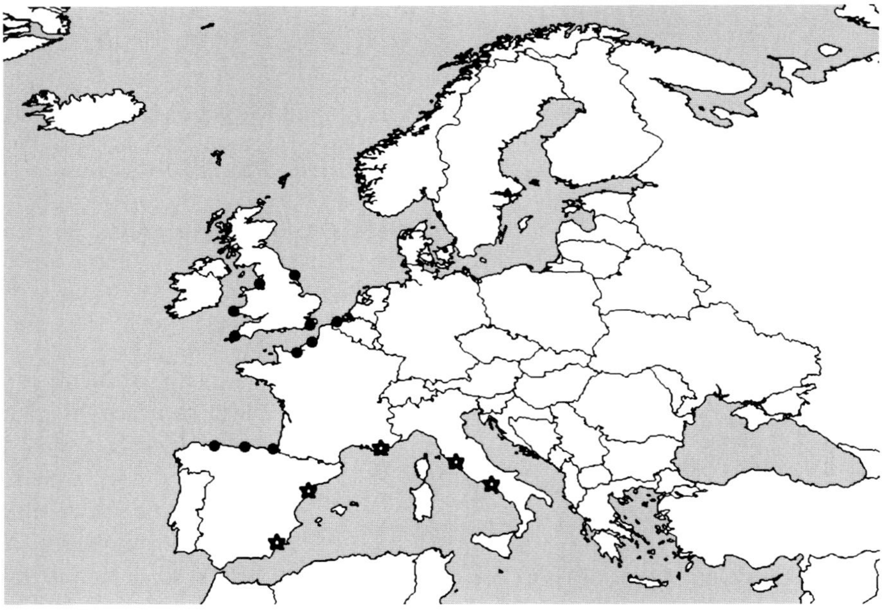

Fig. 1. Distribution of *Brassica oleracea* (marked by *black stars*) and *B. oleracea* ssp. *oleracea* (marked by *black dots*) considered as wild form of cabbages (adopted from Thompson 1976; Tsunoda et al. 1980; Song et al. 1988; Gray 1989; Kalloo and Bergh 1993)

(Kianian 1990). Although minor chromosomal rearrangements were observed, those had little effect on fertility or chromosome pairing in meiosis, as it was observed in a cross of broccoli and cabbage (Slocum et al. 1990). The fertility of the resulting F_1 hybrids is often variable; it is generally high enough to assure the receiving of hybrid progeny through the F_2 and subsequent generations (Snogerup 1980). The frequency of hybridization and outcrossing is greatly enhanced by a high degree of self-incompatibility among the *Brassica* (Snogerup 1980). Natural genetic variation of *B. oleracea* facilitates investigation of the genetic basis of morphological differences. For example, the phenotype with the curd inflorescence is under complex genetic control (Lan and Paterson 2000).

8.1.2 Botanical Description

Brassica oleracea L. belongs to the Brassicaceae (formerly called Cruciferae) family, also known as the mustard family (Fig. 3). The name crucifer comes from the shape of flowers, with four diagonally opposite petals in the form of a cross.

Within the Brassicaceae family there appears a fully sequenced and well-annotated model plant *Arabidopsis thaliana* (AGI 2000). The comparative analysis based on mitochondrial DNA revealed that *A. thaliana* and *Brassica* lineages diverged 14.5 to 24.0 million years ago (Yang et al. 1999; Koch et al. 2000, 2001). Nevertheless, the average similarity of coding sequences at the nucleotide level is 87% in both species (Lydiate et al. 1993; Cavell et al. 1998). Hence, it is expected that the close evolutionary relationship of *Brassica* species and *A. thaliana* will facilitate the exploitation of genome information concerning the control of basic biological processes for genetic analysis of *Brassica* crops.

B. oleracea together with *B. rapa* and *B. nigra* represent three basic diploid species of *Brassica* genus with C, A and B genomes, respectively, which differ in their chromosome number and genome size. The chromosome number of $n = 8$, $n = 9$ and $n = 10$ was found for *B. nigra*, *B. oleracea* and *B. rapa*, respectively, and was determined over 80 years ago (Karpechenko 1922). The genome size of diploid *Brassica* species is characterized by the low span of nuclear DNA contents and it ranges from 0.97 pg/2C (468 Mb/1C; where 1C represents the value in the haploid nucleus) for *B. nigra* to 1.37 pg/2C (600–670 Mb/1C) for *B. oleracea* subspecies (Arumuganathan and Earle 1991). Within the *B. oleracea* species, the size of the genomes varies from 599 Mb/1C in broccoli to 662 Mb/1C in cauliflower. The phylogenetic and genome relationships between diploids and amphidiploids derived from them were illustrated by the well known U's triangle (1935). Later, these relationships between diploid and amphidiploid species were confirmed by cytogenetic

Fig. 2. Distinct vegetable forms of *Brassica oleracea* compared to wild form *B. oleracea* var. *silvestris*. **a** Broccoli (*B. oleracea* ssp. *italica* L.). **b** Cauliflower (*B. oleracea* ssp. *botrytis* L.). **c** Chinese kale (*B. oleracea* var. *alboglabra*). **d** Brussels sprouts (*B. oleracea* ssp. *gemmifera* DC). **e** Wild cabbage (*B. oleracea* var. *silvestris*). **f** Cabbage (*B. oleracea* ssp. *capitata* L.). **g** Curly kale (*B. oleracea* var. *sabelica*). **h** Kohlrabi (*B. oleracea* var. *gongylodes*). **i** Savoy cabbage (*B. oleracea* var. *sabauda*) (**d** and **g** reprinted from http://www.mpiz-koeln.mpg.de/~pr/garten by permission of Wolgang Schuchert; **e** reprinted from http://www.bioimages.org.uk by permission of Malcolm Storey)

studies (Prakash and Hinata 1980), nuclear DNA content (Verma and Rees 1974), isozymes (Quiros et al. 1987), proteins (Vaughan 1977), rRNA genes (Quiros et al. 1987), chloroplast DNA analysis (Erickson et al. 1983; Palmer et al. 1983), artificial resynthesis of hybrids (Ramanujam and Srinivasachar 1943; Mizushima 1950; Olsson 1960), and RFLP markers (Song et al. 1988a,b). The RFLP analysis supported the hypothesis that three *Brassica* diploid species evolved in two pathways: *B. nigra* derived from one pathway, and *B. rapa* and *B. oleracea* derived from another having a common origin (Song et al. 1988a,b, 1990).

Cytological analysis and construction of a karyotype for *B. oleracea* is still rather problematic due to morphological features of chromosomes, characterized by the extremely small size, as well as similar lengths and/or arm ratios. This makes it difficult to identify homologs and distinguish between different pairs in the complement (Olin-Fatih and Heneen 1992). Initially, the mitotic chromosomes of *B. oleracea* based on their size and primary construction were classified into six groups designated as AABBbCDDEF (Richharia 1937a,b). Later, the analysis of the *B. oleracea* chromosome structure was based on the number of satellites and nucleoli (Sikka 1940; Prakash and Hinata 1980). Meaningfully, the morphology of pachytene chromosomes in *Brassica* was described on the basis of structural characteristics of total chromosome length, symmetry of the arms and the shape of the heterochromatic pericentromere regions (Röbbelen 1960). Pachytene chromosomes are useful for their extension compared to mitotic chromosomes and they reveal clear chromosome patterns, particularly at telomeric regions. Röbbelen classified the *B. oleracea* chromosomes on the basis of their absolute length as follows: a single chromosome was described as short (20–25 μm), two medium (25 to 30 μm), five long (30–40 μm) and one very long (more than 40 μm). In the 1980s, chromosome banding techniques to determine C-banding patterns of late prophase chromosomes (Wang and Luo 1987; Wang et al. 1989; Olin-Fatih and Heneen 1992) were developed for *Brassica*. These analyses enable the classification of chromosomes into the following groups: *m* (4 pairs), *sm* (4) and *st* (1). Finally, a partial karyotype for *B. oleracea* ssp. *alboglabra* (A12DH line) has been constructed, based on the combination of chromosome size, arm ratios and fluorescent in situ hybridization (FISH) to

Fig. 3. Taxonomy of the *Brassica oleracea* L. species (from NCBI Taxonomy Browser; http://www.ncbi.nlm.nih.gov/Taxonomy)

Kingdom: Plantae – Plants
Subkingdom: Tracheobionta – Vascular plants
Superdivision: Spermatophyta – Seed plants
Division: Magnoliophyta – Flowering plants
Class: Magnoliopsida – Dicotyledons
Family: *Brassicaceae* – Mustard family
Genus: *Brassica* L. – mustard
Species: *Brassica oleracea* L. – cabbage
Subspecies: *acephala* D.C. – kale
***alboglabra* (L. H. Bailey) Musil** – Chinese kale, Chinese broccoli
***botrytis* L.** – cauliflower
***capitata* L.** – head cabbage
***gemmifera* DC.** – Brussels sprouts
***gongylodes* L.** – kohlrabi
***italica* Plenck** – broccoli
***medullosa* Thell.** – marrow-stem kale
***oleracea* L.** – wild cabbage
***ramosa* DC.** – perrenial kale
***viridis* L.** – collards
***sabellica* L.** – curly kale, feathered cabbage

mitotic and meiotic chromosomes, using 45S rDNA, 5S rDNA and pericentromeric heterochromatin sequences as probes (Armstrong et al. 1998; Howell et al. 2002). This karyotype provides a valuable framework for the physical mapping and integration of the genetic and physical maps of *B. oleracea* (Howell et al. 2002; Fig. 4).

Detailed analyses of the *B. oleracea* genome structure based on isozyme studies and RFLP genetic mapping indicate that many markers/loci are reiterated in the genome and form duplicated chromosomal segments (Quiros et al. 1987; McGrath et al. 1990; Slocum et al. 1990; Lan et al. 2000; Babula et al. 2003). Approximately, over 50% of the identified loci in genome C are duplicated, providing examples for divergence of gene function (Kianian and Quiros 1992a). The duplication level indicates that the genome of *Brassica* is very complex as an outcome of multiple rounds of polyploidization during their ancestry (Kianian and Quiros 1992a; Lagercrantz 1998) and supports the hypothesis that *B. oleracea* is a secondary polyploid (Prakash and Hinata 1980; Quiros et al. 1987, 1988; McGrath et al. 1990; Slocum et al. 1990). It is possible that the modern genome of *B. oleracea* derives from one or more ancestral genomes that have suffered extensive rearrangements during or after the events responsible for their duplication (Kianian and Quiros 1992a). Within duplicated segments found in the genome numerous rearrangements were observed. They may be explained by chromosomal translocations. In addition to translocations, deletions and inversions seem to be another important molding force of the *Brassica* genomes (Song et al. 1991; Babula et al. 2003). Such aberrations are of common occurrence in *Brassica* and have been reported by independent investigators as a widespread event in various species (Snogerup 1980; Quiros et al. 1988). Rearranged duplicated segments resulting in novel syntenic combinations seem to be the rule in the C genome.

8.1.3 Economic Importance

Brassica oleracea species is known mainly as highly nutritious vegetables: cabbage, cauliflower, broccoli, collard and kale. They are an essential part of the

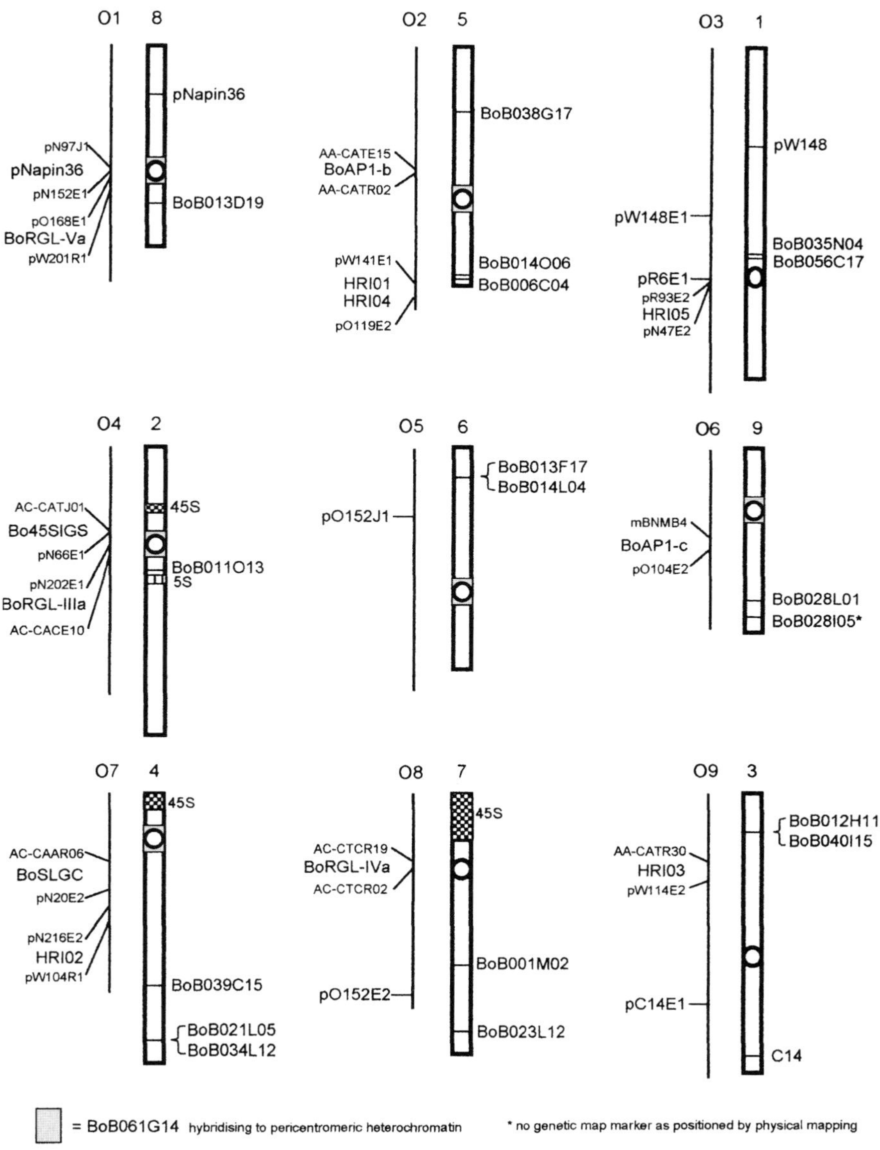

Fig. 4. The haploid idiogram of the *B. oleracea* ssp. *alboglabra* line A12DHd karyotype defined based on cDNA, 5S rDNA and 45S rDNA probes and assigned to nine linkage groups of the *Brassica* C genome. Linkage groups are represented by *vertical bars*, whereas ideograms of chromosomes show relative lengths. *Open circles* represent centromeres. Chromosomes 1, 3, 5, 6, and 8 are inverted to follow the orientation of linkage groups (reprinted from Howell et al. 2002 with permission of Genetics Society of America)

diets of many nations in developing countries and are widely used in the cuisine of many cultures. Vegetables are a rich source of dietary fiber, vitamin A, B and C, minerals and other possible salubrious factors, such as anticancerogenic compounds (Fahey and Talalay 1995). One example shows that the Chinese consume 0.25 kilogram of crucifer vegetables per capita daily and in Korea the consumption is even higher (Williams and Hill 1986). Moreover, leafy forms of cabbage and kales with thickened succulent stems are cultivated for animal fodder in regions too cool for maize, providing grazing for sheep and cattle in Northern Europe and New Zealand (Williams and Hill 1986). Furthermore, the tree forms are grown on the Channel Islands, where the leaves are stripped from the stem as cattle feed in winter and the remaining stalks are cut and dried for manufacture as walking sticks (Parker and Cox 1974). The Food and Agriculture Organization of the United Nations (FAO) in its 2004 statistical data reported that harvested area was 3,242,105 ha and 893,216 ha for cabbages and cauliflowers, respectively. However, production amounted to 68,389,593 Mt and 16,413,583 Mt for cabbages and cauliflowers, respectively (http://apps.fao.org/). Some data concerning harvested area, yield and production of cabbages

Table 1. Specification of area harvested, yield and production of cabbages worldwide and in selected continents on the basis of FAO data (2004)

	Area harvested (ha)	Yield (hg/ha)	Production (Mt)
World	3,242,105	210,942	68,389,593
Africa	84,930	174,465	1,481,727
Asia	2,388,893	209,740	50,104,721
Europe	555,913	233,544	12,983,030
North and Central America	150,321	213,280	3,206,047

worldwide and in selected continents are shown in Table 1. The detailed data are available on the website (http://apps.fao.org/).

8.1.4 Breeding Objectives

The high level of the *Brassica* vegetables crop results from their importance for human and animal consumption, induced nutritional value and their being a potentially important source of anticancerogenic compounds (Beecher 1994). It is a well known fact that cardiovascular disease and cancer are ranked to be main causes of death in most industrialized countries. Over 20 independent dietary epidemiological studies have provided evidence that *Brassica* consumption reduces the risk of cancer (Verhoeven et al. 1997; www.brassica.info). For example, broccoli contains high levels of sulforaphane, a constituent that may provide powerful protection against carcinogenesis, mutagenesis, and other forms of toxicity by electrophiles and reactive forms of oxygen chemoprotection and hence has stimulated much interest in seed production of this crop (Farnham et al. 2003). Sulforaphane probably protects the body from cancer by induction of detoxification enzymes such as quinone reductase (QR). Other studies provide the premise that *Brassica* vegetables are beneficial in the prevention of other major illnesses such as Alzheimer, cataracts, and some of the functional declines associated with aging (Granado et al. 2003). Furthermore, broccoli is characterized by high contents of glucosinolates, secondary metabolites that, upon hydrolysis, release bioactive isothiocyanates (ITCs) and are a good source of bioactive compounds, such as phenolics (flavonoids and hydroxycinnamoyl derivatives), glucosinolates, and vitamin C. *Brassica* are high in dry matter with digestibility at 85 to 95%, as well. This is significant because the digestible energy of most forage crops is too low for high gains and growth in animals. *Brassica* increase the availability of certain minerals and are also high in protein. Leaves contain 18 to 25% crude proteins, while turnip and swede roots contain about 10% crude protein. These quality traits are important reasons why these leaf and root crops have been commonly grown in New Zealand and Europe as nutritional fodder for sheep and cattle. *Brassica* retain their nutritive value well into freezing temperatures and can be expected to be grazed in most years as late as the end of December, or even longer in the coastal regions of Massachusetts.

Besides, *B. oleracea* is characterized by the occurrence of anticancerogenic compounds. It contains some glucosinolates being antinutritive factors, which limits its value as an animal feed. About 10 to 20 individual glucosinolates are usually present in the *Brassica* species, only a few predominant (Fenwick et al. 1989). Problems of the thyroid disorder in animals grazing vegetables were addressed by assaying for thiocyanate ion content.

8.1.5 Classical Mapping Efforts

Cytogenetic studies have provided most of the information on the genome structure in *Brassica*. Genetic studies of *Brassica oleracea* have been limited in part by the long generation time of the biennials, the complex inheritance patterns of some traits, and the difficulty in overcoming self-incompatibility. However, genetic analysis of progeny from crosses between cultivar groups of *B. oleracea* have provided information on the inheritance of some morphological traits, such as annual habit

(Detjen 1926; Dickson 1968; Baggett and Walhert 1975; Pelofske and Baggett 1979), internode distance (Pease 1926; Dickson 1968; Pelofske and Baggett 1979), heading (Kristofferson 1924; Pelofske and Baggett 1979), and leaf characteristics (Kristofferson 1924; Pease 1926). Complex inheritance was often reported for these traits, and in some cases different studies reported conflicting results, probably due in part to differences in the accessions or forms used as parents to determine inheritance. Although *B. oleracea* does not have an extensive genetic map of morphological markers, some linkage associations for genes controlling morphological traits have been detected in segregants of crosses between cultivar groups (Pease 1926; Pelofske and Baggett 1979). Only seven linkage groups containing a few markers each have been recognized (Wills 1977; Sampson 1978).

Several isozyme markers have been analyzed in segregating populations of *Brassica oleracea* (Arus and Orton 1983) and in *B. campestris-oleracea* addition lines, to monitor the presence of specific alien chromosomes (Quiros et al. 1987). Linkage among isozyme loci was not observed and they have not been tested for linkage to morphological markers, although one isozyme locus, 6-PGD-2, was localized to the same chromosome as a ribosomal DNA restriction fragment (Quiros et al. 1987). Duplicated isozyme loci have been observed, supporting the hypothesis that *B. oleracea* is a secondary polyploid, but the chromosomal organization of such duplicated sequences has not been described (Quiros et al. 1987, 1988).

8.1.6 Classical Breeding Achievements

The breeding programs concern improving the quality and increasing the yield of existing cultivars. Conventional approaches are based on crossing individuals, of which one is deficient in one or more traits and the other possesses the desired traits. They cover crossing whole genomes followed by selection of the superior recombinants from among the several segregation products. Therefore, such procedures are time-consuming, have at least a few limitations and hence are difficult to achieve the desired objectives. In *B. oleracea* species, out of the conventional breeding approaches, further improvements of traits are limited by the availability of suitable genes in the germplasm stocks (Metz 2001; Gapper et al. 2002; Radchuk et al. 2002).

One of the major causes of reduction in the yield and quality of *Brassica* vegetables are diseases, such as black rot (bacterial disease) caused by *Xanthomonas campestris* pv. *campestris* (Pam.) Dawson, clubroot caused by *Plasmodiophora brassicae* Wor., white blister caused by *Albugo candida*, downy mildew caused by *Peronospora parasitica* (Pers.:Fr.) Fr., and *turnip mosaic* (= cabbage black ringspot) caused by TuMV (Williams 1980; Buczacki 1983). These diseases are distributed worldwide and can be found on many economically important species in the Brassicaceae family. Other programs concentrate on improving resistance to pests such as the cabbage aphid (*Brevicoryne brassicae*) and the cabbage fly (*Delia brassicae*). The wild *Brassica* species (for example *B. fruticulosa*) are sources of resistance to several biotic and abiotic stresses (Crouch et al. 1994; Ramsey and Ellis 1996) and hence they provide important genetic resources for the improvement of *Brassica* crops (Lázaro and Aguinagalde 1998). These wild species are hybridized with cultivars followed by several generations of backcrossing to recover the cultivar characteristics.

Resistance to black rot has been extensively studied and some genotypes with incomplete resistance have been identified (Bain 1952; Hunter et al. 1987). They have been used to improve resistance in *B. oleracea* but none of them have provided complete resistance to the disease and the quantitative genetic control complicates its use in producing resistance hybrid varieties (Camargo et al. 1995). In case of clubroot, resistant and susceptible varieties have been identified; however, breeding for clubroot resistance has been limited (Crute et al. 1980). Since the introduction of molecular techniques and the association of molecular markers with genes affecting morphological traits the process of selection of desired disease resistant morphotypes has advanced.

In the case of downy mildew, many sources of genetic resistance to this disease have been identified in *B. oleracea* (broccoli) at both the seedling and adult plant stages (Natti and Atkin 1960; Natti et al. 1967; Dickson and Petzoldt 1993; Hoser-Krauze et al. 1995). They were controlled by single dominant genes, but none of these resistance genes have been incorporated into cultivars.

Progress with breeding of *B. oleracea* is expected with the help of molecular strategies and use of marker-assisted selection strategy.

8.2 Construction of Genetic Maps

8.2.1 First Generation Maps

Contemporary molecular tools at the present stage of their development permit detailed studies of "chromosomal archaeology". Stebbins (1966) foresaw these advances nearly four decades ago, when he suggested that "the opportunities for profitable investigations for this sort are by no means at an end, and new techniques may extend them to degrees of clarity and certainty which at present can hardly be imagined."

For a genus such as *Brassica*, where the cultivated species have a close genomic relationship based on interspecific hybridization and polyploidy, examining genome structure and organization is of great interest. This is not only because of locating quantitative trait loci (QTLs) or major genes controlling traits of agricultural interest, but also for clarifying *Brassica* evolution, taxonomy and synteny with related species and genera, and not the least with *Arabidopsis* (Sebastian et al. 2000).

The development of RFLP markers for *Brassica* crops has facilitated genetic and evolutionary studies. Although many morphological and isozyme markers were identified in *Brassica* species, detailed linkage maps had not been developed before the use of DNA markers. Figdore et al. (1988) assessed the feasibility of creating a restriction fragment length polymorphism (RFLP) linkage map in *Brassica* species and found a high degree of polymorphism even among closely related *Brassica* accessions. The work was continued and the first detailed genetic map for *Brassica oleracea* was constructed based on the segregation of 258 restriction fragment length polymorphisms in a broccoli (*B. oleracea* var. *italica*) × cabbage (*B. oleracea* var. *capitata*) F_2 population (Slocum et al. 1990) (Table 2). The genetic markers defined nine linkage groups, covering 820 recombination units. The population used to map the cloned sequences consisted of 96 individuals. Genomic DNA clones used as probes were derived from genomic DNA libraries prepared from three different subspecies of *Brassica* (*B. oleracea* ssp. *botrytis*, *B. oleracea* ssp. *capitata*, *B. oleracea* ssp. *pekinensis*) digested with *Pst*I. A majority of the informative probes hybridized to more than two RFLPs in the population tested. Conservation of the relative linkage arrangement of the loci between linkage groups was observed. This data supported previous cytological evidence for the existence of duplicated regions and the evolution of *B. oleracea* from a lower chromosome number progenitor, no evidence was provided for the current existence of blocks of homoeology spanning entire pairs of linkage groups. Instead, the arrangement of the analyzed duplicated loci suggested that a fairly high degree of genetic rearrangements occurred in the evolution of *B. oleracea*.

The same F_2 individuals (90 out of 96) used previously for RFLP mapping by Slocum et al. (1990) were analyzed by Kennard et al. (1994) (Table 2) for variation in 22 morphological traits, such as flower development, leaf dimension, heading habit, and internode distance. Marker loci were used to localize and estimate the effect of genes controlling these traits. The source for DNA probes was also the same as in Slocum et al. (1990). Seventy-two RFLP loci, which covered the mapped genome at an average of 10 cM intervals on all nine linkage groups, were tested individually for associations to phenotypic measurements by single factor ANOVA, and markers with significant associations ($P < 0.05$) were used to develop multilocus models.

The RFLP marker system has proved to be useful in *Brassica* species, as several RFLP maps with different purpose, have been constructed. Most of them have been coupled with QTL analysis (Landry et al. 1992; Figdore et al. 1993; Ramsay et al. 1996; Camargo et al. 1997; Morigushi et al. 1999). RFLP markers, along with morphological and isozyme markers, have been also used to identify *Brassica oleracea* monosomic alien chromosome addition lines (McGrath et al. 1990). The main goal of the study by Ramsay et al. (1996) was to use molecular markers to introgress defined short regions of chromosome from a donor doubled haploid calabrese line of *B. oleracea* var. *italica* into a recipient short generation variety *B. oleracea* var. *alboglabra* (Fig. 5, Table 2). Linkage maps of nine chromosomes of *Brassica oleracea* constructed by Kearsey et al. (1996) (Table 2) based on 75 informative molecular markers, were compared in first and second backcross progeny from a cross between two doubled haploid lines. It was from a 66% increase of the observed map length from BC_1 to BC_2 population that a hypothesis was made about the higher recombination frequency in female meiosis in this species. Another possible explanation was that there was a significant shift in chiasma location within the mapped intervals.

Table 2. A summary of first generation of *Brassica oleracea* mapping efforts

Population Cross	Type	No. of individuals	Source of probes	Markers Type	No.	Map length (cM)	Av. loci distance	No. of LGs	Authors	Additional information
Broccoli (*B. oleracea* var. *italica*) × cabbage (*B. oleracea* var. *capitata*)	F_2	96	Genomic DNA libraries (*B. oleracea* ssp. *botrytis*, *B. oleracea* ssp. *capitata*, *B. oleracea* ssp. *pekinensis*) digested with *Pst*I	RFLP	258	820	3.2	9	Slocum et al. (1990)	–
See Slocum et al. (1990)	F_2	90	See Slocum et al. (1990)	RFLP	72	820	11.4	9	Kennard et al. (1994)	Variation in twenty-two morphological traits
No. 86-16-5 (originated from a cross between *B. oleracea* ssp. *capitata* × *B. napus* ssp. *rapifera*) and line CrGC No. 85 (rapid-cycling *B. oleracea*)	F_2	90	Library of *B. napus* embryo cDNA clones	RFLP	201	1,112	5.5	9 + 4	Landry et al. (1992)	Two dominant QTLs for resistance to race 2 of the clubroot disease, Leaf morphology and biennial flowering
Collard (B115) × cauliflower (B265), collard (B115) × broccoli (B008), wild kale from USSR (B1661) × cauliflower (B265) and one interspecific hybrid *B. oleracea* var. *kohlrabi* (B255) × *B. insularis* (B364)	F_2	60 60 60 52	cDNA library of *B. napus*, genomic libraries of *B. oleracea* and *B. napus*	RFLP Isozyme Morphological	108	747	6.9	8 + 3	Kianian and Quiros (1992a)	8 LGs were assigned to their respective chromosomes, by alignment with gene synteny groups for *B. campestris-oleracea* addition lines (McGrath et al. 1990)
Clubroot resistant broccoli (*B. oleracea* var. *italica*) × susceptible cauliflower (*B. oleracea* var. *botrytis*)	F_2	180	Genomic library of *B. oleracea* var. *botrytis*	RFLP	58	576	11.8	9	Figdore et al. (1993)	Segregation for disease resistance, morphological and maturity examined
B. oleracea var. *italica* (GD) × *B. oleracea* var. *botrytis* (RMA12)	BC_1 BC_2	296 30	Genomic *B. napus*, and *B. oleracea* libraries *Pst*I digested; Operon RAPD primer; phosphoglucose isomerase	RFLP RAPD Isozyme	75	1,051	16.4	9	Kearsey et al. (1996)	–

Table 2. (continued)

Population Cross	Type	No. of individuals	Source of probes	Markers Type	No.	Map length (cM)	Av. loci distance	No. of LGs	Authors	Additional information
DH line of *B. oleracea* var. *italica* × *B. oleracea* var. *alboglabra*	BC_1	176	*Eco*RI genomic DNA library from *B. rapa* cv. Tobin; *Pst*I genomic DNA library of *B. napus*; *Pst*I *Brassica* genomic DNA library; genomic clones for S-glycoprotein and S-locus related protein; phosphoglucose isomerase	RFLP RAPD Isozyme	115 22 1	747	5.4	9	Ramsay et al. (1996)	–
Cabbage (*B. oleracea* var. *capitata*) × broccoli (*B. oleracea* var. *italica*)	F_2	124	*Pst*I genomic DNA library and a cDNA library for *B. napus* and an *Eco*RI genomic DNA library from *B. rapa* cv. Tobin	RFLP RAPD	112 47	921	5.6	9	Camargo et al. (1997)	SI locus located
Cabbage (*B. oleracea* var. *capitata* × kale (*B. oleracea* var. *acephala*)	F_2	138	*Pst*I genomic *B. napus* library; cDNA *B. napus* library;	RAPD RFLP Isozyme Morphological	99 21 2 2	823.6	7.16	9	Moriguchi et al. (1999)	QTL for clubroot resistance located

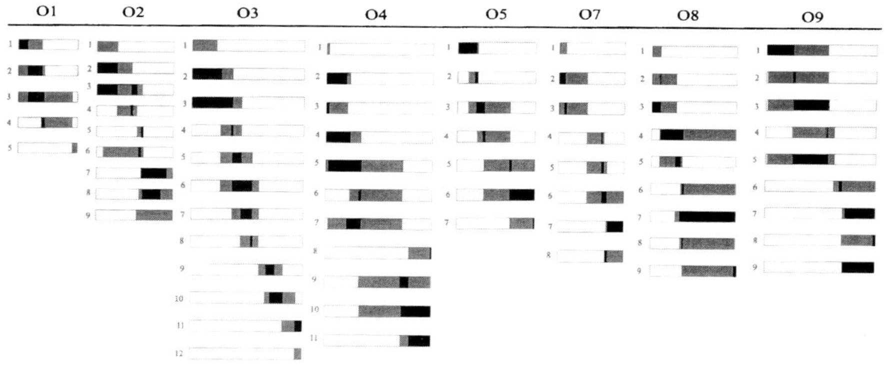

Fig. 5. A set of 71 recombinant backcross lines with determined position and size of donor segments. Each line of the recurrent background genotype (A12DHd, derived from *B. oleracea* ssp. *alboglabra*) is represented by a *white horizontal rectangle*, the size of which corresponds to the size of each linkage group (designed O1-O9) and drawn essentially to scale. *Black shaded* regions represent donor segments (GDDH33, derived from *B. oleracea* ssp. *italica*) and *grey shaded* regions represent tracts containing a crossover. This set of introgressed lines covers almost the entire genome of *B. oleracea*, except for linkage group O6 (reprinted from Ramsay et al. 1996 with permission of NRC Research)

To conclude this section on the first generation of *Brassica oleracea* genetic maps we should mention a paper by Kianian and Quiros (1992a) (Table 2) on the development of a composite linkage map of *Brassica oleracea* from maps of four different populations, derived from 108 DNA, isozyme and morphological loci covering over 747 cM in 11 linkage groups. Of these linkage groups, eight were assigned to their respective chromosomes, by alignment with gene synteny groups for *B. campestris-oleracea* addition lines (McGrath et al. 1990). The four populations used in this study were: three intraspecific hybrids namely collard (B115) × cauliflower (B265), collard (B115) × broccoli (B008), wild kale from the USSR (B1661) × cauliflower (B265) (at least 60 plants each) and one interspecific hybrid *B. oleracea* var. *kohlrabi* (B255) × *B. insularis* (B364) (52 plants). The majority of probes were isolated from a cDNA library of *B. napus* (courtesy of Dr. J Harada). In addition, genome-specific probes isolated from *B. oleracea* and *B. napus* genomic libraries were utilized. Also eight isozyme and morphological markers were used. The average interval between adjacent markers was 8.0 cM. This map could also be treated as a first integrated map. Results of this analysis were in agreement with previous findings that a portion of *B. oleracea* loci is duplicated (Slocum et al. 1990). The distribution of duplicated markers on more than one chromosome and the duplication of chromosomal segments agreed with the hypothesis that *B. oleracea* is, at least partially, a secondary polyploid species (Prakash and Hinata 1980). Consistently with previous reports, authors found evidence for extensive rearrangements that the *B. oleracea* genome have suffered during or after the events responsible for their duplication.

8.2.2 Second Generation Maps

In case of such an important crop species as *Brassica oleracea*, one of the objectives for genetic map construction is marker-assisted breeding, which is facilitated by the presence of dense, high-resolution genetic maps. One way to achieve this purpose was to integrate already existing maps. The other way, used by Bohuon et al. (1996), was to develop a new map, using an immortal population of doubled haploid lines, which could be further saturated (Fig. 6, Table 3). The first dense genetic map for *B. oleracea* was constructed based on the segregation of 303 RFLP-defined loci in a population of 169 microspore-derived DH lines produced from a highly polymorphic cross between *B. oleracea* var. *italica* (GDDH33) and *B. oleracea* var. *alboglabra* (A12DHd). Genetic linkage analysis produced a map with the total length of 875 cM. When the *B. oleracea* linkage map was compared, based on a common set of 129 RFLP markers, to one of *B. napus* (Parkin et al. 1995) good colinearity between the C genome linkage groups of the two species was observed, suggesting that the structures of the C genome chromosomes are essentially identical in the two species. The linkage groups of *B. oleracea* map were numbered and orientated so as to match the corresponding linkage groups of *B. napus*.

Table 3. A summary of high-resolution genetic maps of *Brassica oleracea*

Population Cross	Type	No. of individuals	Source of probes	Markers Type	No.	Map length (cM)	Av. loci distance	No. of LG	Authors	Additional information
B. oleracea var. *italica* (GDDH33) × *B. oleracea* var. *alboglabra* (A12DHd)	DH	169	Genomic *B. napus*, *B. rapa* and *B. oleracea* libraries *Pst*I digested; cDNA *B. napus* library	RFLP	303	879	2.9	9	Bohuon et al. (1996)	*B. oleracea* LGs were numbered and orientated so as to match the corres-ponding linkage groups of *Brassica napus*
I. Kianian and Quiros (1992a) II. Landry et al. (1992) III. Camargo et al. (1997) IV. Ramsay et al. (1996)	F_2	69	Genomic *B. oleracea* and *B.napus* libraries, *B. napus* cDNA library, gene-specific clones	RFLP	167	1,738	10.4	9	Hu et al. (1998)	–
I. (A12 × GD) Bohuon et al.(1996) II. (N × G) cauliflower DH line, *B. oleracea* var. *botrytis* × Brussels sprout hybrid, *B. oleracea* var. *gemmifera*	DH	210 97	*Pst*I genomic *B. oleracea*, *B. rapa* and *B. napus* libraries	RFLP AFLP Microsatelite	547	893	2.6	9	Sebastian et al. (2000)	–
Collard (*B. oleracea* var. *acephala*) × cauliflower (*B. oleracea* var. *botrytis*) Kianian and Quiros (1992a)	F_2	62	*B. napus* genomic library	Microsatelite	31	*	*	8	Saal et al. (2001)	–
Sebastian et al. (2000)	DH	210 97	See Sebastian et al. (2000), nine PCR gene-specific markers	RFLP SNP	23 1	893	♦	9	Howell et al. (2002)	Integration with cytogenetic map
Rapid-cycling *B. oleracea* DH line × *B. oleracea* var. *italica*	F_2	163	AFLP Analysis System I and II; Operon Technologies	AFLP RAPD ISSR SSR	301 55 46 3	731.9	1.64	9	Farinhó et al. (2004)	Gene for downy mildew resistance located

* Marker order and distances as in Hu et al. (1998)
♦ Marker order and distances as in Sebastian et al. (2000)

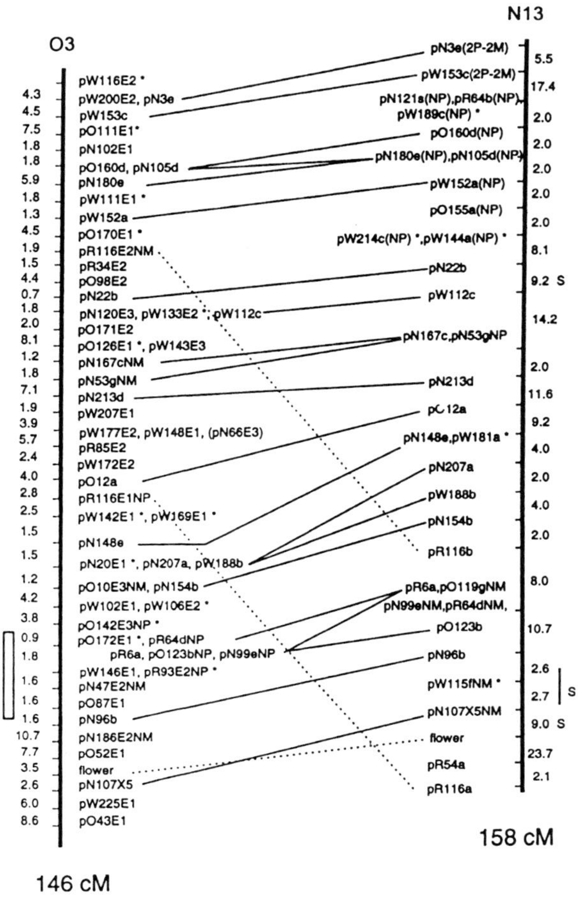

Fig. 6. Genetic linkage maps of the *Brassica* C genome in *Brassica oleracea* and *Brassica napus* (fragment). *Vertical line* on the left represents *B. oleracea* linkage group O3. On the *right* of this pair is *B. napus* N13 linkage group. The flower locus controls flower color and the allele for white flower is dominant to the allele for yellow flowers. Corresponding loci with equivalent positions in both *B. oleracea* and *B. napus* are connected by *solid lines* while the pairs of loci that might be equivalent but which map to different locations are joined by *dotted lines*. (*) equals loci detected in one population by probes that were not used to characterize the other population. *Open bar* indicate region where markers were clustered in the C genome maps of both *B. oleracea* and *B. napus* (reprinted from Bohuon et al. 1996 with permission of Springer)

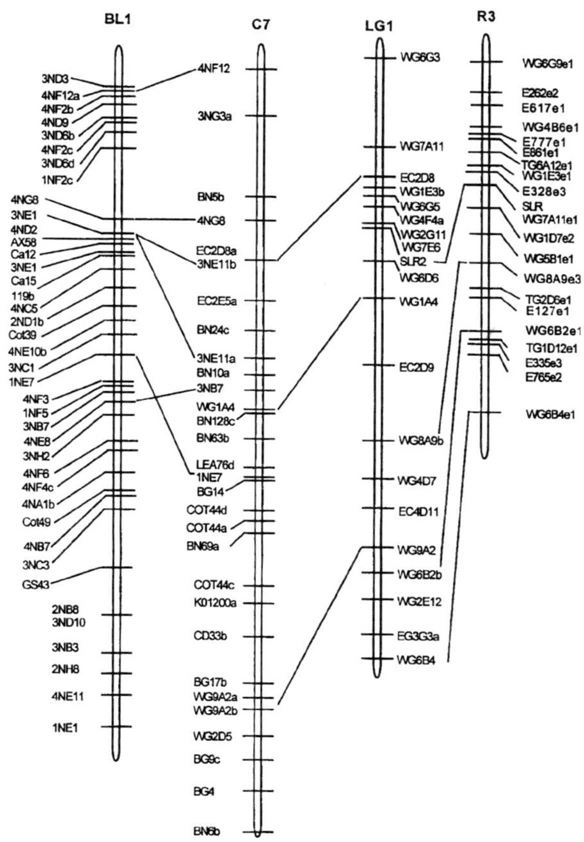

Fig. 7. Alignment of one of *B. oleracea* linkage group from four *B. oleracea* maps. C7 group corresponds to the base map developed by Hu et al. (1998), BL1 group to the map of Landry et al. (1991); LG1 group to the map of Camargo et al. (1997); and R3 group to the map of Ramsay et al. (1996) (reprinted from Hu et al. 1998 with permission of NRC Research)

As it was mentioned above, the alignment of the existing *B. oleracea* maps is necessary for more effective use. The first integrated map was constructed by Hu et al. (1998) from an F_2 population of 69 individuals with sequences previously mapped independently in three linkage maps of this species (Fig. 7, Table 3). These were the maps published by Kianian and Quiros (1992a), Landry et al. (1992) and Camargo et al. (1997) (Table 2). The base map constructed in this study consisted of 167 RFLP loci in nine linkage groups, plus eight markers in four unlinked linkage pairs, covering 1,738 cM. Linkage group alignment was also possible with the fourth map published by Ramsay et al. (1996) (Fig. 5, Table 2) that contained loci found in common with the map of Camargo et al. (1997). Common sequences across the mapping populations served to align most of the linkage groups of the independently developed maps. The probes used in this work included both anonymous genomic clones from *B. napus*, *B. oleracea*, *B. rapa* and *A. thaliana* cDNA library and clones identifying known genes. Similarly to previous reports, a duplication of loci, both intrachromosomal and interchromosomal, could be detected (Truco et al. 1996).

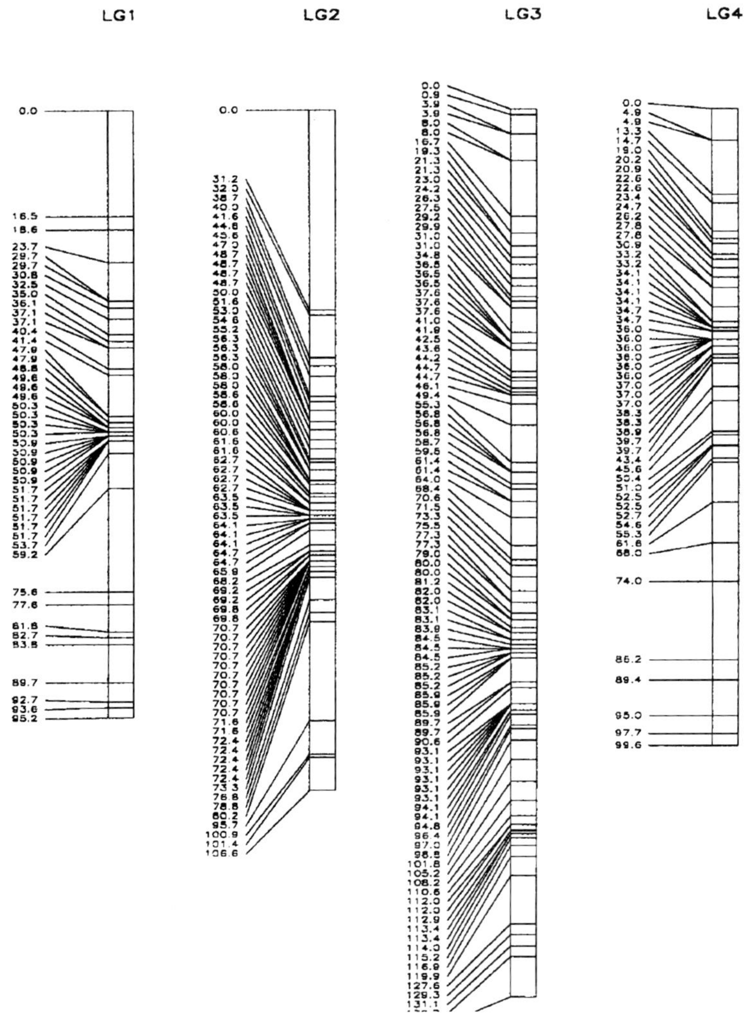

Fig. 8. Integrated linkage map of *B. oleracea*. Linkage groups are numbered LG1 through to LG9, with their order and orientation as defined by Bohuon et al. (1996). Recombination distances are in Kosambi's cM. Only the marker positions are shown here, illustrating the high-marker density and good coverage obtained through map integration (reprinted from Sebastian et al. 2000 with permission of Springer)

The second integrated map of *B. oleracea* was constructed by Sebastian et al. (2000), based on genetic maps generated from the segregation data of two very different F_1-derived doubled haploid mapping populations (Fig. 8, Table 3). The first mapping population (A12 × GD), consisting of 210 DH lines, was produced from a cross between two DH parents, a rapid-cycling Chinese kale line (*B. oleracea* var. *alboglabra*) and a calabrese (*B. oleracea* var. *italica*) (Bohuon et al. 1996) (Fig. 6, Table 3). A second mapping population (N × G) of 97 DH lines was derived from a cross between a cauliflower DH line, *B. oleracea* var. *botrytis*, and a Brussels sprout hybrid, *B. oleracea* var. *gemmifera*. Integration of the two component maps, using JoinMap v. 2.0 was based on 105 common loci including RFLPs, AFLPs and microsatellites. It is worth mentioning that, even though the populations used were derived from very different genetic backgrounds, close conservation of marker order was revealed between these two maps. Based on 547 markers mapping to nine linkage groups, the integrated map covers a total map length of 893 cM, with an average locus interval of 2.6 cM. Comparison back to the component linkage maps revealed similar sequences of common markers, although significant differences in recombination frequency were observed between some pairs of homoeologous markers. Map integration resulted in an increased locus density and effective population size, providing a stronger framework for subsequent physical mapping and for precision mapping of QTLs using substitution lines.

Similarly to Sebastian et al. (2000) (Fig. 8, Table 3), who introduced AFLP markers to *Brassica* genome mapping, Saal et al. (2001) (Table 3) proved the usefulness of microsatellite markers in *Brassica* species.

Fig. 8. (continued)

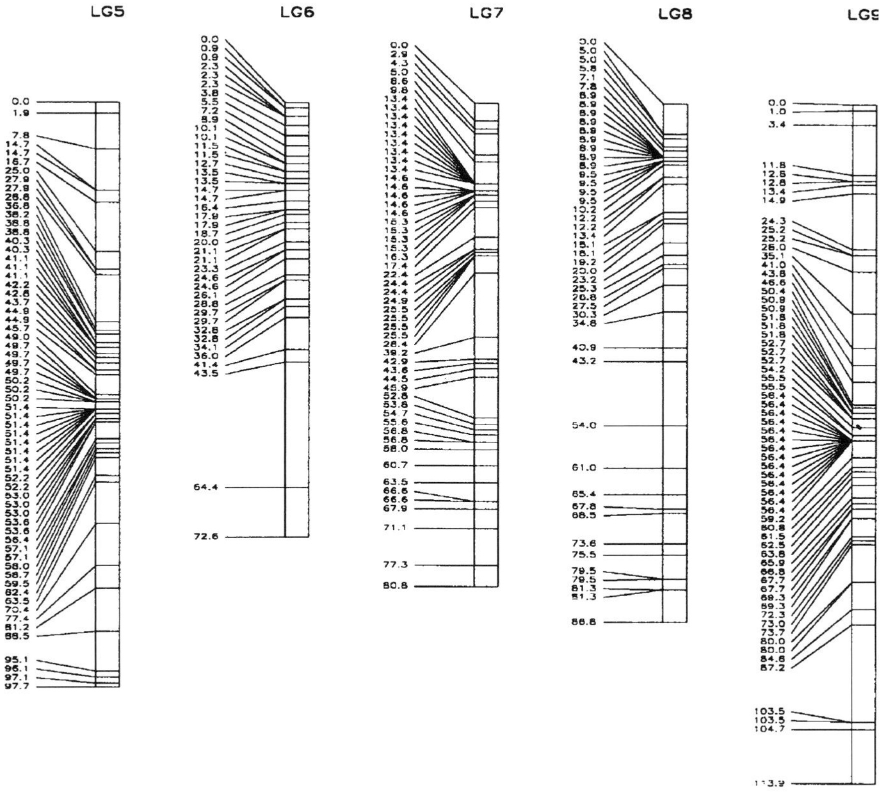

The distribution of rapeseed microsatellites in the C genome was investigated by genetic mapping. Ninety-two dinucleotide microsatellites were screened for polymorphism in an F_2 population derived from a cross between collard and cauliflower, for which an RFLP map had been constructed previously (Kianian and Quiros 1992a; Hu et al. 1998) (Fig. 7, Tables 2 and 3). A total of 29 primer pairs disclosed 34 loci, of which 31 were evenly distributed on 8 of the 9 *B. oleracea* linkage groups. Moreover, the identification of A and C genome-specific microsatellite loci was reported, by using resynthesized rapeseed and its respective parents, as well as *Brassica carinata* (BC genome) and *Brassica juncea* (AB genome).

Every genetic map constructed by different research groups provides more information on genome structure and evolution. The data flow would be more efficient if comparison of corresponding linkage groups would have been possible. The integration of the cytogenetic map with the genetic one provides a framework of such an alignment. The assignment of all nine linkage groups of the *Brassica oleracea* genetic map to each of the nine chromosomes of the karyotype derived from mitotic metaphase spreads of the *B. oleracea* var. *alboglabra* line A12DHd using FISH was done by Howell et al. (2002) (Fig. 4, Table 3). The A12DHd line was previously used as a parent of a doubled haploid line mapping population (Bohuon et al. 1996) (Fig. 6, Table 3), the genetic map of which (A12 × GD) was integrated with another genetic map of *B. oleracea* (N × G; Sebastian et al. 2000) (Fig. 8, Table 3). The majority of probes were BACs, with A12DHd inserts, which gave clear reliable FISH signals. To the existing integrated linkage map nine markers were added, distributed over six linkage groups. A cytogenetic map of *B. oleracea* associated with a genetic map will contribute to a wide range of research. For example, it can provide information complementary to that from physical molecular maps currently being developed (http://brassica.bbsrc.ac.uk), as well as the location of genes in relation to features of chromosomal organization.

Recently, construction of a genetic map of *Brassica oleracea* var. *italica* based on RAPD, AFLP, ISSR and SSR markers with simultaneous mapping of a dominantly inherited downy mildew resistance gene expressed at the adult stage *Pp523* has been reported by Farinhó et al. (2004) (Table 3). Additionally, a locus determining flower color (white vs. yellow) was also located on this map. Since the average loci distance of only 1.64 cM is one of the smallest reported to date, this is probably the densest *B. oleracea* genetic map.

8.2.3 Comparative Maps

One way to study the evolution of plant genomes is to compare the level of conserved gene order among related species or genera (Doebley and Wendel 1989). Using the same set of DNA probes, it is possible to construct linkage maps for each of the species to be compared. Gene order, linkage group conservation, and chromosome distribution can then be determined. Examples of the use of this technique for evolutionary studies in plants are common. These include comparisons of genomes between related species with the same and different chromosome numbers. Among the first to do so was Slocum (1989), who compared *Brassica oleracea* with *Brassica rapa* genome, and Kowalski et al. (1994), the proponent of a whole series of comparisons between *B. oleracea* and *A. thaliana* genomes. The former disclosed extensive genomic regions of conserved homology, while the latter observed, as expected, the lower level of colinearity.

Comparisons of the linkage maps of *Brassica* species have been carried out for *B. oleracea* and *B. rapa* (McGrath and Quiros 1991), for *B. rapa* with *B. oleracea* and *B. napus* (Teutonico and Osborn 1994), and for synthetic *B. napus* with *B. oleracea* and *B. rapa* (Lydiate et al. 1993). These studies have revealed many regions with conserved linkage arrangements between *B. oleracea* and *B. rapa*, in some of them the order of loci being disturbed by deletions and inversions.

First comparative analysis between all the three diploid *Brassica* genomes was performed by Truco et al. (1996). In this study, linkage maps of *B. nigra* (Truco and Quiros 1994), *B. rapa* (McGrath and Quiros 1991) and *B. oleracea* (Kianian and Quiros 1992a) (Table 2) developed by a set of common probes were compared for colinearity. Their primary objective was to ascertain the degree of homology among the genomes of these three species and then relate it to their phylogeny. Altogether, 85 loci were scored after testing the three species with 42 common probes. Intergenomic conserved regions, but with extensive reordering among the genomes, were found. Eighteen linkage groups could be associated on the basis of homologous segments based on at least three common markers. The highest homology was observed for the A and C genomes. Intragenomic homologous conservation was also observed for some of the chromosomes of the A, B and C genomes. The best estimate for intragenomic homology was obtained from *B. oleracea*, since its map was the largest and it corresponded to 40% of the C genome. Based on the analysis performed, at least five and no more than seven ancestral chromosomes could be postulated to explain the existing linkage groups and their homologous relationship. Their results suggested that the *Brassica* genomes differentiated from an originally smaller genome that served as a foundation to build the present-day genomes by duplicating and reshuffling the existing information. Hybridization and chromosomal rearrangements have thus played a substantial role in their evolution.

The genome-wide triplication of *Brassica* species was first suggested by Lagercrantz and Lydiate (1996). They used a linkage map derived from a highly polymorphic cross between *B. nigra* accessions and additional maps of *B. rapa* and *B. oleracea*, generated using the same sets of RFLP probes (Parkin et al. 1995; Sharpe et al. 1995; Bohuon et al. 1996), to perform their own comparative analysis. It was noted previously that intragenomic duplication is a prominent feature of the *B. oleracea* and *B. rapa* genomes (Slocum et al. 1990; Song et al. 1991; Kowalski et al. 1994; Parkin et al. 1995), but not to this extent. The three genomes have distinct chromosomal structures differentiated by a large number of rearrangements, but colinear regions involving virtually the whole of each of the three genomes were identified. The gene contents of *B. nigra*, *B. oleracea* and *B. rapa* were basically equivalent and the differences in chromosome number (8, 9 and 10, respectively) are probably the result of chromosome fusions and/or fissions.

As mentioned above (see Sect. 8.1), the importance of revealing the *Brassica oleracea* genome structure is not only related to the vegetable crops that it represents itself, but also to *B. napus*, oilseed rape. Based on interspecific hybridization and cytogenetic data, *B. oleracea* and *B. rapa* were proposed as the progenitors of the amphidiploid *B. napus* (U 1935). A comparison of *B. napus* and *B. oleracea* linkage maps (Lydiate et al. 1993; Cheung et al. 1997) showed that linkage groups of *B. oleracea* are conserved in the *B. napus* map. These results provided molecular evidence for *B. oleracea*, or a closely related $2n = 18$ *Brassica* species, as the C genome progenitor, and also reflected on the homoeology between the A and C genome in *B. napus*. Parkin et al. (1995) identified the 10 linkage groups of the *B. rapa* A genome progenitor (N1-10) and the 9 linkage groups of *B. oleracea* C genome progenitor (N11-19) of *B. napus*. As described in Bohuon et al. (1996) (Fig. 6, Table 3), the nine C genome linkage groups from the *B. napus* map were aligned with the nine homologous groups of the *B. oleracea* genetic linkage map.

Comparative genome analyses have been carried out not only among members of the *Brassica* genus, but also between species belonging to the Brassicaceae family. A close relative of the *Brassica* species is *A. thaliana* ($n = 5$), an extensively utilized model plant, both in terms of biochemistry, physiology, classical and molecular genetics. The relationship between *Arabidopsis* and *Brassica* was further suggested by extensive conservation of coding sequences, 87% on average (Lydiate et al. 1993; Cavell et al. 1998).

The first global comparison of genomic structure and organization between *B. oleracea* and *A. thaliana* was done by Kowalski et al. (1994). By applying previously mapped *B. oleracea* DNA probes (Slocum et al. 1990) to *A. thaliana* populations, they analyzed the relative organization of chromosomes of genomes of these two species. Extensive rearrangement distinguishes the chromosomes of *A. thaliana* and *B. oleracea*, although numerous regions of locally conserved linkage and/or homoeology were also apparent. The conserved regions (11) between *Arabidopsis* and *Brassica*, identified based on 49 corresponding markers, covered 24.6% and 29.9% of these two genomes, respectively. An average length of such a region was estimated to be 21.3 cM in *A. thaliana*. These data suggested that 26 chromosomal rearrangements have occurred since the divergence of these two species. The results of this study shows evidence of sequence duplication (as 12.5% of probes mapped to more than one locus), which may be a result of an ancient duplication of chromosomes or chromosomes segments in an ancestor of *A. thaliana* (McGrath et al. 1993). Also, consistently with previous reports (Slocum et al. 1990) (Table 2) the triplication of some regions of *B. oleracea* genome was also suggested.

These studies, however, did not provide a complete scope of the genome comparison between *Brassica oleracea* and *Arabidopsis thaliana* because of the limited number of common markers. Another very interesting study is that of Lan et al. (2000), who established a detailed comparative map of *B. oleracea* and *Arabidopsis* based largely on mapping of *Arabidopsis* ESTs in two *Arabidopsis* and four *B. oleracea* F_2 populations. Each of these populations was composed of 78, 78, 56, 247, 250 and 246 individuals. A total of 113 *Brassica PstI* genomic clones, 35 *Arabidopsis* genomic clones, 23 *Arabidopsis* anonymous cDNA clones, four cloned RAPD-PCR products, 198 *Arabidopsis* EST clones and 19 putatively embryo-specific *Arabidopsis* EST clones were used as probes. In this analysis, based on 186 corresponding loci, a much more detailed picture of the comparative genome organization of *B. oleracea* and *A. thaliana* was provided. More than half of comparative loci (57%) have been found in syntenic segments between *Arabidopsis* and *Brassica*, established on the basis of conservative criteria for synteny. At least 19 chromosomal structural rearrangements differentiate *B. oleracea* and *A. thaliana* orthologs. Chromosomal duplication in the *Brassica oleracea* genome was strongly suggested based on parallel arrangements of 41% of loci mapped in *Brassica*. At least 22 rearrangements differentiate *Brassica* homologs from one another. Results of this study also point to the possibility of the *Brassica* genome being triplicated and the *Arabidopsis* genome duplicated. The linkage maps of *B. oleracea* spans on average 863.6 cM. A total of 367 loci were detected in the composite map, with an average distance between loci of 2.35 cM.

The year 2000 ended with the publication of the complete DNA sequence of the *Arabidopsis thaliana* genome, which provided a comprehensive repository of data that can be exploited in many different ways. These analyses (AGI 2000; Blanc et al. 2000; 2003) revealed the genome duplication reported by Lan et al. (2000) in *Arabidopsis* took place before *Arabidopsis-Brassica* split. Hence, the *Brassica* genome should be considered as a more complex polyploid (each pair of duplicated chromosomal segments in *Arabidopsis* has three pairs of counterparts in *Brassica*), although a very high level of gene loss hinders detection of the former rounds of duplication.

The approach that was taken by Ryder et al. (2001) contrasted with previous colinearity studies between *Brassica* and *Arabidopsis*. They used a combination of RFLP and PCR-based markers (previously described in Bohuon et al. 1996 and Parkin et al. 1995; Table 3) from two regions of the *B. oleracea* genome to establish the position of corresponding loci in *A. thaliana*, which was facilitated by hybridizing probes onto the whole *A. thaliana* genome, as represented by BAC contigs. Since they were not limited by the availability of polymorphic *Arabidopsis* markers, their results were much more complete. They observed a contrasting pattern of chromosomal organization between the two regions of the *Brassica* genome sampled from linkage groups O3 and O6. The linear organization of linkage group O6 appeared to be well-conserved with respect to a 25-cM region of *A. thaliana*. On the contrary, the top section of linkage group O6, in common with the section of linkage group O3 sampled, displayed considerably more evidence of rearrangements with respect to *A. thaliana*.

Another genome-wide comparative study between these two species has been conducted in our laboratory. Similarly to Lan et al. (2000), ESTs were used from *A. thaliana* to construct a genetic RFLP map of *B. oleracea*. Since the *A. thaliana* genome sequence has been determined (AGI 2000), it was possible to assign in silico the sequences of EST probes and their homologs to genetic and physical maps of the *A. thaliana* genome and show their homology to known genes. To gain deeper insight into the degree of chromosome collinearity between *Arabidopsis* and *B. oleracea* genomes, the reports on the duplication status of the *A. thaliana* genome (Blanc et al. 2000; Grant et al. 2000; Vision et al. 2000) were particularly considered. The *B. oleracea* genetic map was constructed (Babula et al. 2003; Kaczmarek et al. 2004; Fig. 9) based on the segregation of 280 loci in an F_2 population described previously (Kianian and Quiros 1992a; Hu et al. 1998; Fig. 7, Tables 2 and 3). Based on the pair-wise comparison of the linear organization of loci on the *A. thaliana* and *B. oleracea* maps, it was possible to identify over 20 putative syntenic blocks covering 30% of the *Brassica* genetic map and 40% of the *Arabidopsis* map. The average segment consisted of 3.3 loci corresponding to 12.8 cM in *Arabidopsis* and 36.4 cM in *B. oleracea*. The number of segments appeared to be well-aligned with segmental duplication in the *Arabidopsis* genome. Although the order of genes was preserved in a majority of these segments, in some it was disrupted by inversion and deletion based rearrangements. Consistent with previous reports, evidence was found on the evolution of diploid *Brassica* genomes from a hexaploid ancestor (Lagercrantz and Lydiate 1996; Lagercrantz 1998; O'Neill and Bancroft 2000). The identification of sequence homologs to the *Arabidopsis* genome sequence dispersed in the *Brassica* genome suggested extensive chromosomal repatterning has occurred since the divergence of these two species (Kowalski et al. 1994; Lan et al. 2000).

The results of previously mentioned comparative studies, both between *B. oleracea* and *A. thaliana* and within the *Brassica* genus, lead to contrasting conclusions concerning genome structure and evolution. This discrepancy was likely due to the partial replication of all the studied genomes and to the ambiguous criteria for conserved region identification. Lukens et al. (2003) compared the position of sequenced *Brassica* loci with known position on a *B. oleracea* genetic map (Bohuon et al. 1996) (Fig. 6, Table 3) to the positions of their putative orthologs within the *A. thaliana* genome. They developed an algorithm written in PERL that used explicit criteria to identify orthologous regions and to establish their significance. Consistently with previous reports, evidence was found for substantial genomic replication in *B. oleracea* as compared to *A. thaliana*, along with evidence that multiple chromosomal rearrangements have occurred since the divergence of the two species. They reported that *B. oleracea* and *A. thaliana* shared 34 significant colinear regions. The average segment had 3.3 markers corresponding to 2.1 Mbp in *A. thaliana* and 7.1 cM in *B. oleracea*. In total, the significant regions covered over one-fourth of the *B. oleracea* genome. Finally, they inferred that differences in the genomic arrangements between *B. oleracea* and *A. thaliana* appeared to be due to the recent history of polyploidy in *B. oleracea*.

Particularly interesting results were obtained by Li et al. (2003a) in studies of gene alignment of the *B. oleracea* and *A. thaliana* genomes. They reported on the construction of the first transcriptome map based on *B. oleracea* cDNAs obtained from leaf tissue of 88 F_2 individuals of a mapping population derived from a cross between DH of broccoli and cauliflower. The map consisted of 247 cDNA markers obtained by the SRAP technique (Li and Quiros 2001). Sequencing of 190 of the polymorphic cDNA bands and performing of FASTA searching resulted in the detection of 169 sequences similar to *Arabidopsis* genes. Consistent with previous reports, they found evidence of extensive collinearity between the two genomes for chromosomal segments rather than for whole chromosomes, often disrupted by inversions and dele-

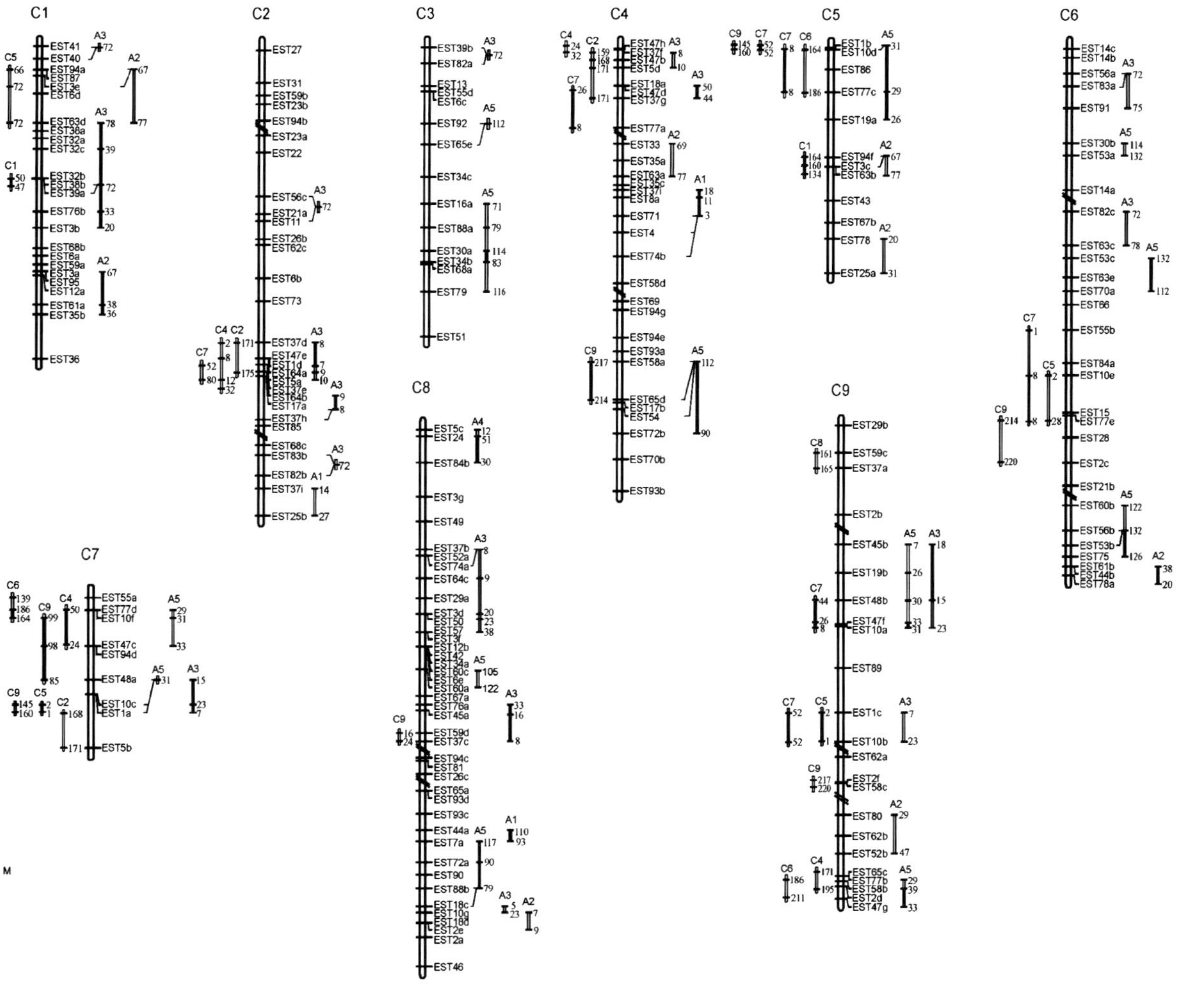

Fig. 9. Comparative map of *B. oleracea* and *A. thaliana*. The nine linkage groups of *B. oleracea* are represented by *vertical bars*. EST loci on the right of each linkage group were identified with EST probes previously assigned to *A. thaliana* BACs. The regions duplicated in *B. oleracea* (intragenomic duplications) are indicated to the left of each linkage group (except C3) as *thin vertical bars* with chromosome designations (C1-C9) above them. The *A. thaliana* conserved regions are shown on the *right* site of the *B. oleracea* linkage groups as *thin vertical bars* with chromosome number (A1-A5) above. *Thin black bars* (or their part) indicate inversions events in the corresponding regions in *B. oleracea* and *A. thaliana* chromosomes as compared to homologous regions in the main linkage groups. The *horizontal lines* indicate positions of EST loci with corresponding cM values from the original chromosome location. The *diagonal breaks* in the linkage groups C2, C4, C6, C8 and C9 stand for genetic distances that are greater than 30 cM (but less than 60 cM) between neighbouring loci (reprinted from Babula et al. 2003 by permission of Springer)

tion/insertions. They have also observed large-scale duplications of *Brassica* genome, but their distribution argued against an ancient triplication of the entire genome. The obtained results are rather consistent with the hypothesis that events of complexity higher than a simple polyploidization have lead to the synthesis of *Brassica* genomes including also aneuploidy and chromosomal rearrangement (Quiros 2001). Clear differences in the similarity score value of related sequences made possible the identification of orthologs. The SRAP technique has one advantage over ESTs used previously in Brassicaceae comparative analysis (Lan et al. 2000; Babula et al. 2003; Fig. 9), in that it might detect some genes with low levels of expression or detect gene expression more evenly.

8.3 Gene Mapping

First genetical analyses of genes responsible for simple-inherited traits in *B. oleracea* come from the 1950s and 1960s (Anstey and Moore 1954; Anstey 1955; Cole 1959; Sampson 1966). These early attempts used in the analysis such characters as white and cream petals, persistent sepals, purple ovary, glossy and pale foliage, male-sterility and self-incompatibility. Although some linkages between genes coding those traits were detected, their number was too low to construct any genetic map. Thus, major improvement in *Brassica* mapping began with developing DNA markers, especially RFLPs. However, only some morphological traits were used in further maps, because they usually had no economical importance and were relatively laborious in comparison with DNA markers. These include the presence of an enlarged stem (bulb *B*), anthocyanin pigmentation (purple color *P*) and glossy foliage (*gl-1*) that were analyzed by Kianian and Quiros (1992a). Unfortunately, only the last from these traits was assigned to a linkage group, the rest being unlinked. Other morphological traits that have been mapped include genes responsible for leaf morphology (Landry et al. 1992; a single codominant locus detected) and a locus responsible for flower color (Farinhó et al. 2004) (see Table 4, for details).

To coordinate reproductive development with seasonal changes, many plant species found in temperate climates have evolved a biennial or winter-annual growth habit. The distinguishing feature of this growth habit is an obligate or facultative requirement for vernalization before flowering can occur. Thus flowering in biennials and winter annuals is blocked before winter, and exposure to winter permits flowering the next spring. In *B. oleracea*, there exist both winter-annual and rapid-flowering types, which flower rapidly without vernalization. As annual habit is a simply inherited trait dependent on a few genes (in a closely related species, *A. thaliana*, there are two genes responsible for this phenomenon (Michaels et al. 2004)), it could be mapped using conventional strategies. The trait was mapped by Kianian and Quiros (1992a) (called annual habit; see Table 4) and by Landry et al. (1992) (called vernalization). In the latter work the corresponding gene was unlinked to other map markers available so far (see also Sect. 8.4, where QTL mapping approaches are described).

Besides morphological traits, metabolic characteristics could also be mapped by a similar approach, although they need additional analyses of crop chemical components. For instance, glucosinolate contents were evaluated in the mapping of genes coding enzymes involved in the biosynthesis of aliphatic glucosinolates. There are three genes, which control the synthesis of aliphatic glucosinolates in *B. oleracea*: *BoGSL-ELONG*, *BoGSL-PRO* and *BoGSL-ALK*. The dominant allele for the *BoGSL-ELONG* gene results in four-carbon glucosinolates, whereas the presence of the dominant allele for the *BoGSL-PRO* gene results in three-carbon glucosinolates. Plants carrying both dominant alleles produce both types of glucosinolates (Li et al. 2001a). The third gene, *BoGSL-ALK*, controls sidechain desaturation of aliphatic glucosinolates. A population with segregating alleles for these genes was used by Li and Quiros (2001) and Li et al. (2003a) for genetic mapping of this gene (Table 4). The *BoGSL-ALK* gene was further map-based cloned (Gao et al. 2004; see Sect. 8.6), whereas *BoGSL-ELONG* was cloned using an *Arabidopsis* candidate gene approach (Li and Quiros 2002).

Another group of genes that are of special interest in *Brassica* breeding are disease resistance genes. Studies on molecular markers linked to these genes in *B. oleracea* have mainly been focused on resistance to *Plasmodiophora brassicae*, the inciting agent of clubroot (Landry et al. 1992; Figdore et al. 1993; Voorrips et al. 1997; Moriguchi et al. 1999; Rocherieux et al. 2004). However, this trait is quantitative under polygenic control, and will be described more comprehensively in the QTL Sect. 8.4. Notwithstanding, the resistance to downy mildew, incited by *Peronospora parasitica*, is controlled by a single dominant or recessive gene depending on the source (Farnham et al. 2002; Coelho and Monteiro 2003). Giovanelli et al. (2002) used an F_2 population of 100 individuals with segregating resistance/susceptibility to downy mildew at the cotyledon stage (a single dominant gene) to perform bulked segregant analysis (BSA) and RAPD screening. The authors identified eight RAPD markers that were linked to the resistance locus and converted two of them into SCARs (see Table 4, for details). Another work aimed at the mapping of resistance to downy mildew at adult plant stages (a single dominant gene) by using similar approaches, but with a construction of a genetic map (the map described in Sect. 8.2) (Farinhó et al. 2004; Table 3). A population of 163 F_2 plants coming

Table 4. A summary of gene mapping efforts for *B. oleracea*

Trait mapped/ gene	Marker type/probe	No. of loci	Linkage group	Flanking marker (with distance)		References
Self-incompatibility, SLG gene	RFLP/cDNA clone BOS5 (Nasrallah et al. 1985)	Three loci	2	pBOS5-1	pBN120-1 (14 cM) pBOS5-2 (3 cM)	(Kianian and Quiros 1992a; Kianian and Quiros 1992b)
				pBOS5-2*	pBOS5-1 (3 cM) pBOS5-3 (0 cM)	
				pBOS5-3*	pBOS5-2 (0 cM) pBSMS1-1 (20 cM)	
		Single locus: cosegregation with SI reaction	2	AA11E (4.7 cM) WG5A1 (2.1 cM)		(Camargo et al. 1997)
	RFLP/genomic clone S29 (Trick and Flavell 1989)	Single locus	6	pWG7E10e2 (34 cM) pWG5A1e1 (35 cM)		(Ramsay et al. 1996)
Napin	cDNA clone pN2 (Crouch et al. 1983)	Five loci	1	pN2-6	pBN24-1 (4 cM) pBN7-1 (0 cM)	(Kianian and Quiros 1992a; Kianian and Quiros 1992b)
				pN2-5	pBN121-3 (2 cM) pLEA76-1 (2 cM)	
				pN2-3	pCOT50-3 (3 cM) pN2-1 (3 cM)	
				pN2-1	pN2-3 (4 cM) pN2-2 (2 cM)	
				pN2-2	pN2-1 (2 cM) pBN11-3 (14 cM)	
		Two loci (one unlinked)	5	pLea76b (3 cM) 4NC2 (1 cM)		(Landry et al. 1992)
Cruciferin	cDNA clone pC1 (Crouch et al. 1983)	Two loci	3	pC1-1	pTA71-2 (26 cM) pBN98-8 (4 cM)	(Kianian and Quiros 1992a; Kianian and Quiros 1992b)
			8	pC1-2	pBN129-2 (12 cM) pB845-2 (5cM)	
		Single locus	2	3NH4 (5 cM) 2NH5 (1 cM)		(Landry et al. 1992)
Annual habit An	Morphological	Single locus	6	pCOT49-1 (8 cM) pB370-1 (10 cM)		(Kianian and Quiros 1992a)
Leaf morphology	Morphological	Single locus	3	2NF2 (21 cM) 3ND12 (2 cM)		(Landry et al. 1992)
Glossy foliage (gl-1)	Morphological	Single locus	B**	– pBN98-5 (11 cM)		(Kianian and Quiros 1992a)
β-carotene accumulation/ *Or* gene	RFLP (converted from AFLPs)***	Single locus	Bulked segregant analysis	RA9 and RA1 (0.5 cM) RA4 (1.6 cM)		(Li and Garvin 2003)
Downy mildew resistance locus (cotyledons)	SCAR (converted from RAPDs)	Single locus	Bulked segregant analysis	UBC359 (6.7 cM) OPM16 (3.3 cM)****		(Giovannelli et al. 2002)
Downy mildew resistance locus (mature plants)	RAPD, AFLP	Single locus	3	AAG.CTA_113 (2.7 cM) AT.CTA_133/134 (3.6 cM)		(Farinhó et al. 2004)

* The order of loci pBOS5-2 and pBOS5-1 is unknown
** An additional linkage group not assigned to any chromosome
*** AFLP markers were converted also into SCARs
**** Both markers map above the resistance locus

Table 4. (continued)

Trait mapped/ gene	Marker type/probe	No. of loci	Linkage group	Flanking marker (with distance)	References
Desaturation of glucosinolates/ *BoGSL-ALK* gene	Phenotypic analysis, SRAP, AFLP	Single locus	1	SRAP133 (1.4 cM♦) AFLP99 (29.3 cM)	(Li and Quiros 2001)
Sidechain elongation of glucosinolates/ *BoGSL-PRO*	Phenotypic analysis, cDNA-based SRAP	Single locus	1	T123 (7.1 cM) T15 (0.0 cM)	(Li et al. 2003a)
Sidechain elongation of glucosinolates/ *BoGSL-ELONG*	Phenotypic analysis, cDNA-based SRAP	Single locus	1	T42 (4.7 cM) T6 (15.8 cM)	(Li et al. 2003a)
Resistance gene homolog *BoRGL*-Va	CAPS	Single locus	1±	ACCTCJ03 (4.2 cM) pW201R1 (1.1 cM)	(Vicente and King 2001)
Resistance gene homolog *BoRGL*-IIIa	CAPS	Single locus	4±	ACCTCJ03 (0.9 cM) ACCATJ05 (0.8 cM)	(Vicente and King 2001)
Resistance gene homolog *BoRGL*-IVa	CAPS	Two loci	8± 9±	pW104E2 (4.9 cM) pW188J1 (5.6 cM) pO119J1 (2.3 cM) pO119E3 (1.6 cM)	(Vicente and King 2001)

♦ The distance overestimated by the residual heterozygosity of the mapping population
± Linkage groups correspond to those described in Bohuon et al. (1996) and Sebastian et al. (2000)

from lines resistant and susceptible to downy mildew was screened with RAPD, AFLP and ISSR markers (DNA bulking). This made it possible to identify five markers closely linked to the resistance gene (see Table 4).

Besides classical mapping strategies, which are based on finding resistance phenotypes, there are approaches utilizing candidate gene analysis. Vicente and King (2001) used high-homology regions from several cloned resistance genes to design primers for PCR amplification of their homologs in *B. oleracea*. Products of amplification were cloned, sequenced, and further hybridized to *B. oleracea* and *A. thaliana* BAC libraries to estimate the copy number for these genes and their relative position in the *Arabidopsis* genome, respectively. This made it possible to confirm the products corresponding to resistance genes in *A. thaliana*. Results of hybridization to the *B. oleracea* linkage map are shown in Table 4. Notwithstanding, no correlation between these genes and resistance QTLs has been found in *Brassica* yet.

Another trait that was widely analyzed on different levels, including genetic (and physical) mapping was self-incompatibility (SI). SI has been utilized in the breeding of hybrid varieties of economically important vegetables of Brassicaceae, within *B. oleracea* including cabbage, broccoli and cauliflower. Thus, SI is considered to be very important as an agricultural trait. SI is based on the ability of the pistil to discern the presence of self-pollen and to inhibit the germination or subsequent development of self-related, but not genetically unrelated, pollen. In the Brassicaceae, the genetic control of SI was deciphered in the early 1950s with Bateman's analysis of *Iberis amara* (Bateman 1955). The essence of this model is that an incompatible response occurs when the same *S*-allele is active in the stigma and pollen. The number of alleles (haplotypes) that occur at the *S*-locus in *B. oleracea* was estimated at 60 (Ockendon and Gates 1975), and the multifunctional gene complex of the *S*-locus may span several hundred kilobases (Boyes et al. 1997). Molecular genetic analyses have demonstrated that among the genes contained at the locus there are three highly polymorphic genes. They encode proteins expressed specifically in the epidermal cells of the stigma, the cell wall-localized glycoprotein (SLG) and the plasma membrane-spanning receptor protein kinase (SRK) (reviewed in Kachroo et al. 2002). The third gene, described more recently (Schopfer et al. 1999), is expressed exclusively in anthers and encodes Cys-rich protein SCR. The *S*-locus was mapped in *B. oleracea* by Kianian and Quiros (1992a,b), Ramsay et al. (1996), and Camargo et al. (1997) (Table 4). The latter team of researchers proved cosegregation of the *S*-locus detected with the SI reaction in plants. Finally, the *S*-locus (or three linked loci in Kianian and Quiros 1992a) was used in the construction of map alignments (Hu et al. 1998) (Table 3).

Other genes that were mapped by using cDNA clones as probes are napins and cruciferins. They code basic storage proteins synthesized in *Brassica* embryos during seed maturation (Crouch et al. 1983) and thus could be interesting from the agricultural point of view. There have been several attempts to map these genes in *B. oleracea*. Song et al. (1988) used RFLPs for napin and cruciferin, together with some other nuclear probes, to analyze phylogenetic relationships within different *B. oleracea* and *B. rapa* cultivars. The same probes were further used by McGrath et al. (1990) for the assignment of genes to chromosomes in *B. oleracea* monosomic addition lines – both genes were assigned to chromosome 1. Napin and cruciferin genes were mapped genetically by Landry et al. (1992), and Kianian and Quiros (1992a,b), and the alignment of the group from independent maps was performed by Hu et al. (1998) (Fig. 7, Tables 2 and 3).

Besides SLG (the *S*-locus), napins and cruciferins, there have been many other genes mapped in *B. oleracea* using cDNA clones. Probes for these experiments consist of *Brassica* cDNA/EST clones (Kianian and Quiros 1992a,b; Landry et al. 1992; Li et al. 2003a; Camargo et al. 1997; Bohuon et al. 1998; Table 4) or *Arabidopsis* cDNA/ESTs (Sadowski et al. 1996; Babula et al. 2000; Lan et al. 2000; Babula et al. 2003). The corresponding genes mapped by these attempts are usually not related to any simply-inherited trait and in many cases their biological function is unknown. Thus, they will not be described here in detail.

Sometimes the main goal of genetic mapping is the isolation of a gene connected with a given trait. This was the case of the cauliflower high β-carotene gene *Or* (*Orange*). The *Or* mutant in cauliflower is a spontaneous, single-locus mutation that causes accumulation of high levels of β-carotene in plastids. As plant carotenoids are precursors for vitamin A synthesis and fulfill essential requirements for human nutrition, their levels in plants may be of economical importance (Mayne 1996). The *Or* gene appears to induce β-carotene accumulation by derepressing a mechanism that suppresses carotenoid accumulation (Li et al. 2001b). In order to clone the *Or* gene, a mapping population of 195 F_2 individuals was produced from a single F_1 plant derived from a cross between a wildtype cauliflower and an *Or* mutant (Li and Garvin 2003). By using the AFLP technique in conjunction with bulked segregant analysis, the authors identified 10 markers closely linked to the *Or* gene. A few of them were further converted into RFLP markers and SCAR markers (the closest linked at 0.5 cM), and used for positional cloning (see Sect. 8.6).

8.4 *Brassica oleracea* Quantitative Trait Loci (QTL)

8.4.1 Resistance to *Plasmodiophora brassicae*

Clubroot, caused by the obligate, biotropic soil-borne pathogen *Plasmodiophora brassicae*, is one of the most damaging diseases of vegetable *Brassica* crops worldwide. Infected plants show symptoms leading to gall formation on the root system. These clubs inhibit nutrient and water transport, and lead to stunt plant growth and increased susceptibility to wilting. In areas where *Brassica* vegetables are cultivated intensively, cultural practices or chemical treatments either have not been successful in keeping these crops healthy, or are too expensive. Therefore, the development of resistant cultivars is the most efficient way to control clubroot for all *Brassica* crops. Breeding of clubroot-resistant varieties is an important goal in cole crop improvement.

Resistance to clubroot has been reported in different cruciferous species, including the three most commonly cultivated *Brassica* species: *B. napus*, *B. rapa* and *B. oleracea* (Crute et al. 1980). Although several sources of resistance have been identified in *B. oleracea* (Crute et al. 1980; Crisp et al. 1989; Monteiro and Williams 1989; Dias et al. 1993; Manzanares-Dauleux et al. 2000), few breeding programs for resistance have been successful. Clubroot resistance in *B. oleracea* is frequently incomplete, rarely expressed at a high level and is largely carried by materials unsuitable for production purposes (Crute et al. 1983). Precise genetic control of resistance to clubroot in the various crops of *B. oleracea* is still unclear and information is relatively limited. In *B. oleracea* this character appears to be determined by several genes with either recessive (Weisaeth 1974; Crute et al. 1983; Voorrips and Visser 1993) or dominant alleles (Laurens and Thomas 1993).

The linkage of clubroot resistance genes with genetic markers in *B. oleracea* was first reported by Landry et al. (1992) (Table 5). Two QTLs for resistance to race 2 of *P. brassicae* were identified. In the case of a cabbage resistant line, alleles for resistance of two QTLs appeared to be dominant over susceptibility and the two QTLs accounted for 61% of the total variation for clubroot resistance.

Similarly, Figdore et al. (1993) detected three QTLs in an F_2 population obtained from a cross between a *P. brassicae*-resistant to race 7 broccoli and a *P. brassicae*-susceptible cauliflower (Table 5). One dominant QTL came from the broccoli line, one QTL came from the cauliflower line and the third was due to a heterozygous state.

The resistance to *P. brassicae* Woronin in an F_2 population of a cross between a clubroot-resistant kale (*B. oleracea* var. *acephala*) and a susceptible cauliflower (*B. oleracea* var. *botrytis*) was examined by Grandclement and Thomas (1996) (Table 5). QTL detection was performed with RAPD markers. Two resistant notations, carried out at different times after inoculation, were used. Three markers were associated with these two notations and three were specific to only one notation. QTL analysis suggested the existence of at least two genetic mechanisms implicated in the resistance phenomenon.

More precise procedures for locating QTLs and estimating their effects were developed by Ramsay et al. (1996). This alternative approach is to use molecular markers to introgress small, defined regions of a chromosome from a donor line into a recurrent recipient line by systematic backcrossing and inbreeding (Fig. 5, Table 5). They described the construction of a library of backcross lines in *B. oleracea* and proved that for *B. oleracea*, the production of some recombinant backcross lines containing a single introgressed tract is possible in the BC_2S_1 generation.

Resistance in the DH line population, obtained from a cross between two DH lines, derived from cabbage and broccoli, was determined by Voorrips et al. (1997) in two ways: by assigning symptom grades to each plant, and by measuring fresh weights of healthy and affected parts of the root system of each plant (Table 5). Using a multiple QTL mapping approach to analyze the fresh weight data, they found two loci for clubroot resistance; these were designated *pb-3* and *pb-4*. The additive effects of these loci were responsible for 68% of the difference between the parents and for 60% of the genetic variance among DH line means. Also, indications for the presence of two additional, minor QTLs were found. Analysis of symptom grades revealed the two QTLs, *pb-3* and *pb-4*, as well as one of the two minor QTLs indicated by the analysis of the fresh weight data. In field evaluations under natural infection, Moriguchi et al. (1999) detected only one QTL controlling clubroot resistance.

Rocherieux et al. (2004) reported on the genetic analysis of resistance to clubroot from C10, an inbred line selected from a French kale landrace showing a high level of resistance to several pathotypes of *P. brassicae* (Manzanares-Dauleux et al. 2000) (Table 5). The location, effects and specificity of loci involved in the control of five different *P. brassicae* isolates (four SSI and one field isolate) were compared in a $F_{2/3}$ population from a cross between the resistant line C10 and a highly susceptible broccoli (HDEM). Of the nine QTLs detected, one was common to all the isolates and the others were specific to one, two or three isolates. Depending on the isolates, the magnitude of the effect of both the common and isolate-specific QTLs was variable.

8.4.2 Transformation and Plant Regeneration

Agrobacterium-mediated transformation is the most widely used method to deliver transgenes into *Brassica* species (reviewed by Poulsen 1996; Puddephat et al. 1996). Efficiency of transformation is limited by genetic constrains imposed either by the plant or bacterial genome. The study by Cogan et al. (2002) (Table 5) is the first report of plant genes that regulate this process, except the strong evidence for a plant genotypic basis to transformation (Gelvin 2000). They identified seven QTLs for transgenic (GFP, green fluorescent protein) and adventitious root production using an *Agrobacterium rhizogenes*-mediated cotransformation system in conjunction with a *Brassica oleracea* doubled haploid mapping population.

Table 5. A summary of QTL mapping efforts for *B. oleracea*

Target trait	Population used	Location	Seasons of study	Marker type	Strategy	Software	QTL	Major/ Minor	Gene action	% Genetic variation; Allele effect	LG/flanking markers with distances (cM)	Authors
Resistance to race 2 of *Plasmodiophora brassicae* Leaf morphology (LMOR)	78 F_3 *B. oleracea* var. *capitata* resistant No. 86-16-5 × rapid-cycling *B. oleracea* CrGC No. 85	Growth cabinet	43 Days after inoculation	RFLP	Interval mapping	Mapmaker/QTL	CR2A CR2B LMOR	Major Major		D (58) D (15)	LG6, 2NF11-2ND3 (22); LG1, 3NE4a-3ND3 (12) LG3, 3ND12, 21	Landry et al. (1992)
Resistance to race 7 of *Plasmodiophora brassicae* (R)	180 F_2 clubroot resistant broccoli *B. oleracea* var. *italica* CR7 × susceptible cauliflower *B. oleracea* var. *botrytis* "Early White"	Greenhouse	35 Days after inoculation	RFLP	Chi-square, R-square and F-tests		R	Major Minor Minor Minor		12.5 7.5 10.2	LG1, 14a LG1, 34-75a (35) LG4, 48 LG9, 177b	Figdore et al. (1993)
Head type (H)							H	Major Minor Minor		16.6 7.3 – 9.8 9.9 – 13.9	LG9, 2 LG3, 96a-188(4) LG9, 2-37a(19)	
Head color (C)							C			12.5 16.2 13.9	LG3, 96a-188(4) LG4,71-59(25) LG9,79-37a(39)	

Gene action: when the effect of the QTL allele followed a Mendelian model, it could either be additive (A), dominant (D), recessive (R), partial dominant (PD), over dominant (OD);
[a] linkage groups on the map published by Sebastian et al. (2000);
[b] linkage groups on the map published by Saal et al. (2001);
[c] separate analysis of the data obtained from block 1 of field trial;
[d] separate analysis of the data obtained from block 2 of field trial; QTL (Sebastian et al. 2002): lamina width (LW), lamina petiole length (LPL), bare petiole length (BPL), auricle petiole length (APL), wing petiole length (WPL), leaf apex shape (LAS), midrib width (MW), lobe number (LN), wing number (WN), leaf type (LT), leaf transition zone (LTZ), vernalization (VN), node number (NN), leaf number (LN), axillary buds (AB), plant height (PH), stem length (SL).

Table 5. (continued)

Target trait	Population used	Location	Seasons of study	Marker type	Strategy	Software	QTL	Major/Minor	Gene action	% Genetic variation; Allele effect	LG/flanking markers with distances (cM)	Authors
Head extrusion (E)							E			12.2 7.6	LG8,6a-82a(6) LG1,64	
Heading data (maturity M)							M			12.7; 15.5 6.6–12.8 4.0–9.4 4.7	LG1, 19-14a(72) LG4, 71-59 (25) LG5, 210b-118(40) LG6, 110	
Resistance to *Plasmodiophora brassicae* Wor.	90 F_2 clubroot resistant kale *B. oleracea* var. *acephala* C10 × susceptible cauliflower *B. oleracea* var. *botrytis* 48.4.7	Greenhouse	6 Weeks 2 years	RAPD	Student's t-test Kruskal and Wallis test One way ANOVA			Major Minor			OPA4-700, OPA16-510; OPB11-740 OPA1-1880 OPL6-780 OPA18-1490 OPE20-1250	Grand-clement and Thomas (1996)
Developmental traits:	70–90 F_2	Greenhouse	37–97 Days	RFLP	Single factor	SAS GLM procedure				6.7–42.7%		Kennard
	cabbage		after sowing		ANOVA	SAS FREQ				(single		et al.
Flowering traits:	*B. oleracea*				Chi-square	SAS UNIVARIATE			–	marker	2C, 152 (18)	(1994)
Annual/biennial	var. *capitata*				tests	SAS CORR			PD	loci),	5C, 67a- 4a (27)	
	"Wisconsin				Pearson				PD	60.1%	7C, 179-131b (35)	
Days to bud	Golden Acre" ×				correlation				A	(multilocus	1C, 29b (78)	
	broccoli				coefficients				A	model)	2C, 152-165 (42)	
	B. oleracea var.								D		3C, 133-92b (21)	
	italica "Packman"								A		7C, 148-131b (53)	
									D		9C, 43b	
Days to flower									–		1C, 84-93a (12)	
									–		1C, 29b (78)	
									–		2C, 152-165 (42)	
									–		3C, 133-113 (3)	
									–		3C, 92b (61)	
									–		7C, 148-131b (53)	
									–		7C, 25b (93)	

Table 5. (continued)

Target trait	Population used	Location	Seasons of study	Marker type	Strategy	Software	QTL	Major/ Minor	Gene action	% Genetic variation; Allele effect	LG/flanking markers with distances (cM)	Authors
Days from bud								–		3C, 133 (40)		
to flower								–		8C, 170 (37)		
Size of cluster								–		2C, 165 (60)		
								–	12.0	4C, 26 (47)		
								–	8.0	5C, 97a (25)		
								–	17.9	9C, 43b		
									19.5			
Leaf traits:									PD	10.3	4C, 10b-26 (15)	
Lamina length									D	11.3	5C, 67[a]	
									PD	11.1	6C, 30	
									A	12.1	6C, 111	
Lamina width									A	15.6	1C, 61a	
									OD	12.1	4C, 128	
									OD	13.5	5C, 174	
									OD	11.6	7C, 25b	
Petiole length									D-PD	16.9	4C, 10b-26 (25)	
									A	19.4	5C, 92a	
Basal petiole width									OD	18.1	7C, 25b	
Basal petiole thickness									PD	6.7	4C, 10b-26 (25)	
									A	7.4	6C, 111	
									D	13.9	4C, 10b	
Leaf length									PD	9.5	4C, 26	
									PD	14.2	5C, 67a	
									OD	15.4	5C, 92a	
									PD	11.5	6C, 30-111	
Lamina length/width									A	7.1	1C, 61a	
									A	8.1	4C, 128	
									PD		4C, 10b	
									D		4C, 26	
									PD		6C, 30	

Table 5. (continued)

Target trait	Population used	Location	Seasons of study	Marker type	Strategy	Software	QTL	Major/ Minor	Gene action	% Genetic variation; Allele effect	LG/flanking markers with distances (cM)	Authors
Lamina/ petiole length									A		5C, 92a	
Stem traits:									A		2C, 8(41)	
Distance between second and third node											3C, 133-95 (35)	
											5C, 97a-9a (43)	
									A		8C, 170-168 (28)	
Distance between third and fourth node											3C, 23-95 (48)	
											5C, 9a	
											7C, 179	
									A		8C, 115-143 (10)	
Distance of twenty internodes									OD		1C, 29b	
									A		2C, 83b	
									A		3C, 23-95 (35)	
									OD		4C, 26	
											9C, 153a	
Resistance to *Plasmodiophora brassicae* Wor.	91 DH-lines cabbage *B. oleracea* var. *capitata* "Bindsachsener" × *B. oleracea* var. *italica* "Greenia"	Phytotron chamber	6 Weeks	RFLP AFLP	Multiple-QTL Model (MQM), Kruskal and Wallis test	GENSTAT5 MAPQTL 3.0	*pb-3*	Major	–	54%	LG3: 4NE11a	Voorrips et al. (1997)
							pb-4	Minor		6%	LG1: 2NA8c	
								Minor			LG6: 2NA1b-A20	
Resistance to *Plasmo-diophora*	Ms6 266 $F_{2/3}$ resistant kale *B. oleracea* var.	Greenhouse	6–8 Weeks	RAPD RFLP ACGM	Interval mapping (IM),	PROC GLM QTL Cartographer	*Pb-Bo1*			34.0	LG1, T2, 20.41	Rochérieux et al. (2004)
							Pb-Bo2			6.8	LG2, S07.1900, 85.41	
							Pb-Bo5a			4.3	LG5, PBB7b, 0.01	

Table 5. (continued)

Target trait		Population used	Location	Seasons of study	Marker type	Strategy	Software	QTL	Major/ Minor	Gene action	% Genetic variation; Allele effect	LG/flanking markers with distances (cM)	Authors
brassicae Wor.		*acephala* C10 ×			(am-	composite		*Pb-Bo5b*			3.0	LG5, ab1.1350, 116.81	
pathotype 1, 2,		susceptible			plified	interval		*Pb-Bo9b*			3.4	LG9, a04.1900, 44.78	
4 and 7 (five	K92	broccoli			con-	mapping		*Pb-Bo1*	Major		77.4	LG1, T2, 20.41	
isolates		*B. oleracea* var.			sensus	(CIM)		*Pb-Bo2*			2.5	LG2, PBB38a, 66.01	
described		*italica* HDEM			genetic	ANOVA		*Pb-Bo8*			2.1	LG8, c01.980, 36.49	
near)	Pb137-				mark-	Pearson		*Pb-Bo1*	Major		80.7	LG1, T2, 20.41	
	522				ers)	coefficient		*Pb-Bo3*			4.7	LG3, aa7.1400, 144.46	
	K92-							*Pb-Bo1*	Major		68	LG1, T2, 20.41	
	16							*Pb-Bo4*			2.8	LG4, aa9.983, 112.89	
								Pb-Bo5a			3.3	LG5, s07.200, 29.27	
								Pb-Bo1			20.7	LG1, ae05.800, 0.01	
	eH							*Pb-Bo2*			21.0	LG2, r10.1200, 88.21	
								Pb-Bo5a			4.2	LG5, PBB7b, 0.01	
								Pb-Bo5b			6.1	LG5, a18.1400, 118.19	
								Pb-Bo9a			5.5	LG9, aj16.570, 10.01	
Resistance to *Plasmodiophora brassicae* Wor. pathotype 1 and 3		138 F_2 cabbage *B. oleracea* var. *capitata* × kale *B. oleracea* var. *acephala*	Infested field	60 Days after trans-planting	RAPD RFLP Isozyme Morpho-logical	Interval mapping	QGENE	–	–	–	30%	LG3: WG6A1-WG1G5 (2.6 cM)	Moriguchi et al. (1999)
Agrobacterium rhizogenes-mediated transformation efficiency		73 DH lines *B. oleracea* var. *alboglabra* (A12DHd) × *B. oleracea* var. *italica* (GDDH33)	Phytotron chamber	35 Days after inoculation	RFLP AFLP micro-satelite	Marker regression, interval mapping ANOVA	QTL CAFE http://web.bham.ac.uk/g.g.seaton/	–	–	–		GFP-fluorescent root production:	Cogan et al. (2002)
											8.8%	LG1, 100 cM	
											5.4%	LG3, 85 cM	
											4.3%	LG7, 62 cM	
												Adventitious root production:	
											29.9%,	LG3, 34 cM and 92 cM	
											8.3%	LG5, 38 cM	
											7.8%	LG7, 66 cM	

Table 5. (continued)

Target trait	Population used	Location	Seasons of study	Marker type	Strategy	Software	QTL	Major/ Minor	Gene action	% Genetic variation; Allele effect	LG/flanking markers with distances (cM)	Authors
Agrobacterium tumefaciens susceptibility	59 DH lines *B. oleracea* var. *alboglabra* (A12DHd) × *B. oleracea* var. *italica* (GDDH33) and 5 substitution lines A12DHd with GDDH33 regions	Phytotron chamber	50 Days after inoculation	RFLP	Interval mapping	MAPQTL v 4.0 Genstat v 5.0	–	–	–	–	LG O9, 24.6-67.9 cM; pW233-40.1 cM	Sparrow et al. (2002)
Regeneration from protoplasts	128 F_2 broccoli *B. oleracea* var. *italica* × rapid-cycling *B. oleracea* (RC)	Phytotron chamber	5, 10 and 15 Weeks after transfer of microcalli	AFLP micro-satelite	Interval mapping, modified bulked segregant analysis	PlabQTL PROC GLM SAS PROC CORR	–	–	–	82.8% (total)	LG O2[a]/LG C8[b], 16 cM LG O9[a]/LG C7[b], 11cM	Holme et al. (2004)
Developmental characteristics Leaf traits	86 DH lines cauliflower *B. oleracea* var. *botrytis* × Brussel sprout *B. oleracea* var. *gemnifera*	Field	78, 142 and 162 Days after sowing	RFLP AFLP SSR	Marker regression ANOVA	http://web. bham.ac.uk/ g.g.seaton/	LW			14.5% 32.4% 44.5%	LG O6, 16 cM LG O7, 16 cM LG O8, 28 cM LG O8, 80 cM	Sebastian et al. (2002)
							LPL			16.9 10.2 10.6	LG O3, 44 cM LG O6, 16 cM LG O7, 26 cM	
							BPL			45.4 11.3	LG O1, 28 cM LG O1, 70 cM LG O2, 6 cM	
							APL			14.3	LG O1, 28 cM	
							WPL			25.6	LG O1, 0 cM	
							LAS			37.8	LG O3, 94 cM	
							MW			10.5	LG O2, 30 cM	
							LN			26.0	LG O6, 4 cM	

Table 5. (continued)

Target trait	Population used	Location	Seasons of study	Marker type	Strategy	Software	QTL	Major/ Minor	Gene action	% Genetic variation; Allele effect	LG/flanking markers with distances (cM)	Authors
							WN			17.7	LG O8, 4 cM	
							LT			8.3	LG O1, 32 cM	
										11.0	LG O5, 28 cM	
							LTZ			11.9	LG O5, 46 cM	
Flowering traits							VN			14.1	LG O7, 18 cM	
										23.0	LG O8, 24 cM	
Axillary bud traits							NN[c]			42.9	LG O8, 22 cM	
							NL[c]			12.8	LG O4, 6 cM	
										14.1	LG O7, 56 cM	
							NN[d]			17.9	LG O8, 14 cM	
										10.9	LG O9, 62 cM	
							AB[d]			10.8	LG O1, 22 cM	
Stem traits							PH			9.1	LG O7, 22 cM	
							SL			20.6	LG O1, 34 cM	
											LG O1, 68 cM	
										6.6	LG O7, 48 cM	
										11.3	LG O8, 24 cM	
Curd-related traits	247 F_2	Field	180 Days	RFLP	Interval	SAS and	dfb		D	20.9	BK, C4, EW4D04w+6	Lan and
Days to	rapid-cycling		after planting		mapping,	Microsoft			A	39.6	BK, C7, EW8C11a+2	Peterson
budding (dfb)	*B. oleracea*		on the first		multiple	Excel			DA	11.0	BK, C1, EST429a+7	(2000)
	(RCB) ×		flowering day		QTL model	MapmakerQTL			DA	1.2	BK, C7, EST151c+0	
	B. oleracea var.								DA	9.9	CAN, C3, EW8F03a+5	
	Cantanese (CAN),								D	65.3	CAN, C4, EW9E10+12	
	250 F_2 RCB ×								DA	6.5	CAN, C5, EST195b+3	
	B. oleracea var.								D	87.6	CAN, C7, WG3D11+14	
	Pusa Katki (PK),								D	44.4	PK, C3, EST411c+0	
	246 F_2RCB ×								A	44.1	PK, C4, EST114a+1	
	B. oleracea var.								A	13.9	PK, C7, EW6B07a+3	
	Bugh Kana (BK)								DA	5.9	PK, C8, EW8D10b+0	
									D	21.9	PK, C9, EST131a+17	
									DA	27.9	PK, C4, EW9B02X+1	
									A	10.0	BK, C4, EW2B12s+4	

Table 5. (continued)

Target trait	Population used	Location	Seasons of study	Marker type	Strategy	Software	QTL	Major/ Minor	Gene action	% Genetic variation; Allele effect	LG/flanking markers with distances (cM)	Authors
Days from budding to flowering (dofdbf)								dofdbf	A	39.7	BK, C7, EW8C11a+1	
									D	7.8	BK, C8, EW5G04b+10	
									A	15.9	CAN, C4, EW9E10+12	
									DA	9.4	CAN, C5, EST195b+3	
									RA	43.8	CAN, C7, EW8C11a+4	
									RA	7.3	CAN, C8, EW8E09c+2	
									DA	4.6	CAN, C3, EW8F03a+1	
									A	11.6	CAN, C4, EST453g+3	
									D	63.6	PK, C3, EST411c+0	
									A	76.6	PK, C4, EST373a+16	
									A	11.4	PK, C7, EW7E01a+1	
									DA	7.4	PK, C8, EW8D10b+1	
									D	24.6	PK, C9, EST131a+13	
									R	5.5	PK, C5, EW8A11a+2	
First-rank branching (rk1)								hrk1	D	15.8	BK, C7, WG3D11+11	
									DA	8.2	CAN, C1, EST566a+2	
									DA	7.2	CAN, C3, EW8F03a+0	
									A	14.3	CAN, C4, EST453g+3	
									DA	5.9	CAN, C5, EW8F11c+0	
									DA	27.2	CAN, C7, WG3D11+8	
									RA	7.6	CAN, C8, EW8E09c+0	
									A	34.8	PK, C4, EST122b+6	
									R	7.6	PK, C5, EW8A11a+0	
									RA	9.2	PK, C6, WR1D12a+0	
									DA	7.9	PK, C7, EW6B07a+3	
									RA	6.2	PK, C1, EST217a+3	
									DA	6.5	PK, C9, EST131a+14	
Side-branching (sidb)								sidb	A	45.9	BK, C7, EW8C11a+1	
									RA	13.6	CAN, C4, EST55b+16	
									RA	30.4	CAN, C7, EW8C11a+5	
									RA	8.1	CAN, C8, EW7C10b+7	

Table 5. (continued)

Target trait	Population used	Location	Seasons of study	Marker type	Strategy	Software	QTL	Major/ Minor	Gene action	% Genetic variation; Allele effect	LG/flanking markers with distances (cM)	Authors
									A	37.8	PK, C4, EST429E+1	
									R	11.4	PK, C5, EW8A11a+1	
									DA	12.5	PK, C7, EW8F06+0	
									DA	3.8	PK, C9, EW1G03+0	
Cluster width (cluw)								cluw	DA	6.1	CAN, C1, EW5H03+0	
									RA	5.9	CAN, C2, EW9F06a+0	
									RA	8.0	CAN, C4, EW9E10+0	
									D	6.9	PK, C1, EST125+0	
									D	33.4	PK, C3, EST411c+0	
									RA	5.6	PK, C4, EW9B02x+0	
Curd width (headw)								headw	DA	14.9	BK, C4, EW4D04w+7	
									DA	10.5	BK. C7. EW8C11a+4	
									DA	11.3	CAN. C1, EW7C10a+0	
									A	11.7	CAN, C4, EST453g+4	
									DA	7.8	CAN, C5, EW7E08a+1	
									A	22.1	CAN, C7, WG3D11+6	
									RA	5.1	CAN, C9, EW1G03a+1	
									DA	3.7	CAN, C3, EST22b+4	
									?	6.2	CAN, C8, EW8E09c+0	
									D	51.2	PK, C3, EST411c+0	
									A	54.3	PK, C4, EST122b+4	
									R	7.6	PK, C5, EW8A11a+0	
									DA	6.4	PK, C7, EW5A12a+0	
									A	4.3	PK, C9, EW8E09b+2	
Apical shoot length (rrl1st)								rrl1st	A	22.7	BK, C1, EW2E07a+12	
									RA	11.3	BK, C4, EW2B12s+10	
									R	15.2	BK, C7, EW8C11a+1	
									DA	7.5	BK, C7, EW6C11a=0	
									D	9.1	CAN, C3, EW9E05a+0	
									DA	11.0	CAN, C4, EST453g+0	
									RA	7.6	CAN, C5, EST195b+7	

Table 5. (continued)

Target trait	Population used	Location	Seasons of study	Marker type	Strategy	Software	QTL	Major/Minor	Gene action	% Genetic variation; Allele effect	LG/flanking markers with distances (cM)	Authors
									D	6.8	CAN, C7, EW6B07a+0	
									RA	12.8	CAN, C9, EW1G03a+1	
									?	11.0	PK, C1, EW8C11x+3	
									R	7.9	PK, C1, WST88c+0	
First branch								rrl2nd	DA	25.5	BK. C1. EW2E07a+15	
length (rrl2nd)									RA	15.4	BK. C4, EW4D04w+23	
									RA	15.6	BK, C7, EW8C11a+2	
									?	7.5	BK. C7. EW3A04b+23	
									RA	4.3	BK. C7, EW5C12f+0	
Plant size	247 F_2	Field	Until the	RFLP	Interval	SAS and	Laml		?	15.0	BK, C4, EW4D04w+43	Lan and
Lamina length	rapid-cycling		data of		mapping,	Microsoft			A	33.0	BK, C7, EW8C11a+0	Paterson
	B. oleracea		appearance		multiple	Excel			R	4.5	BK, C6, EW2C08a+14	(2001)
	(RCB) ×		of the		QTL model	MapmakerQTL			RA	3.6	BK, C9, EST131a+0	
	B. oleracea var.		first flower						RA	13.7	CAN, C4, WG3F04b+4	
	Cantanese (CAN),								RA	30.3	CAN, C7, EW8C11a+4	
	250 F_2 RCB ×								DA	4.0	CAN, C1, EW6F02+1	
Lamina width	*B. oleracea* var.						Lamw		DA	20.0	BK, C1, EW2E07a+16	
	Pusa Katki (PK),								RA	32.4	BK, C7, EW8C11a+0	
	246 F_2RCB ×								RA	9.6	CAN, C4, EST453g+2	
	B. oleracea var.								A	36.7	CAN, C7, EW8C11a+4	
	Bugh Kana (BK)								R	9.3	CAN, C8, EW809c+0	
Petiole length							Petl		A	19.5	BK, C7, EW8C11a+2	
									A	12.9	CAN, C4, EW7B02+2	
									RA	25.7	CAN, C7, EW8C11a+2	
									?	1.5	CAN, C7, EW3A04b+3	
									?	7.3	CAN, C8, WR2F06x+7	
Stem length							Stl		DA	11.1	CAN, C1, EST125+9	
									DA	6.8	CAN, C1, EW5H03+0	
									DA	7.0	CAN, C3, EST165d+0	
									DA	10.8	CAN, C5, EST195b+8	
									RA	9.7	CAN, C6, EW8A09+6	

Table 5. (continued)

Target trait	Population used	Location	Seasons of study	Marker type	Strategy	Software	QTL	Major/ Minor	Gene action	% Genetic variation; Allele effect	LG/flanking markers with distances (cM)	Authors
									D	22.1	CAN, C7, WG3D11+12	
									DA	14.3	PK, C1, EST373c+4	
									DA	5.9	PK, C3, EW5F09a+6	
									A	11.3	PK, C4, EST122b+2	
									RA	9.5	PK, C6, WR1D12a+0	
Stem width							Stw		D	18.7	BK, C2, EW7B04c+16	
									DA	7.1	BK, C4, EW2B12s+0	
									A	32.2	BK, C7, EW8C11a+2	
									?	5.6	BK, C8, EW5G04b+13	
									A	6.9	BK, C9, EW6B07d+0	
									?	5.7	BK, C6, EWc208a+2	
									DA	11.7	CAN, C4, EST55b+14	
									D	11.0	CAN, C5, EST195b+10	
									A	33.3	CAN, C7, Ew8C11a+2	
									A	32.5	PK, C4, EST133b+7	
									RA	8.0	PK, C7, EW5A12a+0	
									DA	12.4	PK, C9, EST131A+15	
									RA	5.3	PK, C5, EST453d+0	
Node number							Non		RA	13.7	BK, C7, EW8C11a+0	
									DA	13.7	CAN, C3, EW8F03a+3	
									RA	16.5	CAN, C7, EW8C11a+4	
									RA	5.8	CAN, C4, EST55b+6	
									?	4.3	CAN, C6, EW5C05c+0	
									DA	17.7	PK, C4, EW9E10+4	
									DA	6.1	PK, C9, EW8E11a+0	
Flowering time (FT)	92 F_3 families cabbage	Field	15 surveys at 5-15 days intervals	RFLP	Interval mapping	MapMaker/ QTL 1.1	FT		AR	19.7	LG2, *ec2h2-slg6*	Camargo and Osborn (1996)
	(*B. oleracea*								AD	29.9	LG6, *wg6f10-wg7g10*	
	var. *capitata*) ×				single-factor ANOVA	Minitab's GLM procedure	PF		AD	19.2	LG8, *ec3b2-ec3e10*	
Annual/biennial (PF)	broccoli								AR	2.9	LG2, *ec2h2-slg6*	
	(*B. oleracea* var. *italica*)								AD	3.4	LG6, *wg7g10-wg3g9a*	

Table 5. (continued)

Target trait	Population used	Location	Seasons of study	Marker type	Strategy	Software	QTL	Major/ Minor	Gene action	% Genetic variation; Allele effect	LG/flanking markers with distances (cM)	Authors
Flowering time	149 DH lines *B. oleracea* var. *alboglabra* (A12DHd) × *B. oleracea* var. *italica* (GDDH33)	Field (two trials)	Until the first flower	RFLP	Marker regression ANOVA SIM function Interval mapping	http://web.bham.ac.uk/g.g.seaton/ MQTL Mapmaker/ QTL	FTO2.1			9 3	O2, 78±9 cM	Bohuon et al. (1998)
							FTO3.1			8 32	O3, 12 cM	
							FTO5.1			8 16	O5, 16±13 cM	
							FTO5.2			– –	O5, 40±10 cM	
							FTO9.1			47 22	O9, 36-46 cM	
							FTO9.2			11 37	O9, 74-94 cM	
Flowering time Late flowering (LF) Early flowering (EF)	79 Substitution lines *B. oleracea* A12DHd × GDDH33	Field (eight trials)	Until the first flower	RFLP AFLP	ANOVA Tukey-Kramer multiple comparison tests		LF				O1, 0-30.3 cM	Rae et al. (1999)
							EF				O1, 30.3-38.1 cM	
							LF				O2, 65-80 cM	
							LF1				O3, 0-35 cM	
							LF2				O3, 35-75 cM	
							LF				O5, 0-34.2 cM	
							LF1				O9, 0.0-43.0 cM	
							LF2				O9, 43.0-64.4 cM	
							LF3				O9, 70.8-106.8 cM	
							EF				O9, 23.1-47.0 cM	
Flowering time	60 F_2 *B. juncea*; 60 F_2 *B. oleracea*; 95 F_2 *B. rapa*	Greenhouse	Until the first fully expanded flower	RFLP	Interval mapping Two-factor analysis of variance	MapMaker/ QTL 1.1 JMP 3.02				14	J2, 43 cM long	Axelsson et al. (2001)
										36	J3, 33 cM	
											J10, 25 cM	
											J12, 27 cM	
											J15, 27 cM	
										14	J18, 20 cM	
											O2, 25 cM	
										13	O3, 13 cM	
										13	O9, 18 cM	
										21	R2, 43 cM	
										23	R3, 18 cM	
											R10, 14 cM	

Table 5. (continued)

Target trait	Population used	Location	Seasons of study	Marker type	Strategy	Software	QTL	Major/ Minor	Gene action	% Genetic variation; Allele effect	LG/flanking markers with distances (cM)	Authors
Germination:	105		Time taken	RFLP	ANOVA	Genstat 5				13.7	O1, AC-CACJ01, 118	Bettey et al.
Unstressed MGT	*B. oleracea*		by shoots to	AFLP	Interval	REML					O3, AC-CAAE02 , 84.2	(2000)
	DH lines		reach 1.5 cm	micro-	mapping	MapQTL 3.0					O5, pN23E2, 5.0	
	A12DHd ×		and by roots	satelites	Kruskal-	QTL Java					O6, pO10E2, 72.6	
Water stress	GDDH33		to reach 4 cm;		Wallis test	Applet V0.70					O3, pW102J1, 88.2	
MGT			up to 15 days		Marker	(http://web.					O9, pO160E1, 71.2	
Heat treatment					regression	bham.ac.uk/				14.5	O6, AC-CATE18, 32.9	
Conductivity						g.g.seaton/					O9, pN3E1, 107.8	
						qtl_ns/					O6, pR64E2, 11.2	
Seedling growth :						brass.html)					O9, AC-CAGJ05, 11.2	
Seed weight											O9, pO160E1, 71.2	
											O4, pO147e1N, 20.8	
Root weight											O9, AC-CAGJ05, 47.3	
										16.4	O6, pR64E2, 11.2	
Shoot weight											O9, AC-CAGJ05, 47.3	
											O1, AC-CACE08, 93.9	
Shoot: root weight											O4, pW188E2N, 37	
ratio											O6, AA-CATJ08, 20.2	
										13.7	O3, LEW6E1, 16.3	
Root growth rate											O6, pR64E2, 11.2	
											O7, pr36E2, 5.9	
											O7, pN64E2, 37	
Shoot growth rate										16.6	O9, AC-CAGJ05, 47.3	
Root length											O3, pN213J2, 61.7	
Shoot length											O9, pC14 (cos), 88	
Shoot: root length											O7, mBN72AJ1, 67.3	
ratio												

Together, they accounted for 26% and 32% of the genetic variation in population, respectively. Two of the QTL regions identified were common to both transgenic and adventitious root production. Regions of the genome that contain enhancing QTLs associated with transgenic root production originate from both parents. Two different methods for QTL analysis were employed (marker regression and interval mapping) and with one exception (the region on linkage group O7 for transgenic root production), both techniques detected the same regions. This work is significant for two reasons: for the first time, the QTL mapping approach was used to identify regions associated with transgenic and adventitious root production and also for the first time the identified regions seem to coregulate these two processes.

The aim of the study performed by Sparrow et al. (2004) was to identify the inheritance patterns and genetic loci associated with susceptibility to *A. tumefaciens* in *B. oleracea* (Table 5). The genetic control and susceptibility was investigated using a double haploid mapping population of *B. oleracea* and the associated RFLP map. Preliminary studies were carried out by the analysis of an 8 × 8 diallel, for which the parental lines were selected to include a range of susceptibilities to *A. tumefaciens*. This analysis showed that both additive and dominant gene effects control this process, with additive gene effects being more important. They identified a highly significant QTL associated with susceptibility to *A. tumefaciens* on linkage group O9. The use of substitution lines covering this region confirmed the localization of this QTL within the central 38% of LG O9. The alignment of linkage groups with chromosomes of *B. oleracea* (Howell et al. 2002) suggested that genes associated with susceptibility to *A. tumefaciens* are located on chromosome 3 of the *B. oleracea* cytogenetic map.

Brassica oleracea contains many commercially important vegetables, which are becoming increasingly popular due to their nutritional value and their anticarcinogenic properties (Beecher 1994). Although both quality and yield are currently improved by conventional breeding, further improvement of traits, such as disease resistance, delayed postharvest senescence and yield, are limited by the availability of suitable genes in the germplasm stocks (Metz 2001; Gapper et al. 2002; Radchuk et al. 2002). Approaches such as protoplast fusion and genetic transformation are now available for the introduction of new genes. A major obstacle to a wide commercial application of these methods in *B. oleracea* is the lack of elite breeding lines with a high capacity for plant regeneration from in vitro cultures. Much effort has been put into improving the culture conditions for plant regeneration, but the genotype of the plant material still has a strong effect on regeneration of *B. oleracea* cell culture systems (Jourdan and Earle 1989; Loudon et al. 1989; Zhao et al. 1995). In the *B. oleracea* genetic study of plant regeneration from protoplast-derived microcalli of F_2 progenies from a cross between two parents with high and low regeneration ability was previously published (Hansen et al. 1999). Results indicated that the plant regeneration ability is controlled by two or three genes.

In their study Holme et al. (2004) (Table 5) used the same plant material and the corresponding evaluation data to identify two major QTLs contributing to the regeneration ability from microcalli of *B. oleracea* protoplasts and locate them to linkage groups O2 and O9 on the map published by Sebastian et al. (2000) (Fig. 8, Table 3) or to chromosomes C8 and C7 on the map published by Saal et al. (2001) (Table 3). In a multiple regression analysis, the two QTLs, both with additive effects from the high-responding parent, explained 83% of the total genetic variation for regeneration recorded 15 weeks after initial transfer of microcalli to regeneration medium. The regions where those QTLs were identified were found to be homologous to a region from the top of chromosome 5 of *A. thaliana* harboring a QTL for the total number of regenerated shoots from root explants and a QTL affecting flowering time (Bohuon et al. 1998) (Table 5). But in material used in this study, the association between flowering time and regeneration ability was not found.

8.4.3 Developmental Characteristics

Morphology

Most analyses of quantitative trait loci have been focused on locating QTLs for disease resistance and flowering time. But the vegetable *Brassica* (*B. oleracea*), which have the most diverse morphological forms, provide not only popular vegetables for human consumption, but also a unique opportunity for scientists who seek to understand the genetic basis of plant growth and development. Early inheritance studies focused on morphological traits, although complex

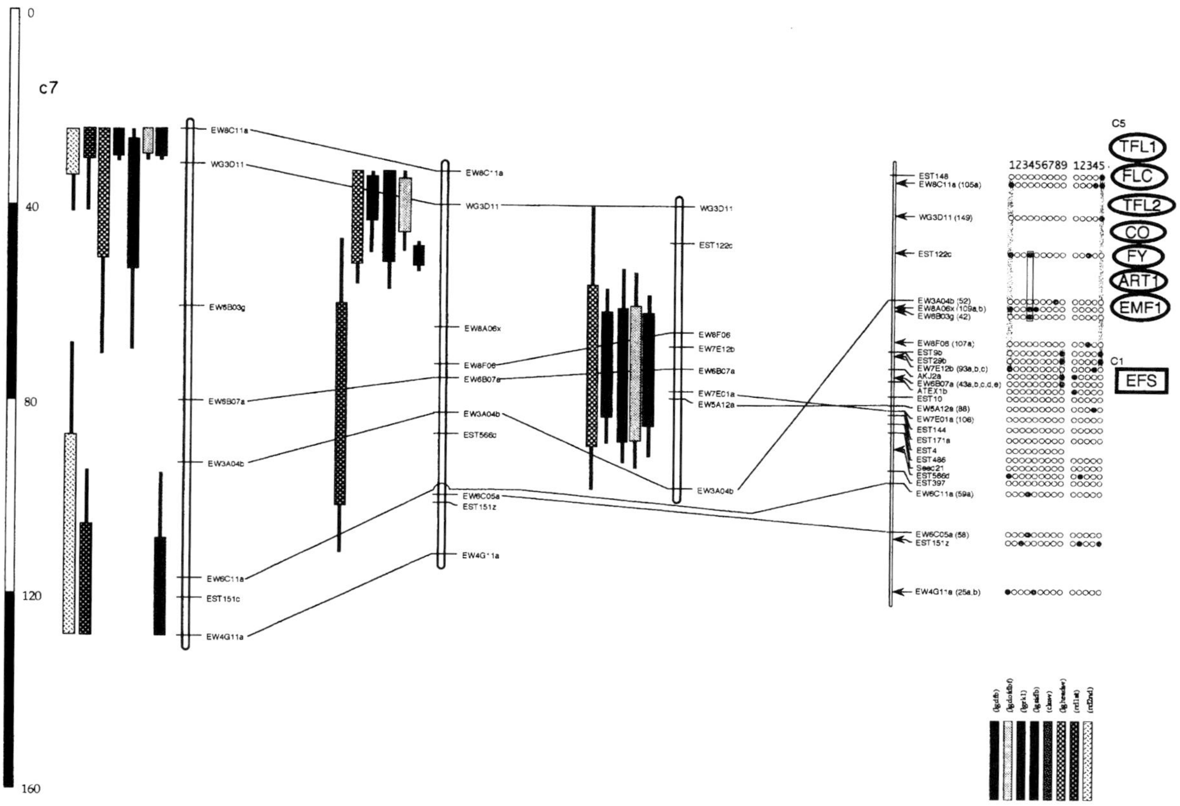

Fig. 10. QTL maps of curd-related traits, plotted on RFLP linkage maps of *B. oleracea* RCB × CAN, RCB × PK, and RCB × BK F_2 populations. Linkage maps of these three crosses used for QTL mapping are also aligned with a more detailed map of RCB × GC, with homoeologous locations in *Brassica* and homologous locations in *Arabidopsis* (as described in more detail elsewhere; Lan et al. 2000). The *solid circles* next to the loci indicate homoeologous *Brassica* loci (chromosomes 1–9, *near right*) or homologous *Arabidopsis* loci (chromosomes 1–5, *far right*) detected by the same probe. All *open circles* indicate that no polymorphism was detected for homoeologous (*Brassica*) or homologous (*Arabidopsis*) loci. The letter "R" next to the probe name indicates that the probe hybridizes to a repetitive DNA sequences in *Arabidopsis*. Chromosome segments that appear to be homoeologous (*Brassica*) or homologous (*Arabidopsis*) where connected by *shaded columns*. *Open columns* indicate possible triplicated (*Brassica*) or duplicated regions (*Arabidopsis*). The likelihood intervals of QTL associated with different phenotypes (as indicated by *fills*; see legend) are plotted as *bars* (1 LOD) and *whiskers* (2 LOD) (reprinted from Lan and Paterson 2000 with permission of Genetics Society of America)

inheritance was often observed, suggesting that the traits were controlled by many genes.

Kennard et al. (1994) (Table 5) attempted to resolve these complex inheritance patterns for a range of morphological traits using a single factor ANOVA on an F_2population of a cabbage × broccoli *B. oleracea* cross. They were able to detect and locate marker-locus associations for 16 of the 22 traits. In some cases, chromosomal regions correlated with different related traits had a common localization with similar gene action and genotypic class ranking. The numbers, action, and linkages of genes controlling traits estimated with marker loci in the population used corresponded to estimates based on classical genetic methods from other studies using similar or similarly wide crosses (Pease 1926; Pelofske and Bagget 1979).

Lan and Paterson (2000) (Table 5) investigated curd-related traits in three F_2 populations of *B. oleracea* derived from crosses of a common parent, the rapid-cycling *Brassica* (RCB), to three *B. oleracea* varieties, Cantonese (CAN), Pusa Katki (PK) and Bugh Kana (BK) (Fig. 10). To survey the genetic divergence that has accumulated during the course of evolution in the *Brassica* genus, they chose genotypes originating from different geographic areas: Italy, India and Thailand, and exhibiting different morphologies, to cross to RCB. A total of 86 QTLs were identified that control eight curd-related traits in *B. oleracea*.

These may reflect allelic variation in as many as 67 different genetic loci and 54 ancestral genes. These results proved that the enlarged curd of cauliflower is under complex genetic control (Watts 1964; Kalloo and Bergh 1993; Kennard et al. 1994). The observation that the BK population showed simpler genetic control of most traits supported the hypothesis that *B. oleracea* var. BK is a progenitor of cauliflower and broccoli. The level of diversity of QTLs among these three populations was also reflected in a comparison to another population in which similar traits have been mapped (Kennard et al. 1994) (Table 5). Nine QTLs corresponded to QTLs reported for these traits in this study. Comparative data helped to identify *Arabidopsis* mutations localized in region homologous to the regions where *Brassica* QTLs were found.

The three previously described *B. oleracea* F_2 populations (Lan and Paterson 2000) were used by the same authors (Lan and Paterson 2001) (Table 5) to detect quantitative trait loci (QTLs) influencing the size of leaves and stems using restriction fragment length polymorphisms (RFLPs). Through comparative mapping they inferred that the 47 QTLs identified might reflect variation in as few as 35 different genetic loci, and 28 ancestral genes, i.e., about one-fifth of the QTLs had possible homoeologs. These data reinforced the picture obtained based on QTLs for curd-related traits (Lan and Paterson 2000), and also previously published data (Kowalski et al. 1994) that the *Brassica* genome may be relatively rapidly evolving, perhaps facilitated by a high level of duplication (Lan et al. 2000). For some QTLs corresponding to ancestral genes they were able to identify putative *Arabidopsis* mutations, which confirm the possibility to use comparative data to point to candidate genes from *Arabidopsis*.

Sebastian et al. (2002) (Table 5) used a population of DH lines derived from a cross between two very different *B. oleracea* crop types, an annual cauliflower and biennial Brussels sprout, to detect and locate QTLs controlling 27 morphological and developmental traits, including leaf, flowering, axillary bud and stem characters. The same population was previously used to construct a detailed linkage map of *B. oleracea* genome based on RFLP, AFLP and simple sequence repeat (SSR) markers (Sebastian et al. 2000). Using the multiple-marker regression approach to QTL mapping, they were able to detect at least 32 QTLs, between zero to four per trait, which individually explained between 6% and 43% of the additive genetic variation for a range of 17 morphological traits. Similarly to Kennard et al. (1994), who found significant marker-trait associations on two linkage groups for three leaf characters, QTLs for two of them were found on the same linkage groups (O6 and O7) in close proximity. All the traits had the same gene action, suggesting that the same putative gene may be involved in controlling these characters. However, because of incomparable marker sets used in these two studies, the homology of identified regions could not be determined. In this study, a putative QTL for vernalization was detected on linkage group O8 at 24 cM, strongly suggesting homology, on the basis of synonymous marker localization, to a QTL detected on N19 linkage group of *B. napus* (Ferreira et al. 1995). Other studies have shown the same region in *B. napus* to have homology with a region in *B. rapa* that also has a putative QTL controlling vernalization (Teutonico and Osborn 1994; Osborn et al. 1997). Similarly, a QTL for petiole length detected on O3 in this study appeared to be homologous to a QTL detected on another *B. oleracea* linkage map (Camargo et al. 1995).

Bettey et al. (2000) (Table 5) measured the traits associated with seed vigor and preemergence seedling growth in a segregating population of 105 DH *Brassica oleracea* lines. They were able to find some correlation, notably among germination traits, and between seed weight and preemergence seedling growth. Collectively the results indicate that germination and preemergence seedling growth are under separate genetic control. Quantitative trait analysis revealed significant loci on linkage groups O1, O3, O6, O7 and O9. It is worth mentioning that two of these QTLs are in a similar position to QTLs for flowering time found previously by Bohuon et al. (1998) (Table 5).

Flowering Time Control

Among many morphological types of *B. oleracea* two flowering habits can be readily distinguished. Some types, such as cabbage and kohlrabi, are biennials and require exposure to low temperatures for flowering initiation (vernalization). Other types, like broccoli and cauliflower, are annuals and will flower during a single growing season. The knowledge on genetic control of timing of transition from the vegetative to reproductive phase is not only of scientific interest because it facilitates the understanding of plant development, but it is also important in agriculture, because its modification may enable the geographical range of *Brassica* crop to be extended.

In early studies, the results obtained by different researchers contradicted one another. Some pointed out the possibility of a single gene control of flowering (Detjen 1926; Horovitz and Perlasca 1954; Walkoff 1963; Landry et al. 1992), whereas others suggested an oligogenic control for time to flower (Baggett and Wahlert 1975; Pelofske and Baggett 1979; Sachan and Singh 1987; Baggett and Kean 1989).

Kennard et al. (1994) (Table 5) used RFLP markers to study genetic control of flowering in single plants of an F_2 population derived from a cabbage by broccoli cross and found four genomic regions associated with the annual/biennial habit and flowering time, and digenic epistasis was detected between two regions controlling flowering time.

Camargo and Osborn (1996) (Table 5) in their study used replicated F_3 families, instead of single F_2 plants, to provide a better estimate of the genotype effects of F_2 plants. Results of marker interval analysis confirmed that flowering time is under oligogenic control, with at least three putative loci involved. Two QTLs on different linkage groups were associated both with the annual/biennial habit and flowering time. The putative locus on LG 2 was linked to *slg6*, a gene that encodes an *S*-locus-specific glycoprotein involved in the self-incompatibility mechanism of *B. oleracea* (Nasrallah et al. 1985), while the putative locus on LG 6 was linked to a QTL that controls petiole length (Camargo et al. 1995). A similar observation was also reported in another *B. oleracea* study (Kennard et al. 1994). A third region on linkage group 8 was associated with flowering time only. Epistasis was detected between two genomic regions controlling flowering time on LG 6 and LG 8, a situation previously reported (Sachan and Singh 1987; Kennard et al. 1994). Authors found no evidence for homology between the flowering time genes identified in *B. napus* (Ferreira et al. 1995) and *B. rapa* (Teutonico and Osborn 1995).

The work reported by Bohuon et al. (1998) (Table 5) was a part of a program developed to explore the extensive genetical variation in the *Brassica* crop genus, capitalizing on the synteny between and within *Brassica* genomes and *A. thaliana* to facilitate the mapping and identification of potential candidate loci. They scored for flowering time in two trials a population of 150 DH lines of rapid-cycling *B. oleracea* derived from a cross between *B. oleracea* var. *alboglabra* × *B. oleracea* var. *italica*. Six QTLs were identified; one each on LG O2 and O3 and two each on LG O5 and O9. In total, these QTLs explained 58 and 93% of the genetic variation in two trials. Three of these QTLs, on O2, O3 and O9, were situated in regions showing considerable homology both with each other and chromosome regions of *B. nigra* LG 2 and LG 8, that have been shown to affect flowering time. The same regions are all homologous to a single region on the *A. thaliana* chromosome 5, which contains a number of the flowering-related genes (*CONSTANS* contig), one or more of which could be candidates for the QTL found in *Brassica*.

The work reported by Rae et al. (1999) (Table 5) reinforced the results of Bohuon et al. (1998) by using the same cross to engineer near-isogenic lines. These lines were grown in field trials for five years and scored for flowering time. Based on the lengths of the substitutions, evidence for 11 QTLs on chromosomes O1, O2, O3, O5 and O9 was found, five of which mapped to similar regions to five of the six found in a previous analysis (Bohuon et al. 1998).

In comparison to two previous reports (Osborn et al. 1997; Bohuon et al. 1998), Lan and Paterson (2000) identified at least six additional *Brassica* QTLs which fall in a homologous region involving the tops of *B. oleracea* chromosome 1, 4 and 7, corresponding to a segment of *Arabidopsis* chromosome 5, that contain seven flower-timing mutations, namely *tfl1, flc, tfl2, co, fy, art1* and *emf1*.

The aim of the analysis performed by Axelsson et al. (2001) (Table 5) was to answer one interesting question related to quantitative traits, i.e., to what extent duplicated genes in polyploid *Brassica* genomes influence the variation in those traits and, whether multiple QTLs for a particular trait are often controlled by homologous genes. Previous comparative mapping revealed that a 1.5-Mb segment at the top of chromosome 5 in *A. thaliana* corresponds to three homologous segments in *B. nigra* (Lagercrantz et al. 1996). Moreover, QTL mapping identified flowering-time loci in two of the three *B. nigra* chromosome segments and homologs to the *Arabidopsis CONSTANS* (*CO*) gene mapped close to both *B. nigra* QTLs. Similar associations between the locations of three flowering-time QTLs in *B. oleracea* and the *CO* gene were found by Bohuon et al. (1998). This work was extended and flowering-related QTLs were mapped in *B. juncea*, *B. oleracea* and *B. rapa*, using rapid-cycling lines of each species as the early-flowering parent. This study focused on chromosomal regions homologous to a segment around the *CO* gene in *A. thaliana*. The results obtained confirmed that segments corresponding to three *B. nigra* segments are present in other diploid *Brassica* species and there are also six such segments present in amphidiploids of *B. juncea*. In all the four species studied, putative QTLs were detected in two of the triplicated segments of each basic diploid (A, B and C) genome, except in the amphidiploid *B. juncea*, in which two QTLs were found in the A genome component, but only one in the B genome component. Based on the close correspondence between several of the likelihood peaks, it seems probable that at least some of the QTLs in the different species are controlled by orthologous and (or) paralogous genes. It is worth mentioning that the QTLs detected on O3 and O9 of the *B. oleracea* genome in this study may be equivalent to those previously identified by Bohuon et al. (1998) (Table 5).

8.5 Marker-Assisted Breeding

One of main goals of breeding programs is the introgression of a gene or genes from a donor parent into the target parent. It is usually achieved through backcrossing or by using molecular markers. These markers have widespread application in plant breeding; they can be used for example to characterize diversity within gene pools and to identify sources of genetic diversity that complement existing breeding populations (Mazur and Tingey 1995). Molecular markers are the basis of the marker-assisted selection (MAS) strategy. This strategy uses molecular markers to infer the presence of a gene from the presence of close linkage between these markers and the gene. If the marker and the gene are located far apart, then the possibility that they will be transmitted together to the progeny is reduced due to double crossover events. Hence, in such selection only these markers should be used, which are tightly linked to the target gene. It is connected with the development of high-density molecular maps or only saturated regions encompassing the locus of interest on these maps. Although for *B. oleracea* several molecular maps have been constructed (Bohuon et al. 1996; Hu et al. 1998; Lan et al. 2000; Babula et al. 2003), not many genes have yet been isolated in this species using the MAS strategy. Till now, studies have been conducted directed towards the construction of a high-density genetic map of *B. oleracea*, which in the future would facilitate the selection and isolation of genes of interest.

The MAS strategy is based on different methods. One of them was associated with developing intervarietal substitution lines, which were used to identify markers tightly linked to a gene. These near-isogenic lines (NILs) differ in the presence or absence of the target gene and a small region flanking the target gene (Muehlbauer et al. 1988).

As mentioned earlier, *B. oleracea* and especially broccoli exhibit anticarcinogenic activity, which is associated with the presence of certain glucosinolates such as 4-methysulphinylbutyl isothiocyanate. Many wild forms of *B. oleracea* have high levels of individual aliphatic glucosinolates (Mithen et al. 1987; Giamoustaris and Mithen 1996). Hence, these wild forms are valuable sources for specific increase of glucosinates contents to enhance anticarcinogenic potential of the *Brassica* vegetables. For this purpose, broccoli cultivars were crossed with selected wild forms of *B. oleracea* (*B. villosa* and *B. drepanensis*). The hybrids obtained by backcrossing with introgressed segments of the genome of *B. villosa* exhibited a 10-fold increase of 4-methylsulphinylbutyl glucosinolate levels (Faulkner et al. 1998). Moreover, the tissue of these plants exhibited a > 100-fold increase in the ability to induce quinone reductase in Hepa 1c1c7 cells over broccoli cultivars. Other studies with the use of wild *Brassica* species as a valuable resistance resource were carried out as well (Happstadius et al. 2003). They concerned the screening for resistance to Verticillium wilt of *B. oleracea* cultivar accessions and wild related species.

Another approach depends on the development of a population of introgression lines (ILs) by systematic backcrossing and inbreeding. Each IL is near-isogenic to the recipient genotype, except for a single homozygous RFLP defined chromosome segment from the donor species. Such ILs are useful not only for identifying markers linked to major genes, but are extremely useful for the precise fine-mapping of QTLs, but it depends upon the size of the introgressed region. In *B. oleracea* the production of introgressed lines is limited. The first experimental work concerned the construction of a substitution library of recombinant backcross lines of doubled haploid plants (derived from a cross between *B. oleracea* ssp. *alboglabra*-A12DHd genotype, the recurrent parent, and *B. oleracea* ssp. *italica*-GDDH33 genotype, the donor parent) by using molecular marker-aided selection (Ramsay et al. 1996) (Fig. 5). Repeated backcrossing and selfing of the progeny from this cross led to the production of the lines with small introgressed segments of donor parent DNA. The analysis of recombinant backcross lines in this study indicated that a full range of recombinant backcross lines with a single introgressed tract and covering the entire genome would be achieved in the BC_3S_1 generation. Such complete genome coverage is a reflection of the power of marker-assisted selection. Ramsay and coworkers derived a set of seventy-one recombinant backcross lines covering almost the whole genome, with the exception of O6 linkage group (Fig. 5). This is caused by the presence of the self-incompatibility locus on O6 linkage group and it hindered the production of homozygous lines containing this region.

Five *B. oleracea* substitution lines derived by Ramsay et al. (1996) (Fig. 5) for linkage group O9 were used to confirm the significance of the QTL locus associated with susceptibility to *Agrobacterium tumefaciens*, which had been detected earlier on the genetic map of *B. oleracea* (Sparrow et al. 2004; see Sect. 8.4). Therefore, for this study five A12DHd lines were selected, which contained the substituted fragments of the GDDH33 genome covering the area with the mapped QTL for *A. tumefaciens* susceptibility. These studies confirmed the presence of this QTL and suggested that genes responsible for susceptibility are located within the central 38% of linkage group O9. Fine-mapping of the genes associated with *A. tumefaciens* susceptibility by further backcrossing of a key substitution line (SL177) to A12DHd could facilitate the isolation and cloning of these genes.

The above cross used by Ramsay et al. (1996) (Fig. 5) to produce a set of NILs, was used to derive another set of NILs as well (Rae et al. 1999). These NILs were employed to identify QTLs for flowering time in *B. oleracea*. In this study 79 lines were selected for flowering behavior testing over a period of five years. Among them, only 23 lines showed significant variation in flowering behavior from the A12DH parent having a short generation time. Ten out of 23 had a single substitution, four had two substitutions, eight had three substitutions and one had four substitutions.

The strategy of bulked segregant analysis (BSA) was developed by Michelmore et al. (1991) for *Dm* genes of *Lactuca sativa* conferring resistance to *Bremia lactucae*. This method is based on comparing two pooled DNA samples of individuals from a segregating population derived from a single cross. Within each pool the individuals are identical for the target trait or gene, but are arbitrary for all the other genes. Two pools contrasting for a trait (resistant vs. susceptible) are analyzed to identify markers that distinguish them. Markers that reveal polymorphism between bulks will be genetically linked to loci determining the target trait. In *B. oleracea*, the BSA method was used, e.g., in the detection of molecular markers closely linked to gene *Pp523*, which confer downy mildew resistance to adult plants of broccoli (Farinhó et al. 2004). Earlier genetic studies demonstrated that resistance to downy mildew is dominantly inherited and controlled by a single major gene (Coehlo et al. 1998; Coehlo and Monteiro 2003). The first step was the construction of a genetic map for a population derived from a cross between the downy mildew-susceptible GK97362 line of rapid-cycling *B. oleracea* and the S4 line derived from the downy mildew adult-stage resistant accession OL87125 of broccoli. This map was based on 430 markers such as RAPD (random amplified polymorphic DNA), AFLP (amplified fragment length polymorphism), ISSR (intersimple sequence repeat) and SSR (simple sequence repeat), and additionally it mapped the *Pp523* gene and a locus determining flower color (white vs. yellow). The analysis of this map made it possible to detect five markers closely linked in coupling to the resistance gene, encompassing a chromosomal region of 15 cM. Two of them, OPK17_980 and ATCTA_133/134, flanked this resistance gene and were located at 3.1 cM and 3.6 cM, respectively. The identified markers can be used as selective markers for marker-assisted selection in breeding programs aiming at the introgression of this gene in susceptible *B. oleracea* genotypes.

In other studies the eight RAPD markers, linked to the cotyledon stage resistance locus *Dm* corresponding to a single dominant gene in broccoli were identified using bulked segregant analysis (Farnham et al. 2001; Giovannelli et al. 2002). Two of these detected RAPD markers, i.e., $UBC359_{620}$ and $OPM16_{750}$, were located relatively close to each other and the same side of *Dm* resistance locus, and they were converted to SCAR markers, for the identification of the downy mildew resistance and susceptibility of F_2 individuals without the need for downy mildew inoculation. These SCAR markers should be more useful for different laboratories due to overcoming problems in reproducibility, with which the RAPD markers are burdened. Furthermore, they can aid breeders in gene pyramiding through incorporating a number of desirable alleles into a single cultivar.

Presently, only a few reports are available on the pyramiding of multiple foreign genes for pest management in the same variety (Cao et al. 2002). A mixture of two or more genes in the same plant is a promising long-term strategy for delaying resistance and is achieved by interplanting two cultivars and by introducing two different resistance genes into the same cultivar (gene pyramiding). The plant crop having two or more pyramided resistance genes with high doses can provide promising insect control and may be useful for resistance management. However, no information is available on how well plants with multiple resistance genes control insect populations resistant to one or more of the pyramided genes (Cao et al. 2002).

One of experiments connected with the analysis of the effect of gene pyramiding concerns the determination of influence of two *Bacillus thuringiensis* (Bt) genes in the same broccoli plant on the production of Bt proteins and the control of diamondback moths resistance (DBM, *Plutella xylostella*) to one or the other protein (Cao et al. 2002). The production of transgenic crops expressing genes from bacterium *B. thuringiensis* has revolutionized research and the ability of farmers to control insect pests. For example, in 1999 the cultivation of transgenic crops containing a transgene for an insecticidal protein derived from Bt in the US accounted for 26% area of corn (US EPA 1999) and 32% of cotton (Carpenter and Giannessi 2000). Broccoli plants with pyramided two Bt genes (*cry1Ac* and *cry1C*) were obtained by sexual crossing between the *cry1Ac*- and *cry1C*-transgenic plants, which had been well-characterized in terms of the number of introduced genes, the levels of gene expression and insect control (Metz et al. 1995; Cao et al. 1999). To increase the chance of obtaining functional *cry1Ac/cry1C* hybrid broccoli, i.e., with active and stable expression of both genes and production of their protein, various cross combinations were made. Molecular analysis of these cross combinations showed that both genes were expressed in all progeny. Interestingly, parents with high levels of expression of one gene produced progeny with high expression of both genes. It indicates that broccoli can successfully pyramid two closely related Bt genes into a single plant for resistance management and is a useful resource for further studies on insect resistance.

For an efficient MAS strategy and gene pyramiding, the conversion of anonymous molecular markers such as RAPD and AFLP into SCARs or STSs may be useful. The examples applying these techniques to *B. oleracea* crops were described above. An additional example concerns the identification of markers closely linked to the *Or* gene that induces the accumulation of high levels of β-carotene in cauliflower (Li and Garvin 2003). In the mapping process based on an F_2 population (derived from a cross between a wildtype, white-curd cauliflower cultivar Stovepipe and an *Or* homozygous line, 1227) 10 AFLP markers linked to the *Or* gene were identified. Four of them, which were most closely linked to this gene, were converted into RFLP markers. Next, three of the RFLP markers were converted into sequence-characterized amplified region (SCAR) markers. The application of two-step conversion gives a guarantee of success in receiving valuable SCAR markers. These SCAR markers based on PCR are simple and they can greatly facilitate the analysis of a large segregating population. Additionally, two RFLP markers were placed on the genetic map at 0.5 cM and 1.6 cM from the *Or* locus on both sides. The identification of *Or* gene and markers closely linked to this gene will be an essential step towards its isolation and gaining insight into the regulation of carotenoid synthesis and accumulation.

In searching for valuable resistance sources in crops germplasm screening is carried out among crop accessions and wild-related forms of a given species. It makes it possible to characterize the genetic diversity within given gene pools and to identify sources of genetic diversity that complement existing breeding populations. Within the limits of *Brassica* species, the wild-related plants still exist in nature and they can be intercrossed. These wild relatives have the potential to provide access to a much wider range of allelic variation, which is limited in crop types. To facilitate the genetic diversity for crop improvement, the *Brassica* gene pools were developed (King et al. 2004; www.brassica.info). Presently, the core collection for *Brassica* spp. within EU GenRes and other programs was generated (Boukema et al. 1997; Maggioni 1998). Seeds in this collection were multiplied and accessions were screened for a number of fungal and other resistance traits. However, these collections consisted of heterogeneous and heterozygous material as well, which limits their use in detailed genetic studies. Furthermore, Diversity fixed foundation sets (DFFS) were developed, covering an informative set of genetically fixed lines representing a structured sampling of diversity across a gene pool. The DFFS are collections of genetically fixed lines that represent a snapshot of diversity within the relevant gene pool (Fig. 11).

The approach of generating fixed lines is particularly powerful as the data obtained are cumulative, allowing long-term comparative analysis. In addition, fixed lines enable experimental trials to be established with replicate plants. As such they may be interpreted to provide insight into the contribution and interaction of genetic, environmental and developmental components of variation (King et al. 2004).

A core collection of the *B. oleracea* gene pool (BOCC) was created for an EU project. In the process of creating the BOCC a new tool was introduced, named the path indicator. This tool allows a good overview of the content of the core collection. It points to groups of material instead of individual accessions and can therefore contribute to a better interpretation of evaluation results.

The production of a DFFS for *B. oleracea* by Horticulture Research International (HRI, Wellesbourne, UK) was first initiated. It comprises a structured sample of 386 accessions representing diversity within the gene pool. These have been selected as nested subsets of 94 (so that DNA samples fit on a 96-well microtiter plate, with 2 controls). Each subset has been selected to represent diversity within species subtaxa and crop types, as well as variation in the ecogeographic origin. Detailed data for the *B. oleracea* DFFS will be published on http://www.brassica.info.

One of the examples of executing germplasm screening in *Brassica* concerns the screening of 307

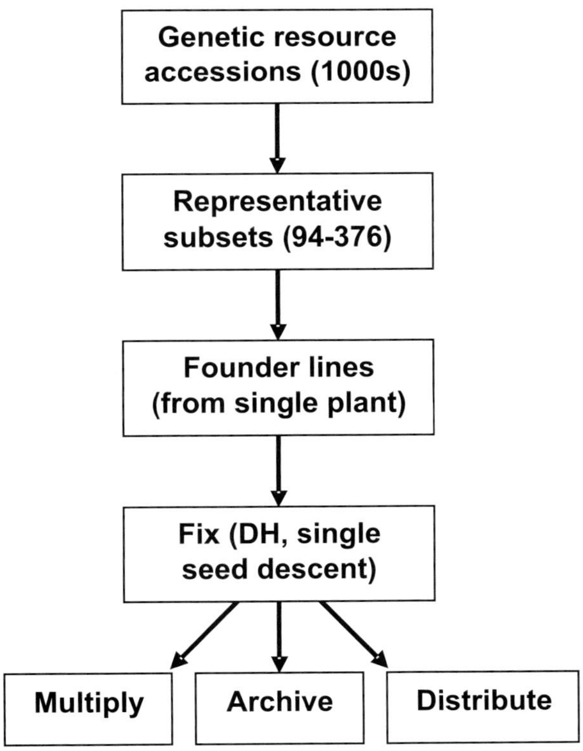

Fig. 11. Derivation scheme of a Diversity Fixed Foundation Set (DFFS) (www.brassica.info)

selected *B. oleracea* accessions, representing 11 of 14 cultivar groups and four wild-related species (*B. cretica*, *B. incana*, *B. insularis* and *B. villosa*) to identify resistance resource to a fungal disease, Verticillium wilt (Happstadius et al. 2003). In this study several accessions with a high level of resistance to Verticillium wilt were identified. The results of this and earlier studies (Zeise and Buchmüller 1997) suggest that the genome C of *Brassica* contains resistance trait(s) to Verticillium wilt.

Although the cost of MAS is high for most applications, the precision of selection offered by DNA markers and the development of newer markers could make MAS more economical and the method of choice for breeding programs in the future.

8.6 Map-Based Cloning

During the last few decades, development of novel and comfortable marker systems resulted in a significant progress in the field of forward genetics. Marker types such as AFLP, CAPS, SSR, ISSR or others enable relatively simple and rapid generation of very dense genetic maps that can be further used for gene cloning. Although map-based cloning is now relatively straightforward in *Arabidopsis*, where a large number of genomic tools are available, in other plant systems there are still some limitations. However, for first steps of map-based cloning, the genome-wide AFLP mapping can confidently be used, because no prior sequence information is required. In the case of a single-gene project, a useful modification of the strategy is first to identify linked markers using bulked segregant analysis (BSA), a technique developed by Michelmore et al. (1991). This could be done by performing a large set of AFLP primer combinations on DNA pools of mutants and wildtypes. As a result, one should be able to identify closely linked markers, enabling subsequent regional fine-mapping, followed by joining BAC clones.

Up to now, BSA has been used with success in isolation of a single gene in *B. oleracea*. This is a cauliflower high β-carotene gene, which was described in detail in Sect. 8.3 (Li et al. 2003b). Li and Garvin (2003) constructed a genetic map of 195 F_2 individuals and found a few markers closely linked with the *Or* loci (Fig. 12a). To increase the resolution of this map, two previously developed SCAR markers (SA4 and SA7) that flank the *Or* gene were used to screen an additional 1437 F_2 individuals to find recombinants between these two markers. The authors identified 54 F_2 individuals which had undergone a recombination event between SA4 and SA7; they were further used for fine-mapping of AFLP markers closer to the *Or* locus. Four additional AFLPs, which resided between SA4 and SA7, were mapped by using this approach (Fig. 12b). One of these markers, A9, was used to screen a cauliflower BAC library that made it possible to identify 27 BAC clones. Those were screened with A9, A1 and A7 markers (A3 marker was not used in this analysis, as it contains a high copy-number sequence). Based on this study, only seven BACs were finally connected with the *Or* region and hybridized with A9, A1 and A7 markers in the Southern hybridization to establish their relative position and orientation. The left end of BAC7 was used in subsequent screening of the BAC library. Selected clones were end-sequenced and the left end of the BAC45 was used for the next round of chromosome walking. Finally, a BAC contig was assembled (Fig. 12c) and end fragments of some BACs were converted into RFLP markers, and mapped in the population of 54 recombinants between SA4 and SA7. This facilitated the identification of BACs encompassing the *Or* locus. For the purpose of gene cloning, Li et al. (2003b) constructed a cauliflower BAC library from *Or* homozygous plants using the pBeloBAC11 vector. The library contains 60,288 clones that provide approximately 10-fold coverage of the genome. The average insert size was calculated to be 110 kb.

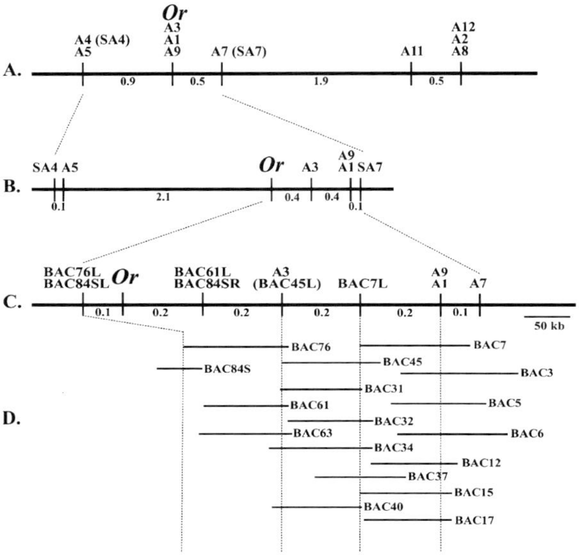

Fig. 12. Genetic and physical mapping of the *Or* region in cauliflower. Map positions of the markers are listed above the lines, whereas genetic distances (cM) are indicated below. Prefixes A, SA, and BAC represent AFLP, SCAR, and BAC end-derived markers, respectively. **a** Low-resolution linkage map of the *Or* region (Li and Garvin 2003). **b** High-resolution genetic map based on the segregation data from 1632 F_2 individuals. **c** Genetic map of *Or* constructed using AFLP and BAC end-derived markers. **d** BAC contig encompassing the *Or* region. Only clones with informative ends are displayed (*horizontal lines*) (reprinted from Li et al. 2003b with permission of Springer)

The approach used in the cloning of the *Or* locus was based on a classical map-based chromosome walking technique, where the main work is related to BAC selection and contig assembly. On the contrary, chromosome landing is an approach where the scientists concentrate on efficient and very precise genetic mapping of a gene of interest that facilitates one-step selection of a desired genomic clone. A modification of the chromosome landing for the isolation of the *GLS-ALK* gene from *B. oleracea* can be found in papers of Li and Quiros (2001, 2003). The *GLS-ALK* gene is responsible for desaturation of glucosinolates and its alleles could be traced in the mapping population by appropriate chemical analyses (see also Sect. 8.3). Li and Quiros (2001) mapped the gene on the genetic map based on AFLP and SRAP markers. The latter marker system enables tagging of coding sequences using GC-rich primers, and one of such markers, SRAP133, appeared to be closely linked (see Table 4), and completely cosegregated with the *BoGLS-ALK* gene in three populations, which have different genetic backgrounds (Li and Quiros 2001, 2003). Sequencing of this marker and homology searching with the *Arabidopsis* genome sequence revealed that it is related to a functional gene for the desaturation of glucosinolates in *A. thaliana*. Thus, the authors concluded SRAP133 tagged the *GLS-ALK* gene in *B. oleracea* and used this marker for BAC library screening (Li and Quiros 2003). This made it possible to identify a single BAC clone, B21H13, harboring 17 genes including gene *BoGSL-ALK* (Li and Quiros 2003; Gao et al. 2004). The library used for BAC selection was constructed for the broccoli doubled haploid line Early Big-10 with 6-fold coverage and 90-kb insert size on average (Carlos F. Quiros, personal communication).

In order to highlight the functionality of the *BoGSL-ALK* gene in *B. oleracea*, its functional allele was introduced by transformation into *A. thaliana* (Columbia) (Li and Quiros 2003). Columbia is homozygous recessive for the null allele of *GS-ALK*, thus the transformation should change glucosinolate contents in this species. The authors received transformants expressing the gene and used them to analyze the glucosinolates using HPLC. They detected three new peaks in leaves and four new peaks in seeds of transformed plants, which corresponded to different specific glucosinolates. As the conversion of some forms of glucosinolates into others was very significant (more than 80% in some cases), it could be concluded that cloning of the *BoGSL-ALK* gene provides the opportunity to engineer *Brassica* crops by modifying their GSL sidechains. It should be emphasized, however, that possible applications such as the elimination of antinutrient progoitrin with a simultaneous accumulation of anticarcinogenic glucoraohanin, would probably be more important in seed crops, e.g., rapeseed. On the other hand, if the objective was to accumulate sinigrin for biological control (i.e., incorporating crop residue in the soil to control pathogenic fungi, weeds or nematodes), then *BoGSL-ALK* could be introduced to *Brassica* crop with high biomass capacity.

The cloning of the *GSL-ALK* gene is a good example of how the information from a model genome project could be transferred to crop species. Besides *GSL-ALK* Quiros's team has recently cloned another important gene for glucosinolate modification, *GSL-ELONG*, based on homology with an *Arabidopsis* candidate gene (Li and Quiros 2002), and are going to clone the *GSL-PRO* gene. With these genes in hands, manipulation of the aliphatic glucosinolate profiles of *Brassica* species could be greatly facilitated.

Although there are no other examples of positional cloning in *B. oleracea*, some other types of gene isolation should be mentioned. One of the first genes cloned in *B. oleracea* were the *S*-locus genes (Nasrallah et al. 1988). The authors firstly selected cDNA clones which maternal mRNAs exhibited high levels of transcript in mature stigma and no expression in seedlings (Nasrallah et al. 1985). These clones were further used as probes in the screening of the *B. oleracea* cosmid library (Nasrallah et al. 1988; Stein et al. 1991). Gene cloning could also be performed by screening a genomic library with a probe from other species or a cDNA clone for a specific gene (Tang et al. 1996; Attieh et al. 2002; Das et al. 2002; Li and Quiros 2002; Henriksson et al. 2003).

8.7 Advanced Works

8.7.1 Comparative Physical Mapping in the *A. thaliana–B. oleracea* System

The cultivated *Brassica* species are most closely related to the *A. thaliana* model plant. This makes it a very attractive and comfortable system for transferring data and information obtained for *Arabidopsis* to *Brassica* crops. Starting from 1994 (Kowalski et al. 1994; Sadowski et al. 1994), a large number of research projects based on this assumption have been reported. Besides comparative genetic maps described in Sect. 8.2 (Kowalski et al. 1994; Lan et al. 2000; Babula et al. 2003; Lukens et al. 2003), some other advanced approaches have been utilized. Sadowski et al. (1996) reported genetic and physical analysis by pulse-field gel electrophoresis (PFGE) in three *Brassica* diploid genomes for a cluster of five genes characterized in a selected segment of 15 kb on chromosome 3 of *A. thaliana*. The analysis on *B. oleracea* revealed that genes from the complex are present at four different positions in this genome, although gene content differs largely among them. The complete complex with all the five genes being conserved was detected at one position only. This result is consistent with the present hypothesis of a hexaploid ancestor of diploid *Brassica* species (Lagercrantz and Lydiate 1996). A more complex study was performed by O'Neill and Bancroft (2000) who screened the *B. oleracea* BAC collection against 19 low-copy genes from a 222 kb region of the *Arabidopsis* chromosome 4. It should be noted that nine out of those 19 genes are present in a homoeologous region on *A. thaliana* chromosome 5. The authors established seven different *Brassica* BAC contigs, from which four corresponded to the *Arabidopsis* chromosome 4, and three to chromosome 5 (Fig. 13). All the seven *B. oleracea* regions showed collinearity with the *Arabidopsis* counterparts, but in all of the triplicated *Brassica* segments several of the genes were missing. Evidence for a translocation and an inversion that took place after *B. oleracea* speciation was detected. Recent continuation of this work on the *B. rapa* genome confirmed that the microstructure of this region is very similar in this species (Rana et al. 2004). This is in good agreement with the fact that these two lineages diverged only ca. 4 Mya (Inaba and Nishio 2002). The microstructures of the triplicated segments

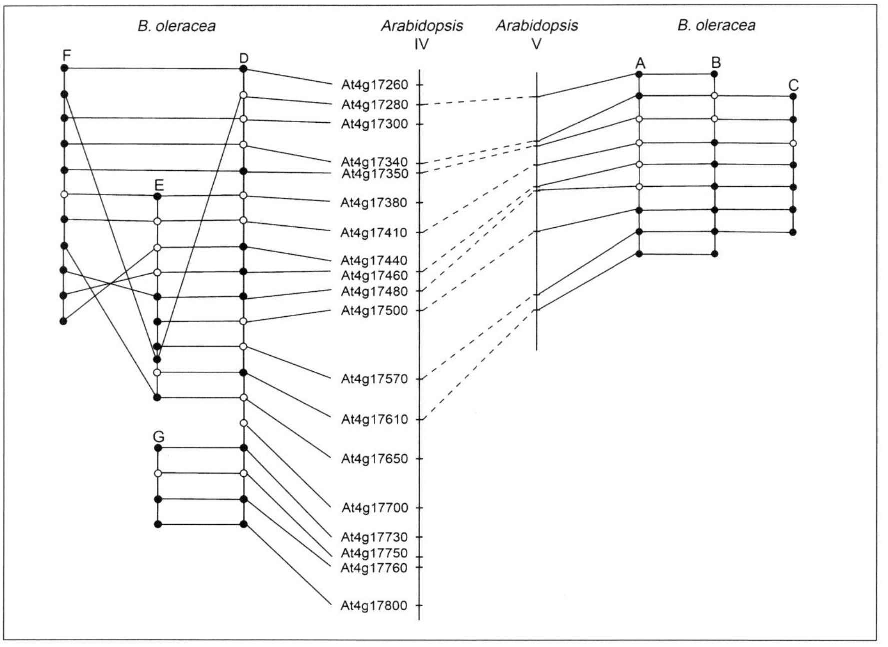

Fig. 13. Comparison of a region of the *A. thaliana* chromosome 4 and its homoeolog on chromosome 5, and corresponding segments in the *B. oleracea* genome. Nineteen gene probes from this region are present (*closed circles*) or absent (*open circles*) in chromosomal regions of *B. oleracea* represented by seven BAC contigs (A to G). *Lines* represent relative positions of homologous genes within *Brassica* segments (O'Neill and Bancroft 2000) (reprinted from Bancroft 2001 with permission of Elsevier)

are primarily diverged from each other and their *Arabidopsis* homoeologs, by an interspersed pattern of gene loss. Moreover, Rana et al. (2004) analyzed the chromosomal organization of this gene complex also in *B. napus*, and found that the microstructure in this amphidiploid species is almost identical to its diploid relatives. This is important information for the prospect of transferring data regarding genetic and physical mapping among cultivated *Brassica* species.

More detailed studies were based on the sequencing of selected *B. oleracea* regions and comparative analyses with their homoeologs in the *Arabidopsis* genome. Quiros et al. (2001) sequenced two *B. oleracea* cosmid contigs corresponding to the *ABI1-Rps2-Ck1* segment on chromosome 4 of *A. thaliana*. Although the regions were relatively short (only seven genes in *B. oleracea*), some changes in gene content were detected in comparison to their *Arabidopsis* counterparts. Sequence data indicate that although coding sequences were very similar, promoters were poorly conserved. Gao et al. (2004) analyzed a 101-kb-long BAC clone B21H13 from *B. oleracea* by sequence comparison with the corresponding region in *Arabidopsis* (119 kb), and detected even more frequent structural changes between both genomes. As was expected, a significant number of genes has been lost in the region in *Brassica* after the split from the *Arabidopsis* lineage: out of 37 genes in *A. thaliana* only 21 were conserved in *B. oleracea* (Fig. 14). This is probably a consequence of triplication of the *Brassica* ancestral genome, which enhanced the process of gene loss within triplicated *Brassica* chromosomal segments. The gene loss was most probably due to larger deletion events, because usually a few neighboring genes were deleted simultaneously, and there was no significant extension of intergenic regions at corresponding sites in *B. oleracea*. However, transposition into new chromosomal locations could not be ruled out. It needs to be emphasized that all genes within the complex in *Brassica* had the same orientation as in *Arabidopsis*.

The establishment of structural similarity between the *B. oleracea* and *A. thaliana* genomes is preliminary to identification and understanding of gene function in the *Brassica* species. However, the triplicated nature of the *Brassica* genomes, which gives rise to the increase of gene redundancy and the number of duplicated segments in relation to simpler *A. thaliana* genome, can impede functional analysis of the particular genes. It can be assumed that a particular known homolog of *A. thaliana* having specific partner(s) in the *Brassica* genome can provide predictable data in terms of their specific function, as has been shown in several reports (e.g., Wiersma et al. 1989; Arondel et al. 1992; Osborn et al. 1997; Schranz et al. 2002). Therefore, studies on gene expression in *Brassica* should be carried out in support

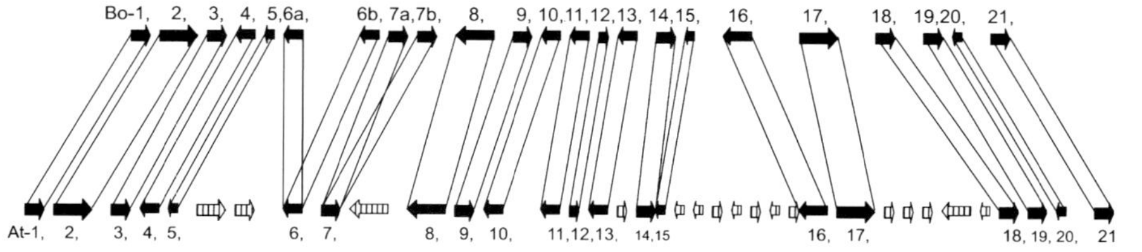

Fig. 14. Comparison of gene contents in the *B. oleracea* BAC B21H13 (Bo) with its homoeologous region in *A. thaliana* (At). *Arrows* represent individual genes with their orientation; conserved genes are indicated with *closed arrows*, whereas genes lost in *Brassica* are indicated with *open arrows*. Two genes (Bo-6 and Bo-7) were duplicated in *B. oleracea* after *Arabidopsis-Brassica* split (reprinted from Gao et al. 2004 with permission of NRC Research)

of the genetic and functional analyses of individual gene homologs. This strategy has been recently employed for specific homologs gene copies involved in the ethylene biosynthesis (*SAM*, *ACS* and *ACO*) and signaling pathways (*ETR1*, *CTR1*, *MKK4*, *MKK5*, *EIN2*, *EIN3*, *EREBP*, *ERF5* and *ERF7*) in the *B. oleracea* genome (Babula et al. 2006). First, these homologs have been recognized through in silico analysis of the *A. thaliana* genome data and their differential ozone- and drought-responsiveness revealed by a DNA microarray experiment (Ludwików et al. 2004). Then, the genome collinearity has been analyzed in the *A. thaliana* and *B. oleracea* genomes on the basis of chromosomal regions carrying the gene homologs studied. The identification of the conserved segments in the both genomes allowed to initially deduce functions of the *B. oleracea* genes mapped. Finally, with the use of the semiquantitative RT-PCR approach the conservation of differential transcriptional induction of the ACC synthases homologs for *B. oleracea* and *A. thaliana* during ozone stress have been confirmed. The results indicate that conservation at the gene sequence and colinearity levels can provide guidelines for the functional analysis of gene homologs from closely related species.

8.7.2 New Genomics Tools Developed for *B. oleracea* Research

In the genomic and postgenomic era new facilities have been developed for *B. oleracea*. One of the most important initiatives is the Multinational *Brassica* Genome Project (MBGP) (the general information and database gateway available at http://www.brassica.info and http://brassica.bbsrc.ac.uk, respectively), which aims at the creation of a platform for the exchange of information and genomic resources among the *Brassica* research community. Physical maps are being constructed for the *Brassica* A genome in Korea and for both the A and C genomes in the UK. Partial physical mapping of the genome of *B. napus* is being conducted in Canada and in the EU. Although such physical maps will be of great value for the identification of specific genome regions of these important crops, they will not permit a detailed analysis of the entire *Brassica* genome, the preparation of microarrays to analyze the transcriptome, or an efficient design of markers associated with the sequences of specific genes to be used in breeding programs. To achieve these things, the complete sequence of at least one of the *Brassica* genomes will be required. It is necessary to sequence only one *Brassica* genome initially as both the macrostructure (chromosome level) and the microstructure (gene-by-gene level) of *Brassica* genomes show extensive, though imperfect, collinearity. For this purpose the *B. rapa* A genome was selected as it is smaller than that of *B. oleracea* or *B. napus* (at 550 Mb relative to 600 Mb and 1,200 Mb, respectively), seems to have less dispersed repetitive elements, and there are freely available BAC libraries providing very deep (>20-fold) coverage of the genome.

The BAC-based physical mapping of *B. oleracea* that is being carried out at the John Innes Center has also made substantial progress during the last two years. Two *B. oleracea* BAC libraries have been used to create the physical map: JBo and BoB (see Table 6, for details). A *B. oleracea* map consists of 21,014 BAC clones arranged in 1365 contigs (contigs are assembled based on fingerprinting) with 1205 gene-specific probes (data from the IGF project; JIC 2005) providing anchors to the genome of *A. thaliana*. As a complement to the GST anchor probe set, 970

Table 6. *B. oleracea* BAC libraries available from M*B*GP

Clones	Average insert size	Coverage	Variety/line used for library construction	Institution	Contact	Comments
33,792	145 kb	8×	var. *alboglabra* A12 line	PBL/JIC	Ian Bancroft (UK)	Described as JBo publicly available from BBSRC IGF (http://brassica.bbsrc.ac.uk/IGF/) being contiged as part of IGF physical map project
24,960	130 kb	6×	var. *alboglabra* A12 line	HRI/TAMU	Graham King (UK)	Described as BoB publicly available from TAMU (http://hbz.tamu.edu) being contiged as part of IGF physical map project
18,000	90 kb	3×	var. *italica* Early Big	UC Davis	Carlos Quiros (USA)	
87,168	107 kb	14×	T01434 rapid-cycling line	U Georgia	Andrew Paterson (USA)	Same line as TIGR *B. oleracea* genomic sequencing
36,864	85 kb	5×	Badger Inbred 16	U Georgia	Andrew Paterson (USA)	

genetically mapped *Brassica* RFLP probes have been hybridized to BAC clones, in order to provide an internal framework for the contig building. Both data regarding the present state of the physical map, and the BAC libraries and individual clones, are simply searchable, and could be ordered from the *Brassica* IGF project at http://brassica.bbsrc.ac.uk/IGF/. Moreover, integration of cytogenetic and genetic linkage maps of *B. oleracea* has been realized using BAC and other chromosome-specific probes (Howell et al. 2002). This effort offers a unique opportunity of assigning BAC contigs to the appropriate physical chromosome arm by genetic mapping of a marker designed for this region. The integration was carried out with the use of a mapping population of DH lines: A12DH line was crossed with GDDH33 (commercial calabrase, *B. oleracea* subsp. *italica*) deriving a reference population AG for the integration of loci detected by different marker systems into a common genetic map (Bohuon et al. 1996). As one parent of the population (A12DH) was a DNA donor for the BAC library BoB, this population should be further used as an aid in the physical mapping approach. The results of integration are shown in Fig. 4. The population is freely distributed by M*B*GO and thus could be used for integration with any other *B. oleracea* genetic map.

Besides the BAC-based physical map project, TIGR (The Institute for Genomic Research) in collaboration with Cold Spring Harbor received funding to perform a whole genome shotgun sequencing of the *B. oleracea* genome, with a goal of reaching 0.5 to 1× coverage (contact person: Chris Town). Actually, almost six hundred thousand (595,321) sequences of 600 bp on average have been deposited in NCBI/GenBank databases. Individual clones are available from the M*B*GP platform (http://www.brassica.info).

On the other hand, relatively small numbers of EST and cDNA clones have so far been annotated within *B. oleracea* in the public domain. There are only 369 EST sequences for *B. oleracea* in the NCBI/GenBank, which is quite little in comparison with other *Brassica* crops (for instance *B. napus*: 45,906 ESTs, *B. rapa* subsp. *pekinensis*: 12,583 ESTs). Notwithstanding, the *B. oleracea* EST sequencing program has been initiated at Warwick HRI, that intends to sequence 15,000 EST clones from at least 10 cDNA libraries (http://grc.warwick.ac.uk/brassica/est.html), from different tissues and for abiotic/biotic stresses. The libraries generated will be available for targeted screening or sequencing in other projects and for public access. To date, a total of 22,000 good sequences (both forward and reverse) have been obtained, and are awaiting assignment of GenBank accession numbers. These represent approximately 16,000 individual ESTs.

Other important resourses have already been or are being developed for *Brassica* crops, including those of *B. oleracea*. This includes mapping populations based on rapid-cycling lines (AG population provided by Graham King, AG substitution lines provided by Mike Kearsey, and TB population provided by Tom Osborn), wild germplasm databases, numerous genetic markers, and mutant stocks. These are fully described in an updated form on the web site of MBGP at www.brassica.info, thus they need not be specified here.

8.8 Future Scope of Works: Comparative Assessment vis-a-vis Model Systems

The *Brassica oleracea* species stands for unique model of morphological diversity within a single species. Equally for basic biologists and breeders, this makes the model an example of the most spectacular illustration of structural evolution in plants under domestication on the one hand, as well as an example of particular genetic variability at the level of many traits on the other hand. Natural genetic variation of *B. oleracea* facilitates investigation of the genetic basis of morphological characters. Studies on wild relatives of the cultivated *B. oleracea* towards introducing the resistance/tolerance genes to several biotic and abiotic stresses appear to be of importance. They provide important genetic resources available for the improvement of *Brassica* crops (Lázaro and Aguinagalde 1998). In case of such an important crop species as *B. oleracea*, one of the objectives for genetic map construction is marker-assisted breeding, which is facilitated by the presence of dense, high-resolution genetic maps. One way to achieve this purpose was to integrate already existing maps. As mentioned above, the alignment of the existing *B. oleracea* maps is necessary for their more effective use. Common sequences across the mapping populations should serve to align most of the linkage groups of the independently developed maps.

Genomes within Brassicaceae show variability in size and chromosome number. Comparative genome analyses proved to be informative not only among members of the *Brassica* genus, but also among species of the Brassicaceae family. The example is *A. thaliana*, an extensively utilized model plant in the studies of biochemistry, physiology, classical and molecular genetics. Although the comparative analysis of sequences homologs to *Arabidopsis* and *Brassica* genomes based on genetic maps, suggested extensive chromosomal repatterning since the divergence of these two species (Kowalski et al. 1994; Lan et al. 2000), the same studies indicated structural conservation in most of the genome regions. Meaningfully, differences in the genomic arrangements between *B. oleracea* and *A. thaliana* appeared to be due to the recent history of polyploidy in *B. oleracea*, mainly connected to two additional cycles of polyplodization and diploidization. Nevertheless, comparative genetic mapping has revealed genome collinearity of related species. Importantly, sequence-based comparisons between *Brassica* and *Arabidopsis* proved the conservation of gene arrangements. However, numerous small rearrangements such as duplications, inversions, translocations and insertions/deletions, have been detected. Therefore, a high degree of gene linear order conservation has been found unambiguously only at the microcollinearity level between *B. oleracea* and *A. thaliana*. This is also in agreement with the general finding that duplications and subsequent gene loss have been identified as a particular important factor in the evolution of plant genomes. Particularly interesting results were obtained by Li et al. (2003a) on global gene for gene alignment of the genomes of *B. oleracea* and *A. thaliana*. Consistent with previous reports, they found evidence of extensive collinearity between the two genomes for chromosomal segments rather than for whole chromosomes, often disrupted by inversions and deletion/insertions. They have also observed large-scale duplications of the *Brassica* genome, but their distribution argued against ancient triplication of the entire genome. The results obtained are rather consistent with the hypothesis that events of higher complexity than simple polyploidization have lead to the synthesis of *Brassica* genomes including also aneuploidy and chromosomal rearrangement (Quiros 2001). Clear differences in the similarity score value of related sequences allowed the identification of orthologs. The SRAP technique has one advantage over ESTs used previously in *B. oleracea/A. thaliana* comparative analysis (Lan et al. 2000; Babula et al. 2003), that it might detect some genes with low level of expression or detect gene expression more evenly.

Finally, comparative genome mapping analysis will serve as an efficient tool for transfer of resources and information from the model genome of *Arabidopsis* to *Brassica* species. The chromosomal mapping of agronomically important loci, as it has been shown, can be performed in the species with larger genome, and cloning can be carried out using information from related model organism for which wide genomic information and resources have been established.

Hampered so far cytological analysis and construction of the informative karyotype for *B. oleracea* should be developed by complementing the classical mitotic metaphase analysis with fine decondensed meiotic chromosomes studies as it was demonstrated for this species by Röbbelen (1960). Applying fluorescence in situ hybridization to pachytene bivalents with highly informative genomic clones, should improve physical mapping resolution (Ziółkowski and Sadowski 2002). Similar approaches applied to *A. thaliana*, in connection with the whole-genome sequencing program, have proved to be of high importance in studies of small genomes (Fransz et al. 1998, 2000; Koornneef et al. 2003).

Acknowledgement. This work was supported by grants 3PO6A 036 24 and 2P06A0 35 28 from the State Committee for Scientific Research (Poland).

References

Anstey TH (1955) Inheritance of white petals in green sprouting broccoli. Can J Agri Sci 35:573–578

Anstey TH, Moore JF (1954) Inheritance of glossy foliage and cream petals in green sprouting broccoli. J Hered 45:39–41

Arabidopsis Genome Initiative (AGI) (2000) Analysis of the genome sequence of the flowering plant *Arabidopsis thaliana*. Nature 408:796–815

Armstrong SJ, Fransz P, Marshall DF, Jones GH (1998) Physical mapping of DNA repetitive sequences to mitotic and meiotic chromosomes of *Brassica oleracea* var. *alboglabra* by fluorescence in situ hybridization. Heredity 81:666–673

Arondel V, Lemieux B, Hwang I, Gibson S, Goodman HM, Somerville CR (1992) Map-based cloning of a gene controlling omega-3 fatty acid desaturation in *Arabidopsis*. Science 258:1353–1355

Arumuganathan K, Earle ED (1991) Nuclear DNA content of some important plant species. Plant Mol Biol Rep 9:208–218

Arus P, Orton TJ (1983) Inheritance and linkage relationships of isozyme loci in *Brassica oleracea*. J Hered 74:405–412

Attieh J, Djiana R, Koonjul P, Etienne C, Sparace SA, Saini HS (2002) Cloning and functional expression of two plant thiol methyltransferases: a new class of enzymes involved in the biosynthesis of sulfur volatiles. Plant Mol Biol 50:511–521

Axelsson T, Shavorskaya O, Lagercrantz U (2001) Multiple flowering time QTLs within several *Brassica* species could be the result of duplicated copies of one ancestral gene. Genome 44:856–864

Babula D, Kaczmarek M, Delseny M, Quiros CF, Sadowski J (2000) Construction of a genetic map for *Brassica oleracea* based on ESTs from the *Arabidopsis thaliana* genome. Acta Hort 539:95–99

Babula D, Kaczmarek M, Barakat A, Delseny M, Quiros CF, Sadowski J (2003) Chromosomal mapping of *Brassica oleracea* based on ESTs from *Arabidopsis thaliana*: complexity of the comparative map. Mol Genet Genom 268:656–665

Babula D, Misztal LH, Jakubowicz M, Kaczmarek M, Nowak W, Sadowski J (2006) Genes involved in biosynthesis and signalization of ethylene in *Brassica oleracea* and *Arabidopsis thaliana*: identification and genome comparative mapping of specific gene homologs. Theor Appl Genet 112:410–420

Baggett JR, Kean D (1989) Inheritance of annual flowering in *Brassica oleracea*. HortScience 24:662–664

Baggett JR, Wahlert W (1975) Annual flowering and growth habit in cabbage-broccoli crosses. HortScience 10:170–172

Bain DC (1952) Reaction of *Brassica* seedlings to black rot. Phytopathology 42:497–500

Bancroft I (2001) Duplicate and diverge: the evolution of plant genome microstructure. Trends Genet 17:89–93

Bateman AJ (1955) Self-incompatibility systems in angiosperms. III. Cruciferae. Heredity 9:52–68

Beecher CWW (1994) Cancer preventative properties of varieties of *Brassica oleracea*. Am J Clin Nutr 59:1166–1170

Bettey M, Finch-Savage WE, King GJ, Lynn JR (2000) Quantitative genetic analysis of seed vigour and preemergence seedling growth traits in *Brassica oleracea*. New Phytol 148:277–286

Blanc G, Barakat A, Guyot R, Cooke R, Delseny M (2000) Extensive duplication and reshuffling in the *Arabidopsis* genome. Plant Cell 12:1523–1540

Blanc G, Hokamp K, Wolfe KH (2003) A recent polyploidy superimposed on older large-scale duplications in the *Arabidopsis* genome. Genome Res 13:137–144

Bohuon EJR, Keith DJ, Parkin IAP, Sharpe AG, Lydiate DJ (1996) Alignment of the conserved C genomes of *Brassica oleracea* and *Brassica napus*. Theor Appl Genet 93:833–839

Bohuon EJR, Ramsay LD, Craft JA, Arthur AE, Marshall DF, Lydiate DJ, Kearsey MJ (1998) The association of flowering time quantitative trait loci with duplicated regions and candidate loci in *Brassica oleracea*. Genetics 150:393–401

Boukema IW, van Hintum ThJL, Astley D (1997) The creation and composition of the *Brassica oleracea* Core Collection. Plant Genet Resour News 111:29–32

Boyes DC, Nasrallah ME, Vrebalov J, Nasrallah JB (1997) The self-incompatibility (S) haplotypes of *Brassica* contain highly divergent and rearranged sequences of ancient origin. Plant Cell 9:237–247

Buczacki ST (1983) *Plasmodiophora*: an interrelationship between biological and practical problems. In: Buczacki ST (ed.) Zoosporic Plant Pathogens: A Modern Perspective. Academic, London, pp 161–191

Camargo LEA, Osborn TC (1996) Mapping loci controlling flowering time in *Brassica oleracea*. Theor Appl Genet 92:610–616

Camargo LEA, Williams PH, Osborn TC (1995) Mapping of quantitative trait loci controlling resistance of *Brassica oleracea* to *Xanthomonas campestris* pv. *campestris* in the field and greenhouse. Phytopathology 85:1296–1300

Camargo LEA, Savides L, Jung G, Nienhuis J, Osborn TC (1997) Location of the self-incompatibility locus in an RFLP and RAPD map of *Brassica oleracea*. J Hered 88:57–60

Cao J, Tang JD, Strizhov N, Shelton AM, Earle ED (1999) Transgenic broccoli with high levels of *Bacillus thuringiensis* Cry1C protein, control diamondback moth larvae resistant to Cry1A or Cry1C. Mol Breed 5:131–141

Cao J, Zhao J-Z, Tang JD, Shelton AM, Earle ED (2002) Broccoli plants with pyramided cry1Ac and cry1C Bt genes control diamondback moths resistant to Cry1A and Cry1C proteins. Theor Appl Genet 105:258–264

Carpenter JE, Gianessi LP (2000) Value of Bt and herbicide-resistant cottons. Proc Beltwide Cotton Conf 1:76–79

Cavell AC, Lydiate DJ, Parkin IAP, Dean C, Trick M (1998) Collinearity between a 30-centiMorgan segment of *Arabidopsis thaliana* chromosome 4 and duplicated regions within the *Brassica napus* genome. Genome 41:62–69

Cheung WY, Champagne G, Hubert N, Landry BS (1997) Comparison of the genetic maps of *Brassica napus* and *Brassica oleracea*. Theor Appl Genet 94:596–582

Coehlo PS, Monteiro A (2003) Inheritance of downy mildew resistance to mature broccoli plants. Euphytica 131:65–69

Coehlo PS, Leckie D, Bahcevandziev K, Valério L, Astley D, Boukema I, Crute IR, Monteiro A (1998) The relationship between cotyledon and adult plant resistance to downy mildew (*Peronospora parasitica*) in *Brassica oleracea*. Acta Hort 459:335–342

Cogan NOI, Lynn JR, King GJ, Kearsey MJ, Newbury, Puddephat IJ (2002) Identification of genetic factors controlling the efficiency of *Agrobacterium rhizogenes*-mediated transformation in *Brassica oleracea* by QTL analysis. Theor Appl Genet 105:568–576

Cole K (1959) Inheritance of male sterility in green sprouting broccoli. Can J Genet Cytol 1:203–207

Crisp P, Crute IR, Sutherland RA, Angeli SM, Bloor K, Burgess H, Gordon P (1989) The exploitation of genetic resources of *Brassica oleracea* in breeding for resistance in clubroot (*Plasmodiophora brassicae*). Euphytica 42:215–226

Crouch ML, Tenbarge KM, Simon AE, Ferl R (1983) cDNA clones for *Brassica napus* seed storage proteins: evidence from nucleotide sequence analysis that both subunits of napin are cleaved from a precursor polypeptide. J Mol Appl Genet 2:273–283

Crouch JH, Lewis BG, Mithen RF (1994) The effect of a genome substitution on the resistance of *Brassica napus* to infection by *Leptosphaeria maculans*. Plant Breed 112:265–278

Crute IR, Gray AR, Crisp P, Buczacki ST (1980) Variation in *Plasmodiophora brassicae* and resistance to clubroot disease in Brassicas and allied crops: a critical review. Plant Breed Abstr 50:91–104

Crute IR, Phelps K, Barnes A, Buczacki ST, Crisp P (1983) The relationship between genotypes of three *Brassica* species and collections of *Plasmodiophora brassicae*. Plant Pathol 32:405–420

Das S, Roscoe TJ, Delseny M, Srivastava PS, Lakshmikumaran M (2002) Cloning and molecular characterization of the Fatty Acid Elongase 1 (FAE 1) gene from high and low erucic acid lines of *Brassica campestris* and *Brassica oleracea*. Plant Sci 162:245–250

Detjen LR (1926) A preliminary report on cabbage breeding. Proc Am Soc Hort Sci 23:325–332

Dias JS, Ferreira ME, Williams PH (1993) Screening of Portuguese cole landraces (*Brassica oleracea* L.) with *Peronospora parasitica* and *Plasmodiophora brassicae*. Euphytica 67:135–141

Dickson MH (1968) Eight newly described genes in broccoli. Proc Am Soc Hort Sci 93:356

Dickson MH, Petzoldt R (1993) Plant age and isolate source affect expression of downy mildew resistance in broccoli. HortScience 28:730–731

Doebley JF, Wendel JF (1989) Application of RFLP to plant systematics. In: Helentjaris T, Burr B (eds.) Development and Application of Molecular Markers to Problems in Plant Genetics. Cold Spring Harbor Lab, Cold Spring Harbor, NY, pp 57–67

Erickson LR, Straus NA, Beversdorf WD (1983) Restriction patterns reveal origins of chroloplast genomes in *Brassica* amphidiploids. Theor Appl Genet 65:201–206

Fahey J, Talalay P (1995) The role of crucifers in cancer chemoprotection. In: Gustine DL, Flores HE, Rockville MD (eds.) Phytochemicals and Health. American Society of Plant Physiologists, Rockville, MD, pp 187–189

Farinhó M, Coelho P, Carlier J, Svetleva D, Monteiro A, Leitão J (2004) Mapping of a locus for adult plant resistance to downy mildew in broccoli (*Brassica oleracea* convar. *italica*). Theor Appl Genet 109:1392–1398

Farnham MW, Wang M, Giovannelli JL (2001) Towards mapping genes for downy mildew resistance in broccoli. In: Proceedings of the Plant and Animal Genome IX Conference, San Diego, CA, 2001, p 18

Farnham MW, Wang M, Thomas CE (2002) A single dominant gene for downy mildew resistance in broccoli. Euphytica 128:405–407

Farnham MW, Stephenson K, Fahey J (2003) Glucoraphanin concentration of broccoli seed is highly influenced by genotype. HortScience 38:677

Faulkner K, Mithen R, Williamson G (1998) Selective increase of the potential anticarcinogen 4-methylsulphinylbutyl glucosinolate in broccoli. Carcinogenesis 19:605–609

Fenwick GR, Heaney RK, Mawson R (1989) Glucosinolates. In: Cheeke PR (ed.) Toxicants of Plant Origin, vol II. CRC, Boca Raton, FL, pp 1–41

Ferreira ME, Satagopan J, Yandell BS, Williams PH, Osborn TC (1995) Mapping loci controlling vernalization requirement and flowering time in *Brassica napus*. Theor Appl Genet 90:727–732

Figdore SS, Kennard WC, Song KM, Slocum MK, Osborn TC (1988) Assessment of the degree of restriction fragment length polymorphism in *Brassica*. Theor Appl Genet 75:833–840

Figdore SS, Fereira ME, Slocum MK, Williams PH (1993) Association of RFLP markers with trait loci affecting clubroot resistance and morphological characters in *Brassica oleracea* L. Euphytica 69:33–44

Fransz P, Armstrong S, Alonso-Blanco C, Fischer TC, Torres-Ruiz RA, Jones G (1998) Cytogenetics for the model system *Arabidopsis thaliana*. Plant J 13:867–876

Fransz PF, Armstrong S, de Jong JH, Parnell LD, van Drunen C, Dean C, Zabel P, Bisseling T, Jones GH (2000) Integrated cytogenetic map of chromosome arm 4S of *A. thaliana*: structural organization of heterochromatic knob and centromere region. Cell 100:367–376

Gao MQ, Li GY, Yang B, McCombie WR, Quiros CF (2004) Comparative analysis of a *Brassica* BAC clone containing several major aliphatic glucosinolate genes with its corresponding *Arabidopsis* sequence. Genome 47:666–679

Gapper NE, McKenzie MJ, Christey MC, Braun RH, Coupe SA, Lill RE, Jameson PE (2002) *Agrobacterium tumefaciens*-mediated transformation to alter ethylene and cytokinin biosynthesis in broccoli. Plant Cell Tiss Org Cult 70:41–50

Gelvin SB (2000) *Agrobacterium* and plant genes involved in T-DNA transfer and integration. Annu Rev Plant Physiol Plant Mol Biol 51:223–256

Giamoustaris A, Mithen R (1996) Genetics of aliphatic glucosinolates IV. Sidechain modification in *B. oleracea*. Theor Appl Genet 93:1006–1010

Giovannelli JL, Farnham MW, Wang M, Strand AE (2002) Development of sequence characterized amplified region markers linked to downy mildew resistance in broccoli. J Am Soc Hort Sci 127:597–601

Granado F, Olmedilla B, Blanco I (2003) Nutritional and clinical relevance of lutein in human health. Br J Nutr 90:487–502

Grandclement C, Thomas G (1996) Detection and analysis of QTLs based on RAPD markers for polygenic resistance to *Plasmodiophora brassicae* Woron. in *Brassica oleracea* L. Theor Appl Genet 93:86–90

Grant D, Cregan P, Shoemaker RC (2000) Genome organization in dicots: genome duplication in *Arabidopsis* and synteny between soybean and *Arabidopsis*. Proc Natl Acad Sci USA 97:4168–4173

Gray AR (1989) Taxonomy and evolution of broccolis and cauliflowers. Baileya 23:28–46

Hansen LN, Rodomiro O, Andersen SB (1999) Genetic analysis of protoplast regeneration ability in *Brassica oleracea*. Plant Cell Tiss Org Cult 58:127–132

Happstadius I, Ljungberg A, Kristiansson B, Dixelius C (2003) Identification of *Brassica oleracea* germplasm with improved resistance to Verticillium wilt. Plant Breed 122:30–34

Helm J (1963) Morphologish-taxonomische Gliederung der Kultursipen von *Brassica oleracea* L. Kulturpflanze 11:92–210

Henriksson H, Denman SE, Ademark P, Master ER, Teeri TT, Brumer IH, Campuzano IDG (2003) N-linked glycosylation of native and recombinant cauliower xyloglucan endotransglycosylase 16A. Biochem J 61–73

Holme IB, Trop AM, Hansen LN, Andersen SB (2004) Quantitative trait loci affecting plant regeneration from protoplasts of *Brassica oleracea*. Theor Appl Genet 108:1513–1520

Horovitz S, Perlasca G (1954) Genetics of normal flowering of cabbage in the torrid tropics. Agron Trop 4:81–93

Hoser-Krauze J, Lakowska-Ryk E, Antosik J (1995) The inheritance of resistance of some *Brassica oleracea* L. cultivars and lines to downy mildew, *Peronospora parasitica* (Pers) ex. Fr. J Appl Genet 36:27–33

Howell EC, Barker GC, Jones GH, Kearsey MJ, King GJ, Kop EP, Ryder CD, Teakle GR, Vicente JG, Armstrong SJ (2002) Integration of the cytogenetic and genetic linkage maps of *Brassica oleracea*. Genetics 161:1225–1234

Hu J, Sadowski J, Osborn TC, Landry TC, Quiros CF (1998) Linkage group alignment from four independent *Brassica oleracea* RFLP maps. Genome 41:226–235

Hunter JE, Dickson MH, Ludwig JW (1987) Sources of resistance to black rot in crucifers and inheritance of resistance. HortScience 26:1545–1547

Inaba R, Nishio T (2002) Phylogenetic analysis of *Brassiceae* based on the nucleotide sequences of the *S*-locus related gene, SLR1. Theor Appl Genet 105:1159–1165

Jourdan PS, Earle ED (1989) Genotypic variability in the freqency of plant regeneration from leaf protoplasts of four *Brassica* ssp. and of *Raphanus sativus*. J Am Soc Hort Sci 114:343–349

Kachroo A, Nasrallah ME, Nasrallah JB (2002) Self-incompatibility in the Brassicaceae: receptor-ligand signaling and cell-to-cell communication. Plant Cell 14:S227–S238

Kaczmarek M, Babula D, Sadowski J (2004) Application of the *Arabidopsis thaliana* sequence data in understanding the *Brassica* genome. In: Sadowski J (ed.) Understanding the Plant Genome. IPG PAS, Poznan, pp 113–122

Kalloo G, Bergh BO (1993) Genetic Improvement of Vegetable Crops. Pergamon, Oxford

Karpechenko GD (1922) The number of chromosomes and the genetic correlation of cultivated cruciferae. Bull Appl Bot Genet Plant Breed 13:3–14

Kearsey MJ, Ramsay LD, Jennings DE, Lydiate DJ, Bohuon EJR, Marshall DF (1996) Higher recombination frequencies in female compared to male meiosis in *Brassica oleracea*. Theor Appl Genet 92:363–367

Kempin S, Savidge S, Yanofsky MF (1995) Molecular basis of the cauliflower phenotype in *Arabidopsis*. Science 267:522–525

Kennard WC, Slocum MK, Figdore SS, Osborn TC (1994) Genetic analysis of morphological variation in *Brassica oleracea* using molecular markers. Theor Appl Genet 87:721–732

Kianian SF (1990) Evolutionary trends in the *Brassica oleracea*-cytodeme: chromosomal and linkage relationships. Dissertation, University of California, Davis, CA

Kianian SF, Quiros CF (1992a) Generation of a *Brassica oleracea* composite RFLP map: linkage arrangements among various populations and evolutionary implications. Theor Appl Genet 84:544–554

Kianian SF, Quiros CF (1992b) Genetic analysis of major multigene families in *Brassica oleracea* and related species. Genome 35:516–527

King GJ, Teakle GR, Allender CJ, Barker GC, Pink DA, Ryder CD (2004) From trait to genome: characterizing *Brassica* diversity. In: Proceedings of the Plant, Animal and Microbe Genome XII Conference, San Diego, CA, 2004, p 46

Koch M, Haubold B, Mitchell-Olds T (2000) Comparative evolutionary analysis of chalcone synthase and alcohol dehydrogenase loci in *Arabidopsis*, *Arabis*, and related genera. Mol Biol Evol 17:1482–1498

Koch M, Haubold B, Mitchell-Olds T (2001) Molecular systematics of the Brassicaceae: evidence from coding plastidic *matK* and nuclear *Chs* sequences. Am J Bot 88:534–544

Kop EP, Teakle GR, McClenaghan RE, Lynn JR, King GJ (2003) Genetic analysis of the bracting trait in cauliflower and broccoli. Plant Sci 164:803–808

Koornneef M, Fransz P, de Jong H (2003) Cytogenetic tools for *Arabidopsis thaliana*. Chrom Res 11:183–194

Kowalski SP, Lan T-H, Feldmann KA, Paterson AH (1994) Comparative mapping of *Arabidopsis thaliana* and *Brassica oleracea* chromosomes reveals islands of conserved organization. Genetics 138:499–510

Kristofferson KB (1924) Contributions to the genetics of *Brassica oleracea*. Hereditas 5:297–364

Lagercrantz U (1998) Comparative mapping between *Arabidopsis thaliana* and *Brassica nigra* indicates that *Brassica* genomes have evolved through extensive genome replication accompanied by chromosome fusion and frequent rearrangments. Genetics 150:1217–1228

Lagercrantz U, Lydiate DJ (1996) Comparative genome mapping in *Brassica*. Genetics 144:1903–1910

Lagercrantz U, Putterill J, Coupland G, Lydiate D (1996) Comparative mapping in *Arabidopsis* and *Brassica*, fine scale genome collinearity and congruence of genes controlling flowering time. Plant J 9:13–20

Lan T-H, Paterson AH (2000) Comparative mapping of quantitative trait loci sculpting the curd of *Brassica oleracea*. Genetics 155:1927–1954

Lan T-H, Paterson AH (2001) Comparative mapping of QTLs determining the plant size of *Brassica oleracea*. Theor Appl Genet 103:383–397

Lan T-H, DelMonte TA, Rieschmann KP, Hyman J, Kowalski SP, McFerson J, Kresovich S, Paterson AH (2000) An EST-enriched comparative map of *Brassica oleracea* and *Arabidopsis thaliana*. Genome Res 10:776–788

Landry BS, Hubert N, Crete R, Chiang MS, Lincoln SE, Etoh T (1992) A genetic map for *Brassica oleracea* based on RFLP markers detected with expressed DNA sequences and mapping of resistance genes to race 2 of *Plasmodiophora brassicae* (Woronin). Genome 35:409–420

Laurens F, Thomas G (1993) Inheritance of resistance to clubroot (*Plasmodiophora brassicae* Wor.) in kale (*Brassica oleracea* ssp. *acephala*). Hereditas 119:253–262

Lázaro A, Aguinagalde I (1998) Genetic diversity in *Brassica oleracea* L. (Cruciferae) and wild relatives (2n = 18) using isozymes. Ann Bot 82:821–828

Li G, Quiros CF (2001) Sequence-related amplified polymorphism (SRAP), a new marker system based on a simple PCR reaction: its application to mapping and gene tagging in *Brassica*. Theor Appl Genet 103:455–461

Li G, Quiros CF (2002) Genetic analysis, expression and molecular characterization of BoGSL-ELONG, a major gene involved in the aliphatic glucosinolate pathway of *Brassica* species. Genetics 162:1937–1943

Li G, Quiros CF (2003) In planta sidechain glucosinolate modification in *Arabidopsis* by introduction of dioxygenase *Brassica* homolog BoGSL-ALK. Theor Appl Genet 106:1116–1121

Li G, Riaz A, Goyal S, Abel S, Quiros CF (2001a) Inheritance of three major genes involved in the synthesis of aliphatic glucosinolates in *Brassica oleracea*. J Am Soc Hort Sci 126:427–431

Li G, Gao M, Yang B, Quiros CF (2003a) Gene for gene alignment between the *Brassica* and *Arabidopsis* genomes by direct transcriptome mapping. Theor Appl Genet 107:168–180

Li L, Garvin DF (2003) Molecular mapping of *Or*, a gene inducing β-carotene accumulation in cauliflower (*Brassica oleracea* L. var. *botrytis*). Genome 46:588–594

Li L, Paolillo DJ, Parthasarathy MV, Dimuzio EM, Garvin DF (2001b) A novel gene mutation that confers abnormal patterns of β-carotene accumulation in cauliflower (*Brassica oleracea* var. *botrytis*). Plant J 26:59–67

Li L, Lu S, O'Halloran DM, Garvin DF, Vrebalov J (2003b) High-resolution genetic and physical mapping of the cauliflower high-β-carotene gene *Or* (Orange). Mol Genet Genom 270:132–138

Loudon PT, Nelson RS, Ingram DS (1989) Studies of protoplast culture and plant regeneration from commercial and rapid-cycling *Brassica* species. Plant Cell Tiss Org Cult 19:213–224

Ludwików A, Gallois P, Sadowski J (2004) Ozone induced oxidative stress response in *Arabidopsis*: transcription profiling by microarray approach. Cell Mol Biol Lett 9:829–842

Lukens L, Zou F, Lydiate D, Parkin I, Osborn T (2003) Comparison of a *Brassica oleracea* genetic map with the genome of *Arabidopsis thaliana*. Genetics 164:359–372

Lydiate D, Sharpe A, Lagercrantz U, Parkin I (1993) Mapping the *Brassica* genome. Outlook Agri 22:85–92

Maggioni L (1998) Activities and achievements of the ECP/GR *Brassica* Working Group. Acta Hort 459:243–248

Mandel MA, Gustafson-Brown C, Savidge B, Yanofsky MF (1992) Molecular characterization of the *Arabidopsis* floral homeotic gene APETALA1. Nature 360:273–277

Manzanares-Dauleux MJ, Divaret I, Baron F, Thomas G (2000) Evaluation of French *Brassica oleracea* landraces for resistance to *Plasmodiophora brassicae*. Euphytica 113:211–218

Mayne ST (1996) β-carotene, carotenoids, and disease prevention in humans. FASEB J 10:690–701

Mazur BJ, Tingey SV (1995) Genetic mapping and introgression of genes of agronomic importance. Curr Opin Biotechnol 6:175–182

McGrath JM, Quiros CF (1991) Inheritance of isozyme and RFLP markers in *Brassica campestris* and comparison with *B. oleracea*. Theor Appl Genet 82:668–673

McGrath JM, Quiros CF, Harada JJ, Landry BS (1990) Identification of *Brassica oleracea* monosomic alien chromosome addition lines with molecular markers reveals extensive gene duplication. Mol Gen Genet 223:198–204

McGrath JM, Jancso MM, Pichersky E (1993) Duplicate sequences with a similarity to expressed genes in the genome of *Arabidopsis thaliana*. Theor Appl Genet 86:880–888

Metz TD (2001) Transgenic broccoli (*Brassica oleracea* var. *italica*) and cabbage (var. *capitata*). In: Bajaj YPS (ed.) Biotechnology in Agriculture and Forestry, vol 47. Transgenic Crops II. Springer, Berlin Heidelberg New York, pp 69–86

Metz TD, Dixit R, Earle ED (1995) *Agrobacterium tumefaciens*-mediated transformation of broccoli (*Brassica oleracea* var. *italica*) and cabbage (*B. oleracea* var. *capitata*). Plant Cell Rep 15:287–292

Michaels SD, Bezerra IC, Amasino RM (2004) FRIGIDA-related genes are required for the winter-annual habit in *Arabidopsis*. Proc Natl Acad Sci USA 101:3281–3285

Michelmore RW, Paran I, Kesseli RV (1991) Identification of markers linked to disease-resistance genes by bulked segregant analysis: a rapid method to detect markers in specific genomic regions by using segregating populations. Proc Natl Acad Sci USA 88:9828–9832

Mithen R, Lewis BG, Heaney R, Fenwick GR (1987) Glucosinolates of wild and cultivated *Brassica* species. Phytochemistry 26:1969–1973

Mizushima U (1950) Karyogenetic studies of species and genus hybrids in the tribe *Brassiceae* of *Cruciferae*. Tohoku J Agri Res 1:15–27

Monteiro AA, Williams PH (1989) The exploration of resources of Portuguese cabbage and kale for resistance to several *Brassica* diseases. Euphytica 41:215–225

Moriguchi K, Kimizuka-Takagi Ch, Ishii K. Nomura K (1999) A genetic map based on RAPD, RFLP, isozyme, morphological markers and QTL analysis for clubroot resistance in *Brassica oleracea*. Breed Sci 49:257–265

Muehlbauer GJ, Specht JE, Thomas-Compton MA, Staswick PE, Bernard RL (1988) Near-isogenic lines: A potential resource in the integration of conventional and molecular marker linkage maps. Crop Sci 28:729–735

Nasrallah JB, Kao TH, Goldberg ML, Nasrallah ME (1985) A cDNA clone encoding an *S*-locus specific glycoprotein from *Brassica oleracea*. Nature 318:262–267

Nasrallah JB, Yu SM, Nasrallah ME (1988) Self-incompatibility genes of *Brassica oleracea*: expression, isolation, and structure. Proc Natl Acad Sci USA 85:5551–5555

Natti JJ, Atkin JD (1960) Inheritance of downy mildew resistance in broccoli. Phytopathology 50:241

Natti JJ, Dickson MH, Atkin JD (1967) Resistance of *Brassica oleracea* varieties to downy mildew. Phytopathology 57:144–147

Ockendon DJ, Gates PJ (1975) Growth of cross and self-pollen tubes in the styles of *Brassica oleracea*. New Phytol 75:155–160

Olin-Fatih M, Heneen WG (1992) C-banded karyotypes of *Brassica campestris*, *B. oleracea* and *B. napus*. Genome 35:283–289

Olsson G (1960) Species crosses within the genus *Brassica*. II. Artificial *Brassica napus* L. Hereditas 46:351–396

O'Neill CM, Bancroft I (2000) Comparative physical mapping of segments of the genome of *Brassica oleracea* var. *alboglabra* that are homoeologous to sequenced regions of chromosomes 4 and 5 of *Arabidopsis thaliana*. Plant J 23:233–243

Osborn TC, Kole C, Parkin IAP, Sharpe AG, Kuiper M, Lydiate DJ, Trick M (1997) Comparison of flowering time genes in *Brassica rapa*, *B. napus* and *Arabidopsis thaliana*. Genetics 146:1123–1129

Palmer JD, Shield CR, Cohen DB, Orton TJ (1983) Chloroplast DNA evolution and the origin of amphidiploid *Brassica* species. Theor Appl Genet 65:181–189

Parker S, Cox GS (1974) The Giant Cabbage of the Channel Islands. Toucan, St. Peter Port Guernsey, UK

Parkin IAP, Sharpe AG, Keith DJ, Lydiate DJ (1995) Identification of the A and C genomes of an amphidiploid *Brassica napus* (oilseed rape). Genome 38:1122–1131

Pease MA (1926) Genetic situation in *Brassica oleracea*. J Genet 16:363–385

Pelofske PJ, Baggett JR (1979) Inheritance of internode lengh, plant form and annual habit in a cross of cabbage and broccoli (*Brassica oleracea* var. *capitata* and var. *italica*). Euphytica 28:189–197

Poulsen GB (1996) Genetic transformation of *Brassica*. Plant Breed 115:209–225

Prakash S, Hinata K (1980) Taxonomy, cytogenetics and origin of crop Brassicas, a review. Opera Bot 55:1–57

Puddephat IJ, Riggs TJ, Fenning TM (1996) Transformation of *Brassica oleracea* L.: a critical review. Mol Breed 2:185–210

Quiros CF (2001) DNA-based marker maps of *Brassica*. In: Phillips RL, Vasil IK (eds.) DNA-Based Markers in Plants. Kluwer Academic, London, pp 201–238

Quiros CF, Ochoa O, Kianian SF, Douches D (1987) Analysis of the *Brassica oleracea* genome by the generation of *B. rapa-oleracea* chromosome addition lines: characterization by isozymes and rDNA genes. Theor Appl Genet 74:758–766

Quiros CF, Ochoa O, Douches DS (1988) Exploring the role of x = 7 species in *Brassica* evolution: hybridization with *B. nigra* and *B. oleracea*. J Hered 79:351–358

Quiros CF, Hu J, Truco MJ (1994) DNA based *Brassica* maps. In: Phillips RL, Vasil IK (eds.) DNA-Based Markers in Plants, Kluwer Academic, London, pp 199–222

Quiros CF, Grellet F, Sadowski J, Suzuki T, Li G, Wroblewski T (2001) *Arabidopsis* and *Brassica* comparative genomics: sequence, structure and gene content in the *ABI1-Rps2-Ck1* chromosomal segment and related regions. Genetics 157:1321–1330

Radchuk VV, Ryschka U, Schumann G, Klocke E (2002) Genetic transformation of cauliflower (*Brassica oleracea* var. *botrytis*) by direct DNA uptake into mesophyll protoplasts. Physiol Plant 114:429–438

Rae AK, Howell EC, Kearsey MJ (1999) More QTL for flowering time reveled by substitution lines in *Brassica oleracea*. Heredity 83:586–596

Ramanujam S, Srinivasachar D (1943) Cytogenetic investigations in the genus *Brassica* and the artificial synthesis of *B. juncea*. Ind J Genet 3:73–88

Ramsay LD, Jennings DE, Bohuon EJR, Arthur AE, Lydiate DJ, Kearsey MJ, Marshall DF (1996) The construction of a substitution library of recombinant backcross lines in *Brassica oleracea* for the precision mapping of quantitative trait loci. Genome 39:558–567

Ramsey AD, Ellis PR (1996) Resistance in wild *Brassica* to the cabbage whitefly, *Aleyrodes proletella*. In: Proceedings of the 9th Crucifer Genet Workshop. Acta Hort 407:507–514

Rana D, Boogaart T, O'Neill CM, Hynes L, Bent E, Macpherson L, Park JY, Lim YP, Bancroft I (2004) Conservation of the microstructure of genome segments in *Brassica napus* and its diploid relatives. Plant J 40:725–733

Richharia RH (1937a) Cytological investigations of *Raphanus sativus*, *Brassica oleracea* and their F_1 hybrids. J Genet 34:19–44

Richharia RH (1937b) Cytological investigations of chromosome species of *Brassica* and their F_1 hybrids. J Genet 34:45–55

Röbbelen G (1960) Beiträge zur analyse des *Brassica* genomes. Chromosoma 11:205–228

Rocherieux J, Glory P, Giboulot A, Boury S, Barbeyron G, Thomas G, Manzanares-Dauleux MJ (2004) Isolate-specific and broad-spectrum QTLs are involved in the control of clubroot in *Brassica oleracea*. Theor Appl Genet 108:1555–1563

Ryder CD, Smith LB, Teakle GR, King GJ (2001) Contrasting genome organization: two regions of the *Brassica oleracea* genome compared with collinear regions of the *Arabidopsis thaliana* genome. Genome 44:808–817

Saal B, Plieske J, Hu J, Quiros CF, Struss D (2001) Microsatellite markers for genome analysis in *Brassica*. II. Assignment of rapeseed microsattelites to the A and C genomes and genetic mapping in *Brassica oleracea* L. Theor Appl Genet 102:695–699

Sachan JS, Singh B (1987) Generation mean analysis for flowering and maturity in Indian mustard [*Brassica juncea* (L.) Czern & Cross]. Theor Appl Genet 73:571–574

Sadowski J, Hu J, Delseny M, Quiros C (1994) Genetical and physical mapping of an *Arabidopsis* gene complex in *Brassica* genomes. Crucifer News 16:47–48

Sadowski J, Gaubier P, Delseny M, Quiros CF (1996) Genetic and physical mapping in *Brassica* diploid species of a gene cluster defined in *Arabidopsis thaliana*. Mol Gen Genet 251:298–306

Sampson DR (1966) Genetic analysis of *Brassica oleracea* using nine genes from sprouting broccoli. Can J Genet Cytol 8:404–413

Sampson DR (1978) A second gene for hairs in *B. oleracea* and its tentative location in linkage group four. Can J Genet Cytol 20:101–109

Schopfer CR, Nasrallah ME, Nasrallah JB (1999) The male determinant of self-incompatibility in *Brassica*. Science 286:1697–1700

Schranz ME, Quijada P, Sung S-B, Lukens L, Amasino R, Osborn TC (2002) Characterization and effects of the replicated flowering time gene FLC in *Brassica rapa*. Genetics 162:1457–1468

Sebastian RL, Howell EC, King GJ, Marshall DF, Kearsey MJ (2000) An integrated AFLP and RFLP *Brassica oleracea* linkage map from two morphologically distinct double haploid mapping populations. Theor Appl Genet 100:75–81

Sebastian RL, Kearsey MJ, King GJ (2002) Identification of quantitaive trait loci controlling developmental characteristics of *Brassica oleracea* L. Theor Appl Genet 104:601–609

Sharpe AG, Parkin I, Keith DJ, Lydiate DJ (1995) Frequent translocations in the amphidiploid genome of oilseed rape (*Brassica napus*). Genome 38:1112–1121

Sikka SM (1940) Cytogenetics of *Brassica* hybrids and species. J Genet 40:441–509

Slocum MK (1989) Analyzing the structure of *Brassica* species using RFLP analysis. In: Helentjaris T, Burr B (eds.) Development and Application of Molecular Markers to Problems in Plant Genetics. Cold Spring Harbor Lab, Cold Spring Harbor, NY, pp 73–80

Slocum MK, Figdore SS, Kennard WC, Suzuki JY, Osborn TC (1990) Linkage arrangement of restriction fragment length polymorphism loci in *Brassica oleracea*. Theor Appl Genet 80:57–64

Snogerup S (1980) The wild forms of the *Brassica oleracea* group and their possible relations to the cultivated ones. In: Gomez-Campo C, Hinata K, Tsunoda S (eds.) *Brassica* Crops and Wild Allies: Biology and Breeding. Japan Scientific Societies, Tokyo, pp 121–132

Song KM, Osborn TC, Williams PH (1988a) *Brassica* taxonomy based on nuclear restriction fragment length polymorhisms (RFLPs). 1. Genome evolution of diploid and amphidiploid species. Theor Appl Genet 75:784–794

Song KM, Osborn TC, Williams PH (1988b) *Brassica* taxonomy based on nuclear restriction fragment length polymorphisms (RFLPs) 2. Preliminary analysis of subspecies within *B. rapa* (syn. *campestris*) and *B. oleracea*. Theor Appl Genet 76:593–600

Song KM, Osborn TC, Williams PH (1990) *Brassica* taxonomy based on nuclear restriction fragment length polymorphisms (RFLPs). 3. Genome relationships in *Brassica* and related genera and the origin of *B. oleracea* and *B. rapa* (syn. *campestris*). Theor Appl Genet 79:497–506

Song KM, Suzuki JY, Slocum MK, Williams PH, Osborn TC (1991) A linkage map of *Brassica rapa* (syn. *campestris*) based on restriction fragment length polymorphism loci. Theor Appl Genet 82:296–304

Sparrow PAC, Townsend TM, Arthur AE, Dale PJ, Irwin JA (2004) Genetic analysis of *Agrobacterium tumefaciens* susceptibility in *Brassica oleracea*. Theor Appl Genet 108:644–650

Stebbins JL (1966) Chromosomal variation and evolution; polyploidy and chromosome size and number shed light on evolutionary processes in higher plants. Science 152:1463–1469

Stein JC, Howlett B, Boyes DC, Nasrallah ME, Nasrallah JB (1991) Molecular cloning of a putative receptor protein kinase gene encoded at the self-incompatibility locus of *Brassica oleracea*. Proc Natl Acad Sci USA 88:8816–8820

Tang GQ, Bai YY, Loo SW (1996) Molecular cloning and primary sequence analysis of a gene encoding a putative chitinase gene in *Brassica oleracea* var. *capitata*. Cell Res 65–73

Teutonico RA, Osborn TC (1994) Mapping of RFLP and qualitative traits loci in *Brassica rapa* and comparison to the linkage maps of *B. napus*, *B.oleracea* and *Arabidopsis thaliana*. Theor Appl Genet 82:296–304

Teutonico RA, Osborn TC (1995) Mapping loci controlling the vernalization requirement in *Brassica rapa*. Theor Appl Genet 91:1279–1283

Thompson KF (1976) Cabbages, kales etc. *Brassica oleracea* (Cruciferae). In: Simmonds NW (ed.) Evolution of Crop Plants. Longman, London, pp 49–52

Trick M, Flavell RB (1989) A homozygous S genotype of *Brassica oleracea* expresses two S-like genes. Mol Gen Genet 218:112–117

Truco MJ, Quiros CF (1994) Structure and organization of the B genome based on a linkage map in *Brassica nigra*. Theor Appl Genet 89:590–598

Truco MJ, Hu J, Sadowski J, Quiros CF (1996) Inter- and intragenomic homology of the *Brassica* genomes: implications for their origin and evolution. Theor Appl Genet 93:1225–1233

Tsunoda S, Hirata K, Gomez-Campo C (1980) *Brassica* Crops and Wild Allies. Japan Scientific Societies, Tokyo

U N (1935) Genome analysis in *Brassica* with special reference to the experimental formation of *B. napus* and peculiar mode of fertilization. Jpn J Genet 7:784–794

US EPA (1999) Proceedings of the EPA/USDA Workshop on Bt Crop Resistance Management.

Vaughan JG (1977) A multidisciplinary study of the taxonomy and origin of *Brassica* crops. BioScience 27:35–40

Verhoeven DT, Verhagen H, Goldbohm RA, van den Brandt PA, van Poppel G (1997) A review of mechanisms underlying anticarcinogenicity by *Brassica* vegetables. Chem Biol Interact 103:79–129

Verma SC, Rees H (1974) Nuclear DNA and the evolution of allotetraploid *Brassicae*. Heredity 33:61–68

Vicente JG, King GJ (2001) Characterization of disease resistance gene-like sequences in *Brassica oleracea* L. Theor Appl Genet 102:555–563

Vision TJ, Brown DG, Tanksley SD (2000) The origins of genomic duplications in *Arabidopsis*. Science 290:2114–2117

Voorrips RE, Visser DL (1993) Examination of resistance to clubroot in accessions of *Brassica oleracea* using a glasshouse seedling test. Neth J Plant Pathol 99:269–276

Voorrips RE, Jongerius MC, Kanne HJ (1997) Mapping of two genes for resistance to clubroot (*Plasmodiophora brassicae*) in a population of double haploid lines of *Brassica oleracea* by means of RFLP and AFLP markers. Theor Appl Genet 94:75–82

Walkoff C (1963) A mutant annual cabbage. Euphytica 12:77–80

Wang X-H, Luo P (1987) Studies on the karyotypes and C-banding patterns of Chinese kale (*Brassica alboglabra*) and cabbage (*B. oleracea* var. *capitata*). Acta Bot Sin 29:149–155

Wang X-H, Luo P, Shu J-J (1989) Giemsa N-banding pattern in cabbage and Chinese kale. Euphytica 41:17–21

Watts LE (1964) Studies of maturity in F_1 and F_2 generation of cauliflowers from crosses between summer, autumn and winter types. J Hort Sci 39:84–91

Weisaeth G (1974) Some problems and results in breeding for clubroot resistance in cole crops. In: Proceedings of the Eucarpia Meetings–Cruciferae, pp 101–107

Williams PH (1980) Black rot: a continuing threat to world crucifers. Plant Dis 64:736–742

Williams PH, Hill CB (1986) Rapid-cycling populations of *Brassica*. Science 232:1385–1389

Wills AB (1977) A preliminary gene list in *Brassica oleracea*. Eucarpia Cruciferae News 2:22–24

Wiersma PA, Schmiemann MG, Condie JA, Crosby WL, Moloney MM (1989) Isolation, expressionand phylogenetic inheritance of an acetolactate synthase gene from *Brassica napus*. Mol Gen Genet 219:413–420

Yang YW, Lai KN, Tai PY, Ma DP, Li WH (1999) Molecular phylogenetic studies of *Brassica*, *Rorippa*, *Arabidopsis* and allied genera based on the internal transcribed spacer region of 18S-25S rDNA. Mol Phylogenet Evol 13:455–462

Zeise K, Buchmüller M (1997) Studies on the susceptibility to *Verticillium dahliae* KLEB. var. *longisporum* STARK of six related *Brassica* species. Z Pflanzenkrankheiten Pflanzenschutz 104:501–505

Zhao K-N, Bittisnich DJ, Halloran GM, Whitecross MI (1995) Studies of cotyledon protoplast cultures from *B. napus*, *B. campestris* and *B. oleracea*. II. Callus formation and plant regeneration. Plant Cell Tiss Org Cult 40:73–84

Ziółkowski PA, Sadowski J (2002) FISH-mapping of rDNAs and *Arabidopsis* BACs on pachytene complements of selected Brassicas. Genome 45:189–197

9 Eggplant

A. Frary[1], S. Doganlar[1], and M. C. Daunay[2]

[1] Izmir Institute of Technology, Department of Biology, Gulbahcekoyu Campus, Urla 35430, Izmir Turkey
e-mail: samidoganlar@iyte.edu.tr
[2] INRA, Unité de Génétique & Amélioration des Fruits et Légumes, Domaine St. Maurice, BP 94, 84143 Montfavet cedex, France

9.1 Introduction

9.1.1 A Brief History of the Crop

Eggplant or brinjal eggplant, *Solanum melongena* L., is an ancient crop in Eastern countries. Domestication of eggplant took place in the area between northeastern India and southwestern China and very early agricultural, botanical, and medicinal documents described its morphology, uses and properties. Wild *S. melongena* forms are still found in the area formerly known as Indo-China and some modern cultivars from this region have fruits that closely resemble wild *Solanum* species fruit types (round green berries, 3–5 cm in diameter). The domestication of eggplant primarily involved dramatic expansion of fruit shape, size and color diversity, as well as decreases in plant prickliness and fruit bitterness.

From its Indo-Chinese center of origin and domestication, eggplant moved even farther east to Japan (where the first records are dated from the eighth century AD) and west probably at the time of Muslim conquests (eighth to eleventh centuries) or even earlier. Progressively, eggplant cultivation spread to the whole Mediterranean basin, Central Europe, Africa and then America and is cultivated worldwide today.

S. incanum L., a complex of several wild species found in East Africa and Middle Eastern countries, has been demonstrated to be the closest taxon to *S. melongena*, and is the most probable progenitor of this latter species (Daunay et al. 2001a). *S. insanum* L. is probably a primitive cultivated form of eggplant which has escaped from cultivated fields and spontaneously returned to the wild state (Lester and Hasan 1991). Chloroplast DNA studies confirm the very close relationship between *S. melongena* and *S. incanum* (Sakata and Lester 1994, 1997; Sakata et al. 1991; Isshiki et al. 1998; Bohs and Olmstead 1999), formerly known for their morphological likeness and relatively easy crossability (Lester and Hasan 1991). Isozymes (Isshiki et al. 1994a,b; Karihaloo and Gottlieb 1995; Isshiki 1996) and nuclear DNA (Karihaloo et al. 1995; Mace et al. 1999) demonstrate little polymorphism within *S. melongena* germplasm and confirm the close relationship to *S. incanum*, which itself displays a much higher level of genetic diversity.

9.1.2 Botanical Description

S. melongena belongs to the Solanaceae family, and to the tribe Solaneae, which comprises several cultivated species including pepper (*Capsicum* sp.), tomato (*Solanum* section *Lycopersicon* sp.), potato (tuberous *Solanum* sp.), *Physalis* and *Cyphomandra* species (Daunay et al. 2001b). Eggplant and its related species belong to subgenus *Leptostemonum*, the largest subgenus of genus *Solanum*. The plant is woody and develops several branches according to a roughly dichotomic ramification pattern. Anthocyanins, prickles and hairiness on vegetative parts vary quantitatively. Inflorescences are 1 to 5 andromonoecious cymes, although most modern cultivars display solitary hermaphrodite flowers. The basic flower type is 5-merous (5 sepals, 5 petals, 5 stamens) but 6, 7, and 8-merous flowers are commonly found in globose and round fruited types. Eggplant is generally considered to be an autogamous species; however, in open fields and warm conditions, flowers are visited by insects and the rate of allogamy can reach 70% or more. The fruits are berries of highly variable shape (round, intermediate, long, snake-like) and size (tens of grams to more than a kilo). The absence or presence as well as the distribution pattern of two kinds of pigments, chlorophylls and anthocyanins, control a wide diversity of fruit colors (Daunay et al. 2004). Eggplant is a diploid species, with a basic chromosome number

Genome Mapping and Molecular Breeding in Plants, Volume 5
Vegetables
C. Kole (Ed.)

of 12 and a genome size of approximately 956 Mbp (Bennett and Leitch 2004).

9.1.3 Economic Importance

Eggplant is a very popular native vegetable in Asia and the Mediterranean basin. In 2003, eggplant world production was 29 million tons (t) from 1.6 million hectares (FAO 2003). Asia is the major eggplant producer, with China (16 million t; 851,000 ha) and India (8 million t; 510,000 ha) as leaders, followed far behind, by Turkey (970,000 t; 37,000 ha), Egypt (703,000 t; 31,000 ha), Japan (420,000 t; 12,000 ha), and Italy (377,000 t; 13,000 ha). The average yield (18 t/ha) is extremely variable, depending on climate, cultural system, crop duration and grower technology. The Netherlands is the absolute champion with yields of 390 t per hectare.

The nutritional energy value of eggplant is limited (Gebhardt and Thomas 2002), but the presence of good fiber and various vitamins and minerals in the fruit is beneficial to human health. Furthermore the fruit contains phenolic compounds such as anthocyanins and phenolic acids which have antioxidant properties (Cao et al. 1996; Stommel and Whitaker 2003) as well as alkaloids (Aubert et al. 1989a,b), which have several beneficial biological and pharmaceutical properties. The beneficial effects of eggplant alkaloids are further supported by the widespread use of eggplant in traditional medicine. Alkaloids can, however, have negative effects as excessive levels can result in extremely bitter, unpalatable fruit. Fortunately, the fruits of cultivated species normally do not contain such high levels of alkaloids unless the plants have been subjected to extreme stress.

9.1.4 Breeding Objectives

Regardless of the country of cultivation, the main breeding objectives concern quantitative and qualitative yield. Thus, a major goal is the adaptation of cultivars to various growing conditions including climate (temperate to tropical), season of growth and production system (open field or greenhouse). Pest and disease resistances are also significant breeding objectives. Quality factors include fruit color and glossiness, fruit palatability and plant prickliness. Other important factors are local aesthetic and culinary preferences. In Turkey, for example, different cultivars of eggplant with distinct morphologies are used for stuffing (small, slightly oblong fruits), roasting (large, fasciated fruits), and frying (elongated fruits). Improved postharvest quality is another important trait as it facilitates transport and extends shelf-life.

9.1.5 Classical Breeding Achievements

Classical breeding efforts in eggplant have included the examination of heterosis (Kakizaki 1931) and the production of F_1 hybrids which have been adopted for cultivation worldwide. In addition, increased earliness, reduced prickliness and decreased fruit bitterness have been important achievements. Dramatic progress has also been made in the last decade by breeders who have worked on the adaptation of eggplant production in the greenhouse in winter. As a result, cultivars adapted to a wide range of agricultural and climatic conditions are available on the seed market and provide good yield and quality.

Despite good progress in the area of quantitative and qualitative yield, the range of pathogens controlled by genetic resistance is limited primarily to bacterial wilt (*Ralstonia solanacearum*), Fusarium wilt (*Fusarium oxysporum* f. sp. *melongena*) and fruit anthracnosis (*Colletotrichum gloeosporiodes*). In some cases, wild relatives of eggplant have been used as sources of these disease resistances, which have been transferred into cultivated eggplant by interspecific hybridization. For example, *S. aethiopicum* has been used as a source of bacterial wilt resistance (Ano et al. 1991) and *S. incanum* has been used as a source of Verticillium wilt resistance (Robinson et al. 2001). Partial resistance to *Verticillium dahliae* has also been introduced into *S. melongena* from *S. linnaeanum* (syn. *S. sodomeum*) (Acciari et al. 2001). Resistances to *Meloidogyne* nematode species and various insects have been described either in *S. melongena* germplasm or related species, but their use poses difficulties such as insufficient levels of resistance, limited crossability and difficulty of evaluation. Interspecific protoplast fusion has yielded various progenies. However, this technique usually only succeeds in producing fertile hybrids for combinations for which sexual hybridization is also successful. This general rule has been observed for crosses between *S. melongena* and *S. aethiopicum* (Ano et al. 1991; Daunay et al. 1993), *S. melongena* and *S. torvum* (McCammon and Honma 1983; Sihachakr et al. 1989), *S. melongena* and *S. khasianum* (Sharma et al. 1980; Sihachakr et al. 1988), and *S. aethiopicum* and *S. violaceum* (Daunay et al. 1991; Tamura et al. 2002). One exception is the case of *S. melongena* and *S. sisymbriifolium*. For this cross, somatic hybridization produced a viable, albeit sterile, hybrid (Gleddie et al. 1986; Collonnier et al. 2003) whereas sexual hybridization was not successful (Sharma et al. 1984). Despite progress in this area, no eggplant cultivar, to date, has been produced via interspecific protoplast fusion.

9.1.6 Classical Mapping Efforts

As compared to other crops in the Solanaceae, knowledge of the genetic control of many morphological traits and abiotic and biotic stress tolerances in eggplant is rather poor. What is known about the genetic control of these traits has been summarized by Daunay et al. (2001b). Very little classical mapping has been performed in eggplant. Genetic control of anthocyanin accumulation is probably the best studied trait. For example, in early work, Nolla (1932) showed that green fruit color was completely linked to white corolla and nonstriped anthers. Linkage among three anthocyanin genes, *D*, *X* and *Puc*, was reported by Janick and Topoleski (1963). Tigchelaar et al. (1968) described the interactions and linkages among nine anthocyanin genes. They found that the three basic color genes, *D*, *P* and *Y*, have different functions and are not linked. In contrast, there was linkage among some modifier genes (*Puc*, *Sa* and *Dil2*) that control anthocyanin distribution and intensity. Linkage was also detected among genes for purple stem color (*Pst*), purple hypocotyl (*Phy*) and inhibition of anthocyanins in shaded fruit (*Ishyf*) (Wanjari and Khapre 1977). It has also been reported that male-sterility (*fms*) is linked to the *X* locus for purple fruit color (Phatak et al. 1991).

9.1.7 Limitations of Classical Endeavors and Utility of Molecular Mapping

Eggplant breeding has been limited by several factors. As with many other crops, most agronomic traits for eggplant are under oligogenic or polygenic control, which complicates their dissection and breeding. Most of the genetic research in solanaceous vegetables has focused on tomato, potato and pepper. As a result, there is very little knowledge about the genetic control of important traits in eggplant. In addition, the difficulty of measuring some traits in eggplant such as fruit glossiness, which depends on fruit developmental stage, epidermis color and light quality has restricted breeding progress. Because trait characterization and breeding are usually performed on adult plants which require ample greenhouse or field space, only limited numbers of genotypes can be screened. Classical breeding techniques also make it difficult to screen several traits at once on a given progeny (for example, to assess several disease resistances). Moreover, related germplasm which may be a rich resource for eggplant improvement (Daunay et al. 2001b; Robinson et al. 2001) has not been sufficiently characterized.

The application of molecular techniques will help improve the efficiency and effectiveness of eggplant breeding in several ways. Molecular analysis of complex quantitative traits helps to simplify breeding by allowing the determination of which loci have the greatest phenotypic effects so that breeding efforts can be targeted to a few selected loci. Molecular analysis also allows the identification of genes with minor phenotypic effects which, likewise, can be tracked with linked markers. Although molecular mapping cannot completely eliminate time-consuming or technically difficult trait characterization, once molecular markers that are tightly linked to the trait of interest (such as a particular disease resistance or fruit glossiness) are identified, plants can be prescreened at the seedling stage to eliminate unwanted types. This limits the number of genotypes grown to the adult stage, thereby, reducing the time required for phenotypic characterization and the space needed for greenhouse/field tests. Thus, marker-assisted selection is ideal for backcross breeding programs and can allow the rapid assessment of many more plants than some types of phenotypic screening. Moreover, linked molecular markers can be used for resistance screening tests for quarantined pathogens (e.g., for *Ralstonia solanacearum* in Europe). Marker-assisted selection also enables the simultaneous screening of many different traits on a single plant. Of course, continued linkage of the marker to the trait of interest must be periodically verified as genetic recombination can render a marker useless. Molecular techniques are also valuable for characterization of the genetic diversity and relationships among related *Solanum* species, which are increasingly being used as a gene pool for eggplant improvement. Despite the many benefits of molecular markers, implementation is often very expensive. As a result, their use for routine eggplant breeding may be limited to the most important traits (e.g., disease resistance) as eggplant, unlike other solanaceous crops like tomato, does not provide private breeding companies with high returns on investment.

9.2 Construction of Genetic Maps

9.2.1 A Brief History of Mapping in Eggplant

Although the related solanaceous species tomato, potato and pepper have been the focus of much molecular research, genetic mapping in eggplant has only initiated in the past decade. Random amplified polymorphic DNA (RAPD) markers were the first markers to be used for the development of a linkage map (Nunome et al. 1998). Since the development of this first map, amplified fragment length polymorphism (AFLP), restriction fragment length polymorphism (RFLP) and microsatellite (or simple sequence repeat, SSR) markers have also been mapped in eggplant populations (Nunome et al. 2001; Doganlar et al. 2002a; Nunome et al. 2003a). These mapping efforts, summarized in Table 1, have used both intraspecific and interspecific populations and have allowed a limited number of genes and quantitative trait loci (QTL) to be localized.

9.2.2 First Generation Maps

The first molecular map of eggplant was published in 1998 by Nunome and colleagues. These researchers used an F_2 population of 168 individuals derived from an intraspecific cross between two *S. melongena* lines: EPL-1, a Japanese commercial-type line, and WCGR112-8, a bacterial wilt-resistant line. The resulting map contained 94 RAPD markers with an average spacing of 8.8 cM. The markers fell into 13 linkage groups covering a total of 716.7 cM. AFLP markers were screened in the same population and integrated into the RAPD map to produce a map containing 181 markers (93 AFLP, 88 RAPD) in 21 linkage groups spanning 779.2 cM with an average interval of 4.9 cM (Nunome et al. 2001). Later, seven SSR markers were mapped in 120 individuals of the same F_2 population (Nunome et al. 2003a). This most recent intraspecific map contains a total of 162 markers (97 RAPD, 58 AFLP, 7 SSR) in 17 linkage groups encompassing 716.9 cM at an average spacing of 4.9 cM. In these studies, the existence of linkage groups in ex-

Table 1. Summary of currently available eggplant molecular genetic linkage maps

Population type	Parents	Number of individuals	Numbers and types of markers	Total map distance (cM)	Average interval distance (cM)	Number of linkage groups	Reference
Intraspecific F_2	*S. melongena* cv. EPL-1 × *S. melongena* cv. WCGR112-8	168	94 RAPD	716.7	8.8	13	Nunome 1998
Intraspecific F_2	*S. melongena* cv. EPL-1 × *S. melongena* cv. WCGR112-8	168	88 RAPD 93 AFLP	779.2	4.9	21	Nunome 2001
Interspecific F_2	*S. linnaeanum* MM195 × *S. melongena* MM738	58	233 RFLP	1480	7.6	12	Doganlar et al. 2002a
Intraspecific F_2	*S. melongena* cv. EPL-1 × *S. melongena* cv. WCGR112-8	120	97 RAPD 58 AFLP 7 SSR	716.9	4.9	17	Nunome et al. 2003
Interspecific F_2	*S. sodomeum*[1] PI1767 × *S. melongena* cv. Buia	48	117 RAPD 156 AFLP	736	2.7	12	Sunseri et al. 2003

[1] *S. sodomeum* = *S. linnaeanum*

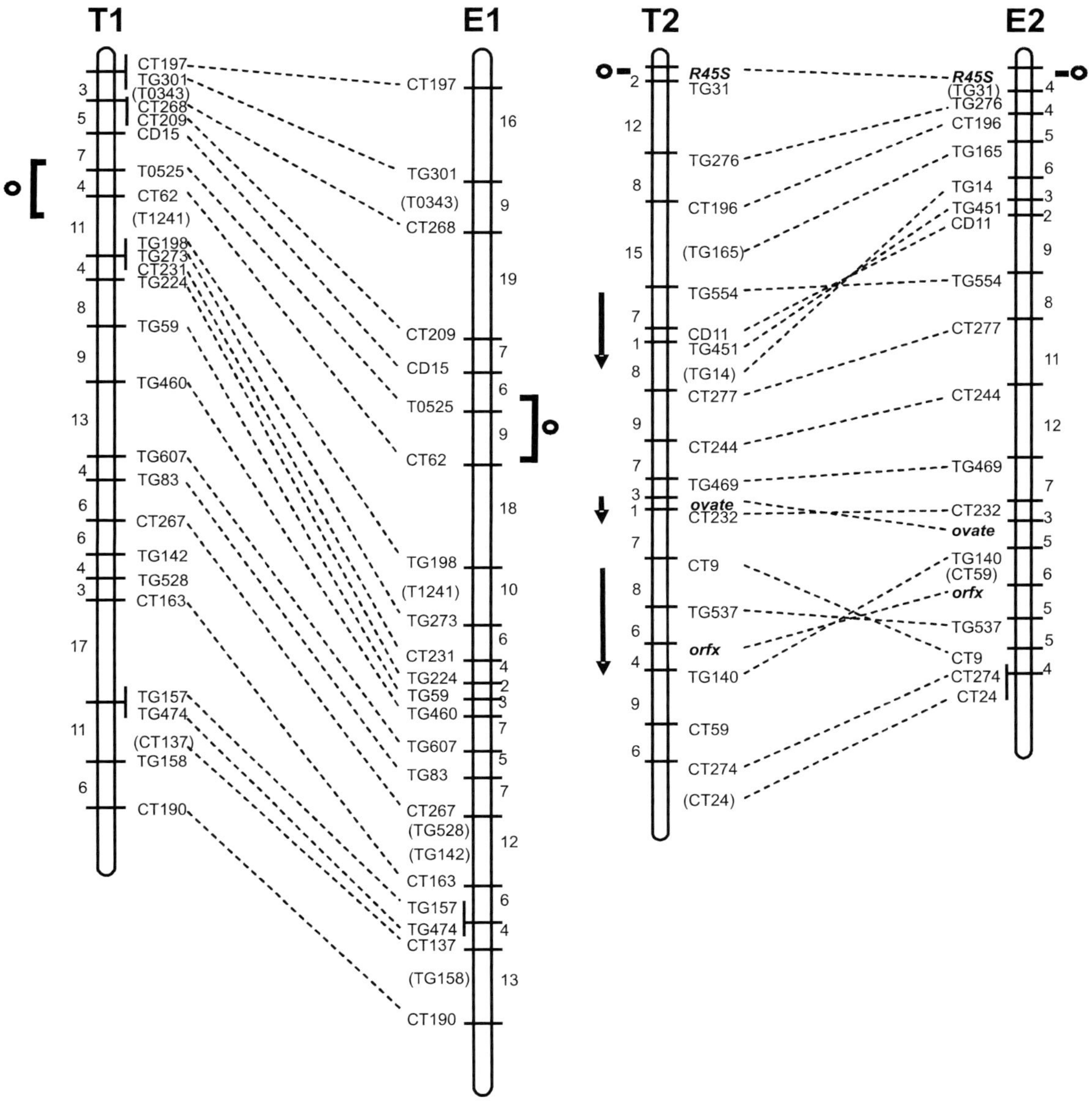

Fig. 1. Molecular linkage map of the eggplant genome constructed using an interspecific *S. linnaeanum* × *S. melongena* F_2 population and comparison with homeologous regions of tomato. Eggplant linkage groups are labeled E1 to E12, tomato chromosomes are labeled T1 to T12 with S and L used to differentiate short and long arms of tomato chromosomes. Markers by *tick marks* (framework markers) on eggplant linkage groups ordered at LOD $\geq$ 3.0, markers in *parentheses* ordered at 2.0 < LOD < 3.0. Map distances are in cM and approximate positions of centromeres are indicated by *solid bars* and *open circles*. *Dashed lines* connect each eggplant framework marker to its tomato counterpart. *Arrows* next to tomato chromosomes indicate the locations of inversions that distinguish the two genomes (adapted with permission from Doganlar et al. 2002a)

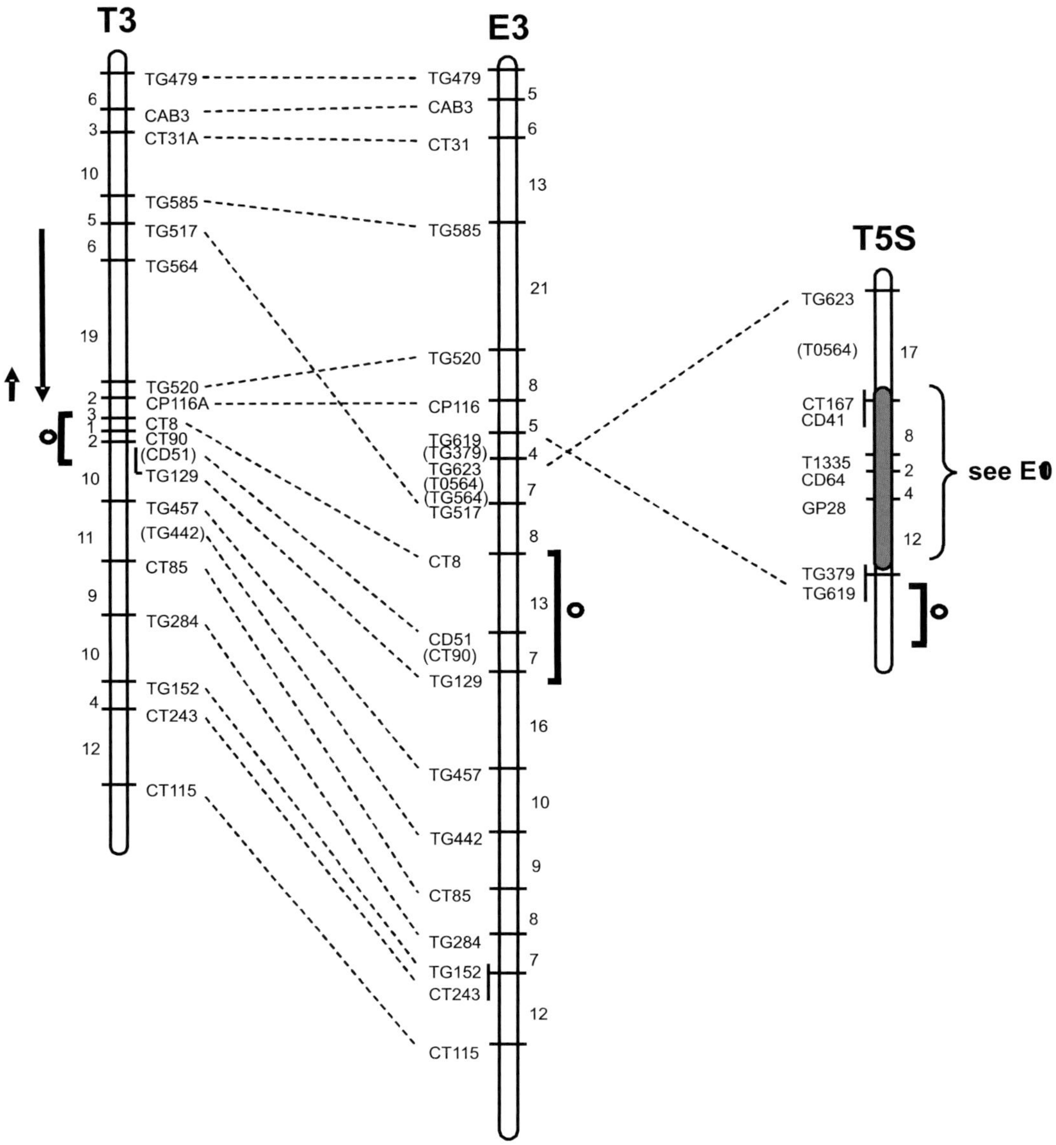

Fig. 1. (continued)

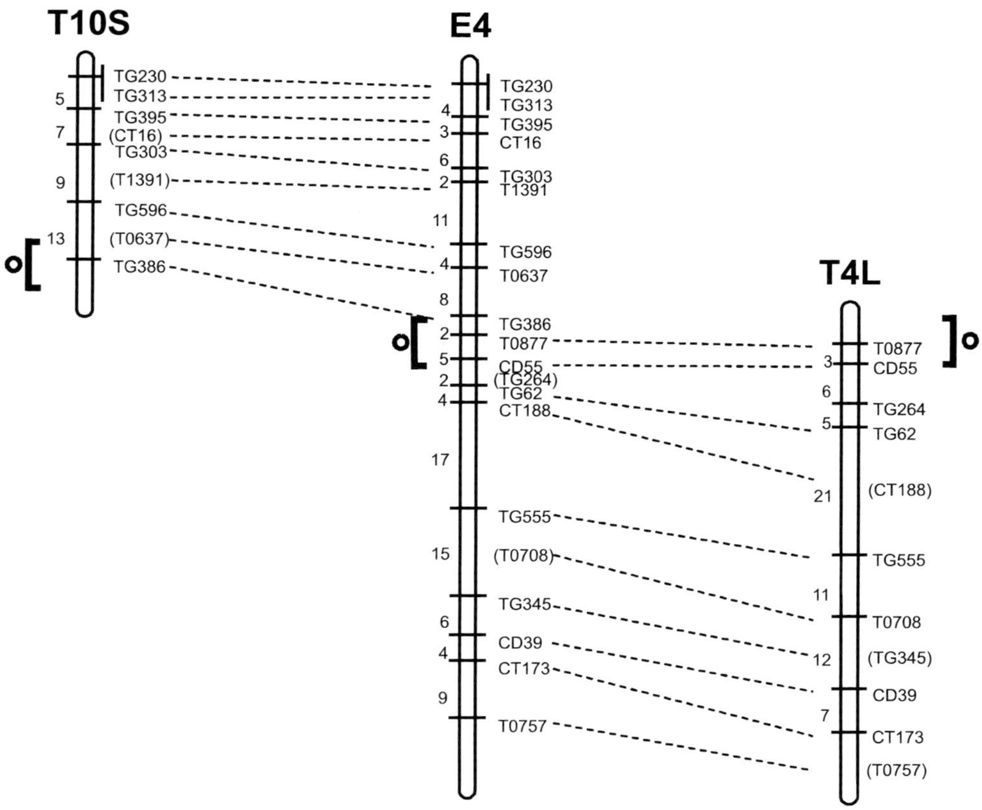

Fig. 1. (continued)

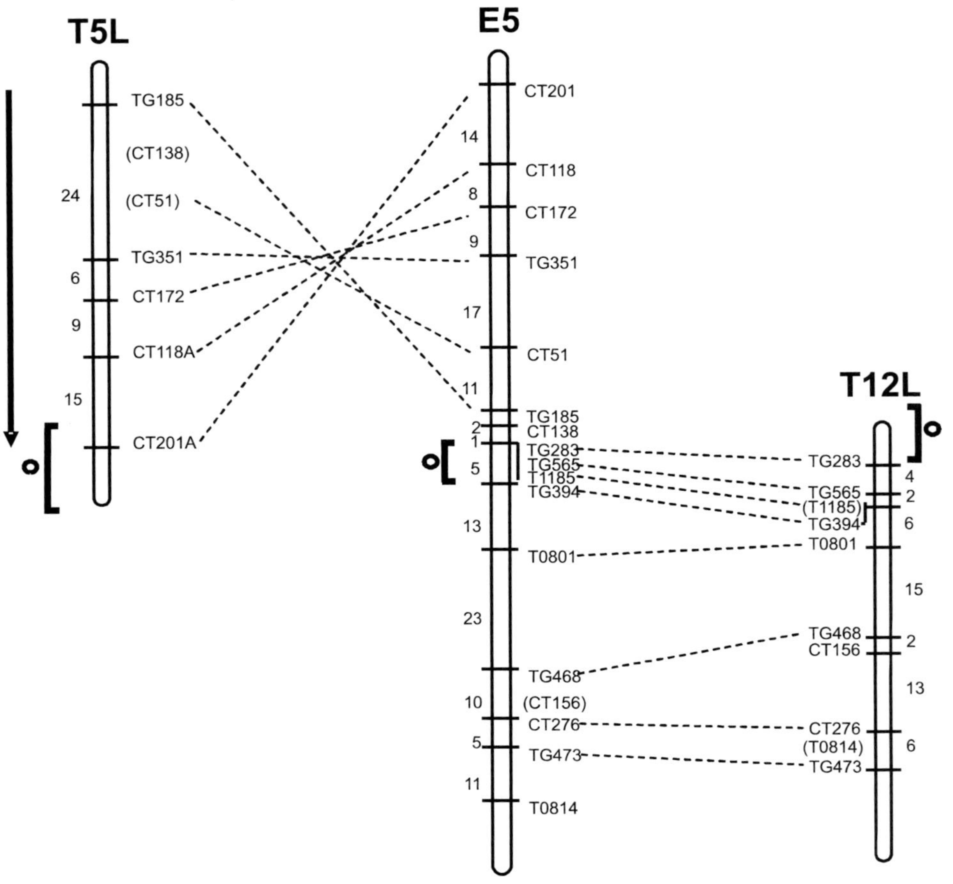

Fig. 1. (continued)

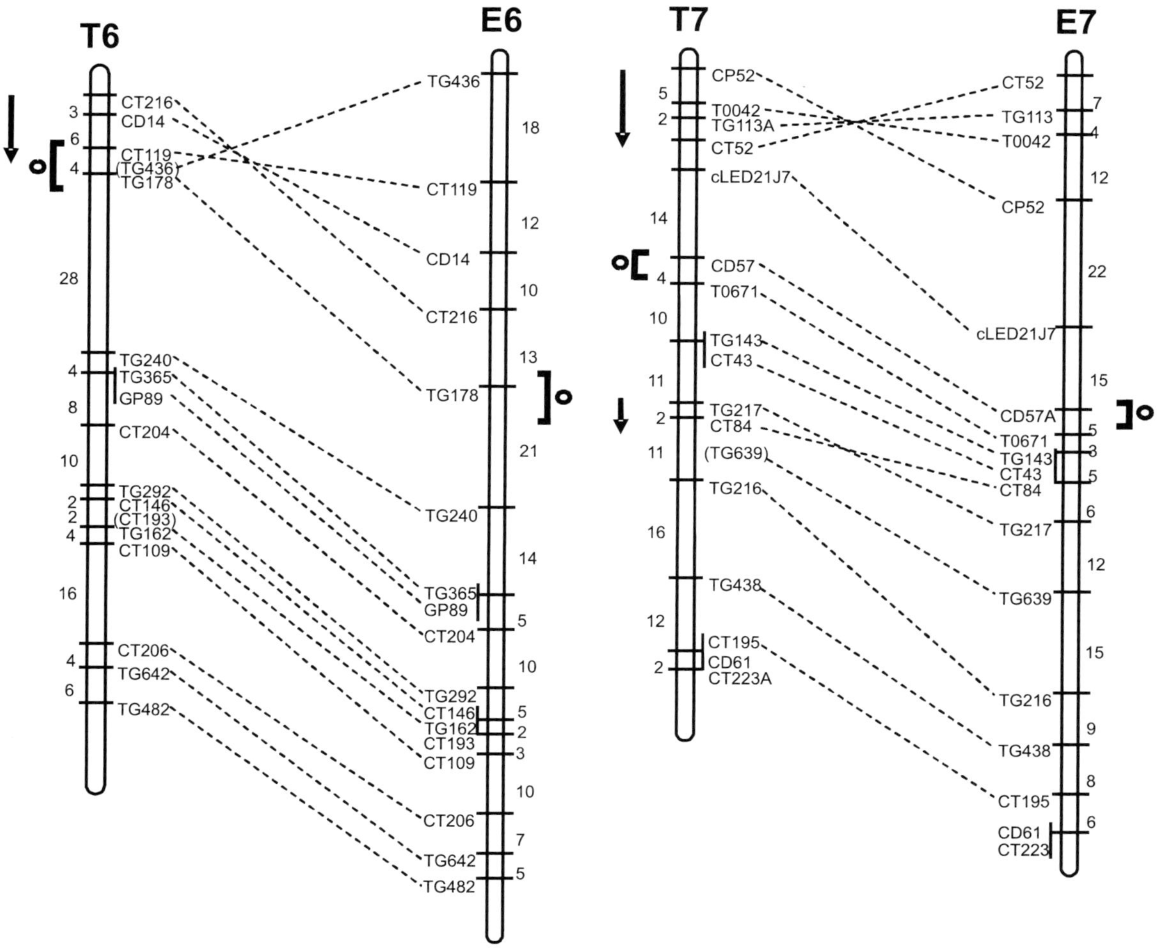

Fig. 1. (continued)

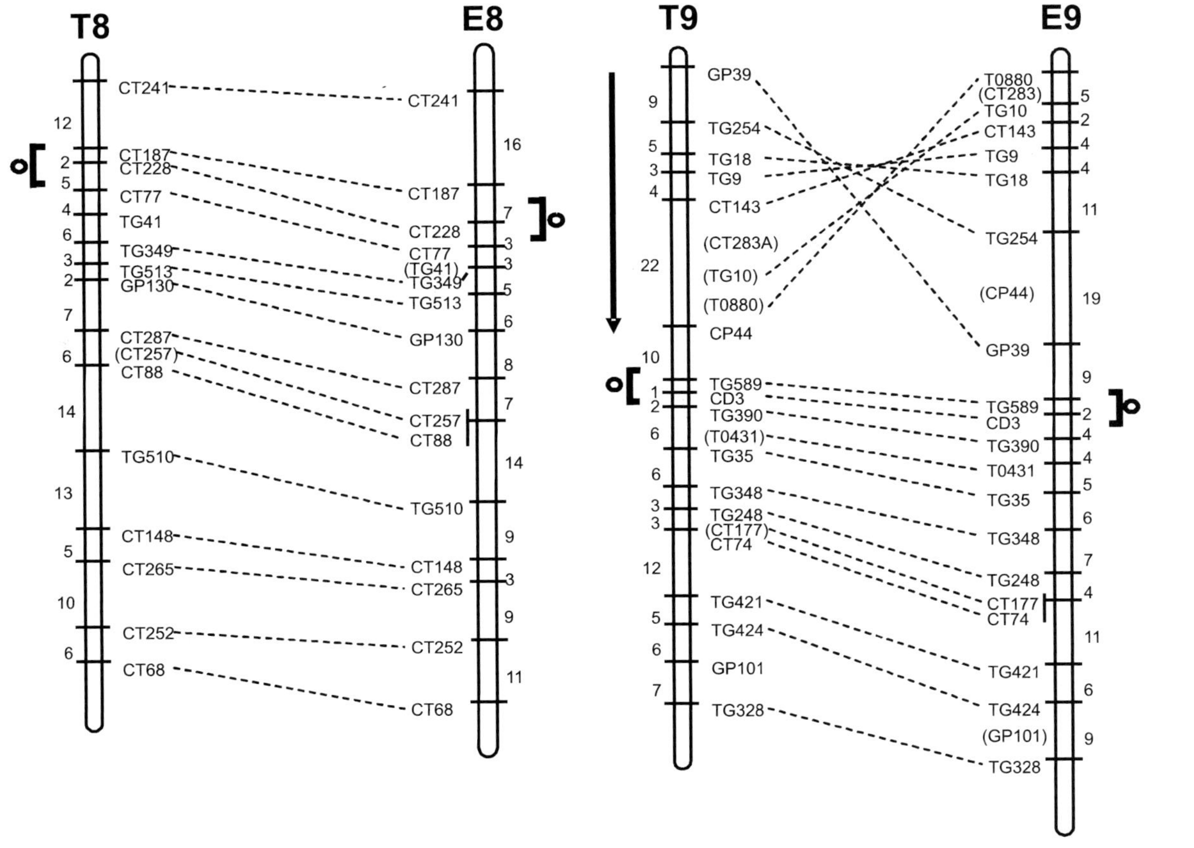

Fig. 1. (continued)

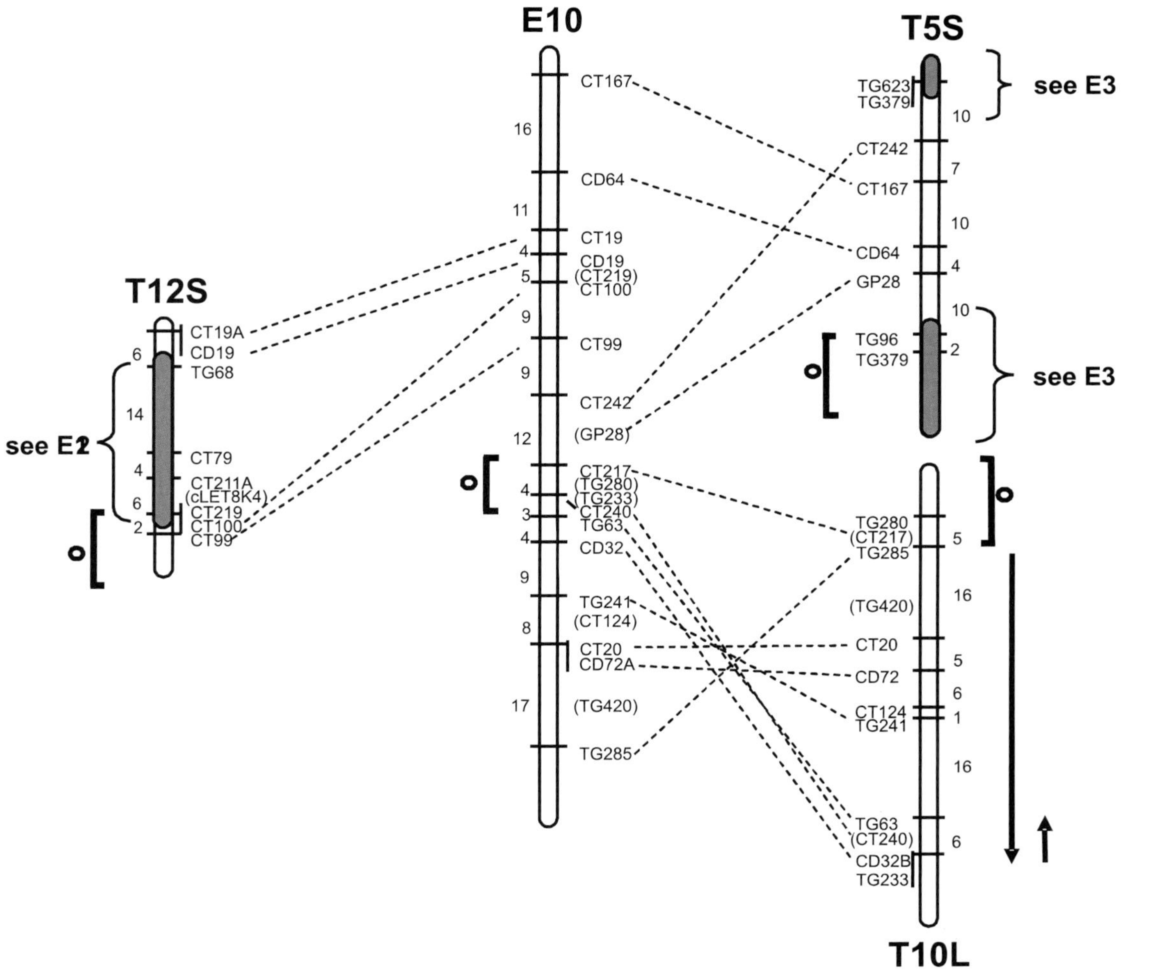

Fig. 1. (continued)

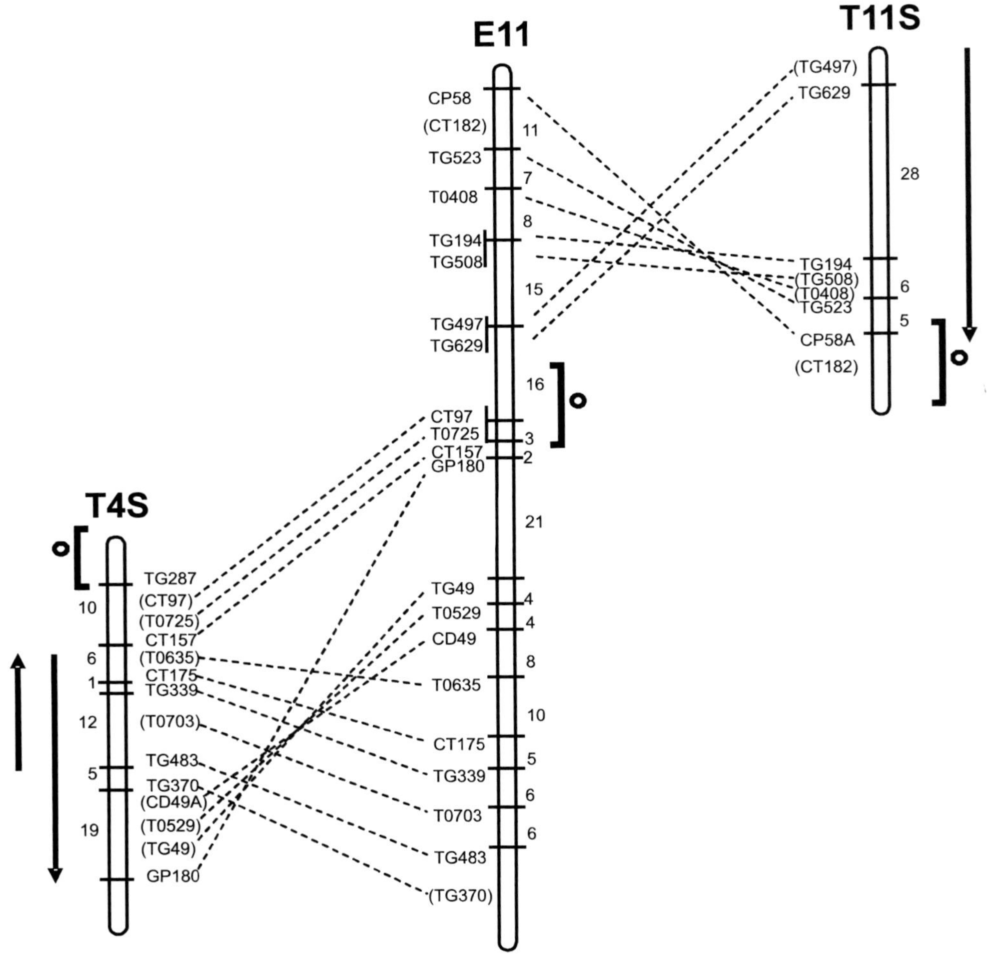

Fig. 1. (continued)

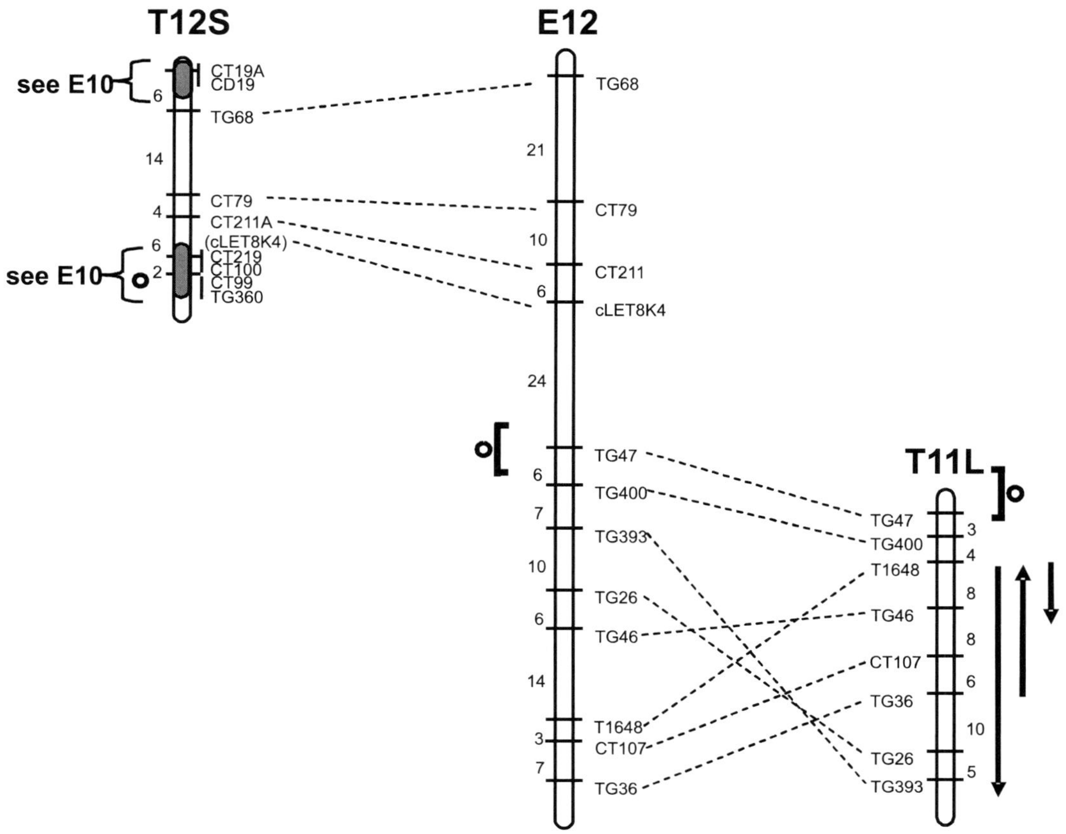

Fig. 1. (continued)

cess of the haploid chromosome number of eggplant ($n = 12$) indicates that whole genome coverage was not obtained, perhaps due to the limited DNA polymorphism characteristic of many intraspecific populations of eggplant or the limited number of markers that were mapped.

Molecular genetic linkage maps of eggplant have also been constructed using interspecific populations (Doganlar et al. 2002a; Sunseri et al. 2003). In the first study, an F_2 population of 58 individuals was derived from a cross between the wild species *S. linnaeanum* (MM195) and cultivated eggplant *S. melongena* (MM738) (Doganlar et al. 2002a). The map developed from this population contains 12 linkage groups and spans 1,480 cM with 233 RFLP markers at an average distance of 7.6 cM (Fig. 1). All of the mapped RFLPs which consisted of cDNA, genomic DNA and EST markers were previously mapped in tomato, thus allowing comparison of the genome organization of eggplant with tomato as described in the next section.

A RAPD-AFLP map was also constructed using an F_2 population derived from a cross between a Verticillium wilt resistant line of *S. linnaeanum* (also called *S. sodomeum*) and *S. melongena* cv. Buia (Sunseri et al. 2003). The map was based on 48 individuals and contains 273 markers (156 AFLP, 117 RAPD) encompassing 736 cM in 12 linkage groups with an average spacing of 2.7 cM. Current progress is being made toward higher density, saturated maps with the addition of more AFLP, SSR and other PCR-based markers to the existing eggplant maps (Nunome et al. 2003a; personal communication, Doganlar, 2006).

9.2.3 Comparative Maps

To date, only one comparative map has been developed for eggplant (Doganlar et al. 2002a; Fig. 1) and was constructed from tomato markers as described above. This map shows that large regions of the genome have remained colinear since the divergence of eggplant, tomato and potato from their last common ancestor. It was found that 28 chromosomal rearrangements including five translocations and 23 paracentric inversions differentiate the eggplant and tomato genomes. Eggplant and potato are also differentiated by five translocations but only 19 inversions as four of the inversions detected in the eggplant/tomato comparison are also present in potato. Thus, paracentric inversion, rather than pericentric inversion or translocation, seems to be the primary mechanism for genome rearrangements in the Solanaceae. Inversions, specifically paracentric inversions, are also known to be a major cause of chromosomal rearrangements in the Brassicaceae and *Drosophila* (Lagercrantz 1998; Ranz et al. 2001). Comparison of the eggplant map with those of tomato, potato, and pepper also indicated that the portions of the genome corresponding to tomato chromosomes 5, 9, 11 and 12 underwent the most rearrangements during evolution of the Solanaceae.

The strong conservation in marker order and chromosome arrangement between eggplant and tomato suggests that the genomic resources that are actively being developed for tomato will be useful for the more rapid advancement of eggplant genetic research. Moreover, the addition of a comparative eggplant genetic map to the repertoire of other resources available in the family has made the Solanaceae a preeminent system for comparative genomics studies in plants.

9.3 Gene and QTL Mapping

Because of the dearth of molecular genetic maps, relatively few genes and QTLs have been mapped in eggplant. The first report of trait localization on an eggplant molecular map described linkage of a fruit shape gene to RAPD markers on linkage group (LG) 10 of the *S. melongena* EPL-1 × WCGR112-8 F_2 map (Nunome et al. 1998). Interval analysis as implemented by Mapmaker QTL (Lincoln et al. 1992) was used to identify this QTL and the result was confirmed with the addition of AFLP markers to the map. In the later map, however, the linkage group was renamed as LG 2 (Nunome et al. 2001). In the same study, fruit, stem and calyx color mapped to LGs 7, 11, 12 and 16 with the strongest and most significant associations for all three traits with markers on LG 7 (Table 2, Fig. 2; Nunome et al. 2001).

More extensive QTL mapping was done in the interspecific *S. linnaeanum* MM 195 × *S. melongena* MM738 F_2 mapping population using single point linear regression analysis (Doganlar et al. 2002b; Frary et al. 2003) as implemented by QGENE (Nelson 1997). One study identified QTL related to domestication including those controlling fruit size and shape; fruit,

Table 2. QTL identified in the intraspecific and interspecific eggplant mapping populations. Significant markers column includes the most significant marker linked to each trait (*underlined*) as well as flanking markers that were also significant ($P \leq 0.05$). P-value, magnitude of phenotypic effect (% PVE) and gene action (*d/a*) are for the underlined marker

Population type	Cross (reference)	Trait	QTL name	Linkage group	Significant marker(s)	P-value	% PVE	*d/a*
Intraspecific F_2	*S. melongena* EPL-1 ×	Fruit shape	None	1	eUBC242-wAEM53a	<0.0001	31	nd
	S. melongena WCGR112-8	Fruit color	None	1	eAEM64a-eAEM26b	<0.0001	27	nd
	(Nunome et al. 2001)	Stem color	None	1	eAEM64a-eAEM26b	<0.0001	54	nd
		Calyx color	None	1	eAEM64a-eAEM26b	<0.0001	68	nd
Interspecific F_2	*S. linnaeanum* MM195 ×	Fruit weight	*fw2.1*	1	TG469-*orfx*	0.002	23	−0.8
	S. melongena MM738		*fw9.1*	1	TG589-TG35	0.005	44	−0.1
	(Doganlar et al. 2002b)		*fw11.1*	11	TG523	0.009	19	−7.3
		Fruit length	*fl2.1*	1	TG469-*orfx*-CT24	0.0004	29	−0.4
			fl9.1	1	TG589-CD3	0.0008	27	−0.1
			fl11.1	11	TG523-T408	0.004	22	−3.5
		Fruit diameter	*fd1.1*	1	TG273-TG607	0.002	23	0.8
			fd11.1	11	TG523	0.01	19	−6.0
		Fruit shape	*fs2.1*	1	TG140-CT9-CT274	0.002	22	−1.0
			fs7.1	1	TG639-TG216-CT195	0.0008	29	1.0
		Ovary length	*ovl1.1*	1	TG224-TG83-CT163	0.0008	35	−0.3
		Ovary diameter	*ovd4.1*	1	CD55-TG62	0.003	29	−7.0
			ovd9.1	1	TG390	0.003	31	−0.8
			ovd9.2	1	TG35-CT74	0.0008	35	−2.3
		Ovary shape	*ovs4.1*	1	TG596-TG386-TG62	0.0009	36	−3.0
		Ovary locule number	*oln5.1*	1	CT118-CT172	0.003	29	−1.2
		Fruit anthocyanin presence	*fap10.1*	10	CT100-CT240-TG285	<0.0001	93	1.0
		Fruit anthocyanin intensity	*fai1.1*	1	TG607-TG83	0.01	34	−0.3
			fai10.1	10	CT242-TG241	0.003	42	−0.1
			fai12.1	12	CT79	0.001	48	−5.4
		Fruit color	*fc8.1*	1	TG510-CT148	0.006	28	−0.1
			fc10.1	10	CT99-CT240-TG285	<0.0001	81	0.7
		Leaf lamina anthocyanin	*lla2.1*	1	*ovate*-*orfx*	0.005	20	−4.3
			lla6.1	1	TG178-CT204-CT109	0.0003	27	−0.2
			lla9.1	1	TG421-TG424	0.002	22	76.0
			lla10.1	10	CT99-CT240-TG285	<0.0001	60	0.4
		Leaf rib anthocyanin	*lra10.1*	10	CT99-CT240-TG285	<0.0001	76	0.8

Table 2. (continued)

Population type	Cross (reference)	Trait	QTL name	Linkage group	Significant marker(s)	P-value	% PVE	d/a
		Stem anthocyanin	*sa6.1*	1	TG178-TG240-CT204	0.0001	33	−0.2
			sa6.2	1	CT193-CT109	0.007	19	0.2
			sa10.1	10	CT217-CT240-TG241	<0.0001	39	0.9
			sa12.1	12	CT79-CT211-cLET8K4	0.01	24	3.0
		Prickle anthocyanin	*pa10.1*	10	CT100-CT240-TG285	<0.0001	93	0.8
		Corolla anthocyanin	*ca3.1*	1	CT115	0.0007	26	2.0
			ca5.1	1	TG351-TG565-T801	0.0003	31	−5.5
			ca6.1	1	TG240-GP89-CT109	0.0002	30	−0.7
			ca10.1	10	CT240-CD72A-TG285	0.0009	26	0.3
		Fruit stripe	*fst4.1*	1	TG230-TG303-TG62	<0.0001	67	−0.9
			fst10.1	10	CD19-CT99-CT242	0.002	25	1.3
		Leaf prickle	*lp6.1*	1	TG240-CT109-TG482	<0.0001	79	−0.7
			lp10.1	10	CT20-TG285	0.006	23	1.8
		Stem prickle	*sp6.1*	1	TG365-CT109-TG482	<0.0001	64	−0.5
		Flower calyx prickle	*flcp6.1*	1	TG240-CT193-TG642	<0.0001	64	−0.8
		Fruit calyx prickle	*ftcp6.1*	1	TG240-TG162-TG642	<0.0001	65	−0.7
			ftcp9.1	1	TG328	0.006	29	1.0
			ftcp11.1	11	T408-TG194	0.008	19	9.0
			ftcp11.2	11	CT175-TG339	0.003	34	−10.0
		Petiole prickle	*pp6.1*	1	TG240-CT109-TG482	<0.0001	71	−0.4
Interspecific F_2	*S. linnaeanum* MM195 × *S. melongena* MM738 (Frary et al. 2003)	Leaf length	*ll11.1*	11	CT175-T0703	0.007	19	−0.3
			ll12.1	12	cLET8K4	0.006	20	−10.2
		Leaf width	*lw1.1*	1	TG273-TG607	0.004	21	1.4
			lw3.1	1	TG585-TG520	0.002	22	0.0
			lw3.2	1	TG619-TG623-CT8	0.007	20	0.8
			lw7.1	1	TG639-TG216-CT223	0.004	23	1.4
		Leaf shape	*lsh1.1*	1	CT209-TG607-CT163	<0.0001	40	0.6
			lsh5.1	1	CT51-TG185-T0801	<0.0001	36	1.3
			lsh7.1	1	TG639-TG216-CT223	0.0002	33	0.5
			lsh8.1	1	TG349-CT287	0.007	20	−0.4
		Leaf lobing	*llob6.1*	1	TG240-CT109-TG482	<0.0001	76	−0.5
			llob10.1	10	TG241-CD72A	0.006	19	1.4

Table 2. (continued)

Population type	Cross (reference)	Trait	QTL name	Linkage group	Significant marker(s)	P-value	% PVE	*d/a*
		Leaf surface	*lsur4.1*	1	TG395-TG303-T1391	0.001	25	−3.1
		Flower diameter	*fld7.1*	1	CT52-TG113	0.005	23	0.0
		Flower shape	*fls1.1*	1	TG83-CT137	0.008	19	0.0
			fls7.1	1	CT52	0.002	27	−2.0
			fls9.1	1	TG35-TG424	0.01	19	−2.3
		Days to flowering	*dtf1.1*	1	T0525-TG198	0.01	20	0.9
			dtf2.1	1	TG451-CT277-*fw2.2*	0.0006	28	−0.2
			dtf3.1	1	TG517-CT8	0.004	24	1.5
			dtf5.1	1	CT172	0.005	20	−0.5
			dtf5.2	1	TG394-T0801-TG468	0.002	23	0.3
			dtf8.1	1	CT187-CT228-CT77	0.003	24	−1.5
			dtf9.1	1	TG348	0.007	23	1.1
			dtf11.1	11	T0408	0.005	20	−1.0
		Flowers/inflorescence	*fln3.1*	1	TG284-TG152-CT115	0.009	19	−2.7
			fln4.1	1	T0637-CT188	0.002	25	−0.5
		Fruit/infructescence	*ftn3.1*	1	TG152	0.002	24	4.2
			ftn4.1	1	TG303-T1391	0.004	21	−1.2
			ftn7.1	1	TG216-TG438-CT223	0.003	23	1.8
			ftn10.1	10	TG241-CD72A-TG285	0.0007	26	−29.5
		Fruit set	*fset4.1*	1	CT16-TG303	0.004	21	−1.2
			fset7.1	1	TG216-TG438-CT223	0.003	23	−0.3
		Fruit calyx size	*cs2.1*	1	*R45S*-TG276	0.009	19	−0.5
			cs2.2	1	TG469-*ovate*-*fw2.2*	0.01	18	−0.2
			cs9.1	1	TG10-TG18	0.003	23	1.4
		Fruit glossiness	*fglo1.1*	1	CT268-CD15-CT267	0.0002	30	0.0
			fglo6.1	1	TG365-CT204-CT206	<0.0001	40	−1.2
			fglo8.1	1	CT187-TG349	0.007	20	−0.7
			fglo9.1	1	TG421-TG424	0.005	22	2.5
			fglo12.1	12	TG68	0.007	23	−38.5
		Plant height	*ht2.1*	1	CT196-TG14-CD11	0.008	18	0.1
			ht5.1	1	CT51-TG565-TG394	0.0006	28	−2.1
			ht10.1	10	CT217-TG285	0.001	26	1.4
			ht12.1	12	TG36	0.008	18	2.3

Table 2. (continued)

Population type	Cross (reference)	Trait	QTL name	Linkage group	Significant marker(s)	P-value	% PVE	*d/a*
		Apex hairs	*ah3.1*	1	TG585-CP116A	0.0004	28	−0.1
			ah4.1	1	TG555	0.008	21	−2.5
			ah5.1	1	CT172-TG351	0.002	25	0.3
			ah10.1	10	CT242	0.01	17	−3.3
		Leaf hairs	*lh1.1*	1	CT268	0.009	18	0.1
			lh2.1	1	CT24	0.009	19	1.7
			lh3.1	1	TG585-TG520	0.0005	28	−0.3
			lh5.1	1	TG351	0.004	22	0.1
			lh6.1	1	CT119-CT216	0.006	19	0.6
			lh8.1	1	TG513-GP130-CT287	0.006	23	−0.1
			lh10.1	10	CT242	0.007	18	−5.9
		Stem hairs	*sh2.1*	1	CT232-*ovate*	0.006	20	0.2
			sh3.1	1	TG585-TG619	0.0004	28	−0.1
			sh10.1	10	CT242	0.003	21	−1.9
		Ovary hairs	*ovh1.1*	1	CT137	0.007	28	−0.5
			ovh3.1	1	TG442-CT85	0.001	35	3.0
			ovh6.1	1	CT146-CT109	0.009	25	−0.6
			ovh10.1	10	TG241-TG285	0.0004	40	1.6

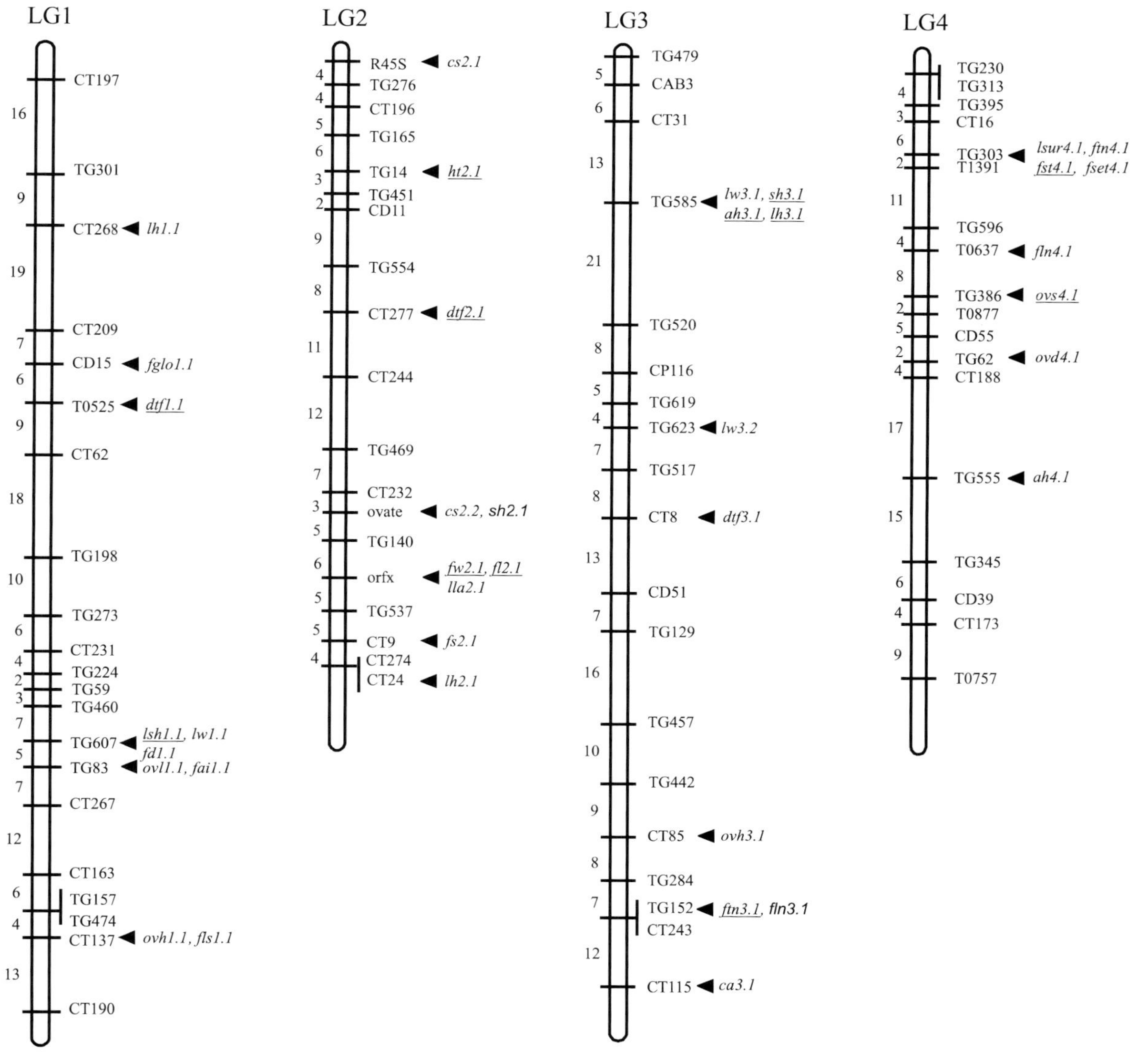

Fig. 2. Eggplant linkage map showing the locations of domestication and morphological QTL identified in Doganlar et al. (2002b) and Frary et al. (2003). *Triangles* are used to indicate the most significant marker for each locus. Trait abbreviations correspond to those used in Table 2. QTL with putative orthologs in other solanaceous species (see Table 3) are *underlined*

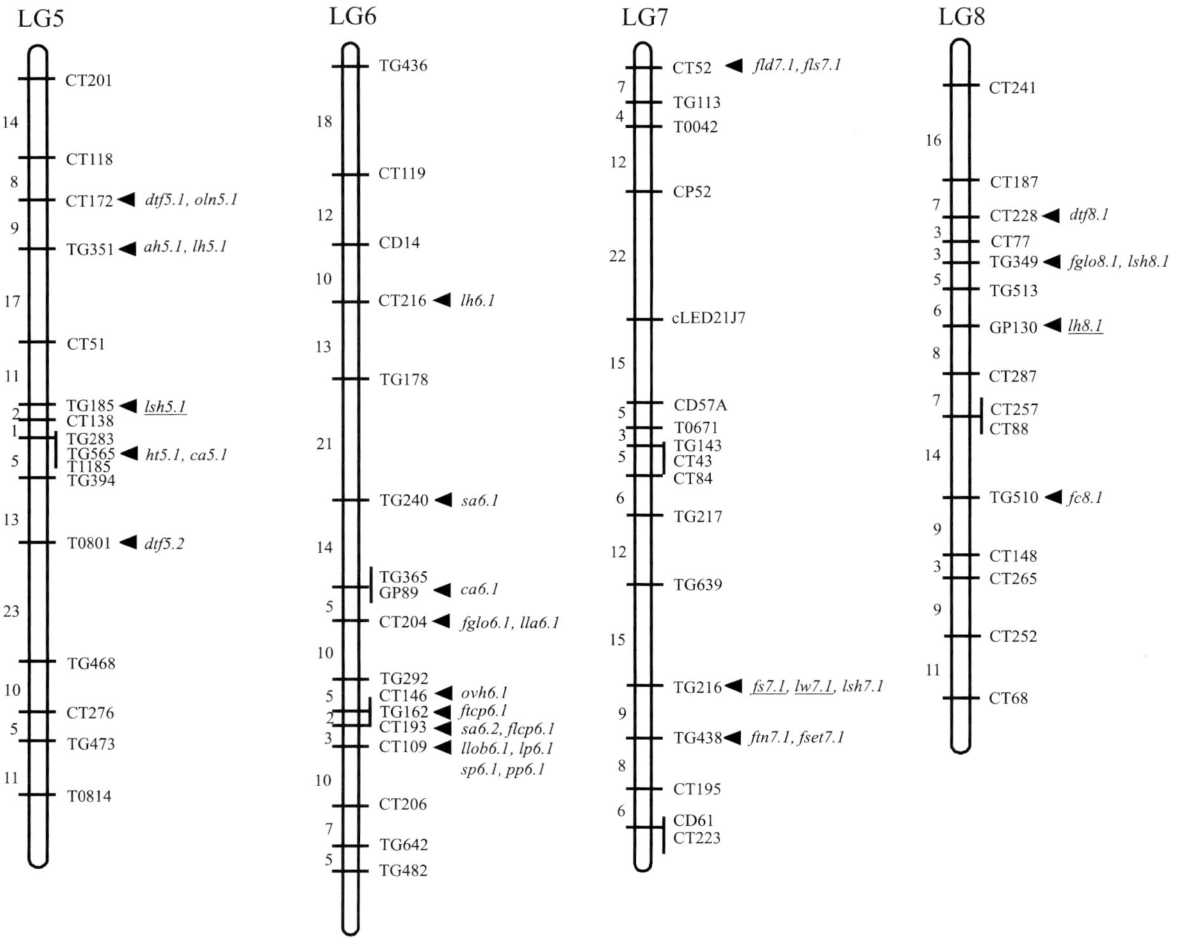

Fig. 2. (continued)

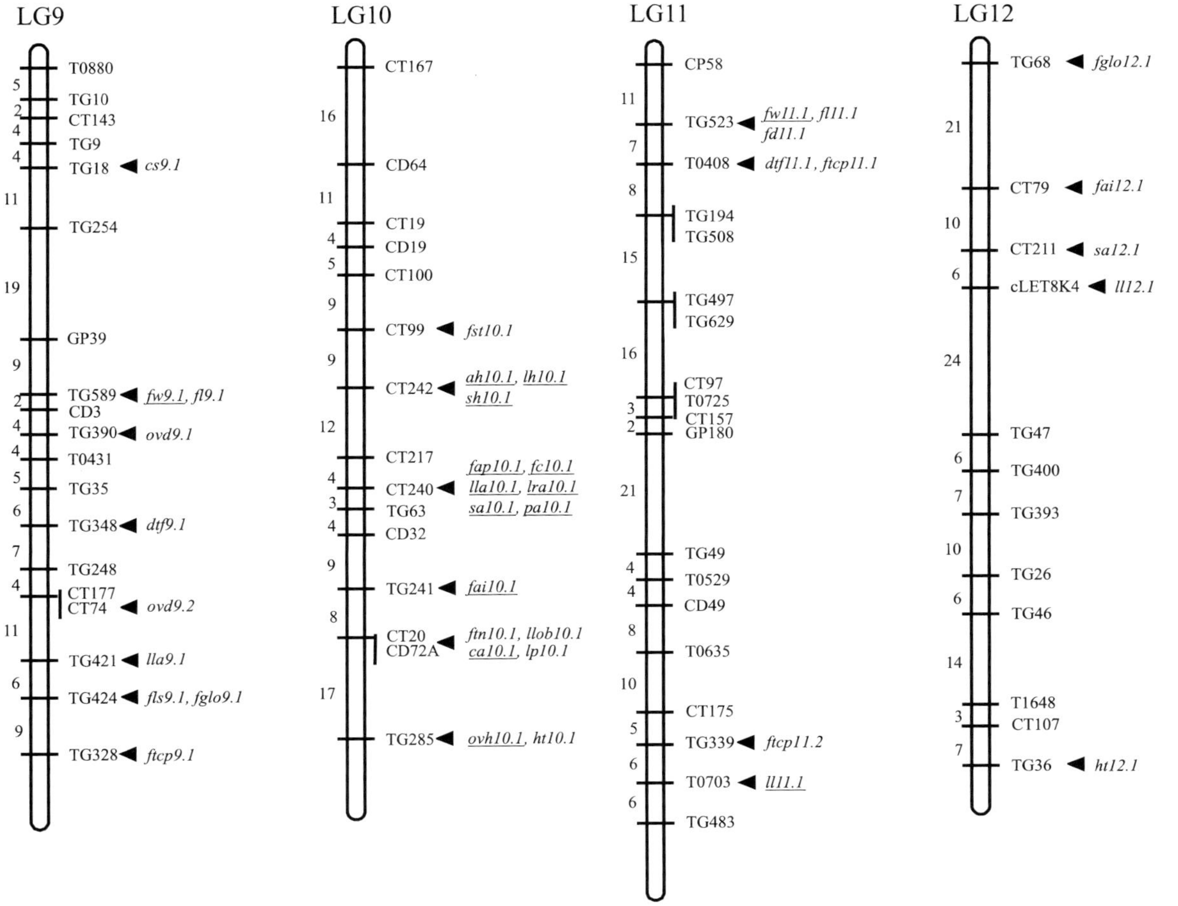

Fig. 2. (continued)

Table 3. Domestication-related and morphological QTL with putative conservation in the Solanaceae (compiled from Doganlar et al. 2002b and Frary et al. 2003)

Trait	Eggplant Locus name	Linkage group	Putative ortholog Locus name	Chromo-some	Crop	Locus type	Reference
Fruit weight	*fw2.1*	2	*fw2.2*	2	Tomato	QTL	Frary et al. 2000
			fw2.1	2	Pepper	QTL	Ben Chaim et al. 2001
	fw9.1	9	*fw9.2*	9	Tomato	QTL	Grandillo et al. 1999
	fw11.1	11	*fw11.1*	11	Tomato	QTL	Grandillo et al. 1999
Fruit shape	*fl2.1*	2	*ovate*	2	Tomato	Morphological	Ku et al. 1999
	fs7.1	7	*fs7.b*	7	Tomato	QTL	Grandillo et al. 1999
	ovs4.1	4	*fs10.1*	10	Tomato	QTL	Grandillo et al. 1999
			fs10.1	10	Pepper	QTL	Ben Chaim et al. 2001
Anthocyanins	*fap10.1*[a]	10	*af*	5	Tomato	Morphological	Tanksley et al. 1992
			ag	10	Tomato	Morphological	Tanksley et al. 1992
			an2	10	Tomato	Known gene	De Jong et al. 2004
			F	10	Potato	Morphological	van Eck et al. 1993
			I_{ep}, I_{co}	10	Potato	Morphological	van Eck et al. 1994
Fruit stripe	*fst4.1*	4	*Fs*	10	Tomato	Morphological	Tanksley et al. 1992
			u	10	Tomato	Morphological	Grandillo and Tanksley 1996
Leaf length	*ll11.1*	11	Leaf length QTL	4	Tomato	QTL	Paran et al. 1997
			lflr4.1	4	Tomato	QTL	Frary et al. 2004
Leaf width	*lw7.1*	7	*lfw7.1*	7	Tomato	QTL	Frary et al. 2004
Leaf shape	*lsh1.1*	1	*lra1a*	1	Tomato	QTL	de Vicente and Tanksley 1993
			lra1b	1	Tomato	QTL	de Vicente and Tanksley 1993
	lsh5.1	5	*lr5*	5	Tomato	QTL	de Vicente and Tanksley 1993
			lfw12.1	12	Tomato	QTL	Frary et al. 2004
			lfl12.1	12	Tomato	QTL	Frary et al. 2004
Leaf lobing	*llob6.1*	6	*c*	6	Tomato	Morphological	Liharska et al. 1997
			Pts	6	Tomato	Morphological	Tanksley et al. 1992
Flowering time	*dtf1.1*	1	*dtf1.1*	1	Tomato	QTL	de Vicente and Tanksley 1993
			dfl1.1	1	Tomato	QTL	Grandillo and Tanksley 1996
	dtf2.1	2	*dftf2.1*	2	Tomato	QTL	de Vicente and Tanksley 1993
			dfl2.1	2	Tomato	QTL	Grandillo and Tanksley 1996
Flower number	*fln3.1*	3	*nflt3.1*	3	Tomato	QTL	Grandillo and Tanksley 1996
Plant height	*ht2.1*	2	Height QTL	2	Tomato	QTL	Paran et al. 1997

[a] Other anthocyanin QTL at this position: *fai10.1, fc10.1, lla10.1, lra10.1, sa10.1, pa10.1, ca10.1*

Table 3. (continued)

Trait	Eggplant Locus name	Linkage group	Putative ortholog Locus name	Chromosome	Crop	Locus type	Reference
Hairs	*ah3.1*[b]	3	*Ln*	3	Tomato	Morphological	Tanksley et al. 1992
	lh8.1	8	*Hr*	8	Tomato	Morphological	Tanksley et al. 1992
	ah10.1[c]	10	Type B trichome QTL	5	Potato	QTL	Bonierbale et al. 1994
			TriV-1	5	Tomato	QTL	Maliepaard et al. 1995
	ovh10.1	10	*h*	10	Tomato	Morphological	Tanksley et al. 1992

[b] Other hairiness QTL at this position: *lh3.1*, *sh3.1*
[c] Other hairiness QTL at this position: *lh10.1*, *sh10.1*

stem and leaf color; and plant prickliness (Doganlar et al. 2002b). The other study examined the population for morphological traits such as leaf and flower size and shape, numbers of flowers and fruit per inflorescence, plant height and hairiness (Frary et al. 2003). Both studies used data gathered from plants grown in the greenhouse in Ithaca, NY, USA, and plants grown in the field in Montfavet, France.

In the domestication study, 62 QTLs were identified for the 22 traits measured (Table 2); however, it was found that only six loci were responsible for most of the variation in fruit weight, fruit shape, fruit and plant color, and prickliness. These results suggest that the domestication of eggplant, like that of maize and other cereals, involved mutations at a limited number of major loci. In the plant morphology study, 63 QTLs were identified for 18 traits (Table 2). In both studies, several of the identified loci had putative orthologs in at least one other solanaceous species. Thus, 40% of the domestication-related loci and 27% of the morphological QTLs had putative orthologs in tomato, pepper or potato suggesting that the functions of these loci have been conserved since the divergence of these different crop species from their last common ancestor (Table 3). Since publication of the eggplant results, more QTL and candidate gene mapping studies in tomato have identified additional loci with putative orthology to eggplant loci. The eggplant loci with putative orthologs in the Solanaceae are listed in Table 3. These results reinforce the idea that domestication has been driven by only a few, conserved genes and that the morphological diversity seen in the Solanaceae may be the result of different mutations in these conserved loci. Lester and Daunay (2003), on the basis of several examples from cultivated eggplant and other crops, also state that relatively few loci are concerned in the domestication process, since generally the molecular diversity within the crops is much narrower than the diversity in their wild counterparts, although the morphological diversity within the crops is much greater. Further, they note that the mutations at these loci controlling domestication traits concern in most cases a loss of function.

9.4 Marker-Assisted Breeding

To date, marker-assisted breeding in eggplant has been limited to screening germplasm of *S. melon-*

gena and related species and assessing the hybrid nature and cytoplasmic parentage of hybrids produced from interspecific pollinations and somatic hybridization. The earliest studies used chloroplast DNA and isozyme markers to examine genetic relationships within and among species. In one of the first studies, Sakata et al. (1991) compared the restriction fragment patterns of chloroplast DNA from 10 species of eggplant including the three cultivated species, *S. melongena*, *S. macrocarpon* and *S. aethiopicum*, with the intention of using this method to confirm the hybrid nature of progeny from interspecific crosses and somatic hybridization. RFLP analysis of specific amplified regions of chloroplast DNA was also used to study taxonomy and determine the cytoplasmic origin of hybrids produced by crosses between *S. melongena* and *S. aethiopicum* (Isshiki et al. 1998). Other studies have used chloroplast DNA to determine genetic relationships among eggplant and wild species accessions that are morphologically similar (Sakata and Lester 1994) and as a potential predictor of the success of interspecific crosses (Sakata and Lester 1997). Isozyme markers have also been applied to examine genetic relationships in cultivated, weedy and wild eggplants and for cultivar differentiation (Isshiki et al. 1994a–c; Karihaloo and Gottlieb 1995); however, in general, isozymes exhibit low polymorphism. Despite this limitation, isozymes have been used to study segregation in self progeny from *S. melongena* × *S. integrifolium* (synonym *S. aethiopicum* Aculeatum group) amphidiploids (Isshiki et al. 2000). RFLP analysis of mitochondrial DNA has also been tested and found to be appropriate for phylogenetic studies of eggplant and related species (Isshiki et al. 2003).

More recently, PCR-based markers have become increasingly popular for studying genetic relationships and diversity. Karihaloo et al. (1995) used data from RAPD marker analysis to support their claim that *S. melongena* and *S. insanum* (weedy eggplant, also called *S. melongena* var. *insanum*) should not be considered as a distinct species. RAPD and RFLP markers have also been used to confirm the hybridity of a *S. integrifolium-S. violaceum* protoplast fusion product with bacterial wilt resistance (Tamura et al. 2002). Mace et al. (1999) used AFLP markers to examine genetic diversity and found that the results agreed with previous studies that employed morphological, isozyme and ITS-1 ribosomal sequence data. AFLP marker data has also been used in combination with morphological data, to assign or correct the species names of *Solanum* germplasm and differentiate between wild and cultivated forms of eggplant (Furini and Wunder 2004). Accessions of cultivated eggplant and related species have also been assessed for SSR polymorphism using both dinucleotide (Nunome et al. 2003a) and trinucleotide (Nunome et al. 2003b) repeats. In this work it was found that both types of repeat had low levels of polymorphism within *S. melongena* and that, although all of the trinucleotide repeat primers could be used in eggplant's closest wild relative, *S. incanum*, only about half of the primers could be used to amplify products in other related species (Nunome et al. 2003b). Intersimple sequence repeat (I-SSR) markers have been used with isozymes and RAPDs to confirm the origin of dihaploid plants produced from *S. melongena-S. aethiopicum* somatic hybrids carrying Fusarium wilt resistance derived from the *S. aethiopicum* parent (Rizza et al. 2002).

9.5 Other Resources

As a close relative, eggplant has benefited from the recent explosion of genomics research in tomato. Currently, 3181 expressed sequence tags (ESTs) of eggplant are available via GenBank and the Solanaceae Genomics Network (SGN, http://www.sgn.cornell.edu). The sequences were obtained from a cDNA library developed from buds and flowers of four to eight-week old eggplant plants. In addition, a bacterial artificial chromosome (BAC) library has been made from *S. melongena* MM738. This *Hind*III library of approximately 50,000 BACs in the pBeloBACII vector provides six-fold coverage of the eggplant genome with an average insert size of 115 kb (personal communication, Vrebalov, 2006).

9.6 Future Scope of Work

Molecular genetics research in eggplant to date provides a sound basis for future advances in the species and in comparative studies with other solanaceous species. Currently, a set of PCR-based markers that can be applied to eggplant, tomato, potato and pepper is being developed (personal communication, Tanksley, 2006). These markers will be invaluable for comparative analyses. The current international effort to sequence the tomato genome

(SOL, http://www.sgn.cornell.edu) will also provide a wealth of information applicable to eggplant. One of the goals of the project is to develop deep EST databases from various solanaceous species. These databases will be used to identify orthologous genes in the Solanaceae and to answer questions about gene and genome evolution in the family. Another aspect of the project is the completion of high-resolution comparative genetic maps for eggplant and related species using a common set of markers. These maps will be useful for detailed analyses of chromosomal evolution in Solanaceae and will hasten the identification of orthologous genes in the family. Moreover, interspecific backcross inbred lines will be developed for eggplant to provide the necessary populations for efficient gene and QTL discovery and mapping. With all of these resources, the future of eggplant molecular genetics research promises to be exceptionally fruitful.

Acknowledgement. We are grateful to several individuals for valuable comments on the manuscript: RN Lester, University of Birmingham, UK; RW Robinson, Cornell University, USA; T Allersma, P Verlaan and D Vreugdenhil, De Ruiter Seeds, The Netherlands; J Haanstra, Rijk Zwaan, The Netherlands; JW Hennart, Vilmorin, France; and R Mank, Keygene, The Netherlands. This work was supported by a Career Project from the Scientific and Technical Research Council of Turkey to Sami Doganlar (104T224).

References

Acciarri N, Rotino GL, Valentino D, Vitelli G, Sunseri F, Martelli G, Tamietti G (2001) Genetic improvement of eggplant for resistance to *Verticillium* through interspecific hybridization with *Solanum sodomeum* L. In: Proceedings of the 11th Eucarpia Meeting on Genetics and Breeding of *Capsicum* and Eggplant, Antalya, Turkey, 2001, pp 298–302

Ano G, Hebert Y, Prior P, Messiaen CM (1991) A new source of resistance to bacterial wilt of eggplants obtained from a cross *Solanum aethiopicum* L. × *Solanum melongena* L. Agronomie 11:555–560

Aubert S, Daunay MC, Pochard E (1989a) Saponosides stéroïdiques de l'aubergine (*Solanum melongena* L.). I-Intérêt alimentaire, méthodologie d'analyse, localization dans le fruit. Agronomie 9:641–651

Aubert S, Daunay MC, Pochard E (1989b). Saponosides stéroidiques de l'aubergine (*Solanum melongena* L.). II. Variations des teneurs liées aux conditions de récolte, aux génotypes et à la quantité de graines des fruits. Agronomie 9:751–758

Ben Chaim A, Paran I, Grube RC, Jahn M, van Wijk R, Peleman J (2001) QTL mapping of fruit-related traits in pepper (*Capsicum annuum*). Theor Appl Genet 102:1016–1028

Bennett MD, Leitch IJ (2004) Plant DNA C-values database (release 3.0, Dec 2004). http://www.rbgkew.org.uk/cval/homepage.html

Bohs L, Olmstead RG (1999) *Solanum* phylogeny inferred from chloroplast DNA sequence data. In: Nee M, Symon, DE, Lester RN, Jessop JP (eds.) Solanaceae IV: Advances in Biology and Utilization. Royal Botanic Gardens, Kew, UK, pp 97–110

Bonierbale MW, Plaisted RL, Pineda O, and Tanksley SD (1994) QTL analysis of trichome-mediated insect resistance in potato. Theor Appl Genet 87:973–987

Cao G, Sofic E, Prior RL (1996) Antioxidant capacity of tea and common vegetables. J Agri Food Chem 44:3426–3431

Collonnier C, Fock I, Daunay MC, Servaes A, Vedel F, Siljak-Yakovlev S, Souvannavong V, Sihachakr D (2003) Somatic hybrids between *Solanum melongena* and *S. sisymbrifolium*, as a useful source of resistance against bacterial and fungal wilts. Plant Sci 164:849–861

Daunay MC, Lester RN, Laterrot H (1991) The use of wild species for the genetic improvement of brinjal-eggplant (*Solanum melongena*) and tomato (*Lycopersicon esculentum*). In: Hawkes JG, Lester RN, Nee M, Estrada RN (eds.) Solanaceae III. Taxonomy, Chemistry, Evolution. Royal Botanic Gardens, Kew, UK, pp 389–412

Daunay MC, Chaput MH, Sihachakr D, Allot M, Vedel F, Ducreux G (1993) Production and characterization of fertile somatic hybrid plants of eggplant (*Solanum melongena* L.) with *S. aethiopicum* L. Theor Appl Genet 85:841–850

Daunay MC, Lester RN, Ano G (2001a) Eggplant. In: Charrier A, Jacquot M, Hamon S, Nicolas D (eds.) Tropical Plant Breeding. Science, Enfield, NH, pp 199–222

Daunay MC, Lester RN, Gebhardt C, Hennart JW, Jahn M, Frary A, Doganlar S (2001b) Genetic resources of eggplant (*Solanum melongena* L.) and allied species: a new challenge for molecular geneticists and eggplant breeders. In: van den Berg RG, Barendse GWM, van der Weerden GM, Mariani C (eds.) Solanaceae V. Advances in Taxonomy and Utilization. Nijmegen University Press, Nijmegen, The Netherlands, pp 251–274

Daunay MC, Aubert S, Frary A, Doganlar S, Lester RN, Barendse G, van der Weerden G, Hennart JW, Haanstra J, Dauphin F, Jullian E (2004) Eggplant (*Solanum melongena*) fruit color: pigments, measurements and genetics. In: Proceedings of the 12th Eucarpia Meeting on Genetics and Breeding of *Capsicum* and Eggplant, Noordwijkerhout, The Netherlands, 2004, pp 108–116

De Jong WS, Eannetta NT, De Jong DM, Bodis M (2004) Candidate gene analysis of anthocyanin pigmentation loci in the Solanaceae. Theor Appl Genet 108:423–432 de Vicente MC, Tanksley SD (1993) QTL analysis of transgressive segregation in an interspecific tomato cross. Genetics 134:585–596

Doganlar S, Frary A, Daunay MC, Lester RN, Tanksley SD (2002a) A comparative genetic linkage map of eggplant (*Solanum melongena*) and its implications for genome evolution in the Solanaceae. Genetics 161:1697–1711

Doganlar S, Frary A, Daunay MC, Lester RN, Tanksley SD (2002b) Conservation of gene function in the Solanaceae as revealed by comparative mapping of domestication traits in eggplant. Genetics 161:1713–1726

FAO Statistics (2003) http://apps.fao.org/faostat. Cited 2003

Frary A, Nesbitt TC, Frary A, Grandillo S, Van der Knaap E, Cong B, Lui J, Meller J, Elber R, Alpert KB, Tanksley SD (2000) *fw2.2*: a quantitative trait locus key to the evolution of tomato fruit size. Science 289:85–88

Frary A, Doganlar S, Daunay MC, Tanksley SD (2003) QTL analysis of morphological traits in eggplant and implications for conservation of gene function during evolution of solanaceous species. Theor Appl Genet 107:359–370

Frary A, Fritz LA, Tanksley SD (2004) A comparative study of the genetic bases of natural variation in tomato leaf, sepal and petal morphology. Theor Appl Genet 109:523–533

Furini A, Wunder J (2004) Analysis of eggplant (*Solanum melongena*)-related germplasm: morphological and AFLP data contribute to phylogenetic interpretations and germplasm utilization. Theor Appl Genet 108:197–208

Gebhardt SE, Thomas RG (2002) Nutritive value of foods. Home Garden Bull (USDA) 72:80–81

Gleddie S, Keller WA, Setterfield G (1986) Production and characterization of somatic hybrids between *S. melongena* L. and *S. sisymbriifolium* Lam. Theor Appl Genet 71:613–621

Grandillo S, Tanksley SD (1996) Analysis of horticultural traits differentiating the cultivated tomato from the closely related species *Lycopersicon pimpinellifolium*. Theor Appl Genet 92:935–951

Grandillo S, Ku HM, Tanksley SD (1999) Identifying the loci responsible for natural variation in fruit size and shape in tomato. Theor Appl Genet 99:978–987

Isshiki S, Okubo H, Fujieda K (1994a) Phylogeny of eggplant and related *Solanum* species constructed by allozyme variation. Sci Hort 59:171–176

Isshiki S, Okubo H, Oda N, Fujieda K (1994b) Isozyme variation in eggplant (*Solanum melongena* L.). J Jpn Soc Hort Sci 63:115–120

Isshiki S, Okubo H, Fujieda K (1994c) Genetic control of isozymes in eggplant and its wild relatives. Euphytica 80:145–150

Isshiki S (1996) Genetic variation of isozymes and its application to breeding in eggplant and its wild relatives. Bull Faculty Agri Saga University Japan 80:1–46

Isshiki S, Uchiyama T, Tashiro Y, Miyazaki S (1998) RFLP analysis of a PCR amplified region of chloroplast DNA in eggplant and related *Solanum* species. Euphytica 102:295–299

Isshiki S, Okubo H, Fujieda K (2000) Segregation of isozymes in selfed progenies of a synthetic amphidiploid between *Solanum integrifolium* and *S. melongena*. Euphytica 112:9–14

Isshiki S, Suzuki S, Yamashita K (2003) RFLP analysis of mitochondrial DNA in eggplant and related *Solanum* species. Genet Resour Crop Evol 50:133–137

Janick J, Topoleski LD (1963) Inheritance of fruit color in eggplant (*Solanum melongena*). Proc Am Soc Hort Sci 83:547–558

Kakizaki Y (1931) Hybrid vigor in eggplants and its practical utilization. Genetics 16:1–25

Karihaloo JL, Gottlieb LD (1995) Allozyme variation in the eggplant, *Solanum melongena* L. (Solanaceae). Theor Appl Genet 90:578–583

Karihaloo JL, Brauner S, Gottlieb LD (1995) Random amplified polymorphic DNA variation in the eggplant *Solanum melongena* L. Theor Appl Genet 90:767–770

Ku HM, Doganlar S, Chen KY, Tanksley SD (1999) Genetic basis of pear-shaped fruit in tomato. Theor Appl Genet 99:844–850

Lagercrantz U (1998) Comparative mapping between *Arabidopsis thaliana* and *Brassica nigra* indicates that *Brassica* genomes have evolved through extensive genome replication accompanied by chromosome fusions and frequent rearrangements. Genetics 150:1217–1228

Lester RN, Daunay MC (2003) Diversity of African vegetable *Solanum* species and its implications for a better understanding of plant domestication. In: Knüpffer H, Ochsmann J (eds.) Rudolf Mansfeld and Plant Genetic Resources. Schriften Genet Ressourcen 22:137–152

Lester RN, Hasan SMZ (1991) Origin and domestication of the brinjal eggplant, *Solanum melongena*, from *S. incanum*, in Africa and Asia. In: Hawkes JG, Lester RN, Nee M, Estrada RN (eds.) Solanaceae III. Taxonomy, Chemistry, Evolution. Royal Botanic Gardens, Kew, UK, pp 369–387

Liharska TB, Hontelez J, van Kammen A, Zabel P, Koornneef M (1997) Molecular mapping around the centromere of tomato chromosome 6 using irradiation-induced deletions. Theor Appl Genet 95:969–974

Lincoln S, Daly M, Lander E (1992) Mapping genes controlling quantitative traits with Mapmaker/QTL1.1. Whitehead Institute Tech Rep 2

McCammon KR, Honma S (1983) Morphological and cytogenetic analyses of an interspecific hybrid eggplant, *Solanum melongena* × *Solanum torvum*. HortScience 18:894–895

Mace ES, Lester RN, Gebhardt CG (1999) AFLP analysis of genetic relationships among the cultivated eggplant, *Solanum melongena* L., and wild relatives (Solanaceae). Theor Appl Genet 99:626–633

Maliepaard C, Bas N, van Heusden S, Kos J, Pet G, Verkerk R, Vrielink R, Zabel P, Lindhout P (1995) Mapping of QTLs for glandular trichome densities and *Trialeurodes vaporariorum* (greenhouse whitefly) resistance in an F_2 from *Lycop-*

ersicon esculentum × *L. hirsutum* f. *glabratum*. Heredity 75:425–433

Nelson JC (1997) QGENE: software for marker-based genomic analysis and breeding. Mol Breed 3:229–235

Nolla JAB (1932) Inheritance of color in eggplant. J Dept Agri Puerto Rico 16:19–30

Nunome T, Yoshida T, Hirai M (1998) Genetic linkage map of eggplant. In: Proceedings of the 10th Eucarpia Meeting on Genetics and Breeding of *Capsicum* and Eggplant, Avignon, France, 1998, pp 239–242

Nunome T, Ishiguro K, Yoshida T, Hirai M (2001) Mapping of fruit shape and color development traits in eggplant (*Solanum melongena* L.) based on RAPD and AFLP markers. Breed Sci 51:19–26

Nunome T, Suwabe K, Iketani H, Hirai M (2003a) Identification and characterization of microsatellites in eggplant. Plant Breed 122:256–262

Nunome T, Suwabe K, Ohyama A, Fukuoka H (2003b) Characterization of trinucleotide microsatellites in eggplant. Breed Sci 53:77–83

Paran I, Goldman I, Zamir D (1997) QTL analysis of morphological traits in a tomato recombinant inbred line population. Genome 40:242–24

Phatak SC, Liu J, Jaworski CA, Sultanbawa AF (1991) Functional male sterility in eggplant: inheritance and linkage to the purple fruit color gene. J Hered 82:81–83

Ranz JM, Casals F, Ruiz A (2001) How malleable is the eukaryotic genome? Extreme rates of chromosomal rearrangement in the genus *Drosophila*. Genome Res 11:230–239

Rizza F, Mennella G, Collonnier C, Sihachakr D, Kashyap V, Rajam MV, Prestera M, Rotino GL (2002) Androgenic dihaploids from somatic hybrids between *Solanum melongena* and *S. aethiopicum* group *gilo* as a source of resistance to *Fusarium oxysporum* f. sp. *melongenae*. Plant Cell Rep 20:1022–1032

Robinson RW, Shail JW, Gao Y (2001) Interspecific hybridization of eggplant for Verticillium wilt resistance and other useful traits. In: van den Berg RG, Barendse GWM, van der Weerden GM, Mariani C (eds.) Solanaceae V. Advances in Taxonomy and Utilization. Nijmegen University Press, Nijmegen, The Netherlands, pp 279–292

Sakata Y, Nishio T, Matthews PJ (1991) Chloroplast DNA analysis of eggplant (*Solanum melongena*) and related species for their taxonomic affinity. Euphytica 55:21–26

Sakata Y, Lester RN (1994) Chloroplast DNA diversity in eggplant (*Solanum melongena*) and its related species *S. incanum* and *S. marginatum*. Euphytica 80:1–4

Sakata Y, Lester RN (1997) Chloroplast DNA diversity in brinjal eggplant (*Solanum melongena* L.) and related species. Euphytica 97:295–301

Sharma DR, Chowdhury JB, Ahuja U, Dhankhar BS (1980) Interspecific hybridization in genus *Solanum*. A cross between *S. melongena* and *S. khasianum* through embryo culture. Z Pflanzenzücht 85:248–253

Sharma DR, Sareen PK and Chowdhury JB (1984) Crossability and pollination in some nontuberous *Solanum* species. Ind J Agri Sci 54:514–517

Sihachakr D, Haicour R, Serraf I, Barrientos E, Herbreteau C, Ducreux G, Rossignol L, Souvannavong V (1988) Electrofusion for the production of somatic hybrid plants of *Solanum melongena* L. and *S. khasianum* CB Clark. Plant Sci 57:215–223

Sihachakr D, Haicour R, Chaput MH, Barrientos E, Ducreux G, Rossignol L (1989) Somatic hybrid plants produced by electrofusion between *S. melongena* L. and *S. torvum* Sw. Theor Appl Genet 77:1–6

Stommel JR, Whitaker BD (2003) Phenolic acid content and composition of eggplant fruit in a germplasm core subset. J Am Soc Hort Sci 128:704–710

Sunseri F, Sciancalepore A, Martelli G, Acciarri N, Rotino GL, Valentino D, Tamietti G (2003) Development of RAPD-AFLP map of eggplant and improvement of tolerance to Verticillium wilt. In: Hammerschlag FA, Saxena P (eds.) Proceedings of the 26th IHC Biotechnology in Hort Crop Improvement. Acta Hort 625, ISHS, pp 107–115

Tamura N, Murata Y, Mukaihara T (2002) A somatic hybrid between *Solanum integrifolium* and *S. violaceum* that is resistant to bacterial wilt caused by *Ralstonia solanacearum*. Plant Cell Rep 21:353–358

Tanksley SD, Ganal MW, Prince JP, de Vicente MC, Bonierbale MW, Broun P, Fulton TM, Giovannoni JJ, Grandillo S, Martin GB, Messeguer R, Miller JC, Miller L, Paterson AH, Pineda O, Roder S, Wing RA, Wu W, Young ND (1992) High-density molecular linkage maps of the tomato and potato genomes. Genetics 132:1141–1160

Tigchelaar EC, Janick J, Erickson HT (1968) The genetics of anthocyanin coloration in eggplant (*Solanum melongena* L.). Genetics 60:475–491

van Eck HJ, Jacobs JM, van Dijk J, Stiekema WJ, Jacobsen E (1993) Identification and mapping of three flower color loci of potato (*S. tuberosum* L.) by RFLP analysis. Theor Appl Genet 86:295–300

van Eck HJ, Jacobs JM, van den Berg PM, Stiekema WJ, Jacobsen E (1994) The inheritance of anthocyanin pigmentation in potato (*S. tuberosum* L.) and mapping of tuber skin color loci using RFLPs. Heredity 73:410–421

Wanjari KB, Khapre PR (1977) Inheritance of pigmentation in *Solanum melongena* × *S. indicum* L. Genetica Agaria 31:327–332

10 Cucurbits

Yi-Hong Wang[1], Tarek Joobeur[2], Ralph A. Dean[2], and Jack E. Staub[3]

[1] School of Science, Penn State University Behrend College, Erie, PA 16563, USA
e-mail: yxw23@psu.edu
[2] Fungal Genomics Laboratory, North Carolina University, Campus Box, 7251, Raleigh, NC 27695, USA
[3] USDA-ARS, 1575 Linden Drive, Department of Horticulture, University of Wisconsin, Madison, WI 53703, USA

10.1 Introduction

Cucurbits (the Cucurbitaceae family) include 118 genera and 825 species distributed primarily in tropical and subtropical regions of the world (Jeffrey 1990). The most economically important cucurbit crops in terms of world total production (FAO 2006) are watermelon [*Citrullus lanatus* (Thunb.) Matsum. & Nakai], cucumber (*Cucumis sativus* var. *sativus* L.), and melon (*Cucumis melo* L.) and are reviewed in this chapter. Other reviews on cucurbit crops have been provided by Robinson and Decker-Walters (1997), Andres (2004), Fehér (1993), McCreight et al. (1993) and Tatlioglu (1993). A comprehensive coverage on cucurbit diseases can be found in Blancard et al. (1994) and Zitter et al. (1996). During the past decade, considerable efforts have been made to expand the knowledge of genetic mapping, marker development/molecular breeding, and genetic engineering in cucurbit species. Although progress has been made in melon and cucumber (e.g., resistance genes cloned in melon; Joobeur et al. 2004; Pauquet et al. 2004; Taler et al. 2004), comparatively little is known about the genomes of other cucurbit species (Brown and Myers 2002; Levi et al. 2002, 2004).

10.1.1 Origin and Botany

Wild ancestors of cultivated melon and watermelon originated in Africa, and wild, free-living cucurbit species are found on that continent today (Pitrat et al. 1999). Both the species were domesticated in multiple areas of secondary diversity such as the Middle and Near East, and India (Esquinas-Alcazar 1977; Staub et al. 1992; Robinson and Decker-Walters 1997; Mliki et al. 2001). In contrast, cucumber is of southwest Asian origin, and was domesticated around 1500 BC (Whitaker and Davis 1962; Jeffrey 1980). Melon was first domesticated in Egypt around 3000 BC, slightly earlier than watermelon. Its cultivation was first spread to the Mediterranean, Middle East, and then to Asia. It was introduced to Europe, and then to the Americas from the Mediterranean. While dispersal of watermelon may have been similar to that of melon, cucumber was introduced from India to China, North Africa and Southern Europe, and then from Europe to the New World (i.e., Haiti) by Columbus in 1494 (Staub et al. 1997).

Cucurbits are typically tendril-bearing indeterminant plants with simple and palmately veined leaves that can be shallow to deeply lobed and are borne singly on each node (Jeffery 1980). Melon leaves are subcordate and less lobed than cucumber while the leaves of watermelon are pinnatified. Although most commercial cucumber and watermelon varieties are monoecious, in recent years gynoecious phenotypes have been developed for open-field and greenhouse production. Likewise, while commercial melon varieties are typically andromonoecious, and produce round or oval fruits, monoecious phenotypes do exist which develop elongated fruits. Compared to cucumber and watermelon, melon fruits are extremely diverse and are often used as criteria to partition varieties into horticultural groups based on shape, color, netting, and other external and internal characters (see below). For instance, rind color can vary from white to dark green, where sutures can be white, yellow, or tan in color (López-Sesé et al. 2003). Interior flesh color may be orange, pink, green or white.

The genus *Cucumis* has two major species, cucumber (*C. sativus*) and melon (*C. melo*) (Jeffery 1980). Within *C. melo*, Munger and Robinson (1991) proposed six cultivar groups based on fruit characteristics which were expanded into 16 groups by Pitrat et al. (2000). However, a more rigorous taxonomic treat-

Genome Mapping and Molecular Breeding in Plants, Volume 5
Vegetables
C. Kole (Ed.)

ment will likely allow for the classification of these morphotypes more specifically as cultivar groups as defined by Spooner et al. (2003). Phylogenetic analysis using molecular markers has separated these into two major groups: the North American and European cantalupensis and inodorus morphotypes as one group and the more exotic morphotypes (conomon, chito, dudaim, agrestis and momordica) in a second group (Stepansky et al. 1999). These six groups are:

1. Cantalupensis (syn. cantaloupe): medium-sized fruits with netted, warty or scaly surface; flesh usually orange or green; flavor aromatic or musky. Consumed as a fresh market product, and is typically andromonoecious.
2. Inodorus (syn. winter melon): fruits usually larger, maturing later and longer keeping than Cantalupensis. Rind not netted; flesh white or green, lacking a musky odor. Consumed as a fresh market product. Mostly andromonoecious.
3. Flexuosus (syn. snake melon): long, slender and often ribbed fruits. Consumed as a fresh market product. Mostly monoecious.
4. Conomon (syn. pickling melon): small fruits with smooth, tender skin, white flesh, early maturity. Mainly consumed as a processed product and is andromonoecious.
5. Dudaim (syn. pomegranate melon): fruits small, round to oval with white flesh and thin rind. Consumed as a fresh market product, and is typically andromonoecious.
6. Momordica (syn. snap melon): fruits ovoid to cylindrical, flesh white or pale orange, low in sugar content, mealy, insipid or sour tasting. Fruit cracks or bursts as it approaches maturity and degenerates when barely ripe. Consumed as a fresh market product, and is typically monoecious in sex expression.

Most cultivated or edible watermelons belong to *Citrullus lanatus*, which include two morphological subspecies: *C. lanatus* subsp. *lanatus* (Thunb.) Matsum. & Nakai (although *Citrullus vulgaris* Schrad. is still found in the literature) and *C. lanatus* subsp. *vulgaris* Schrad. The latter contains various botanical varieties such as citron which is a source of Fusarium wilt resistance in watermelon (Robinson and Decker-Walters 1997). Although different ploidy levels exist in cultivated watermelons, diploids (may contain 0.5–1% polyploids), triploids and tetraploids have been commercially important (Fehér 1993). Tetraploids may produce smaller fruits than diploids, but their fruit quality is excellent and germination is typically better than their diploid counterparts.

Cytological studies of cucurbits have lagged behind other crop plants primarily because their chromosomes are relatively small and stain poorly. Chen et al. (1998) karyotyped cucumber chromosomes ($2n = 2x = 14$) using a C-banding technique and found that the length of mitotic chromosomes ranges from 1.48 to 2.31 μm, which was similar to that reported by Ramachandran and Seshadri (1.25–2.72 μm; 1986). Melon chromosomes ($2n = 2x = 24$) were karyotyped by Ramachandran and Seshadri (1986) who found chromosomes ranged from 1.06 to 1.88 μm, and more recently by Ma et al. (1994) who reported slightly smaller chromosome lengths (0.91–1.65 μm). Cucurbit chromosomes are small because their haploid DNA content is low. For example, melon has a genome size of 4.5×10^8 base pairs (bp) and the genomes of cucumber and watermelon are 3.67×10^8 and 4.25×10^8 bp, respectively (Arumuganathan and Earle 1991).

10.1.2 Cucurbit Production

Cucurbits are an important part of the human diet. Historical data (FAO 2006) indicate that total production of cucumber, melon and watermelon has increased more than fourfold in the last 40 years due to increased yield. According to the Food and Agriculture Organization (FAO) 2005 data (Table 1), watermelon is the most popular cucurbit, followed by cucumber and melon. Although China, Turkey, Iran and the United States produce the highest quantity of cucurbit crops, there is a gap in cucurbit production between developing and developed countries. For example, Cameroon has the second largest cucumber growing areas (100,000 ha) in the world (next to China), but its total production is only 115,000 tons compared to 440,000 tons produced on 700 ha in the Netherlands (FAO 2006). This probably reflects a difference in the sophistication of production systems.

Delicious as dessert (melon and watermelon) or vegetable (cucumber), cucurbits are a rich source of vitamin A, β-carotene and potassium (Robinson and Decker-Walters 1997). However, cucurbit fruit also contains in its placenta the bitter cucurbitacins which are most abundant in colocynth (*Citrullus colocynthis*) and various wild species of *Cucumis*. Cucurbitacins can serve to deter insects such as aphids

Table 1. World cucurbit production in 2005 based on FAO (2006) (ha = hectare; Mt = metric ton; countries are ranked based on their total productions; cucumber figures also include those of gherkins)

Melons Country	Area (ha)	Production (Mt)	Cucumbers Country	Area (ha)	Production (Mt)	Watermelon Country	Area (ha)	Production (Mt)
China	578,500	15,138,000	China	1,553,100	26,559,600	China	2,014,500	69,315,000
Turkey	103,000	1,700,000	Turkey	60,000	1,725,000	Turkey	137,000	3,800,000
Iran	80,000	1,230,000	Iran	80,000	1,400,000	Iran	100,000	2,150,000
Spain	35,200	1,176,900	US	68,660	969,400	US	57,140	1,669,940
US	45,180	1,150,440	Japan	14,500	680,000	Egypt	62,000	1,500,000
Romania	36,000	755,000	Egypt	28,000	600,000	Mexico	42,979	970,055
Italy	28,254	665,398	Mexico	17,000	475,000	S. Korea	23,000	850,000
India	31,500	645,000	Netherlands	600	435,000	Brazil	75,000	622,000
Egypt	24,000	565,000	Indonesia	50,064	423,333	Syria	25,000	620,000
Mexico	31,500	510,000	S. Korea	6,000	400,000	Kazakhstan	45,000	600,000
World	1,308,018	28,321,159	World	2,488,600	41,743,840	World	3,424,507	95,292,051

and spider mites from attacking foliage and fruits. But some insects like cucumber beetles are attracted to plants with high cucurbitacins. Although at least five genes regulate cucurbitacin biosynthesis, a single recessive mutation in cucumber produced a nonbitter phenotype, and this mutation has been bred into many cucumber cultivars for sustained pest resistance (Robinson and Decker-Walters 1997).

10.2 Genetic Mapping of Cucurbit Genomes

Earlier efforts to map cucurbit genomes used morphological and protein markers such as in cucumber (Fanourakis and Simon 1987; Pierce and Wehner 1990; Knerr and Staub 1992; Vakalounakis 1992; Meglic and Staub 1996). Some of these markers are listed in the cucumber gene list (Xie and Wehner 2001) curated by TC Wehner, North Carolina State University, Raliegh, NC. Early melon mapping work employing classical markers was, however, limited to varietal surveys (Esquinas-Alcazar 1977; Perl-Treves et al. 1985). Pitrat (1991) constructed the first melon genetic map that consisted of eight linkage groups with 23 disease resistance, flower biology and vegetative characters. The first watermelon map was constructed using isozyme markers that contained seven linkage groups spanning 354 cM (Navot et al. 1990). Updated information regarding mapped genes has been provided by Pitrat (2002) and Guner and Wehner (2003) for melon and watermelon, respectively.

It is clear from earlier works that the use of phenotypic markers alone is not sufficient for the construction of saturated maps in cucurbit species. This is due to the paucity of such markers and the difficulty inherent in identifying and/or producing such markers. Moreover, unlike DNA-based markers, the expression of morphological and certain isozyme marker systems may be affected by environment factors (Staub et al. 1982).

DNA markers (mostly RFLP and RAPD) were first used to map the cucumber genome by Kennard et al. (1994). Because of the comparatively small number of markers included in these maps, the average distance between markers is dramatically less than the recent maps of Serquen et al. (1997), Park et al. (2000), Staub and Serquen (2000), Bradeen et al. (2001), and Fazio et al. (2003a). The map by Park et al. (2000) contains the most markers (347) spanning 816 cM, which is within the range of the length of the cucumber genome (750–1,000 cM) estimated by Staub and Meglic (1993).

RAPD and RFLP markers were also applied by Baudarcco-Arnas and Pitrat (1996) to produce the first genetic map of melon that consisted of 102 markers and a second map with 188 predominantly AFLP markers was constructed by Wang et al. (1997). Oliver et al. (2001) mapped the melon genome with nearly 400 markers. A high-density composite map was constructed by Périn et al. (2002a) consisting of 668 AFLP, IMA and phenotypic markers. The latter includes both disease resistance and fruit quality genes and

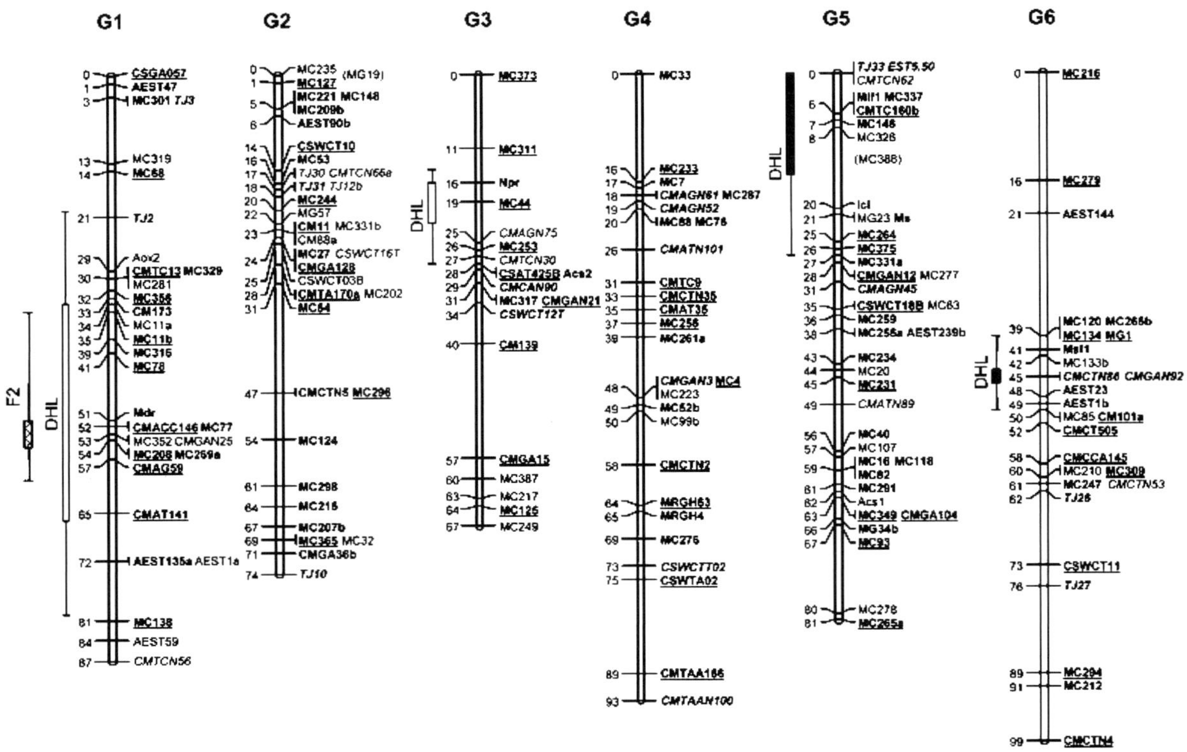

Fig. 1. A composite linkage map of melon (*Cucumis melo* L.) after merging the F_2 and double haploid line (DHL) maps. Linkage groups are represented by *vertical double lines* and labeled G1–G12. Loci are listed to the *right*, and recombination distances (in centiMorgans) to the *left*, of each linkage group. Loci in *boldface* define the framework of each linkage group. *Underlined* loci were scored in both populations; those in *italics* were scored only in the DHL population and the remainder only in the F_2 population. Loci in *brackets* could not be mapped after map merging; these markers are in their most likely map positions, according to those calculated with the individual populations. *Boxes* on the *left* of the linkage groups indicate regions with distorted segregation ($P < 0.005$) in the population indicated. *Solid boxes* indicate excess of Piel de Sapo alleles, *white boxes* excess of PI 161375 alleles, and *dashed boxes* excess of heterozygous genotypic class and lack of PI 161375 homozygous. The confidence interval of the position of the putative locus involved in the distorted segregation is shown with *bars*, including markers with distortion at $P < 0.05$ (adapted from Gonzalo et al. 2005 with permission of Springer Science and Business Media)

will be useful for identifying markers linked to these traits. Another composite map has been generated of melon by Gonzalo et al. (2005) using SSR markers (Fig. 1). Maps using different parents were also generated (Liou et al. 1998; Brotman et al. 2000; Danin-Poleg et al. 2002; Fukino et al. 2002; Silberstein et al. 2003). Even though there was a considerable lag in mapping of the watermelon genome after the construction of the initial map (Hashizume et al. 1996), efforts in this species are increasing (Fan et al. 2000; Hawkins et al. 2001; Levi et al. 2001, 2004; Zhang et al. 2004) and have resulted in a map consisting of 554 molecular markers (Hashizume et al. 2003).

Given that maps of increasing density are becoming available for these cucurbits species, experiments designed to investigate the synteny (comparative mapping) between species genomes may be possible in the near future. The fact that cucurbits are similar to each other at the DNA level increases the potential usefulness of such studies for gene identification and marker development. The homology at DNA level between cucumber and watermelon is 78% (Pasha and Sen 1998). Likewise, the homology between watermelon and melon and between cucumber and melon is 84% and 90%, respectively. Furthermore, SCAR primers (Wang et al. 2000) designed for melon amplified products of similar size and sequence in cucumber (Wang 1999). A homology of 76 to 94.8% was found in six PCR fragments amplified by SSR primers between cucumber and melon (Danin-Poleg et al. 2001) and 20 of 31 melon SSR primers can be used in cucumber (Chiba et al. 2003). Similarly, 33 of 67 melon SSR markers was found to amplify DNA sequences in the cucumber genome and 16 ampli-

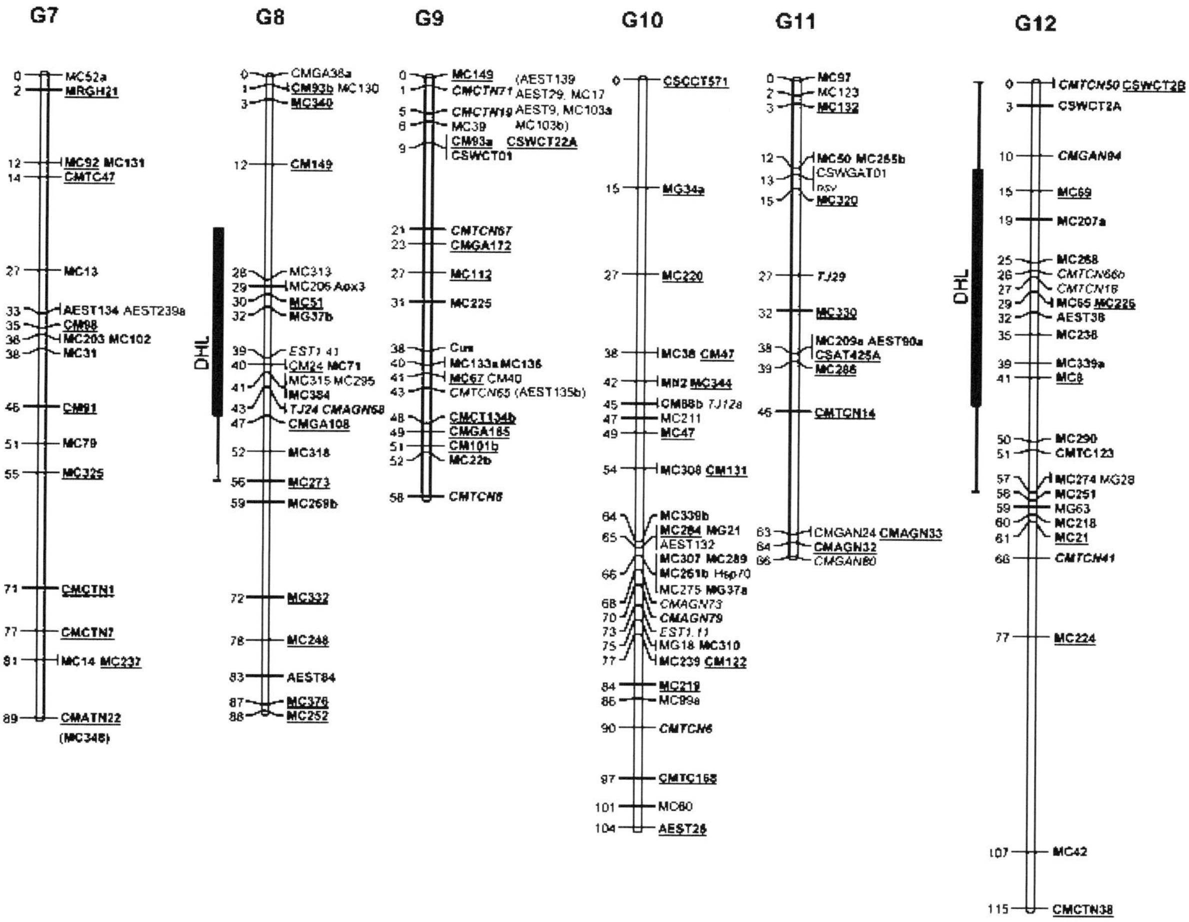

Fig. 1. (continued)

fied watermelon genomic DNA sequences (Ritschel et al. 2004). These explain the utility of SSR markers (Danin-Poleg et al. 2000) and cDNA and chloroplast DNA (ccpSSR) markers (Neuhausen 1992; Silberstein et al. 1999; Brotman et al. 2000; Klingler et al. 2001; Oliver et al. 2001; Chung and Staub 2003; Chung et al. 2003; Silberstein et al. 2003) for genome analysis in both melon and cucumber.

10.3 Genetic Mapping of Mendelian Traits

Genes conditioning economically important traits such as disease resistance have been intensively mapped in melon and cucumber by a number of research groups. Most melon research has been focused on identifying markers linked to disease resistance, especially Fusarium wilt and virus diseases, and yield and quality traits. Markers tightly linked to a trait (<5 cM) have potential applications in genotypic selection of the trait.

10.3.1 Fusarium Wilt Resistance

The first disease resistance gene (*Fom-2*) tagged using molecular (RAPD) markers in cucurbits was Fusarium wilt of melon caused by *Fusarium oxysporum* f. sp. *melonis* by Wechter et al. (1995; 1998) and Wang et al. (2000). Two AFLP markers were also found to flank *Fom-2* at 1.7 and 3.3 cM apart, where one additional marker cosegregated with the gene (Wang et al. 2000). *Fom-2* was also mapped by Baudarcco-Arnas and Pitrat (1996) with flanking RAPD markers E07-1.3 and G17-1.0. Subsequently, using disease resistance gene homologs, Brotman et al. (2002) mapped *Fom-2* at 0.7 cM from marker NBS3. More recently, Joobeur et al. (2004) identified three markers (STS411, SSR430 and SSR138) with STS411 and SSR430 flanking *Fom-2* in a 6-kb interval during the process of cloning the gene.

Another gene conferring resistance to Fusarium wilt pathogen races 0 and 2, *Fom-1*, was mapped by Brotman et al. (2002) with flanking RFLP markers at 2.6 cM and 7.0 cM apart. Brotman et al. (2005) reported that *Fom-1* was 1.4 cM from *Prv*, the *potyvirus* resistance gene in melon (see Sect. 10.3.2), and that both genes were flanked by AFLP (ACA/CAT90 and ATC/CAT497) and RFLP/CAPS (NBS47-3, NBS1-CAPS and 62-CAPS) markers at 1.7 and 1.4 cM, respectively. These markers are most closely linked to *Fom-1* in genetic maps constructed to date. The watermelon Fusarium wilt (caused by *Fusarium oxysporum* f. sp. *niveum*) race 1 resistance gene (*Fom-1*) was also mapped (Xu et al. 1999; Fan et al. 2000; Levi et al. 2002).

10.3.2 Mapping of Other Resistance Genes

Several virus resistance genes have also been identified by molecular markers. The gene *nsv*, conferring resistance to *melon necrotic spot virus*, was first mapped by Morales et al. (2002). *nsv* was subsequently fine-mapped using 408 F_2 plants and 2727 backcross progeny to a single BAC clone with cosegregating marker 52K20sp6 (Morales et al. 2005). Since the single BAC clone physically contains the *nsv* gene, it is a matter of sequencing the BAC to identify a candidate *nsv* gene.

The gene, *Prv*, controls resistance to *potyvirus* which is a debilitating disease in melon. *Prv* was tightly linked to NBS47-3 (Brotman et al. 2002) which is flanked at 1.4 cM by *Fom-1* and NBS47-3 on each side (Brotman et al. 2005). The zucchini yellow mosaic virus (ZYMV) resistance gene, *Zym-1*, was mapped in both melon (Danin-Poleg et al. 2002) and cucumber (Park et al. 2000). However, using the *zym* cosegregating marker Park et al. (2004) did not find this region to be syntenic between these *Cucumis* species.

Other disease resistance genes mapped include an aphid (*Aphis gossypii* Glover) resistance gene (*Vat*) in melon (Klingler et al. 2001; Silberstein et al. 2003), a scab resistance gene (*Ccu*) in cucumber (Bradeen et al. 2001), and a downy mildew resistance gene (*dm*) (Horejsi et al. 2000; Bradeen et al. 2001).

10.3.3 Fruit/Flower Traits

Several horticultural traits have been mapped in *Cucumis* species using molecular markers. In melon, these include the *pH* locus, which conditions flesh acidity (Danin-Poleg et al. 2002), the *p* locus conferring five carpels (*pp* plants have five carpels) (Oliver et al. 2001), a "nonyellowing" long shelf-life trait (Touyama et al. 2000), a male-sterility gene (*ms-3*) (Park and Crosby 2004) and the *andromonoecious* (*a*) gene (Silberstein et al. 2003; Noguera et al. 2005). The dominant *a* allele confers monoecious phenotype while recessive *aa* plants are andromonoecious (Noguera et al. 2005).

Economically important horticultural traits have also been mapped in cucumber. Gynoecious sex expression (i.e., pistillate flowers) is conditioned by a single major dominant locus *F*, which is involved with the regulation of bud primordial ethylene levels (higher ethylene levels support pistillate flower production) and mapped by Trebitsh et al. (1997). Hybridization of the linked marker (CS-ACS1G) to genomic DNA of monoecious and gynoecious cucumber lines indicated that highly inbred gynoecious lines (*FF*) uniformly contained an additional copy of the ACC synthase gene, which is the key enzyme in ethylene synthesis (Trebitsh et al. 1997; Mibus and Tatlioglu 2004). Other markers cosegregating with *F* (Bradeen et al. 2001), and genes that modify its action (Serquen et al. 1997; Fazio et al. 2003a) have also been identified. However, it was not known whether these are associated with genes in the ethylene synthesis pathway. The map position of traits that tend to increase female sex stability in gynoecious cucumber phenotypes (*FF*, *Ff*) (Lower et al. 1983) are known (Serquen et al. 1997; Fazio et al. 2003a). For instance, the gene for determinate plant habit (*de*) was found to be linked at 3.1 cM to AFLP marker (Bradeen et al. 2001) and 0.8 cM from an RFLP marker (Fazio et al. 2003a).

10.4 Genetic Mapping of Quantitative Traits

Mapping of quantitative trait loci (QTLs) in cucurbits was first initiated in cucumber fruit traits by Kennard and Havey (1995). Among these traits, fruit weight was found to be in the vicinity of fruit diameter (Kennard and Havey 1995; Sequen et al. 1997), and multiple lateral branching (MLB) and fruit diameter were located in the same genomic region (Fazio et al. 2003a). Four QTLs (LOD > 3) for sex expression were detected, as well as four for MLB, two for earliness, and two for fruit length (Serquen et al. 1997). Cumulative QTL effects explained at least 25% (~45% for gynoecy) of the observed variation for any particular trait. In a more rigorous analysis (RILs examined in multiple locations), Fazio et al. (2003a) confirmed the results of Serquen et al. (1997) and detected additional QTLs for the traits examined. For instance, four location-independent factors that cumulatively explained 42% of the observed phenotypic variation were detected for MLB. This and other QTL/marker yield component associations were found to be reliable and consistent over multiple environments and in different cross-progeny types (Dijkhuizen and Staub 2002).

In melon, Monforte et al. (2004) found that the heritability of fruit shape (FS) was comparatively higher than fruit weight (FW) and sugar content (SSC) in agreement with Périn et al. (2002b). Eight QTLs for FS were reported and among them were *fs1.1*, *fs9.1*, and *fs11.1*. Eduardo et al. (2004) generated three near-isogenic lines (NILs) from PI 161375 × Pinyonet Piel de Sapo to verify the effect of *fs1.1*, *fs9.1* and *fs11.1* on fruit shape. They found that NILs carrying PI 161375 alleles at *fs1.1* and *fs9.1* produced more elongated fruits than NILs in a Pinyonet Piel de Sapo background possessing the same allelic configuration. It was further demonstrated that alleles at *fs1.1* acted to produce growth only in the longitudinal direction, while *fs9.1* alleles caused development in the transversal direction. NILs with PI 161375 alleles at *fs11.1* did not affect fruit growth (shape) when compared to alleles at the same locus in a Pinyonet Piel de Sapo background (Eduardo et al. 2004). Traits that are highly correlated with each other such as fruit shape (FS) and ovary shape (OVS) tend to map to the same region. For example, QTLs for FS and OVS overlapped and were linked to the *p* locus in melon (Périn et al. 2002b).

QTLs conferring disease resistance in melon have also been identified. Dogimont et al. (2000) detected seven QTLs for resistance to *cucumber mosaic cucumovirus* (*CMV*). More recently, Perchepied et al. (2005) mapped the *Fom 1.2* gene that confer polygenic partial resistance to the Fusarium wilt race 1.2 using Isabelle (resistant) and Védrantais (susceptible) as parents. Nine QTLs were identified in five linkage groups which explained 41–64% of the observed phenotypic variation.

10.5 Marker-Assisted Selection (MAS)

Molecular marker systems used in cucurbit genome mapping (i.e., RFLP, RAPD, or AFLP) can be amenable to high-throughput genotype screening, which is important for designing breeding strategies which incorporate MAS (Zheng and Wolff 2000; Fazio et al. 2003b). Converting such dominant or codominant markers to sequence characterized amplified regions (SCAR) or sequence-tagged site (STS) marker equivalents will facilitate multiplexing for increased MAS efficiency (Robbins et al. 2002, 2004). Marker conversion usually involves sequencing the marker sequences from both parents, identifying polymorphic segments, and designing PCR primers according to the observed parental sequence differences (Horejsi et al. 1999; Zheng et al. 1999; Wang et al. 2000; Xu et al. 2000). These converted markers can either be dominant (Wechter et al. 1998), codominant (Wang et al. 2000) or as CAPS (cleaved amplified polymorphic sequence) markers (Zheng et al. 1999).

When the codominant FM and AM markers (Wang et al. 2000) were used to screen 45 melon genotypes from different sources for Fusarium wilt resistance, it was found that FM correctly predicted disease phenotypes in 37 genotypes and AM in 41 of the genotypes examined (Wang et al. 2000). More recently, Burger et al. (2003) tested 24 genotypes with FM and AM markers, and found that FM could correctly predict disease phenotype in 22 of the genotypes examined and AM correctly predicted reaction phenotypes for all of the genotypes tested, differentiating heterozygotes from homozygotes, which shows their potential utility for MAS. On the other hand, unconverted RAPD markers cannot be relied on to screen for target phenotypes (Zheng and Wolff 2000).

Markers linked to monogenic traits could be effective in MAS. Wang et al. (2000) and Burger et al. (2003) demonstrated that melon *Fom-2* markers were robust and effective for phenotypic screening in a breeding program. MAS has been successfully used in conventional commercial breeding programs in the recent past (<10 years) for the incorporation of monogenic traits using backcross (BC) and recurrent selection.

However, there are few documented reports of the relative success of MAS when compared to phenotypic selection. One notable example is found in cucumber where MAS for markers linked to multiple lateral branching (MLB) QTLs was compared to phenotypic (PHE; selection for MLB in open-field environments), and no selection (random, RAN; intermating without selection) (Fazio et al. 2003b). Five markers (SSR (2), RAPD (2) and SNP (1)) were used in backcross MAS during three cycles of greenhouse selection and PHE and RAN selection in an H-19 × GY-7 population (GY-7 = recurrent parent). No significant differences ($p < 0.001$) were detected between the means of PHE and MAS.

10.6 Map-Based Cloning

Because of their economic importance, disease resistance genes in cucurbits were among the most studied. Thus, it is not surprising that the first genes cloned in cucurbits were disease resistance genes. Taler et al. (2004) cloned two partially dominant complementary genes that confer resistance to downy mildew in melon using a biochemical method (Balaas et al. 1992). The first gene in cucurbit cloned by map-based cloning approach was melon *Fom-2* (Joobeur et al. 2004), followed by *Vat* (Pauquet et al. 2004), and *nsv* (Morales et al. 2005; Garcia-Mas et al. 2004).

The cloned *Fom-2* shared a high similarity to the previously characterized NBS-LRR class of resistance gene *I2* (Ori et al. 1997; Simons et al. 1998) in tomato (Joobeur et al. 2004). The *I2* locus confers resistance to tomato vascular wilt caused by *Fusarium oxysporum* f. sp. *lycopersici*. Sequence comparisons indicated that the domain of the *Fom-2* LRR was identical in the two resistant lines (MR-1 and PI 161375) tested, except for three nucleotides which resulted in the substitution of two residues V and K in MR-1 with M and E in PI 161375, respectively. LRR domain sequences from the susceptible genotypes (Védrantais, AY and Durango) were identical but 25 amino acids out of 541 were different from those of the resistant sequences (Joobeur et al. 2004). Additional varieties need to be screened to test the consistency of the PCR primer sequences of the *Fom-2* codominant marker. If such comparisons prove consistent, then this primer sequence could be used as an effective selection tool for *Fom-2*-resistant genotypes.

Another resistance gene, *Vat*, has also been cloned (Pauquet et al. 2004). *Vat* confers resistance to colonization by the melon/cotton aphid, and resistance to transmission of unrelated *cucumber mosaic virus* (*CMV*) and *potyviruses*. The cloned *Vat* gene belongs to the coiled-coil (CC) NBS/LRR class of plant disease resistance genes (Pauquet et al. 2004). The gene is 6 kb in size with three introns and encodes a protein of 1473 amino acids. The *Vat* locus contains resistance gene homologs including other *Vat*-like sequences that do not confer any known resistance. Function of this gene was confirmed by complementation of susceptible melon varieties (Pauquet et al. 2004).

Using genetic and physical mapping, Garcia-Mas et al. (2004) delineated the location of *nsv* gene that confers resistance to *melon necrotic spot virus*. From a 15-BAC clone contig spanning 1.2 cM, one BAC clone (1-21-10) was identified which contained the gene (Garcia-Mas et al. 2004) with cosegregating marker 52K20sp6 (Morales et al. 2005). The delineation was carried out using 408 F_2 plants and 2727 backcross progeny (Morales et al. 2005). The single BAC clone now can be sequenced to identify the candidate *nsv* gene.

Cloned disease resistance genes can be transferred to other susceptible melon lines to enhance their resistance to damaging pathogens either through transformation or traditional breeding aided by MAS via markers developed from these genes. They can also be introduced into other cucurbits to determine if heterologous expression of these genes affects host resistance. The sequences of cloned genes could allow for the development of a unique marker that could aid the selection of functional resistance gene during plant improvement.

10.7 Cucurbit Genetic Engineering

Traditionally, new cucurbit varieties are released after years of gene transfer by conventional breeding and evaluation. Transfer of target traits is usually limited within a species depending on cross-compatibilities. Development and utilization of molecular markers may greatly improve the efficiency of the process depending on the traits that are being selected. In contrast, modern technologies incorporating direct gene transfer (i.e., genetic engineering) are attractive because a gene(s) conferring a target trait does not necessarily need to originate in the same species and, theoretically, trait integration can be completed after transgenic plants have been regenerated and tested for their integrity. The incorporation of multiple traits (e.g., genes for fruit yield and quality) is much more complex and requires extensive testing and validation of requisite improvements since traits may be antagonistic (i.e., negatively correlated). Despite unique possibilities of such dramatic additions to conventional breeding programs, most transgenic cucurbit plants have been generated to manipulate only the fruit ripening process and virus resistance. The latter was mostly carried out on melon and squash (e.g., Freedom and Destiny series) because conventional breeding for virus resistance was much more successful in cucumber (see Gaba et al. 2004 for review on squash genetic engineering).

Ethylene is known to promote the ripening of climacteric fruit (Giovannoni 2004). If a gene encoding one key enzyme in the ethylene synthesis pathway, ACC oxidase, is "knocked out" by antisense, ethylene production can drop below 1% of nontransgenic control fruit, and the ripening process is subsequently blocked (either on or off the vine; Ayub et al. 1996). In cases of prolonged ripening melon fruit quality is improved because a higher amount of sucrose is accumulated (Guis et al. 1997). Knocking out the ACC oxidase gene also reduces total volatiles (the aroma compounds) by 60–85% because ethylene preferentially stimulates the synthesis of strong aroma compounds (Bauchot et al. 1998). Similarly, transgenic fruits are also firmer and contain higher levels of chlorophyll and titratable acidity than their nontransgenic counterparts (Silva et al. 2004).

Genetic engineering has also been successfully exploited to manipulate cucurbit virus resistance. Fang and Grumet (1993) introduced the full-length *zucchini yellow mosaic potyvirus* (*ZYMV*) coat protein into melon plants, and transgenic plants expressing the coat protein were found to exhibit apparent immunity to ZYMV infection (i.e., there was neither observed symptom nor virus accumulation in infected transgenic plants). Similarly, when CMV coat proteins were expressed in melon, transgenic plants did not develop symptoms when inoculated with 1 μg/ml virus and showed delayed symptoms when inoculum was 10 μg/ml (Yoshioka et al. 1993). Field evaluation of transgenic melon plants harboring coat protein genes of CMV, ZYMV and WMV-2 (watermelon mosaic 2 *potyvirus*) showed that homozygous transgenic plants never developed symptoms while hemizygous transgenic plants exhibited a significant delay (2–3 weeks) compared to control plants (Fuchs et al. 1997). Overall, only 8% of the homozygous and 33% of the hemizygous plants were infected by more than two viruses, while 99% of nontransgenic plants were infected. However, the incidence of any, unmixed, viral infection was 32–36% in homozygous plants, 70% in hemizygous plants, and 100% in nontransgenic controls. The hemizygous plants had a better yield than the homozygotes because of hybrid vigor despite reduced protection against viruses (Fuchs et al. 1997).

The increasing reluctance of consumers to accept transgenic plants (Aldhous 2003) should be considered prior to their introduction. Currently, among transgenic cucurbits with various enhanced traits, 54% of the trait and risk-assessment field trials have been dedicated to melons and 32% to squash (Gaba et al. 2004). In these field trials, 84% of the transgenic plants were enhanced with virus resistance, and 77% of these field trials were in the US (Gaba et al. 2004).

To ease public concern regarding genetically modified organisms (GMOs), a direct gene transfer without any use of antibiotics was developed by Chinese scientists who initially conducted their experiments in watermelon (Tao et al. 1996; Chen et al. 1998; Xiao et al. 1999a–c). This technology involves simply injecting the source DNA into target plant tissues, and subsequently identifying transgenic progeny based on phenotype. This was first reported by Tao et al. (1996) who injected cashew DNA into watermelon plants. In another experiment, Xiao et al. (1999b) soaked watermelon embryos in bottle gourd total genomic DNA. Watermelon plants susceptible to Fusarium wilt were utilized, while DNA was extracted from resistant bottle gourd plants (Xiao et al. 1999a). The regenerated plants (D_1) showed variations in chromosome arm lengths and ratios. In 22% of D_2 plants, fruit skin color changed from dark green to white or white with green netting, similar to the donor. Variations in fruit shape, seed shape and color each occurred in about one third of the D_2 plants. Two lines from D_3 showed high resistance to the wilt pathogen (Xiao et al. 1999b). Injecting source DNA directly into the ovary produced the same results (Xiao et al. 1999c).

Chen et al. (1998) took an approach resembling traditional transgene transformation in terms of construct preparation. They digested the source DNA from squash plants resistant to the Fusarium wilt pathogen, and then ligated DNA fragments onto a vector containing a GUS reporter and the CaMV 35S promoter. This mix was injected into the ovaries of watermelon plants. Two hundred transgenic plants were identified by GUS staining, of which 10 were resistant to the wilt pathogen (Chen et al. 1998).

10.8 Future Prospects

Cucurbit genomics will play a more significant role in cucurbit genome mapping, gene discovery and molecular marker development. The first step toward genomic analysis of cucurbit genomes is construction of large-insert genomic libraries and generation of expressed sequence tags (ESTs). Large-insert genomic libraries are available in melon (Luo et al. 2001; van Leeuwen et al. 2003) and cucumber (Nam et al. 2005). Several groups are generating ESTs in cucumber as seen by sequence cataloging in GenBank. Melon ESTs are also being generated by Garcia-Mas et al. (2002), and likely other research groups. These ESTs are useful in genetic mapping, gene discovery and construction of a genome array (Yariv et al. 2002, 2004).

Information gathered from other crop plants should be exploited for use in cucurbits. For example, tomato fruit size (Frary et al. 2000) and shape (Liu et al. 2002) genes have been isolated. Theoretically, it should be possible to find cucurbit homologs and examine their association with similar traits in cucurbits. In order to utilize information generated from model plant systems, a comparison of cucurbit genome with that of *Arabidopsis thaliana* has been performed by Monforte et al. (2004) and van Leeuwen et al. (2003). Tools developed for *Arabidopsis* can be readily adapted for cucurbit research, especially genomics tools for comparative studies of different cucurbit crops, the identification of novel genes associated with a trait of interest, and the genotyping progenies in a breeding program as an aid to phenotypic selection.

References

Aldhous P (2003) Time to choose. Nature 425:655

Andres TC (2004) Website for the plant family Cucurbitaceae and home of the Cucurbit Network. http://www.cucurbit.org/family.html. Cited 2004

Arumuganathan K, Earle ED (1991) Nuclear DNA content of some important plant species. Plant Mol Biol Rep 9:211–215

Ayub R, Guis M, Ben-Amor M, Gillot L, Roustan JP, Latche A, Bouzayen M, Pech JC (1996) Expression of ACC oxidase antisense gene inhibits ripening of cantaloupe melon fruits. Nat Biotechnol 14:862–866

Balass M, Cohen Y, Bar-Joseph M (1992) Identification of a constitutive 45 kD soluble protein associated with resistance to downy mildew in muskmelon (*Cucumis melo* L.) line PI 124111F. Physiol Mol Plant Pathol 41:387–396

Bauchot AD, Mottram DS, Dodson AT, John P (1998) Effect of aminocyclopropane-1-carboxylic acid oxidase antisense gene on the formation of volatile esters in Cantaloupe Charentais melon (cv. Vedrandais). J Agri Food Chem 46:4787–4792

Baudracco-Arnas S, Pitrat M (1996) A genetic map of melon (*Cucumis melo* L.) with RFLP, RAPD, isozyme, disease resistance and morphological markers. Theor Appl Genet 93:57–64

Blancard D, Lecoq H, Pitrat M (1994) A Color Atlas of Cucurbit Disease: Observation, Identification & Control. Wiley, New York, 299 p

Bradeen JM, Staub JE, Wye C, Antonise R, Peleman J (2001) Towards an expanded and integrated linkage map of cucumber (*Cucumis sativus* L.). Genome 44:111–119

Brotman Y, Kovalski I, Dogimont C, Pitrat M, Portnoy V, Katzir N, Perl-Treves R (2005) Molecular markers linked to papaya ring spot virus resistance and *Fusarium* race 2 resistance in melon. Theor Appl Genet 110:337–345

Brotman Y, Silberstein L, Kovalski I, Klinger J, Thompson G, Katzir N, Perl-Treves R (2000) Linkage groups of *Cucumis melo*, including resistance gene homologs and known genes. Acta Hort 510:441–448

Brotman Y, Silberstein L, Kovalski I, Perin C, Dogimont C, Pitrat M, Klingler J, Thompson GA, Perl-Treves R (2002) Resistance gene homologs in melon are linked to genetic loci conferring disease and pest resistance. Theor Appl Genet 104:1055–1063

Brown RN, Myers JR (2002) A genetic map of squash (*Cucurbita* spp.) with randomly amplified polymorphic DNA markers and morphological markers. J Amer Soc Hort Sci 127:568–575

Burger Y, Katzir N, Tzuri G, Portnoy V, Saar U, Shriber S, Perl-Treves R, Cohen R (2003) Variation in the response of melon genotypes to *Fusarium oxysporum* f.sp. *melonis* race 1 determined by inoculation tests and molecular markers. Plant Pathol 52:204–211

Chen JF, Staub JE, Jiang JM (1998) A reevaluation of karyotype in cucumber (*Cucumis sativus* L.) Genet Resour Crop Evol 45:301–305

Chen WS, Chiu CC, Liu HY, Lee TL, Cheng JT, Lin CC, Wu YJ, Chang HY (1998) Gene transfer via pollen-tube pathway for antiFusarium wilt in watermelon. Biochem Mol Biol Intl 46:1201–1209

Chiba N, Suwabe K, Nunome T, Hirai M (2003) Development of microsatellite markers in melon (*Cucumis melo* L.) and their application to major cucurbit crops. Breed Sci 53:21–27

Chung SM, Staub JE (2003) The development and evaluation of concensus chloroplast SSRs for chloroplast genetic analysis. Theor Appl Genet 107:757–767

Chung SM, Decker-Walters DS, Staub JE (2003) Genetic relationships within the Cucurbitaceae as assessed by ccSSR marker and sequence analysis. Can J Bot 81:814–832

Danin-Poleg Y, Reis N, Baudracco-Arnas S, Pitrat M, Staub JE, Oliver M, Arus P, deVicente CM, Katzir N (2000) Simple sequence repeats in *Cucumis* mapping and map merging. Genome 43:963–974

Danin-Poleg Y, Reis N, Tzuri G, Katzir N (2001) Development and characterization of microsatellite markers in *Cucumis*. Theor Appl Genet 102:61–72

Danin-Poleg Y, Tadmor Y, Tzuri G, Reis N, Hirschberg J, Katzir N (2002) Construction of a genetic map of melon with molecular markers and horticultural traits, and localization of genes associated with ZYMV resistance. Euphytica 125:373–384

Dijkhuizen A, Staub JE (2002) QTL conditioning yield and fruit quality traits in cucumber (*Cucumis sativus* L.): effects of environment and genetic background. J New Seeds 4:1–30

Dogimont C, Leconte L, Perin C, Thabuis A, Lecoq H, Pitrat M (2000) Identification of QTLs contributing to resistance to different strains of cucumber mosaic cucumovirus in melon. Acta Hort 510:391–398

Eduardo I, Arús P, Monforte AJ (2004) Genetics of fruit quality in melon. Verification of QTLs involved in fruit shape with near-isogenic lines (NILs). In: Lebeda A, Paris HS (eds.) Progress in Cucurbit Genetics and Breeding Research. Proceedings of Cucurbitaceae 2004, 8th Eucarpia Meeting on Cucurbit Genetics and Breeding, Olomouc, Czech Republic, 12–17 July 2004, pp 499–502

Esquinas-Alcazar JT (1977) Alloenzyme variation and relationships in the genus *Cucumis*. Dissertation, University of California, Davis, CA, USA

Fan M, Xu Y, Zhang H-Y, Ren H-Z, Kang G-B, Wang Y-J, Chen H (2000) Identification of quantitative trait loci associated with fruit traits in watermelon (*Citrullus lanatus*) and analysis of their genetic effects. Acta Genet Sin 27:902–910

Fang GW, Grumet R (1993) Genetic engineering of potyvirus resistance using constructs derived from the zucchini yellow mosaic virus coat protein gene. Mol Plant Micr Interact 6:358–367

Fanourakis NE, Simon PW (1987) Analysis of genetic linkage in cucumber. J Hered 78:238–242

FAO Statistics (2006) http://faostat.fao.org/faostat/. Accessed March 2006

Fazio G, Staub JE, Stevens MR (2003a) Genetic mapping and QTL analysis of horticultural traits in cucumber (*Cucumis sativus* L.) using recombinant inbred lines. Theor Appl Genet 107:864–874

Fazio G, Chung SM, Staub JE (2003b) Comparative analysis of response to phenotypic and marker-assisted selection for multiple lateral branching in cucumber (*Cucumis sativus* L.). Theor Appl Genet 107:875–883

Fehér T (1993) Watermelon *Citrullus lanatus* (Thunb.) Matsum. & Nakai. In: Kalloo G, Bergh BO (eds.) Genetic Improvement of Vegetable Crops. Pergamon, New York, pp 295–311

Frary A, Nesbitt TC, Grandillo S, Knaap E, Cong B, Liu J, Meller J, Elber R, Alpert KB, Tanksley SD (2000) *fw2.2*: a quantitative trait locus key to the evolution of tomato fruit size. Science 289:85–88

Fuchs M, McFerson JR, Tricoli DM, McMaster JR, Deng RZ, Boeshore ML (1997) Cantaloupe line CZW-30 containing coat protein genes of cucumber mosaic virus, zucchini yellow mosaic virus, and watermelon mosaic virus-2 is resistant to these three viruses in the field. Mol Breed 3:279–290

Fukino N, Taneishi M, Saito T, Nishijima T, Hirai M (2002) Construction of a linkage map and genetic analysis for resistance to cotton aphid and powdery mildew in melon. Acta Hort 588:283–286

Gaba V, Zelcer A, Gal-On A (2004) Cucurbit biotechnology: the importance of virus resistance. In Vitro Cell Dev Biol Plant 40:346–358

Garcia-Mas J, Morales M, Monforte MA, Arús P, Monfort A, Sánchez N, Puigdomènech P (2002) EST sequencing and SNP discovery in melon. Cucurbitaceae. ASHS, Alexandria, VA, USA, pp 20–24

Garcia-Mas J, Morales M, van Leeuwen H, Monfort A, Puigdomenech P, Arús P, Nieto C, Aranda MA, Dogimont C, Orjeda G, Caboche M, Bendahmane A (2004) A physical map covering the *Nsv* locus in melon. In: Lebeda A, Paris HS (eds.) Progress in Cucurbit Genetics and Breeding Research. Proceedings of Cucurbitaceae 2004, 8th Eucarpia Meeting on Cucurbit Genetics and Breeding, Olomouc, Czech Republic, 12–17 July 2004, pp 209–212

Giovannoni JJ (2004) Genetic regulation of fruit development and ripening. Plant Cell 16:S170–S180

Gonzalo MJ, Oliver M, Garcia-Mas J, Monfort A, Dolcet-Sanjuan R, Katzir N, Arús P, Monforte AJ (2005) Simple-sequence repeat markers used in merging linkage maps of melon (*Cucumis melo* L.). Theor Appl Genet 110:802–811

Guis M, Botondi R, Ben-Amor M, Ayub R, Bouzayen M, Pech JC, Latche A (1997) Ripening-associated biochemical traits of Cantaloupe Charentais melons expressing an antisense ACC oxidase transgene. J Am Soc Hort Sci 122:748–751

Guner N, Wehner TC (2003) Gene list for watermelon. Cucurbit Genet Coop Rep 26:76–92

Hashizume T, Shimamoto I, Harushima Y, Yui M, Sato T, Imai T, Hirai M (1996) Construction of a linkage map for watermelon (*Citrullus lanatus* (Thumb.) Matsum & Nakai) using random amplified polymorphic DNA (RAPD). Euphytica 90:265–273

Hashizume T, Shimamoto I, Hirai M (2003) Construction of a linkage map and QTL analysis of horticultural traits for watermelon [*Citrullus lanatus* (Thunb.) Matsum & Nakai] using RAPD, RFLP and ISSR markers. Theor Appl Genet 106:779–785

Hawkins LK, Dane F, Kubisiak TL, Rhodes BB, Jarret RL (2001) Linkage mapping in a watermelon population segregating for Fusarium wilt resistance. J Am Soc Hort Sci 126:344–350

Horejsi T, Box JM, Staub JE (1999) Efficiency of randomly amplified polymorphic DNA to sequence characterized amplified region marker conversion and their comparative polymerase chain reaction sensitivity in cucumber. J Am Soc Hort Sci 124:128–135

Horejsi T, Staub JE, Thomas CE (2000) Linkage of random amplified polymorphic DNA markers to downy mildew resistance in cucumber (*Cucumis sativus* L.). Euphytica 115:105–113

Jeffrey C (1980) A review of the Cucurbitaceae. Bot J Linn Soc 81:233–247

Jeffrey C (1990) Systematics of the Cucurbitaceae: an overview. In: Bates DM, Robinson RW, Jeffrey C (eds.) Biology and Utilization of the Cucurbitaceae. Cornell University Press, Ithaca, NY, USA, pp 3–9

Joobeur T, King JJ, Nolin SJ, Thomas CE, Dean RA (2004) The Fusarium wilt resistance locus *Fom-2* of melon contains a single resistance gene with complex features. Plant J 39:283–297

Klingler J, Kovalski I, Silberstein L, Thompson GA, Perl-Treves R (2001) Mapping of cotton-melon aphid resistance in melon. J Am Soc Hort Sci 126:56–63

Knerr LD, Staub JE (1992) Inheritance and linkage relationships of isozyme loci in cucumber (*Cucumis sativus* L.). Theor Appl Genet 84:217–224

Kennard WC, Havey MJ (1995) Quantitative trait analysis of fruit quality in cucumber: QTL detection, confirmation, and comparison with mating-design variation. Theor Appl Genet 91:53–61

Kennard WC, Poetter K, Dijkhuizen A, Meglic V, Staub JE, Havey MJ (1994) Linkages among RFLP, RAPD, isozyme, disease-resistance, and morphological markers in narrow and wide crosses of cucumber. Theor Appl Genet 89:42–48

Levi A, Thomas CE, Thies J, Simmons A, Xu Y, Zhang X, Reddy OUK, Tadmor Y, Katzir N, Trebitsh T, King S, Davis A, Fauve J, Wehner T (2004) Developing a genetic linkage map for watermelon: polymorphism, segregation and distribution of markers. In: Lebeda A, Paris HS (eds.) Progress in Cucurbit Genetics and Breeding Research. Proceedings of Cucurbitaceae 2004, 8th EUCARPIA Meeting on Cucurbit Genetics and Breeding, Olomouc, Czech Republic, 12–17 July 2004, pp 515–523

Levi A, Thomas CE, Joobeur T, Zhang X, Davis A (2002) A genetic linkage map for watermelon derived from a test cross population: (*Citrullus lanatus* var. *citroides* × *C. lanatus* var. *lanatus*) × *Citrullus colocynthis*. Theor Appl Genet 105:555–563

Levi A, Thomas CE, Zhang X, Joobeur T, Dean RA, Wehner TC, Carle BR (2001) A genetic linkage map for watermelon based on Randomly Amplified polymorphic DNA (RAPD) markers. J Am Soc Hort Sci 126:730–737

Liou PC, Chang YM, Hsu WS, Cheng YH, Chang HR, Hsiao CH, Drew RA (1998) Construction of a linkage map in *Cucumis melo* (L.) using random amplified polymorphic DNA markers. Acta Hort 461:123–131

Liu J, Van Eck J, Cong B, Tanksley SD (2002) A new class of regulatory genes underlying the cause of pear-shaped tomato fruit. Proc Natl Acad Sci USA 99:13302–13306

López-Sesé AI, Staub JE, Gomez-Guillamon ML (2003) Genetic analysis of Spanish melon (*Cucumis melo* L.) germplasm using a standardized molecular marker array and reference accessions. Theor Appl Genet 108:41–52

Lower RL, Smith OS, Ghaderi A (1983) Effects of plant density, arrangement, and genotype on stability of sex expression in cucumber. HortScience 18:737–738

Luo M, Wang Y-H, Frisch D, Joobeur T, Wing R, Dean RA (2001) Melon BAC library construction using improved methods and identification of clones linked to the locus conferring resistance to melon Fusarium wilt (*Fom-2*). Genome 44:154–162

Ma D, Guo Z, Zhang C, Gao S, Wang M (1994) Chromosome number and karyotype of melons (*Cucumis melo* L.). Cucurbit Genet Coop Rep 17:61–65

McCreight JD, Nerson H, Grumet R (1993) Melon *Cucumis melo* L. In: Kalloo G, Bergh BO (eds.) Genetic Improvement of Vegetable Crops. Pergamon, New York, USA, pp 267–294

Meglic V, Staub JE (1996) Inheritance and linkage relationships of allozyme and morphological loci in cucumber (*Cucumis sativus* L.). Theor Appl Genet 92:865–872

Mibus H, Tatlioglu T (2004) Molecular characterization and isolation of the *F/f* gene for femaleness in cucumber (*Cucumis sativus* L.). Theor Appl Genet 109:1669–1676

Mliki A, Staub JE, Sun Z, Ghorbel A (2001) Genetic diversity in melon (*Cucumis melo* L.): an evaluation of African germaplasm. Genet Resour Crop Evol 48:587–597

Monforte AJ, Oliver M, Gonzalo MJ, Alvarez JM, Dolcet-Sanjuan R, Arús P (2004) Identification of quantitative trait loci involved in fruit quality traits in melon (*Cucumis melo* L.). Theor Appl Genet 108:750–758

Morales M, Luis-Arteaga M, Alvarez JM, Colcet-Sanjuan R, Monfort A, Arus P, Garcia-Mas J (2002) Marker saturation of the region flanking the gene *NSV* conferring resistance to the melon necrotic spot Carmovirus (MNSV) in melon. J Am Soc Hort Sci 127:540–544

Morales M, Orjeda G, Nieto C, van Leeuwen H, Monfort A, Charpentier M, Caboche M, Arús P, Puigdomènech P, Aranda MA, Dogimont C, Bendahmane A, Garcia-Mas J (2005) A physical map covering the *nsv* locus that confers resistance to melon necrotic spot virus in melon (*Cucumis melo* L.). Theor Appl Genet 111:914–922

Munger HM, Robinson RW (1991) Nomenclature of *Cucumis melo* L. Cucurbit Genet Coop Rpt 14:43–44

Nam YW, Lee JR, Song KH, Lee MK, Robbins MD, Chung SM, Staub JE, Zhang HB (2005) Construction of two BAC libraries from cucumber (*Cucumis sativus* L.) and identification of clones linked to yield component quantitative trait loci. Theor Appl Genet 111:150–161

Navot N, Sarfatti M, Zamir D (1990) Linkage relationships of genes affecting bitterness and flesh color in watermelon. J Hered 81:162–165

Neuhausen SL (1992) Evaluation of restriction fragment length polymorphism in *Cucumis melo*. Theor Appl Genet 83:379–384

Noguera FJ, Capel J, Alvarez JI, Lozano R (2005) Development and mapping of a codominant SCAR marker linked to the *andromonoecious* gene of melon. Theor Appl Genet 110:714–720

Oliver M, Garcia-Mas J, Cardús M, Pueyo N, López-Sesé AI, Arroyo M, Gómez-Paniagua H, Arús P, de Vicente MC (2001) Construction of a reference linkage map for melon. Genome 44:836–845

Ori N, Eshed Y, Paran I, Presting G, Aviv D, Tanksley S, Zamir D, Fluhr R (1997) The *I2C* family from the wilt disease resistance locus *I2* belongs to the nucleotide binding, leucine-rich repeat superfamily of plant resistance genes. Plant Cell 9:521–532

Park SO, Crosby KM (2004) Development of RAPD markers linked to the male-sterile *ms-3* gene in melon. Acta Hort 637:243–249

Park Y, Katzir N, Brotman Y, King J, Bertrand F, Havey M (2004) Comparative mapping of ZYMV resistances in cucumber (*Cucumis sativus* L.) and melon (*Cucumis melo* L.). Theor Appl Genet 109:707–712

Park YH, Sensoy S, Wye C, Antonise R, Peleman J, Havey MJ (2000) A genetic map of cucumber composed of RAPDs, RFLPs, AFLPs, and loci conditioning resistance to papaya ringspot and zucchini yellow mosaic viruses. Genome 43:1003–1010

Pasha MK, Sen SP (1998) Numerical and chemical approaches to taxonomy and relationships. In: Nayar NM, More TA (eds.) Cucurbits. Science, Enfield, NH, USA, pp 19–32

Pauquet J, Burget E, Hagen L, Chovelon V, Menn A, Valot N, Desloire S, Caboche M, Rousselle P, Pitrat M, Bendahmane A, Dogimont C (2004) Map-based cloning of the *Vat* gene from melon conferring resistance to both aphid colonization and aphid transmission of several viruses. In: Lebeda A, Paris HS (eds.) Progress in Cucurbit Genetics and Breeding Research. Proceedings of Cucurbitaceae 2004, 8th Eucarpia Meeting on Cucurbit Genetics and Breeding, Olomouc, Czech Republic, 12–17 July 2004, pp 325–329

Perchepied L, Dogimont C, Pitrat M (2005) Strain-specific and recessive QTLs involved in the control of partial resistance to *Fusarium oxysporum* f. sp. *melonis* race 1.2 in a recombinant inbred line population of melon. Theor Appl Genet 111:65–74

Périn C, Hagen LS, De Conto V, Katzir N, Danin-Poleg Y, Portnoy V, Baudracco-Arnas S, Chadoeuf J, Dogimont C, Pitrat M (2002a) A reference map of *Cucumis melo* based on two recombinant inbred line populations. Theor Appl Genet 104:1017–1034

Périn C, Hagen LS, Giovinazzo N, Besombes D, Dogimont C, Pitrat M (2002b) Genetic control of fruit shape acts prior to anthesis in melon (*Cucumis melo* L.). Mol Genet Genom 266:933–941

Perl-Treves T, Zamir D, Navot N, Galun E (1985) Phylogeny of *Cucumis* based on isozyme variability and its comparison with plastome phylogeny. Theor Appl Genet 71:430–436

Pierce LK, Wehner TC (1990) Review of genes and linkage groups in cucumber. HortScience 25:605–615

Pitrat M (1991) Linkage groups in *Cucumis melo* L. J Hered 82:406–411

Pitrat M (2002) 2002 gene list for melon. Cucurbit Genet Coop Rep 25:76–93

Pitrat M, Chauvet M, Foury C (1999) Diversity, history and production of cultivated cucurbits. Acta Hort 492:21–28

Pitrat M, Hanelt P, Hanner K (2000) Some comments on intraspecific classification of cultivars of melon. Acta Hort 510:29–36

Ramachandran C, Seshadri VS (1986) Cytological analysis of the genome of cucumber (*Cucumis sativus* L.) and muskmelon (*Cucumis melo* L.). Z Pflanzensucht 96:25–38

Ritschel PS, de Lima Lins TC, Tristan RL, Buso GSC, Buso JA, Ferreira ME (2004) Development of microsatellite markers from an enriched genomic library for genetic analysis of melon (*Cucumis melo* L.). BMC Plant Biol 4:9

Robbins MD, Staub JE, Fazio G (2002) Deployment of molecular markers for multitrait selection in cucumber. In: Proceedings of the Cucurbitaceae 2002, 8–12 Dec 2002, Naples, FL, USA, ASHS, Alexandria, VA, pp 41–47

Robbins MD, Staub JE (2004) Strategies for selection of multiple quantitatively inherited yield components in cucumber. In: Lebeda A, Paris HS (eds.) Progress in Cucurbit Genetics and Breeding Research. Proceedings of Cucurbitaceae 2004, 8th Eucarpia Meeting on Cucurbit Genetics and Breeding, Olomouc, Czech Republic, 12–17 July 2004, pp 401–410

Robinson RW, Decker-Walters DS (1997) Cucurbits. CABI, New York, 226 p

Serquen FC, Bacher J, Staub JE (1997) Mapping and QTL analysis of a narrow cross in cucumber (*Cucumis sativus* L.) using random amplified polymorphic DNA markers. Mol Breed 3:257–268

Silberstein L, Kovalski I, Brotman Y, Perin C, Dogimont C, Pitrat M, Klingler J, Thompson G, Portnoy V, Katzir N, Perl-Treves R (2003) Linkage map of *Cucumis melo* including phenotypic traits and sequence-characterized genes. Genome 46:761–773

Silberstein L, Kovalski I, Huang R, Anagnostou K, Jahn MMK, Perl-Treves R (1999) Molecular variation in melon (*Cucumis melo* L.) as revealed by RFLP and RAPD markers. Sci Hort 79:101–111

Silva JA, da Costa TS, Lucchetta L, Marini LJ, Zanuzo MR, Nora L, Nora FR, Twyman RM, Rombaldi CV (2004) Characterization of ripening behavior in transgenic melons expressing an antisense 1-aminocyclopropane-1-carboxylate (ACC) oxidase gene from apple. Post Harvest Biol Technol 32:263–268

Simons G, Groenendijk J, Wijbrandi J, Reijans M, Groenen J, Diergaarde P, Van der Lee T, Bleeker M, Onstenk J, de Both M, Haring M, Mes J, Cornelissen B, Zabeau M, Vos P (1998) Dissection of the *Fusarium I2* gene cluster in tomato reveals six homologs and one active gene copy. Plant Cell 10:1055–1068

Spooner DM, van den Berg RG, Hetterscheid WLA, Brandenburg WA (2003) Plant nomemclature and taxonomy: an horticultural and agronomic perspective. Hort Rev 28:1–60

Staub JE, Knerr LD, Holder DJ, May B (1992) Phylogenetic relationships among several African *Cucumis* species. Can J Bot 70:509–517

Staub JE, Kuhns LJ, May B, Grun P (1982) Stability of potato tuber isozymes under different storage regimes. J Am Soc Hort Sci 107:405–408

Staub JE, Meglic V (1993) Molecular genetic markers and their legal relevance for cultivar discrimination: a case study in cucumber. HortTechnology 3:291–300

Staub JE, Serquen FC (2000) Towards an integrated linkage map of cucumber: map merging. Acta Hort 510:357–366

Staub JE, Serquen FC, McCreight JD (1997) Genetic diversity in cucumber (*Cucumis sativus* L): III. An evaluation of Indian germplasm. Genet Resour Crop Evol 44:315–326

Stepansky A, Kovalski I, Perl-Treves R, Naudin CV (1999) Intraspecific classification of melons (*Cucumis melo* L.) in view of their phenotypic and molecular variation. Plant Syst Evol 217:313–332

Taler D, Galperin M, Benjamin I, Cohen Y, Kenigsbuch D (2004) Plant *eR* genes that encode photorespiratory enzymes confer resistance against disease. Plant Cell 16:172–184

Tao L, Xie WJ, Yang WX, Gong ZZ, Liu JP, Xu YY, Xu ZH (1996) RAPD analysis of introduced exogenous cashew DNA in watermelon cultivars. J Fruit Sci (China) 13:175–177 (in Chinese)

Tatlioglu T (1993) Cucumber *Cucumis sativus* L. In: Kalloo G, Bergh BO (eds.) Genetic Improvement of Vegetable Crops. Pergamon, New York, pp 197–234

Touyama T, Asami I, Oyabu T, Yabe K, Sugahara S, Kanbe M (2000) Development of PCR-based markers linked to a long-shelf-life-character "nonyellowing" in muskmelon. Res Bull Aichi Ken Agri Res Center 32:47–52

Trebitsh T, Staub JE, O'Neill SD (1997) Identification of a 1-aminocyclopropane-1-carboxylic acid synthase gene linked to the female (*F*) locus that enhances female sex expression in cucumber. Plant Physiol 113:987–995

Vakalounakis DJ (1992) Heart leaf; a recessive leaf shape marker in cucumber: linkage with disease resistance and other traits. J Hered 83:217–221

van Leeuwen H, Monfort A, Zhang H-B, Puigdomènech P (2003) Identification and characterization of a melon genomic region containing a resistance gene cluster from a constructed BAC library. Microcolinearity between *Cucumis melo* and *Arabidopsis thaliana*. Plant Mol Biol 51:703–718

Wang Y-H (1999) Marker development and genome mapping for localizing disease resistance genes in melon (*Cucumis melo* L.). Dissertation, Clemson University, Clemson, SC, USA

Wang Y-H, Thomas CE, Dean RA (1997) A genetic map of melon (*Cucumis melo* L.) based on amplified fragment length polymorphism (AFLP) markers. Theor Appl Genet 95:791–798

Wang Y-H, Thomas CE, Dean RA (2000) Genetic mapping of a Fusarium wilt resistance gene (*Fom-2*) in melon (*Cucumis melo* L.). Mol Breed 6:379–389

Wechter WP, Whitehead MP, Thomas CE, Dean RA (1995) Identification of a randomly amplified polymorphic DNA marker linked to the *Fom-2* Fusarium wilt resistance gene in muskmelon MR-1. Phytopathology 85:1245–1249

Wechter WP, Dean RA, Thomas CE (1998) Development of sequence-specific primers that amplify a 1.5-kb DNA marker for race 1 Fusarium wilt resistance in *Cucumis melo* L. HortScience 33:291–292

Whitaker TW, Davis GN (1962) Cucurbits: Botany, Cultivation, and Utilization. Interscience, New York

Xiao GH, Wu DX, Liu JX (1999a) Introduction of bottle gourd DNA to watermelon by soaking embryo for resistance to wilt disease. China Vegetable 2:36–37 (in Chinese with English summary)

Xiao GH, Wu DX, Liu JX, Zheng SQ, Wang SW, Fu F (1999c) Variations in the characters of watermelon offspring induced by exogenous bottle gourd DNA introduction. Hunan Agri Res News 6:5–9 (in Chinese)

Xiao GH, We DX, Liu JX, Zheng SQ, Xiao LY (1999b) Resistance to wilt disease of watermelon germplasm resources by exogenous DNA introduction. J Hunan Agri Univ 25:453–457 (in Chinese with English summary)

Xie JH, Wehner TC (2001) Gene list 2001 for cucumber. Cucurbit Genet Coop Rep 24:110–136

Xu Y, Ouyang X-X, Zhang H-Y, Kang G-B, Wang Y-J, Chen H (1999) Identification of a RAPD marker linked to Fusarium wilt resistance gene in wild watermelon germplasm (*Citrullus lanatus* var. *citroides*). Acta Bot Sin 41:952–955

Xu Y, Zhang HY, Kang GB, Wang YJ, Chen H (2000) Studies of molecular marker-assisted selection for resistance to Fusarium wilt in watermelon (*Citrullus lanatus*) breeding. Acta Genet Sin 27:151–157

Yariv Y, Portnoy V, Burger Y, Debbi H, Tzuri G, Mozes-Daube N, Lewinsohn E, Katzir N, Giovannoni J, White R, Schaffer A (2002) The isolation and characterization of fruit ripening-related genes in melon (*Cucumis melo*): a geneomic approach. Cucurbitaceae 2002. ASHS, Alexandria, VA, USA, pp 421–426

Yariv Y, Portnoy V, Burger Y, Benyamini Y, Lewinsohn E, Tadmor Y, Ravid U, White R, Giovannoni J, Schaffer AA, Katzir N (2004) Isolation and characterization of fruit-related genes in melon (*Cucumis melo*) using SSH and macroarray techniques. In: Lebeda A, Paris HS (eds.) Progress in Cucurbit Genetics and Breeding Research. Proceedings of Cucurbitaceae 2004, 8th Eucarpia Meeting on Cucurbit Genetics and Breeding, Olomouc, Czech Republic, 12–17 July 2004, pp 491–497

Yoshioka K, Hanada K, Harada T, Minobe Y, Oosawa K (1993) Virus resistance in transgenic melon plants that express the cucumber mosaic virus coat protein gene and in their progeny. Jpn J Breed 43:629–634

Zhang RB, Xu Y, Yi K, Zhang HY, Liu LG; Yi GG, Levi A (2004) A genetic linkage map for watermelon derived from recombinant inbred lines. J Am Soc Hort Sci 129:237–243

Zheng XY, Wolff DW (2000) Randomly amplified polymorphic DNA markers linked to Fusarium wilt resistance in diverse melons. HortScience 35:716–721

Zheng XY, Wolff DW, Baudracco-Arnas S, Pitrat M (1999) Development and utility of cleaved amplified polymorphic sequences (CAPS) and restriction fragment length polymorphisms (RFLPs) linked to the *Fom-2* Fusarium wilt resistance gene in melon (*Cucumis melo* L.). Theor Appl Genet 99:453–463

Zitter TA, Hopkins D, Thomas C (1996) Compendium of Cucurbit Diseases. Am Phytopathol Soc, APS, St. Paul, MN, USA, 87 p

11 Onion

John McCallum

New Zealand Institute for Crop & Food Research, Private Bag 4704, Christchurch New Zealand
e-mail: mccallumj@crop.cri.nz

11.1 Introduction

Onion (*Allium cepa* L.) has been valued as a food and medicine since antiquity. Although it is a widely cultivated crop and second only to tomato in value among the vegetables (FAO 2005), the genome and genetic resources of onion are relatively poorly characterized. This chapter surveys aspects of genetic, genomic and horticultural onion and closely related *Allium* species relevant to the genetic analysis and improvement of onion using genomic approaches. It focuses on the nuclear genome and readers are referred to the review of Havey (2002) for a more detailed review of *Allium* organellar genomes.

11.1.1 Taxonomy and History

Onion has been domesticated for over 4000 years, as shown by records dating from Old Kingdom Egypt (2700 BC) and Sumeria (3000 BC). Even earlier records appear in Indian literature dating to ~4000 BC (Hanelt 1990).

A. cepa is a member of section *Cepa* (Mill.) Prokh. of genus *Allium*, which includes wild and cultivated species important to the origin and improvement of domesticated *Allium*. *A. cepa* does not exist in the wild and the wild species *A. vavilovii* has been identified as the closest relative (Havey 1992; Fritsch et al. 2001). The distribution of *A. vavilovii* and related species of the Cepa alliance within section *Cepa* in Turkmenistan/Iran suggests that domestication of a wild relative occurred during emergence of ancient civilizations in this area (Hanelt 1990). The Galanthum alliance within section *Cepa* includes *A. galanthum*, now used as source of novel cytoplasm (Havey 1999; Sulistyaningsih et al. 2002). The Altaicum alliance includes *A. fistulosum* (bunching onion or Welsh onion) the major cultivated *Allium* of North Asia and the widely distributed wild species *A. altaicum*.

The species *A. cepa* includes groupings distinguished by bulb phenotypes and by dormancy and photoperiod requirements (Fritsch and Friesen 2002). The most phenotypically diverse is the common onion group, which includes numerous traditional and commercial open-pollinated varieties as well as modern F_1 hybrid derivatives. These are characterized by larger bulbs, generally enclosed by a single skin and biennial reproduction by seed in an appropriate environment. The aggregatum group, which includes shallot, is distinguished by predominantly asexual production via smaller, clustered bulbs (Rabinowitch and Kamenetsky 2002). However since fertile shallots and highly multicentered onion varieties exist, and molecular marker patterns have shown strong similarity between these groups (Klaas 1998) they are now accepted as horticultural rather than taxonomic groupings.

11.1.2 Genetic Resources and Breeding

Resources and Breeding Systems

Bulb onion is an outcrossing diploid ($2n = 2x = 16$). Populations are maintained through insect cross pollination and may harbor significant phenotypic variation, as well as deleterious alleles (Berninger and Buret 1967). Onion exhibits strong inbreeding depression, with substantial loss of fitness following two or more rounds of self-pollinations (Jones and Davis 1944). The key characteristic constraining breeding, maintenance and exchange of germplasm is meeting the adaptation requirements for varieties to produce bulbs in a given environment (Brewster 1994). *Allium cepa* is relatively poorly represented in public genetic resource collections due to the expense of mother bulb production and storage, and of mass pollinations. Asexually-propagated shallot-types by

Genome Mapping and Molecular Breeding in Plants, Volume 5
Vegetables
C. Kole (Ed.)

contrast are poorly represented because of requirements for virus elimination via tissue culture (Rabinowitch and Kamenetsky 2002). Major sources of breeding stocks in the twentieth century were traditional, widely adapted open-pollinated varieties and public releases from the USDA-ARS and US land-grant universities (Goldman et al. 2000).

Historically, onion improvement has been conducted principally by mass selection of open-pollinated populations or segregating families (Dowker 1990), although studies have shown improved gains can be made using recurrent selection strategies (Dowker et al. 1984; Havey and Randle 1996). Commercial plant breeding of onion is now increasingly targeted toward the release of F_1 hybrids based on cytoplasmic male-sterility, since these provide protection of breeding investment to seed companies and possible gains in yield and uniformity for growers. This is based primarily on S-cytoplasm, which was derived from a single sterile plant arising in Italian Red 13-53 (Jones and Mann 1963). Although other sources are used, including T-cytoplasm (Berninger 1965) and *A. galanthum* (Havey 1999; Yamashita et al. 2005), the wide adoption of S-cytoplasm from US public programs has probably increased the cytoplasmic uniformity of cultivated onion. Reduced production and maintenance of open-pollinated varieties may have serious implications for conservation of genetic diversity in onion. Use of S-cytoplasm requires development of lines homozygous recessive (i.e., *msms*) at the nuclear restorer locus, which generally requires substantial inbreeding.

Hybridization and Introgression

Breeders have long sought to improve the pest and disease resistance of bulb onion by intercrossing with other cultivated *Allium*. The most recent and systematic survey of crossability in cultivated *Allium* was reported by van Raamsdonk et al. (2003). The main focus of introgression efforts has been *A. fistulosum*, which harbors many desirable resistance traits. The top onion and wakegi onion are minor asexually propagated *Alliums* that have been shown to be diploid hybrids between *A. cepa* and *A. fistulosum* on the basis of molecular data (Maass 1997; Sugiharto Arifin et al. 2000). A seed-propagated amphidiploid from the same parent species is known as Beltsville Bunching (Jones and Mann 1963). Despite more than 70 years of efforts by breeders, developing useful hybrids between *A. cepa* and *A fistulosum* has proved challenging and has yet to provide commercial releases. Putative introgressions of *A. fistulosum* into *A. cepa* have been reported on the basis of cytogenetic evidence (Hou and Peffley 2000) and molecular markers (Peffley and Hou 2000). In contrast to the difficulties of achieving sexual hybridization between *A. cepa* and *A. fistulosum*, the development of asexually-propagated *A. fistulosum* lines containing alien additions of *A. cepa* chromosomes has proved highly productive for *Allium* research. The series of shallot monosomic addition lines developed by Shigyo et al. (1996) has enabled chromosomal assignment of many marker and major morphological loci. A partial set of onion alien addition lines was developed by Barthes and Ricroch (2001) and these authors reported homeologous crossovers between the genomes.

The wild accession *A. roylei* Baker has proven to be a key species enabling useful introgressions from wild and cultivated genetic resources of other members of section *Cepa* into onion (Meer and Vries 1990). Since it can cross easily with both *A. cepa* and *A. fistulosum*, *A. roylei* has enabled the construction of "bridge crosses" (Khrustaleva and Kik 1998) allowing disease resistance and other useful characteristics to be introgressed from *A. roylei* and *A. fistulosum* into onion, as well as providing valuable resources for genetic analysis. Molecular analyses have shown that although *A. roylei* shows affiliation with section *Cepa* on the basis of nuclear genome markers, chloroplast markers show similarity with section *Schoenoprasum* (van Raamsdonk et al. 2000, 2003), suggesting a hybrid origin.

Other Technologies

Doubled haploid onions are appealing for development of breeding and genetic stocks due to the high levels of genetic variation and inbreeding depression observed in onion populations (Bohanec 2002). Although several groups have successfully produced double haploid onion lines (Campion and Alloni 1990; Campion et al. 1995; Javornik et al. 1998; Alan et al. 2004), major technical challenges remain. Wide variation is observed among populations in rates of haploid induction and a genetic analysis has revealed evidence for quantitative inheritance and dominance of low induction (Bohanec et al. 2003). In addition to variability in induction, genome-doubling remains inefficient, with over 90% of regenerants remaining haploid (Bohanec 2002).

Development of novel genetic and breeding stocks through genetic transformation has proven challenging in onion and other *Alliums*, and is not yet a routine procedure (Zheng et al. 2001; Eady 2002; Eady et al. 2003; Aswath et al. 2006).

11.1.3 Cytology and Genome Structure

The organization of *Allium* nuclear and organellar genomes has been reviewed by Havey (2002) and many common features shared by onion and garlic genomes are discussed by Cavagnaro and Galmarini in Chapter 12 of this volume.

Genome Size and Composition

The most conspicuous feature of *Allium* nuclear genomes is their great size, which, along with a modest chromosome number, has made *Allium* a favored genus for cytological research and education but complicates most molecular genetic analyses. The onion genome is 17 pg or 15 Mbp per 1C (Arumuganathan and Earle 1991; Bennett and Leitch 1995), one of the largest among cultivated plants, and has a G+C content of 32% (Stack and Comings 1979), among the lowest of the angiosperms. Extreme genome size may influence the biology and fitness of *Allium*. Onion has a slow growth rate relative to most crops and recent surveys suggest that larger genome size is associated with reduced maximum photosynthetic rates (Knight et al. 2005).

Repetitive Elements

Reassociation kinetic studies (Stack and Comings 1979) showed that high, medium and single copy DNA comprise 41, 36 and 6% of the onion genome, respectively. Sequencing of onion BACs completed to date (Do et al. 2004) has provided a similar estimate of about 5% coding regions and reveal diverse transposable element-related sequences in the flanking regions, as seen in other plant genomes. FISH analysis using whole onion BACs revealed that most of these hybridized to whole chromosomes (Suzuki et al. 2001). GISH studies in the *A. cepa* × (*A. fistulosum* × *A. roylei*) bridge cross showed that the repetitive element composition of these species is significantly different (Khrustaleva and Kik 1998). Although sequence variation in onion Ty1-*copia*-like retrotransposons has been characterized and their concentration in terminal heterochromatic regions has been reported (Pearce et al. 1996; Kumar et al. 1997) there is little other published information on the sequence variation and biology of mobile repetitive elements that have shaped the evolution of the onion genome. Bioinformatic analysis of onion expressed sequences has suggested that DNA and RNA elements and/or their products comprise a significant proportion of the transcriptome (Kuhl et al. 2004).

Instead of the telomeric repeats which stabilize chromosome ends in most higher plants such as *Arabidopsis* (Fajkus et al. 2005), this is replaced in onion and other *Allium* species by a GC-rich satellite (Barnes et al. 1985; Pich et al. 1996).

Recombination

Comparative studies of the chromosomal distribution of crossovers have shown highly significant differences in the distribution of chiasmata among genera of section *Cepa*. While chiasmata are randomly distributed in onion (Emsweller and Jones 1935) and *A. roylei* (Vries et al. 1992a), in *A. fistulosum* they are concentrated near the centromere (Emsweller and Jones 1935; Albini and Jones 1987). This suggests that *A. fistulosum* may be a more efficient target for genomic cloning since pericentromeric regions may contain a higher gene density. Khrustaleva et al. (2005) used GISH and AFLP to construct an integrated map of two chromosomes in the *Allium cepa* × (*A. roylei* × *A. fistulosum*) bridge cross population and also observed that recombination predominantly occurred in the proximal half of chromosome arms.

11.2 Molecular Markers and Other Genome Resources

Marker, sequence and other genomic resources in onion are very limited compared to other vegetable taxa such as the solanaceous and cruciferous vegetables. A variety of molecular markers have been successfully used to characterize nuclear and organellar variation between *Allium* species and interspecific hybrids (Klaas and Friesen 2002). However, development of markers to provide useful discrimination among cultivated *A. cepa* has been much more challenging.

11.2.1 Protein Markers

Morphological and biochemical markers of onion were most recently reviewed by Cramer and Havey (1999). Due to the lack of portable codominant markers, *Allium* researchers have continued to use isozyme markers until quite recently for a variety of applications (Rouamba et al. 2001; Shigyo et al. 2003; Mangum and Peffley 2005). Their use within cultivated onion has been restricted by low levels of polymorphism (Peffley and Orozco-Castillo 1987). The use of seed protein markers has been reported for cultivar discrimination purposes (Mennella et al. 2005) but these markers have yet to be genetically characterized.

11.2.2 RAPD and AFLP Markers

Although they have proven informative for analysis of interspecific comparisons, additions or introgressions, random genomic markers have yet to find major application within cultivated *A. cepa*. The informativeness of such markers for interspecific comparisons is to be expected since GISH and other genomic analyses indicate considerable divergence in the repetitive DNA content of species within section *Cepa* (Khrustaleva and Kik 1998). RAPD markers have been used for fingerprinting (Wilkie et al. 1993; Tanikawa et al. 2002) and introgression (Peffley and Hou 2000) applications, and assigned to shallot chromosomes by Shigyo et al. (1997a). More critical evaluation of RAPDs during the development of the onion linkage map (Bradeen and Havey 1995; King et al. 1998a) showed that caution is required in using genetically uncharacterized RAPDs in onion and that relatively few could be placed on the map. AFLP markers have been used successfully to construct a map in an intraspecific onion cross (personal communication, van Heusden, Kik, 2006) as well to add low-density dominant markers to intraspecific onion maps (Havey et al. 2001; McCallum et al. 2006).

11.2.3 Expressed Sequence, SSR and SNP Markers

The most useful marker development resource to date has been the collection of 20,000 onion EST sequences from immature bulb, callus and roots of four cultivars developed by Kuhl et al. (2004), and smaller sets from earlier projects (King et al. 1998a; McCallum et al. 2002). This represents a total of 11,726 unique ESTs curated in the Onion Gene index version 1.0 (TIGR 2003). Markers derived from this resource include RFLP (King et al. 1998a), SSCP (McCallum et al. 2001) and SNP and SSR markers (Kuhl et al. 2004; Martin et al. 2005). Due to the gigabase genome size of onion, RFLP markers are technically demanding, but have provided a crucial source of codominant markers for framework map development and comparative studies. Importantly, when used in segregating populations or DH lines they provide information about copy number that is useful for development and interpretation of PCR-based markers in the large and complex onion genome. Conversion of RFLP to PCR-based markers is in many cases difficult due to the prevalence of highly expressed representatives of multigene families in the nonnormalized cDNA libraries used to develop first-generation RFLP markers. Use of standard mutation-scanning methods of SSCP and heteroduplex analysis has permitted development of polymorphic markers from amplicons of 3′ and 5′ UTR regions of genes (McCallum et al. 2001). The SNP markers published to date have been based on direct sequencing of the products from nested PCR reactions and are therefore not particularly user-friendly. By contrast, mining the nonredundant EST resource has provided over 300 sequences containing simple sequence repeat motifs and evaluations to date have provided over 100 polymorphic markers from these (Kuhl et al. 2004; Martin et al. 2005).

Comparisons among inbred onion populations using RFLP (King et al. 1998b) showed that these could discriminate inbreds within market classes and revealed the range of heterozygosity within inbreds. Similar comparisons of inbreds using EST-SNP and SSR markers confirmed that the SSR markers could discriminate among these and reveal relationships consistent with known pedigrees (Jakse et al. 2005). Unpublished work in our laboratory surveying allelic diversity in a broader sample of cultivated onion germplasm using these markers has revealed substantial polymorphism within populations and relatively fewer polymorphisms between these, confirming earlier observations (Bark and Havey 1995). Although these markers revealed relatively few common alleles on the basis of size, sequencing reveals additional haplotypes as SNPs in the majority of these. This suggests that more general mutation-scanning methods may be used to survey allelic variation in low-copy amplicons from 5′ UTR and coding regions of unigene homologs (Parida et al. 2006).

11.2.4 BAC Resources

The only onion BAC resource developed to date is a 0.3× genome library (Suzuki et al. 2001, 2002). Although limited by partial coverage, this resource has provided some valuable insights into onion genome composition (Do et al. 2004) and confirmed the inefficiency of BAC-based genome sequencing approaches in onion.

11.2.5 Genomic SSR Markers

A set of genomic SSR markers was developed for onion by standard enrichment methods (Fischer and Bachmann 1998) and used to assess genetic diversity in a set of shallot and onion accessions (Fischer and Bachmann 2000). To date these have not been mapped or used in published studies due to complex PCR conditions, though .Masuzaki et al. (2006a) have developed simplified conditions and assigned 21 of these to *A. cepa* chromosomes. By contrast, genomic SSRs developed in *A. fistulosum* (Song et al. 2004) have provided a highly polymorphic set of framework markers for mapping (Ohara et al. 2005). It is likely that methods for expressed genome enrichment and emerging technologies for massively parallel nanoliter sequencing will soon enable economical genome sequencing and less biased discovery of genomic SSRs and other polymorphisms in *Allium*. This approach is already being applied in *Pinus* genomes of similar size (Peterson et al. 2006).

11.3 Genetic Mapping

11.3.1 Inter- and Intraspecific Maps

The first public linkage map in *Allium* was a low-density map of onion developed by King et al. (1998a) in the cross between partially inbred lines Brigham Yellow Globe 15-23 (BGY15-23) and Alisa Craig 43 (AC43). Although based on only 58 F_3 families derived by selfing F_2 from a single F_1 plant, this family provided sufficient seed for RFLP mapping and robust replicated phenotyping of segregating traits, including fertility restoration, bulb carbohydrate content, pungency and platelet aggregation. A genetic map was developed in an F_2 family ($N = 65$) originating from the interspecific cross *A. cepa* var Jumbo × *A. roylei* by van Heusden et al. (2000b) based on *EcoR*1/*Mse*1 AFLP markers. The population size did not permit integration of maternal (15 linkage groups, 626 cM) and paternal (8 linkage groups, 694 cM) maps due to the population size. Subsequent use of *A. fistulosum*-shallot alien addition lines permitted the assignment of 51 *A. roylei* AFLPs, 7 isozymes and 3 CAPs markers to chromosomes (van Heusden et al. 2000a). As in the RFLP map development, ∼10% of the *cepa*-specific markers were amplified in more than one addition line, suggesting unlinked duplications. The authors also noted that onion and shallot parents shared ∼90% of AFLP fragments.

A medium-density linkage map based on the BGY15-23 × AC43 population was subsequently published by Martin et al. (2005), which was similarly anchored to chromosomes by use of alien addition lines. This map spans 1,907 cM in 14 linkage groups, only two of which are not assigned to chromosomes. Comparisons with rice in this study revealed no evidence for synteny at the recombinational level. More detailed analyses by Jakše et al. (2006) in asparagus showed no microsynteny with either onion or rice, suggesting that genomic resources will need to be developed independently for *Alliaceae* and *Asparagaceae*.

11.3.2 Other Map and Marker Resources

Another partial intraspecific onion linkage map has been developed in a single F_2 population ($N = 81$) from the cross W202A × Texas Grano 438 (McCallum et al. 2006). To date this map, based on SSR and SSCP markers, spans 450 cM on all eight chromosomes and is very well-aligned with the BYG15-23 × AC43 and *A. cepa* × *A. roylei* interspecific maps. Importantly, the parental populations are representative of long-day storage and short-day sweet germplasm widely utilized by breeders. During evaluation of PCR-based markers used to develop both this map and the medium-density map of Martin et al. (2005), a considerable number of markers failed to segregate due to heterozygosity in parental populations, as observed previously using RFLPs (King et al. 1998a). An intraspecific map of onion has been developed in the cross Bessanovski × Sweet Onion using AFLP (personal communication, van Heusden, Kik, 2006), suggesting that with appropriate optimization these markers will provide useful medium to high-density maps in the future.

In addition to the above applications in anchoring genetic maps, the alien addition line stocks have been used for chromosomal assignment of isozyme (Shigyo et al. 1995), RAPD (Shigyo et al. 1997a) and genomic SSR (Masuzaki et al. 2006a) markers. These will continue to be valuable stocks for efficient evaluation and chromosomal assignment of codominant markers in onion.

Medium-density linkage maps have now been developed in *A. fistulosum* using AFLP and genomic SSR (Ohara et al. 2005). Although *A. fistulosum* genomic SSR markers evaluated in onion to date have not shown promise for comparative mapping, onion EST-SSR markers seem to provide a useful source of orthologous polymorphic SNPs in *A. fistulosum* (personal communication, McCallum, Tsukazaki 2006).

11.4 Phenotypic Variation and Economic Traits

11.4.1 Adaptation and Bulb Quality

The trait that is of primary concern to breeders is adaptation to the day length and temperature conditions in the target production environment(s). A major division in onion germplasm is that of short-day (i.e., produce bulbs in shortening days) versus long-day (i.e., produce bulbs in lengthening days) (Brewster 1990b). Selection for bienniality is a critical requirement, since bolting seriously compromises bulb crop yield and quality. In some cases, early planting is used to encourage bolting to achieve annual breeding cycles for seed production, although this poses risks in terms of bolting susceptibility. Selection for uniformity and vigor are closely related but have potentially conflicting requirements, since onions exhibit serious inbreeding depression (Jones and Davis 1944).

Bulb storability (Brewster 1990a) is a complex trait that is not only of great practical but also of fundamental interest, since storage selection has undoubtedly strongly shaped population genetic structure during domestication and breeding. Storage life can range from a month to two years, depending on genetic and agronomic factors, temperature and humidity of postharvest storage and preharvest exposure to insect, bacterial and fungal pathogens. Storage life is critical not only for export of crops but also for continuity of supply in small local markets of the tropics and semitropics.

Bulb appearance, morphology and composition of onion vary widely and define market niches. Dry skin color, number and thickness have a major impact on appearance and durability during handling and storage. In many commercial hybrid onion breeding programs strong selection is carried out for single-centeredness, whereas multicentered and multibulbed phenotypes are common and expected in traditional and shallot varieties. Multicenteredness may provide some yield advantages at the expense of quality.

11.4.2 Carbohydrate Composition

The carbohydrate composition of bulbs has a major impact on storage and culinary qualities. The total dry matter content (dry matter % or DM%) of onion varies from around 5% in fresh salad-style varieties to over 30% in dehydrator onions and shallots (Jones and Mann 1963). Around 65% of the dry matter content is made up of the nonstructural carbohydrates fructose, glucose, sucrose and fructooligosaccharides (fructans) of the kestose and neokestose series (Darbyshire and Henry 1978; Shiomi et al. 2005). Low dry matter onion varieties contain mainly monosaccharides and only small amounts of short-chain fructooligosaccharides. By contrast dehydrator varieties and shallots contain predominantly fructans of longer chain lengths. Quantitatively fructose dominates sweetness. Onions and other *Alliums* are a major source of dietary fructan (Moshfegh et al. 1999), which confer benefits as a prebiotic (Roberfroid and Delzenne 1998) .

11.4.3 Organosulfur Compounds

The most distinctive properties of onion and other *Allium* vegetables are the aromas and medicinal properties imparted by organosulfur compounds. In contrast to garlic (as discussed by Cavagnaro and Galmarini in Chapter 12 of this volume), *A. cepa* accumulates predominantly the 1-propenyl cysteine sulfoxide (1-PECSO) flavor precursor, along with lesser amounts of methyl- and propyl- CSOs (Randle and Lancaster 2002). Substantial quantities of the gamma-glutamyl forms of these precursors may also be present (Lancaster and Shaw 1991). Following maceration of tissue, sulfoxides are hydrolyzed by the enzyme alliinase, producing unstable sulfenic acids. A molar equivalent of pyruvic acid is also produced, which is used as a measure of gross pungency (Schwimmer and Weston 1961). The sulfenic acid produced from 1-PECSO is acted on by the lachrymatory factor synthase (LFS) enzyme to form lachrymatory factor (LF), which causes the tearing characteristic of onion (Imai et al. 2002). LF can also be used as a measure of flavor intensity (Kopsell et al. 2002; McCallum et al. 2005b). A much wider range of transient and bioactive compounds are produced in onion compared to garlic (Block 1992; Block et al. 1997). Gross pungency and precursor levels exhibit wide environmental and genetic variation (Randle and Lancaster 2002). Physiological studies indicate genetic differences in S assimilation and partitioning (Randle et al. 1999; McCallum et al. 2002, 2005b).

11.4.4 Pigmentation

The flavonoid composition of dry and fleshy scales provides distinctive variation in appearance and also a significant dietary source of these antioxidant pigments. Red color is a valuable trait and is erroneously associated with milder flavor by many consumers, although no such correlation exists. The major pigments present in all colored onions are quercetin mono-, di- and triglucosides (Price and Rhodes 1997; Fossen et al. 1998). Conjugates and oxidation products of these also contribute to dry scale color (Ly et al. 2005). In addition, red onions also contain anthocyanins, predominantly cyanidin 3-glucosides and malonylglucosides (Fossen et al. 1996).

11.4.5 Human Health Benefits

Although epidemiological studies have shown strong links between onion and other *Allium* vegetable intake and reduced risks of cancer (Hsing et al. 2002) and other ailments (Griffiths et al. 2002), onion poses several challenges for genetic dissection and breeding for health benefits. Due to the outbreeding nature of onion, populations may harbor substantial bulb-to-bulb variation for phytochemicals compared to asexually propagated *Alliums* such as garlic. Since the onion is a significant dietary source of fructans, flavonoids and organosulfur compounds, as well as other bioactive phytochemicals such as saponins (Corea et al. 2005), resolving health benefits into individual compounds or groups of compounds will pose analytical and methodological challenges. Benefits include labile ones, such as antiplatelet activity (Morimitsu and Kawakishi 1990; Ali et al. 1999), and stable ones, such as inhibition of bone density loss (Wetli et al. 2005) and anticoagulative properties (Kim et al. 2002). The recent identification of putative molecular modes of action of *Allium* organosulfur compounds (Cooper and Pinto 2006) may facilitate more rational targeting and analysis of specific compounds for quantitation and improvement of onion health benefits.

11.4.6 Pest and Disease Susceptibility

Onion is susceptible to a wide variety of serious pests and diseases, which compromise the yield, quality and storage life of bulbs, as well as seed yield and quality. Production systems over the past 50 years have come to rely on substantial agrichemical use to stabilize yield and storage, but environmental and consumer health concerns are providing new impetus to reduce these through improved genetics. With few exceptions, sources of resistance or tolerance to major pests and diseases are wanting or poorly characterized in cultivated germplasm. US public breeding programs have characterized (Nichols et al. 1965; Cramer 2000) and released useful sources of resistance to Fusarium basal rot and pink root (*Pyrenochaeta terrestris*) (Goldman 1996; Goldman et al. 2000). The other major soil-borne fungal pathogen afflicting onion and other *Allium* crops is white rot (*Sclerotium cepivorum*), the persistent resting sclerotes of which are stimulated to germinate by *Allium* organosulfur compounds (Coley-Smith and Holt 1966; King and Coley-Smith 1969). Hovius and Goldman (2005) confirmed that field epidemic severity is related to levels of organosulfur leaching, suggesting that selection for less stimulatory lines may be one strategy to reduce susceptibility. The use of transformation technologies to remove toxic oxalic acid secreted by the fungus through overexpression of transgenic or endogenous oxalate oxidases is also being evaluated (Eady 2002). A major foliar fungal pathogen of most temperate production regions is downy mildew (*Perenospora destructor*) (Maude 1990). Although there is wide variation in susceptibility to this pathogen, much of it probably related to leaf morphology and composition, none is sufficient to remove the need for substantial fungicide usage in many temperate regions. Introgression of the resistance from *A. roylei* (Meer and Vries 1990) from backcross material into commercial breeding lines is well-advanced and similar introgression of Botrytis leaf blight (Vries et al. 1992b) has been initiated. The shallot and *A. fistulosum* gene pools may also provide useful sources of tolerance to many foliar and soil pathogens. Onion thrips (*Thrips tabaci*) are the most serious insect pest of intensive onion production, causing yield reduction, storage losses and transmission of viral diseases. Although variation in susceptibility has been reported (Coudriet et al. 1979; Hamilton et al. 1999; Jones et al. 1934), efforts to exploit these sources have not led to demonstrable gains.

Future gains in onion resistance genetics and agrichemical reductions will rely on a combination of these strategies, including large-scale multidisciplinary surveys of germplasm collections, introgression from shallot and other species in section *Cepa* and transgenic methods.

11.5 Genetic and QTL Analysis of Economic Traits

The first economic traits to be subjected to classical genetic analyses were bulb color and S-cytoplasm fertility restoration (Jones et al. 1944; Jones and Mann 1963).

11.5.1 Bulb Pigmentation

Color inheritance studies (Clarke et al. 1944; El-Shafie and Davis 1967) suggested that at least five loci were involved: a dominant locus *C* conditioning colored versus white, dominant *L* and *R* giving red in presence of *C*, a partially dominant locus *I* conditioning white bulbs, and a dominant locus *G* giving yellow. Complementary factors conditioning light red colors in crosses between yellow onions were first described by Jones and Peterson (1952).

The first genome-level analysis of flavonoid pigmentation was the assignment of loci controlling the characteristic red scale color of shallot to chromosome 5 (Shigyo et al. 1997b). A tentative assignment of the *crb* locus conditioning a complementary recessive red to chromosome 7 was made by mapping in BYG15-23 × AC43(King et al. 1998a). Masuzaki et al. (2006b) assigned the anthocyanin biosynthetic genes flavonoid 3′-hydroxylase to this chromosome and flavonol synthase to chromosome 4.

Comparative studies of flavonoid biosynthetic gene expression activity in differently colored onions carried out by Kim and coworkers have suggested the possible molecular nature of the mutations underlying these loci. Transcription of two chalcone synthase (CHS) genes was shown to be reduced in white onions (Kim et al. 2005b), suggesting that the *C* locus may be a regulator of expression of this early step in flavonoid biosynthesis. Transcription of the dihydroflavonol 4-reductase (DFR) gene later in the pathway was shown to occur only in red onions, suggesting that this gene was a candidate for *L* or *R* locus (Kim et al. 2004a, 2005c). Similar studies of another locus conditioning pink color, provisionally named *P*, demonstrated that certain alleles were associated with reduced expression of the anthocyanidin synthase gene (Kim et al. 2004b, 2005c). This gene was subsequently implicated as the *L* locus by co-segregation analysis of complementary red in yellow × yellow crosses (Kim et al. 2005a). Although primer sets were presented for these candidate genes, color trait cosegregation data was only presented for small progeny sets and none have yet been placed on the onion genetic map.

11.5.2 Male-Fertility

Genetic analysis of the inheritance of male sterility deriving from Italian Red 13-53 by Jones and Clarke (1943) suggested that it arose from interaction between the cytotype and a recessive nuclear gene *Ms*. Molecular genetic analyses of organellar mutations associated with S-cytoplasm led to development of PCR-based assays for distinguishing N and S-cytoplasms on the basis of chloroplast (Havey 1995) and mitochondrial .(Sato 1998) polymorphisms. The *Ms* locus was initially mapped to a 28-cM interval flanked by the RFLP markers *API65-E5-10.0* and *AOB210-H3-6.6/5.5* in the BYG15-23 × Alisa Craig population (King et al. 1998a), a region since assigned to chromosome 2 (Martin et al. 2005). Due to the small population size ($N = 58$), subsequent improvements in phenotyping of test cross families and marker coverage led to revision of linkage order and the RFLP locus *AOB272-E1-10.0/12.0* was shown to lie ~0.9 cM from *Ms* (Gokce et al. 2002). Studies in two open-pollinated populations heterozygous at the marker locus and at *Ms* showed that these were at linkage equilibrium (Gokce and Havey 2002). Sequencing of *AOB272* alleles showed that SNP markers can be developed to aid the selection of N *msms* maintainer lines (Gokce et al. 2002). The linkage associations of *Ms* reported by Havey and coworkers were confirmed by QTL analysis of fertility restoration assessed as F_3 seed yield from 80 selfed F_2 plants from the cross W202A × Texas Grano 438, which were segregating for fertility restoration at *Ms* in an S-cytoplasm background (personal communication, McCallum 2006; Fig. 1). Significant disequilibrium of SSR markers in this population with seed yield that were assigned by Martin et al. (2005) to chromosome 6 suggests that chromosomal assignment of additional anchor loci and/or further comparative mapping is required to resolve the chromosomal locations of these linkage groups.

Other cultivated *Allium* species may provide models for future genetic and functional studies of nuclear-cytoplasmic interactions underlying CMS systems. The genetic basis of male-sterility in *A. fistulosum* is poorly characterized, although markers linked to the *Rf* gene restoring pollen fertility in backcross progenies with *A. galanthum* cytoplasm have been reported (Yamashita et al. 1999), and these workers have also assigned this locus to *A. fistulosum* chromosome 5 (Yamashita et al. 2005). The most appealing model in *Allium*, however, is chive (*Allium schoenoprasum* L.), since the CMS systems of this species is particularly well-characterized (Engelke et al. 2004).

11.5.3 Bulb Carbohydrate Composition

The development of the linkage map in BYG15-23 × AC43 enabled the first quantitative genetic analysis of loci underlying variation in bulb composition (Galmarini et al. 2001). This provided evidence for QTLs on chromosomes 3, 5 and 8, affecting bulb dry matter and solids content, as assessed by oven drying and refractometry (Mann and Hoyle 1944), the methods traditionally used to assess and select for gross bulb carbohydrate content. Interestingly, QTLs reported on chromosomes 3 and 5 were associated with RFLPs revealed by cDNAs encoding an acid invertase (API89) and a phloem-unloading sucrose transporter (API66), both candidate genes for carbohydrate metabolism. More detailed phenotyping of bulb carbohydrate in a subset ($N = 48$) of F_3 families from this population in two environments confirmed the effect of the chromosome 3 locus on bulb sucrose levels (Havey et al. 2004). QTL analysis of bulb carbohydrate composition in the storage × sweet onion populations W202A × Texas Grano 438 and Colossal Grano × Early Longkeeper revealed that a major dominant gene, *Frc*, located on chromosome 8 conditioned the large differences in fructan content in these populations and also in BYG15-23 × AC43 (McCallum et al. 2006). This was distal to the QTL affecting soluble solids content detected by Galmarini et al. (2001) using a shorter linkage map of this chromosome. QTL analysis in W202A × Texas Grano 438 also confirmed that the region of chromosome 3 reported by (Galmarini et al. 2001) also had a significant effect on bulb dry matter content (Fig. 2). Although *Frc* conditions a large shift in the osmotic potential, it did not exert a significant effect on dry matter content in W202A × Texas Grano 438 and studies in much larger F_2 or backcross progenies are desirable to better resolve the pleiotropic and epistatic effects of loci mapped to date. During these studies, genes encoding the two enzymes responsible for fructan biosynthesis, 1-sucrose-sucrose fructosyltransferase (1-SST; Vijn et al. 1998) and fructan: fructan 6(G)-fructosyltransferase (6G-FFT; Fujishima et al. 2005), as well as a sucrose synthase gene were mapped to chromosome 6. The key role for loci on chromosome 8 in the regulation of *A. cepa* carbohydrate metabolism was independently demonstrated by Hang et al. (2004),who showed that the *A. fistulosum* monosomic addition lines carrying this chromosome from shallot also accumulated nonreducing carbohydrates in winter. This suggests that these addition lines may be useful tools for functional analysis to identify genes underlying *Frc* and other loci affecting onion carbohydrate metabolism.

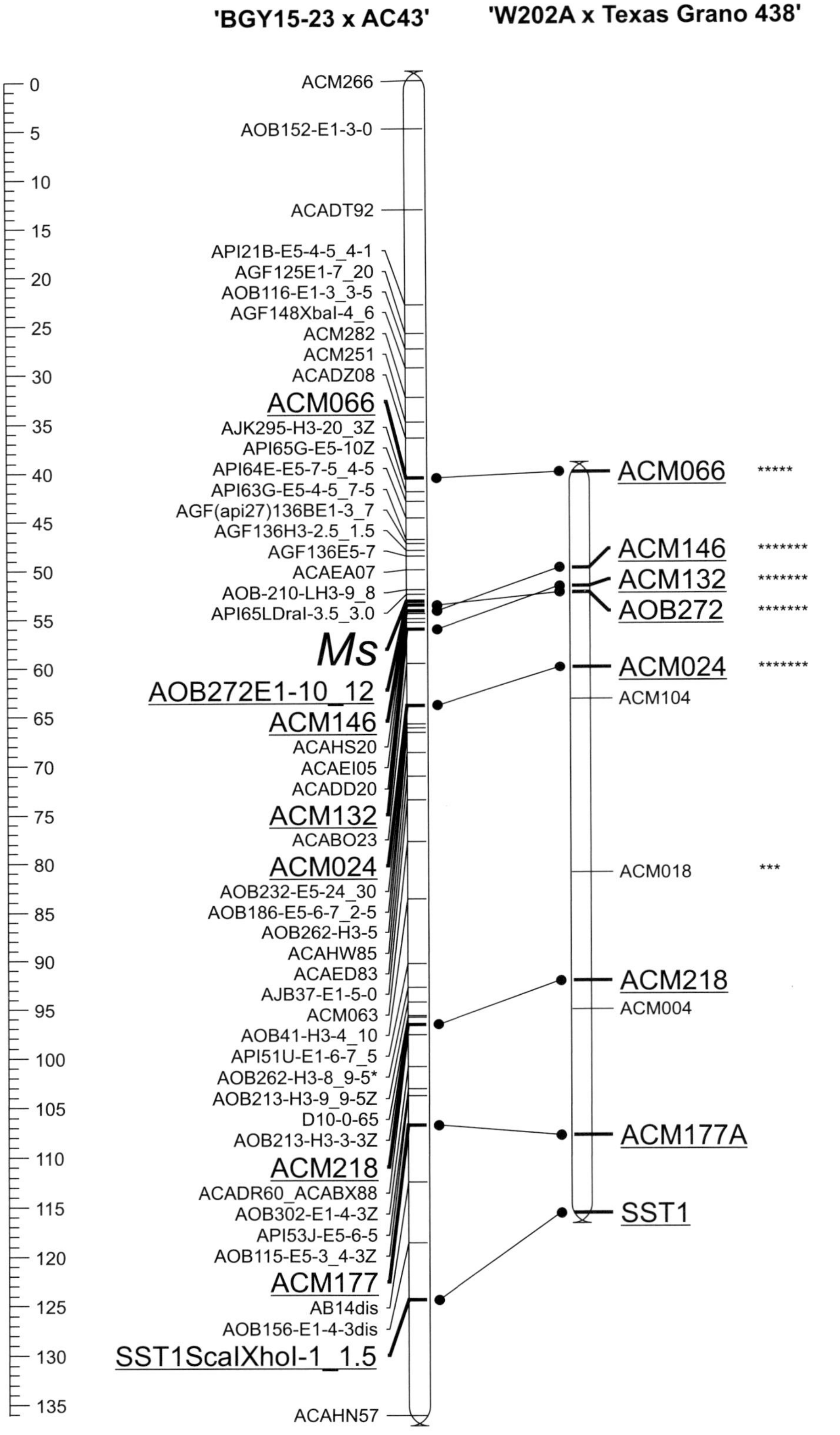

Fig. 1. Alignment of linkage groups associated with male-fertility restoration in S-cytoplasm from two onion mapping populations using portable PCR-based markers. SSR markers (prefixed ACM) were described by Kuhl et al. (2004) and Martin et al. (2005). SSCP markers for AOB272 and SST-1 were described previously (McCallum et al. 2001; Gokce et al. 2002). Linkage mapping was performed using JoinMap 3.0 (van Ooijen and Voorips 2002) using a LOD 6 criteria for forming linkage groups using data of Martin et al. (2005) and data from the W202A × Texas Grano 438 population (McCallum et al. 2006). *Scale* denotes Kosambi distance in centiMorgans. *Asterisks* adjacent to markers in the W202A × Texas Grano 438 group denote significance levels for Kruskal-Wallis single-marker tests for association of genotypes with seed yield from $N = 80$ selfed F_2 plants (***: 0.01, ******: 0.0005, *******: 0.0001)

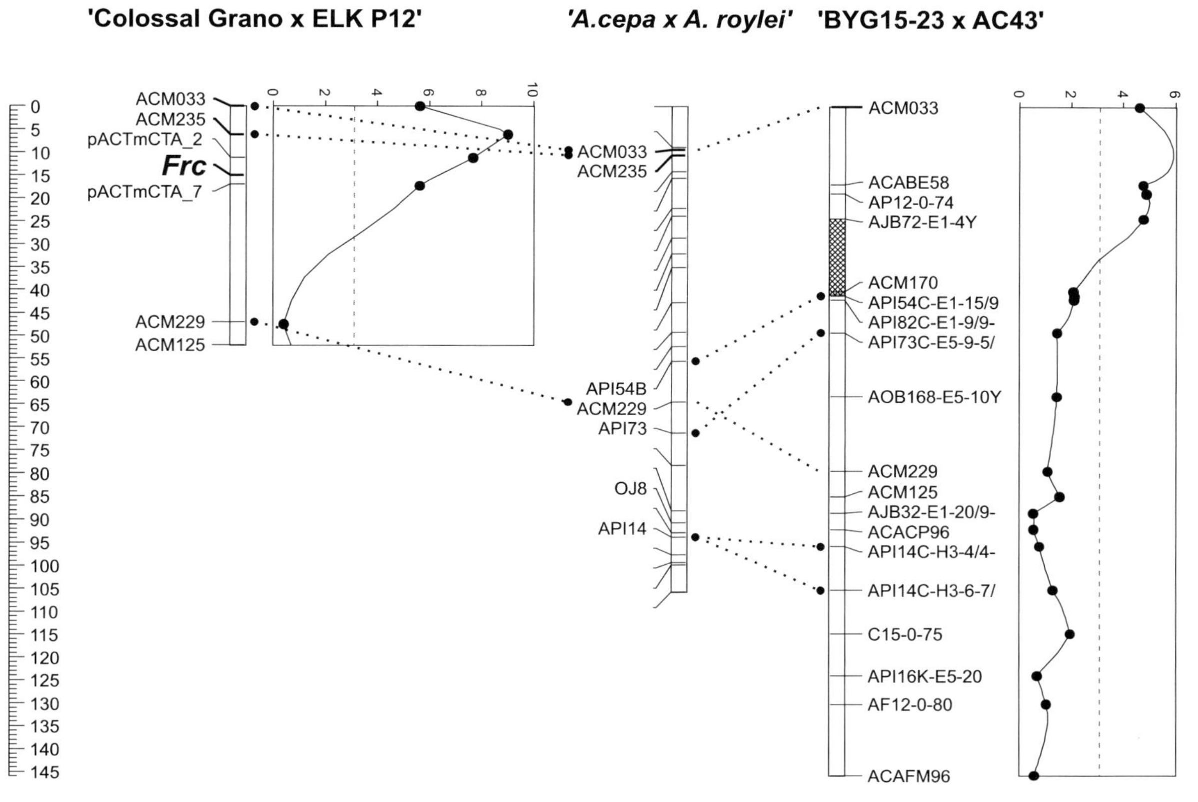

Fig. 2. Assignment of the *Frc* locus to chromosome 8 of onion by QTL and linkage analysis in inter- and intraspecific *Allium cepa* mapping populations (McCallum et al. 2006). Names of AFLP loci in the *A. cepa* × *A. roylei* interspecific map are omitted for clarity. Location shown for *Frc* in Colossal Grano × ELK population was estimated by scoring fructan content as a Mendelian character in F_2 bulbs ($N = 46$). *Plots* denote results from interval mapping of sucrose content in Colossal Grano × ELK and total fructan content in BYG15-23 × AC43 populations and *dashed lines* denote 95% significance levels derived from permutation testing. The *crosshatched* interval in BYG15-23 × AC43 was reported by Galmarini et al. (2001) to affect bulb dry matter. Scale denotes Kosambi distance in centiMorgans

11.5.4 Bulb Pungency

In contrast to the successes in mapping loci affecting bulb carbohydrate content it has proved more difficult to identify QTL for bulb pungency. The biosynthesis of the flavor precursors is as yet poorly understood (Jones et al. 2004), limiting the use of candidate gene approaches. Galmarini et al. (2001) reported that the interval on chromosome 5 in BYG15-23 × AC43 also affecting bulb dry matter influenced bulb and antiplatelet activity, suggesting that this may reflect a pleiotropic effect of this locus. The support and chromosomal locations of QTL for pungency and antiplatelet activity reported on linkage groups B and G in this study are uncertain given the current uncertainty regarding chromosomes 2 and 6 noted previously in this review (see Sect. 11.5.2). QTL analysis for bulb pungency has also been conducted in W202A × Texas Grano 438 F_3 families following phenotypic evaluation in two environments. It has been possible to map several candidate genes involved in sulfur assimilation in this population, including a serine acetyl-transferase (McManus et al. 2005), ATP sulfurylase (ATPS) and sulfite reductase (McCallum et al. 2005a) and align these genome regions well with the BYG15-23 × AC43 linkage map. QTL analysis showed a significant association between the regions of chromosome 3 containing the ATPS and SiR loci and this marker association has been verified in two independently derived F_2 families from the same pedigree (McCallum et al. 2006 in press).

11.6 Conclusion

The recent progress in mapping loci affecting onion bulb carbohydrate content highlights the value of collaborative genetic map development and application of diverse genetic resources to genetically analyze economic traits in *Allium cepa*. Resolving relevant genetic loci and useful sources of variation in sexual *A. cepa* for bulb health and quality traits, tolerance to major pests and diseases and fertility restoration will probably require collaborative development and characterization of diverse genetic and genomic resources and standardized methodologies for phenotyping. Advances in functional characterization of genes underlying fitness, composition and adaptation in model plant systems will facilitate increasing use of linkage disequilibrium methods in onion. Currently the major issues facing *Allium* genetic and genomics research are a decline in the already small public research community and limited public resources for enabling future research compared to other crops of comparable value. Investment in precompetitive genetic research by seed companies, similar to that provided in tomato, would assist closing this gap. A priority for map development is the establishment of substantially larger mapping populations to enable resolution of framework marker order. Collaborative phenotyping and population development from double haploid parent lines would be to the advantage of all parties involved in onion research and breeding. The emergence of robust technologies for whole genome amplification means that DNA templates from intercross or backcross progenies could now be readily replicated from original stocks for distribution. Development of a much larger set of public, portable, EST-derived markers for genetic mapping, diversity and identity analyses would also greatly advance commercial and research interests. Rapid changes in the technology and economics of genomic sequencing suggest that, in the short term, limited resources within the *Allium* research community are best applied to developing genetic stocks and EST resources to leverage future opportunities for exploration and exploitation of variation in the *Allium cepa* genome.

Acknowledgement. The author wishes to thank Colin Eady (Crop & Food Research) for critical reading of this manuscript and Mike Havey (USDA-ARS) for sharing linkage data from the BYG15-23 × AC43 populations. The author acknowledges funding support from the New Zealand Foundation for Research, Science and Technology.

References

Alan AR, Brants A, Cobb E, Goldschmied PA, Mutschler MA, Earle ED (2004) Fecund gynogenic lines from onion (*Allium cepa* L.) breeding materials. Plant Sci 167:1055–1066

Albini SM, Jones GH (1987) Synaptonemal complex spreading in *Allium cepa* and *A. fistulosum*. Chromosoma 95:324–338

Ali M, Bordia T, Mustafa T (1999) Effect of raw versus boiled aqueous extract of garlic and onion on platelet aggregation. Prostaglandins Leukot Essent Fatty Acids 1999:60:43–47

Arumuganathan K, Earle ED (1991) Nuclear DNA content of some important plant species. Plant Mol Biol Rep 9:208–218

Aswath CR, Mo SY, Kim DH, Park SW (2006) Agrobacterium and biolistic transformation of onion using nonantibiotic selection marker phosphomannose isomerase. Plant Cell Rep 25:92–99

Bark OH, Havey MJ (1995) Similarities and relationships among populations of the bulb onion as estimated by nuclear RFLPs. Theor Appl Genet 90:407–414

Barnes SR, James AM, Jamieson G (1985) The organization, nucleotide sequence, and chromosomal distribution of a satellite DNA from *Allium cepa*. Chromosoma 92:185–192

Barthes L, Ricroch A (2001) Interspecific chromosomal rearrangements in monosomic addition lines of *Allium*. Genome 44:929–935

Bennett MD, Leitch IJ (1995) Nuclear DNA amounts in angiosperms. Ann Bot 76:113–176

Berninger E (1965) Contribution a L'Etude de la Sterilitemale de L'Oignon(*Allium cepa* L.). Ann Amelior Plantes 15:183–199

Berninger E, Buret P (1967) Etudes des defcients chlorophyllines chez deux especes cultivees du genre *Allium*: l'oignon (*A. cepa* L.) et le poireau (*A. porrum* L.). Ann Amelior Plantes 17:175–194

Block E (1992) The organosulfur chemistry of the genus *Allium*: implications for the organic chemistry of sulfur. Agnew Chem Int Ed Engl 31:1135–1178

Block E, Gulati H, Putman D, Sha DY, You N, Zhao SH (1997) *Allium* chemistry: synthesis of 1-[Alk(en)ylsulphinylpropyl Alk(en)yl disulphides (cepaenes), antithrombotic flavorants from homogenates of onion (*Allium cepa*). J Agri Food Chem 45:4414–4422

Bohanec B (2002) Double haploid onions. In: Rabinowich HD, Currah L (eds.) Allium Crop Science: Recent Advances. CABI, New York, pp 145–158

Bohanec B, Jakse M, Havey MJ (2003) Genetic analyses of gynogenetic haploid production in onion. J Am Soc Hort Sci 128:571–574

Bradeen JM, Havey MJ (1995) Randomly amplified polymorphic DNA in bulb onion and its use to assess inbred integrity. J Am Soc Hort Sci 120:752–758

Brewster JL (1990a) Cultural systems and agronomic practices in temperate climates. In: Rabinowitch HD, Brewster JL (eds.) Onions and Allied Crops 1. CRC, Boca Raton, FL, pp 1–30

Brewster JL (1990b) Physiology of crop growth and bulbing. In: Rabinowitch HD, Brewster JL (eds.) Onions and Allied Crops 1. CRC, Boca Raton, FL

Brewster JL (1994) Onions and other vegetable *Alliums*. CABI, Wallingford, Oxon, UK

Campion B, Alloni C (1990) Induction of haploid plants in onion (*Allium cepa* L) by in vitro culture of unpollinated ovules. Plant Cell Tiss Org Cult 20:1–6

Campion B, Perri E, Azzimonti MT, Vicini E, Schiavi M (1995) Spontaneous and induced chromosome doubling in gynogenic lines of onion (*Allium cepa* L). Plant Breed 114:243–246

Clarke AE, Jones HA, Little TM (1944) Inheritance of bulb color in the onion. Genetics 29:569–575

Coley-Smith JR, Holt RW (1966) The effects of species of *Allium* on germination in soil of sclerotia of *Sclerotium cepivorum* Berk. Ann Appl Biol 58:273–278

Cooper AJL, Pinto JT (2006) Cysteine S-conjugate beta-lyases. Amino Acids 30:1–15

Corea G, Fattorusso E, Lanzotti V, Capasso R, Izzo AA (2005) Antispasmodic saponins from bulbs of red onion, *Allium cepa* L. var. Tropea. J Agri Food Chem 53:935–940

Coudriet DL, Kishaba AN, McCreight JD, Bohn GW (1979) Varietal resistance in onions to thrips. J Econ Entomol 72:614–615

Cramer CS (2000) Breeding and genetics of Fusarium basal rot resistance in onion. Euphytica 115:159–166

Cramer CS, Havey MJ (1999) Morphological, biochemical, and molecular markers in onion. HortScience 34:589–593

Darbyshire B, Henry RJ (1978) The distribution of fructans in onions. New Phytol 81:29–34

Do GS, Suzuki G, Mukai Y (2004) Genomic organization of a novel root alliinase gene, ALL1, in onion. Gene 325:17–24

Dowker BD (1990) Onion breeding. In: Rabinowitch H, Brewster JL (eds.) Onions and Allied Crops 1. CRC, Boca Raton, FL, pp 215–232

Dowker BD, Horobin JF, Crowther TC, Fennell JFM (1984) Breeding of improved open-pollinated populations of spring-sown onions. J Agri Sci (Cambridge) 102:615–623

Eady C (2002) Genetic transformation of onions. In: Currah L (ed.) Allium Crop Science: Recent Advances. CABI, New York, pp 119–144

Eady C, Davis S, Farrant J, Reader J, Kenel F (2003) *Agrobacterium tumefaciens*-mediated transformation and regeneration of herbicide resistant onion (*Allium cepa*) plants. Ann Appl Biol 142:213–217

El-Shafie MW, Davis GN (1967) Inheritance of bulb color in the onion (*Allium cepa* L.). Hilgardia 38:607–622

Emsweller SL, Jones HA (1935) An interspecific hybrid in *Allium*. Hilgardia 9:265–273

Engelke T, Gera D, Tatlioglu T (2004) Determination of the frequencies of restorer and maintainer-alleles involved in CMS1 and CMS2 in German chive varieties. Plant Breed 123:51–59

Fajkus J, Sakarova E, Leitch AR (2005) Telomeres in evolution and evolution of telomeres. Chrom Res 13:469–479

FAO Statistics (2005) FAOSTAT data. http://faostat.fao.org/faostat/. Cited 2006

Fischer D, Bachmann K (1998) Microsatellite enrichment in organisms with large genomes (*Allium Cepa* L.). Biotechniques 24:796–802

Fischer D, Bachmann K (2000) Onion microsatellites for germplasm analysis and their use in assessing intra- and interspecific relatedness within the subgenus *Rhizirideum*. Theor App Genet 101:153–164

Fossen T, Andersen OM, Ovstedal DO, Pedersen AT, Raknes A (1996) Characteristic anthocyanin pattern from onions and other *Allium* spp. J Food Sci 61:703–706

Fossen T, Pedersen AT, Andersen OM (1998) Flavonoids from red onion (*Allium cepa*). Phytochemistry 47:281–285

Fritsch RM, Friesen N (2002) Evolution, domestication and taxonomy. In: Currah L (ed.) Allium Crop Science: Recent Advances. CABI, Wallingford, Oxon, UK, pp 5–31

Fritsch RM, Matin F, Klaas M (2001) *Allium vavilovii* M. Popov et Vved. and a new Iranian species are the closest among the known relatives of the common onion *A. cepa* L. (Alliaceae). Genet Resour Crop Evol 48:401–408

Fujishima M, Sakai H, Ueno K, Takahashi N, Onodera S, Benkeblia N, Shiomi N (2005) Purification and characterization of a fructosyltransferase from onion bulbs and its key role in the synthesis of fructooligosaccharides in vivo. New Phytol 165:513–524

Galmarini CR, Goldman IL, Havey MJ (2001) Genetic analyses of correlated solids, flavor, and health-enhancing traits in onion (*Allium cepa* L.). Mol Genet Genom 265:543–551

Gokce AF, Havey MJ (2002) Linkage equilibrium among tightly linked RFLPs and the *Ms* locus in open-pollinated onion populations. J Am Soc Hort Sci 127:944–946

Gokce AF, McCallum J, Sato Y, Havey MJ (2002) Molecular tagging of the *Ms* locus in onion. J Am Soc Hort Sci 127:576–582

Goldman IL (1996) A list of germplasm releases from the University of Wisconsin Onion Breeding Program, 1957–1993. HortScience 31:878–879

Goldman IL, Schroek G, Havey MJ (2000) History of public onion breeding programs in the United States. Plant Breed Rev 20:68–103

Griffiths G, Trueman L, Crowther T, Thomas B, Smith B (2002) Onions: a global benefit to health. Phytother Res 16:603–615

Hamilton BK, Pike LM, Sparks AN, Bender DA, Jones RW, Candeia J, de Franca G (1999) Heritability of thrips resistance in the IPA-3 onion cultivar in South Texas. Euphytica 109:117–122

Hanelt P (1990) Taxonomy, evolution, and history. In: Rabinowitch H, Brewster J (eds.) Onions and Allied Crops. CRC, Boca Raton, FL, pp 1–26

Hang TT, Shigyo M, Yaguchi S, Yamauchi N, Tashiro Y (2004) Effect of single alien chromosome from shallot (*Allium cepa* L. Aggregatum group) on carbohydrate production in leaf blade of bunching onion (*A. fistulosum* L.). Genes Genet Syst 79:345–350

Havey M (2002) Genome organization in *Allium*. In: Rabinowitch HD, Currah L (eds.) Allium Crop Science: Recent Advances. CABI, Wallingford, Oxon, UK, pp 59–80

Havey M, King JJ, Bradeen JM, Bark O, Sato Y, Gökçe AF (2001) A low-density genetic map of onion (*Allium cepa* L.) and its use for marker-facilitated selection of maintainer lines. Acta Hort 555:87–89

Havey MJ (1992) Restriction enzyme analysis of the chloroplast and nuclear 45s ribosomal DNA of *Allium* sections *Cepa* and *Phyllodolon* (*Alliaceae*). Plant Syst Evol 183:17–31

Havey MJ (1995) Identification of cytoplasms using the polymerase chain reaction to aid in the extraction of maintainer lines from open-pollinated populations of onion. Theor Appl Genet 90:263–268

Havey MJ (1999) Seed yield, floral morphology, and lack of male fertility restoration of male-sterile onion (*Allium cepa*) populations possessing the cytoplasm of *Allium galanthum*. J Am Soc Hort Sci 124:626–629

Havey MJ, Galmarini CR, Gokce AF, Henson C (2004) QTL affecting soluble carbohydrate concentrations in stored onion bulbs and their association with flavor and health-enhancing attributes. Genome 47:463–468

Havey MJ, Randle WM (1996) Combining abilities for yield and bulb quality among long and intermediate-day open-pollinated onion populations. J Am Soc Hort Sci 121:604–608

Hou A, Peffley EB (2000) Recombinant chromosomes of advanced backcross plants between *Allium cepa* L. and *A. fistulosum* L. revealed by in situ hybridization. Theor Appl Genet 100:1190–1196

Hovius MHY, Goldman IL (2005) Flavor precursor S-alk(en)yl-L-cysteine sulfoxidel concentration and composition in onion plant organs and predictability of field white rot reaction of onions. J Am Soc Hort Sci 130:196–202

Hsing AW, Chokkalingam AP, Gao YT, Madigan M, Deng J, Gridley G, Fraumeni JF (2002) *Allium* vegetables and risk of prostate cancer: a population-based study. J Natl Cancer Inst 94:1648–1651

Imai S, Tsuge N, Tomotake M, Nagatome Y, Sawada H, Nagata T, Kumagai H (2002) Plant biochemistry: an onion enzyme that makes the eyes water. Nature 419:685

Jakše J, Telgman A, Jung C, Cheung F, Town CD, Havey MJ (2006) Comparative sequence and genetic analyses of Asparagus BACs reveal no microsynteny with onion or rice. Theo Appl Genet, in press

Jakse M, Martin W, McCallum J, Havey M (2005) Single nucleotide polymorphisms, indels, and simple sequence repeats for onion cultivar identification. J Am Soc Hort Sci 130:912–917

Javornik B, Bohanec B, Campion B (1998) Second cycle gynogenesis in onion, *Allium cepa* L, and genetic analysis of the plants. Plant Breed 117:275–278

Jones HA, Bailey SF, Emsweller SL (1934) Thrips resistance in the onion. Hilgardia 8:215–232

Jones HA, Clarke AE (1943) Inheritance of male sterility in the onion and the production of hybrid seed. Proc Am Soc Hort Sci 43:189–194

Jones HA, Clarke AE, Stevenson FJ (1944) Studies in the genetics of the onion (*Allium cepa* L.). Proc Am Soc Hort Sci 44:479–484

Jones HA, Davis GN (1944) Inbreeding and heterosis and their relation to the development of new varieties of onions. USDA Tech Rep No 28

Jones HA, Mann LK (1963) Onions and Their Allies 1st edn. Interscience, London

Jones MG, Hughes J, Tregova A, Milne J, Tomsett AB, Collin HA (2004) Biosynthesis of the flavor precursors of onion and garlic. J Exp Bot 55:1903–1918

Khrustaleva LI, de Melo PE, van Heusden AW, Kik C (2005) The integration of recombination and physical maps in a large-genome monocot using haploid genome analysis in a trihybrid *Allium* population. Genetics 169:1673–1685

Khrustaleva LI, Kik C (1998) Cytogenetical studies in the bridge cross *Allium cepa* × (*A. fistulosum* × *A. roylei*). Theor Appl Genet 96:8–14

Kim GN, Lee MK, Park I (2002) Acid and heat stability of the anticoagulative activity of an onion extract. Biosci Biotechnol Biochem 66:859–861

Kim S, Binzel M, Park S, Yoo K-S, Pike L (2004a) Inactivation of DFR (dihydroflavonol 4-reductase) gene transcription results in blockage of anthocyanin production in yellow onions (*Allium cepa*). Mol Breed 14:253–263

Kim S, Binzel ML, Yoo KS, Park S, Pike LM (2004b) Pink (P), a new locus responsible for a pink trait in onions (*Allium cepa*) resulting from natural mutations of anthocyanidin synthase. Mol Genet Genom 272:18–27 (epub 2004 Jul 28)

Kim S, Jones R, Yoo K-S, Pike LM (2005a) The *L*locus, one of complementary genes required for anthocyanin production in onions (*Allium cepa*), encodes anthocyanidin synthase. Theor App Genet 111:120–127

Kim S, Yoo K-S, Pike LM (2005b) The basic color factor, the *C*locus, encodes a regulatory gene controlling transcription of chalcone synthase genes in onions (*Allium cepa*). Euphytica 142:273–282

Kim S, Yoo K-S, Pike LM (2005c) Development of a codominant, PCR-based marker for allelic selection of the pink trait in onions (*Allium cepa*), based on the insertion mutation in the promoter of the anthocyanidin synthase gene. Theor Appl Genet 110:628–633

King J, Coley-Smith J (1969) Production of volatile alkyl sulphides by microbial degradation of synthetic alliin and alliin-like compounds, in relation to germination of sclerotia of *Sclerotium cepivorum* Berk. Ann Appl Biol 64:303–314

King JJ, Bradeen JM, Bark O, McCallum JA, Havey MJ (1998a) A low-density genetic map of onion reveals a role for tandem duplication in the evolution of an extremely large diploid genome. Theor Appl Genet 96:52–62

King JJ, Bradeen JM, Havey MJ (1998b) Variability for restriction fragment-length polymorphisms (RFLP) and relationships among elite commercial inbred and virtual hybrid onion populations. J Am Soc Hort Sci 123:1034–1037

Klaas M (1998) Applications and impact of molecular markers on evolutionary and diversity studies in the genus *Allium*. Plant Breed 117:297–308

Klaas M, Friesen N (2002) Molecular markers in *Allium*. In: Currah L (ed.) Allium Crop Science: Recent Advances. CABI, New York, pp 159–186

Knight CA, Molinari NA, Petrov DA (2005) The large-genome constraint hypothesis: evolution, ecology and phenotype. Ann Bot 95:177–190

Kopsell DE, Randle WM, Schmidt NE (2002) Incubation time, cultivar, and storage duration affect onion lachrymatory factor quantification. HortScience 37:567–570

Kuhl JC, Cheung F, Yuan Q, Martin W, Zewdie Y, McCallum J, Catanach A, Rutherford P, Sink KC, Jenderek M, Prince JP, Town CD, Havey MJ (2004) A unique set of 11,008 onion expressed sequence tags reveals expressed sequence and genomic differences between the monocot orders Asparagales and Poales. Plant Cell 16:114–125

Kumar A, Pearce SR, McLean K, Harrison G, Heslop-Harrison JS, Waugh R, Flavell AJ (1997) The Ty1-copia group of retrotransposons in plants: genomic organization, evolution, and use as molecular markers. Genetica 100:205–217

Lancaster JE, Shaw ML (1991) Metabolism of gamma-glutamyl peptides during development, storage and sprouting of onion bulbs. Phytochemistry 30:2857–2859

Ly TN, Hazama C, Shimoyamada M, Ando H, Kato K, Yamauchi R (2005) Antioxidative compounds from the outer scales of onion. J Agri Food Chem 53:8183–8189

Maass HI (1997) Genetic diversity in the top onion, *Allium* × *Proliferum* (Alliaceae), analyzed by isozymes. Plant Syst Evol 208:35–44

Mangum PD, Peffley EB (2005) Central cell nuclear-cytoplasmic incongruity: a mechanism for segregation distortion in advanced backcross and selfed generations of (*Allium cepa* L. × *Allium fistulosum* L.) × *A. cepa* interspecific hybrid derivatives. Cytogenet Genome Res 109:400–407

Mann L, Hoyle BJ (1944) Use of the refractometer for selecting onion bulbs high in dry matter for breeding. Proc Am Soc Hort Sci 46:285–292

Martin W, McCallum J, Shigyo M, Jakse J, Kuhl J, Yamane N, Pither-Joyce M, Gokce A, Sink K, Town C, Havey M (2005) Genetic mapping of expressed sequences in onion and in silico comparisons with rice show scant colinearity. Mol Genet Genom 274:197–204

Masuzaki S, Araki N, Yamane N, Yamauchi N, Wako T, Kojima A, Shigyo M (2006a) Chromosomal locations of microsatellites in onion. HortScience 41:315–318

Masuzaki S, Shigyo M, Yamauchi N (2006b) Direct comparison between genomic constitution and flavonoid contents in *Allium* multiple alien addition lines reveals chromosomal locations of genes related to biosynthesis from dihydrokaempferol to quercetin glucosides in scaly leaf of shallot (*Allium cepa* L.). Theor Appl Genet 112:607–617

Maude R (1990) Leaf diseases of onions. In: Rabinowitch H, Brewster J (eds.) Onions and Allied Crops II. CRC, Boca Raton, FL, pp 173–190

McCallum J, Clarke A, Pither-Joyce M, Shaw M, Butler R, Brash D, Scheffer J, Sims I, van Heusden S, Shigyo M, Havey M (2006) Genetic mapping of a major gene affecting onion bulb fructan content. Theor Appl Genet 112:958–967

McCallum J, Pither-Joyce M, Shaw M, Porter N, Searle B, McManus M, Havey MJ (2005a) Genomic and biochemical analysis of onion (*Allium cepa* L.) sulfur metabolism. Acta Hort 688:75–84

McCallum J, Pither-Joyce M, Shaw M, Kenel F, Davis S, Butler R, Scheffer J, Jakse J, Havey MJ (2007) Genetic Mapping of Sulfur Assimilation Genes Reveals a QTL for Onion Bulb Pungency Theor App Genet, in press

McCallum J, Porter N, Searle B, Shaw M, Bettjeman B, McManus M (2005b) Sulfur and nitrogen fertility affects flavor of field-grown onions. Plant Soil 274:151–158

McCallum JA, Leite D, Pither-Joyce MD, Havey MJ (2001) Expressed sequence markers for genetic analysis of bulb onion (*Allium cepa*. L.). Theor Appl Genet 103:979–991

McCallum JA, Pither-Joyce MD, Shaw ML (2002) Sulfur deprivation and genotype affect gene expression and metabolism of onion roots. J Am Soc Hort Sci 127:583–589

McManus MT, Leung S, Lambert A, Scott RW, Pither-Joyce M, Chen B, McCallum J (2005) Molecular and biochemical characterization of a serine acetyltransferase of onion, *Allium cepa* (L.). Phytochemistry 66:1407–1416

Meer QP, Vries JN (1990) An interspecific cross between *Allium roylei* Stearn and *Allium cepa* L., and its backcross to *A. cepa*. Euphytica 47:29–31

Mennella G, Sanaja' V, D'Alessandro A, Desiderio A (2005) Biochemical characterization of white onion landraces (*Allium cepa* L.) through HPLC analysis of endosperm seed proteins. Euphytica 141:169–180

Morimitsu Y, Kawakishi S (1990) Inhibitors of platelet aggregation from onion. Phytochemistry 29:343–3439

Moshfegh AJ, Friday JE, Goldman JP, Ahuja JKC (1999) Presence of inulin and oligofructose in the diets of Americans. J Nutr 129:1407–1411S

Nichols CG, Gabelman WH, Larson RH, Walker JC (1965) The expression and inheritance of resistance to pink root in onion seedlings. Phytopathology 55:752–756

Ohara T, Song YS, Tsukazaki H, Wako T, Nunome T, Kojima A (2005) Genetic mapping of AFLP markers in

Japanese bunching onion (*Allium fistulosum*). Euphytica 144:255–263

Parida S, Anand Raj Kumar K, Dalal V, Singh N, Mohapatra T (2006) Unigene derived microsatellite markers for the cereal genomes. Theor Appl Genet 112:808–817 (epub 2006 Jan 21)

Pearce SR, Pich U, Harrison G, Flavell AJ, Heslopharrison JSP, Schubert I, Kumar A (1996) The Ty1-copia group retrotransposons of *Allium Cepa* are distributed throughout the chromosomes but are enriched in the terminal heterochromatin. Chrom Res 4:357–364

Peffley E, Orozco-Castillo C (1987) Polymorphism of isozymes within plant introductions of *Allium cepa* L. HortScience 22:956–957

Peffley EB, Hou A (2000) Bulb-type onion introgressants possessing *Allium fistulosum* L. genes recovered from interspecific hybrid backcrosses between *A. cepa* L. and *A. fistulosum* L. Theor Appl Genet 100:528–534

Peterson DG, Chouvarine P, Thummasuwan S, Saha S, Mukherjee D, Carlson JE (2006) Exploring the pine genome using Cot Filtration and 454 Life Sciences Massively Parallel Shotgun Sequencing. In: Proceedings of the Plant and Animal Genomes XIV Conference, San Diego, CA

Pich U, Fritsch R, Schubert I (1996) Closely related *Allium* species (*Alliaceae*) share a very similar satellite sequence. Plant Syst Evol 202:255–264

Price KR, Rhodes MJC (1997) Analysis of the major flavonol glycosides present in four varieties of onion (*Allium Cepa*) and changes in composition resulting from autolysis. J Sci Food Agri 74:331–339

Rabinowitch HD, Kamenetsky R (2002) Shallot (*Allium cepa*, Aggregatum Group). In: Rabinowitch HD, Currah L (eds.) Allium Crop Science: Recent Advances. CABI, New York, pp 409–430

Randle WM, Kopsell DE, Kopsell DA, Snyder RL (1999) Total sulfur and sulfate accumulation in onion is affected by sulfur fertility. J Plant Nutr 22:45–51

Randle WM, Lancaster JE (2002) Sulphur compounds in *Alliums* in relation to flavor quality. In: Rabinowitch H, Currah L (eds.) Allium Crop Science: Recent Advances. CABI, Oxford, UK, pp 1–62

Roberfroid MB, Delzenne NM (1998) Dietary fructans. Annu Rev Nutr 18:117–143

Rouamba A, Sandmeier M, Sarr A, Ricroch A (2001) Allozyme variation within and among populations of onion (*Allium cepa* L.) from West Africa. Theor Appl Genet 103:855–861

Sato Y (1998) PCR amplification of CMS-specific mitochondrial nucleotide sequences to identify cytoplasmic genotypes of onion (*Allium cepa* L.). Theor Appl Genet 96:367–370

Schwimmer S, Weston WJ (1961) Enzymatic development of pyruvic acid in onion as a measure of pungency. J Agri Food Chem 9:301–304

Shigyo M, Miyazaki T, Isshiki S, Tashiro Y (1997a) Assignment of randomly amplified polymorphic DNA markers to all chromosomes of shallot (*Allium cepa* L. Aggregatum group). Genes Genet Syst 72:249–252

Shigyo M, Tashiro Y, Iino M, Terahara N, Ishimaru K, Isshiki S (1997b) Chromosomal locations of genes related to flavonoid and anthocyanin production in leaf sheath of shallot (*Allium cepa* L. Aggregatum group). Genes Genet Syst 72:149–152

Shigyo M, Tashiro Y, Isshiki S, Miyazaki S (1995) Genetic analyses of glutamate dehydrogenase isozymes in cultivated and wild species of section Cepa in *Allium*. J Jpn Soc Hort Sci 64:121–127

Shigyo M, Tashiro Y, Isshiki S, Miyazaki S (1996) Establishment of a series of alien monosomic addition lines of Japanese bunching onion (*Allium fistulosum* L) with extra chromosomes from shallot (*A. cepa* L. Aggregatum group). Genes Genet Syst 71:363–371

Shigyo M, Wako T, Kojima A, Yamauchi N, Tashiro Y (2003) Transmission of alien chromosomes from selfed progenies of a complete set of *Allium* monosomic additions: the development of a reliable method for the maintenance of a monosomic addition set. Genome 46:1098–1103

Shiomi N, Benkeblia N, Onodera S (2005) The metabolism of the fructooligosaccharides in onion bulbs: a comprehensive review. J Appl Glycosci 52:121–127

Song YS, Suwabe K, Wako T, Ohara T, Nunome T, Kojima A (2004) Development of microsatellite markers in bunching onion (*Allium fistulosum* L.). Breed Sci 54:361–365

Stack SM, Comings DE (1979) The chromosomes and DNA of *Allium cepa*. Chromosoma 70:161–181

Sugiharto Arifin N, Ozaki Y, Okubo H (2000) Genetic diversity in Indonesian shallot (*Allium cepa* var. *ascalonicum*) and *Allium wakegi* revealed by RAPD markers and origin of *A. wakegi* identified by RFLP analyses of amplified chloroplast genes. Euphytica 111:23–31

Sulistyaningsih E, Yamashita K, Tashiro Y (2002) Haploid induction from F-1 hybrids between CMS shallot with *Allium galanthum* cytoplasm and common onion by unpollinated flower culture. Euphytica 125:139–144

Suzuki G, Do GS, Mukai Y (2002) Efficient storage and screening system for onion BAC clones. Breed Sci 52:57–159

Suzuki G, Ura A, Saito N, Do GS, Seo BB, Yamamoto M, Mukai Y (2001) BAC FISH analysis in *Allium cepa*. Genes Genet Syst 76:251–255

Tanikawa T, Takagi M, Ichii M (2002) Cultivar identification and genetic diversity in onion (*Allium cepa* L.) as evaluated by random amplified polymorphic DNA (RAPD) analysis. J Jpn Soc Hort Sci 71:249–251

TIGR (2003) Onion Gene Index (OnGI). v 1.0. http://www.tigr.org. Cited 2003

van Heusden AW, Shigyo M, Tashiro Y, Vrielink-van Ginkel R, Kik C (2000a) AFLP linkage group assignment to the chromosomes of *Allium cepa* L. via monosomic addition lines. Theor Appl Genet 100:480–486

van Heusden AW, van Ooijen JW, Vrielink-van Ginkel R, Verbeek WHJ, Wietsma WA, Kik C (2000b) A genetic map of an interspecific cross in *Allium* based on amplified fragment length polymorphism (AFLP) markers. Theor Appl Genet 100:118–126

van Ooijen JW, Voorips RE (2002) JoinMap 3.0: software for the calculation of genetic linkage maps. Plant Research International BV, Wageningen, The Netherlands

van Raamsdonk LWD, Ensink W, van Heusden AW, Vrielink-van Ginkel M, Kik C (2003) Biodiversity assessment based on cpDNA and crossability analysis in selected species of *Allium* subgenus Rhizirideum. Theor Appl Genet 107:1048–1058

van Raamsdonk LWD, Vrielink-van Glinkel M, Kik C (2000) Phylogeny reconstruction and hybrid analysis in *Allium* subgenus Rhizirideum. Theor Appl Genet 100:1000–1009

Vijn I, Vandijken A, Luscher M, Bos A, Smeets E, Weisbeek P, Wiemken A, Smeekens S (1998) Cloning of sucrose-sucrose 1-fructosyltransferase from onion and synthesis of structurally defined fructan molecules from sucrose. Plant Physiol 117:1507–1513

Vries JN, Wietsma WA, Jongerius MC (1992a) Linkage of downy mildew resistance genes Pd1 and Pd2 from *Allium roylei* Stearn in progeny of its interspecific hybrid with onion (*A. cepa* L.). Euphytica 64:131–137

Vries JN, Wietsma WA, Vries T (1992b) Introgression of leaf blight resistance from *Allium roylei* Stearn into onion (*A. cepa* L.). Euphytica 62:127–133

Wetli HA, Brenneisen R, Tschudi I, Langos M, Bigler P, Sprang T, Schürch S, Mühlbauer RC (2005) A glutamyl peptide isolated from onion (*Allium cepa* L.) by bioassay-guided fractionation inhibits resorption activity of osteoclasts. J Agri Food Chem 53:3408–3414

Wilkie SE, Isaac PG, Slater RJ (1993) Random amplified polymorphic DNA (RAPD) markers for genetic analysis in *Allium*. Theor Appl Genet 86:497–504

Yamashita K, Arita H, Tashiro Y (1999) Isozyme and RAPD markers linked to fertility restoring gene for cytoplasmic male sterile *Allium fistulosum* L. with cytoplasm of *A. galanthum* Kar. et Kit. J Jpn Soc Hort Sci 68:954–959

Yamashita K, Takatori Y, Tashiro Y (2005) Chromosomal location of a pollen fertility-restoring gene, *Rf*, for CMS in Japanese bunching onion (*Allium fistulosum*L.) possessing the cytoplasm of *A. galanthum* Kar. et Kir. revealed by genomic in situ hybridization. Theor Appl Genet 111:15–22

Zheng S-J, Khrustaleva L, Henken B, Sofiari E, Jacobsen E, Kik C, Krens FA (2001) *Agrobacterium tumefaciens*-mediated transformation of *Allium cepa* L.: the production of transgenic onions and shallots. Mol Breed 7:101–115

12 Garlic

P. F. Cavagnaro[1,2] and C. R. Galmarini[1,2,3]

[1] INTA, EEA La Consulta, CC8 (5567), La Consulta, San Carlos, Mendoza, Argentina
e-mail: crgalmarini@mendoza.inta.gov.ar
[2] Consejo Nacional de Investigaciones Científicas y Tecnológicas (CONICET)
[3] FCA, UNCuyo, Alte. Brown 500, Chacras de Coria (5505), Mendoza Argentina

12.1 Introduction

12.1.1 Center of Origin, Spread and Domestication

Origin

The center of origin of garlic has been unclear for a long time although garlic has been widely cultivated since ancient civilizations. Early taxonomists like Linnaeus (1973) and Don (1827) considered garlic to be a Mediterranean species. Later, the works of Regel (1887), DeCandolle (1886), Vavilov (1951) and Kazakova (1971) indicated that garlic was originally from Central Asia, and Vavilov (1951) proposed the Mediterranean area as a secondary center of origin. More recently, a number of fertile plants of primitive type were discovered in the northwest side of the Tien Shan Mountains of Central Asia by Etoh (1986) and Kotlinska et al. (1991); this fact supports the idea that this area was the primary center of origin of garlic. Pooler (1991) and Maaß and Klaas (1995) also suggested a Central Asian origin based on the existence of fertile plants and the most primitive morphotypes in this region and also because biochemical and molecular markers were most variable among Central Asian clones. Etoh and Simon (2002) have recently reviewed the origin of garlic and the history of its cultivation and the readers are referred to their article for a more comprehensive text.

It has been proposed that *Allium longicuspis* Regel is the progenitor species of garlic, because both species are morphologically and karyologically similar and *A. longicuspis* can be found in the western side of the Tien Shan Mountains, the center of origin of garlic (Vvedensky 1935; Etoh and Ogura 1984; Mathew 1996; Etoh and Simon 2002). Morphologically, these species are generally distinguished because garlic filaments are typically shorter than the perianths, while *A. longicuspis* has exserted anthers (Regel 1875; Fedschenko 1876; Vvedensky 1935). However, exserted anthers have also been observed in garlic clones with fertile flowers (Konokov 1953; Konvicka et al. 1978; Etoh 1983; Kotlinska et al. 1991), while *A. longicuspis* may not always have open flowers in which case anthers are not exserted from the perianths (Vvedensky 1935; Kazakova 1978). Etoh (1986) stated that it is difficult to distinguish between *A. sativum* and *A. longicuspis* and Etoh and Simon (2002) doubted that this feature should be used as a key for separating the two species. According to Etoh and Simon (2002), the results from the karyotype, isozyme, RAPD and AFLP analysis done by Etoh (1984), Pooler and Simon (1993a), Maaß and Klaas (1995), Hong (1999) and Ipek et al. (2003), demonstrate that variation for *A. longicuspis* lies within the range of variation observed among garlic accessions. These results have led some researchers to believe that *A. longicuspis* is not a separate species but rather a subgroup or subspecies of *Allium sativum* (Etoh and Simon 2002). Nonetheless, *A. longicuspis* is considered either the closest wild relative of garlic or its wild ancestor.

The possibility that *A. longicuspis* and garlic have a common wild ancestor has been considered. Mathew (1996), based on morphological similarities and the presence of a garlic odor, suggested that *Allium tuncelianum*, a fully fertile Turkish species could be the wild ancestor of garlic and *A. longicuspis*. So far, solid evidence supporting this hypothesis has not been presented.

Spread

Vvedensky (1935), Wendelbo (1971) and Kazakova (1978) considered the natural habitat of *A. longicuspis* as the region in Central Asia that extends from the Kopet Dag Mountains in the west to the Tien

Genome Mapping and Molecular Breeding in Plants, Volume 5
Vegetables
C. Kole (Ed.)

Shan Mountains in the east. Later, Mathew (1996) added eastern Turkey and the Caucasus to this region. Engeland (1991) referred to these areas as "the garlic crescent" or "the extended garlic crescent" for the broader region. He suggested that *A. longicuspis* was cultivated in the garlic crescent, or the extended garlic crescent region, by seminomads more than 10,000 years ago. Presumably, wild garlic was also extensively spread in this region (Engeland 1991). Engeland hypothesized that, since this area served as a trading route between China and the Mediterranean, it is possible that garlic was passed to more southern tribes and later introduced to Egypt, the Mediterranean and India. There are evidences indicating that garlic was known in Egypt (Tackholm and Drar 1954) and India (Engeland 1991) more than five thousands years ago. According to Burkill (1966), garlic has been consumed in India since ancient times, and from India it spread to the south east of Asia.

Engeland (1991) remarked on the importance of the Caucasus region regarding the spread of garlic to the North and the West. He suggested that from the Caucasus, garlic was introduced into Russia, the Ukraine and Eastern Europe. Through Turkey it reached the Mediterranean and southeastern Europe. According to evidence from writings, garlic was extensively consumed during Greek and Roman civilizations (Sturtevant 1919; McCollum 1976). It is believed that garlic was introduced to North and South America as well as to the subSaharan Africa from the Mediterranean region since most of the clones cultivated in Africa, North and South America belong to the Mediterranean type.

Apparently, neither garlic nor *A. longicuspis* was introduced early in China. Although Chia (530 – 550) reported that garlic was first introduced in this country during the second century BC, Laufer (1919) and Kitamura (1950) considered this hypothesis unlikely. Because the Chinese name of garlic indicates a western Chinese origin, it seems possible that garlic introduction was done from Central Asia through the western desert (Etoh and Simon 2002). According to Etoh and Simon (2002) after being introduced in China, garlic was introduced to Korea and from there to Japan. Interestingly, garlic rapidly became widely used in Korea, while it was little used in Japan for a long time (Etoh and Simon 2002).

12.1.2 Taxonomy

Garlic, *Allium sativum* L. is a diploid species in the *Allium* genus of the Alliaceae family, order Asparagales. Besides garlic, other economically important *Allium* species include onion (*A. cepa*), chives (*A. schoenoprasum*) and the tetraploid leek (*A. ampeloprasum*). Garlic intraspecific taxonomy has followed different classification criteria. Based on morphological data, Helm (1956) considered three botanical varieties (*A. sativum* L. var. *sativum*, var. *ophioscorodon* and var. *pekinense*) but Jones and Mann (1963) argued that garlic accessions could be grouped in just two of Helm's varieties (var. *sativum* and var. *ophioscorodon*) since many garlic clones showed a combination of features of the later varieties. They further proposed that these should be designated as horticultural groups rather than botanical taxa. One important feature frequently included in several garlic classifications is the bolting phenotype. Zagorodskij (1935), Kuznetsov (1954), Alekseeva (1960) and Komissarov (1964, 1965) separated garlic cultivars into two subspecific groups: bolting and nonbolting. Komissarov (1964, 1965) further classified the bolting types according to their geographical distribution (Mediterranean, Central Asian and Sino-Mongolian) and suggested that the nonbolters evolved from the corresponding bolting types. Nonbolting clones from the Mediterranean would have evolved from garlics cultivated for many centuries in a region including the Caucasus, the Balkans and the Crimea. Some researchers observed that the bolting habit of some cultivars was not a stable trait under their climate and field conditions (Kazakova 1978; Pooler and Simon 1993b) and disagreed with classifications based on the bolting phenotype criteria (Kazakova 1978). However, according to Etoh (1985), complete-bolting types always produce scapes and non-bolting types never form scapes, whereas incomplete bolters show an intermediate behavior and can vary, to some extent, in their bolting response. It is possible, therefore, that the garlic cultivars varying in the bolting habit, observed by Kazakova (1978) and Pooler and Simon (1993b) correspond to the intermediate (incomplete-bolting) type of garlic. Moreover, diversity assessments of garlic collections using RAPD (Bradley et al. 1996; Alzahim et al. 1997) and AFLP (Ipek et al. 2003) markers have separated garlic accessions according to their bolting phenotype, thus supporting the early classifications (Zagorodskij

1935; Kuznetsov 1954; Alekseeva 1960; Komissarov 1964, 1965) into two distinct genetic groups.

Engeland (1991) proposed a classification based on cultivar bolting phenotype combined with morphological features of the scape, inflorescence, bulbils and maturation of bulbs. Burba (1993) described Argentinean garlic as non- or incomplete-bolting types, like the Mediterranean type of garlic.

Messiaen et al. (1993) and Lallemand et al. (1997) used morphophysiological characters and isozyme profiles to classify garlic clones. Six groups of the western world were distinguished: one group comprising Eastern European bolting clones and five Mediterranean groups that included one bolting, two incomplete-bolting and two nonbolting types. Asian clones were not clearly classified but they noted different isozyme profiles for this group of cultivars.

Maaß and Klaas (1995) assessed garlic intraspecific diversity using isozyme and RAPD analysis. They classified garlic clones from the old world into four subgroups; the sativum group from the Mediterranean, the ophioscorodon group from middle and eastern Europe, the longicuspis group from Central Asia and the subtropical group from southern Asia. The longicuspis group includes *A. longicuspis* and the East Asian subgroup pekinense. Clones from this group are bolting type and have coiling scapes and some produce, to a varying extent, fertile flowers. This is the most primitive group and it is believed that the other three groups originated from the longicuspis group. In the ophioscorodon group, all the plants corresponded to bolting type and possessed coiling scape but the flowers were sterile. The sativum group included bolting, non- and incomplete-bolting phenotypes and these subgroups were discriminated by the isozyme and RAPD marker analysis. It should be mentioned that only the *longicuspis* group from Central Asia had any fertile accessions.

12.1.3 Habitat

According to Vvedensky (1935) and Mathew (1996), garlic's natural habit is in "gullies shaded during the day" or "rocky valleys and river flats", 1,350–2,100 m, respectively. Etoh and Simon (2002) point out that since garlic and its presumed closest wild relative *A. longicuspis* develop bulbs during the summer, it is likely that hot and dry summers were typical in the centers of origin of both species. The weather in Central Asia supports these conditions. Moreover, today's cultivated garlic varieties, as well as *A. longicuspis* plants, grow well under relatively dry conditions and intense sunlight. However, some regions in Central Asia (e.g., the Dzungaria Basin desert, described by Regel (1875) as part of the natural habitat of garlic) as we know them today, seem too dry to support garlic growth. Hence, it is possible that the natural habitat of garlic in Central Asia was restricted to gullies, rocky valleys or riverbeds that contained enough soil moisture for the plant growth. Alternatively, if in earlier times this region received more rainfall, a more extensive distribution of wild garlic in these mountains can be imagined (Etoh and Simon 2002).

12.1.4 Economic Importance

Garlic is a widely cultivated crop with a current worldwide production of 14 million tons cultivated in 1.13 million hectares (FAOSTAT data 2005). Most of the garlic is produced in Asia (87%), China and India being the largest producers, collectively accounting for 78% of the production. It is interesting to point out that the yield in China (16.6 ton/ha) is four times higher than the yield observed in India (4.16 ton/ha) and higher than the world mean (12.4 ton/ha). The yield differences exhibited by the two countries most likely reflect differences in technological resources and aspects related to the management of the crop, rather than to differences in genetic background and performance of the cultivars used. Europe, America and Africa account for 5.4, 4.8 and 2.8% of the production, respectively. Since 1970, world garlic production has increased more than four times while the production area increased slightly less than three times, thus, also indicating an increment in yield (Fig. 1). Worldwide, the average yield almost doubled during this period (from 6.9 to 12.5 t/ha) and this was mainly due to the increase exhibited by the major producers; China went from 9.2 to 16.6 t/ha. Although garlic world production history shows a steady increase during the last 35 years, the most dramatic increment in the total production and yield has been observed in recent years (Fig. 1). Considering the time period from 1999 to 2004, total garlic production and yield increased 45 and 27%, respectively (FAOSTAT data 2005). In the same period, garlic consumption raised from 1.62 to 2.21 kg/capita representing a 36% increment. Currently, there is great variation among

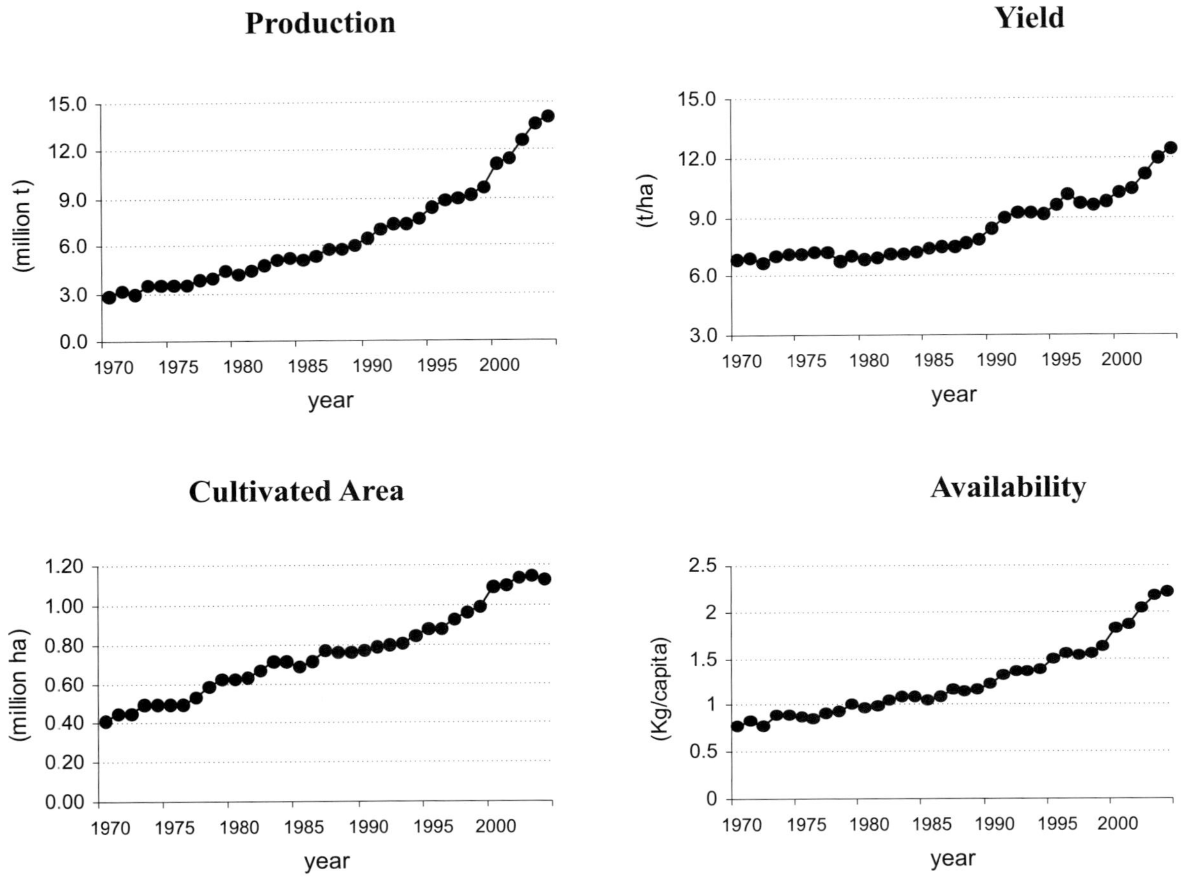

Fig. 1. World garlic production, cultivated area, yield and per capita availability from 1970 to 2004 (from FAO Yearbook Production Statistics)

countries for garlic yield, ranging from 1 t/ha in the case of Lithuania to almost 50 t/ha for the Netherlands (values are means of the years 2002, 2003 and 2004).

Different types of garlic are grown and consumed around the word. Growers' and consumers' preferences vary with geographic region, cultural background and end-use. Although fresh intact bulbs is the usual form of garlic traded in commerce, fresh garlic preparations (e.g., chopped, sliced or minced) as well as dehydrated forms of condiments are common in several countries. For the latter, clones with high dry matter are preferred. In parts of Asia and North Africa garlic leaves are marketed, whereas colored garlic bulbs are appreciated in some specialty markets (Simon and Jenderek 2003). Another specialty market resulted from the nutraceutical industry where a number of garlic preparations (e.g., powder, tablets, water, oil or ethanol-based formulations) are marketed, as a significant source of organosulfur compounds, such us allicin, that are often correlated with blood antiplatelet activity.

12.1.5 Morphology

Garlic morphology and factors affecting the plant growth and bulb development have been extensively discussed (Mann 1952; Mann and Minges 1958; Takagi 1990). As most edible *Allium*, this species is typically cultivated to produce underground bulbs for human consumption. Garlic bulbs consist of cloves, usually in numbers from 4 – 12, surrounded by a few dried sheaths that correspond to the basal part of the plant foliage. Botanically, a clove is a modified axillary bud originated in the basal part of inner foliage leaves. Several cloves can develop in the axis of a foliage leaf. A garlic clove consists of an outer thin protective, often colored, leaf surrounding an inner thickened storage leaf. The basal central part of the storage leaf contains a vegetative bud originated from a modified (flattened) very short stem called basal plate. The vegetative bud carries a larger leaf (sprout leaf) and several leaf primordia that surround the apical meristem. The sprout leaf and inner foliage leaves will elongate through the storage leaf to finally protrude above it (sprouting), after growth inhibitors are broken down, typically as a consequence of the exposure of the garlic cloves to cold temperatures. The length ratio between the sprout leaf and the storage leaf has been used as an index for monitoring rupture of dormancy in garlic cloves (Burba et al. 1993). The inner leaves will produce the plant foliage. Leaves have opposite and alternate orientation and their basal part overlap, forming a pseudostem, whereas the leaf blades are not overlapped. Roots originate from the basal plate.

Some garlic clones have a tendency to produce a flower stalk (referred to as bolting, stalking or hard neck type) whereas others have lost this ability (nonbolting, nonstalking or soft neck) and some exhibit an intermediate response (incomplete-bolting). For the bolting types, the apical meristem of the scape differentiates floral initials interspersed with bulbils primordia (Kothari and Shah 1974; Etoh and Ogura 1977; Etoh 1985; Kamenetsky and Rabinowitch 2001). There is great variation among garlic clones for scape length and morphology, as well as for number, size, color and shape of bulbils (Simon and Jenderek 2003). Like onion and other *Allium*, garlic flowers are perfect and consist of 6 petals, 6 anthers and 3 locules of 2 ovules each (Simon and Jenderek 2003). However, Etoh and Simon (2002) point out that even garlic bolting cultivars may not develop fully mature flowers.

12.1.6 Genome Size and Karyotype

The nuclear genomes of many *Allium* are among the largest known among all eukaryotes (Labani and Elkington 1987; Ori et al. 1998). Garlic has a DNA content of 32.7 pg per 2C nucleus (Ranjekar et al. 1978). Comparably, this is slightly smaller than the onion and hexaploid wheat genome and ~32 and 6 times larger than rice and maize, respectively (Bennet and Smith 1976). No evidence of polyploidy as a means for explaining the extremely large garlic genome has been presented. The garlic genome contains a high amount of repetitive DNA and an overall GC content of 36.9% (Kirk et al. 1970). Considering only the coding regions, GC content was 30.5 and 46% for introns and exons, respectively (Kuhl et al. 2004). Interestingly, the GC contents of garlic and onion (32%; Kirk et al. 1970) nuclear genomes are among the lowest among all angiosperms (Kirk et al. 1970; Matassi et al. 1989).

Garlic is a diploid species ($2n = 2x = 16$). Extensive variation has been observed at the structural level of chromosomes in the garlic karyotype. Battaglia (1963) described significant differences in overall chromosome length, arm ratio and position of secondary constrictions for a collection of European clones. In some cases variation was associated with the geographic origin of the cultivars. For example, the European clones from Battaglia (1963) were remarkably uniform for the number of nucleolar chromosomes per haploid set, whereas Verma and Mittal (1978) and Konvicka and Levan (1972) observed variation, for this aspect, within Indian and Swedish garlic collections, respectively. According to Simon and Jenderek (2003), a preponderance of the "basic" karyotype (2 sets of chromosomes) was demonstrated for bolting Central Asian clones by Hong et al. (2000a) whereas incomplete-bolting cultivars from the Iberian Peninsula usually exhibited "nonbasic" karyotypes (e.g., heteromorphic pairs) and also variations in centromere location and/or satellites.

12.1.7 Nutritional Composition and Nutraceutical Value

Fenwick and Hanley (1990) reviewed the nutritional composition of garlic and other edible *Allium*. The composition of a garlic bulb varies greatly depending on many factors: cultivars, agronomic practices, climate, soil fertility and postharvest storage conditions determine the quality and intensity of garlic flavor as well as its nutritional and nutraceutical value. Thus, values given herein are only reference values. Treutner et al. (1978) found dry matter contents of 31–56% (w/w) among 20 garlic clones while Cantwell et al. (2005) reported values of 32–45% in a collection of 198 accessions grown under California weather. According to Fenwick and Hanley (1990), the protein content was 3.5–6.2, 2.6 and 1.3% for bulbs, leaves and flower and scapes, respectively, while the mean carbohydrate contents were 29.1, 9.5 and 20.1%, respectively. Free sugars and polysaccharides of the fructan type account for part of garlic carbohydrates. Significant amounts of pectin are also present in the skins of garlic and onion. The fat content was 0.2–0.3% for garlic bulbs, scapes and flowers, while leaves had twice that concentration (0.5%). Of the ash fraction (1.3%) of bulbs, potassium, phosphorous and calcium were the most abundant elements. According to Lupea and Vranceanu (1972), garlic was found to contain 42 µg of molybdenum per 100 g fresh weight, ranking below only common bean as a dietary source of this mineral.

12.1.8 Organosulfur Constituents

Organosulfur compounds are responsible for *Allium* flavor, the lachrymatory effect of crushed onions and many of the health benefits attributed to garlic and onion consumption. Several authors have extensively studied the biochemistry of *Allium* flavor over the last 33 years since alliin was first reported. For a comprehensive account on *Allium* organosulfur chemistry see Block (1992). Briefly, cells of the *Alliaceae* contain, in their cytoplasm, the odorless flavor precursors *S*-alk(en)yl cysteine sulfoxides (ACSOs) which, when hydrolyzed by the enzyme alliinase, give rise to the flavor and pungency characteristic of *Allium* plants. Alliinase is a piridoxal-5′-phosphate-dependent enzyme located in the vacuole. Thus, when tissues are disrupted, the ACSOs are broken down by the enzyme to produce thiosulfinates, ammonia and pyruvic acid. The production of thiosulfinates takes place through intermediate compounds, which are highly reactive sulfenic acids that rapidly condense, in pairs, to form thiosulfinates. The thiosulfinates are responsible for the flavor and pungency of fresh onions, garlic and other alliums. There is variation for pungency in garlic germplasm (Cavagnaro et al. 2005b; Natale et al. 2005). Four different ACSOs have been found: *S*-methyl-L-cysteine sulfoxide (MCSO), *S*-propyl-L-cysteine sulfoxide (PCSO), S-(2-propenyl)-L-cysteine sulfoxide (alliin or 2-PECSO), S-(1-propenyl)-L-cysteine sulfoxide (1-PECSO). *Allium* species vary in the type and concentration of flavor precursors that they contain, thus giving rise to different combinations and concentrations of thiosulfinates, which determines, to a great extent, the characteristic flavor of the different species. In garlic bulbs, alliin is the predominant flavor precursor while allicin is, consequently, the major thiosulfinate accounting for 60–95% of the total thiosulfinates (Block et al. 1992). Many bioactive properties have been attributed to allicin, which are mostly related with plant defenses against pathogens, e.g., antibiotic (Cavallito and Bailey 1954) and antifungal (Singh et al. 1990), and its nutraceutical properties (Briggs et al. 2000). It should be pointed out that, although thiosulfinates are the predominant organosulfur compounds in freshly crushed garlic, they can, after some time at room temperature, participate in a cascade of nonenzymatic rearrangements to produce a wide range of other organosulfur compounds (Randle and Lancaster 2002). More than 70 sulfur compounds have been identified in *Allium* (Randle and Lancaster 2002).

A "functional food" or "nutraceutic" can be defined as the food that encompasses potentially healthful compounds that may provide a health benefit beyond the traditional nutrients it contains. A wide range of health benefits has been attributed to garlic consumption and many of these have been substantiated by modern research: antihypertensive (Suetsuna 1998), hypolipidemic (Augusti and Mathew 1973), hepatoprotective (Wang et al. 1998), antidiabetic (Kumari et al. 1995), antitumor (Kyo et al. 1998), cholesterol reducing (Bordia et al. 1975) and platelet antiaggregatory activity (Bordia et al. 1996). Although it has not been fully substantiated, it is believed that organosulfur compounds are responsible for most of garlic's nutraceutical effects. There is variation among germplasm (Araniti et al. 2005). A significant positive correlation among garlic-induced in vitro antiplatet activity (IVAA), allicin and pyruvate content (pungency) was observed in garlic F_2 segregating families (Cavagnaro et al. 2005a) and between allicin and piruvate in a collection of relatively unrelated garlic clones (Cavagnaro et al. 2005b), indicating that it will be difficult to develop garlic populations with low pungency and high IVAA. Similar results were obtained in onion varieties and onion F_3 families where pungency was positively correlated with IVAA (Goldman et al. 1996; Galmarini et al. 2001; Havey et al. 2004).

12.2 Garlic Breeding and Construction of the First Genetic Maps

Garlic as a crop, has a long history of asexual reproduction. Most cultivated garlic clones do not flower and those flowering are nearly or completely sterile, since very few fully mature flowers are produced (Simon and Jenderek 2003). The lack of sexual reproduction in garlic has restricted the improvement of the crop through conventional breeding. In the last decades, researchers have made progress towards the understanding of the factors involved in garlic sterility and fertility and recently, systems for garlic seed production have been developed (see for review Etoh and Simon 2002; Simon and Jenderek 2003). With routine seed production underway, inheritance studies and the construction of linkage maps for this crop, are now possible.

Very recently, Ipek et al. (2005) and Zewdie et al. (2005) published the first garlic genetic maps, using different marker systems on F_2 families with different genetic background.

Ipek et al. (2005) used two garlic populations generated by self-pollination of relatively unrelated plants to construct two low-density linkage maps, based on AFLP markers.

AFLPs have proved to be useful markers for the construction and saturation of genetic maps in many crops, since they provide a large number of scorable bands with relatively few reactions. Due to the huge garlic genome, the original AFLP technique (Vos et al. 1995) yielded too many bands per gel, making the data analysis ambiguous (Ipek et al. 2003). Therefore, for the construction of garlic AFLP-based linkage maps, a fourth selective nucleotide was added to all *Eco*RI selective primers to reduce the number of amplified bands (Ipek et al. 2005). Similar modifications were made for the construction of AFLP-based maps in two other *Allium* with big genomes: onion (van Heusden et al. 2000) and leek (Smilde et al. 1999).

According to Ipek et al. (2005), of the total 681 polymorphic AFLPs (combined data from the two families), 548 (80.5%) segregated according to the expected 3:1 ratio and 126 (18.5%) showed a 15:1 ratio, typical for duplicated regions. Only the bands fitting a 3:1 segregation ratio were used for the linkage analysis excluding those showing segregation distortions (~20%). Thus, the maps were constructed upon 305 and 243 polymorphic AFLP bands from the two F_2 families (designated as MP1 and MP2), which consisted of 53 and 54 individuals, respectively. For MP1, 175 markers (57.4%) were mapped to 19 major linkage groups containing 5 to 21 markers (Fig. 2) while 37 markers (12.1%) were assigned to 12 minor groups of less than 5 markers and the remaining 93 AFLPs (30.5%) were unlinked. For MP2, 101 (41.6%) and 40 (16.5%) markers were assigned to 12 major groups consisting of 5 to 25 markers (Fig. 3) and 12 minor groups (<5 markers), respectively. One hundred and two (41.9%) markers remained unlinked. MP1 and MP2 maps spanned 1166 and 862 cM of the garlic genome, respectively, and the average distance among markers was 5.4 cM for MP1 and 6.0 cM for MP2. A minimum LOD score of 3.0 (α = 0.5) was considered for linkage analysis of both families.

According to Ipek et al. (2005) the relatively small population size used, most likely account for the large number of linkage groups and the high percentage of unlinked markers observed in both families. The authors point out that large garlic populations are difficult to attain since a self-pollinated garlic plant typically yields 10 to 50 seeds. For comparison of both maps, comigrating AFLP bands that segregated in both families were identified in the polyacrylamide gels and in the linkage maps. Only a few common segregating bands were identified between MP1 and MP2, reflecting important differences regarding the genetic background of the self-pollinated plants that originated both families. In other words, the plants that originated both families were distantly related. Among the markers in common those that were linked in one family were always linked in the other family (Figures 2 and 3).

Also recently, Zewdie et al. (2005) published a garlic genetic map based on several PCR-based markers mainly derived from onion expressed DNA regions reported previously (King et al. 1998; Kuhl et al. 2004). The authors used an S_1 progeny of 84 plants generated by self-pollination of a single fertile plant from garlic clone PI 540316 (USDA-ARS, Western Regional Plant Introduction Station, Pullman, WA, USA) to map 36 molecular markers and a male-fertility (*Mf*) trait into 9 linkage groups covering 415 cM of the genome.

Their approach for identifying garlic molecular polymorphism consisted in designing nested primers based on sequences from onion ESTs (Kuhl et al. 2004) and onion cDNAs (King et al. 1998) and using these primers to amplify garlic genomic regions from the parental plant to search for polymorphisms. Thus, single PCR amplicons from the parental line were excised from the agarose gels, cloned, and several colonies with inserts were sequenced. Sequences were aligned and compared, revealing a total of 83 SNPs and 8 indels that were further tested among S_1 plants. Also a number of decamers and onion SSR primers (Kuhl et al. 2004) were tested on the S_1 progeny. Male-fertility was evaluated based on pollen viability (Jenderek and Hannan 2004) from clonally propagated S1 plants in three growing seasons. Plants with no vi-

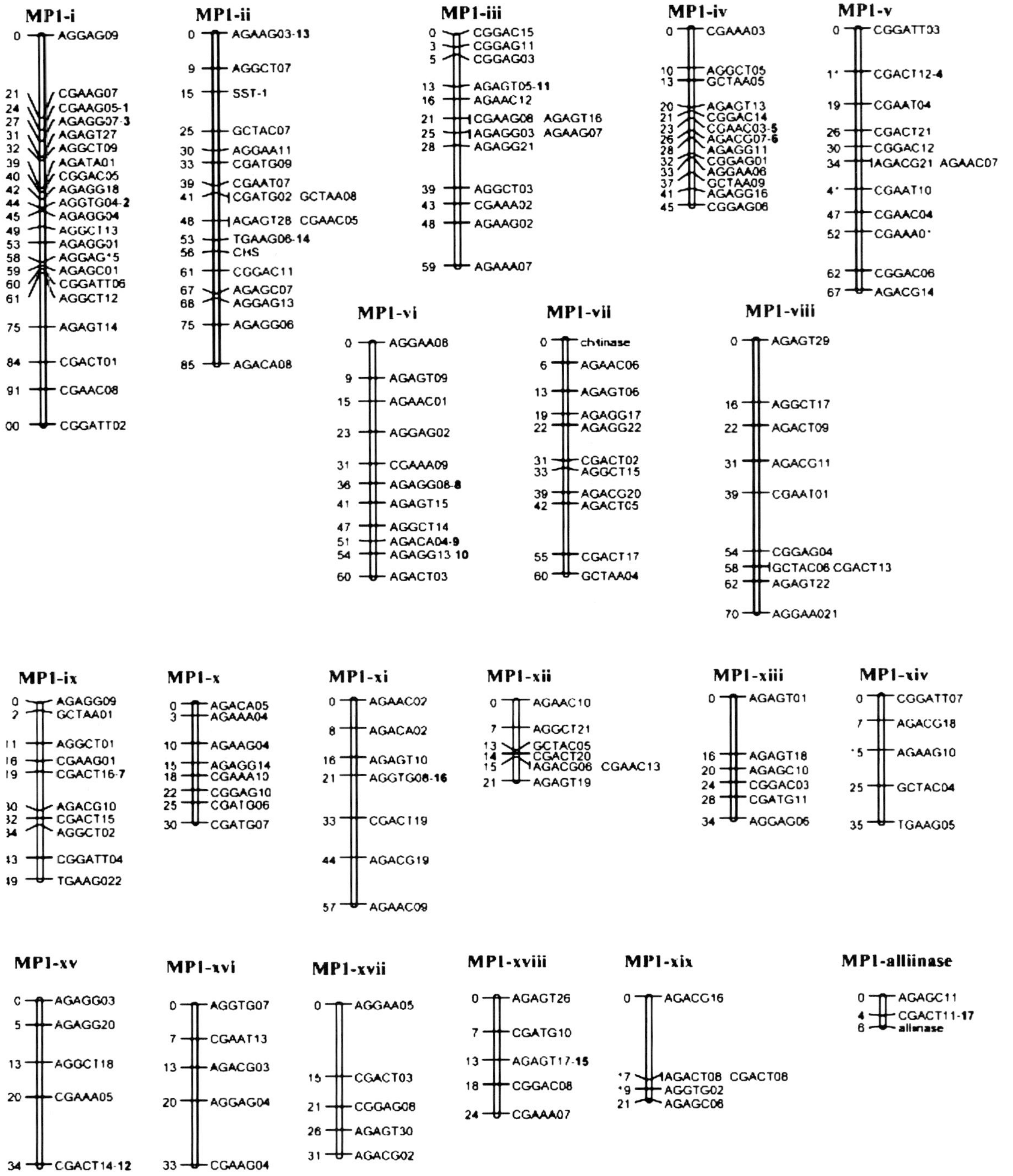

Fig. 2. Linkage groups in the genetic map of *Allium sativum*, based on the MP1 family. Map distances are in centiMorgans. Numbers in *boldface* are common markers segregating in both MP1 and MP2 (from Ipek et al. 2005)

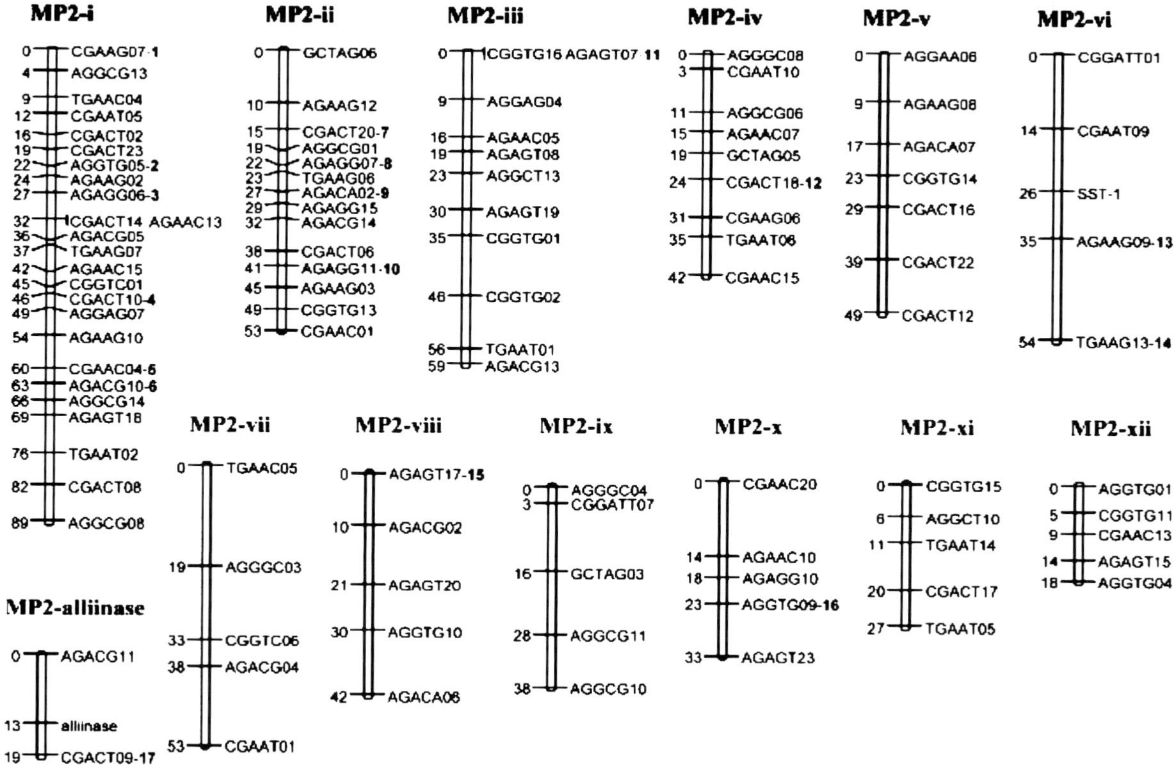

Fig. 3. Linkage groups in the genetic map of *A. sativum* based on the MP2 family. Numbers in *boldface* are common markers segregating in both MP1 and MP2 (from Ipek et al. 2005)

able pollen were scored as male-sterile whereas plants with one or more viable pollen grains were considered male-fertile.

According to Zewdie et al. (2005), among the characters evaluated, only 52 molecular markers (namely, 30 SNPs, 18 dominant amplicons, 1 SSR and 2 RAPDs) and the *Mf* trait segregated among S_1 progenies. Of these, 42 (79%) exhibited significant segregation distortions from the expected 1:2:1 and 3:1 ratios, for codominant and dominant loci, respectively. Male-fertility segregation (52 fertile: 32 sterile) also deviated from a normal segregation considering a single-gene model. Linkage analysis at logarithm of the odds (LOD) 5 and maximum distance of 30 cM, included all 53 loci. The resulting map consisted of 9 linkage groups of 2 to 8 markers with an average distance between loci of 15 cM (Fig. 4). Sixteen (30%) loci remained unlinked. The *Mf* locus was loosely linked (~23 cM) to one SSR and one SNP and the three loci conformed linkage group E (Fig. 4).

It is remarkable the relatively low percentage of markers showing segregation distortion reported by Ipek et al. (2005) (~20%) as compared with the almost 80% found by Zewdie et al. (2005). In the former study, most of the markers with distorted segregations fitted a 15:1 ratio, typical of duplicated loci and therefore they did not attempt to map these markers. Whether the markers exhibiting segregation distortion found by Zewdie et al. (2005) fitted a ratio for duplicated genes was not indicated. They attributed the high percentage of segregation-distorted-loci to the loss of progeny plants due to linkage with deleterious alleles. According to the authors, since garlic has been asexually propagated for millennia, it is possible that mutations have produced deleterious alleles that are maintained in the heterozygous state, thus after sexual reproduction, plants homozygous for deleterious mutations would show reduced vigor or perish. Although this hypothesis is reasonably possible, in this case, comparable percentages of markers with and without segregation distortions could be expected in the progenies used by Ipek et al. (2005). However, one can speculate that garlic clones have accumulated varying amounts and types of deleterious mutations through history and since the families used in both studies were generated from different garlic

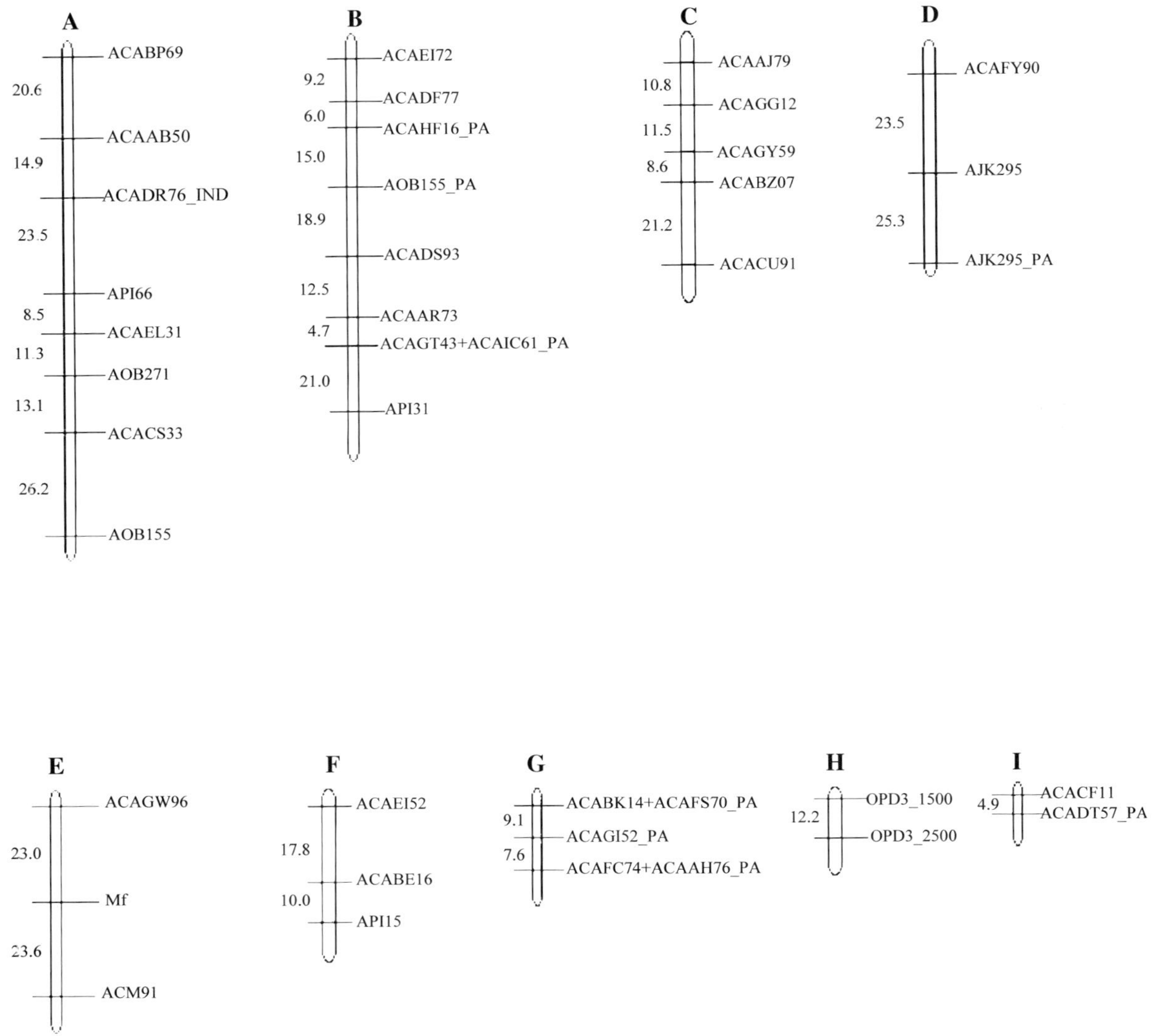

Fig. 4. Garlic genetic map by Zewdie et al. (2005). Map distances are in cM and locus designation was after their respective primers name and corresponds to onion ESTs (AC prefixes) or SSRs (ACM91) as described by Kuhl et al. (2005) or previously mapped cDNAs (AOB or API prefixes) as described by King et al. (1998). The indel marker has extension of IND and presence/absence of band markers have extensions of PA (from Zewdie et al. 2005)

genotypes it is possible that meiotic recombination in PI 540316 (Zewdie et al. 2005) resulted in a family with a higher number of lethal recombinants.

12.3 Mapping Gene-Specific Markers

Besides the anonymous AFLPs, gene-specific markers for alliinase, chitinase, sucrose 1-fructosyltransferase (*SST*-1) and chalcone synthase (*CHS*) were mapped (Ipek et al. 2005). Since in garlic and other species these enzymes are encoded by gene families, rather than by single genes, the authors selected and scored one dominant marker from each gene (to avoid scoring ambiguity), thus mapping one copy of the respective gene families. In MP1 markers for the four genes were mapped, being *SST*-1 and *CHS* in the same linkage group, whereas in MP2 only alliinase and *SST*-1 could be mapped (Figures 2 and 3). Interestingly, two AFLP markers in common for both families were linked to gene-specific markers: one marker tightly linked (<10 cM) to alliinase and one marker loosely linked (10–30 cM) to *SST*-1 (Figures 2 and 3).

The map of Zewdie et al. (2005) was mostly based on PCR-based markers developed from onion ESTs and cDNAs. Thus, many of the mapped loci represent putative expressed regions of the garlic genome. Sequences from onion ESTs and cDNAs (King et al. 1998; Kuhl et al. 2005) used by Zewdie et al. (2005) were compared with the National Center for Biotechnology Information (NCBI) sequence databases. Significant sequence homology with known genes was observed for a number of the markers mapped by Zewdie et al. (2005). For example, loci *ACADR76_IND*, *ACABE16* and *ACACF11* in linkage groups A, F and I showed significant hits with a putative α-glucosidase, a regulatory subunit of a phosphate protein and an adenylate kinase gene, respectively. The loci *ACAAJ79* and *ACAGG12*, linked at 10.8 cM in group C (Fig. 4), showed high homology with a sulfate reductase and a glycerol-3-phosphate dehydrogenase gene, respectively. However, a number of the sequences blasted to the NCBI database either had no significant homology with previously characterized sequences or had homology with hypothetical proteins with unknown function.

The *Mf* locus is the first phenotypic trait mapped in garlic. As more gene-specific markers and traits of agronomic interest are included in garlic genetic maps, their utility in marker-assisted selection and map-based cloning will be enhanced, becoming a powerful tool for breeding of this species.

12.4 BAC Libraries and Physical Mapping

With two garlic maps including anonymous and gene-specific markers recently published, it is of interest to anchor linkage groups to specific chromosomes. One way of approaching this is through in situ hybridization (ISH) techniques using as probes detecting markers in each linkage group. However, physical mapping of unique sequences is limited to the size of the probe used. According to Jiang and Gill (1994) probes shorter than 10 kb targeting single copy genomic regions are difficult to map in plants since they give weak or undetectable signals. The construction of a garlic BAC library would provide plenty DNA probes suitable for ISH analysis. Lee et al. (2003) constructed two low coverage (∼0.08×) BAC libraries from a South Korean cultivar as a mean for developing fluorescent in situ hybridization (FISH) probes for garlic chromosomes. The construction of a full-coverage library in garlic is hampered by its huge genome size. Nevertheless, if a few clones carrying markers from the genetic garlic map were identified in the BAC library constructed by Lee et al. (2003) they could be used to anchor linkage groups to specific chromosomes. Moreover, clones from an onion partial BAC library (0.32× coverage) (Suzuki et al. 2001) may also be useful for physical mapping in garlic since they are close relatives and some degree of synteny is expected between both species.

12.5 Genome Organization in Garlic

The relatively large number of AFLP markers (18.5%) segregating into a 15:1 ratio as observed by Ipek et al. (2005) suggests that duplicated regions may be common in the garlic genome. These results are in agreement with those obtained by King et al. (1998) in onion, where 21% of the RFLP loci were duplicated, and those by Jones and Rees (1968) and Ranjekar et al. (1978) suggesting that duplications account for a large part of the *Allium* genome. It appears that duplications of genes are, as well, frequent in the garlic genome. Cavagnaro et al. (2003) observed 4–8 bands in garlic alliinase Southern blot hybridizations and, later Cavagnaro et al. (2004) sequence-characterized the 5′ end of 12 putative alliinase alleles from one garlic accession, indicating extensive duplications for this gene family. Ipek et al. (2005), based on sequence data from cloned amplicons of alliinase, chitinase and *SST-1* genes, found six, five and three-sequence variants, respectively.

Stack and Comings (1979) based on Cot reassociation kinetics data concluded that the onion genome consists of middle-repetitive sequences interspersed among single copy regions. King et al. (1998) proposed that intrachromosomal tandem duplications, and not polyploidization followed by diploidization (paleopolyploidy), played an important role in the evolution of large-genome *Allium* species. In onion, a fair amount of cytological and genetic evidence from RFLP-based linkage analysis supports this conjecture. In garlic sufficient evidence supporting either hypothesis is lacking. Mapping different members of selected gene families may provide insights into the structure of the garlic genome. For this, different copies must be previously characterized and their sequence compared to identify DNA polymorphisms (e.g., in noncoding regions) that could be used as mapping markers in segregating populations.

King et al. (1998) proposed two possible mechanisms that can account for the tandemly duplicated RFLPs observed in onion: unequal crossing over during meiotic recombination (Smith 1976) and duplication of coding and noncoding regions through a retroviral-like mechanism (Vanin 1985). In the latter mechanism, newly duplicated genes, or at least some of them, must represent intronless copies since, presumably, processing of introns takes place before reinsertion in nearby genomic regions. The fact that the 12 alliinase sequence-variants found by Cavagnaro et al. (2004) in one garlic clone contained introns contradicts the retroviral-like mechanism hypothesis as the major evolutionary force responsible for gene duplications in the garlic genome. However, more evidence is necessary to determine if either of these mechanisms have a preponderant roll in the process of gene duplication and genome expansion in garlic. Hence, for now, retropositional duplications cannot be ruled out in garlic. Moreover, active retrotransposons are ubiquitous elements in the genomes of other monocots (e.g., in the Gramineae family) (Vicient et al. 2001) and *Copia*-like retroelements have been reported in onion (Pearce et al. 1996), a close relative of garlic.

12.6 Interspecific Hybrids

The opportunity of producing interspecific hybrids between garlic and other *Allium* became feasible with the discovery of fertile garlic clones. Hybrids between onion (*Allium cepa* L.) and fertile garlic were obtained by Ohsumi et al. (1993) by conventional crossing followed by embryo rescue. They reported a high level of sterility in the hybrids as indicated by the only 2% of pollen viability obtained and the complete lack of seed set. Etoh (1984) was able to cross a sterile *A. longicuspis* with pollen from a fertile garlic. The few interspecific hybrids obtained were also sterile. Sugimoto et al. (1991) succeeded in obtaining interspecific hybrids between the tetraploid ($2n = 2x = 32$) leek (*A. ampeloprasum*) and the diploid garlic. Several interspecific triploids and aneuploids close to triploids were recovered. Although some tetraploids and diploids were also obtained from the same cross, it is possible that these are not hybrids (Etoh and Simon 2002).

The utility of the interspecific hybrids obtained so far has been restricted due to the lack of seed production. However, the presence of viable pollen in onion-garlic hybrids, even in such small amount, raises the possibility of introgressing part of the garlic genome into the onion genome by backcrossing these hybrids with onion (Etoh and Simon 2002).

12.7 Genetic Transformation

Genetic transformation is, potentially, a valuable tool for improving garlic, a crop that, only until recently, has been closed to improvement by classical breeding methods due to its lack of sexual reproduction. However, stable *Allium* transformation is not straightforward and garlic, onion and leek are considered recalcitrant to genetic transformation (Eady et al. 1996). A limited number of embryogenic calli attained, its extremely large genome and the presence of strong endogenous nucleases appear to affect garlic transformation and regeneration systems. Nonetheless, stable genetic transformation of garlic has been achieved by particle bombardment (Sawahel 2002) and *Agrobacterium*-mediated (Kondo et al. 2000) gene transfer systems. In both studies, the reporter gene *uid*A was used to transform garlic plants.

12.8 Progress of Garlic Breeding and Future Scope of Works

The development of systems for the production of relatively large amount of seeds is a major accomplishment towards the domestication of this species. In fact, there is no evidence indicating that sexual reproduction and selection were ever used by garlic growers throughout history, thus, although garlic has been cultivated since ancient times, breeding the crop has just begun (Simon and Jenderek 2003). Breeding progress, in aspects related to the seed production system, has already been reported and future goals for garlic improvement are being considered. According to Simon and Jenderek (2003), garlic responds to selection for reduced numbers of small bulbils and, after a few seed-propagated generations, bulbil-free strains can be obtained in some genetic backgrounds. Likewise, they report improved pollen fertility in the first progeny derived from sexual reproduction. Marker-assisted selection (MAS) for this trait is possible since molecular markers associated with pollen fertility have been identified (Etoh 1985; Hong et al. 1997, 2000b). Seeds vigor and size and seed germination rates can be dramatically improved after two or three cycles. Concerning the latter, Inaba et al. (1995) and Jenderek (1998) found that after two or three generations of selection for improved seed germination, germination rate increased from 10–35% to 65–93%. Simon and Jenderek (2003) indicated that seed yield can also be significantly improved and suggested that up to 1200 seeds can be produced per plant, if all the flowers in a garlic inflorescence set the maximum possible number of seeds.

With routine seed production underway, combining desired traits found in different garlic clones is now feasible. Previously evaluated and characterized garlic collections, by means of morphological, isozyme and molecular markers, as well as regarding specific traits of interest (e.g., flavor, nutraceutical properties, resistance to pathogens) will provide a source of classified genetic variability, useful in the planning of garlic breeding programs. Among the breeding goals for this crop, besides those related to the improvement of the seed production system, are typical bulb traits, including clove number and color, bulb size and shape, color of outer and inner bulb scales, time of bulb maturity, yield and bulb storage parameters. Genetic resistance to viruses, fungi and nematodes are also very important breeding objectives. Flavor and health-enhancing attributes are of particular interest for the consumer and the processing and the nutraceutical industry.

According to Simon and Jenderek (2003), as garlic breeding proceeds, two different objectives can be pursued: development of new clones for asexual reproduction or development of seed-propagated cultivars. Although the latter is tempting since virus-free seed-propagated cultivars, and even hybrids, can be imagined, its likelihood of success is uncertain and, according to Simon and Jenderek (2003) it will likely be this way for several years. Instead, the development of new clones, by crossing plants with specific traits of interest, seems more feasible in the near future. By this means, a large number of recombinant phenotypes can be obtained, considering the broad diversity observed in garlic collections. If more gene markers and traits of agronomic importance are added to the linkage map of Ipek et al. (2005) and/or new, more detailed maps are constructed, molecular markers linked to traits or genes of interest can assist with the selection process of garlic breeding programs.

Genetic transformation of selected garlic cultivars, with genes conferring specific traits of interest may create new phenotypes useful for different purposes of the grower, the consumer and the nutraceutical industry. For example, if the positive association between IVAA and pungency observed in garlic and onion is due to linkage, and not pleiotropy, transgene technology can be used to "break" this linkage, for developing mild (low pungency) varieties with high antiplatelet activity.

References

Alekseeva MV (1960) Cultivated onion. Kolos, Moscow

Alzahim M, Newbury HJ, Ford-Lloyd BV (1997) Classification of genetic variation in garlic (*Allium sativum* L.) revealed by RAPD. HortScience 32:1102–1104

Araniti E, Galmarini CR, Sance MM, Burba JL (2005) Diferenciación de cultivares de ajo por su aptitud para la elaboración de extractos añejados y por sus propiedades nutracéuticas: Actividad Antiplaquetaria. IX Curso Taller sobre Prodcucción, Comercialización e Industrialización de Ajo. Mendoza, 29 June 2005, pp 191–192

Augusti KT, Mathew PT (1973) Effect of long term feeding of aqueous extracts of onion (*Allium cepa*) and garlic (*Allium sativum*) on normal rats. Ind J Exp Biol 11:239–241

Battaglia E (1963) Mutazione chromosomica e cariotipe fondamentale in *Allium sativum* L. Caryologia 26:1–46

Bennett MD, Smith JB (1976) Nuclear DNA amounts in angiosperms. Phil Trans R Soc London Series B Biol Sci 274:227–274

Block E (1992) The organosulfur chemistry of the genus *Allium*: implications for the organic chemistry of sulfur. Angewandte Chemie Intl Edn Engl 31:1135–1178

Block E, Haganathan S, Putman D, Zhao S-H (1992) *Allium* chemistry: HPLC analysis of thiosulfinates from onion, garlic, wild garlic (ramsoms), leek, scallion, shallot, elephant (great-headed) garlic, chive, and Chinese chive. Uniquely high allyl to methyl ratios in some garlic samples. J Agri Food Chem 40:2418–2430

Bordia A, Bansal HC, Arora SK, Singh SV (1975) Effect of essential oils of garlic and onion on alimentary hyperlipidemia. Atherosclerosis 21:15–19

Bordia A, Verma SK, Srivastava KC (1996) Effect of garlic on platelet aggregation in humans: a study in healthy subjects and patients with coronary artery disease. Prostaglandins Leukot Essent Fatty Acids 55(3):201–205

Bradley KF, Rieger MA, Collins GG (1996) Classification of garlic cultivars by DNA fingerprinting. Aust J Exp Agri 36:613–618

Briggs WH, Xiao H, Parkin KL, Shen C, Goldman IL (2000) Differential inhibition of human platelet aggregation by selected *Allium* thiosulfinates. J Agri Food Chem 48:5731–5735

Burkill IH (1966) A dictionary of the economic products of the Malay Peninsula, vol. 1, reprinted edn. Ministry of Agriculture and Co-operatives, Kuala Lumpur, Malaysia, pp 102–103

Burba JL (1993) Produccion de "semilla" de ajo. Asociación Cooperadora INTA EEA La Consulta, Argentina, p 163

Burba JL, Casali VW, Buteler MI (1993) Intensidad de la dormición como parámetro fisiológico para agrupar cultivares de ajo (*Allium sativum* L.). Horticultura Argentina 12(32):47–52

Cavallito CJ, Bailey JH (1944) Allicin, the antibacterial principle of *Allium sativum*. I. Isolation, physical properties and antimicrobial action. J Am Chem Soc 66:1950–1951

Cantwell M, Hong G, Voss R (2005) Compositional characterization of a garlic germplasm collection. Am Soc Hort Sci 2005 Annu Conf HortScience 40(4):1039

Cavagnaro PF, Masuelli RW, Simon PW (2003) Molecular data suggest multiple members comprising the alliinase gene family in garlic. Am Soc Hort Sci 2003 Centennial Conf HortScience 38(5):804

Cavagnaro PF, Galmarini CR, Masuelli RW, Simon PW (2004) 5' sequence characterization of alliinase gene family members in garlic. In: Proceedings of the Plant and Animal Genome XII Conference, San Diego, CA, p 112

Cavagnaro PF, Senalik D, Galmarini CR, Simon PW (2005a) Correlation of pungency, thiosulfinates, antiplatelet activity and total soluble solids in two garlic families. Am Soc Hort Sci 2005 Annu Conf HortScience 40(4):1019

Cavagnaro PF, Galmarini CR, Simon PW (2005b) Evaluación de caracteres ligados al sabor y a características nutracéuticas de una colección de 162 clones de ajo. Resúmenes del XII Congreso Latinoamericano y XXVIII Congreso Argentino de Horticultura, General Roca, Río Negro, 6–8 September 2005, p 267

Chia S-H (530–550) Ch'imin Yaoshu (Essential Arts for the People): Chinese Book of Husbandry. Translated into Japanese by Nishiyama B, Kumashiro Y (1969) Asian Economic, Tokyo

De Candolle A (1886) Origin of cultivated plants, reprinted from English edn. (1967) Hafner, New York

Don G (1827) A monograph of the genus *Allium*. Mem Wernerian Nat History Soc (1832) 1:1–102

Eady CC, Lister CE, Suo Y, Schaper D (1996) Transient expression of *uid* A constructs in in vitro onion (*Allium cepa* L.) cultures following particle bombardment and *Agrobacterium*-mediated DNA delivery. Plant Cell Rep 15:958–962

Engeland RL (1991) Growing great garlic. Filaree Productions, Okanogan, WA

Etoh T (1983) Accomplishment of microsporogenesis in a garlic clone. Mem Fac Agri Kagoshima Univ 19:55–63

Etoh T (1984) Hybrids between wild garlic (*Allium longicuspis* Regel) and garlic (*A. sativum* L.). In: Proceedings of the Eucarpia 3rd Allium Symposium, IVT, Wagenigen, The Netherlands, pp 78–82

Etoh T (1985) Studies on the sterility in garlic, *Allium sativum* L. Mem Fac Agri Kagoshima Univ 21:77–132

Etoh T (1986) Fertility of the garlic clones collected in Soviet Central Asia. J Jpn Soc Hort Sci 55:312–319

Etoh T, Ogura H (1977) A morphological observation on the formation of abnormal flowers in garlic (*Allium sativum* L.). Mem Fac Agri Kagoshima Univ 13:77–88

Etoh T, Ogura H (1984) Comparison between *Allium longicuspis* Regel and a fertile garlic clone and their hybrid seeds. 1984 Abstr Jpn Soc Hort Sci Spring Meeting, pp 170–171

Etoh T, Simon PW (2002) Diversity, fertility and seed production of garlic. In: Rabinowitch HD, Currah L (eds.) *Allium* Crop Science: Recent Advances. CABI, New York, pp 101–107

Fedschenko AP (1876) Flora Turkestans, vol. 3. St. Petersburg, Moskva, USSR, pp 38–40

Fenwick GR, Hanley AB (1990) Chemical composition. In: Brewster JL, Rabinowitch HD (eds.) Onion and Allied Crops, vol. 3, Biochemistry, Food Science and Minor Crops. CRC, Boca Raton, FL, pp 73–92

Galmarini CR, Goldman IL, Havey MJ (2001) Genetic analyses of correlated solids, flavor, and health-enhancing traits in onion (*Allium cepa* L.). Mol Genet Genom 265:543–551

Goldman IL, Kopelberg M, Debaene JP, Schwartz BS (1996) Antiplatelet activity in onion (*Allium cepa* L) is sulfur dependent. Thromb Haemost 76:450–452

Havey MJ, Galmarini CR, Gökçe AF, Henson C (2004) QTL affecting soluble carbohydrate concentrations in stored onion bulbs and their association with flavor and health-enhancing attributes. Genome 47:463–468

Helm J (1956) Die zuWurz- und Speisezwecken kultivierten Arten der Gattun *Allium* L. Kulturpflanze 4:130–180

Hong C-J (1999) Fundamental studies on the crossbreeding in garlic, *Allium sativum* L. Dissertation, Kagoshima University, Kagoshima, Japan

Hong C-J, Etoh T, Laundry B, Matsuzoe N (1997) RAPD markers related to pollen fertility in garlic (*Allium sativum* L.) Breed Sci 47:359–362

Hong C-J, Wattanabe H, Etoh T, Iwai S (2000a) Morphological and karyological comparison of garlic clones between the center of origin and westernmost area of distribution. Mem Fac Agri Kagoshima Univ 36:1–10

Hong C-J, Wattanabe H, Etoh T, Iwai S (2000b) A search of pollen fertile clones in Iberian garlic by RAPD markers. Mem Fac Agri Kagoshima Univ 36:1–16

Inaba A, Ujiie T, Etoh T (1995) Seed productivity and germinability of garlic. Breed Sci 45(2):310

Ipek M, Ipek A, Simon PW (2003) Comparison of AFLPs, RAPD markers, and isozymes for diversity assessment of garlic and detection of putative duplicates in germplasm collections. J Am Soc Hort Sci 128:24–252

Ipek M, Ipek A, Almquist SG, Simon PW (2005) Demonstration of linkage and development of the first low-density genetic map of garlic, based on AFLP markers. Theor Appl Genet 110:22–236

Jenderek MM (1998) Generative reproduction of garlic (*Allium sativum*). Zeszyty Naukowe Akademii Rolniczej im. H. Kolltaja w Krakowie 57:14–145

Jenderek MM, Hannan RM (2004) Variation in reproductive characteristics and seed production in the USDA garlic germplasm collections. HortScience 39:48–488

Jiang J, Gill B (1994) Nonisotopic in situ hybridization and plant genome mapping: the first 10 years. Genome 37:71–725

Jones HA, Mann LK (1963) Onions and Their Allies. Leonard Hill Books, London

Jones R, Rees H (1968) Nuclear DNA variation in *Allium*. Heredity 23:59–605

Kamenetsky R, Rabinowitch DH (2001) Floral development in bolting garlic. Sexual Plant Reprod 13:23–241

Kazakova AA (1978) Allium. In: Brezhnev DD (ed.) Flora of Cultivated Plants, vol. 10. Kolos, Leningrad, USSR

King JJ, Bradeen JM, Bark O, McCallum JA, Havey MJ (1998) A low-density genetic map of onion reveals a role for tandem duplication in the evolution of an extremely large diploid genome. Theor Appl Genet 96:5–62

Kirk JT, Rees OH, Evans G (1970) Base composition of nuclear DNA within the genus *Allium*. Heredity 25:50–512

Kitamura S (1950) Origins of Chinese cultivated plants. Toho-Gakuho Kyoto 19:76–101

Komissarov VA (1964) Evolution of the cultivated garlic, *A. sativum* L. News Timirijazev Agri Acad 4:70–73

Komissarov VA (1965) Classification of garlic, A. sativum L. Rep Timirjazev Agri Acad 108:351–357

Kondo T, Hasegawa H, Suszuki M (2000) Transformation and regeneration of garlic (*Allium sativum* L.) by *Agrobacterium*-mediated gene transfer. Plant Cell Rep 19:989–993

Konokov PF (1953) The question of obtaining garlic seed. Sad i Ogorod (Fruit Vegetable Gardens) 8:38–40

Kothari IL, Shah JJ (1974) Histogenesis of seed stalk and inflorescence in garlic. Phytomorphology 24:42–48

Kotlinska T, Havranek P, Navratill M, Gerasimova L, Pimakhov A, Neikov S (1991) Collecting onion, garlic and wild species of *Allium* in central Asia, USSR. FAO/IBPGR Plant Genet Resour News 83/84:31–32

Konvicka O, Levan A (1972) Chromosome studies in *Allium sativum*. Hereditas 72:12–148

Konvicka O, Nienhaus F, Fischbeck G (1978) Untersuchungen uber die Ursachen der Pollensterilitat bei *Allium sativum* L. Z Pflanzenzuchtung 80(4):265–276

Kuhl JC, Cheung F, Yuan Q, Martin W, Zewdie Y, McCallum J, Catanach A, Rutherford P, Sink KC, Jenderek M, Prince JP, Town CD, Havey MJ (2004) A unique set of 11,008 onion expressed sequence tags reveals expressed sequence and genomic differences between the monocot orders Asparagales and Poales. Plant Cell 16:114–125

Kumari K, Mathew BC, Augusti KT (1995) Antidiabetic hypolipidemic effects of S-methylcysteine sulfoxide isolated from *Allium cepa* L. Ind J Biochem Biophys 32(1):49–54

Kuznetsov AV (1954) Cultivated Garlic. Selkhozgiz, Moscow, USSR

Kyo E, Uda N, Suzuki A, Kakimoto M, Ushijima M, Kasuga S, Itakura Y (1998) Immunomodulation and antitumor activities of aged garlic extract. Phytomedicine 5:259–267

Labani R, Elkington T (1987) Nuclear DNA variation in the genus *Allium* L. (Liliaceae). Heredity 59:119–128

Lallemand J, Messiaen CM, Briand F, Etoh T (1997) Delimitation of varietal groups in garlic (*Allium sativum* L.) by morphological, physiological and biochemical characters. In: Burba JL, Galmarini CR (eds.) Proceedings of the First International Symposium on Edible Alliaceae. ISHS, Leuven, Belgium, pp 123–132

Laufer B (1919) Sino-Iranica, reprinted from original edn. (1973) Ch'eng Wen, Taipei, Taiwan

Lee HR, Eom EM, Lim YP, Bang JW, Lee DH (2003) Construction of a garlic BAC library and chromosomal assignment of BAC clones using the FISH technique. Genome 46:514–520

Linnaeus C (1753) Species Plantarum, vol. 1. R Society, London

Lupea V, Vranceanu C (1972) Beitrag zur untersuchung des molybdangehaltes einiger Lebensmittel and Gebieten mit endemischer Nephritis. Nahrung 16:637–639

Maaß HI, Klaas M (1995) Intraspecific differentiation of garlic (*Allium sativum* L.) by isozyme and RAPD markers. Theor Appl Genet 91:89–97

Mann LK (1952) Anatomy of the garlic bulb and factors affecting bulb development. Hilgardia 21:195–251

Mann LK, Minges PA (1958) Growth and bulbing of garlic (*Allium sativum* L.) in response to storage temperature of planting stocks, day length, and planting date. Hilgardia 27:385–419

Matassi G, Montero L, Montero LM, Salinas J, Bernardi G (1989) The isochore organization and the compositional distribution of homlogous coding sequences in the nuclear genome of plants. Nucl Acids Res 17:5273–5290

Mathew B (1996) A review of *Allium* Section *Allium*. Royal Botanic Gardens Kew, Richmond, UK

McCollum GD (1976) Onion and allies. In: Simmonds NW (ed.) Evolution of Crop Plants. Longman, London, pp 186–190

Messiaen CM, Cohat J, Leroux JP, Pichon M, Beyries A (1993) Les Allium alimentaires reproduits par voie vegetative. INRA, Paris

Natale PJ, Camargo A, Galmarini CR (2005) Characterization of argentine garlic cultivars by their pungency. Acta Hort 688:313–316

Ori D, Fritsch RM, Hanelt P (1998) Evolution of genome size in *Allium* (Alliaceae). Plant Syst Evol 210:57–86

Ohsumi C, Kojima A, Hinata K, Etoh T, Hayashi T (1993) Interspecific hybrid between *Allium cepa* and *Allium sativum*. Theor Appl Genet 85:969–975

Pearce SR, Pich U, Harrison G, Flavell AJ, Hesiop-Harrison JS, Schubert I, Kumar A (1996) The Ty1-*copia* group retrotransposons of *Allium cepa* are distributed throughout the chromosomes but are enriched in the terminal heterochromatin. Chrom Res 4:357–364

Pooler MR (1991) Sexual reproduction in garlic (*Allium sativum* L.). Dissertation, University of Wisconsin, Madison, WI

Pooler MR, Simon PW (1993a) Characterization and classification of isozyme and morphological variation in a diverse collection of garlic clones. Euphytica 68:121–130

Pooler MR, Simon PW (1993b) Garlic flowering in response to clone, photoperiod, growth temperature and cold storage. HortScience 28:1085–1086

Randle WM, Lancaster JE (2002) Sulphur compounds in *Allium* in relation to flavor quality. In: Rabinowitch HD, Currah L (eds.) *Allium* Crop Science: Recent Advances. CABI, New York, pp 329–356

Ranjekar PK, Pallota D, Lafontaine JG (1978) Analysis of plant genomes. V. Comparative study of molecular properties of DNAs of seven *Allium* species. Biochem Genet 16:957–970

Regel E (1875) Alliorum adhuc cognitorum monographia. Acta Hort Petropolitani 3:1–266

Sawahel WA (2002) Stable genetic transformation of garlic plants using particle bombardment. Cell Mol Biol Lett 7:49–59

Simon PW, Jenderek MM (2003) Flowering, seed production and the genesis of garlic breeding. In: Janick J (ed.) Plant Breeding Reviews, vol. 2. Wiley, New York, pp 211–244

Singh UP, Pandey VN, Wagner KG, Singh KP (1990) Antifungal activity of ajoene, a constituent of garlic (*Allium sativum* L.) Can J Bot 68:1354–1359

Smilde WD, van Heusden AW, Kik C (1999) AFLPs in leek (*Allium porrum*) are not inherited in large linkage blocks. Euphytica 110:127–132

Smith GP (1976) Evolution of repeated DNA sequences by unequal crossover. Science 191:524–529

Stack SM, Comings DE (1979) The chromosomes and DNA of *Alllium cepa*. Chromosoma 70:161–181

Sturtevant EL (1919) Sturtevant's Notes on Edible Plants. JB Lyon, Albany, NY

Suetsuna K, (1998) Isolation and characterization of angiotesin converting enzyme inhibitor dipeptides derived from *Allium sativum* (garlic). J Nutr Biochem 9:415

Sugimoto H, Tsuneyoshi T, Tsukamoto M, Uragami Y, Etoh T (1991) Embryocultured hybrids between garlic and leek. Allium Improv Newsl 1:67–68

Suzuki G, Ura A, Saito N, Do GS, Seo BB, Yamamoto M, Mukai Y (2001) BAC FISH analysis in *Allium cepa*. Genes Genet Syst 76(4):251–255

Tackholm V, Drar M (1954) Flora of Egypt, vol. 3. Cairo Univ Bulletin Fac Sci 30:86–113

Takagi H (1990) Garlic *Allium sativum* L. In: Brewster JL, Rabinowitch HD (eds.) Onion and Allied Crops, vol. III. Biochemistry, Food Science and Minor Crops. CRC, Boca Raton, FL, pp 109–146

Treutner R, Jankovsky M, Hubacek J (1978) The content of some substances in selected varieties of garlic (*Allium sativum* L.). Rostl Vyroba 24:1003

van Heusden AW, van Ooijen JW, Vrielink-van Ginkel R, Verbeek WHJ, Wiestma WA, Kik C (2000) A genetic map of an interspecific cross in *Allium* based on amplified fragment length polymorphism (AFLP) markers. Theor Appl Genet 100:118–126

Vanin EF (1985) Processed pseudogenes: characteristics and evolution. Annu Rev Genet 19:253–272

Vavilov NI (1951) The origin, variation, immunity and breeding of cultivated plants. Chron Bot 13:1–364

Verma SC, Mittal RK (1976) Chromosome variation in the common garlic, *Allium sativum* L. Cytologia 43:383–396

Vicient CM, Jaaskelainen MJ, Kalendar R, Schulman AH (2001) Active retrotransposons are a common feature of grass genomes. Plant Physiol 125:1283–1292

Vos P, Hogers R, Blecker M, Reijans M, van de Lee T, Hornes M, Frijters A, Pot J, Peleman J, Kuipe M, Zabeau M (1995) AFLP: a new technique for DNA fingerprinting. Nucl Acids Res 23:4407–4414

Vvedensky AI (1935) *Allium* L. In: Komarov VL (ed.) Flora SSSR, vol. 4. Translated into English by Airy-Shaw HK (1944) as The genus *Allium* in the USSR. Herbetia 11:65–218

Wang BH, Zuzel KA, Rahman K, Billington D (1998) Protective effects of aged garlic extract against bromobenzene toxicity to precision cut rat liver slices. Toxicology 126:213–222

Wendelbo P (1971) Alliaceae. In: Rechinger KH (ed.) Flora Iranica 76. Acad Druck-u Verragsanstalt, Graz, pp 1–100

Zagorodkij PF (1935) Garlic. Rukovodstvo po Aprobatsii Sejlskokhozyastvennikh Kultur 5:253–264

Zewdie Y, Havey MJ, Prince JP, Jenderek MM (2005) The first genetic linkages among expressed regions of the garlic genome. J Am Soc Hort Sci 130(4):569–574

Subject Index